Principles of NMR
Spectroscopy

An Illustrated Guide

David P. Goldenberg
UNIVERSITY OF UTAH

University Science Books
Mill Valley, California

University Science Books
Mill Valley, California
www.uscibooks.com

PRODUCTION MANAGER   Paul C. Anagnostopoulos
MANUSCRIPT EDITOR   John Murdzek
DESIGNER   Melissa Ehn
ILLUSTRATORS   Cohographics
COMPOSITOR   Windfall Software, using ZzTEX
PRINTER AND BINDER   Bang Printing

This book is printed on acid-free paper.

Print ISBN: 978-1-891389-88-7
eBook ISBN: 978-1-938787-86-7

LIBRARY OF CONGRESS CATALOGING-IN-PUBLICATION DATA

Names: Goldenberg, David P., 1954–
Title: Principles of NMR spectroscopy : an illustrated guide / David P. Goldenberg, University of Utah.
Other titles: Principles of nuclear magnetic resonance spectroscopy
Description: Mill Valley, California : University Science Books, [2016] | Includes index.
Identifiers: LCCN 2015032333 | ISBN 9781891389887 (alk. paper)
Subjects: LCSH: Nuclear magnetic resonance spectroscopy.
Classification: LCC QD96.N8 G625 2016 | DDC 543/.66—dc23
LC record available at http://lccn.loc.gov/2015032333

Printed in the United States of America
10 9 8 7 6 5 4 3 2 1

For Karen, Benjamin, and Jeremy, who are unlikely to read much of this book, but inspire with their own passions.

Brief Contents

Contents

CHAPTER 8   A More Mathematical Look at Molecular Motion and NMR Relaxation   195

CHAPTER 15   Two-Dimensional Spectra Based on Scalar Coupling: COSY and TOCSY Experiments   425

CHAPTER 16   Heteronuclear NMR Techniques   459

Preface

In the seven decades since the demonstration of nuclear magnetic resonance in liquid and solid samples, spectroscopic techniques based on this phenomenon have evolved to a degree that few could have imagined in the 1940s. In contrast to the early experiments of Felix Bloch and Edward Purcell, who surprised many with their ability to detect resonance signals at all from pure water or paraffin wax, it is now routine to record and interpret spectra from molecules containing thousands of atoms in dilute solutions. This remarkable progress has been made possible by advances in several technologies, including electronics, computing, superconducting magnets and isotopic labeling, as well as developments in the underlying theory of NMR. Steady progress in these areas have made magnetic resonance one of the most powerful techniques presently available for studying molecular structure, dynamics and function, as well as a widely used tool for medical imaging.

As NMR techniques have become dramatically more powerful and more widely adopted, they have also become much more complicated and difficult to understand. When the applications of NMR were largely limited to one-dimensional spectra of small molecules, it was possible to gain a working understanding using relatively simple physics, and most instruction focused on the practical aspects of recording and interpreting spectra. In contrast, modern multidimensional NMR experiments are based on quite subtle quantum physics, and their execution relies on complex equipment and computer programs that many users come to view as mysterious black boxes.

The challenges in learning and teaching modern NMR spectroscopy became painfully apparent to me about 20 years ago, when my students and I began using the technique to study protein structure and dynamics. While there were many textbooks available, they seemed to fall into two large categories. In one group, there were relatively accessible books that covered important basics, but did not offer much insight into mysteries such as the "magnetization transfer" that gives rise to the cross peaks in multidimensional spectra. Books in the second group covered the multidimensional techniques in varying degrees of detail, but typically relied on a great deal of mathematics and rather abstract constructs, such as density matrices and product operators. There seemed to be no clear path from the elementary treatments to the ones that would enable me to really understand what was going on. For better or worse, I felt compelled to find such a path, and this book can be viewed as a record of that journey, though extensively rewritten to eliminate, I hope, evidence of the numerous side trips, dead ends and other wanderings.

For many students, at all levels, the most difficult challenge along the way to mastering modern NMR is encountered when trying to understand the magnetic interactions of nuclei linked by covalent bonds, the class of interaction referred to as scalar coupling. This interaction provides the link between NMR spectra and the covalent structures of molecules, and it is the basis of most multidimensional experiments, especially heteronuclear experiments (i.e., those involving two or more types of nuclei). Although some of the consequences of scalar coupling, such as the splitting of spectral peaks, can be accounted for using simple pictures, with the magnetic dipoles pointing either up or down, these pictures quickly fail when applied to more complex experiments. The underlying difficulty is that nuclei linked by scalar coupling behave in ways that can only be described quantum mechanically and that often defy our intuition based on experience in the macroscopic world.

Because of the central role of quantum mechanics in NMR, especially as currently practiced, one of the challenges in teaching or writing about the subject lies in deciding how much quantum physics to include, and when. Indeed, the chapters of this book have been reorganized several times in trying to deal with this question. Though early drafts presented much of the quantum mechanics early on, the current version reflects my growing appreciation of the traditional visual description of NMR, in the form of vector diagrams, which not only excel in describing the basics but can serve as a reference when moving on to the aspects for which a quantum treatment is necessary. At the same time, however, I feel that it is useful to include an early qualitative introduction to spin quantum mechanics, so that students begin with an appreciation for the relationship between the macroscopic magnetization that is detected in a spectrometer and the microscopic magnetic dipoles.

A related decision in teaching NMR, especially to students with different backgrounds, is how much mathematics to include. Most of the mathematics used here should be familiar to advanced undergraduate and graduate students in the sciences, who have typically taken a year or more of calculus, but my experience has been that many students taking my classes are a bit rusty. For this reason, the text includes brief reviews of some important math concepts, such as complex numbers, Fourier transformations, vectors, and matrices. These topics are introduced as appropriate in the development of the subject, perhaps breaking the flow of some sections but ensuring that the tools are at hand when needed. These subjects are also summarized, and in some cases extended, in the appendices.

In the resulting organization of this book, the chapters are divided roughly in half. Except for Chapter 2, which is a qualitative introduction to quantum mechanics and spin, the first 10 chapters cover topics that are well described using a classical approach, including chemical shifts, dipolar coupling, pulse methods, relaxation, and the nuclear Overhauser effect. The final chapter in this group (Chapter 10) introduces two-dimensional spectroscopy in the form of the NOESY experiment, which can be described without a quantum mechanical treatment. The second part of the book relies much more heavily on quantum mechanics and begins with three chapters that introduce the mathematical formalism and the quantum description of first isolated spins and then scalar-coupled pairs. The special properties of coupled spins are emphasized, and a new form of vector diagram is introduced to help visualize quantum correlations (or entanglement) that lie at the heart of the matter.[1]

The next chapters then use the quantum mechanical treatment and diagrams to discuss a variety of NMR experiments based on scalar coupling, including COSY, HSQC, HMQC and three-dimensional experiments based on these. Finally, in Chapters 17 and 18, the density matrix and product-operator formalism are introduced.

Leaving the density matrix and product operators to the end of the book represents a rather unconventional approach and calls for some explanation, especially for instructors considering this text for their classes. NMR depends on both quantum mechanics, which describes the behavior of individual spins and coupled groups of spins, and statistical mechanics, which determines the distribution of molecules in different quantum states. For uncoupled spin-1/2 nuclei, the statistical aspect can be dealt with by considering only the small excess population of spins that are in the lower energy state at equilibrium. For two or more scalar-coupled spins, things are not so simple, and the density matrix provides an efficient means of handling the statistical and quantum mechanical aspects together. The product-operator formalism, in turn, greatly simplifies the density matrix calculations. I believe, however, that this simplification comes at a pedagogical cost, because the quantum and statistical aspects are merged in the mathematical manipulations. In particular, the more compact treatments make it difficult to recognize the special quantum properties that give rise to the peaks in multidimensional spectra.

Rather than first developing the product-operator formalism and then using it to describe different NMR experiments, as most texts do, the approach followed here is to describe a set of important experiments in a way that treats explicitly both the quantum and statistical aspects of a scalar-coupled system. For a system of two spin-1/2 nuclei, this requires following, through the various manipulations of an experiment, the quantum states of the four subpopulations that begin in the eigenstates at equilibrium. Finally, the signals generated by the total population are calculated by combining the signals from the subpopulations according to the initial distribution of the eigenstates. Although this treatment is more laborious than using the product-operator formalism, I believe that it provides a much clearer picture of the physical processes underlying the NMR experiments, as mysterious as some of the quantum effects may still be. When the density matrix and product-operator formalism are presented in the final chapters, the reader will, I hope, both appreciate

---

[1] A description of the correlation diagrams and the underlying ideas they represent has been published in a more condensed form in Goldenberg, D. P. (2010) *Concepts Magn. Reson. Part A.*, 36A, 49–83. Portions of Chapters 13, 14, 17, and 18 have been adapted from this article, with permission of John Wiley and Sons.

the utility of these mathematical tools and have a greater physical intuition of the phenomena that they represent.

The material presented here would likely fill a full semester course that emphasizes the theoretical and practical aspects of modern NMR experiments, with perhaps only limited time left for applications in chemistry or structural biology. However, the organization of the book should also make it suitable for other kinds of courses. For instance, a course with a greater emphasis on applications might include the following:

- An introduction to the basics of NMR, including the pulse method (Chapters 1–6).
- An introduction to relaxation (Chapter 7, but probably not the more mathematical treatment in Chapter 8).
- The nuclear Overhauser effect (Chapter 9).
- An introduction to two-dimensional spectroscopy, using the NOESY experiment as an example (Chapter 10).
- A qualitative description of multidimensional experiments based on scalar coupling, drawing on the descriptions of these experiments in Chapters 15 and 16, but without the quantum mechanics needed for a full understanding.
- Applications chosen for the target audience, with examples drawn from the relevant literature.

A more advanced course, for students already familiar with the material listed above, could follow the sequence below:

- The quantum mechanics of isolated and scalar-coupled spins (Chapters 11–14).
- Multidimensional experiments based on scalar coupling (Chapters 15 and 16).
- The density matrix and product-operator formalism (Chapters 17 and 18).
- Treatments of more advanced experiments, using the product-operator formalism and material from the literature and other texts.
- Relaxation and the use of relaxation measurements to study molecular motions (Chapters 7 and 8).
- Applications, with examples drawn from the relevant literature.

As an alternative, for those who wish to go straight to the density matrix and product operators before discussing specific experiments, Chapters 15 and 16 can be omitted without a loss of continuity.

Electronic supplements to this book will be made available for free download through links at http://uscibooks.com/goldenberg.htm. These will include all of the illustrations (with color used in many of the vector diagrams representing two-spin systems) and a set of files for use with the program Maxima, a free computer algebra system. The Maxima files are intended as aids for exploring some of the more mathematical topics, especially the quantum-mechanical treatments introduced in the second half of the book.

Because my own practical experience with NMR has focused on studies of proteins, nearly all of the examples in this text are based on this class of molecules. I apologize for this rather limited perspective, but hope that readers will find the treatment of NMR general enough that it can be readily extended to other kinds of molecules.

An important convention that should be mentioned at the outset, because there is no universal agreement, concerns the "left-hand" and "right-hand" rules used by different authors to describe the precession of magnetization and the effects of pulses. Because these motions cannot really be seen, the choice is arbitrary, but it is a source of ongoing confusion. From my perusal, it seems that most texts based on classical descriptions and vector diagrams, especially earlier ones, have used the left-hand convention to describe precession and the effects of pulses. More advanced texts and much of the current literature, especially texts and papers that employ the product-operator formalism, generally describe the rotations relative to a right-handed coordinate system. Since a major goal of this book is to serve as a bridge between the two approaches, the choice was especially problematic. In the end, it seemed most important to maintain consistency with the more recent literature, and the right-hand convention has been used here. I have also attempted to abide consistently with the more modern practice of treating Larmor frequencies as signed quantities reflecting the gyromagnetic ratio. These conventions are discussed in some detail in Chapter 5.

Acknowledgments

I am grateful to numerous friends and colleagues who have aided my study of NMR and offered suggestions for this book. The use of NMR in my laboratory began with an outstanding graduate student, Scott Beeser, and a wonderful collaborator, Terry Oas, of Duke University. Scott's talent and enthusiasm and Terry's rich background in NMR made for a delightful and productive collaboration, and my first real immersion in NMR. The next major step in my NMR education was a sabbatical visit in Gerhard Wagner's laboratory at Harvard Medical School, supported by a fellowship from the John Simon Guggenheim Memorial Foundation. I remain indebted to Gerhard and the members of his research group, especially Dara Gilbert, Kwaku Dayie, and Volker Dotsch, for their hospitality and guidance during that year. A few years later, I began to teach a section on NMR as part of a graduate course in biomolecular structural methods at the University of Utah, from which this book emerged. The numerous students who have enrolled in that course over the past decade have been critical to the refinement of the material and its presentation, and I thank them for their probing questions, suggestions and patience. Other students, working in my research group, have made additional contributions, especially W. Miachel Hanson and Gourab Bhattacharje. The University of Utah is fortunate to have two excellent scientists to direct its NMR facilities, Peter F. Flynn and Jack Skalicky. I have benefited from numerous illuminating discussions with them both, and both contributed directly to this book by recording some of the spectra used for illustrations. Also at the University of Utah, Julio Facelli carefully read an early version of this book and provided numerous helpful suggestions, Brian Saam helped with the material on NMR imaging in Chapter 10, and Aaron Fogelson carefully reviewed Appendix C, on Fourier series and transforms. Special thanks also goes to Michael Summers, who helped clarify for me the special problems associated with the behavior of coupled spins, especially the phenomenon of multiple-quantum coherence. Russel Hopson and Lawrence McIntosh carefully reviewed the book and made numerous suggestions that have greatly improved it. The final production of this book reflects the dedicated efforts of Jane Ellis of University Science Books, John Murdzek, and Paul Anagnostopoulos. In spite of all of this help, errors and omissions are certain to remain, for which I bear full responsibility. Corrections and suggestions from readers will be greatly appreciated.

Finally, I thank my wife, Karen, and sons, Benjamin and Jeremy, for their steadfast support of this and other projects that likely defy common sense. They make it all worthwhile.

CHAPTER 1

An Overview of Modern Solution NMR

1.1 The General Nature of NMR Spectroscopy

Like other spectroscopic methods, NMR spectroscopy is based on the absorption or emission of electromagnetic radiation. For many readers of this book, the most familiar form of spectroscopy is likely to be absorbance spectrophotometry using ultraviolet (UV) or visible light, a technique that is widely used in chemistry and biochemistry laboratories to determine the concentrations of a wide variety of compounds, especially those containing systems of conjugated double bonds or liganded metal ions. It may be useful, therefore, to briefly consider this spectroscopic technique and discuss its similarities to and differences from NMR spectroscopy. The source of radiation in a UV–visible spectrometer is usually an ordinary tungsten lamp, for visible light, or a deuterium lamp for UV wavelengths. The light from the lamp passes through a monochromator, essentially an adjustable filter that selects a relatively narrow range of wavelengths, and then enters the sample. Some fraction of the light is absorbed by the molecules in the sample, and the fraction that is not is measured by a detector. The amount of light absorbed depends on the wavelength of light used, the concentration of molecules and the properties of those molecules. An absorbance spectrum is generated by measuring the absorbance as a function of wavelength. Typically, a UV–visible absorbance spectrum contains one or a few rather broad and overlapping peaks that correspond to functional groups that absorb light of different wavelengths.

The absorbance of UV or visible light is coupled to a transition of the molecule from one quantum mechanical state to another. The two states differ from one another in the distribution of the electrons about the nuclei and in their energies. This is often depicted in an energy-level diagram such as shown in Fig. 1.1.

The transition and the associated absorption of light only occur when the energy of the photons matches the difference in the energies of the two states, ΔE. The energy of a photon is related to the frequency of the radiation according to

$$E = h\nu$$

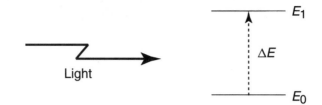

Figure 1.1
Energy diagram for the
absorption of radiation.

where h is Planck's constant ($\approx 6.626 \times 10^{-34}$ J·s). The frequency is also related to the wavelength of the radiation, λ, according to

$$\nu = c/\lambda$$

where c is the velocity of light ($\approx 3 \times 10^{8}$ m/s). Combining these two relationships gives

$$E = hc/\lambda$$

For UV and visible radiation, the wavelength ranges from about 200 to 800 nm, corresponding to energies of about 10^{-18} J per photon, or about 10^{6} J per mole of molecules undergoing the transition. This is a very large energy relative to the thermal energies of molecules at ordinary temperatures, which means that virtually all of the molecules will be in the lower-energy, or ground, state in the absence of radiation to stimulate the transition to the higher-energy state. Once a molecule absorbs a quantum of energy this large, there is a very high probability that it will spontaneously return to the ground state, usually emitting a photon of a lower energy, in a process known as fluorescence. While some of this emitted radiation will reach the detector, much of it is emitted in other directions, and some of it is converted to heat rather than emitted. As a consequence, the amount of light reaching the detector is less than that which entered the sample, and we say that the difference is "absorbed," usually without worrying too much about what happened to it.

In NMR spectroscopy, the quantum states that are exploited arise from a property, the nuclear "spin," of certain nuclei that causes them to take on discrete energy levels when placed in a magnetic field. The differences in these energies depend on the kind of nucleus involved and the strength of the field. In the simplest, and most commonly used, systems, there are only two energy levels, and the energy difference is given by

$$\Delta E = \frac{h}{2\pi} \gamma B \qquad (1.1)$$

where B is the strength of the magnetic field, in units of tesla (T), and γ is a constant, for a specific nucleus type, called the gyromagnetic ratio (or, slightly more properly but less commonly, the magnetogyric ratio). The gyromagnetic ratios of some of the nuclei most commonly used in NMR spectroscopy of organic and biological molecules are listed in Table 1.1. (Don't worry for now about the negative gyromagnetic ratio for ^{15}N; the sign of the constant reflects other properties of the signal arising from the nucleus.) Unlike the other isotopes listed in the table, which

Table 1.1   Gyromagnetic ratios of nuclei commonly used in NMR.[a]

Nucleus	Gyromagnetic ratio ($T^{-1} s^{-1}$)
^1H	2.675221×10^8
^2H	4.106628×10^7
^{13}C	6.728284×10^7
^{15}N	-2.712618×10^7
^{19}F	2.518148×10^8
^{31}P	1.083940×10^8

a. From Harris, R. K., Becker, E. D., Cabral de Menezes, S. M., Goodfellow, R. & Granger, P. (2001). NMR nomenclature. Nuclear spin properties and conventions for chemical shifts. *Pure Appl. Chem.*, 73, 1795–1818. http://dx .doi.org/10.1351/pac200173111795

Table 1.2   Energies, wavelengths and frequencies associated with commonly used nuclei in an 11.7-T magnetic field

Nucleus	ΔE (J/particle)	Wavelength (m)	Frequency (MHz)
^1H	3.31×10^{-25}	0.60	500
^2H	5.09×10^{-26}	3.91	77
^{13}C	8.33×10^{-26}	2.38	126
^{15}N	3.36×10^{-26}	5.91	51
^{19}F	3.12×10^{-25}	0.64	471
^{31}P	1.34×10^{-25}	1.48	203

have two energy levels, ^2H has three. Fluorine is not normally found in biological molecules, but can be incorporated using synthetic or biosynthetic methods and has quite useful NMR properties, including sensitivity similar to that obtained with ^1H. In modern NMR spectrometers, the magnetic field strength generally lies in the range of 7 to 20 T. A field strength of 11.7 T is one of the most common and yields energy differences, wavelengths and frequencies as listed in Table 1.2. Even in the strongest magnetic fields currently available, these energy differences are very small. For comparison, thermal energy at room temperature is about 4×10^{-21} J/particle. The very low energy differences associated with NMR are perhaps the single factor that is most important in defining the method's general character, its strengths and its limitations.

One important consequence of the small energy differences is that the radiation used in NMR spectroscopy has a much longer wavelength, and correspondingly lower frequency, than those used for most other spectroscopic techniques. The frequencies used in NMR are similar to those used in radio broadcasting, and much of the electronic technology originally used for NMR instruments was derived from radio and radar technology. Rather than a light bulb, as used for visible-light spectroscopy, the radiation source for NMR is an electronic circuit that generates an electric current that alters direction, or oscillates, at a frequency chosen to match the energy difference

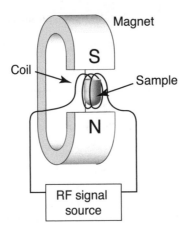

Figure 1.2

Schematic representation of an NMR spectrometer.

of the nuclei of interest. In an NMR spectrometer, the sample is placed in the field of a very strong magnet and is also surrounded by a coil of wire connected to the signal-generating circuit, often referred to as an "rf," for radio frequency, source. A highly schematic representation of an NMR spectrometer is shown in Fig. 1.2.

The oscillating current in the coil generates an oscillating magnetic field, which can be thought of as a bath of photons, with the very low energies indicated in Table 1.2. A small fraction of these photons are absorbed by the nuclei in the sample, causing transitions in the quantum mechanical states of these nuclei, much as the electrons of a molecule undergo a transition when they absorb the much higher energy photons of visible or UV light. Because of the very low energies involved, detecting the absorption is much more difficult than for optical spectroscopy, and we will ignore for now exactly how this is done.

While they present large technical challenges, the low energies associated with NMR spectroscopy are also the origin of some of the technique's unique powers. One important consequence of the small energy differences concerns the processes by which the nuclei return to their equilibrium states after absorbing photons. The rates of these processes depend greatly on the energy difference between the states, so the excited states generated by UV and visible light typically have lifetimes on the order of nanoseconds or less, while the excited states generated in NMR spectroscopy can persist for seconds or longer. These long lifetimes make it possible to carry out elaborate manipulations of the quantum states, by applying successive pulses of rf energy, and these manipulations are the basis of advanced techniques such as multidimensional spectroscopy.

The rates at which the excited nuclei return to equilibrium, their *relaxation rates*, are quite sensitive to molecular motion, both overall tumbling in solution and internal motions. An important application of NMR spectroscopy involves special experiments to measure these relaxation rates, from which information about the molecular motions can be derived. Relaxation processes are also the basis of methods for measuring the distance between nuclei. The sensitivity of relaxation rates to molecular motion can also present a challenge to NMR spectroscopists, because the tumbling rates of large molecules can lead to very short lifetimes, sometimes eliminating a detectable signal altogether.

Although the low energy differences inherent in NMR spectroscopy are the source of much of the technique's power, in most situations increasing the energy differences by increasing the magnetic field strength pays large dividends in increased sensitivity and resolution. As a consequence, there is a constant demand for ever more powerful magnets. It is common to refer to the strength of a spectrometer's magnetic field in terms of the corresponding resonance frequency for ^{1}H. Thus, the 11.7-T magnet referred to above would be called a "500-MHz magnet". Today (i.e., in 2015), most spectrometers used for biomolecular NMR have magnets that yield ^{1}H resonance frequencies of 500–900 MHz, and a few 1-GHz instruments are just becoming available. Although the differences within this range may seem modest, the higher fields are often essential for some projects, especially those involving larger proteins or nucleic acids.

1.2   Examples of Simple Spectra of a Small Protein

Before delving into the technical details of NMR spectroscopy, it is useful to briefly examine a simple spectrum of a small protein to illustrate the great potential of the technique as well as the challenges in its application. The ^{1}H spectra shown in Fig. 1.3 are of the protein bovine pancreatic trypsin inhibitor (BPTI), composed of 58 amino acid residues.

The spectrum in panel A was recorded from a sample of BPTI in its native, folded conformation and shows that the individual ^{1}H nuclei in the protein have a range of resonance frequencies, giving rise to peaks that are *almost* resolvable in this spectrum. The absolute differences in resonance frequencies are actually quite small and are given as "parts per million" (ppm), relative to a standard. The energy differences that correspond to these small differences in frequency are truly minuscule. (Note also that the frequency scale increases from right to left, a convention that is explained in Chapter 4.)

That the resonance frequencies are very sensitive to the physical environment of the nuclei can be seen by looking at the spectrum of the same protein after disrupting its folded conformation (Fig. 1.3B). The folded structure of BPTI depends, in part, on three disulfide bonds between the sulfur atoms of Cys residues. The sample used to record the spectrum in panel B was treated with a reducing agent, and the resulting thiol groups were irreversibly blocked to create a chemically stable form that is extensively unfolded. In this spectrum, there are still protons with significantly differing resonance frequencies. This variation arises primarily from the different functional groups to which the protons are attached. For instance, the resonances clustered at about 8 ppm are those that arise from the backbone amide groups of the protein, while those in the range from about 2 to 0 ppm come from protons attached to carbon atoms in aliphatic side chains. Within the groups of protons associated with different chemical groups, however, there is much less variation in the resonance frequencies than is seen for the native protein. This is especially apparent for the amide protons.

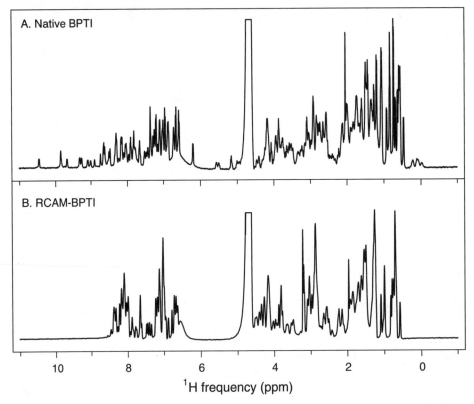

Figure 1.3 One-dimensional ^{1}H NMR spectra of bovine pancreatic trypsin inhibitor in its native conformation (A) and after reducing its three disulfide bonds and reacting the resulting thiols with iodoacetamide, to produce the reduced and carboxyamidomethylated (RCAM) form (B). In each sample, the protein concentration was approximately 0.25 mM and the pH was 4. The spectra were recorded in a spectrometer with a magnetic field strength of 11.7 T and ^{1}H resonance frequency of 500 MHz. The large truncated peaks near the center of the spectra are due to the hydrogen nuclei of water. (Spectra generously provided by Dr. Jack Skalicky.)

Thus, the resonance frequency is sensitive to both covalent structure and to three-dimensional conformation. Collectively, these differences are refereed to as *chemical shifts*, which are discussed further in Chapter 4. The extent of variation in observed chemical shifts for nuclei in a given functional group is commonly described as the *chemical shift dispersion* and is often used as a qualitative measure of the extent of ordered structure in a large molecule, such as a protein.

Although these spectra suggest that information can be obtained for individual atoms within a very complex molecule, the resolution is not quite high enough to distinguish all the individual signals. In addition, it is not obvious how to assign the resonance peaks to specific nuclei in the molecule, or how to interpret the spectrum in terms of the structural, chemical or physical properties of the molecule. Although the resonance frequencies are clearly sensitive to the chemical environment of the nuclei, deriving structural information from the frequencies alone is very difficult. Over the past five decades, tremendous technical and theoretical advances have been made in NMR spectroscopy, and these challenges have largely been met, at least for moderately sized proteins and nucleic acids, as well as other complex molecules. These advances are described briefly in the following section.

1.3   NMR Technology, Circa 2015

When the first NMR experiments were performed in the mid 1940s, as discussed in Chapter 3, it was thought astounding that signals from the ^1H nuclei of pure paraffin or water could be detected at all. Today, we routinely detect the signals from solute molecules at concentrations of 1 mM or less. Furthermore, we are able to resolve the signals from hundreds of nuclei in a molecule, assign the signals to specific atoms and derive detailed structural information from the spectra. The following are some aspects of the technology that has made this possible.

1.3.1   Instrumentation

As noted earlier, the sensitivity of NMR is intrinsically limited by the strength of the magnetic field employed. In addition, the magnetic field must be extremely uniform. Today, the magnets used for most forms of NMR are solenoidal electromagnets that are maintained in a superconducting state by using very low temperatures and special materials for the coil. The geometry of this type of magnet is shown in Fig. 1.4.

The sample is placed within the coils of the electromagnet, where the field is strongest and most uniform. In a superconducting NMR magnet, the wire is made of a niobium alloy and is surrounded by liquid helium to maintain a temperature of about 4 K. Once an electric current is established in the coil, the power source is disconnected, and the ends of the coil are connected to one another. Due to the lack of electrical resistance, the current is then maintained indefinitely, unless the magnet is somehow perturbed.

To isolate the liquid helium from the environment and slow its evaporation, it is surrounded by a vacuum chamber, which in turn is surrounded by liquid nitrogen, at 77 K, and another vacuum chamber. The result is a quite large vessel that dominates the NMR laboratory. In 1995, engineers at JEOL, a manufacturer of NMR instruments, carefully cut away sections of a decommissioned NMR magnet to provide a

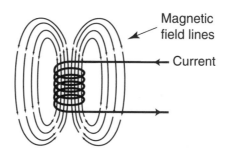

Magnetic field lines

Current

Figure 1.4
Diagram of a solenoidal electromagnet.

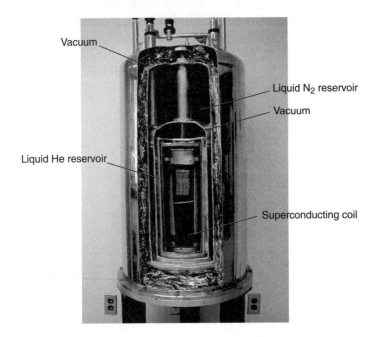

Figure 1.5 Photograph of the interior of a partially dissected superconducting NMR magnet. (Photo courtesy of JEOL USA, http://www.jeolusa.com/RESOURCES/AnalyticalInstruments/ NMRMagnetDestruction/tabid/390/Default.aspx)

rare inside view (Fig. 1.5). When still intact, this magnet generated a field of 6.3 T in its core, corresponding to a ^{1}H frequency of 270 MHz. The coil contained approximately 19 km of wire, and the full assembly weighed about 190 kg. These dimensions increase dramatically as the field strength is increased, as does the price.

In addition to the main solenoid, the magnet contains additional coils that are adjusted to make the field as uniform as possible. Some of these are superconducting coils that are adjusted when the magnet is first installed in a laboratory, and others are conventional electromagnets that are adjusted by the user each time a new sample is placed in the instrument. These coils are called *shims*, a term derived from the use of thin pieces of metal to adjust the fit of mechanical devices, and the processes of adjusting these coils is called *shimming*. Historically, shimming has been one of the aspects of NMR spectroscopy that requires the most skill and patience. In the past several years, however, even this process has become largely automated.

As large and impressive as it is, the cryogenic magnet is only one component of an NMR instrument. The sample and the electrical coils for stimulating the nuclei and detecting the NMR signal are housed in a *probe* that fits within the central core of the magnet, as shown schematically in Fig. 1.6. Notice that the probe coil is placed so that its axis is perpendicular to that of the much larger superconducting magnet. As we will see later, this geometry is critical to the operation of the instrument. The probe coil is part of the electronic circuitry of the spectrometer and it must be tuned to the appropriate frequency. As a consequence, the probes are highly specialized devices, designed for particular nuclei (e.g., ^{1}H or ^{13}C) and field strengths. Experiments that utilize multiple nuclei types often require probes with multiple coils, each designed

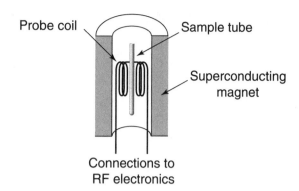

Figure 1.6
Diagram of an NMR probe.

for the appropriate frequency. In addition, modern probes often include coils for modifying the external field felt by the molecules in the sample tube, leading to quite complex designs. The probe also typically contains mechanisms for regulating the temperature of the sample and for spinning the sample tube about its axis, which is sometimes useful for averaging out inhomogeneities. Fig. 1.7 is a photograph of a typical probe, in this case for a 500-MHz spectrometer.

The probe is inserted into the core of the cryogenic magnet from the bottom, and the sample is inserted into the core of the probe from the top of the entire assembly. A relatively recent advance is the development of probes with cryogenically cooled coils and signal amplifiers. The low temperature greatly reduces the background electronic noise in the probe and provides significantly improved sensitivity.

The coils in the probe are connected to a bank of electronic circuitry that generates the radio-frequency signals that stimulate the nuclear transitions and, in turn, detects and amplifies the signals generated by the transitions. Because of the low inherent sensitivity of NMR, the electronics must be capable of both applying strong stimulating signals and detecting very weak signals. In addition, the frequency of the signals must be maintained with very high precision and, as we will discuss later, the circuitry must apply precisely timed pulses of radiation.

The overall operation of the instrument is controlled by a computer, usually a UNIX workstation, which also stores the data. The data can be processed and analyzed on this computer or, more commonly, they are transferred via a network to another computer for offline analysis.

Figure 1.7
Photograph of an NMR probe.
(Courtesy of Peter F. Flynn)

1.3.2   Continuous Wave and Pulse Techniques

In UV–visible spectroscopy, a spectrum is usually recorded by systematically changing the wavelength passed by the monochromator and measuring the absorbance of light at each wavelength. During the first two decades or so of NMR spectroscopy, a similar approach was used to measure the absorption of energies by nuclei with different resonance frequencies. Although this could, in principle, be done by adjusting the frequency of the stimulating electrical circuit, it is more practical to keep this frequency constant and then make small adjustments to the strength of the external magnetic field. Nuclei in different chemical environments then satisfy the resonance condition at different external field strengths, giving rise to a spectrum that is equivalent to one that would be generated by varying the signal frequency while maintaing a constant magnetic field strength. This classic mode of collecting data is referred to as *continuous wave* spectroscopy. In the continuous wave mode, the external field strength must be adjusted very slowly in order to take advantage of the very high resolution intrinsic to NMR. In addition, it is often necessary to collect data for an extended time in order to attain reasonable sensitivity.

In the mid 1960s, a radically different approach for collecting NMR data was developed. In the "pulse" method, the sample is initially irradiated with a strong burst of energy with a frequency band broad enough to stimulate all of the nuclei of a given type. Immediately afterwards, the stimulated nuclei continue to resonate and generate a composite signal with all of their frequencies superimposed. This is somewhat analogous to striking all of the keys on a piano at once and recording the not very musical superposition of notes. Mathematical methods are then used to determine the strengths of the signals with different frequencies, resulting in a spectrum that is equivalent to one generated by the continuous wave method. Spectra can be recorded much more rapidly in this way, which can be translated into greatly increased sensitivity because the signals from multiple recordings can be summed. Furthermore, the introduction of the pulse method allowed the development of more sophisticated experiments involving multiple pulses and multidimensional spectroscopy, discussed below.

The development of the pulse method as a practical method for NMR depended on advances in electronics and computer technology, as well as applied mathematics. In the early twenty-first century, it is just now becoming feasible to apply analogous techniques to infrared spectroscopy, with its much higher frequencies and shorter time scales.

1.3.3   Multidimensional Spectroscopy

As illustrated in the spectra shown in the previous section, the resonances from ^{1}H nuclei of a large molecule are almost, but not quite, resolvable in a simple spectrum. In the early 1970s methods were devised to introduce additional dimensions in an NMR spectrum. In two-dimensional spectra, each of the axes represents a resonance frequency, and peaks on the two-dimensional surface identify *pairs* of nuclei that interact in some way. The interaction may be based either on covalent connectivity between the nuclei or their proximity in three-dimensional space. The additional dimension greatly enhances the ability to resolve individual resonances and, of equal or greater importance, provides information about covalent linkages between nuclei or the distances that separate them. Three- and even four-dimensional NMR experiments have also become widely used. Multidimensional spectra are discussed further in Section 1.4.

1.3.4   Isotopic Labeling

Of the elements that are found in most biological macromolecules (i.e., hydrogen, carbon, nitrogen, oxygen, phosphorous and sulfur), the only ones for which the most abundant isotopes are suitable for solution NMR spectroscopy are hydrogen (^{1}H) and phosphorous (^{31}P). There are, however, relatively rare but stable (non-radioactive) isotopes of carbon and nitrogen, ^{13}C and ^{15}N, that give rise to excellent signals. In natural samples, ^{13}C and ^{15}N represent approximately 1% and 0.4% of the total carbon and nitrogen, respectively. For relatively simple compounds that can be prepared in large quantities, useful spectra can be recorded from the small fraction of molecules that contain these isotopes, and this approach is widely used in identifying and characterizing natural and synthetic organic compounds. For larger molecules, however, the signals from the natural abundance ^{13}C and ^{15}N are usually too weak to be useful.

The rare isotopes of carbon and nitrogen can be enriched artificially and the resulting material used as nutrients to grow microorganisms such as bacteria and yeast, resulting in biological molecules similarly enriched with the corresponding isotopes. Genetic engineering methods developed over the last few decades have made it possible to produce nearly any protein, and many nucleic acids, in microorganisms. As a consequence, samples of many biological macromolecules enriched with ^{13}C or ^{15}N can now be prepared, allowing the use of powerful new NMR techniques. This represents a very important technical advance that significantly increases the range of molecules that can be studied by NMR. Although the signals from ^{13}C and ^{15}N are intrinsically weaker than those from ^{1}H, they provide a means of introducing selectivity into complex spectra, for instance to look only at those ^{1}H nuclei that are covalently attached to ^{15}N nuclei. In addition, some multidimensional techniques combine the signals from different nuclei types, as illustrated in the next section.

NMR experiments also commonly utilize a rare isotope of hydrogen, ^{2}H, or deuterium. This isotope gives rise to an NMR signal, but with a frequency very different from that of ^{1}H (76.7 MHz in a "500-MHz" spectrometer). In almost all solution NMR experiments, a small amount of deuterated water, or chloroform for non-aqueous solutions, is added to the sample and is used as an internal frequency reference to "lock" the spectrometer frequency. In some situations, it is also useful to replace a fraction of the natural ^{1}H nuclei in a sample with ^{2}H in order to make other signals more easily detectable. Another application of ^{2}H is based on the observation that the hydrogen atoms of some functional groups, especially amides, undergo spontaneous exchange with the hydrogens of water. This process can be monitored by

NMR if either the molecule of interest or the solvent water is initially labeled with ^{2}H. The rate at which this exchange occurs is sensitive to the extent to which the amide group is accessible to the solvent and can provide useful information about the structures of macromolecules and their transient fluctuations.

1.4   Some Examples of Multidimensional Spectra

Of the various technical advances discussed above, multidimensional spectroscopy is perhaps the one that has made the biggest impact on macromolecular NMR. In this section, a few examples of two-dimensional spectra are shown to introduce the general concept and the power of these methods.

1.4.1   The HSQC (Heteronuclear Single-Quantum Coherence) Spectrum

An example of a particularly useful kind of multidimensional spectrum is shown in Fig. 1.8, again recorded from a sample of BPTI.

This spectrum, called an HSQC spectrum (for heteronuclear single-quantum coherence), uses both ^{1}H and ^{15}N and selectively detects only pairs of covalently attached nuclei. The protein used for this spectrum was uniformly labeled with ^{15}N as described above. To create a two-dimensional representation, peaks in the spectrum are represented as contours, much like those used in topographical maps. Each peak (or spot in the contour plot) represents an ^{1}H–^{15}N pair. The position on the horizontal axis indicates the ^{1}H resonance frequency and the position on the vertical axis indicates the ^{15}N frequency. Because each amino acid residue (except the amino-terminal and prolyl residues) contains a backbone amide group, this spec-

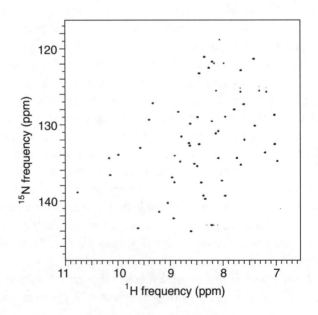

Figure 1.8
HSQC spectrum of BPTI.

trum represents a "fingerprint" in which nearly every residue gives rise to a single peak. (Residues with N–H groups in their side chains usually give rise to additional peaks.)

The spectrum illustrates the two major advantages of multidimensional spectra. First, introducing a second dimension greatly enhances the separation of signals from individual nuclei. Nearly every ^{1}H–^{15}N pair is well resolved, and even more peaks could be resolved if necessary for a larger protein. Second, the spectrum provides direct information about how some of the nuclei in the molecule are covalently linked. We know, for instance, that the ^{1}H nucleus with a resonance frequency of about 10.8 ppm is covalently attached to the ^{15}N nucleus with a resonance frequency of about 138 ppm. This is the sort of information that is necessary to begin determining which resonance is associated with a specific atom in the molecule.

In this spectrum, the spectral dimensions are correlated by the direct covalent connectivity between nuclei. There are other multidimensional spectra based on other types of covalent connectivities, such as between two ^{1}H nuclei separated by multiple bonds. Spectra of this type are especially useful for identifying different residue types in proteins, as illustrated by the TOCSY spectrum shown in Fig. 1.9. In still other spectra, cross peaks appear when pairs of nuclei are close together in three-dimensional space, irrespective of covalent attachment.

As we will discuss in later chapters, the way in which the resonances are correlated in a particular spectrum depends on the details of the series of rf pulses that was applied to the sample, which distinguishes the many types of multidimensional experiments.

1.4.2   The TOCSY (TOtal Correlation SpectroscopY) Spectrum

The example shown in Fig. 1.9 is a portion of a spectrum for BPTI. In this spectrum, both dimensions represent ^{1}H resonance frequencies. Cross peaks appear when two nuclei are linked together by two or three covalent bonds, or when the nuclei are connected by a series of linkages, with no more than three bonds separating any pair of ^{1}H nuclei. In the region shown, the horizontal dimension corresponds to resonances from backbone amide groups, and the vertical dimension corresponds to protons on α-carbon and side-chain atoms. Each vertical line of spots corresponds to the protons that lie within the same amino acid residue as the amide proton with the resonance frequency indicated on the horizontal axis. Different amino acid residue types give rise to distinct patterns in this type of spectrum. The TOCSY and related spectra are very important for assigning resonances to atoms.

1.4.3   The NOESY (Nuclear Overhauser Effect SpectroscopY) Spectrum

This example, shown in Fig. 1.10, is from the same region of a spectrum of BPTI. Like the TOCSY spectrum, both dimensions represent ^{1}H resonance frequencies, but cross peaks appear in this spectrum when two nuclei are within about 5 Å of one another,

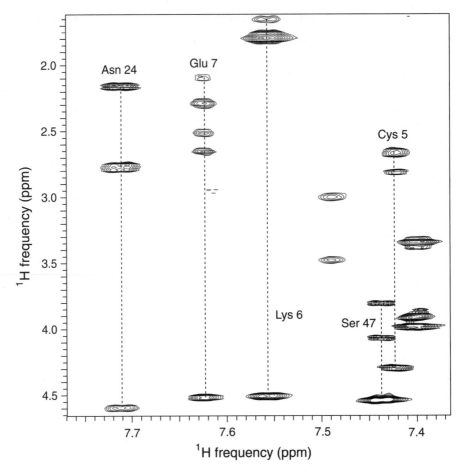

Figure 1.9 Portion of a TOCSY spectrum of BPTI, illustrating cross-peaks between amide and side-chain ^{1}H-nuclei. The labels indicate the amino-acid residue types and sequence numbers.

irrespective of their covalent linkage. The physical phenomenon that gives rise to the cross peaks is called the nuclear Overhauser effect (NOE) and is discussed in detail in Chapter 9. Since most or all of the protons of a particular amino acid residue will lie within relatively close proximity of one another, the NOESY spectrum shows nearly all of the peaks found in the TOCSY. In addition, however, there are cross peaks that arise from protons in different residues.

NOESY spectra have historically been very important for assigning resonances in proteins, because they can be used to identify sequentially adjacent residues. But, newer methods using heteronuclear spectra are making the NOESY less important for assignments. The most important application of NOESY is in deducing spatial arrangements of atoms, forming the basis of structure determination.

There are many other variations on the theme of two-dimensional spectra. In addition, these kinds of spectra can be combined to generate three- and even four-dimensional spectra. A three-dimensional spectrum can be thought of as a stack

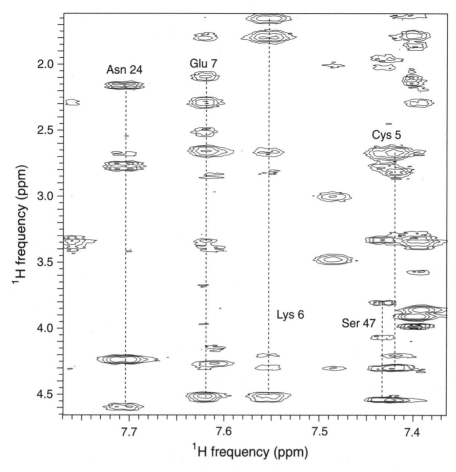

Figure 1.10 Portion of a NOESY spectrum of BPTI, illustrating spatial proximities between amide and side-chain ^{1}H-nuclei. The labels indicate the amino-acid residue types and sequence numbers associated with the amide protons.

of two-dimensional spectra, each associated with a third frequency value, as diagrammed in Fig 1.11.

In this example, each of the planes might be a two-dimensional NOESY spectrum, and the third dimension represents an ^{15}N frequency. A peak appears in the plane only if one or both of the protons is covalently attached to an ^{15}N nucleus with that resonance frequency. Thus, the peaks in the two-dimensional NOESY are spread out among the individual planes. If you were to look down the stack of ^{1}H–^{1}H planes from the top, the superimposed peaks would look very similar to the two-dimensional NOESY. The view looking straight on to the front face of the spectrum, as oriented in the figure, corresponds to the ^{1}H–^{15}N HSQC spectrum.

The addition of a third dimension further enhances the ability to resolve individual signals and provides additional information about relationships among individual resonances. In the hypothetical example shown here, we know that one or the other of the protons is covalently attached to a nitrogen with the indicated resonance

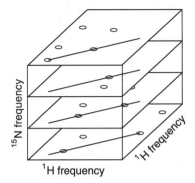

Figure 1.11

Diagrammatic representation of a three-dimensional NMR spectrum.

frequency. The concept can be extended even further to generate four-dimensional spectra. A four-dimensional spectrum is composed of a series of three-dimensional spectra, each of which is associated with yet another resonance frequency. For instance, the fourth dimension might represent ^{13}C resonance frequencies, so that each three-dimensional spectrum contains only peaks in which one of the nuclei is covalently attached to a ^{13}C nucleus with a given frequency. Multidimensional spectra have become extremely important as NMR has been applied to larger and larger molecules. The major limitation arises from the extra time required to collect spectra with more dimensions. Generally, each additional dimension increases the time by a factor of 10 to 100.

1.5   Structural, Dynamic and Chemical Information from NMR

In the past 40 years it has become possible to determine the three-dimensional structures of proteins and nucleic acids from NMR data alone. Although other NMR data are almost always used, the primary source of information is the nuclear Overhauser effect mentioned above. Determining a structure requires identifying a large number of NOESY cross peaks, each representing a pair of atoms that lie within about 5 Å of each other. Computer methods are then used to calculate structures that are compatible with these constraints. If there are enough constraints, all of the calculated structures will be very similar to one another and will represent an accurate model of the protein's structure in solution. As a rough rule of thumb, about 10 distance restraints per amino acid residue are required to calculate a high-quality structure. An example of an ensemble of calculated structures, for a fragment of a eukaryotic transcription factor (the ETS domain of murine ETV6), is shown in Fig. 1.12.

A major advantage of this approach to structure determination, relative to x-ray crystallography, is that it does not require crystals of the molecule. Growing crystals of some macromolecules can be difficult and unpredictable, sometimes making crystallography impossible. Also, NMR of liquid samples precludes the possibility that formation of the crystal might perturb the average structure of the molecule or its

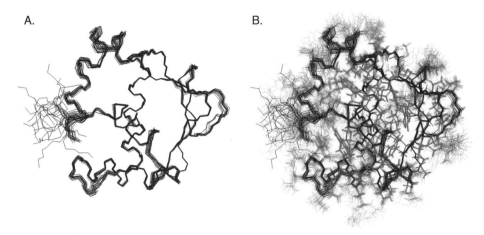

A.   B.

Figure 1.12   An ensemble of 20 calculated structures of the ETS domain of the murine ETV6 transcription factor, derived from NMR constraints. In panel A, only the bonds between non-hydrogen backbone atoms are represented as line segments. In panel B, bonds between backbone atoms are shown in black, and those between side-chain atoms are shown in gray. (The diagrams were drawn using the program PyMOL, from the coordinates of Protein Data Bank entry 2LF7. The structure is described in Coyne III, H. J., De, S., Okon, M., Green, S. M., Bhachech, N., Graves, B. J. & McIntosh, L. P. (2012). *J. Mol. Biol.*, 421, 67–84.)

dynamics. On the other hand, NMR is currently limited to relatively small macro-molecules, such as proteins of a few hundred amino acid residues or less.

It is important to emphasize that the two methods of structure determination are fundamentally different. NMR provides information about specific atoms and pairs of atoms that are combined to calculate a global structure. This can be an advantage, since it is often possible to deduce features of some regions of the molecule before the entire structure is determined. On the other hand, because signals from some atoms may be missing or hard to resolve, the calculated structures may be more heavily biased by information from some regions of the molecule than from others. In x-ray crystallography, every portion of the diffraction pattern contains information from every part of the molecule. As a consequence, crystal structures are usually, but not always, an all or nothing proposition. The final structure, however, is relatively free of biases due to specific features or regions of the molecule. These differences make it very hard to say whether one method is "better" than the other. Often, complementary information can be obtained by the use of both methods.

In addition to NOE distance constraints, NMR can provide other useful structural information, especially estimates of dihedral angles of rotatable bonds, which are typically incorporated into structure calculations. Also, it is sometimes possible to deduce the presence of hydrogen bonds. These types of information are often combined with the NOE constraints in structure calculations such as the one shown in Fig. 1.12.

As mentioned earlier, the non-equilibrium states that are induced by the absorption of radiation in an NMR experiment can be very long lived. The rate at which the system returns to equilibrium is generally sensitive to the rates and nature of molecular motions. While this feature of NMR is the cause of some of its limitations, it can also be a rich source of information about molecular motions.

Although much emphasis is often placed on determining structures by NMR, especially those of biological macromolecules, the technique can also be used to gain a variety of chemical information, often with great specificity. It is the ability of NMR to selectively monitor a single atom in a huge molecule that really gives it its power. Even if the system under study isn't amenable to overall structure determination, it is often possible to learn something about specific portions of the molecule. Examples include measuring pK_a values of specific functional groups and measuring the kinetics of hydrogen exchange between amide groups and water. Hydrogen exchange kinetics is especially useful in studies of protein stability and folding.

Summary Points for Chapter 1

- NMR spectroscopy is based on the absorption of radiation by certain nuclei that take on discrete energy levels when placed in an external magnetic field.
- The energy differences, and frequencies, associated with nuclear magnetism are determined by the properties of the nuclei, expressed as the gyromagnetic ratio, and the strength of the external field, according to Eq. 1.1 (p. 2). These energy differences are much smaller than those associated with most other spectroscopic techniques.
- Resonance frequencies are very sensitive to the local environments of nuclei and are influenced by both covalent chemical structure and three-dimensional conformation.
- Modern NMR spectroscopy relies on advances in several important technologies, including superconducting magnets, electronics, computing and isotopic labeling.
- Multidimensional NMR techniques have made it possible to resolve the signals from molecules containing thousands of atoms and to assign these signals to individual nuclei.
- NMR can be used to determine the three-dimensional structures of complex molecules and to characterize their dynamics on multiple time scales.

Exercises for Chapter 1

1.1. One day while browsing eBay, I came across a used "200-MHz" NMR spectrometer. Assuming that the specification referred to the ^{1}H resonance frequency, calculate the field strength of the magnet, in tesla units. Also, calculate the resonance frequencies for ^{15}N, ^{13}C and ^{2}H nuclei.

1.2. Like the ^{1}H, ^{15}N and ^{13}C nuclei widely used in NMR, the electron is a spin-1/2 particle. In most atomic and molecular structures, all of the electrons are paired, hiding their magnetic properties. But, free-radicals contain an unpaired electron, and electron spin resonance (ESR; also called electron paramagnetic resonance or EPR) spectroscopy is widely used to study such species. The gyromagnetic ratio of an isolated electron is approximately $1.761 \times 10^{11}\,\mathrm{T^{-1}s^{-1}}$, much greater than that of any nucleus. The magnetic field strengths typically used for ESR are about 0.35 T.

(a) Calculate the resonance frequency of an isolated electron in a 0.35-T field. What region of the electromagnetic spectrum does this correspond to?

(b) Calculate the corresponding wavelength for an isolated electron in a 0.35-T field.

(c) Calculate the energy difference associated with the quantum transition for an electron in a 0.35-T field.

1.3. When operational, the NMR magnet shown in Fig. 1.5 produced a magnetic field of 6.3 T from a solenoidal coil made up of approximately 19 km of niobium–titanium wire. The magnetic field strength (B) in the core of an ideal solenoid depends on the number of turns of wire (N), the length of the solenoid (l) and the current (I), according to the relationship

$$B = \mu_0 \frac{N \cdot I}{l}$$

where μ_0 is a constant, called the permeability of free space, with the value $4\pi \times 10^{-7}\,\mathrm{T \cdot m/A}$. Assume for the following that the length of the solenoid is 0.2 m and the average radius of the coil turns (not the wire itself) is 0.1 m.

(a) How many turns of wire are there in the coil?

(b) How much current must pass through the coil to generate a field of 6.3 T?

(c) Because the coil is made of superconducting wire, no voltage is required to maintain the current. Suppose, however, that you wanted to build an electromagnet with the same field strength, but using ordinary 1 mm diameter copper wire, which has a resistance of about 0.02 ohms per meter of length. What voltage would be required to maintain the current you calculated above?

(d) How much power would be consumed by the 6.3-T electromagnet with a copper coil? What are the practical implications of this power requirement?

1.4. NMR spectroscopy is now routinely applied to protein molecules of moderate size. This exercise is meant to give you a sense of the number of nuclei of different types that might be involved. You may find it useful to use a database, such as the Protein Data Bank (www.pdb.org) or UniProt (www.uniprot.org), and a tool such as the online ProtParam calculator (expasy.org/protparam/).

(a) Identify a known protein composed of 100–200 amino acids.

(b) Calculate the number of $^1\mathrm{H}$, $^{15}\mathrm{N}$ and $^{13}\mathrm{C}$ nuclei in this protein molecule, assuming that it has been fully labeled with $^{15}\mathrm{N}$ and $^{13}\mathrm{C}$.

Further Reading: Other Books on NMR spectroscopy

No single text can cover all of modern NMR spectroscopy or even present the basics in a way that is suitable for all readers. The following is a list of just a few of the many books that cover this topic from different perspectives. The titles are listed roughly in order of their likely accessibility to beginning students.

◇ Derome, A. E. (1987). *Modern NMR Techniques for Chemistry Research.* Tetrahedron Organic Chemistry Series. Pergamon Press, Oxford.

A very good, if older, introduction to the basics of pulse NMR, using vector diagrams almost exclusively. The author is very careful to point out where this approach breaks down.

◇ Claridge, T. (1999). *High-Resolution NMR Techniques in Organic Chemistry*. Tetrahedron Organic Chemistry Series. Pergamon Press, Amsterdam.

This book follows in the tradition of Derome's, but includes more on multidimensional techniques and some more modern developments such as composite and shaped pulses.

◇ Homans, S. (1992). *A Dictionary of Concepts in NMR*, revised ed. Biophysical Techniques Series. Oxford University Press, Oxford.

A very useful book with short, clear discussions of the major concepts of modern NMR. It is a good place to learn the vocabulary and to go back to for explanations of important ideas.

◇ *eMagRes*. http://dx.doi.org/10.1002/9780470034590

This reference work was first published in 1996 as a multivolume print edition called the *Encyclopedia of Magnetic Resonance*. Subsequently, it was converted to an online resource that continues to be updated. It includes excellent articles on both specialized topics and more general ones, covering NMR spectroscopy as well as magnetic resonance imaging (MRI).

◇ Fukushima, E. & Roeder, S. B. W. (1981). *Experimental Pulse NMR: A Nuts and Bolts Approach.* Addison-Wesley, Reading, MA.

Though written before the more recent advances in NMR, this is a great book for those who really like to know how things work. It includes excellent explanations of the basic mathematics and the electronics used in pulse NMR, as well as very useful discussions of relaxation.

◇ Keeler, J. (2010). *Understanding NMR Spectroscopy*, 2nd ed. Wiley, Chichester, U.K.

This book covers much of the same material as in the one you are reading, but with a somewhat different perspective and more of an emphasis on practical aspects.

◇ Levitt, M. H. (2008). *Spin Dynamics: Basics of Nuclear Magnetic Resonance*, 2nd ed. Wiley, Chichester, U.K.

An excellent book that develops the theory of NMR in a clear and methodical way, as well as covering many of the practical aspects of the technique.

◇ Freeman, R. (1998). *Spin Choreography: Basic Steps in High Resolution NMR.* Oxford University Press, Oxford.

Ray Freeman and his colleagues have been responsible for the development of many important techniques in solution NMR, especially those involving heteronuclei. In this book, rather than trying to cover all aspects of NMR, he presents his distinct perspective on some of the ways to make the spins dance on command.

⬦ Cavanagh, J., Fairbrother, W. J., Palmer III, A. G., Rance, M. & Skelton, N. J. (2007). *Protein NMR Spectroscopy*, 2nd ed. Elsevier Academic Press, Amsterdam.

Probably the best current text covering the major NMR experiments now being applied to biological macromolecules. This book relies heavily on the product-operator formalism and includes a thorough theoretical chapter in which this approach is developed. The treatment is likely to be rather challenging for most beginners.

⬦ Ernst, R. R., Bodenhausen, G. & Wokaun, A. (1987). *Principles of Nuclear Magnetic Resonance in One and Two Dimensions*. Oxford University Press, Oxford.

Widely considered the definitive text, but aimed at an audience well trained in mathematical physics. Ernst and his colleagues played major roles in the development of pulse methods, multidimensional spectroscopy and magnetic resonance imaging.

CHAPTER 2

An Introduction to Spin and Nuclear Magnetism

Spectroscopy, viewed generally as the interaction of radiation with matter, provides some of the clearest evidence for quantum mechanical behavior. Indeed, much of the early development of quantum theory was motivated by the desire to understand the discrete pattern of frequencies of light emitted from atoms when they are heated. For some forms of spectroscopy, such as UV–visible absorbance, a relatively simple view of quantum mechanics is adequate; it is usually sufficient to think of molecules taking on discrete states that can interconvert through the absorption or emission of energy, without too much concern for how the transitions occur. In NMR, however, these processes take place over relatively long time scales, and much of the power of the technique comes from the ability to monitor, and even manipulate, the time-dependent changes in the quantum states. To exploit this ability requires a more subtle description of quantum mechanics than what is usually implied by simple energy diagrams.

In this chapter, we will consider one of the landmark experiments that stimulated the development of quantum theory, particularly the idea of spin, which gives rise to the magnetic properties of nuclei such as ^{1}H. Some of the general features of quantum mechanics and its application to NMR will also be described in a very qualitative way, so as to provide a visual picture of the behavior of magnetic nuclei when placed in an external field, as in an NMR spectrometer. In the second half of the book, we will return to quantum mechanics, with much more mathematics, in order to treat the interactions between nuclei upon which most multidimensional experiments are based.

2.1 The Stern–Gerlach Experiment

One of the most dramatic demonstrations of quantum mechanics is a conceptually simple experiment first described in 1922 by Otto Stern and Walther Gerlach. The experiment was designed to measure the magnetic properties of free atoms and was

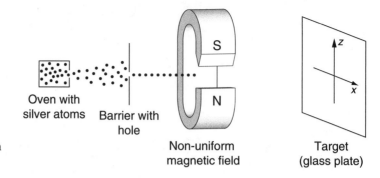

Figure 2.1
Diagram of the Stern–Gerlach
experiment.

set up as shown in Fig. 2.1. In this experiment, silver metal was evaporated by heating in an oven, and the free atoms were allowed to pass through multiple small openings, so as to generate a narrow beam of atoms. The atoms were then aimed between the poles of a special magnet designed to create a non-uniform magnetic field. The magnetic field is illustrated in Fig. 2.2, looking edge-on as the silver atoms would approach it. The field lines are closer together near the pointed **S** pole than near the flat **N** pole, indicating that the field is stronger near the **S** pole. If the silver atoms have a net magnetic dipole, they will be deflected toward one pole or the other, depending on the orientation of the dipole, as illustrated in Fig. 2.3. Atoms with their **S** pole pointed straight up will be repelled by the stronger interaction with the **S** pole of the magnet and deflected downward. Those with the **N** pole pointed straight up will be deflected upward. In each case, the interaction with the **S** pole of the external field is stronger than the interaction with the **N** pole. If the atomic magnetic dipole is perpendicular to the field, then there will be no deflection.

After passing through the magnetic field, the atoms strike a target, which was just a glass plate in the original experiment. The silver atoms accumulated on the glass, leaving a record of their direction when they left the magnet. In the figures, the target is drawn with two axes on it. Following the convention to be used throughout this book, the vertical axis is labeled z and the horizontal axis is labeled x. There is also a y-axis that, in these experiments, is parallel to the direction of the initial beam.

Figure 2.2 Non-uniform field generated by the special magnet in the Stern–Gerlach apparatus.

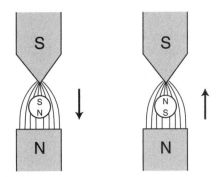

Figure 2.3

Deflection of magnetic dipoles in the non-uniform magnetic field of the Stern–Gerlach apparatus.

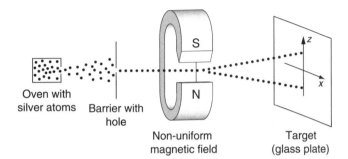

Figure 2.4

Result of the Stern–Gerlach experiment.

If the atoms were "classical" magnets, their orientations would be random as they entered the magnetic field, and they would be spread out along the z-axis of the target. What Stern and Gerlach observed, however, was a concentration of the atoms on two spots on the target, as shown in Fig. 2.4.

The magnet is acting as a filter, separating the atoms into two classes with different magnetic orientations. After passing through the filter, the particles are said to be *polarized*, a state that is similar to that of polarized light, as discussed in Chapter 11. But why are there only two classes of particles? There is no obvious reason that the atoms leaving the oven would be oriented in any particular direction. It seems as though the magnet is imposing an orientation on the atoms, as well as separating them.

This early experiment was tremendously influential in the development of quantum mechanics. While quantization of energy had been proposed much earlier, on the basis of somewhat subtle arguments, the observation of spatial quantization in a direct experiment left little doubt that the universe was quantized, if on a very small scale.

2.2 Spin

When Stern and Gerlach first devised their experiment, they viewed it as a test of the early quantum theory first proposed by Niels Bohr and modified by Arnold Sommerfeld, in which the motions of electrons around a nucleus were assumed to have quantized orientations or "orbits," and the results of the experiment were widely interpreted as a validation of the theory. Further refinement of the theory, however, indicated that there should be an odd number of quantized orbits rather than the two observed. At the same time, there were other observations, involving the fine structure in the optical spectra of samples placed in magnetic fields, that suggested that there must be some additional source of atomic magnetization.

In about 1925, two graduate students, Samuel Goudsmit and George Uhlenbeck, proposed a simple explanation for some of the previously unexplained magnetization effects. They suggested that if an electron were spinning about its axis, the motion of charge on the surface of the electron would give rise to a magnetic dipole along the axis. Unknown to them, another young physicist, Ralph Kronig, had suggested this idea previously, but had rejected it because of a difficulty: In order to account for the strength of the observed magnetization, the surface of the electron would have to move faster than the speed of light! Uhlenbeck and Goudsmit did not realize this initially, and they quickly wrote a paper. By the time they were aware of this problem, the paper had been submitted, and *spin* quickly became entrenched in the language of quantum mechanics. Though they did not point this out initially, the Goudsmit and Uhlenbeck spin theory provided the correct explanation for the quantization observed in the Stern–Gerlach experiment. As it turned out, the quantization observed in that experiment was due to the spin of an unpaired electron in each silver atom.

Particles, such as an electron, that can take on two discrete magnetization values are said to have a spin quantum number, S, of 1/2. Other examples of spin-1/2 particles include the nuclei that are most commonly used in NMR experiments—namely, ^{1}H, ^{13}C and ^{15}N. The half-integer value for the spin quantum number was chosen so that the energies arising from spin quantum levels of the electron would be consistent with the energies associated with orbital angular momentum. More generally, particles can have integer or half-integer values of S, and the number of discrete magnetization values observed is $2S + 1$. For instance, the ^{2}H nucleus is an $S = 1$ particle and would give rise to three beams in a Stern–Gerlach experiment.

2.3   The Gyromagnetic Ratio and Angular Momentum

The relationship between the spinning of a charged object and its magnetic dipole is quantified by the gyromagnetic ratio introduced in the previous chapter. A classical object spinning about an axis is described by a property called *angular momentum*, **I**. This is a vector quantity, indicated by the bold type, with the vector pointing along the axis of rotation and with units of kg·m²/s, or equivalently J·s. The length, or magnitude of this vector reflects the speed of rotation, the mass of the object and the distribution of this mass about the axis.

If the spinning object has a net charge, then a magnetic dipole will be generated that is also aligned with the axis of rotation. The strength and direction of the magnet are given by another vector, **μ**, called the *magnetic moment*. The two vectors are parallel, as shown in Fig. 2.5. Moreover, they are related by a constant of proportionality called the gyromagnetic ratio, γ:

$$\boldsymbol{\mu} = \gamma \mathbf{I} \tag{2.1}$$

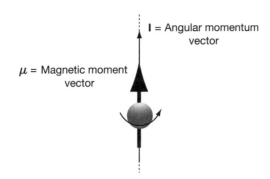

Figure 2.5

Angular momentum and magnetic moment vector for a spinning, charged sphere.

As noted in the previous chapter, the gyromagnetic ratio is a property of a given type of particle and has the units of $T^{-1}s^{-1}$. The units of the magnetic moment are then J/T. The two vectors are drawn as pointing in the same direction in the figure above, as is the case if γ is positive. However, some nuclei, including ^{15}N, have negative gyromagnetic ratios, meaning that the two vectors point in opposite directions.

   Like any vector in a three-dimensional coordinate system, the classical angular momentum, **I**, can be considered the sum, or resultant, of three vectors, each pointed along one of the axes. The lengths and directions of these component vectors determine the net direction and length of the resultant vector. Mathematically, this can be written as

$$\mathbf{I} = I_x\mathbf{x} + I_y\mathbf{y} + I_z\mathbf{z}$$

where **x**, **y** and **z** are each vectors one unit long and pointing in the direction of the corresponding axis. I_x, I_y and I_z are real numbers that represent the contribution of each of the unit vectors to the resultant. In other words, these numbers can be thought of as representing the extent to which the vector points in the three directions of the coordinate system (Fig. 2.6).

   Because the magnetic moment and angular momentum vectors point in the same direction (for spins with positive gyromagnetic ratios), the terms I_x, I_y and I_z also indicate the relative extent to which the magnetic dipole points in the three directions. For this reason, it is common to describe the direction of the magnetic dipole in terms of the angular momentum components. We will often speak, for instance, of I_z as the magnetization component in the z-direction. This, in fact, is what is measured by the deflection of the beams in the Stern–Gerlach experiment with the axes defined as above. However, the meaning of the angular momentum components is subtly different when applied to quantum mechanical spin, as discussed below.

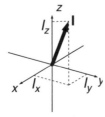

Figure 2.6

Projections of an angular momentum vector, **I**, onto the coordinate axes.

2.4   Quantized Magnetization

NMR spectroscopy is based upon the ability of electromagnetic radiation to alter the quantum states of nuclei with spin. Although a full description of these phenomena requires the mathematical formalism of quantum mechanics, the essential features of many aspects of NMR can be illustrated using a relatively simple physical picture based on the analogy to a spinning object, as originally suggested by Goudsmit and Uhlenbeck. Before adopting this approach, however, it is instructive to consider a bit further the differences between classical and quantum physics and the meaning of the angular momentum vectors.

As an example of a classical object on the macroscopic scale, the behavior of a baseball can be described by its position and instantaneous motion. If the baseball is spinning about an axis, we can also describe that motion. Using Newtonian mechanics, the future position, orientation and motion of the baseball can be reliably predicted. Importantly, the parameters that we use to describe the state of the object can all be measured with as much precision as we like, at least in principle.

In quantum mechanics the situation is quite different, in that there are intrinsic limits to the precision of individual experimental measurements, and we can often only predict the average outcome of measurements applied to many particles. In the case of NMR, the properties that we are interested in are the orientations of the magnetic dipoles associated with the "spinning" nuclei. The important lesson from the Stern–Gerlach experiment is that if we try to measure the orientation of the dipole for an individual spin-1/2 nucleus along a particular direction, we will obtain one of two possible results.

From this, it may be tempting to conclude that there are only two states that are available to the particles, as is often suggested in elementary treatments. Indeed, a common misconception about quantum mechanics is the belief that systems subject to its rules are only allowed to exist in discrete, or quantized, states. The more correct statement is that the results of experimental observations are quantized, in this case the magnetization along the z-axis. In fact, the magnetic dipole of a nucleus can point in *any* direction, even in the presence of the uniform external magnetic field of an NMR spectrometer. Importantly, however, at any instant the orientation can only be measured along a single (arbitrary) direction, as in the Stern–Gerlach experiment, and these measurements will yield quantized results.

In quantum mechanics, the state of a particular particle is defined by a mathematical object known as a *wavefunction*, usually represented by the Greek letter Ψ.[1] The wavefunction cannot, in general, be determined by experimental means, but if it is known (or assumed), the *average* values of experimental observations, such as the magnetization along a particular direction, can be predicted using a set of mathematical rules.

---

[1] The term "wavefunction" is actually rather misleading. As we will see, the wavefunctions that describe the spin states of particles are not really functions in the commonly used sense, where, for instance, $f(x)$ is a formula or rule for calculating a value, say y, for each value of x.

Figure 2.7
Resonance structures of
benzene.

For a spin-1/2 nucleus with its magnetic dipole aligned with the positive z-axis, the wavefunction is, by convention, designated α, while a spin aligned in the opposite direction has the wavefunction β. In a commonly used system of notation, devised by Paul Dirac, the symbols for wavefunctions are placed between a vertical bar and an angle bracket, such as $|\alpha\rangle$ and $|\beta\rangle$. A symbol of this type is referred to as a "ket" (half of a bracket). The usefulness of this notation is not likely to be apparent until the later chapters of this book, but it is introduced here for consistency and to help distinguish symbols for wavefunctions from other mathematical symbols.

As noted above, the magnetic dipole of a spin-1/2 nucleus can be pointed in any direction, and the wavefunction for a spin with an arbitrary orientation can be written as a linear combination of $|\alpha\rangle$ and $|\beta\rangle$:

$$|\Psi\rangle = c_\alpha |\alpha\rangle + c_\beta |\beta\rangle$$

where c_α and c_β represent the relative contributions of $|\alpha\rangle$ and $|\beta\rangle$ to $|\Psi\rangle$. The coefficients c_α and c_β are generally complex numbers; that is, they can have real and imaginary parts, but we will ignore this important point for the moment.

A state that contains non-zero components of both $|\alpha\rangle$ and $|\beta\rangle$ is said to be a *superposition* of $|\alpha\rangle$ and $|\beta\rangle$. The superposition of $|\alpha\rangle$ and $|\beta\rangle$ to generate other wavefunctions is closely related to the concept of resonance structures that is so important in organic chemistry. The classic example is benzene, which is often drawn as shown in Fig. 2.7. Importantly, this representation does not imply that there are two forms of the molecule that are interconverting, but rather that the best description of a single molecule is a superposition of the states represented by the two drawings. Similarly, a superposition of $|\alpha\rangle$ and $|\beta\rangle$ is distinct from both of these wavefunctions, but can be described mathematically as a linear combination of them.

The mathematical formalism of quantum mechanics provides a set of rules from which the *average* value of an experimental measurement can be calculated from the wavefunction. As an example, a spin-1/2 nucleus can be in a state with the wavefunction

$$|\Psi\rangle = \frac{1}{\sqrt{2}}|\alpha\rangle + \frac{1}{\sqrt{2}}|\beta\rangle$$

In this case, $|\alpha\rangle$ and $|\beta\rangle$ make equal contributions.[2] The mathematical rules predict that if a pure population of nuclei with this wavefunction is passed through the Stern–Gerlach filter shown in Fig. 2.1, the two possible outcomes, 1/2 and $-1/2$,

---

[2] The coefficients in this example have been chosen to satisfy the condition that $c_\alpha^2 + c_\beta^2 = 1$, which is a requirement for a properly formed wavefunction. For the moment we are restricting ourselves to wavefunctions for which the coefficients, c_α and c_β, are real numbers. The requirement has a slightly different form when complex coefficients are considered.

Figure 2.8
A Stern–Gerlach filter arranged to measure the x-magnetization component, applied to spins with the wavefunction $|\Psi\rangle = \frac{1}{\sqrt{2}}|\alpha\rangle + \frac{1}{\sqrt{2}}|\beta\rangle$.

$$\frac{1}{\sqrt{2}}|\alpha\rangle + \frac{1}{\sqrt{2}}|\beta\rangle$$

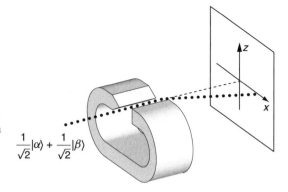

will be observed with equal probability. Thus, the average value of the measurement, which we will write as $\langle I_z \rangle$, is zero. It is impossible, however, to predict the outcome for any individual particle.

From this result, it might appear that the magnetic dipoles of the individual particles in the superposition state have no defined direction until they interact with the filter. If, however, the filter is rotated by 90°, as illustrated in Fig. 2.8, then all of the spins in this particular superposition state will be deflected in the same direction along the x-axis.

For this wavefunction, then, the magnetization is aligned with the positive x-axis. If, on the other hand, we were to apply this "x-filter" to a population of spins in the $|\alpha\rangle$ state, then we would find that half of the spins would be deflected in the positive x-direction and half in the negative x-direction. The same result would be obtained for a population of spins in the $|\beta\rangle$ state. For these spins, the average magnetization along the x-axis will be zero. For any wavefunction, there will be a particular orientation for which a measurement of the magnetization will give a single predictable result. It is in this sense that we can say that the magnetic dipole has a defined direction. In the following section we consider a graphical representation of the average magnetization of a population of spins.

2.5   Vector Representations of Average Magnetization

By analogy with the angular momentum vector of a classical spinning object, we can use a vector representation to describe the average magnetization state of a spin population. For instance, the magnetization arising from a population of spins with the $|\alpha\rangle$ wavefunction is represented by the diagram in Fig. 2.9A: By convention, the stationary magnetic field of an NMR spectrometer is defined to be along the z-axis, which is also referred to as the *longitudinal axis*. The x- and y-axes are said to define the *transverse plane*.

In addition to showing that the $|\alpha\rangle$ wavefunction always gives rise to a positive z-magnetization, the drawing also conveys the information that the *average* magnetization along the x- or y-axis is zero. For a population of particles with the $|\beta\rangle$ wave-

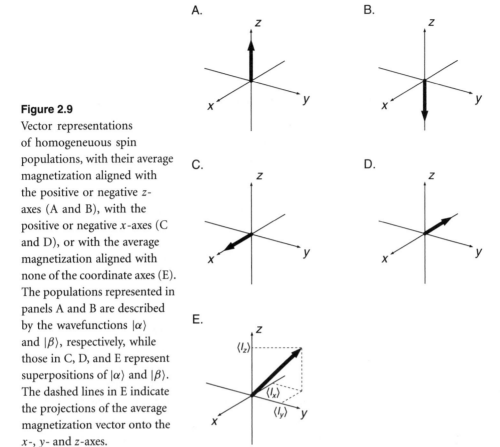

Figure 2.9
Vector representations
of homogeneuous spin
populations, with their average
magnetization aligned with
the positive or negative z-
axes (A and B), with the
positive or negative x-axes (C
and D), or with the average
magnetization aligned with
none of the coordinate axes (E).
The populations represented in
panels A and B are described
by the wavefunctions $|\alpha\rangle$
and $|\beta\rangle$, respectively, while
those in C, D, and E represent
superpositions of $|\alpha\rangle$ and $|\beta\rangle$.
The dashed lines in E indicate
the projections of the average
magnetization vector onto the
x-, y- and z-axes.

function, the vector representation is shown in Fig. 2.9B. In this case, there is a net
magnetization in only the negative z-direction. Similarly, the drawings in Fig. 2.9C
and D represent the average magnetization for populations of particles with wavefunctions that give rise to net magnetization along the positive or negative x-axes.
In these cases, there is no net magnetization along the z- or y-axes, meaning that
individual measurements of I_z or I_y will give values of 1/2 and $-1/2$ with equal probability.

There is no requirement that the net average magnetization lie along one of the
axes of the coordinate system. For instance, there is a wavefunction for the vector representation in Fig. 2.9E. The dashed lines represent the projection of the vector onto
the x-, y-, and z-axes. The interpretation of this vector diagram requires some care.
If we were to take any individual particle from a population with this wavefunction
and measure the z-magnetization using a Stern–Gerlach device, we would, as usual,
obtain only the values 1/2 and $-1/2$. In this case, however, the positive outcome is
somewhat more likely than the negative outcome. This relationship is represented in
the diagram by the projection on the z-axis, which lies between 0 and the full length of
the vector. The projection represents the average value of the measurement, $\langle I_z \rangle$, for

this population. This situation arises when the wavefunction includes contributions from both $|\alpha\rangle$ and $|\beta\rangle$, but that from $|\alpha\rangle$ is larger. Individual measurements of I_x or I_y would give values of 1/2 or $-1/2$ for individual particles, and what the diagram tells us is that a value of 1/2 is somewhat more likely than $-1/2$ for the measurement of I_y, while a value of $-1/2$ is more likely if we were to measure I_x. While each particle has the potential to behave differently in any one measurement, we can think of the macroscopic sample as if it were made up of a population of little magnets all aligned along the direction of the vector. This simplified representation provides a powerful tool for describing many important aspects of NMR spectroscopy.

2.6   Average Magnetization of Bulk Samples

So far, we have considered an imaginary situation in which we have somehow managed to get all of the spins in a sample into the same quantum state simultaneously. This ideal can be approached in the case of the Stern–Gerlach type experiments, but not for liquid or solid samples (except at extremely low temperatures). However, if the population of spin states is such that there is an excess with a net magnetization in a particular direction, then we can still use the vector diagram, with the understanding that the arrow represents the direction of the net magnetization averaged over the entire population.

In the absence of an external magnetic field, the individual spin-1/2 particles in a sample will be oriented randomly. When the nuclei are placed in the field of an NMR spectrometer, however, their individual energies will depend on their orientation and the external field strength. The energy of the nucleus is, itself, a quantized value which, for $S = 1/2$ and a positive gyromagnetic ratio, takes on the values of

$$E_\alpha = -\frac{h\gamma B}{4\pi} \quad \text{or} \quad E_\beta = \frac{h\gamma B}{4\pi}$$

If the spin is in the $|\alpha\rangle$ state, a measurement of its energy will always yield E_α, while the energy of the $|\beta\rangle$ state is always E_β. For superposition states, the energy of an individual spin will be either E_α or E_β, with the relative probability being proportional to the relative contributions of $|\alpha\rangle$ and $|\beta\rangle$ to the superposition state, which is also proportional to the average magnetization along the z-axis. Through a quite complex process we will discuss later, the spin states in a bulk sample establish an equilibrium determined by the Boltzmann distribution and the sample temperature.

At any instant, an individual nucleus might exist in either the $|\alpha\rangle$ or $|\beta\rangle$ state, or any of the innumerable superposition states, as suggested by the drawing shown in Fig. 2.10A. For the two states $|\alpha\rangle$ and $|\beta\rangle$, the relative populations will be

$$\frac{N_\alpha}{N_\beta} = e^{(E_\beta - E_\alpha)/kT} \tag{2.2}$$

where N_α and N_β are the number of nuclei in the two states, k is the Boltzmann constant and T is the temperature. For the various superposition states, the relative

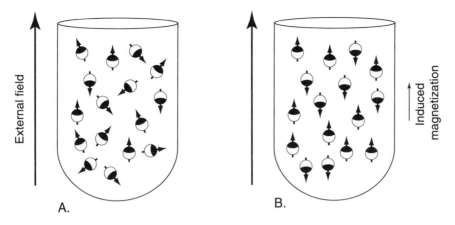

External field

Induced magnetization

A.

B.

Figure 2.10   Diagrammatic representation of the orientation of individual spins at equilibrium in the presence of an external magnetic field. The magnetization moment of an individual spin can point in any direction, as suggested in A, but the net magnetization of the sample can be described by assuming that the individual spins are aligned with either the positive or negative z-axis, as in B, with relative probabilities determined by a Boltzmann distribution.

concentrations will depend on the relative contributions of $|\alpha\rangle$ and $|\beta\rangle$ to the wave-functions for those nuclei.

Note that the definitions of the energies of the $|\alpha\rangle$ and $|\beta\rangle$ states reflect the sign of the gyromagnetic ratio. For nuclei with positive gyromagnetic ratios (including ^{1}H and ^{13}C), the $|\alpha\rangle$ state has the lower energy and is more highly populated at thermal equilibrium. For nuclei with negative γ, however, the $|\beta\rangle$ state has the lower energy.

It might appear that it would be hopelessly difficult to calculate the relative populations of all of the possible states and follow their behavior in an NMR experiment. It turns out, however, that the average properties observed in NMR experiments can be accurately described by treating the equilibrium state *as if* it were composed entirely of spins in either the $|\alpha\rangle$ or $|\beta\rangle$ state, with the relative numbers determined by the Boltzmann distribution, as illustrated in Fig. 2.10B. In fact, the behavior of the sample can be described by considering only the very small excess of spins associated with the lower-energy state.

For nuclei with positive γ, the slight excess of nuclei in the $|\alpha\rangle$ state leads to a net magnetization in the z-direction, which we can represent as a vector along the positive z-axis, just as we did for a pure population of spins in the $|\alpha\rangle$ state. At thermal equilibrium, the magnetization components along the x- and y-axes average to zero. However, the direction of magnetization of the excess $|\alpha\rangle$ population can be perturbed by irradiating the sample—that is by exposing it to fluctuating magnetic fields of the proper frequency, and the behavior of the perturbed system gives rise to the NMR signal. The vector diagrams allow us to visualize these effects, as described in the following sections.

2.7   Time-dependent Changes in Magnetization

The properties of a quantum mechanical system can change with time, and these changes are reflected in changes in the wavefunction. The change over a time interval can be predicted by a set of mathematical rules, and the new wavefunction can be used to predict the new average values for any appropriate experimental measurement. We will postpone until Chapter 12 a detailed discussion of the mathematics, and present here a more qualitative description of the time-dependent changes of the magnetization of a population of a spin-1/2 nuclei. These effects can be divided into two general cases—namely, the change in the average nuclear magnetization in the presence of a stationary external magnetic field and those that are promoted by electromagnetic radiation of an appropriate frequency (i.e., by an oscillating magnetic field). Both types of effects are central to NMR spectroscopy.

In the following discussion, we will ignore the tendency of the distribution of spin states to move toward the thermal equilibrium described in the previous section, with the net magnetization aligned with the positive z-axis. This simplification is not intended to minimize the importance of relaxation processes, which we will return to repeatedly. Rather, the goal is to consider idealized cases initially to illustrate some general rules, and then address relaxation in subsequent chapters.

2.7.1   Motion in a Stationary Magnetic Field

One of the special properties of the $|\alpha\rangle$ and $|\beta\rangle$ wavefunctions is that they remain unchanged in the presence of a uniform and stationary magnetic field in the z-direction. For this reason, $|\alpha\rangle$ and $|\beta\rangle$ are often referred to as *stationary states* or, for reasons discussed in Chapter 12, *eigenstates*. As a consequence, the net z-magnetization of a population composed of particles with either of these wavefunctions will remain constant with time (in the absence of any perturbation), as shown in Fig. 2.11.

In contrast, wavefunctions that represent superpositions of $|\alpha\rangle$ and $|\beta\rangle$ *do* change with time in a uniform field. In Chapter 12, the time-dependent change for a superposition state will be derived from quantum mechanical principles. However, the same relationship can be derived by analogy to an example from classical mechanics. Though it was noted earlier that the notion of a charged particle spinning about its axis to generate a magnetic moment is deeply problematic, the analogy of quantum mechanical spin to that of macroscopic objects does have the important virtue of providing a simple connection between the quantum mechanical effects and the frequencies observed in magnetic resonance experiments.

Figure 2.11
Time-dependent change of z-magnetization in a static magnetic field.

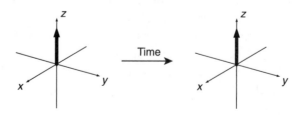

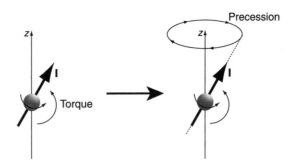

Figure 2.12
Precession of a spinning object due to the interaction between angular momentum and an applied torque.

In the realm of classical mechanics, if an already spinning object is subjected to a force that acts to change the orientation of the object (i.e., a torque), this force will cause either the magnitude or direction of the angular momentum vector to change. If the axis about which the torque acts is the same as that of the already existing rotational motion, the angular velocity, and therefore the angular momentum, will either increase or decrease. If, however, the torque is oriented so as to rotate the object around a different axis, then the axis of rotation will begin to move in a circular path, a motion called *precession*. In the situation depicted in Fig. 2.12, the initial axis of rotation is at an angle of about 30° from the vertical axis, and the torque is applied about an axis perpendicular to the page. The resulting precessional motion is about the z-axis.

In an NMR spectrometer, the stationary magnetic field, represented by the vector **B**, generates a torque on the "spinning" nucleus through its interaction with the magnetic dipole. Just as in the mechanical example, this torque acts to rotate the angular momentum vector about an axis perpendicular to the z-axis and results in a precessional motion, as shown in Fig. 2.13. The rate or frequency of this precessional motion depends on the torque exerted on the magnetic moment by the external field, which depends in turn on the magnitude of the external field and the gyromagnetic ratio. The rate of precession is given by

$$\omega = -\gamma B \tag{2.3}$$

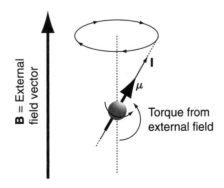

Figure 2.13
Precession of a quantum spin in a static magnetic field.

where B is the magnitude of the external field vector, or the field strength measured in the z-direction. (Actually, B is more correctly defined as the *magnetic flux density*. For most purposes, this distinction is not important, but it leads to some confusion when units are defined.) The value ω is an angular velocity (or angular frequency), with units of radians/second (rad/s, where 2π radians $= 360°$). Note that ω is a signed quantity, and the negative sign in Eq. 2.3 indicates that the direction of precession and the direction of spin are opposite for a nucleus with a positive gyromagnetic ratio.

The frequency of the motion in Hz—that is, the number of times the cycle is completed per second, is given by the Larmor equation:

$$\nu = -\frac{\gamma B}{2\pi} \tag{2.4}$$

The frequency ν is referred to as the *Larmor frequency*. In many contexts, frequencies are treated as unsigned quantities, and the negative sign is often dropped in the Larmor equation. In NMR, however, the direction of precession can be important, and we will follow the more recent practice of treating ν as a signed quantity. We will return to sign conventions and the direction of precession when we discuss pulse NMR in Chapter 5. For now, just note that the direction of precession for a spin with a positive gyromagnetic ratio, such as the ^{1}H and ^{13}C nuclei, is clockwise when looking down the positive z-axis. The direction is reversed for a nucleus with a negative gyromagnetic ratio, such as ^{15}N. By convention, the Greek letter ω is used to indicate angular velocities, with units such as rad/s, and ν is used for frequencies that imply a complete cycle per unit time, with the unit Hz for one cycle per second.

2.7.2 The Gyroscope Demonstration

No introduction to NMR would be complete without a demonstration of angular momentum and precession with a toy gyroscope. This device consists of a disk suspended by freely rotating bearings in a cage, so that the orientation of the rotating disk (or rotor) can be changed without interfering with the internal rotation. Angular momentum is imparted on the rotor by quickly pulling on a string wrapped around the shaft. The string can then be used to suspend the gyroscope from one of its ends, as shown in panel A of Fig. 2.14.

Rather remarkably, the gyroscope remains balanced while supported at only one end. Naively, we might have expected that the force of gravity would exert a torque that would cause the whole device to rotate downward until it fell off of the string. This is just what would happen if the rotor were not spinning. With the rotor spinning, however, the gyroscope moves sideways and precesses about the point held by the string. In fact, the gyroscope does begin to fall, but when this motion is combined with that of the rotating disk, the forces are balanced in such a way that the gyroscope finds an equilibrium position, pointed slightly downward, with a resultant sideways force that causes precession. On the other hand, if the gyroscope is suspended by both ends (as in panel B of Fig. 2.14), so that there is no torque exerted by gravity, then there is no precession.

Figure 2.14
A demonstration of precession by a toy gyroscope. In A, the gyroscope is suspended by one end, leading to a torque and precession. In B, there is no torque and the direction of the gyroscope axis remains stationary. (Photographs by Jeremy Goldenberg.)

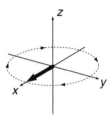

Figure 2.15   Precession of bulk magnetization of a population of spins initially aligned with the x-axis.

2.7.3   Precession of Spins in a Bulk Sample

If we were to have a large sample of spins in the field that all begin with the same wavefunction, so that initially $\langle I_x \rangle = 1/2$, the wavefunctions would all change cyclically in synchrony. As a result, the individual magnetic moments would all precess together and the average magnetization would follow the path illustrated in Fig. 2.15. Even though the outcome of measuring the direction of magnetization for any individual particle would generally not be predictable, the net magnetization observed in any direction would oscillate at the Larmor frequency. It would be as if there were a single macroscopic magnetic revolving about the z-axis. This motion of the net magnetization can be detected by placing a coil of wire around the sample and then connecting the wire to an oscilloscope, which displays the voltage generated in the wire as a function of time. The voltage will fluctuate as the field from the net magne-

Figure 2.16
Electrical signal induced by a magnetic moment rotating within a wire coil.

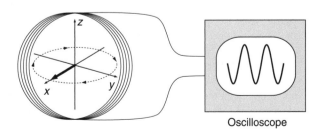

Oscilloscope

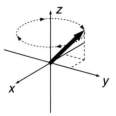

Figure 2.17

Precession of bulk magnetization partially aligned with the
positive z-axis.

tization repeatedly cuts across the wire (Fig. 2.16). The signal, $s(t)$, generated in this
way has the same mathematical form as the projection of a rotating vector onto one
of the coordinate axes and can be written as

$$s(t) = A \sin(\phi + 2\pi \nu t)$$

where ν is the frequency, ϕ is an initial angular offset and A is the amplitude of the
sine function.

The situation we have just considered is, in fact, the one that is most important
in an NMR experiment. The basic trick in nearly all NMR experiments is to generate
the type of non-stationary state described above and then monitor the fluctuation in
the average magnetization, which reflects the average fluctuation of the wavefunction
with time. The resulting magnetic fluctuation is detected by the induction of an
electric current in a surrounding coil.

Importantly, the average z-magnetization of a spin population does *not* change
with time, unless the spins are somehow perturbed by fluctuating magnetic fields.
This is true whether the magnetization initially lies entirely along the z-axis, entirely
in the $x–y$ plane or has some intermediate orientation. For instance, the vector diagram in Fig. 2.17 describes the motion of the average magnetization of spins that
begin with projections along all three axes.

In the quantum mechanical description, the constant z-magnetization reflects the
fact that the relative contributions of $|\alpha\rangle$ and $|\beta\rangle$ to the superposition wavefunction
remains constant with time (in the absence of a fluctuating magnetic field).[3] A very
important result from the full mathematical treatment of this situation is that the
rate of circular motion is the same whether the net magnetization lies entirely in
the $x–y$ plane or is partially aligned with the z-axis. Thus, the frequency detected by
the spectrometer coils is the same in either case. However, the strength of the signal
does depend on the length of the projection onto the $x–y$ plane, with the strongest
signal arising when the average magnetization vector lies entirely in the plane. This
simple geometrical relationship proves to be extremely important in the design of
many NMR experiments, including all of the multidimensional ones.

---

[3] Although the net contributions of $|\alpha\rangle$ and $|\beta\rangle$ remain constant with time, the relative magni-
tudes of the real and imaginary parts of the complex coefficients, c_α and c_β do change. In mathematical terminology, the moduli of the coefficients remain constant while the phases change cyclically,
as reflected in the cyclical motion of the average magnetization. These effects are discussed in detail
in Chapter 12.

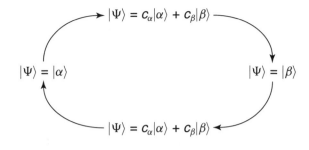

Figure 2.18
Rabi oscillation: cyclic interconversion among quantum states of a spin-1/2 nucleus in the presence of an oscillating magnetic field.

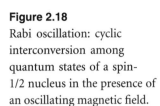

2.7.4   Changes in Magnetization in an Oscillating Magnetic Field

Transitions between quantum states with different energies can be stimulated by electromagnetic radiation if the frequency of that radiation, ν, satisfies the following condition:

$$\nu = \frac{\Delta E}{h}$$

Although quantum transitions are often described as if they were instantaneous, the change in wavefunction is actually a continuous process. If a nucleus is initially in the $|\alpha\rangle$ state, radiation of the proper frequency will convert the wavefunction to a super-position of $|\alpha\rangle$ and $|\beta\rangle$, with the contribution of $|\beta\rangle$ increasing continuously until $|\alpha\rangle$ no longer makes a contribution. At that point, the contribution of $|\alpha\rangle$ increases again until the $|\beta\rangle$ contribution disappears. This process will continue cyclically for as long as the nucleus is irradiated, as diagrammed in Fig. 2.18.

As before, the coefficients c_α and c_β represent the relative contributions of $|\alpha\rangle$ and $|\beta\rangle$ to the wavefunction, which change through the cycle. It is important to emphasize that the intermediate points in the cycle represent superposition states, not a mixture of spins with pure $|\alpha\rangle$ or $|\beta\rangle$ wavefunctions. This cycle is referred to as *Rabi oscillation*, for the physicist Isidore I. Rabi, whose pioneering contributions to magnetic resonance are discussed in the next chapter.

If we begin with a pure population of spins in the $|\alpha\rangle$ state, the shift to a super-position state will be associated with a change in the average magnetization, such that the vector is rotated toward the transverse plane, as shown in Fig.2.19. As soon as there is a transverse component to the magnetization, however, the stationary mag-netic field acts on the magnetization so that it begins to precess about the z-axis at the

Figure 2.19
Rotation of bulk z-magnetization due to an oscillating magnetic field with frequency $\nu = \Delta E / h$.

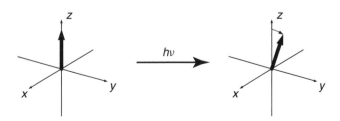

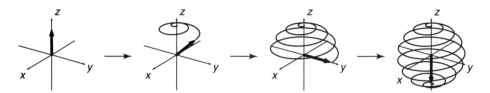

Figure 2.20 Nutation of bulk magnetization due to the combined influence of stationary and oscillating magnetic fields.

same time that it is moving away from that axis. The combination of the two types of motion causes the vector to trace out an elegant spiral on the surface of a sphere (Fig. 2.20). Once the vector points along the negative z-axis, its spiral path continues, bringing it back to the transverse plane and, eventually to the positive z-axis. This motion of the magnetization vector during a period of irradiation, called *nutation*, is discussed further in Chapter 5, using a rotating reference frame to simplify the analysis.

2.8   The "Laws of Motion" for Bulk Magnetization

The discussion of the previous sections can be summarized in a few simple rules that describe the behavior of the net magnetization of a spin population under different conditions. It should be emphasized, however, that these rules do not account for the effects of relaxation, which will eventually return the system to thermal equilibrium, with the net magnetization aligned with the z-axis. With this important caveat, the laws of motion are as follows:

1. In the presence of a uniform and stationary magnetic field aligned with the z-axis:
   (a) If the bulk magnetization is fully aligned with the positive or negative z-axis, it will remain in this orientation as long as the magnetic field is stationary (i.e., in the absence of electromagnetic radiation).
   (b) If the bulk magnetization has a transverse component (i.e., a projection onto the x–y plane), then the transverse component will precess at the Larmor frequency, $\nu = -\gamma B/(2\pi)$.
   (c) The precession frequency does not depend on the projection of the magnetization onto the z-axis.
   (d) The projection along the z-axis will remain constant with time.
2. In the presence of a stationary magnetic field *and* electromagnetic radiation with frequency ν that matches the energy difference between the $|\alpha\rangle$ and $|\beta\rangle$ states:
   (a) The projection of the magnetization vector along the z-axis will change cyclically.

(b) When it is not aligned exactly with the positive or negative z-axis, the magnetization vector will also precess about the z-axis, giving rise to a nutational motion.

The vector picture of bulk magnetization and the rules summarized above are remarkably powerful in describing many important aspects of NMR, and we will rely on them for the first half of this book, rarely needing to think about the underlying quantum mechanical phenomena. There are important limitations to this approach, however, particularly when two or more spins interact so intimately that they must be considered a single quantum mechanical system with a single wavefunction. Systems of this type have fascinating and important properties that are not at all easy to visualize on the basis of our everyday physical experience and can really only be understood using quantum mechanics. A much more mathematical introduction to spin quantum mechanics is presented in Chapters 11–13, as preparation for the discussion of NMR methods that are based on quantum coupling between nuclei.

Summary Points for Chapter 2

- Nuclei with the property of spin display small magnetic dipoles that can be detected by their interactions with external magnetic fields, as demonstrated in the Stern–Gerlach experiment.
- Nuclei and other particles with spin don't really spin! But, much of their behavior can be described by analogy to a charged spinning object.
- Spin is a quantum mechanical property and follows the special rules of the quantum world. Measurements of individual particles, such as the orientation of the magnetic dipole, always yield discrete results, with probabilities determined by the wavefunction that describes the particle's state, represented by the symbol Ψ, or, using Dirac notation, $|\Psi\rangle$.
- The individual magnetic dipoles in bulk samples point in all possible directions, but an external magnetic field favors orientations aligned with the field, inducing a bulk magnetization.
- For many purposes, we can treat the equilibrium distribution of states as if it were composed only of the states fully aligned with or against the field, the $|\alpha\rangle$ and $|\beta\rangle$ states, with relative populations determined by the Boltzmann distribution.
- If the magnetic dipole of a particle is not parallel to the external field, it will undergo a cyclical motion, precession, analogous to that of a spinning top under the influence of a torque.
- The direction of the net magnetization present at equilibrium can be manipulated by the absorption of electromagnetic radiation of the proper energy. The subsequent precession of the individual particles leads to motion of the bulk magnetization, which can be detected by the induction of a current in the coils of an NMR spectrometer.

Exercises for Chapter 2

2.1. As noted in the text, Goudsmit and Uhlenbeck initially imagined that the magnetic moment of an electron was due to the spinning of its charge. The following is a version of the calculation that argues against this simple interpretation. For a spinning sphere, the magnitude of the angular momentum vector, **I**, is given by

$$I = \omega \frac{2}{5} m r^2$$

where ω is the angular velocity (in units of rad/s), m is the mass of the sphere and r is the radius.

(a) For an electron, the mass is 9.11×10^{-31} kg, the magnetic moment is -9.27×10^{-24} J·T^{-1} and the gyromagnetic ratio is 1.76×10^{11} T^{-1}s^{-1}. The radius of an electron is not so easy to define, but its "classical" radius can be defined as approximately 2.82×10^{-15} m. Using these numerical values, the equation given above, and Eq. 2.1 (p. 26), calculate the angular velocity at which an electron would have to rotate in order to account for its magnetic moment. Calculate the maximum velocity of a point on the surface of the electron. Compare this velocity with the speed of light.

(b) The classical electron radius is actually a large overestimate of its radius. Would a smaller radius require a smaller or greater surface velocity to account for the magnetic moment?

(c) For a proton, the mass is 1.67×10^{-27} kg, the magnetic moment is 1.41×10^{-26} J·T^{-1}, the gyromagnetic ratio is 2.68×10^8 T^{-1}s^{-1} and an estimate for the radius is 8.88×10^{-16} m. Calculate the speed at which the surface of a proton would have to move in order to account for the proton's magnetic moment.

2.2. Each of the five vector diagrams in Fig. 2.21 represents a pure population of spin-1/2 particles with a single wavefunction.

In answering the questions below, be sure to briefly explain your reasoning.

(a) Assuming that the vectors represent populations of gaseous molecules or atoms, suppose that you were to take each of the spin populations and subject them to a Stern–Gerlach experiment, with the magnet oriented along the z-direction. For each case illustrated in the figure, describe the expected outcome of the experiment.

(b) Suppose that you now rotate the Stern–Gerlach filter so that its field is aligned with the x-axis, as illustrated in Fig. 2.8. Again, predict the result of passing each of the populations through this filter.

(c) If you were to place each of the populations in a static external magnetic field aligned along the z axis, for which samples would the direction of magnetization change with time (assuming that there is no relaxation)?

(d) Suppose, now, that a static external field is applied along the y-axis. For which of the populations would you expect the direction of magnetization to change with time?

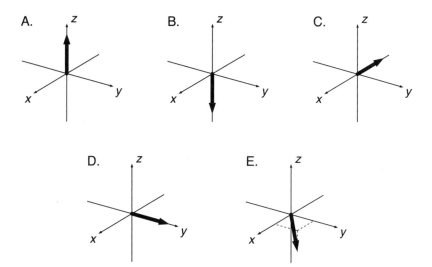

Figure 2.21 Vector diagrams representing the average magnetization due to a pure population of spin-1/2 particles with a single wavefunction, for Exercise 2.2.

Further Reading

An illuminating and entertaining account of the Stern–Gerlach experiment and its historical context:

◇ Friedrich, B. & Herschbach, D. (2003). Stern and Gerlach: How a bad cigar helped reorient atomic physics. *Phys. Today*, 56, 53–59. http://dx.doi.org/10.1063/1.1650229

A first-hand account of the conceptualization of spin:

◇ Goudsmit, S. A. (1998). The discovery of the electron spin. In *Foundations of Modern EPR* (Eaton, G., Eaton, S. & Salikhov, K., eds.). World Scientific, Singapore. http://www.lorentz .leidenuniv.nl/history/spin/goudsmit.html

A much more complete discussion of the precession of a gyroscope:

◇ Feynman, R. P., Leighton, R. B. & Sands, M. (2013). *The Feynman Lectures on Physics*, Vol. I, pp. 20-5–20-7. Basic Books, New York. http://www.feynmanlectures.caltech.edu/ I_20.html#Ch20-S3

The New Millennial Edition of the Feynman Lectures has been completely re-typeset, with extensive corrections, and is available to read (but not download or reproduce) for free on the internet.

CHAPTER 3

Early NMR Experiments and Some Practical Considerations

In the previous chapter, we considered the behavior of spin-1/2 particles in a uniform magnetic field and in the presence of electromagnetic radiation. Now, we will consider the physical manifestation of these behaviors in the context of three historic experiments that defined the NMR phenomenon. These experiments serve to illustrate points that remain important in all modern NMR experiments, including issues of sensitivity and resolution.

3.1 The Rabi Molecular Beam Experiment

The first observation of nuclear magnetic resonance involved a molecular beam experiment, as opposed to the liquid- or solid-phase experiments we usually think of. These experiments were carried out by I. I. Rabi, who earlier had worked with Stern. During the 1930s, Rabi and his co-workers became the masters of spin manipulation in beams of atoms and molecules, and the resonance experiment was their most sophisticated. The actual experimental setup was somewhat more intricate than indicated here, but the following description captures the essential elements.

First imagine an apparatus with three magnets, as shown in Fig. 3.1. The first magnet generates an inhomogeneous field along the z-axis, as in the original Stern–Gerlach experiment; the second creates a uniform field in the z-direction and the third is exactly like the first. With just these components, not much new happens. As before, the first filter splits the atoms into the two states with wavefunctions $|\alpha\rangle$ and $|\beta\rangle$. As the upper beam passes through the uniform field, nothing changes, because the dipoles are now aligned with the z-axis, and the uniform field does not deflect them further. The third magnet just deflects the beam upward again.

The important addition to the experiment is a coil of wire within the gap of the central magnet, as shown in Fig. 3.2. When an electric current passes through the coil, it will create an additional magnetic field that is perpendicular to the z-axis. This coil is connected to a signal generator that produces a high frequency alternating

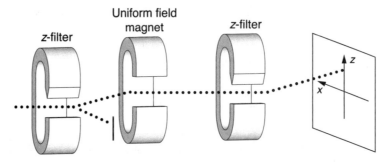

Figure 3.1 Three magnets of the Rabi molecular beam resonance experiment.

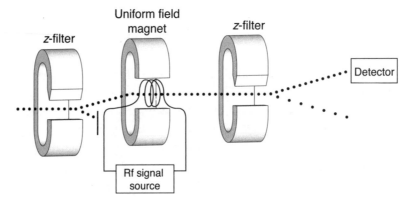

Figure 3.2 The Rabi resonance apparatus, with the rf coil placed in the uniform field of the central magnet.

current—that is, a radio frequency (rf) signal. In addition, Rabi replaced the simple glass-plate target with an electronic detector that could quantify the number of particles that were deflected upward by the second filter.

As the high-frequency signal passes through the coil, electromagnetic radiation is created. The energy of this radiation is given by the relationship

$$E = h\nu$$

where h is Planck's constant and ν is the frequency. If the energy of the radiation is equal to (or close to) the difference in the energies of the $|\alpha\rangle$ and $|\beta\rangle$ states, there will be an interaction between the radiation and the atoms. This occurs when

$$\nu = \frac{|E_\beta - E_\alpha|}{h} = |\gamma|\, B/2\pi$$

This is the same as the Larmor equation (Eq. 2.4, ignoring the sign of γ), which defines of the precession frequency of a spin in an external field. While it isn't obvious that the frequency of radiation required for absorption is the same as the precession

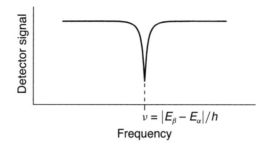

Figure 3.3

Result of the Rabi resonance experiment.

$$\nu = |E_\beta - E_\alpha|/h$$

frequency, it isn't a coincidence either. In both cases, the frequency is defined by the difference between the two allowable energies.

This interaction of radiation with the particles will cause a change in the wavefunctions. As discussed in Chapter 2, the nature of this change is fairly complex, because it is influenced by both the stationary field and the oscillating field. But, the net result is that the particles go from a pure $|\alpha\rangle$ state to a superposition state made up of a combination of $|\alpha\rangle$ and $|\beta\rangle$ components and is changing with time. The particles will then leave the second magnetic field in a state that is generally not pure $|\alpha\rangle$.

When the particles enter the second z-filter, they will, as usual, be deflected either upward or downward. Any particle that is not in the pure $|\alpha\rangle$ state will have a finite probability of being deflected downward. This will cause a decrease in the number of particles reaching the detector. If the frequency of the fluctuating field is varied, the intensity of the detected beam will drop sharply as the frequency approaches that corresponding to the energy difference and then rise again (Fig. 3.3).

Because both the field strength and the frequency can be measured with great accuracy, this method was very important in determining the gyromagnetic ratios of atomic nuclei. For his work on the magnetic properties of nuclei, Rabi won the Nobel Prize in Physics in 1944, one year after Otto Stern received the same award.

3.2   Magnetic Resonance in Solids and Liquids: The Purcell and Bloch Experiments

The atomic beam experiments described above depend on the ability to separate physically the particles in the two states. Once that is done, they can be manipulated and analyzed individually. In condensed phases, we still expect particles with spin to be influenced by a magnetic field, but there is no easy way to separate particles with different orientations. Shortly after World War II, two groups independently devised very similar methods to detect the magnetic resonance of hydrogen nuclei—in one case in solid paraffin and in the other in liquid water. One of the groups, at Harvard, was led by Edward Purcell and the other by Felix Bloch, at Stanford. Both Purcell and Bloch had spent much of the war period working on radar, and this experience was extremely important for their subsequent work on NMR. Even today, much of the electronic technology used in NMR instruments is derived from radio and radar technology. In addition, Rabi played a key role in leading the wartime radar effort, and

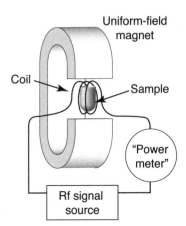

Uniform-field magnet

Coil

Sample

"Power meter"

Rf signal source

Figure 3.4
The Purcell NMR apparatus.

both Purcell and Bloch were greatly influenced by him during that period. In spite of the similarities of their interests, Purcell and Bloch did not know each other well and they approached their experiments with quite different perspectives, initially unaware of the other's efforts. In fact, it took considerable thought and discussion before it was clear that their experiments were based on the same phenomenon.

The Purcell and Bloch NMR experiments both involved placing a sample rich in hydrogen atoms in the field of a strong magnet and within a coil of wire, conceptually analogous to the central magnet of the Rabi experiment (Fig. 3.4). As in the Rabi experiment, the coil was connected to an rf signal generator. In Purcell's experiment there was also a very sensitive circuit designed to measure any changes in the electrical power absorbed by the coil.

Placing particles with spin 1/2, in this case the hydrogen nuclei, in a uniform magnetic field causes them to take on a distribution of states determined by the Boltzmann distribution, provided that there is some mechanism for the states to interconvert. As mentioned in Chapter 2, this distribution can be thought of as being a mixture of spins in either the $|\alpha\rangle$ or $|\beta\rangle$ states, with their relative populations, N_α and N_β, given by the Boltzmann distribution. With the magnets Purcell and Bloch used, the energy difference was about 2×10^{-26} J, as compared to the thermal energy at room temperature, $kT = 4 \times 10^{-21}$ J. Thus, the energy difference due to the interaction with the magnetic field was much smaller than the thermal energies of the molecules, and the ratio of N_α and N_β was only about 1.00001. Although very small, this difference proved sufficient to generate an NMR signal. (Conveniently, the magnets that were readily available then gave rise to ^{1}H Larmor frequencies of about 30 MHz. This enabled Bloch and Purcell to use the very sensitive and "clean" radio electronics that had recently been developed for communications purposes.)

As in the Rabi resonance experiment, the coil surrounding the sample was connected to a radio-frequency signal source. This generated photons which could interact with the nuclei, provided that their energy matched the energy difference between the two states. At this point, the approaches of the two groups diverged, and the techniques they used to detect the energy absorption reflected their different ways of looking at what happens in an NMR experiment. Because these two outlooks are both still important, it is useful to consider both forms of the experiment.

From Purcell's perspective at the time, the key thing that happens when the nuclei interact with electromagnetic radiation is that transitions between the $|\alpha\rangle$ and $|\beta\rangle$ states are stimulated. Interactions with photons of the proper energy ($E = h\nu$) can stimulate the transition in either direction and with equal probability. As a consequence, the radiation tends to equalize the populations. At the same time, however, the system tries to restore the Boltzmann distribution, with a transfer of energy to the surrounding molecules. The net effect is that energy is transferred from the electrical circuit to the sample. In fact, in the 1930s, another scientist, C. J. Gorter, attempted to detect magnetic resonance from the expected increase in the temperature of the sample, but the relatively weak magnetic field used led to effects that were too small to detect. In contrast, Purcell detected the energy transfer by a change in the amount of power absorbed by the coil when the frequency matched the energy difference between the two states. In the actual experiment, the frequency of the oscillator was kept constant, and the strength of the external field was varied, thereby changing the energy difference between the two states. When the energy difference matched the signal frequency, there was a sharp increase in the power absorbed.

In Bloch's approach, the focus was on the average magnetization after the nuclei had absorbed electromagnetic radiation. At the beginning of the experiment, there are a few more nuclei in the $|\alpha\rangle$ state than in the $|\beta\rangle$ state. Thus, if one were to somehow measure the average magnetization in the z-direction, there would be a small net magnetization in the "up" direction. If the nuclei absorb radiation, the population difference will decrease, as discussed above, and the net z-magnetization will decrease.

In the Bloch picture, however, the loss of z-magnetization takes place because the net magnetization vector is tilted toward a horizontal position and then precesses about the z-axis. In terms of the quantum mechanical picture, the absorption of radiation creates a superposition state with a wavefunction made up of both $|\alpha\rangle$ and $|\beta\rangle$ components. As described in Chapter 2, the wavefunction for this superposition state will fluctuate with time, and the net magnetization along the x- or y-axis will fluctuate at the Larmor frequency. Nuclei starting out in either the $|\alpha\rangle$ or $|\beta\rangle$ state will be converted to superposition states, but their average magnetization in the x- or y-direction will exactly cancel out, except for the small excess of nuclei that began in the $|\alpha\rangle$ state. The net result is a small magnetization precessing about the z-axis. The fluctuating x- and y-components of the magnetization were detected in the Bloch experiment by a second coil surrounding the sample, as shown in Fig. 3.5

The second coil, often referred to as the *receive coil*, was placed so that its axis was perpendicular to both the stationary magnetic field and the coil connected to the rf signal source, the *transmit coil*. Because the two coils were perpendicular to one another, there was no direct induction of the signal from the transmit coil to the receive coil. When the frequency of the rf signal source matched the Larmor frequency, the precession of the magnetization was detected by the signal induced in the receive coil. As in the Purcell experiment, the signal frequency was actually kept constant and the strength of the magnetic field was varied.

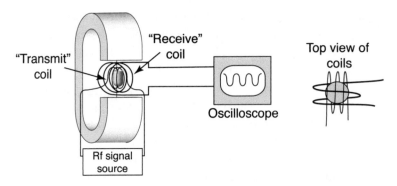

Figure 3.5   The Bloch NMR apparatus.

Bloch referred to the phenomenon he observed as "nuclear magnetic induction," partly in deference to Rabi's earlier demonstration of nuclear magnetic resonance in atomic and molecular beam experiments. In addition, this phrase emphasized that the magnetization of the sample was transferring the signal from the transmit coil to the receive coil. In his paper laying out the theory for the experiment, Bloch introduced what became known as the *Bloch equations*, a set of differential equations that describe the change in bulk magnetization with time. These equations describe the precessional motion of magnetization, in both fixed and fluctuating magnetic fields, and the two types of relaxation processes discussed in the sections below, and in much more detail in Chapters 7 and 12. Although Bloch did not explicitly use vector diagrams like those introduced in Chapter 2, the bulk magnetization was described by a vector quantity, and this paper can be viewed as the origin of such diagrams, which were used for the first time a few years later by Erwin Hahn working in Purcell's laboratory.

Purcell succeeded in detecting absorption by magnetic resonance in December 1945, and Bloch's induction experiment succeeded the next month. In subsequent years the two groups made important contributions to understanding such fundamental aspects of NMR as relaxation and the effects of chemical environment on the resonance frequency. They thus laid the groundwork for the extensive use of NMR spectroscopy in chemistry, which blossomed over the next decades. Purcell and Bloch shared the 1952 Nobel Prize in Physics.

3.3   The Importance of Relaxing

As discussed above, detecting a signal in either the Purcell or Bloch experiments depended on the presence of a small excess of spins in the $|\alpha\rangle$ state. If the populations of the two states had been equal, absorption of the electromagnetic radiation would have still promoted transitions between the two states. But, there would be no net change in the populations and no transfer of energy to the sample to be detected in the Purcell experiment. In the Bloch experiment, detecting a signal also depends on the small excess of spins that begin in the $|\alpha\rangle$ state, because it is this excess that leads to a net magnetization in the $x-y$ plane and thereby induces a current in the receive coil.

How do the nuclei initially reach the equilibrium distribution? This is actually an important question, and the answer is not so obvious. As we saw previously, the $|\alpha\rangle$ and $|\beta\rangle$ states are stationary in the constant field. In principle, a nucleus in the higher energy state, $|\beta\rangle$, could be converted to the $|\alpha\rangle$ state with the release of a photon of the proper energy. This is the process of fluorescence that can be observed with UV, visible or IR radiation, with their much higher energies. However, the probability of this occurring spontaneously is proportional to the cube of the energy difference between the two states, which in this case is extremely small. As a consequence, there is no significant conversion of $|\beta\rangle$ to $|\alpha\rangle$ by this mechanism. In this situation, transitions in either direction require a coupling with radiation of the proper frequency.

However, the molecules in the sample are constantly undergoing thermal motions, and, because the molecules contain magnetic dipoles, these motions generate fluctuations in the magnetic field that each nucleus feels. These fluctuations can cause an interconversion among the quantum states, just as the oscillating field in the Rabi experiment did. In this case, the fluctuating field comes from the interactions among the dipoles of all of the molecules (as well as their interaction with the static field), and energy is transferred among the quantum states of the various molecules. The total energy of the system must remain constant (unless the temperature changes), however, and eventually the distribution between the spin states must satisfy the Boltzmann equation. This process is called *spin–lattice relaxation,* because it results in a transfer of energy between the nuclei and the surrounding molecules, and it is most efficient when the frequencies of the molecular motions are close to the Larmor precession frequency. The rate at which thermal equilibrium is reached is specified by a first-order rate constant, R_1, and depends on many factors, including the frequencies of the thermal motions and the strengths of the moving dipoles.

When Purcell and Bloch began their experiments, they both realized that generating a population difference between the two states was essential, but had no idea of how long it would take. They therefore left the samples in the magnetic field for several hours before trying to make any measurements, though it later turned out that the relaxation took about a second. (It is rumored that Bloch went skiing while waiting for equilibration.) As we will see, relaxation processes play a central role in virtually all NMR experiments, often defining the practical limitations of the experiments and providing some of the most useful information about molecular structure and motions.

3.4 Sensitivity

As noted in Chapter 1, NMR is inherently a rather insensitive method. At the time Purcell and Bloch performed their original experiments, there was good reason to doubt whether the instrumentation then available would be sensitive enough to detect the effect, even though Purcell, for instance, used about 750 g of solid paraffin as his sample. The strengths of the signals generated in the two original experiments, and all subsequent NMR experiments, depend on the following major factors:

1. The magnitudes of the individual magnetic moments in the sample.
2. The net number of moments aligned with the static magnetic field at the beginning of the experiment.
3. The rate of precession of the magnetic moments.

All three of these factors depend on the gyromagnetic ratio, as discussed further below, making the overall sensitivity highly dependent on the type of nucleus used.

In both types of experiment, the signal intensity depends on the net equilibrium magnetization arising from the sample. This is perhaps easiest to see in the Bloch experiment, because the current induced in a wire by the motion of a magnetic field depends on the strength of that field. In this experiment, the field is generated by the aggregate effect of many small magnetic moments, so that the total strength depends on both the magnitudes of the individual moments and the net number of spins aligned in the appropriate direction. In the Purcell experiment, also, the sensitivity depends on the net equilibrium magnetization, because this determines the maximum energy that can be absorbed by the sample. The individual magnetic moments are given directly by the gyromagnetic ratio, and both Purcell and Bloch chose to use the ^{1}H nucleus because of its relatively large value of γ.

The net number of spins that precess about the stationary field corresponds to the initial equilibrium difference between the number of nuclei in the $|\alpha\rangle$ and $|\beta\rangle$ states, as given by the Boltzmann distribution:

$$\frac{N_\alpha}{N_\beta} = e^{\Delta E/kT}$$

where $\Delta E = E_\beta - E_\alpha$. To simplify the expression for the net difference between N_α and N_β, the exponential can be approximated as

$$e^{\Delta E/kT} \approx 1 + \Delta E/kT$$

This approximation is valid provided that the absolute value of ΔE is much smaller than kT, which is the case for NMR, except at extremely low temperatures. The population difference can then be written as

$$N_\alpha - N_\beta = N_\beta \Delta E/kT$$

Because N_α and N_β differ only slightly, $N_\beta \approx N_T/2$, where N_T is the total number of nuclei. Therefore,

$$N_\alpha - N_\beta = \frac{N_T \Delta E}{2kT}$$

Thus, the population difference is proportional to both the gyromagnetic ratio and the static field strength. Note that if the gyromagnetic ratio is negative, then the $|\alpha\rangle$ state has the higher energy, and the fractional population difference defined above is also negative.

Finally, the strength of the induced current is proportional to the rate at which the net magnetic field crosses the wire of the receive coil. This effect may be less obvious than the first two, but it is analogous to the effect of changing the speed of rotation of an electrical dynamo. Anyone who has ridden a bicycle with a dynamo-powered headlight knows that the light is almost always brighter when riding downhill than up. In the NMR experiment, the rotation rate is given by the Larmor equation, and is proportional to the gyromagnetic ratio and the static field strength. In the Purcell-type experiment, the signal strength also depends on the Larmor frequency, because this determines the rate at which energy is absorbed by the sample.

When combined together, the three effects listed above predict that the signal intensity, S, should depend on the following proportionality:

$$S \propto N_T \gamma^3 B^2$$

In practice, however, the sensitivity of the NMR experiment also depends on the noise present in the system, which in turn depends on a number of details involving the construction of the probe coil and its associated electronics. For an ideal solenoidal coil, the signal noise can be shown to be approximately proportional to the square root of the signal frequency. Thus, the signal-to-noise ratio, SNR, is generally assumed to follow the proportionality:

$$\text{SNR} \propto N_T \gamma^{5/2} B^{3/2} \tag{3.1}$$

From these considerations, it is apparent that the signal from ^{1}H nuclei, with $\gamma \approx 2.7 \times 10^8 \, \text{T}^{-1} \, \text{s}^{-1}$, will be much stronger than from an equivalent number of ^{13}C nuclei, for instance, with $\gamma \approx 6.7 \times 10^7 \, \text{T}^{-1} \, \text{s}^{-1}$. Indeed, a sample of ^{13}C nuclei would need to be approximately 30-fold larger than a sample of ^{1}H nuclei in order to generate the same signal to noise ratio. Alternatively, a 10-fold larger magnetic field would be necessary to give rise to an equal SNR from ^{13}C if the number of nuclei in the two samples were the same. Finally, ^{13}C, as well as ^{15}N, are relatively rare isotopes in natural samples, while ^{1}H represents over 99% of natural hydrogen.

For all of these reasons, ^{1}H remains one of the most useful nuclei for NMR spectroscopy. Other nuclei, such as ^{13}C and ^{15}N, however, are also extremely important in modern NMR. In studies of small organic molecules, the resonance frequencies of ^{13}C nuclei, for instance, can provide important structural information. In addition, a variety of clever experiments have been designed to take advantage of the interactions between nuclei of different types, including multidimensional experiments in which the resonances of covalently linked heteronuclear pairs are correlated. Some of these experiments are discussed in Chapter 16.

3.5   Peak Shape and Width

From the discussion so far, it might seem that the peaks in an NMR spectrum would have infinitesimally small widths, since absorption should only be observed when the frequency of the oscillating magnetic field matches the energy difference between the quantum spin states. However, there will always be experimental limitations, such as variations in the frequency generated by the electronic oscillator or variations in the strength of the static magnetic field at different positions within the sample. On top of these effects, it turns out that the shape of an NMR peak is strongly influenced by the macroscopic and molecular properties of the sample.

As discussed in Chapter 4, the actual magnetic field that a nucleus experiences when placed in an external field can be influenced significantly by its chemical environment, including the distribution of electrons and the presence of other nuclei with magnetic moments. The magnitudes of these effects also depend on the orientation of the molecules with respect to the external field. With a solid sample composed of a single crystal, one can observe large changes in resonance frequency as the orientation of the crystal is changed. With powders composed of microcrystals, on the other hand, the individual molecules take on many different orientations, leading to a distribution of effective fields and typically very broad NMR peaks. In liquid samples, molecules typically tumble very rapidly, so that the resonance frequencies are effectively averaged. However, these motions of the molecules generate fluctuating magnetic fields, which, as discussed earlier, contribute to relaxation and establishment of the equilibrium net magnetization. The randomly fluctuating fields also combine with the oscillating field of the spectrometer and contribute, in a complicated way, to the absorption of energy when the spectrometer frequency does not exactly match the Larmor frequency.

In the context of the original Bloch and Purcell experiments, in which the external field strength was slowly varied and continuous spectra were recorded, the relationship between relaxation and peak shape is not so easy to explain at a microscopic level. This relationship becomes much clearer, however, in the context of pulse NMR experiments, and so we will defer a detailed discussion for the later chapters. For now, it is sufficient to note that the shape of an NMR peak from a liquid sample is usually described by an *absorptive Lorentzian* function of the following form:

$$S(f) = \frac{R_2}{R_2^2 + 4\pi^2(f - f_0)^2} \tag{3.2}$$

where f is the frequency at which absorption is being observed, f_0 is the Larmor frequency and R_2 is a relaxation rate constant that is distinct from (and usually larger than) the rate constant for the establishment of thermal equilibrium, R_1, discussed earlier. (The relationship between R_1 and R_2 is discussed in detail in Chapters 7 and 8.)

The absorptive Lorentzian peak has the shape illustrated in Fig 3.6. Note that the denominator of the Lorentzian function is smallest when $f = f_0$, leading to a maximum in the spectrum, and the intensity of the signal at this point is inversely proportional to R_2. It can also be shown that the width of the peak where the intensity is one-half of the maximum is given by R_2/π. Thus, the sharpest peaks are associated with slow relaxation.

The value of R_2 is determined by several factors, but, in liquid samples, one of the most important is the rate at which a molecule tumbles. In general, faster tum-

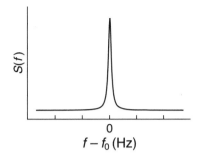

Figure 3.6
The Lorentzian peak-shape function.

bling is associated with lower values of R_2 and, therefore, sharper NMR peaks. As a consequence, larger molecules typically give rise to wider peaks, and this represents one of the fundamental challenges to applying NMR to macromolecules such as proteins and nucleic acids. This effect can be minimized by increasing the temperature of a sample or using a solvent with low viscosity, though there are clearly limits to applying these approaches to biological molecules or other relatively fragile samples. For reasons that will not become fully apparent until later chapters, the relative line widths for large molecules can also be reduced by increasing the strength of the static magnetic field.

Summary Points for Chapter 3

- The first demonstration of nuclear magnetic resonance (i.e., the stimulation by radiation of transitions between spin states) was a molecular beam experiment using particles prepared with a defined state.
- NMR experiments with bulk samples depend on the small excess of spins in the lower energy state at thermal equilibrium.
- Thermal equilibrium of the spin states is established by relaxation processes that depend on molecular motions, which generate random fluctuations of the local magnetic fields experienced by the spins.
- In the Purcell NMR experiment, the stimulation of spin transitions was detected by an increase in the power absorbed by the sample as it was shifted away from thermal equilibrium.
- In the Bloch experiment, and in most modern spectrometers, the absorption of radiation was detected by the induction of an electrical current by the precession of spins.
- Sensitivity in NMR strongly depends on both the strengths of the nuclear dipoles (the gyromagnetic ratio, γ) and the strength of the external field, B.
- The peaks in NMR spectra have finite line widths that reflect relaxation processes that occur as the spectrum is recorded. For liquid samples, larger molecules, which tumble more slowly, generally give rise to wider peaks.

Exercises for Chapter 3

3.1. The following paper describes a method for performing NMR spectroscopy using Earth's magnetic field: Appelt, S., Kühn, H., Häsing, F. & Blümich, B. (2006). Chemical analysis by ultrahigh-resolution nuclear magnetic resonance in the Earth's magnetic field. *Nat. Phys.*, 2, 105–109. http://dx.doi.org/10.1038/nphys211

Although this seems rather remarkable, since much stronger magnetic fields are typically used for NMR, this type of experiment was first described in the 1950s. In fact, this effect is widely used for precision measurements of the strength of Earth's field. The calculations below will give you some feel for the magnitudes involved.

(a) The strength of Earth's magnetic field is approximate 5×10^{-5} T. Calculate the resonance frequencies in this field for ^{1}H, ^{15}N and ^{13}C nuclei.

(b) Calculate the ratio of ^{1}H nuclei in the $|\alpha\rangle$ and $|\beta\rangle$ states (i.e., N_α/N_β) at room temperature in Earth's magnetic field and in a 12-T magnetic field.

(c) Suppose that you could cool your sample to liquid nitrogen temperatures, approximately 77 K. Calculate the ratio of ^{1}H nuclei in the two states in Earth's magnetic field and in a 12-T magnetic field at this temperature.

(d) Now, suppose that you could cool the sample to liquid helium temperature, approximately 3 K. Calculate the ratio N_α/N_β in Earth's magnetic field and in a 12-T magnetic field at this temperature.

3.2. In a modern solution NMR experiment, the sample concentration might be about 1 mM and the volume 0.6 mL. Suppose that you have such a sample in the probe of a 900-MHz NMR spectrometer at 25°C.

(a) Calculate the total number of sample molecules.

(b) Calculate the excess number of ^{1}H nuclei (with a given chemical identity) in the $|\alpha\rangle$ state in the spectrometer, assuming that all of the nuclei are in either the $|\alpha\rangle$ or $|\beta\rangle$ state.

(c) Calculate the relative sensitivity of this spectrometer compared to one based on Earth's magnetic field, assuming that the sensitivity of the electronics for detecting the signals was the same in both.

3.3. Show, as claimed on page 54, that the width of a Lorentzian peak, measured at half of the maximum height, is equal to R_2/π. Hint: Find the values of $(f - f_0)$ in Eq. 3.2 where the signal intensity is one-half of the maximum.

Further Reading

The original papers describing the three classic experiments discussed in this chapter are:

⋄ Rabi, I. I., Millman, S., Kusch, P. & Zacharias, J. R. (1939). The molecular beam resonance method for measuring nuclear magnetic moments. The magnetic moments of $_3$Li^6, $_3$Li^7 and $_9$Li^{19}. *Phys. Rev.*, 55, 526–535. http://dx.doi.org/10.1103/PhysRev.55.526

◇ Purcell, E. M., Torrey, H. C. & Pound, R. V. (1946). Resonance absorption by nuclear magnetic moments in a solid. *Phys. Rev.*, 69, 37–38. http://dx.doi.org/10.1103/PhysRev .69.37

◇ Bloch, F., Hansen, W. W. & Packard, M. (1946). The nuclear induction experiment. *Phys. Rev.*, 70, 474–485. http://dx.doi.org/10.1103/PhysRev.70.474

For Historical perspectives on the discovery of nuclear magnetic resonance and the roles of Rabi, Purcell and Bloch, see the following:

◇ Ramsey, N. F. (1990). The Rabi school. *Eur. J. Phys.*, 11, 137–141. http://dx.doi.org/10.1088/ 0143-0807/11/3/001

◇ Rigden, J. S. (1986). Quantum states and precession: The two discoveries of NMR. *Rev. Mod. Phys.*, 58, 433–448. http://dx.doi.org/10.1103/RevModPhys.58.433

◇ Gerstein, M. (1994). Purcell's role in the discovery of nuclear magnetic resonance: Contingency versus inevitability. *Am. J. Phys.*, 62, 596–601. http://dx.doi.org/10.1119/ 1.17533

The equations now known by his name were laid out by Felix Bloch in a paper accompanying the one describing the nuclear induction experiment:

◇ Bloch, F. (1946). Nuclear induction. *Phys. Rev.*, 70, 460–474. http://dx.doi.org/10.1103/ PhysRev.70.460

The Bloch equations are not used explicitly in this book, but interested readers can find a clear derivation and examples of their application, especially in continuous wave (CW) NMR, in:

◇ Roberts, J. D. (2000). *ABCs of FT-NMR*, Chapter 4. The Bloch equations. Calculating what happens in NMR experiments, pp. 54–78. University Science Books, Sausalito, CA.

Chemical Information from Resonance Frequencies: Chemical Shifts, Dipolar Coupling, and Scalar Coupling

One of the first things that was discovered following the pioneering NMR experiments of Bloch and Purcell was that hydrogen nuclei in different compounds, or in different positions in the same compound, displayed small but measurable differences in their resonance frequencies. The same thing was observed with other nuclei as they were studied. As discussed previously, the resonance frequency, ν, is determined by the gyromagnetic ratio, γ, and the external field strength, B, according to

$$\nu = -\gamma B / 2\pi$$

The observed differences in resonance frequency associated with different chemical environments are *not* due to differences in the gyromagnetic ratios of nuclei of a given type (e.g., ^1H, ^{13}C or ^{15}N), but rather arise from differences in the magnetic field that various nuclei actually "feel" when they are part of a molecule. The field experienced by a given nucleus is influenced by three major effects:

1. *Shielding*, which arises from the effect of the external field on the motion of electrons in the vicinity of the nucleus. The changes in resonance frequency due to shielding are usually referred to as *chemical shifts*, because they reflect the local chemical environment of a nucleus in a molecule.
2. *Dipolar coupling*, in which the spin state of one nucleus (or unpaired electron) influences another nucleus via a direct "through-space" magnetic force.
3. *Scalar coupling*, in which the spin state of one nucleus influences another, but through indirect effects mediated by the electrons involved in the covalent bonds linking the nuclei.

As discussed below, these three types of effects each have distinct features. All are dependent on the orientation of the molecule with respect to the external field, so that they influence the spectra of solid and liquid samples quite differently. The effects of dipolar coupling, for instance, are averaged out completely (or nearly so) by the rapid tumbling of molecules in liquid samples, so that there is usually no effect on the resonance frequency. Although the effects of shielding and scalar coupling are also averaged in liquid samples, they do not disappear altogether.

In studies of simple organic molecules, chemical shift and scalar coupling effects are often sufficient to deduce detailed structures, and measurements of these effects were the mainstay of classical NMR analysis. With larger and more complicated molecules, these effects are also extremely important, in part because they lead to the dispersion of resonance frequencies that makes it possible to resolve the signals of individual nuclei. In addition, dipolar and scalar coupling form the very foundation of multidimensional NMR experiments, which exploit the influence of one nucleus on another.

4.1   Shielding and Chemical Shifts

In solution, the primary effect of chemical structure on resonance frequencies arises from the presence of electrons in the molecule. The orbital motions of the electrons are influenced by the external field in such a way that they induce local magnetic fields. This idea is illustrated very schematically in Fig. 4.1. In this figure, the distribution of electrons is shown as a torus surrounding a nucleus. The presence of an external field increases the probability that an electron will move in a particular path, and this motion induces a magnetic field that usually, but not always, opposes the external field. As a consequence, the field that the nucleus feels is usually smaller than that of the magnet, and the resonance frequency is lower. This effect is called *shielding* and depends on several factors, but it is especially sensitive to the extent to which covalently attached atoms withdraw electron density away from the nucleus of interest.

Because the shielding effect is induced by the external field, the change in effective field is proportional to the applied field strength, and the same *relative change* in resonance frequency is observed regardless of the field strength. One way of expressing the effect is as a shielding coefficient, σ. If the applied external field is B_0, then the

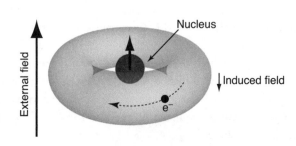

Figure 4.1
Shielding of a nucleus by local magnetic fields induced by the stationary external field.

local field experienced by nucleus i is given as

$$B_i = (1 - \sigma_i) B_0$$

Note that the shielding coefficient is dimensionless and is defined so that a positive value leads to a reduction in the local magnetic field. Strictly speaking, the shielding coefficient is expressed relative to the field felt by a bare nucleus and can only be determined from quantum mechanical calculations, but the symbol σ_i is also used to express the shielding relative to that observed in a specified reference compound, particularly by practitioners of solid-state NMR.

In solution NMR, the shielding effect is commonly expressed in terms of the observed resonance frequency relative to that of a nucleus in a reference compound. The *chemical shift* for nucleus i, usually designated δ_i, is conveniently expressed in units of parts per million (ppm) and is defined as

$$\delta_i = \frac{\nu_i - \nu_{\text{ref}}}{\nu_{\text{ref}}} \times 10^6$$

where ν_{ref} is the resonance frequency for a reference nucleus. Thus, in a 500 MHz spectrometer, a proton chemical shift of 1 ppm represents a frequency shift of 500 Hz. The chemical shift is related to the shielding coefficient, also expressed in parts per million, according to:

$$\delta_i = \frac{\sigma_{\text{ref}} - \sigma_i}{1 - \sigma_{\text{ref}}}$$

$$\approx (\sigma_{\text{ref}} - \sigma_i)$$

where σ_{ref} is the shielding coefficient for the reference nucleus. The chemical shift is also sometimes referred to as an *antishielding* coefficient, because a larger value represents less shielding.

4.1.1 Chemical Shift References

In order to compare resonance frequencies obtained using different spectrometers, or even the same spectrometer under slightly different conditions, it is necessary to use standard compounds that give rise to resonance frequencies that can be used as references. It is most convenient if the reference frequency lies at one end or the other of the spectrum of interest.

For hydrogen nuclei in organic molecules, the largest shielding effects are usually seen in methyl groups, because the carbon atom does not withdraw electron density as much as oxygen and nitrogen atoms, for instance. Among methyl groups, those attached to silicon are particularly well shielded because this element is highly electropositive, virtually guaranteeing that the protons will have a lower resonance frequency than just about any other found in organic compounds. For this reason the standard reference is tetramethylsilane (TMS, Fig. 4.2). By convention, the chemical shift of the hydrogen nuclei in this compound is set to 0.

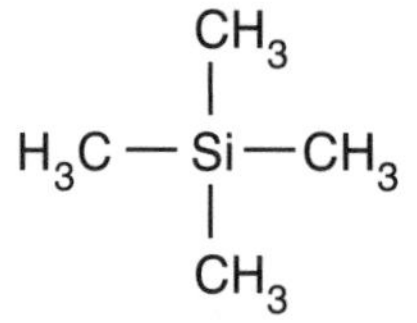

Figure 4.2

Chemical structure of tetramethylsilane (TMS), the standard chemical shift reference for ^{1}H nuclei.

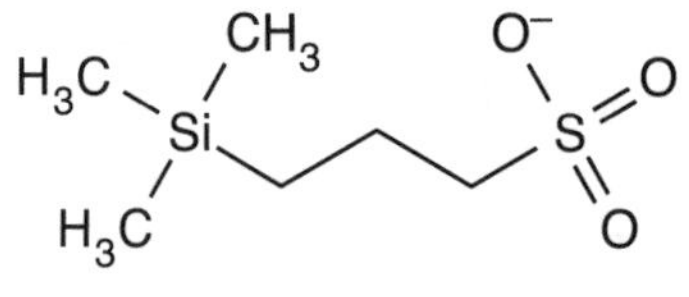

Figure 4.3

Chemical structure of 4,4-dimethyl-4-silapentane-1-sulfonic acid (DSS), commonly used as a chemical shift reference in aqueous solutions.

Unfortunately, TMS is not soluble in water, making it unsuitable for use in most experiments with biological molecules. No one has yet found an ideal reference molecule for aqueous solutions, but the most commonly used one is the closely related compound 4,4-dimethyl-4-silapentane-1-sulfonic acid (DSS, Fig. 4.3). The hydrogens that are not part of the trimethylsilyl group are usually replaced with deuterium, so that they do not contribute to the proton spectrum. DSS is also commonly used as a reference for ^{13}C chemical shifts. For the other nucleus commonly used in biomolecular NMR, ^{15}N, finding a direct chemical shift reference is more problematic, but the resonance frequency of this nucleus can be referenced to that of ^{1}H using the relative gyromagnetic ratios. Further discussion of chemical shift references can be found in the references listed at the end of this chapter.

4.1.2   Upfield and Downfield in an NMR Spectrum

Because of the way chemical shifts are defined and the way in which early NMR experiments were performed, the conventions used to plot and label NMR spectra are rather confusing. If, as is usually the case, the reference compound is more highly shielded than the nuclei of interest, the chemical shift as defined above will always be positive. You might, therefore, expect spectra to be plotted with chemical shift increasing from left to right, and this seems sensible enough. Until the development of pulse NMR methods, however, spectra were determined by maintaining a constant frequency signal in the transmit coil of the spectrometer and slowly varying the strength of the external field. It was thus natural to plot the absorption as a function of the applied field strength. When plotted this way, the absorption peaks from the most highly shielded nuclei appear at the highest field strength, since a larger external field is necessary to overcome the shielding. Thus, the smaller chemical shifts appear on the right of the spectrum, and the largest shifts on the left. To this day, the left of a spectrum is referred to as *downfield* and the right as *upfield*, with the chemical shift decreasing from left to right. For instance, a typical ^{1}H spectrum of a protein (BPTI) is shown in Fig. 4.4. With the spectrum plotted in this way, the absolute value of the Larmor frequency, $|\nu|$, also increases from right to left. For a nucleus with a positive

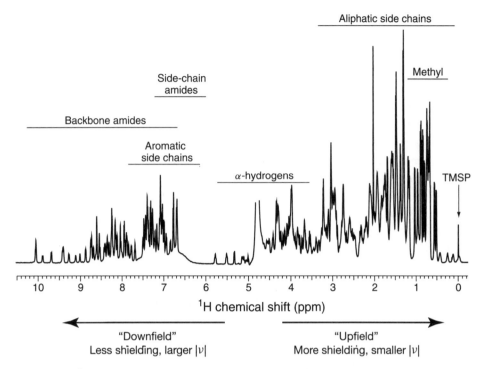

Figure 4.4 ^{1}H NMR spectrum of a small protein, bovine pancreatic trypsin inhibitor (BPTI). The peak labeled "TMSP" corresponds to the ^{1}H nuclei of 3-(trimethylsilyl) propionate, another reference compound used for aqueous solutions.

gyromagnetic ratio, larger chemical shifts represent *more negative* values of ν, so that the signed frequency does actually increases from left to right, following mathematical convention. For a negative gyromagnetic ratio, ν is positive and increases from right to left.

The chemical shifts for the protons in a protein typically have chemical shifts ranging from about 11 to 0 ppm. (The large truncated peak at about 4.75 arises from water.) The chemical shifts of protons found in different functional groups usually fall into characteristic regions of the spectrum, as indicated in the figure. However, the boundaries of these regions are very approximate, and there are frequent exceptions. Generally, the backbone amide protons display the largest chemical shifts, because the covalently attached nitrogen atoms tend to withdraw electron density, while the protons of methyl groups have the smallest shifts. The protons of aromatic side chains also have very characteristic chemical shifts, typically in the region from 6.5 to 7.5 ppm.

The significant range of chemical shifts seen for a given type of proton (e.g., the backbone amides) clearly shows that the resonance frequency is strongly influenced by the environment of a nucleus in the three-dimensional structure, as well as by its covalent neighbor. Usually, however, the shifts due to covalent structure are larger than those due to three-dimensional structure, making the classification shown in Figure 4.4 at least moderately reliable.

Some of the largest effects on chemical shifts in organic and biological molecules arise from aromatic rings, such as those in the amino acids tyrosine, phenylalanine and tryptophan. These shifts are due to effects of the external field on the motions of the delocalized electrons in the aromatic rings. These motions are called *ring currents* and they alter the local magnetic field of nearby nuclei in a way that depends on both the distance from the ring and the orientation of the ring. Nuclei located near the edges of the ring, including those covalently attached to the ring, are deshielded, so they experience stronger magnetic fields and have increased chemical shifts, while nuclei above or below the ring are more highly shielded and display smaller chemical shifts. The magnitude of the shifts for ^{1}H nuclei can be as large as 3 or 4 ppm, sometimes causing a resonance to appear in a completely unexpected portion of the spectrum. A particularly dramatic example is the amide proton of glycine 37 in BPTI, which is located very close to the face of a tyrosine side chain and has a chemical shift of 4.3 ppm, versus the range of 7 to 11 ppm typically seen.

4.1.3   Structural Information from Chemical Shifts

As suggested by the simple one-dimensional spectrum shown in Fig. 4.4, NMR resonance frequencies are exquisitely sensitive to chemical environment, and the observed chemical shifts should represent a rich source of information about the molecule's three-dimensional structure. At present, however, it is quite difficult to extract detailed structural information from chemical shifts alone.

In proteins, the chemical shifts of backbone ^{1}H, ^{13}C and ^{15}N nuclei, as well as the side-chain β-carbon, (illustrated in Fig. 4.5) are influenced significantly by whether or not the atoms are included in regular secondary structures. Table 4.1 lists the average chemical shifts for the different residue types when found in irregular structures (i.e., regions that form neither α-helices nor β-strands) within folded proteins. These averages were compiled from a database of approximately 300 proteins with known three-dimensional structures (http://refdb.wishartlab.com).

Chemical shift values similar to those shown in Table 4.1 are also observed in disordered peptides and proteins, and these values (and those from similar compilations) are often referred to as *random-coil shifts*.

Figure 4.5
Covalent structure of an alanine residue in a peptide or protein, identifying the amide nitrogen (N) and armide hydrogen (H) atoms, the α-carbon (C$_\alpha$) and α-hydrogen (H$_\alpha$) atoms, the carbonyl carbon (C) and oxygen (O) atoms and the side-chain β-carbon (C$_\beta$) atom.

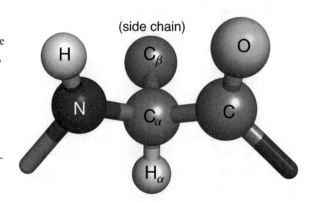

Table 4.1   Random coil chemical shifts (ppm): Average chemical shifts of protein backbone nuclei in conformations other than α-helices or β-strands.

Residue[a]	α-^{1}H	Amide ^{1}H	α-^{13}C	Carbonyl ^{13}C	Amide ^{15}N
Ala	4.32	8.24	52.5	177.8	123.8
Cys (SH)	4.55	8.32	58.2	174.6	118.8
Cys (SS)	4.71	8.43	55.4	174.6	118.6
Asp	4.64	8.34	54.2	176.3	120.4
Glu	4.35	8.42	56.6	176.6	120.2
Phe	4.62	8.30	57.7	175.8	120.3
Gly	3.96	8.33	45.1	174.9	108.8
His	4.73	8.42	55.0	174.1	118.2
Ile	4.17	8.00	61.1	176.4	119.9
Lys	4.32	8.29	56.2	176.6	120.4
Leu	4.34	8.16	55.1	177.6	121.8
Met	4.48	8.28	55.4	176.3	119.6
Asn	4.73	8.40	53.1	175.2	120.4
Pro (trans)	4.42	—	63.3	177.3	—
Pro (cis)	4.42	—	62.8	177.3	—
Gln	4.34	8.32	55.7	176.0	119.8
Arg	4.34	8.23	56.0	176.3	120.5
Ser	4.47	8.31	58.3	174.6	115.7
Thr	4.35	8.15	61.8	174.7	113.6
Val	4.12	8.03	62.2	176.3	119.2
Trp	4.66	8.25	57.5	176.1	121.3
Tyr	4.55	8.12	57.9	175.9	120.3

a. Amino acid residue types are indicated by the standard three-letter code. Separate values are given for the thiol (SH) and disulfide (SS) forms of cysteine and the trans and cis conformations of prolyl residues. From Wishart, D. S. (2011) *Prog. Nucl. Mag. Res. Sp.*, 58, 62–87.

Deviations from the random-coil chemical shifts can be used to infer the presence of specific types of secondary structure, although other environmental effects also influence the chemical shifts of specific nuclei in a molecule. Table 4.2 lists average *secondary shifts* (in ppm) for nuclei in α-helices or β-strands in folded proteins. The numbers in the table represent the average differences between observed shifts of nuclei in helices or strands and the average for the same nucleus and residue type found in unfolded or irregular structures, as listed in Table 4.1.

As an example, a chemical shift of an α-^{13}C nucleus that is greater than the value listed in Table 4.1 is indicative of α-helix structure, while a shift smaller than the random-coil value is indicative of a β-strand. Of these changes, those for the α- and β-carbon nuclei appear to be most reliable. Notice, however, that the variations in the

Table 4.2   Secondary chemical shifts (ppm): Average chemical shift differences of protein nuclei in regular secondary structure elements, relative to the random-coil chemical shifts listed in Table 4.1.

Nucleus	α-Helix[a]	β-Strand[a]
α-^{1}H	−0.29 (0.3)	0.44 (0.5)
Amide ^{1}H	−0.08 (0.6)	0.43 (0.7)
α-^{13}C	2.70 (1.6)	−1.04 (1.5)
β-^{13}C	−0.71 (2.0)	1.57 (2.1)
Carbonyl ^{13}C	1.93 (2.0)	−0.96 (1.8)
Amide ^{15}N	−1.11 (3.1)	0.43 (0.7)

a. The numbers in parentheses represent the sample standard deviations. Calculated from the entries in the RefDB database of protein chemical shifts (Zhang H., Neal, S. & Wishart, D. S.. (2003) *J. Biomol. NMR*, 25, 173–195. http://refdb.wishartlab.com)

secondary shifts for nuclei within a given structural type, indicated by the standard deviations shown in parentheses in Table 4.2, are often as large as the differences observed between structure types. A portion of this variation is due to the identities of neighboring residues in the sequence, and these effects have also been tabulated and can be used to refine the calculation of secondary shifts, as discussed in some of the references provided at the end of this chapter. Even with these corrections, however, it is usually necessary to average the secondary shifts over several amino acid residues in a sequence in order to obtain convincing evidence of regular secondary structure.

The secondary chemical shifts have been used in two general ways. In the absence of information about what resonance frequencies are associated with individual residues in the sequence, the average contents of helix and strand can be estimated from the number of resonances in different regions of the spectrum. This is roughly analogous to estimating secondary structure content from circular dichroism spectra and is reported to be at least as reliable. On the other hand, if the resonances have been assigned to individual residues, then the probable secondary structure of specific polypeptide segments can be derived. Although this approach is not as reliable as direct structure determination by NMR or crystallography, it is much more so than predictions based only on statistical correlations between residue type and secondary structure propensity.

While these methods are very useful, much more detailed structural information could, in principle, be derived from chemical shifts. Presently there is a great deal of activity in developing methods to accurately predict chemical shifts from a known or hypothesized three-dimensional structure. Calculations of this type could, for instance, be used to help rule out possible structures or refine structures based on other information. For relatively small molecules (containing 10–20 non-hydrogen atoms), it is now possible to carry out quite rigorous chemical-shift calculations on a desktop computer using readily available software, such as the program GAUSSIAN. For macromolecules, however, these calculations remain very challenging, but they may represent one of the most exciting avenues for gaining additional structural information from NMR spectroscopy.

4.1.4   Chemical Shift Anisotropy

For liquid samples, it is customary, and almost always adequate, to express the chemical shift as a single number. In reality, however, the chemical shift depends on the orientation of the molecule with respect to the external field. The term *chemical shift anisotropy* (CSA) is used both to refer to the general phenomenon of the orientation dependence and as a specific parameter describing the magnitude of the effect, as described below. The anisotropy is not directly detectable from the resonance frequencies of nuclei in liquid samples because the molecules are rapidly sampling all possible orientations and the effective magnetic fields in the different directions are averaged. In solid samples, however, chemical shift anisotropy is important and contributes to the very broad absorbance bands characteristic of solid-state NMR (unless special techniques are used).

While the average chemical shift is adequate to describe the effects of shielding on the resonance frequencies in liquid samples, anisotropy can still have important consequences for the NMR of liquids. In particular, the field experienced by a nucleus will fluctuate as the molecule tumbles in solution or undergoes other motions, and this can provide a mechanism for relaxation. Over the past several years, relaxation measurements have become an important tool for studying molecular dynamics and, for this reason and others, chemical shift anisotropy has become an increasingly important aspect of solution NMR.

Unfortunately, several different conventions for mathematically describing the chemical shift anisotropy experienced by a nucleus in a given chemical environment have been used in different contexts, and the subject can be a source of considerable confusion and, sometimes, ambiguity. The fundamental mathematical object describing the effect is the *chemical shift tensor*, a matrix composed of nine elements. Together, these elements define a local molecular coordinate system and the local magnetic fields along each coordinate axis. Confusion arises because the axes can be defined in different ways, and a full description requires knowing the relationships among the coordinate system, the external magnetic field and the molecular structure. In addition, the effect can be defined in terms of chemical shifts (δ) or shielding coefficients (σ), and it is not always clear which is being used.

Perhaps the most accessible and widely used representation of the chemical shift tensor is based on a Cartesian coordinate system referred to as the *principal axes frame*, or PAF. This representation can be illustrated graphically by introducing a set of three perpendicular axes, labeled 1, 2 and 3. The orientation of these axes are defined relative to the molecular structure, but are not, in general, aligned with the

Figure 4.6

Axis 1 in the principal axis frame. With the molecule oriented to yield the largest chemical shift (and smallest shielding), axis 1 is defined as the direction of the external field.

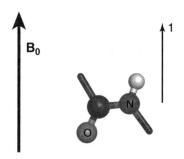

coordinate system used to define the atomic coordinates of the molecule, which has an arbitrary orientation. As the orientation of a molecule changes, by random tumbling in a liquid or by deliberate rotation of a crystal, the principal axes frame rotates with the molecule.

To illustrate the rules by which the PAF axes for a particular nucleus are defined, we will consider the case of a ^{15}N nucleus in a peptide unit of a protein, a case that is particularly important in biomolecular NMR. Conceptually, the process can be visualized in the following steps:

1. The molecule is rotated with respect to the external field until the chemical shift is *maximal*, corresponding to the smallest shielding. With the molecule in this orientation, the direction of the external field, B_0, defines the direction of principal axis 1, and the chemical shift in this orientation is labeled δ_{11}. For the amide nitrogen in our example, this axis is very nearly aligned with the N–H bond, being rotated from this direction by about 18°, as illustrated in Fig. 4.6. The magnitude of the chemical shift in this orientation, δ_{11}, is about 230 ppm for the amide ^{15}N nucleus.

2. Next, the molecule, along with the newly defined axis 1, is rotated so that the chemical shift is at its *minimal* value, as illustrated in Fig. 4.7. In this orientation, the direction of the external field now defines axis 3, corresponding to the chemical shift, δ_{33}. For the amide nitrogen, δ_{33} is about 50 ppm, and the lengths of the two arrows representing axes 1 and 3 in Fig. 4.7. are scaled to represent the relative values of δ_{11} and δ_{33}. Axes 1 and 3 are always perpendicular to each other and define a plane. For the amide nitrogen, this plane is very nearly parallel to the plane defined by the atoms of the amide group—that is, the plane of the page in Figs. 4.6 and 4.7.

3. A final axis is drawn perpendicular to the first two and labeled axis 2. When the molecule is rotated so that this axis is aligned with the external field, the observed chemical shift defines δ_{22}. For the amide nitrogen, this axis is perpendicular to the plane of the peptide bond and the chemical shift is about 80 ppm. In Fig. 4.8, the amide group is rotated so that axis 2 is aligned with the external field.

This set of manipulations can only really be carried out for a solid sample composed of a single crystal, but indirect methods, based on relaxation measurements, can be

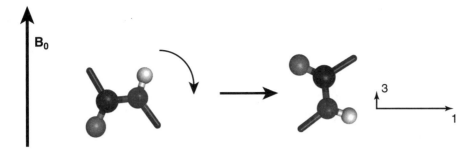

Figure 4.7   Axis 3 in the principal axis frame. With the molecule oriented to yield the smallest chemical shift (and largest shielding), axis 3 is defined as the direction of the external field. The lengths of the arrows are drawn to represent the relative chemical shifts when the external field is oriented along the corresponding axes.

Figure 4.8
Axis 2 in the principal axis frame, defined so as to be perpendicular to axes 1 and 3.

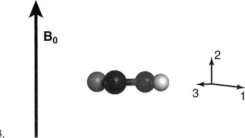

used to define the direction of the axes and the associated chemical shifts for solution samples.

The isotropic chemical shift is simply the average of the three directional chemical shifts.

$$\delta_{iso} = (\delta_{11} + \delta_{22} + \delta_{33})/3$$

and corresponds to the chemical shift observed when the molecule tumbles freely in solution. For an amide ^{15}N nucleus, a typical value of δ_{iso} is about 120 ppm, but, as shown in Tables 4.1 and 4.2, this chemical shift can vary by several ppm depending on the amino acid residue and surrounding environment. However, the directions of the three principal axes are generally very similar for different amide groups, as are the differences in the chemical shifts: $(\delta_{11} - \delta_{iso})$, $(\delta_{22} - \delta_{iso})$ and $(\delta_{33} - \delta_{iso})$. The differences between the chemical shift components are, in fact, usually the properties of the greatest interest, and two other commonly used parameters, described below, are defined in terms of differences.

The chemical shift anisotropy, $\Delta\delta$, is defined as

$$\Delta\delta = \delta_{11} - (\delta_{22} + \delta_{33})/2$$

Note that if δ_{11}, δ_{22} and δ_{33} are all equal, then $\Delta\delta = 0$. For the amide nitrogen, $\Delta\delta \approx$ 160 ppm.

If δ_{22} and δ_{33} are equal to one another, the chemical shift tensor is said to be symmetrical. Deviations from symmetry are described by the parameter η, defined as

$$\eta = \frac{3(\delta_{22} - \delta_{33})}{2\Delta\delta}$$

For the amide nitrogen $\eta \approx 0.28$. This is a relatively small value, reflecting the fact that, for this case, the major influence on the chemical shift is the orientation of axis 1 (which is nearly parallel to the N–H bond vector) with respect to the external field.

The chemical shift anisotropy parameters can also be expressed in terms of shielding coefficients. For instance, the shielding coefficient associated with axis 1, expressed in parts per million, is given by

$$\sigma_{11} = \sigma_{ref} - \delta_{11}$$

where σ_{ref} is the shielding coefficient for the reference compound used to define the chemical shifts. The isotropic shielding coefficient is given by:

$$\sigma_{iso} = (\sigma_{11} + \sigma_{22} + \sigma_{33})/3 = \sigma_{ref} - \delta_{iso}$$

and the shielding anisotropy is

$$\Delta\sigma = \sigma_{11} - (\sigma_{22} + \sigma_{33})/2 = -\Delta\delta \tag{4.1}$$

The terms σ_{11} and $(\sigma_{22} + \sigma_{33})/2$ are often designated $\sigma_\parallel$ and $\sigma_\perp$, respectively, so that $\Delta\sigma = \sigma_\parallel - \sigma_\perp$. For many, but not all, purposes it is the square of the anisotropy parameter that is used in calculations, so that only the absolute value is important. As a consequence there is a tendency to use $\Delta\sigma$ and $\Delta\delta$ interchangeably, a practice that can lead to confusion when the sign does play a role, such as in calculations of cross-relaxation effects, as discussed in Chapter 8.[1]

4.2   Dipolar Coupling

The energy difference between the stationary states of a spin (the eigenstates) can also be influenced by the spin states of nearby nuclei. In classical terms, the effect of one nucleus on another can be viewed as arising from the presence of a nearby small magnet, or dipole. If a molecule containing two spins is placed in an external field, each of the dipoles will be aligned with or against the external field (accepting the simplifying assumption that all of the spins are in either the $|\alpha\rangle$ or $|\beta\rangle$ state). The net magnetic field that one nucleus feels will then depend on the state of the other nucleus, as illustrated in Fig. 4.9.

---

[1] Another source of confusion is the common use of the symbol σ in reporting chemical shift anisotropy data, implying that absolute shielding coefficients are being reported. In general, absolute shielding coefficients can only be determined by theoretical calculations. In some cases, the numbers reported as σ are simply chemical shifts, and in others the reported values are the chemical shifts with the signs reversed. In still others, the individual coefficients may be expressed relative to the average shielding coefficient, σ_{iso}.

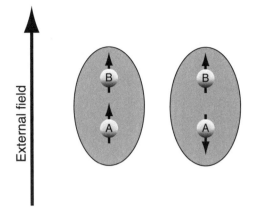

Figure 4.9
Dipolar coupling. The magnetic dipole of each spin can add positively or negatively to the magnetic field experienced by the other.

In the molecule on the left in Fig. 4.9, the field experienced by the B nucleus is enhanced by the A nucleus, while on the right, the field from the A nucleus acts against the external field. We might expect, then, that each nucleus would give rise to two peaks in the spectrum. Importantly, however, the actual effect of one nuclear spin on another depends on the relative orientations of the two nuclei and the external field, which depends, in turn, on the orientation of the molecule with respect to the external field.

To understand how the orientation of the molecule influences the interaction between dipoles, consider the diagram in Fig. 4.10, which shows the field lines generated by one nucleus. Note that the direction of the field is different at different positions around the dipole. Even if the molecule changes its overall orientation, the individual dipoles remain aligned with the external field, as illustrated in Fig. 4.11, (which focuses on the field from the A nucleus).

If the vector between the two nuclei is parallel to the external field, as shown on the left of Fig 4.11, then the local field from nucleus A reinforces the external field, and the absolute value of the resonance frequency of nucleus B is slightly greater than it would be in isolation. If the inter-nucleus vector is perpendicular to the field, then the local field from A slightly decreases the total effective field and the absolute value of the resonance frequency of B decreases. Nucleus B has exactly the same effect on nucleus A. Other things being equal, the strongest effects are seen when the internuclear vector is parallel to the external field.

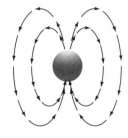

Figure 4.10
Magnetic field lines from a dipole.

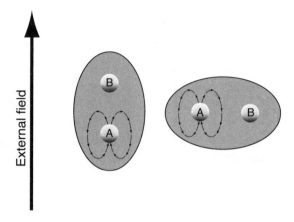

Figure 4.11

The effect of molecular orientation on the influence of one dipole on another. Rotating the molecule leaves the dipole orientations unchanged, but the field experienced by the other nucleus changes.

The total magnitude of the effect depends on the gyromagnetic ratios of the two nuclei (γ_A and γ_B), the distance between them (r) and the angle (θ) between the external field and the vector linking the two nuclei. The gyromagnetic ratios and the internuclear distance are incorporated into the *dipolar coupling constant*, defined as

$$d = \frac{\mu_0 \gamma_A \gamma_B h}{8\pi^2 r^3}$$

The constant μ_0 is called the *magnetic permeability of free space* (and is equal to $4\pi \times 10^{-7}$ T^2· m^3· J^{-1}. Alternatively, the units can be expressed as T·m/A, or as H·m^{-1}, where A and H are the units of electrical current (ampere) and inductance (henry), respectively. This constant is unique to the SI system of measurements and does not appear in many older texts and papers that use the centimeter-gram-second (cgs) system (or, more properly, the Gaussian-cgs system). This reflects a fundamental difference in how the various systems treat the units of electricity and magnetism and is an endless source of confusion.

To calculate the actual perturbation of the resonance frequency requires a quantum mechanical analysis that incorporates the angle θ, and the treatment has a slightly different form for a homonuclear case, such as two protons, and a heteronuclear case, such as a proton and a ^{15}N nucleus. For the homonuclear case, the difference in Larmor frequency, for the case where a nearby nucleus is in the "up" or "down" state, is given by

$$\Delta v = d\frac{3}{4\pi}(3\cos^2\theta - 1) \qquad (4.2)$$

For heteronuclei, where there is a substantial difference between γ_A and γ_B, the frequency difference is

$$\Delta v = \frac{d}{2\pi}(3\cos^2\theta - 1) \qquad (4.3)$$

The change in resonance frequency caused by dipolar coupling can be very large. The plot in Fig. 4.12 shows the value of Δv as a function of the angle θ for the case of two ^{1}H nuclei separated by 2 Å. For this particular case, the maximum value

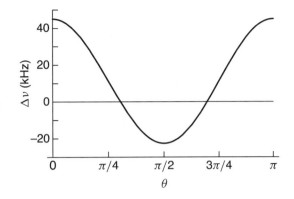

Figure 4.12
The change in resonance frequency due to dipolar coupling as a function of θ, the angle between the internuclear vector and the direction of the external field. Calculated from Eq. 4.2 for two ^{1}H nuclei separated by 2 Å.

is greater than 40 kHz. In a 500-MHz spectrometer, this difference corresponds to approximately 80 ppm, much larger than the entire range of proton chemical shifts seen in a protein spectrum! Note that the effect is maximal when the angle between the external field and the internuclear vector is 0 or π radians (i.e., when the vector is vertical), and changes sign when the vector is perpendicular to the field. The change in dipolar coupling with orientation can be observed directly when a sample composed of a single crystal is rotated in a spectrometer. If the molecules in a solid sample are randomly oriented (e.g., in a powder), then a broad range of resonance frequencies will be observed, making it very difficult to resolve the signals from individual atoms in a molecule. For a molecule tumbling rapidly and isotropically in solution, the dipolar coupling effect averages out to 0, so that there is no effect on the observed resonance frequency.

The magnitude of the dipolar coupling interaction also strongly depends on the distance between the two nuclei, as shown in Fig. 4.13, again for the case of two protons, with $\theta = 0$. Note that the effect largely disappears for distances greater than about 5 Å, so that it is only the closest neighbors in a molecule that influence one another in this way. This has great practical importance for the use of the nuclear Overhauser effect to estimate intramolecular distances, as is discussed in Chapter 9.

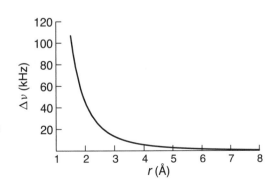

Figure 4.13
The change in resonance frequency due to dipolar coupling as a function of the distance between two spins. Calculated from Eq. 4.3 for two ^{1}H nuclei with the internuclear vector parallel to the external field ($\theta = 0$).

4.2.1   Consequences of Dipolar Coupling for Relaxation, Peak Widths and Resolution

While dipolar coupling does not affect the average resonance frequency of a nucleus in a rapidly tumbling molecule, it does lead to rapid fluctuations in the strength of the magnetic field that a nucleus experiences. This is often the major mechanism for relaxation, contributing both to the establishment of an equilibrium population in a static magnetic field and the peak widths in an NMR spectrum. As discussed briefly in Chapter 3, the width of a peak is directly proportional to the relaxation rate constant, R_2. Examination of Eqs. 4.2 and 4.3 (p. 72) will show that the magnitude of the dipolar coupling interaction is independent of the external field strength. Under conditions where this type of interaction dominates the relaxation mechanisms, the value of R_2 and, therefore, the peak width is also nearly independent of field strength. This has important practical implications.

Remember that chemical shift differences, as expressed in frequency units, *are* proportional to field strength. Thus, if a spectrum is plotted on an absolute frequency scale (Hz), then the separation between peaks due to chemical shifts will increase if the static field strength is increased. Under the conditions stipulated above, however, the peak width, again expressed in frequency units, will remain nearly constant. As a consequence, the resolution between closely spaced peaks increases with field strength. Because spectra are usually plotted on a relative frequency scale (i.e., ppm), this effect is seen as a narrowing of the apparent peak width. Remember, also, that the peak height is inversely proportional to R_2. When these effects are combined with the other effects of field strength on signal intensity, the total change in the spectrum can be quite dramatic. Fig. 4.14 compares idealized spectra for two nuclei that differ by a fixed chemical shift but have the same R_2 relaxation rate, when the static field strength is increased by 60%—that is, the difference between 500- and 800-MHz spectrometers.

The increased field strength enhances both the intensity and resolution of the spectrum. These effects provide much of the motivation for building spectrometers with ever greater field strengths. It should be noted, however, that the benefits of higher fields do eventually begin to diminish, as the cost increases greatly. One reason for the diminishing returns is that other relaxation mechanisms, especially those that depend on chemical shift anisotropy, become more significant at higher fields, so that the relaxation rates and the line widths (in absolute frequency units) begin to increase.

4.2.2   Magic-angle Spinning

The broad signals due to dipolar coupling and chemical shift anisotropy in a solid sample can be sharpened using a clever trick. From Eqs. 4.2 and 4.3 (p. 72), note that the frequency shift from dipolar coupling will disappear when the following condition is satisfied:

$$1 - 3 \cos^2 \theta = 0$$

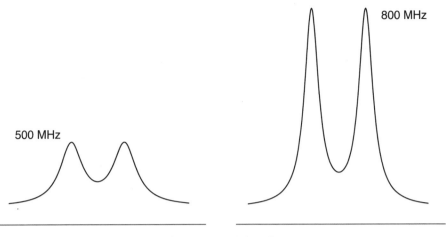

Chemical shift (ppm)

Figure 4.14   The effect of external field strength on resolution and intensity in a solution NMR spectrum, for an idealized case where relaxation is due entirely to dipolar coupling interactions. The peak width, in absolute frequency units, is assumed to be independent of field strength, and the total signal-to-noise ratio is assumed to depend on the field strength raised to the 3/2 power. (Eq. 3.1, p. 53)

This condition is satisfied when θ, the angle between the external field and the inter-nuclear vector, equals approximately 54.7°. If a solid sample is rapidly rotated about an axis, then the *average* orientation of *all* of the bonds will be parallel to this axis, as shown in Fig. 4.15.

   If the rotation axis is oriented 54.7° from the external field, as shown in Fig 4.15, then the dipolar shift for all of the molecules will disappear! This effect was first described by E. R. Andrews and his colleagues and independently by I. J. Lowe in 1959, and the angle of 54.7° quickly became known as the *magic angle*. In order for the spectrum to be completely narrowed, the frequency of rotation must exceed the range of resonance frequencies caused by the inter-nuclear interaction. For the case of the dipolar interaction between two ^{1}H nuclei, the frequency range can be as large

Figure 4.15
Magic-angle spinning. A solid sample is rotated about an axis 54.7° from the direction of the external field. Irrespective of their relative orientations, the average angles of the internuclear vectors with respect to the field are also 54.7°.

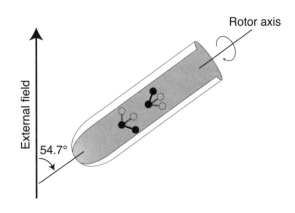

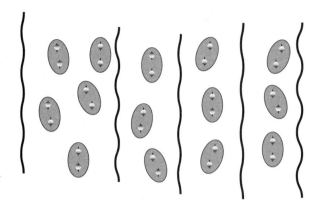

Figure 4.16
Schematic representation of molecules partially oriented by a gel-like medium. The degree of orientation is greatly exaggerated for illustration.

as 50 kHz, requiring a spinning speed of 3,000,000 rpm! Using turbines driven by compressed air or nitrogen, such speeds are now possible.

The same mathematical term, $(1 - 3\cos^2\theta)$, also appears in the expression describing the angular dependence of chemical shifts and scalar coupling interactions (described in Section 4.3). As a consequence, magic-angle spinning also serves to reduce or eliminate line broadening in solids due to these interactions.

4.2.3   Residual Dipolar Coupling Interactions

The magic angle spinning technique allows solid samples to be treated, in some respects, as if they were liquids. Conversely, there are advantages to be gained under some circumstances by causing a liquid sample to take on, to a very limited extent, the properties of a solid. If the rotational freedom of a molecule in solution is restricted slightly, the dipolar coupling interaction is still averaged, but the average value may be different from zero. The observed effects can then be used to derive information about the orientation of internuclear vectors with respect to the average orientation of the molecule.

Macromolecules, such as proteins or nucleic acids, can be partially oriented in solution by including in the sample some larger structure that can be oriented mechanically. Materials used for this purpose include gels formed with polyacrylamide, filamentous virus particles and bicelles (large disc-shaped aggregates of lipid molecules). The fibers of a gel can be oriented by physically stretching the gel, whereas large asymmetric particles often take on preferential orientation in the strong magnetic field of an NMR spectrometer. If the molecules of interest also have some degree of structural or electrostatic asymmetry, the surrounding fibers or particles will favor particular orientations. This effect is illustrated, and greatly exaggerated, in Fig. 4.16.

In this figure, the orienting medium is represented as long thin fibers and the molecules of interest as ellipsoids. Because of simple steric exclusion, the molecules that are close to the fibers tend to be oriented so that their long axes are roughly parallel to the fibers. If we consider a pair of nuclei that lie along a line parallel to the longer molecule axis, as shown in the figure, we can see that the fibers will tend to orient the internuclear vector with the external field. Other pairs of nuclei will tend to be oriented so that the vector is perpendicular to the field, or somewhere between parallel and perpendicular. Because each spin can be oriented either with or against the field, the signal from a particular nucleus will be split into two components, much like the splitting due to scalar coupling described in the following section. The extent of splitting will depend on the average orientation of the internuclear vector with respect to the external field and, therefore, the long axis of the molecule. This effect is referred to as a *residual dipolar coupling interaction* (RDC).

It should be emphasized that the degree of orientation suggested in the figure above is greatly exaggerated. In practice, the density of the orienting medium is chosen so that the orientation of only a small fraction, perhaps 0.1%, of the molecules is biased, and this small population exchanges rapidly with the majority of molecules that are not oriented. This allows each molecule to tumble rapidly so that, for most purposes, the NMR properties are the same as for any other liquid sample. Because of the very large potential effect of dipolar interactions, however, the small degree of net orientation leads to measurable effects.

Measurements of residual dipolar interactions can be combined with other information to help define the conformation of a protein or other large molecule. They provide a particularly useful complement to measurements of interatomic distances based on the nuclear Overhauser effect (Chapter 9), because the distance measurements provide information about local structures, whereas the residual dipolar interaction is influenced by the orientation of a local vector with respect to the entire molecule. One way in which this information can be used is to compare the observed dipolar interactions with those predicted by a model based on other information and then adjust the model to minimize the differences.

4.3   Scalar Coupling

In addition to the direct effects one nucleus can have on another, coupling between nuclei can take place via the electrons that are shared in chemical bonds. In a sense, this is a combination of the two effects discussed above, with the spin state of one nucleus slightly altering the electron shielding that another nucleus experiences.

This effect is usually called *scalar coupling*, a term that has a slightly obscure origin. In general, this type of interaction is sensitive to the orientation of the molecule with respect to the external field, just as chemical shifts and dipolar interactions are. A complete mathematical description, therefore, requires the use of a tensor quantity. For a molecule tumbling isotropically, however, the average effect can be described by a single number, referred to as a *scalar*. As explained further below, this number is usually labeled J, and the effect is also sometimes called J-coupling or indirect coupling.

The effects from scalar coupling are usually much smaller than those of either shielding or dipolar coupling. Unlike the change in resonance frequency due to shielding, this effect does not depend on the external field strength. Scalar coupling is also different from dipolar coupling in that the former does not average out to zero as

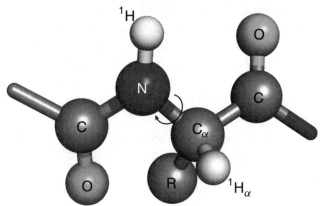

Figure 4.17

Covalent connectivity between the amide and α-hydrogen atoms of an amino acid residue. The bond between the amide-nitrogen and the α-carbon atoms rotates. "R" indicates the β-carbon of the amino acid side chain.

a molecule tumbles in solution, just as the chemical shift does not disappear. Thus, scalar coupling does lead to differences in resonance frequencies in liquid samples.

A simple view of scalar coupling can be gained by considering the equilibrium distributions of nuclei in the "spin-up" and "spin-down" states. In a population of molecules, the spin state of one nucleus will define two subpopulations. The state of this spin will then influence the local magnetic field felt by another nearby nucleus. As a consequence, the resonance frequency for a nucleus within a molecule with a spin-up neighbor will be slightly different than for the chemically equivalent nucleus in another molecule where the neighbor is in the spin-down state. In order for this effect to be detectable, the two nuclei must usually be linked by three or fewer covalent bonds (though four-bond couplings are sometimes observed), rather than just being in spatial proximity. Scalar coupling mediated by hydrogen bonds can, under some circumstances, also be detected.

As an example, the amide and α-protons (^1HN and ^1HA) of an amino acid residue in a polypeptide are linked by three bonds (Fig. 4.17). Because of the covalent linkage, the resonance frequency of the amide proton is influenced by the spin state of the α-proton, and vice versa. Thus, if the unperturbed chemical shifts of the two protons, were for instance, 8 and 5 ppm, the spectrum for just these two protons would contain two split peaks, as shown schematically in Fig. 4.18. The two components of each peak are separated by about 5 Hz, or 0.01 ppm in a 500-MHz spectrometer. Splittings of this magnitude are often resolvable, depending on the peak widths. (Because the splitting, in absolute frequency units, is not proportional to field strength, the relative effect of scalar coupling, in ppm, is actually greater in a low magnetic field than in a high one.) Scalar coupling can take place between different types of nuclei, such as between a proton and an ^{15}N nucleus, as well as between homonuclei. Also, a nucleus can be influenced by more than one other nucleus, leading to quite complex multiplet structures in the spectrum.

The magnitude of the splitting is called the coupling constant and is usually represented by the symbol ^nJ, with n indicating the number of covalent bonds separating the two nuclei. A subscript following the J is often used to identify the particular atoms involved. For instance, the coupling between the amide and α-protons in an

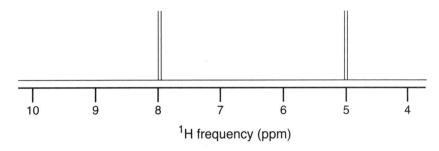

Figure 4.18 Schematic representation of the split ^{1}H NMR peaks corresponding to the amide and α-protons of an amino acid residue.

Table 4.3 Typical scalar coupling constants for nuclei pairs commonly found in organic and biological molecules.

Type	Structure	J (Hz) [a]
^1J	^{1}H $-^{13}$C	140
^1J	^{1}H $-^{15}$N	-90
^1J	^{13}C $-^{13}$C	40
^1J	^{13}C $-^{15}$N	-10
^2J	^{1}H $-$C$-^1$H	-10
^3J	^{1}H $-$C$-$C$-^1$H	2–10

a. The values listed here are adapted from the references listed on page 86 and should be taken only as approximate guides.

amino acid residue can be identified as $^3J_{\text{HN}-\text{HA}}$. The value of the coupling constant can be thought of as a measure of how much the spin state of one nucleus influences the energy of its neighbor.

The value of the coupling constant depends on the nuclei involved and the nature of the covalent linkage between them. In general, the coupling between directly bonded nuclei, ^1J, are larger than those between nuclei separated by two bonds (referred to as geminal coupling constants, ^2J) or three bonds (vicinal coupling constants, ^3J). Table 4.3 lists approximate values for the coupling constants for some of the nuclei pairs that are common in organic and biological molecules.

The exact values of the coupling constants are sensitive to the other constituents in the molecule and, for ^3J, the dihedral angle of the central bond, as discussed further below. Note that the coupling constant is a signed quantity, though the sign cannot be determined from simple spectra. When $J > 0$, the resonance frequencies of each of the nuclei are greater when their dipoles point in the same direction, while a negative coupling constant indicates that the frequencies are larger when the dipoles are oriented in opposite directions.

An important requirement for scalar coupling to be observed is that the two nuclei must have distinct resonance frequencies. This means, for instance, that the three

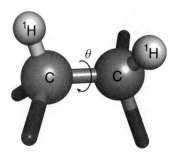

Figure 4.19

Two ^{1}H nuclei linked by a rotatable carbon–carbon bond. The coupling constant, ^3J, is sensitive to the dihedral angle of the central bond.

protons of a methyl group will generally not lead to splitting of one another's resonance lines. Thus, there will be only a single peak with an intensity three times that of a single proton. (The three protons of a methyl group usually have indistinguishable chemical shifts because the methyl group rotates freely, exposing the three protons to the same average environment.) The requirement for different resonance frequencies is an example of something that is not at all obvious from a simple "classical" view of NMR and emerges only from a quantum mechanical treatment (Chapter 13).

4.3.1   The Karplus Equation

When two nuclei are separated by three covalent bonds, the ^3J coupling constant depends on the dihedral angle, θ, of the central bond. For instance, for two protons on adjacent carbon atoms, the coupling constant will depend on the rotation about the carbon–carbon bond (Fig. 4.19).

The maximum coupling constants are usually found when the two nuclei are either aligned with one another ($\theta = 0$) or directly opposite one another ($\theta = 180°$). (By convention, a dihedral angle increases when an observer looks along the direction of the bond and the atom furthest from the viewer is rotated clockwise, as indicated by the arrow in Fig. 4.19.) In these orientations, the intervening electrons are optimally positioned to convey the spin state of the nuclei.

This relationship is expressed mathematically in an equation first proposed by Martin Karplus:

$$^3J = A \cos^2 \theta + B \cos \theta + C \tag{4.4}$$

Where A, B, and C are constants that depend on the nuclei involved and the chemical structure. These parameters are usually determined empirically. As an example, the graph in Fig. 4.20 shows the predicted Karplus relationship between $^3J_{\text{HN–HA}}$ and ϕ, the dihedral angle linking the amide nitrogen and the α-carbon (Fig. 4.17).[2]

---

[2] Note that the Karplus equation is expressed in terms of the dihedral angle defined by the nuclei of interest, in this case the amide and α-protons, whereas the value of ϕ is conventionally defined in terms of the heavy backbone atoms—namely the carbonyl carbons of the residue of interest and the preceding residue. As a consequence, the angles shown on the plot in Fig. 4.20 are shifted by 60° from the values used in the Karplus equation.

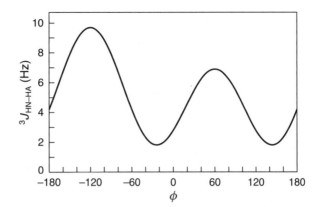

Figure 4.20
Dependence of the $^3J_{\text{HN}-\text{HA}}$ coupling constant on the ϕ dihedral angle, calculated from the Karplus equation.

Table 4.4 Some important ^3J scalar coupling constants for nuclei pairs in proteins.

Coupling constant	Nuclei	Dihedral angle	Atoms of central bond
$^3J_{\text{HN}-\text{HA}}$	Amide ^{1}H	ϕ	Amide nitrogen
	$\alpha\text{-}^1$H		α-carbon
$^3J_{\text{HA}-\text{HB}}$	$\alpha\text{-}^1$H	χ_1	α-carbon
	$\beta\text{-}^1$H		β-carbon
$^3J_{\text{N}-\text{HB}}$	Amide ^{15}N	χ_1	α-carbon
	$\beta\text{-}^1$H		β-carbon

If the bond rotates freely, the observed coupling constant will be an average based on the sampled dihedral angles. If, however, the bond is part of a constrained ring structure or a folded macromolecule, the dihedral angle may be fixed, and the observed coupling constant may provide valuable information about the value of the dihedral angle. In proteins, the most useful coupling constants include those listed in Table 4.4.

In many cases, the measured coupling constant will not be adequate to definitively determine the dihedral angle. For instance, as shown in Fig. 4.20, a $^3J_{\text{HN}-\text{HA}}$ of 3 Hz could be consistent with four values of the ϕ dihedral angle. However, if the residue is not a Gly, the steric restrictions shown on a Ramachandran plot exclude all of the values except those near $-60°$. In other cases, additional information, such as NOE constraints or other coupling constants, may help eliminate possible conformations. As an example, measuring both $^3J_{\text{HA}-\text{HB}}$ and $^3J_{\text{N}-\text{HB}}$ is especially useful in determining the first side-chain dihedral angle in amino-acid residues, χ_1. In practice, coupling constants can only be measured with modest precision and are used to establish relatively wide boundaries on possible dihedral angles (e.g., $\pm 30°$).

Coupling constants can be measured in a variety of ways, including a direct measurement of the frequency difference for the multiple peaks. Directly measuring the splitting requires high resolution spectra and good separation among the peaks from different nuclei. For measuring coupling constants in more complex molecules, several sophisticated multidimensional experiments have been devised.

4.4 Final Comments

In discussing both dipolar and scalar interactions, we have relied on a picture in which the nuclear spins are oriented either "up" or "down," a picture that is pervasive in elementary treatments of NMR. This picture serves us well in accounting for the splitting seen in simple spectra, whether due to residual dipolar coupling or scalar coupling. In Chapter 2, however, it was stressed that individual nuclei can exist in any of a limitless number of superposition states, which give rise to the full range of average directions of magnetization. When attempting to explain more complex experiments, such as multidimensional experiments based on scalar coupling, the simple picture becomes much more problematic, and it is necessary to develop a more complete quantum description, as presented in Chapter 13.

Summary Points for Chapter 4

- Nuclei of the same type (e.g., protons), but in different molecular environments, often display different resonance frequencies, reflecting differences in the local magnetic field.
- The major factors influencing the resonance frequency of a given nucleus are the local electronic structure, the fields of other nuclear dipoles acting through space and the effects of other nuclei acting via bonding electrons.
- Shielding is the effect of the electronic structure on the local magnetic field experienced by a nucleus and is the major cause of resonance frequency differences in liquid samples. Differences in resonance frequency due to shielding are referred to as chemical shifts.
  - The magnitude of the shielding effect is proportional to the external field strength, but chemical shifts are expressed in relative units (ppm) that are independent of field strength.
  - The magnitude of the shielding effect depends on the orientation of the molecule with respect to the external field, but is rapidly averaged in liquid samples.
- Dipolar coupling is the effect of one nuclear (or electron) dipole on another, exerted through space.
  - The magnitude of the dipolar coupling effect is independent of the external field strength, but depends strongly on the orientation of the molecule with respect to the external field and on the distance between the two dipoles.
  - In liquid samples, the dipolar coupling effect averages to zero, but still plays an important role in relaxation.
  - In solid samples, dipolar coupling leads to very broad spectra, but the effect can be reduced by magic angle spinning.

- Scalar coupling is the effect of one nuclear dipole on another, exerted via the bonding electrons linking the atoms.
  - Scalar coupling is usually only observed between nuclei separated by one to three covalent bonds.
  - Scalar coupling gives rise to "splitting" of NMR peaks.
  - The magnitudes of scalar coupling interactions, expressed as the absolute frequency differences (J) between split peaks, is independent of the external field strength.
  - The magnitude of J depends on the number and types of bonds linking the nuclei.
  - For nuclei linked by three bonds, J is sensitive to the dihedral angle of the central bond.

Exercises for Chapter 4

4.1. Suppose, for a particular NMR spectrometer, that the resonance frequency of TMS has been determined to be 600.001336 MHz. The chemical shifts of two ^{1}H nuclei of interest are 3.11 and 6.279 ppm.

(a) Which of the two ^{1}H nuclei has the larger resonance frequency?
(b) For which of the two ^{1}H nuclei is the absolute value of the resonance frequency larger?
(c) Calculate the difference between the resonance frequencies (in Hz) for the two nuclei.
(d) Suppose that the same sample were placed in a spectrometer with a field strength of 12 T. What will the difference between the resonance frequencies for the two ^{1}H nuclei be now?

4.2. Like the amide ^{15}N nucleus, a carbonyl ^{13}C nucleus in a peptide bond displays a significant chemical shift anisotropy. The chemical shift components are approximately

$$\delta_{11} = 250 \text{ ppm} \qquad \delta_{22} = 190 \text{ ppm} \qquad \delta_{33} = 100 \text{ ppm}$$

(a) Suppose that a molecule containing a carbonyl ^{13}C nucleus is placed in a magnetic field of 15 T. What will be the maximum difference in resonance frequencies (in Hz) for this nucleus with the molecule in different orientations?
(b) From the values given above, calculate the isotropic chemical shift (δ_{iso}), the chemical shift anisotropy ($\Delta\delta$), the asymmetry parameter (η) and the shielding anisotropy ($\Delta\sigma$).

4.3. A vivid and classic demonstration of the effect of internuclear dipolar interactions on an NMR spectrum was published by George Pake in 1948: Pake, G. (1948). Nuclear resonance absorption in hydrated crystals: Fine structure of the proton line. *J. Chem. Phys.*, 16, 327–336. http://dx.doi.org/10.1063/1.1746878

Pake, a graduate student of the NMR pioneer Edward Purcell, recorded spectra of ^{1}H nuclei in a single crystal of gypsum oriented at various directions in the static field. Gypsum is the stuff blackboard chalk and drywall are made of and is a dihydrate crystal of $CaSO_4$. The ^{1}H NMR spectrum arises from the water molecules in the crystal. As predicted, each hydrogen atom gave rise to two peaks in the spectrum, due to the dipolar coupling from the other hydrogen atom in the water molecule, and the separation between the peaks changed as the crystal was rotated in the field. From his data, Pake was able to estimate the distance between the two hydrogen atoms in a water molecule and the length of the O-H bond. This is probably the very first example of the use of NMR to derive a molecular structure, even if the structure was already pretty well known. Be sure to show all of the steps in your calculations and use the proper units. If you seek out the original paper, be aware that the units used there are different from those that we use today, and this can be very confusing.

(a) As was standard practice at the time, the spectrum was recorded by slowly varying the strength of the external field and measuring the signal intensity at a constant frequency, in this case 29 MHz. At what approximate static magnetic field strength would you expect the peak to occur for ^{1}H nuclei?

(b) Pake reported his results in terms of the difference in the static field associated with the two peaks. When the crystal was oriented so that the internuclear vector was perpendicular to the external field, the peaks were separated by 1.08×10^{-3} T. Following modern convention, this difference can be expressed, relative to the total static field, in units of parts per million. Calculate the field difference in ppm.

(c) The field difference in ppm is equal to the relative frequency difference (in ppm) in a fixed field strength. Calculate the frequency difference in Hz, assuming the fixed field strength calculated in part (a).

(d) From the frequency difference and Eq. 4.2, calculate the distance, r, between the two nuclei.

(e) What would the frequency difference be if the internuclear vector is parallel to the external field?

(f) The angle between the two bonds of a water molecule is approximately 108°. What is the length of the O-H bond? Hint: This is geometry not NMR!

Just for fun, find out what George Pake worked on later in life.

4.4. Scalar coupling constants are frequently used to help determine the χ_1 side-chain dihedral angle in peptides and proteins. This angle is defined as illustrated in Fig. 4.21.

One of the coupling constants that can be used for this purpose is the one linking the α- and β-protons. For some residue types, there are two β-hydrogens, labeled $\beta 2$ and $\beta 3$, as indicated in the figure, whereas the β-branched residues (Thr, Val and Ile) have only a single β-hydrogen. The Karplus equation coefficients for $^3J_{HA-HB}$ have been estimated[3] as follows:

[3] Pérez, C., Löhr, F., Rüterjans, H. & Schmidt, J. M. (2001). *J. Am. Chem. Soc.*, 123, 7081–7093. http://dx.doi.org/10.1021/ja003724j

$$A = 7.23 \qquad B = -1.37 \qquad C = 2.22$$

Note that θ in the Karplus equation (Eq. 4.4) refers to the angle defined by the two nuclei involved in the scalar coupling interaction, as indicated for $^1H_\alpha$ and $^1H_{\beta 3}$ in Fig. 4.21.

(a) Use the coefficients given above to make a plot of the expected $^1H_\alpha$–$^1H_{\beta 3}$ coupling constant as a function of χ_1. Note that $\theta = \chi_1$ for this coupling constant.

(b) Make a similar plot for the $^1H_\alpha$–$^1H_{\beta 2}$ coupling constant as a function of χ_1, for which $\theta = \chi_1 - 120°$.

(c) What other coupling constants could be used to place constraints on χ_1?

Further Reading

A more rigorous and formal discussion of the factors that influence nuclear resonance frequencies:

◇ Levitt, M. H. (2008). *Spin Dynamics: Basics of Nuclear Magnetic Resonance*, 2nd ed., pp. 195–228. Wiley, Chichester, U.K.

References defining conventions and recommendations for defining chemical shifts and their anisotropies:

◇ Harris, R. K., Becker, E. D., Cabral de Menezes, S. M., Goodfellow, R. & Granger, P. (2001). NMR nomenclature. Nuclear spin properties and conventions for chemical shifts. *Pure Appl. Chem.*, 73, 1795–1818. http://dx.doi.org/10.1351/pac200173111795

◇ Harris, R. K., Becker, E. D., Cabral de Menezes, S. M., Granger, P., Hoffman, R. E. & Zilm, K. W. (2008). Further conventions for NMR shielding and chemical shifts (IUPAC recommendations 2008). *Pure Appl. Chem.*, 80, 59–84. http://dx.doi.org/10.1351/pac200880010059

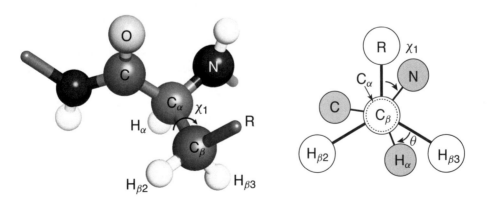

Figure 4.21 Ball-and-stick and Newman projection representations of the χ_1 side chain dihedral angle for an amino acid residue. Positive dihedral angles are defined by clockwise rotation of the atom distal from the viewer, as shown in the figure for $\chi_1 \approx 45°$. For Exercise 4.4.

A review of the factors influencing chemical shifts and methods for calculating them, with an emphasis on quantum mechanical calculations for small molecules:

◇ Facelli, J. C. (2004). Calculations of chemical shieldings: Theory and applications. *Concepts Magn. Reson.*, 20A, 42–69. http://dx.doi.org/10.1002/cmr.a.10096

A short review of the use of chemical shift information in the study of biological macromolecules and methods for predicting shifts in these molecules:

◇ Case, D. A. (2013). Chemical shifts in biomolecules. *Curr. Op. Struct. Biol.*, 23, 172–176. http://dx.doi.org/10.1016/j.sbi.2013.01.007

An extensive review of chemical shift data for proteins and their use in structural analysis:

◇ Wishart, D. S. (2011). Interpreting protein chemical shift data. *Prog. Nucl. Mag. Res. Sp.*, 58, 62–87. http://dx.doi.org/10.1016/j.pnmrs.2010.07.004

A detailed discussion of chemical shift anisotropy and the various conventions used to define the chemical shift tensor, as well as experimental methods for its measurement in solids:

◇ Grant, D. M. (2007). Chemical shift tensors. *eMagRes*. http://dx.doi.org/10.1002/978047 0034590.emrstm0074

A discussion of the theoretical and practical aspects of magic angle spinning:

◇ Andrew, E. R. (2007). Magic Angle Spinning. *eMagRes*. http://dx.doi.org/10.1002/978047 0034590.emrstm0283

A review on the measurement and use of residual dipolar interactions in the study of proteins, by one of the pioneers in this technique:

◇ Bax, A. (2003). Weak alignment offers new NMR opportunities to study protein structure and dynamics. *Protein Sci.*, 12, 1–16. http://dx.doi.org/10.1110/ps.0233303

Compilations of coupling constants for different functional groups:

◇ Becker, E. D. (2000). *High Resolution NMR: Theory and Chemical Applications*, pp. 123–127. Academic Press, San Diego, CA. http://dx.doi.org/10.1016/B978-012084662-7/50049-1

◇ Lambert, J. B. & Mazzola, E. P. (2004). *Nuclear Magnetic Resonance Spectroscopy: An Introduction to Principles, Applications and Experimental Methods*, pp. 119–121. Pearson Education, Upper Saddle River, NJ.

CHAPTER 5

The Pulse NMR Method: The Pulse

For many years after the initial experiments of Purcell and Bloch, the same basic method was used to obtain NMR spectra. Usually, this involved using a constant frequency of radiation and slowly sweeping the strength of the external magnetic field as the absorption of energy was monitored. Both the Purcell single-coil and Bloch double-coil arrangements were used in commercial instruments. This technique is generally referred to as *continuous wave* or CW spectroscopy.

The basic limitation of the CW method is related to the inherently weak signals associated with any NMR experiment, combined with the relatively long time required to collect a spectrum. Because the absorption lines in an NMR spectrum are so sharp, it is natural to want to have as much resolution as possible. This requires a slow sweep of the external field. At the same time, the signal is generally weak compared to the noise, even when extreme measures are used to eliminate electronic noise. Ultimately, the electronic noise represents thermal motions of the electrons in the circuits. Other than increasing the sample concentration, the most straight forward way to enhance the signal-to-noise ratio is to collect more data and average them. As more data points are added together, the signal increases in proportion to the number of points, while the total contribution from random noise increases only in proportion to the square root of the number of points. Thus, the signal-to-noise ratio can be cut in half by averaging four spectra. While this is good, it means that it takes a lot of spectra to obtain more substantial increases. If each spectrum takes 15 minutes, a 10-fold increase in signal-to-noise translates into 25 hours of data collection.

In 1965, Richard Ernst and Weston Anderson introduced a radically new method for collecting NMR spectra. (Others had pursued similar ideas previously, but Ernst and Anderson were the first to make it work.) Instead of collecting the spectrum piecemeal as the frequency or field strength was adjusted, they stimulated all of the nuclei at once with a pulse of many frequencies, and then detected the signal generated as all of the nuclei precessed and induced an electric current in the surrounding coil.

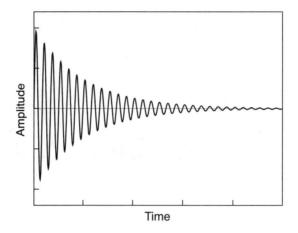

Figure 5.1

A damped sine-wave function.

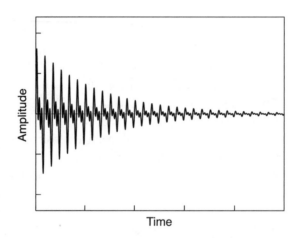

Figure 5.2

Three superimposed damped
sine-wave functions with
different frequencies.

A useful analogy might be the tuning of a piano. Normally, a piano tuner listens
to the tone generated by striking each key individually, and then adjusts the tension
on the string until he or she hears the correct note. An alternative would be to
simultaneously strike all 88 keys, provided that there were some way to sort out all of
the resulting frequencies. In principle, this might be faster, because the information
would be collected for all of the strings at once. That this might be possible can be
shown by looking at the signals that would be generated. Fig. 5.1 shows a plot of
amplitude versus time such as might be observed from striking a single string with
a pure resonance frequency.

The function plotted in Fig. 5.1 is a damped sine wave, in which the amplitude
of the signal decays exponentially. The decay represents the loss of energy from the
vibrating string. The graph in Fig. 5.2 shows the signal that would be generated by
striking three strings, with different frequencies, simultaneously.

Now, the three signals are superimposed to generate a more complicated function.
The numerous local maxima and minima in the graph represent the times at which

all three of the signals either reinforce or cancel one another. Although it isn't obvious how to separate the contributions, it should be apparent that the information for all three is present. While this scheme isn't generally used to tune all 88 notes on a piano, Derome (1987) reports that musical bells really are tuned in just this way. Here the key is having someone with a trained ear who can sort out the superimposed tones.

While the idea of pulse spectroscopy seems fairly straightforward, actually implementing it is quite demanding. The requirements include the following:

1. It must be possible to stimulate all of the absorbing nuclei simultaneously.
2. The resonant state must be relatively long-lived, so that there is time to "listen" to it and resolve the contributions of different frequencies.
3. It must be possible to collect and analyze the data so that the contributions from different frequencies are resolved.

The first two conditions are relatively easy to meet in the case of NMR spectroscopy. The radio signals used to stimulate the nuclei are coherent (i.e., in phase so that all of the nuclei are affected simultaneously), and the low energies associated with NMR transitions leads to relatively long relaxation times, as we have mentioned previously. In addition, the mathematics necessary to separate the combined signals was developed in the 1800s. However, implementing this mathematical technique (the Fourier transform) on the scale needed for NMR requires a digital computer, and the signal must be converted to a digital form. Until the mid 1960s, the requirements for digitizing, storing and manipulating the data exceeded the available technologies. In addition, the algorithm then used to implement the Fourier transform was very inefficient. At about this time, however, computer technology became powerful enough and cheap enough to make it possible to collect and store the digitized data. Of equal importance, there was a major advance in the algorithms for calculating Fourier transforms, which, for large data sets, speeded up the process by several orders of magnitude. (Actually, the mathematical method, now called the fast Fourier transform, FFT, had been invented much earlier but was not widely known until it was rediscovered in 1965 by James Cooley and John Tukey.) Together, these advances made pulse NMR, or Fourier transform NMR, a practical method with much greater sensitivity than the older CW method. In addition, the pulse method forms the basis of multidimensional techniques.

5.1   Some Preliminaries

When considering what happens to the spins and the net magnetization they create during a pulse experiment, things quickly become complicated as we try to keep track of the precession of magnetization arising from nuclei with different Larmor frequencies. In experiments involving multiple pulses, the problem becomes even more difficult unless systematic rules are followed. So, before going on to discuss the details of pulse experiments, it is worthwhile to spend a bit of time on some conventions, especially those used to describe rotational motion.

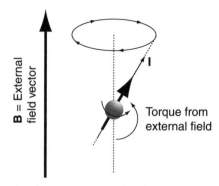

Figure 5.3   Precession of a quantum spin with a positive gyromagnetic ratio in a static magnetic field.

5.1.1   Of Right and Left Hands: Choosing Directions

In Chapter 2, the idea of precessional motion of a spin was introduced, using the drawing shown again in Fig. 5.3. In this drawing, the direction of precession is opposite that of the (fictional) spinning motion of the particle, as indicated by the negative sign in the Larmor equation:

$$\nu = -\frac{\omega}{2\pi} = -\frac{\gamma B}{2\pi}$$

where ν is the Larmor frequency, and ω is the angular velocity. For a nucleus with a positive gyromagnetic ratio (γ), the direction of precession is clockwise when looking down on the vector representing the external magnetic field. Unfortunately, different conventions for defining the direction of precession have been used by different authors over the decades, usually with reference to left- and right-hand rules, leading to a good deal of confusion.

In the hope of, at a minimum, not adding to this confusion, this text abides with conventions that have emerged as a consensus over the past two decades, particularly since the publication of a landmark text by Ernst and colleagues.[1] These conventions are based on the mathematical definitions for a right-handed coordinate system and the right-hand rule employed in the field of electricity and magnetism. One way to visualize the handedness of the coordinate system is illustrated in Fig. 5.4, in which a right hand is shown oriented with the thumb pointed along the positive direction of the z-axis. In this orientation, the fingers curl in the direction of progression from the positive x-axis to the positive y-axis.

The orientation of the right hand is also used to define the direction of rotation. For rotation about the z-axis, positive rotation is in the direction of the fingers (e.g., from the positive x-axis to the positive y-axis). Negative rotation is in the opposite direction, from the positive x-axis to the *negative* y-axis.

---

[1] Ernst, R. R., Bodenhausen, G. & Wokaun, A. (1987). *Principles of Nuclear Magnetic Resonance in One and Two Dimensions*. Oxford University Press, Oxford.

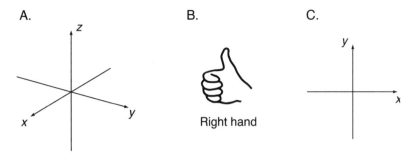

Figure 5.4 Convention for defining a right-handed coordinate system. (A) Perspective drawing of the x,y,z coordinate system in the right-hand orientation. (B) With the thumb pointed along the z-axis, the fingers of the right hand curl from the x-axis to the y-axis. (C) If the x-y plane is rotated to lie in the plane of the page and in the orientation usually used to describe a two-dimensional coordinate system, the positive z-axis projects toward the viewer.

The directions of rotation about the other two axes are defined in the same way: The thumb of the right hand is pointed along the positive axis, and the curl of the fingers defines the direction of positive rotation. The directions for positive and negative rotation about each of the axes are summarized in Table 5.1. It is important to emphasize that the rules summarized in Table 5.1 are defined in terms of the coordinate system, not the forces that give rise to a particular motion.

The standard convention for describing the magnetic field generated by a moving charge is also described as a right-hand rule. The application of this rule to the case of the spinning charge is illustrated in Fig. 5.5. The fingers of the hand are curled in the direction in which the ball is spinning, and the thumb points in the direction of the angular momentum vector, which, for a nucleus with positive γ, is also the direction of the induced magnetic field. If the direction of the induced field is also aligned with the z-axis of a right-handed coordinate system, as in Fig. 5.5, the rotation of the spinning ball is in the direction progressing from the positive x-axis to the positive y-axis and is defined to be positive with respect to the z-axis.

As illustrated in Fig. 5.3, the direction of precession is opposite that of the angular momentum. With the external magnetic field aligned with the positive z-axis, precession of a spin with a positive gyromagnetic ratio is in the negative direction, as

Table 5.1 Rules defining the directions of rotation in a right-handed coordinate system

Axis	Positive rotation	Negative rotation
x	$z \rightarrow -y \rightarrow -z \rightarrow y$	$z \rightarrow y \rightarrow -z \rightarrow -y$
y	$z \rightarrow x \rightarrow -z \rightarrow -x$	$z \rightarrow -x \rightarrow -z \rightarrow x$
z	$x \rightarrow y \rightarrow -x \rightarrow -y$	$x \rightarrow -y \rightarrow -x \rightarrow y$

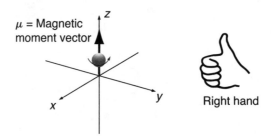

Figure 5.5 The direction of magnetization from a spinning charge as defined by the right-hand rule, for a positive gyromagnetic ratio. With the induced magnetic field aligned with the positive z-axis, the direction of rotation is positive in the right-handed coordinate system.

illustrated in Fig. 5.6. The direction of precession illustrated in Fig. 5.6 can also be described as being left-handed, as shown in the figure: With the thumb of the left hand pointed in the direction of the external field, precession follows the curl of the fingers.

Note that the direction of precession is determined by the direction of the external field and the gyromagnetic ratio, whereas the orientation of the coordinate axes is arbitrary. We could, for instance, rotate the coordinate axes so that the positive x-axis is aligned with the external field (while maintaining the handedness of the coordinates), as illustrated in Fig. 5.7. The direction of precession is still left-handed and follows the direction from the positive z-axis to the positive y-axis, which is still a negative rotation in the coordinate system, now about the x-axis.

In each example above, the external magnetic field was aligned with a positive coordinate axis, leading to a negative direction of precession with respect to the coordinate system (for a positive gyromagnetic ratio). What if we align the magnetic field with a *negative* axis, as shown in Fig. 5.8 for the z-axis? Notice that reversing the direction of the z-axis changes the relationship of the x- and y-axes with respect to the page (and the precessing spin), but the handedness of the coordinate system is maintained (as you should be able to confirm with your right hand.) Now, however,

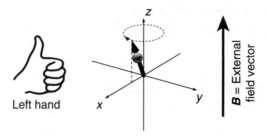

Figure 5.6 Left-handed precession for a spin with positive gyromagnetic ratio. With the external magnetic field aligned with the positive z-axis, precession is in the negative direction relative to the right-handed coordinate axes.

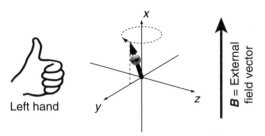

Figure 5.7 Left-handed precession for a spin with positive gyromagnetic ratio, with the coordinate axes rotated so that the x-axis is aligned with the external magnetic field. In this orientation the rotation is defined to be negative about the x-axis

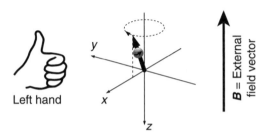

Figure 5.8 Left-handed precession for a spin with positive gyromagnetic ratio, with the external magnetic field aligned with the *negative z*-axis. Precession is now in the positive direction relative to the right-handed coordinate axes.

the precession is in the direction from the positive x-axis to the positive y-axis, which represents *positive* rotation about the (positive) z-axis.

To summarize, the direction of precession relative to the external magnetic field is determined by the sign of the gyromagnetic ratio (left-handed for positive γ and right-handed for negative γ), and is independent of the orientation of the external field with respect to the coordinate system. But, the sign of the precessional motion in the coordinate system *does* depend on the orientation of the external field with respect to the coordinate system. With the field aligned with a positive coordinate axis, the sign of rotation is the same as that of the frequency (ν) and opposite that of the gyromagnetic ratio. By convention, the direction of the stationary magnetic field of an NMR spectrometer is used to define the positive z-axis.

5.1.2   Mathematics of Rotational Motion

Next, we consider the description of rotational motion in terms of projections on the coordinate system. First imagine a vector initially aligned with the x-axis of the coordinate system defined earlier and its motion about the z-axis, on a circular path

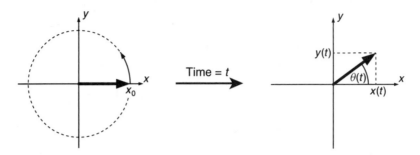

Figure 5.9   Rotational motion of a vector in the x–y plane.

in the x–y plane, as illustrated in Fig. 5.9. In this figure, we are viewing the x–y plane from the positive z-axis, which we will ignore for now. At the beginning, we can define the position of the vector in the x–y plane by its projections on the x and y axis, which we will call x_0 and y_0. In this case, $y_0 = 0$. We can describe the position of the vector after time t either by the new projections, $x(t)$ and $y(t)$, or by the angle $\theta(t)$ formed between the vector and the x-axis, as shown in the right panel of Fig. 5.9.

   If the motion is uniform, then $\theta(t)$ changes linearly with time, according to

$$\theta(t) = \omega t$$

where ω is the angular velocity. If ω is positive, then the motion is from the positive x-axis toward the positive y-axis, as defined above for a positive rotation. $\theta(t)$ can also be expressed in terms of the frequency, ν:

$$\theta(t) = 2\pi \nu t$$

Again, ν should be treated as a signed quantity to indicate the direction of rotation.

   Because the tip of the vector follows a circular path, the values of $x(t)$ and $y(t)$ must always satisfy the equation for a circle, with the radius defined by the initial x-projection:

$$x(t)^2 + y(t)^2 = x_0^2$$

As t and $\theta(t)$ increase (for positive motion), the x-projection initially decreases and the y-projection increases, until $\theta(t) = \pi/2$. At this point, $x(t)$ becomes negative, and $y(t)$ begins decreasing. These changes are described by the cosine and sine trigonometric functions:

$$x(t) = \cos\big(\theta(t)\big)$$

$$y(t) = \sin\big(\theta(t)\big)$$

The time-dependent projections can also be expressed in terms of ω or ν:

$$x(t) = x_0 \cos(\omega t) = x_0 \cos(2\pi \nu t)$$

$$y(t) = x_0 \sin(\omega t) = x_0 \sin(2\pi \nu t) \tag{5.1}$$

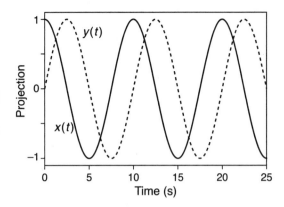

Figure 5.10

Cyclical changes in the x- and y-projections (solid and dashed curves, respectively) of a vector rotating about the z-axis, in the positive direction, as illustrated in Fig. 5.20 and described by Eq. 5.1 with $\nu = 0.1$ Hz and $x_0 = 1$.

The changes in the projections are plotted in Fig. 5.10, for the case where $\nu = 0.1$ Hz and $x_0 = 1$.

If the frequency, ν, is negative, then the motion will proceed from the positive x-axis to the negative y-axis. The corresponding changes in the projections are plotted in Fig. 5.11, without any change to the equations. Note that the curve describing the x-projection is unchanged, while the curve for the y-projection is inverted. These patterns reflect the following identities:

$$\cos(-\theta) = \cos(\theta)$$

$$\sin(-\theta) = -\sin(\theta)$$

These relationships are sometimes expressed by saying that the cosine function is "even" and the sine function is "odd."

Rotational motion in the $x-y$ plane does not have to start at the x-axis, and a more general description defines the starting position in terms of an angle, θ_0, between the x-axis and the vector in its initial position, as shown in Fig. 5.12. Now, the angle at time t is given by:

$$\theta(t) = \theta_0 + \omega t = \theta_0 + 2\pi \nu t$$

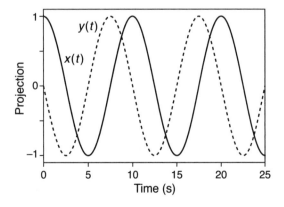

Figure 5.11

Cyclical changes in the x- and y-projections (solid and dashed curves, respectively) of a vector rotating about the z-axis, in the negative direction, as described by Eq. 5.1 with $\nu = -0.1$ Hz and $x_0 = 1$.

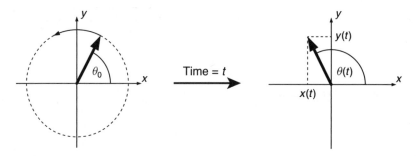

Figure 5.12 Rotational motion of a vector in the $x-y$ plane, starting at a position defined by the angle θ_0.

and the x- and y-projections are

$$x(t) = r \cos(\theta_0 + \omega t) = r \cos(\theta_0 + 2\pi \nu t)$$
$$y(t) = r \sin(\theta_0 + \omega t) = r \sin(\theta_0 + 2\pi \nu t)$$

(5.2)

where r is the radius of the circle swept out by the tip of the vector and can be calculated from the starting projections as

$$r = \sqrt{x_0^2 + y_0^2}$$

The starting angle, θ_0, is commonly referred to the *phase angle*, or just the *phase*, of the cosine and sine functions. The resulting time-dependent changes in x and y are plotted in Fig. 5.13 for the case where $\theta_0 = 0.3\pi$ rad. Comparison of Figs. 5.11 and 5.13 shows that the time axis of the latter is simply shifted to the right by the time required for the vector to rotate from the x-axis to the starting position, θ_0. The time-axis offset is given by θ_0/ω, which is 1.5 s in the case illustrated in Figs. 5.12 and 5.13.

Figure 5.13
Cyclical changes in the x- and y-projections (solid and dashed curves, respectively) of a vector rotating, in the positive direction, about the z-axis, starting at an initial position θ_0 from the x-axis, as illustrated in Fig. 5.23 and described by Eq. 5.2 with $\nu = 0.1$ Hz, $r = 1$ and $\theta_0 = 0.3\pi$ rad.

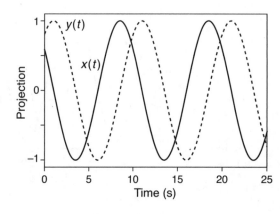

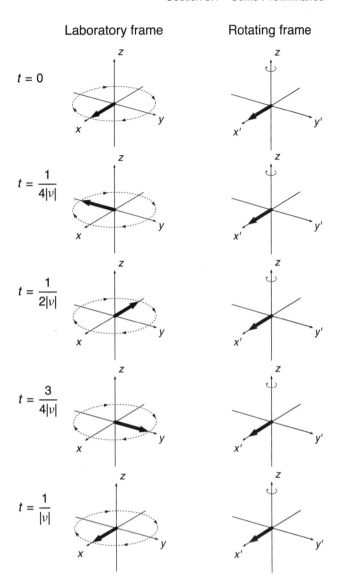

Figure 5.14
Precession as viewed in
the laboratory and rotating
reference frames.

5.1.3 The Rotating Frame

As a final preliminary to discussing pulse NMR, we introduce the concept of a rotating frame of reference. So far, we have followed things from the laboratory frame of reference, from which it appears that the magnetization is spinning around the z-axis at typically 500,000,000 revolutions per second (for protons). Now, we will place ourselves, conceptually, on a coordinate system that is itself revolving at the precession frequency of one of the nuclei we are considering. The diagrams in Fig. 5.14 show how the precession of a single spin population, with a single Larmor frequency, v, is depicted in the two frames of reference.

To remind us that we are looking at a rotating coordinate system, the axes are labeled x' and y', and a circular arrow is drawn at the top of the z-axis, which is not affected by the transformation. Nothing is really changed by this manipulation, and the relationship between the laboratory reference frame and the rotating frame is changing constantly at the precession frequency, ν.

Notice also that the time values in Fig. 5.14 are expressed relative to the frequency, ν, a practice that is often convenient for representing general cases. Because the frequency can be either positive or negative, however, we use the absolute value of the frequency, $|\nu|$, to make sure that time progresses in the right direction.

The usefulness of the rotating frame becomes most apparent when there are two types of spins precessing. Suppose that we now have two spin populations, one with a Larmor frequency, ν_0, and the other with a larger chemical shift, such that the absolute value of the frequency is greater by δ ppm.[2] The frequency for this second spin population will be

$$\nu = \nu_0 + \delta\nu_0 \cdot 10^{-6}$$

If the frequency of the rotating frame is set to ν_0, the spin with this Larmor frequency will be stationary in the rotating frame, and the apparent frequency of the other spin will be

$$\nu_{\text{rot}} = \nu - \nu_0 = \delta\nu_0 \cdot 10^{-6}$$

The magnetization from the first population remains aligned with the x'-axis in the rotating frame, and that of the second will precess, as shown in Fig. 5.15. For clarity, the vectors representing the spins with the two different frequencies are displaced above and below the x–y plane. Keeping one of the magnetization vectors stationary in the rotating frame makes it much easier to follow the difference in magnetization arising from the two populations. Suppose, for instance, that ν_0 is -500 MHz and δ is 1 ppm. We would have to wait $1/1000$ s before seeing the two magnetization vectors pointing in opposite directions. This may not sound very long, but if we were following this process in the laboratory frame, we would have to watch 500,000 cycles of the first magnetization precessing before this happened! In the rotating frame, we watch leisurely and wait for differences to become apparent.

The transformation to the rotating frame can be viewed as a (large) reduction in the effective strength of the magnetic field along the z-axis. Remember that the frequency is related to the strength of the magnetic field, B, according to

$$\nu = -\frac{\gamma B}{2\pi}$$

For the population with frequency, ν_0, in the laboratory frame, the frequency in the rotating frame is 0. Thus, the effective field in the rotating frame is 0. In general, the

---

[2] For a spin with positive gyromagnetic ratio, as indicated in the illustrations, keep in mind that this means that ν is more negative than ν_0.

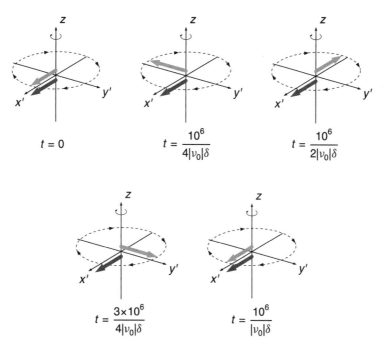

Figure 5.15 Precession of two spins with different Larmor frequencies, visualized in the rotating frame.

field in the rotating frame, B_{rot}, can be calculated as[3]

$$B_{rot} = B + \frac{2\pi v_0}{\gamma}$$

For the spins with Larmor frequency shifted by δ ppm, the *effective* field, B', in the *laboratory* frame is:

$$B' = -\frac{2\pi \left(v_0 + \delta v_0 \cdot 10^{-6}\right)}{\gamma}$$

In the rotating frame, the effective field is

$$B'_{rot} = -\frac{2\pi \left(v_0 + \delta v_0 \cdot 10^{-6}\right)}{\gamma} + \frac{2\pi v_0}{\gamma}$$

$$= -\frac{2\pi \delta v_0 \cdot 10^{-6}}{\gamma}$$

consistent with an apparent precession frequency of $\delta v_0 \cdot 10^{-6}$.

Another important consequence of the transformation to the rotating frame is that the direction of precession in the rotating coordinate system can be either positive or negative. In the example illustrated in Fig. 5.15, the gyromagnetic ratio was

---

[3] Because γ and v_0 have opposite signs, the rotating frame field is always reduced.

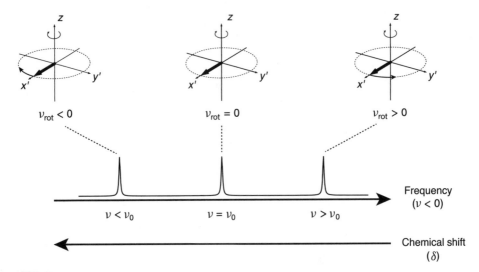

Figure 5.16 Relationships among laboratory-frame Larmor frequencies (ν), the rotating frame frequency (ν_0) and the precession frequencies in the rotating frame (ν_{rot}) for nuclei with positive gyromagnetic ratios. The rotating frame rotates in the *negative* direction relative to the laboratory frame.

assumed to be positive, leading to negative rotation in the laboratory frame, and the frequency of interest was more negative than the rotating frame frequency. Thus, the precession in the rotating frame was also in the negative direction. If the frequency is less negative than the rotating frame frequency, precession in the rotating frame is in the positive direction. The relationships among the various frequencies and the position of the peaks in an NMR spectrum for nuclei with a positive gyromagnetic ratio are illustrated in Fig. 5.16.

Following standard NMR practice, the absolute value of the laboratory-frame Larmor frequency ($|\nu|$) is plotted from right to left in Fig. 5.16, as is the chemical shift (δ). The rotating frame rotates in a negative sense, and the direction of precession in the rotating frame correctly reflects the sign of ν_{rot}.

For a nucleus with a negative gyromagnetic ratio, the Larmor frequency is positive, as is the direction of rotation for the rotating frame, as illustrated in Fig. 5.17. Again, the chemical shifts and absolute value of the laboratory-frame Larmor frequencies are plotted from right to left, but for $\gamma < 0$, this also means that ν increases from right to left in the spectrum.

What remains the same for positive and negative gyromagnetic ratios is the relationship between the direction of precession in the rotating frame and the sign of the rotating frame frequency, ν_{rot}. In both cases, the sign of ν_{rot} matches the convention for the direction of rotation in a right-handed coordinate system. Also, for both positive and negative gyromagnetic ratios, the absolute value of the Larmor frequency plotted in a spectrum increases from right to left (from upfield to downfield). As we will see in Chapter 6, the signal that is recorded in a pulse NMR experiment reflects ν_{rot}, and we will return then to the issue of positive and negative frequencies.

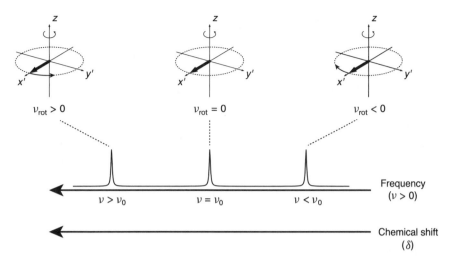

Figure 5.17 Relationships among laboratory-frame Larmor frequencies (ν), the rotating frame frequency (ν_0) and the precession frequencies in the rotating frame (ν_{rot}) for nuclei with negative gyromagnetic ratios. The rotating frame rotates in the *positive* direction, relative to the laboratory frame.

5.2 A Description of Electromagnetic Radiation in the Rotating Frame

As discussed earlier, the goal in a pulse NMR experiment is to simultaneously stimulate all of the nuclei (of a particular type) in a sample and then monitor the precession of the resulting non-eigenstates. This requires a source of electromagnetic radiation covering the entire frequency range of interest. In practice, this is accomplished using an rf source with a single frequency that is turned on for a very short time, typically a few microseconds for liquid samples. The short duration of the pulse leads to an uncertainty in the frequency, so that a quite broad band of radiation frequencies is actually produced.

The notion that a short irradiation time leads to a wide frequency range is critical to the implementation of pulse NMR, and a physical analogy may help illustrate this principle. Suppose that you are pushing a child on a playground swing, which, like any pendulum, has a natural frequency that reflects the length of the ropes and the force of gravity. In order to keep the swing moving or increase the amplitudes of its motion, you give the child a carefully timed push during each oscillation. If the time interval between pushes matches the period of the pendulum, each push will be effective in imparting additional energy to the swing. It takes only a little practice to match the frequency of the swing. Suppose, though, that instead of an adaptable human, the swing is being pushed by a rigidly timed machine that has a frequency that is a little bit different from the natural frequency of the swing. For the first few cycles, the pushes will come just a bit earlier or later than the first one, relative to the swing's cycle, but will still be able to add energy to the swing. After a longer time, however, the machine will be trying to push the swing when it is on the opposite side of its cycle, making it completely ineffective. In an extreme case, a single very strong push will be effective for a swing with any frequency. On the other hand, gentle pushes over many cycles are only effective as long as the machine and the swing are roughly in phase.

The same principle applies to the stimulation of transitions in NMR, or any other form of spectroscopy. For ^{1}H spectra, for instance, the pulse typically has a frequency of about 500 MHz and a duration of a few microseconds. Thus, there are only about a thousand cycles during the pulse. If the natural frequency for the transition differs from the frequency of the radiation by 1 part in 1000, the two will be 180° out of phase at the end of the pulse, but during most of the pulse the radiation will still have been effective in interacting with the nucleus. On the other hand, if the frequency difference is much greater than 1 part in 1000, the phase difference will accumulate more rapidly, and the effectiveness of the pulse will be greatly reduced. In this case, the pulse is able to stimulate nuclei with Larmor frequencies that cover a range of about 1 part in 1000, or 1000 ppm. This range of frequencies is more than adequate to stimulate protons with the normal range of resonance frequencies.

Although pulses with broad frequency ranges are central to the basic pulse NMR method, there are circumstances when it is desirable to selectively stimulate nuclei with a narrow range of Larmor frequencies. The general approach to creating more selective pulses is to reduce the power of the radiation and apply it for a longer period of time. This, however, can introduce additional complications, and further refinements involve shaping the pulse so that the intensity of the radiation varies during the time it is applied. The distribution of frequencies generated by pulses of differing duration and shape can be described quantitatively using Fourier analysis, and we will return briefly to this subject in Section 6.9, after introducing Fourier transforms.

In order to have a net effect on the sample, the electromagnetic radiation must be directed, or polarized, along a single direction. We will consider first the case of the electromagnetic radiation propagating along the y-axis in our laboratory-frame coordinate system. At a given point in the coordinate system, the resulting magnetization from the radiation will oscillate back and forth along the x-axis. This oscillation is generated by placing a coil around the sample with its axis aligned with the x-axis, as shown in Fig. 5.18.

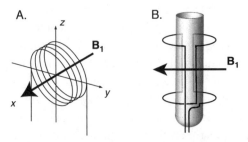

Figure 5.18 Coil used to generate an oscillating magnetic field. A. Schematic representation of a coil with its axis aligned with the x-axis in the laboratory frame. B. A slightly more realistic representation of the saddle coils actually used in NMR spectrometers, showing only a single loop of the coil surrounding the sample tube. Real probes usually contain multiple loops arranged approximately as shown here.

The coil is connected to a oscillating electrical circuit, which generates a fluctuating magnetic field, as shown by the thicker arrows in column A of Fig. 5.19. Note that the frequency of the radiation is $|v_0|$, the absolute value of the frequency we used earlier to define the rotating frame. (In the context of electronic signals, frequencies are not usually signed.)

The oscillating magnetization can also be represented as the sum of two magnetic dipoles rotating about the z-axis in opposite directions, as in column B of Fig. 5.19. Note that the y-components of the two dipoles always cancel out exactly, while the net x-component oscillates just as in the representations in column A. If we now change our view to the rotating frame, the vector rotating clockwise (looking down on the z-axis) appears to be stationary, while the other vector (drawn slightly below the x'–y' plane) rotates counterclockwise at twice the frequency observed in the laboratory frame (column C of Fig. 5.19).

If the effect of the dipole that is rotating at $2v_0$ in the rotating frame is ignored (because $2v_0$ is too far from the Larmor frequency to influence the spins), then the oscillating magnetic field can be interpreted as a static field along the x'-axis in the rotating frame. The effect of the oscillating field can be described using either the classical analogy with precession of a spinning object or quantum mechanically. In this chapter, the focus is on the classical, or vector, representation, and a quantum mechanical treatment is presented in Chapter 12. In both treatments, we will assume that the equilibrium population is composed of only the $|\alpha\rangle$ and $|\beta\rangle$ states, with a slight excess in the $|\alpha\rangle$ state. Thus, net magnetization initially lies in the positive z-direction.

5.3   The Vector Description of a Pulse

In the classical treatment, we consider the effects on a spinning charge that is initially aligned with its magnetic moment along the z-axis. Remember that in the rotating frame, there is no effective field along the z-axis (provided that the Larmor frequency is the same as the rotating-frame frequency.) When the electromagnetic radiation is turned on, the only important magnetic field is the one that is static in the rotating frame. When this field interacts with the angular momentum of the spinning particle, the particle precesses, but now the precession is around the axis defined by the constant field in the rotating frame. If this field lies along the x'-axis, then the magnetization precesses in the y'–z plane, as shown in Fig. 5.20. For a spin with a positive gyromagnetic ratio, the rotation of the magnetization is left handed with respect to the direction of the stationary component of the magnetic field (in the rotating frame), as illustrated in the figure and consistent with the description of precession about the z-axis.

In this case, we describe the pulse as a "negative x-pulse." Importantly, the sign of the pulse indicates the direction of rotation in the right-handed coordinate system, not the orientation of the magnetic field. By following this convention, the directions of pulse rotations are defined in the same way for both positive and negative gyromagnetic ratios, using the right-hand rule and the directions listed in Table 5.1

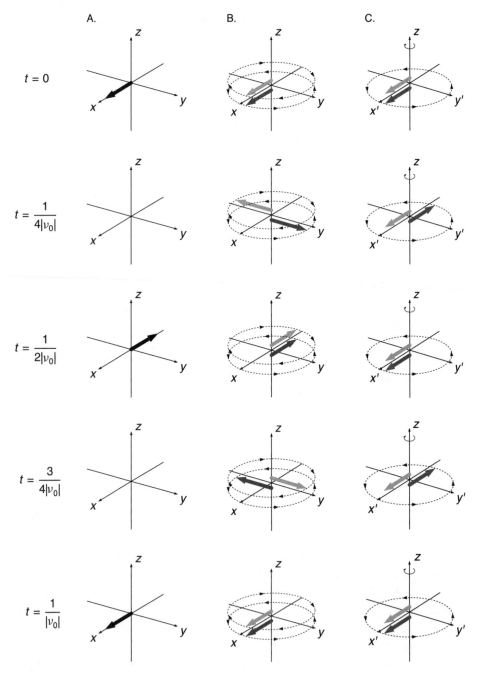

Figure 5.19 Three reprsentations of an oscillating magnetic field aligned with the x-axis of the laboratory reference frame. The thicker arrows in column A represent the direction of the oscillating field. In columns B and C, the oscillating field is represented as two dipoles rotating in opposite directions and viewed in the laboratory and rotating reference fields, respectively. The rotating reference frame is shown for the case of a positive gyromagnetic ratio and negative rotating-frame frequency, ν_0.

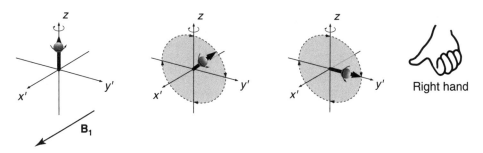

Figure 5.20 Vector description of the effect of a pulse of electromagnetic radiation on the equilibrium z-magnetization. For a positive gyromagnetic ratio The rotation is in the negative direction in the right-handed coordinate system.

(p. 91). When the pulses are actually implemented in the spectrometer, the software controlling the instrument must take into account the sign of the gyromagnetic ratio.[4]

The strength of the oscillating magnetic field is designated B_1. If we stop the radiation at exactly the moment that the magnetic moment of the particle is pointing along the y'-axis, then we will have moved magnetization from pointing in the z-direction to a position lying in the x'–y' plane. After the pulse, if the precession frequency matches the rotating-frame frequency, the magnetization will remain pointed along the y'-axis in the rotating frame. In the laboratory frame, however, the magnetization will be precessing at the Larmor frequency, and this motion can be detected as the current induced in a coil. (The coil is stationary in the laboratory frame.)

To help visualize what happens during a pulse, it may be helpful to think about the motion of the magnetization vector in the fixed laboratory frame of reference. Just before the pulse, when the magnetization is at its equilibrium position aligned with the z-axis, there is no precessional motion, because there is no transverse magnetization component. As soon as the pulse begins, however, a transverse component is generated, and this component begins to precess about the z-axis. At the same time, the oscillating magnetic field continues to act on the nuclear spins and moves the magnetization further away from the z-axis. The combination of these forces causes the average magnetization vector to follow a spiral path from the z-axis to the transverse plane, as described previously in Chapter 2 and illustrated again in the top row of drawings in Fig. 5.21. In these drawings, the spiral curve represents the path of the tip of the magnetization vector on the surface of a sphere.

The drawings in the upper part of Fig. 5.21 greatly exaggerates the vector's rate of motion toward the transverse plane, relative to its rate of precession. Typically, the vector might process around the z-axis approximately 1000 times (in the laboratory frame) before reaching the transverse plane. If we move back to the rotating frame, however, we find that this frame of reference is revolving about the z-axis at the same

---

[4] Unfortunately, spectrometer manufacturers have not yet settled on a universal syntax for defining pulse directions, but the convention used here appears to now be widely adopted in the literature describing pulse sequences.

Laboratory frame

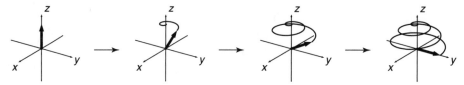

Rotating frame

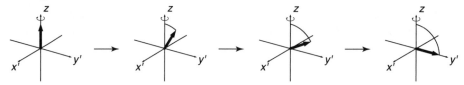

Figure 5.21 The path of magnetization during a pulse, as viewed in the laboratory and rotating reference frames.

rate as the magnetization vector, so the only apparent motion of the vector is its direct rotation from the z-axis to the y'-axis, as shown in the lower set of drawings in Fig. 5.21. The two sets of drawings in Fig. 5.21 should help convince you of the convenience afforded by using the rotating frame for most descriptions of NMR.

The rate at which the magnetic moment is tilted from the z-axis to the y'-axis is given by the Larmor equation, now using the strength of the oscillating field, B_1. If the frequency is expressed as an angular velocity, the Larmor equation is

$$\omega = -B_1\gamma$$

After time, t, the angle of the rotation, a, is

$$a = t\omega = -tB_1\gamma$$

For the case illustrated above, the magnetization is aligned with the y'-axis when $a = -\pi/2$ (because the direction of rotation is negative for positive γ). This will occur when

$$t = \frac{\pi}{2B_1\gamma}$$

This is commonly referred to as a "90-pulse" or a "$\pi/2$-pulse." The direction of the pulse is designated by a subscript, so that the pulse illustrated above is called a "$(\pi/2)_{-x}$-pulse," with the negative sign indicating negative rotation in the coordinate system. A positive $(\pi/2)_x$-pulse is illustrated in Fig. 5.22. A $\pi/2$-pulse serves the important role of generating magnetization in the x-y plane, which then precesses and generates the NMR signal.

A "y-pulse" is one in which the component of the oscillating magnetic field that is stationary lies along the y'-axis. If applied to the equilibrium magnetization state, a positive $(\pi/2)_y$-pulse rotates the z-magnetization about the y'-axis in the x'–z

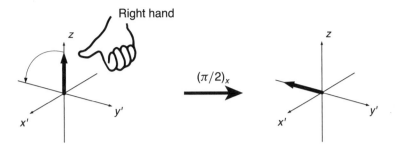

Figure 5.22   A positive $(\pi/2)_x$-pulse. The direction of a positive pulse follows the right-hand rule, with the thumb pointed along the positive x'-axis.

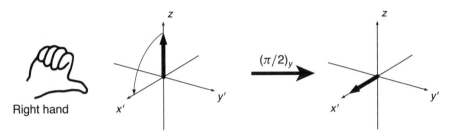

Figure 5.23   Rotation of equilibrium z-magnetization to the x'-axis by a $(\pi/2)_y$-pulse. The pulse follows the right-hand rule for rotation about the y'-axis.

plane, as illustrated in Fig. 5.23. The direction of the y-pulse is defined so that it tilts magnetization from the positive z-axis toward the *positive x'-axis*. While this may, at first glance, seem contrary to the direction of the positive $(\pi/2)_x$-pulse, this is the direction specified by the right-hand rule, as illustrated in the figure.

Pulses can be applied when the net magnetization lies in any direction, and multiple pulses can be applied sequentially. For instance, the diagram in Fig. 5.24 shows the effects of three sequential $(\pi/2)_y$-pulses, starting with the equilibrium z-magnetization. A fourth $(\pi/2)_y$-pulse would return the magnetization to its starting position along the z-axis.

Consider next what would happen if the first $(\pi/2)_y$-pulse were followed by a pulse along the x'-axis, as illustrated in Fig. 5.25. In this case, the second pulse has no effect, because the direction of the pulse is the same as that of the net magnetization. This situation is analogous to the absence of precession in the presence of a uniform

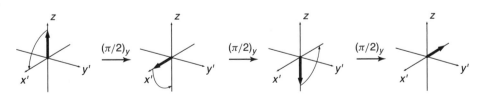

Figure 5.24   Three sequential $(\pi/2)_y$-pulses applied to initial z-magnetization.

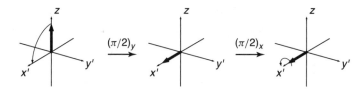

Figure 5.25   A $(\pi/2)_y$-pulse applied to z-magnetization, immediately followed by an x-pulse.

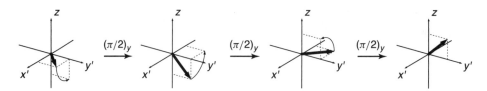

Figure 5.26   Three successive $(\pi/2)_y$-pulses applied to initial magnetization positioned between the x'- and y'-axes.

magnetic field along the z-axis when the net magnetization from the nuclei also lies along the z-axis.

The diagrams in Fig. 5.26 show the effects of three sequential $(\pi/2)_y$-pulses, this time beginning with the magnetization in the x'–y' plane, midway between the x'- and y'-axes. In this case, the y'-component of the magnetization remains constant through the pulses, and the non-y'-component alternates between projections on the x'- and z-axes. As before, a fourth $\pi/2$-pulse will return the magnetization to its starting position (assuming that nothing else happens during the periods between the pulses).

The examples shown above demonstrate the following important generalizations

1. The pulse does not affect any component of the magnetization that lies along the direction of the pulse.
2. Applying four $\pi/2$-pulses in the same direction always returns the magnetization to the original orientation.
3. With a combination of x- and y-pulses of appropriate duration, the magnetization can be made to point in any direction.

After a pulse, the resulting magnetization will evolve with time. Any component lying along the z-axis will remain stationary, and any component with a projection onto the x–y plane will precess at the Larmor frequency.

5.4   Pulses in Different Directions: Timing is Everything!

So far, we have ignored an important practical question: How do we define the direction of a pulse as being along the x'- or y'-axis of the rotating frame? This is really only an issue if an experiment involves more than one pulse, as in multidimensional experiments, because it is the relative directions of the pulses that turn out to be im-

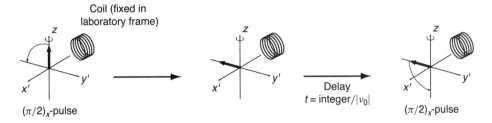

Figure 5.27 Two successive $(\pi/2)_x$-pulses separated by a delay interval, as viewed in the laboratory frame. The time interval between the pulses determines the relative orientation of the rotating frame and the coil.

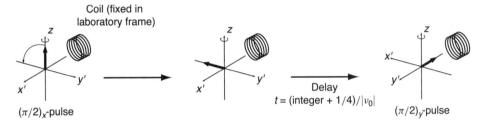

Figure 5.28 A $(\pi/2)_x$-pulse followed by a delay interval and a $(\pi/2)_y$-pulse. Increasing the delay time by $0.25/|\nu_0|$ aligns the y'-axis of the rotating frame with the pulse coil.

portant. If the goal were to create pulses along two perpendicular directions in the laboratory frame, then we would need two perpendicular coils surrounding the sample, as in the Bloch continuous-wave spectrometer. In pulse experiments, however, what is important is the relationship between the directions of different pulses in the *rotating* frame. Pulses in any direction in the x'–y' plane of the rotating frame can be created using a single coil, and all that is required is some careful timing.

We can define the axes of the rotating frame at the time of the first pulse, and we can define them however we like. So, if we call the first $\pi/2$-pulse of an experiment an x-pulse, then the $-y'$-axis of the rotating frame becomes the direction of the magnetization immediately after the pulse, by definition (Fig. 5.27). If, after some period of time, we want to make a second x-pulse, then we will have to make sure that the x'-axis is again aligned with the coil of the spectrometer at the beginning of the pulse. This will happen if the rotating frame has made an integer number of cycles about the laboratory frame—that is, if the time between the pulses is an integer divided by the absolute value of the frequency of the rotating frame, as shown in Fig. 5.27.

On the other hand, if we want to make our second pulse along the y'-axis of the rotating frame, then we increase the delay time so that the rotating frame moves exactly one-quarter of a cycle more, so that now the y'-axis is aligned with the coil in the laboratory frame (Fig. 5.28).

Note in the examples shown in Figs. 5.27 and 5.28 that the magnetization is stationary in the rotating frame (i.e., $\nu = \nu_0$). The result of two $(\pi/2)_x$-pulses is the inversion of the z-magnetization, irrespective of how long we wait between the pulses, the same as a single π-pulse. On the other hand, a y-pulse following an x-pulse has no further effect, because the pulse is along the direction of the existing magnetization. Consider what would happen, however, if there were populations with different Larmor frequencies, so that they move with different rates in the $x'-y'$ plane of the rotating frame. The different populations would then be affected differently by two sequential pulses, depending on how much the directions of their magnetization had diverged during the delay. Effects such as this are key to multidimensional experiments.

Because the frequency of the rotating frame is quite high (e.g., 500 MHz for protons), the timing of multiple pulses must be remarkably precise. As a practical matter, a single oscillator circuit is used to generate the pulses in different directions. This circuit is kept on throughout the experiment and defines the rotating-frame frequency. Pulses are then generated by connecting this ongoing signal to the probe coil at the correct time and then disconnecting it when the pulse is over, a process described as *gating*.

The very small differences in the relative timing of the pulses are established by shifting the phase of the rf signal that is applied to the coil. An increase in the phase angle is equivalent to delaying the pulse by a fraction of a cycle, so that the rf signal to generate a y-pulse is $\pi/2$ rad out of phase from an x-pulse signal, as illustrated in Fig. 5.29. The direction of a pulse is often referred to as its phase, and pulses are sometimes specified by the phase angle, ϕ, rather than by the direction in the rotating-frame coordinate. For pulses along the x'-, y'-, $-x'$- and $-y'$-directions of the rotating frame, the phase angles are 0, $\pi/2$, π and $3\pi/2$ rad, respectively. This phase angle should not be confused with the rotation angle, a, which is determined by the duration of the pulse.

Experiments involving multiple pulses are often described using diagrams such as the one shown in Fig. 5.30. The experiment is divided into a preparation phase, in

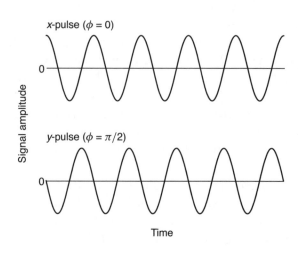

x-pulse ($\phi = 0$)

Signal amplitude

y-pulse ($\phi = \pi/2$)

Time

Figure 5.29
Rf signals used to generate
x- and y-pulses, indicating
the phase offset, ϕ, used to
determine the pulse direction.

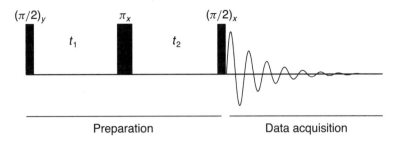

Figure 5.30 Example of a pulse-sequence diagram used to describe a multi-pulse NMR experiment.

which one or more pulses are applied to the sample, followed by the data acquisition phase, which we will discuss in Chapter 6. The pulses are represented with vertical bars, with widths indicating the duration of the pulse and the resulting rotation angle. The rotation angles and directions are usually indicated above the vertical bars, and the time delays between the pulses (t_1 and t_2 in the example) are shown between the bars. The squiggle on the right represents the signal recorded during the data acquisition period.

5.5   Off-resonance Pulses

In the discussions above, we have assumed that the frequency of the pulse is identical to that of the precession frequency of the nuclei. Strictly speaking, the results of the analysis are valid only for those spins with Larmor frequencies identical to that of the rotating frame. The same assumption is made in the quantum mechanical analysis described in Chapter 12. In real applications, there are generally spins with a range of Larmor frequencies, and only one frequency can be exactly stationary in the rotating frame. In addition, the timing of the pulse can only be exactly right for a single effective field.

When the frequency of the rotating frame is not exactly the same as the Larmor frequency, the pulse is said to be *off-resonance*, and there are two important consequences. First, whenever the frequency of the radiation does not match the energy difference between the $|\alpha\rangle$ and $|\beta\rangle$ states, the absorption of radiation will be less efficient than when the match is exact. As a result, the average magnetization does not rotate quite as far as it would for an on-resonance pulse. Second, if the precession frequency does not match the rotating-frame frequency, then the average magnetization will either lead or follow the rotating frame during the pulse.

A complete description of these effects requires yet another transformation of coordinates, and the mathematics becomes a bit involved. Qualitatively, however, the effects can be illustrated with vector diagrams, as shown in Fig. 5.31 for the case of a $(\pi/2)_x$-pulse, which rotates the magnetization from the z-axis to the $-y$-axis.

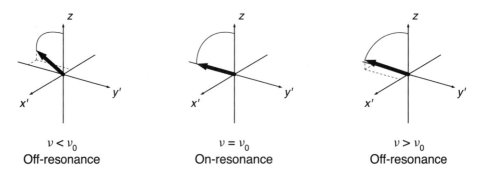

Figure 5.31   Effects of off-resonance pulses, in which the Larmor frequency is greater than or less than the frequency of the pulse irradiation.

Notice that when the Larmor frequency, ν is either larger or smaller than the rotating frame frequency, ν_0, the average magnetization does not quite reach the transverse plane. When ν is less than ν_0, the magnetization also ends up a little bit behind the $-y'$-axis, as viewed in the figure, whereas it ends in front of the $-y'$-axis if the Larmor frequency is greater than the rotating frame frequency.

The deviations from ideal behavior can be minimized by using a strong magnetic field for the pulse, resulting in a short pulse time. For some purposes, such as the $\pi/2$-pulse used to generate $x-y$ magnetization, this is often adequate. For other purposes, even a small residual of magnetization along the z-axis, for instance, may be a significant problem. In these cases, more advanced techniques, often involving *composite pulses*, are employed. As the name suggests, these pulses are made up of multiple shorter pulses of different lengths and phase relationships, and are designed so that the off-resonance effects of the individual pulses largely compensate one another.

Summary Points for Chapter 5

- In a pulse NMR experiment, the signals from all of the nuclei (of a given type) in a sample are recorded simultaneously and are then separated out mathematically.
- The rotating reference frame is used to describe NMR experiments because it simplifies the representation of frequency differences.
- The frequency of the rotating frame is defined to be close to the Larmor frequencies of the nuclei of interest.
- In order to stimulate nuclei with a broad range of resonance frequencies, short, intense pulses of electromagnetic radiation are used.
- Pulses are generated by oscillating electric currents in a coil surrounding the sample.
- A pulse leads to rotation of the bulk magnetization about an axis in the rotating frame.
- The extent of rotation of magnetization, expressed as an angle, is determined by the strength and duration of the pulse.

- The axis of the first pulse (in the rotating frame) in an experiment is defined arbitrarily. The axes of subsequent pulses are set by their phase relationship with the initial pulse.
- If the frequency of the radiation does not match exactly the Larmor frequency, the angle of rotation will be slightly reduced, and the magnetization will lead or follow the rotating frame.

Exercises for Chapter 5

5.1. Consider a case in which a sample containing ^1H nuclei is placed in a 600-MHz spectrometer. Assume that the rotating frame frequency is exactly $-600\,MHz$ and is 5 ppm more negative than the ^1H resonance frequency of TMS.

(a) Calculate the precession frequency (in Hz) for the TMS ^1H nuclei in both the laboratory and rotating frames of reference.

(b) During a period of 100 ms, how many times will the TMS ^1H nuclei precess around the z-axis, in the laboratory and rotating frames?

(c) In which direction, relative to the rotating frame, will the TMS ^1H nuclei precess? (i.e., will magnetization precess from the x'-axis toward the positive or negative y'-axis?)

(d) Suppose that the sample also contains ^1H nuclei with a chemical shift of 10 ppm, relative to TMS. Calculate the precession frequency (in Hz) for these nuclei, in both the laboratory and rotating frames.

(e) For the nuclei with a chemical shift of 10 ppm, in which direction will the magnetization precess?

5.2. Each of the drawings in Fig. 5.32 includes two vector diagrams representing the net magnetization arising from a population of ^1H nuclei in an external field (aligned with the z-axis) before and after either a pulse or a delay period. In each drawing, the axes represent the rotating frame. For the following, assume that the rotating frame frequency is $-400\,MHz$ and that the precession frequency for the spins is 5 ppm greater (more positive) than the rotating frame frequency. For each drawing, indicate whether or not the change in magnetization is due to a simple delay period or a pulse. If the change is due to a delay, calculate the minimum delay that would give rise to the observed change. Assume that there is no significant relaxation of the magnetization during the delay. If the change is due to a pulse, indicate the axis along which the pulse is applied, the rotation angle and whether it is a positive or negative pulse.

5.3. The following questions are also based on the drawings shown in Fig. 5.32. For each question, consider the four vector diagrams representing the situation *after* the pulse or delay period. Each question may have more than one answer.

(a) If an FID were recorded from each of the magnetization states depicted in the vector diagrams, which would give rise to the strongest signal in a coil surrounding the sample?

(b) Which vector would give rise to the weakest signal?

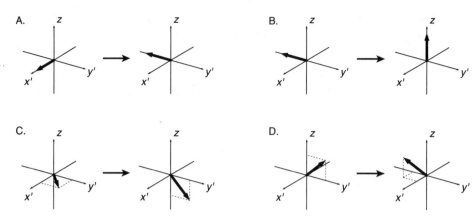

Figure 5.32   Vector diagrams for Exercise 5.2.

(c) In the absence of relaxation, which of the vector diagrams would be expected to change with time?

5.4. Suppose that you have a sample of a (very) simple molecule with only two ^{1}H nuclei, with chemical shifts of 1 and 2 ppm (relative to TMS). You place the samples in a "750-MHz" spectrometer and apply a $\pi/2$-pulse along the y'-axis. For the following, assume that the reference frequency of your spectrometer, which determines the rotating frame frequency, is the same as the resonance frequency for TMS.

(a) If the field strength (B_1) of your pulse is 0.002 T, what should the pulse duration be for the $\pi/2$-pulse?

(b) During the pulse, how many times will the rotating frame rotate, relative to the laboratory frame?

(c) Draw a vector diagram showing the orientation of the two vectors (in the rotating frame) immediately after the pulse.

(d) Draw a vector diagram showing the orientation of the two vectors (in the rotating frame) 1 ms after the pulse. During this 1 ms, how many times will the two vectors precess in the *laboratory* frame?

5.5. For the same molecule in the same spectrometer as described in problem 5.4 above, devise a two-pulse sequence that would enable you to create a spectrum containing only a single peak, corresponding to the larger chemical shift.

(a) Describe this sequence using a series of vector diagrams representing the situation before and after each pulse. Draw your figures to represent the rotating frame, and be sure to show both vectors in each diagram.

(b) What are the orientations of your two pulses, in the rotating frame?

(c) What is the time interval between the two pulses? (Note: You can ignore the "fine" timing necessary to define the orientation of the second pulse. Consider only the timing necessary to define the relative orientations of the magnetization vectors from the two spin populations.)

Further Reading

For further discussion of the practical details of the pulse method, these two older books offer very clear treatments:

⋄ Derome, A. E. (1987). *Modern NMR Techniques for Chemistry Research*. Tetrahedron Organic Chemistry Series. Pergamon Press, Oxford.

⋄ Fukushima, E. & Roeder, S. B. W. (1981). *Experimental Pulse NMR: A Nuts and Bolts Approach*. Addison-Wesley, Reading, MA.

Additional information about the practical aspects of pulse NMR, including the use of composite pulses to minimize off-resonance effects:

⋄ Cavanagh, J., Fairbrother, W. J., Palmer III, A. G., Rance, M. & Skelton, N. J. (2007). *Protein NMR Spectroscopy*, 2nd ed., pp. 165–201. Elsevier Academic Press, Amsterdam, 2nd edition.

CHAPTER 6

The Pulse NMR Method: The Signal and Spectrum

In this chapter, we consider the events following the pulse—namely, the creation of an electrical signal in the spectrometer, the digitization of that signal and its conversion into a spectrum with a familiar form, in which intensity is plotted as a function of frequency. Describing the data processing steps requires a good deal of mathematics, and this chapter is rather rich in equations. The later sections, in particular, deal with some of the practical aspects of data acquisition and processing (such as quadrature detection, phasing the spectrum, zero-filling and apodization), and this material is not absolutely essential for understanding the later chapters. What is important, however, is the general relationship between the signal from the spectrometer and the spectrum generated by Fourier transformation, and how properties of the spectrum, such as resolution and frequency range, are related to properties of the digitized signal. Fourier analysis also provides important insights into the relationship between the duration and shape of an rf pulse and the distribution of the frequencies generated, and this subject is briefly discussed in the final section of this chapter. On a first reading of this chapter, you may want to focus on the drawings of sample signals and spectra, without getting too bogged down with the mathematics. When you begin recording and processing your own data in the lab, the mathematics will become more immediately relevant. For those looking for additional mathematics, a more general and extensive discussion of Fourier transforms and related topics is provided in Appendix C.

6.1   The Signal

In a simple one-dimensional pulse NMR experiment, the $\pi/2$-pulse is followed by a period in which the precession of the resulting magnetization is detected by the induction of an electrical current in a surrounding coil. In fact, the same coil used to make the pulse is usually used to detect the resulting signal. As discussed earlier, the result of the precession in the x–y plane is a sinusoidal signal. If there is only one

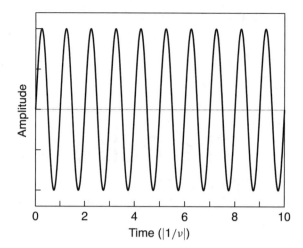

Figure 6.1

The initial signal following a pulse, from a sample with a single Larmor frequency.

population with a single Larmor frequency, the signal will initially look like the curve plotted in Fig. 6.1.

This won't go on forever, and the amplitude of the signal will decay with time. We will discuss later the physical basis of this decay, or relaxation, but for now we will assume that it follows an exponential function:

$$A = A_0 e^{-rt}$$

where A is the amplitude at time, t, A_0 is the initial amplitude, and r is a rate constant for the decay. The signal, s, at time, t, is then given by

$$s(t) = A \sin(2\pi \nu t) = A_0 e^{-rt} \sin(2\pi \nu t)$$

This function is called an *exponentially damped sine wave* and is plotted in Fig. 6.2.

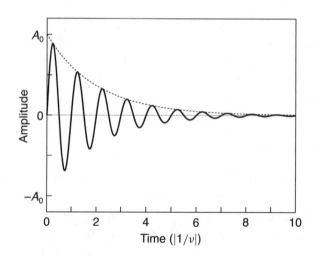

Figure 6.2

An exponentially damped sine function. The dashed line represents the exponential decay of the amplitude of the sine function.

The signal is usually called a *free induction decay*, or "FID." The dashed line shows the exponential decay function. In this plot, the rate constant for the decay is set to one-half of the frequency of the sine wave. After two cycles, the amplitude has been reduced to $1/e$ ($\approx 37\%$) of its initial value. In real NMR experiments, the decay rate constant is much smaller than the frequency. For instance, the decay might occur over a period of hundreds of milliseconds, during which there would be several million cycles.

In real experiments, we want to follow the signals from many different spin populations with different Larmor frequencies. The signals from the different populations will be superimposed, resulting in a much more complicated FID. The trick is to sort out these superimposed signals.

The first challenge in dealing with the signal is that it has a very high frequency. If we want to digitize a 500-MHz signal, we would need to generate at least one byte of computer data for each cycle. Thus, we would need at least 500 megabytes per second of data acquired. Even by today's standards, this would be a huge amount of data—about one CD-ROM per second. (In contrast, to digitize audio frequency signals, of say 20 kHz, we would need a minimum of 20 kilobytes per second, or about 1 CD per 40 min.) A related problem is that the signal is made up of multiple components with very small relative frequency differences, on the order of parts per million. In the case of a 500-MHz ^{1}H spectrum, the range of frequencies would likely cover about 10 ppm, or 5000 Hz, a small fraction of the total frequency. Conceptually, at least, both problems can be solved by subtracting a constant frequency from all of the signals. Thus, if the range of frequencies is from 500,001,000 Hz to 500,006,000 Hz and we subtract 500 MHz, we will have a spectrum with frequencies from 1000 to 6000 Hz, a much more manageable range.

The problem of resolving high-frequency signals that differ by only small amounts was actually solved well before the invention of NMR spectroscopy, during the development of AM (amplitude modulated) radio broadcasting. In an AM radio signal, the vibrations of an audio signal are superimposed on a much higher *carrier* frequency. In the radio receiver, the high-frequency signal must be converted back to an audio frequency. Electronically, this is done by *mixing* the radio frequency signal of interest with a reference signal of similar frequency, ν_0, as illustrated diagrammatically in Fig. 6.3. The mixer does not simply add the two voltages present at a given instant, which would actually complicate the situation by adding another frequency, very close to that of the FID, to the signal. Instead, the mixer *multiplies* the signal intensities at

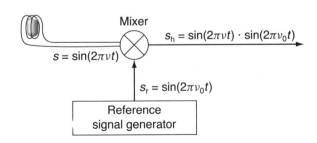

Figure 6.3

Block diagram of the mixer circuit used to generate a heterodyne signal in an NMR spectrometer.

A.
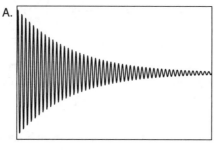

B.
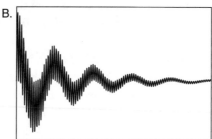

C.
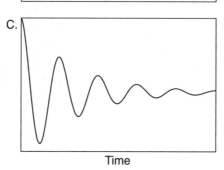

Time

Figure 6.4

An example of generating a heterodyne signal. The function graphed in panel A represents an idealized FID composed of a single damped sine function. In panel B, the FID has been multiplied with a reference sine function, with a frequency slightly lower than that of the FID. Panel C shows the result of filtering out the high-frequency component of the heterodyne signal.

each time point, leading to the product shown in Fig. 6.3. The result is called a *heterodyne signal*, and can be shown, using trigonometric identities, to be the sum of two signals:

$$s_h = \sin(2\pi \nu t) \cdot \sin(2\pi \nu_0 t)$$

$$= \frac{1}{2}\cos\big(2\pi(\nu - \nu_0)t\big) - \frac{1}{2}\cos\big(2\pi(\nu + \nu_0)t\big) \qquad (6.1)$$

The frequencies of the two components are the sum and difference of the original signal frequency and the reference frequency. The reference frequency is typically set to be close to the Larmor frequency, ν, so one of the components has a frequency of about twice ν, and the other has a very low frequency. It is relatively easy to remove the high-frequency component with an electronic filter, leaving just the low-frequency component. A simulated example of a heterodyne signal is shown in Fig. 6.4. Panel A of the figure shows a single damped sine function with a relatively high frequency. The function plotted in panel B was generated by multiplying the idealized FID by a sine function with a frequency equal to 0.9 times that of the FID. There are now two frequencies present, one that is quite low and one that is about twice that of the original signal. If we remove the high-frequency component using a filter, the result shown in panel C is obtained. Now, there is just a single component, with a frequency that is much lower than that of the original FID. Sampling this signal to determine its frequency is much easier than sampling the original one.

In practice there are many superimposed signals with different frequencies, corresponding to the nuclei with different Larmor frequencies. Each of these frequencies will be shifted downward by the same amount, ν_0. It is this frequency-shifted signal that is converted to a digital form.

As a practical matter, the oscillator used to generate the reference signal is usually the same one used to generate the pulses, and the same coils are used for stimulating the sample and collecting the FID. In addition, the same oscillator is used to set the timing between pulses, which effectively determines the frequency of the rotating frame. Thus, the rotating-frame frequency defined in Chapter 5 and the reference frequency are the same. The actual conversion of the FID to a lower frequency is equivalent to the conceptual conversion of the laboratory-frame precession frequency to that of the rotating frame. In other words, the frequency-shifted signal represents the cyclically fluctuating magnetization as it would be observed in the rotating frame. If the Larmor frequency exactly matches that of the rotating frame, the shifted signal has a frequency of zero—that is, it does not change with time. But if the Larmor frequency is different from the rotating frame frequency, then the shifted signal will have a frequency equal to this difference.

Unfortunately, the use of a heterodyne signal also introduces a significant complication. As noted briefly in Chapter 5, the cosine function has the property of being "even," meaning that $\cos(-x)$ is equal to $\cos(x)$. A consequence is that the heterodyne signal that would result from an original signal with a frequency Δ Hz greater than ν_0 is exactly the same as if the frequency were Δ Hz less than ν_0. If we were to use a reference frequency that is either greater or lower than any possible Larmor frequency, this would not be a problem, because there would be no question about the sign of the frequency difference. For a variety of other practical reasons, however, there are significant advantages to placing the reference frequency close to the center of the frequency range of interest. As a cost of this choice, we have to be able to deal with frequency differences that can be either positive or negative, and we must be able to tell the difference.

The physical origin of the frequency ambiguity can be appreciated by imagining that we are sitting on the rotating frame and looking down the y'-axis. From this position, all that we can see is the projection of the magnetization along the x'-axis, as indicated by the dashed line segments in the vector drawings in Fig. 6.5. The time-dependent change in the projection on the x'-axis is exactly the same if the magnetization is in front of the x'-axis ($y' > 0$) and rotating counterclockwise (as viewed from above), or is behind the axis and rotating clockwise.

The solution to the problem is also apparent from this diagram: We need a friend to sit on the x'-axis and watch the projection on the y'-axis while we watch the x'-projection. If the y'-projection is increasing (becoming more positive) while the x'-projection is decreasing, as in the upper part of the diagram, then we can conclude that the precession frequency is greater than that of the rotating frame. On the other

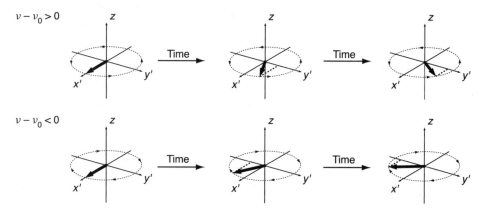

Figure 6.5 Precession of two spins in the rotating frame, one with frequency Δ Hz greater than the rotating frame frequency, ν_0, and the other Δ Hz less than ν_0. The dashed lines indicate the lengths of the projections onto the x'-axis.

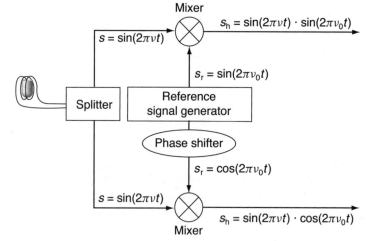

Figure 6.6
Block diagram of the circuit used for quadrature detection of the signal from a single probe coil.

hand, if the y'-projection is becoming more negative, as in the bottom part, then we conclude that the precession frequency is less than the reference frequency.

Detecting the two components of the changing magnetization simultaneously is called *quadrature detection*, and could, in principle, be implemented by having two perpendicular coils surrounding the sample. There is an electronic alternative, however, that is easier and cheaper. This involves splitting the signal from the single coil and mixing the two components with reference signals that are out of phase by 90°. Usually, a single oscillator is used and part of the signal is phase-shifted, as diagrammed in Fig. 6.6.

We now have two heterodyne signals. Like the first one, the second can be expressed as a sum of a high and a low frequency:

$$s_h = \sin(2\pi \nu t) \cdot \cos(2\pi \nu_0 t)$$

$$= \frac{1}{2}\sin\big(2\pi(\nu - \nu_0)t\big) + \frac{1}{2}\sin\big(2\pi(\nu + \nu_0)t\big)$$

If we again filter out the high-frequency component, the resulting low-frequency term will be out of phase from the original heterodyne signal by 90°, exactly as if we had collected a signal from a second coil. We will come back to the question of how to use this second signal later, when we deal with the more general issue of deconvoluting the signals coming from the different spin populations, with their different Larmor frequencies.

6.2 From FID to Spectrum: The Fourier Transform

6.2.1 A Qualitative Overview

The signal from the detector coil is said to be a *time-domain* signal, in that it represents a voltage as a function of time. A spectrum, on the other hand, is a *frequency-domain* signal, in which the intensity is represented as a function of frequency. For the simplest case, a constant sine function, the two representations are shown in Fig. 6.7. If the signal is composed of two superimposed sine curves with different frequencies, the two representations look like Fig. 6.8. In this case, there are two signals of equal amplitude, with the frequency of the second being 20% greater than that of the first. Note that one of the consequences of combining the signals is that the pattern in the time domain doesn't repeat itself until five time units have passed.

As more frequency components are added, both representations become increasingly complex, as shown in Fig. 6.9 for a signal made up of five frequencies. Here, additional components have been squeezed into the same frequency range, with uneven spacings. The result is a much more complex time-domain signal that doesn't repeat itself for a much longer time.

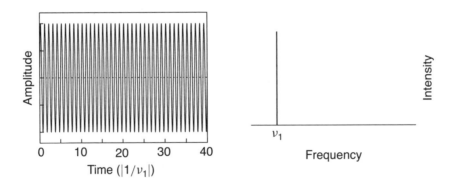

Figure 6.7 Time- and frequency-domain representations of a signal composed of a single frequency.

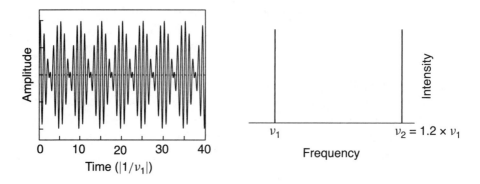

Figure 6.8 Time- and frequency-domain representations of a signal composed of two frequencies.

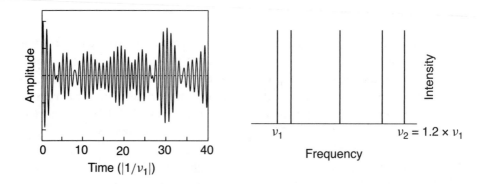

Figure 6.9 Time- and frequency-domain representations of a signal composed of five frequencies.

Because the time-domain and frequency-domain signals represent essentially the same information, there should be some mathematical method for converting one to the other. There are, in fact, several ways in which this can be done, but the Fourier transform is, by far, the one most commonly used. This method is based on the idea that almost any mathematical function (or at least those that are likely to represent physical phenomena) can be approximated as a sum of cosine and sine functions of increasing frequencies. This is just one way in which an arbitrary function can be represented as a sum of other functions. In many applications, series of polynomials are especially useful. The Fourier series is often appropriate when dealing with either periodic functions or functions that are of interest over only a finite range.

6.2.2 A Brief Diversion into Complex Numbers

In addition to being based on the sine and cosine functions, the mathematics of the Fourier transform is made more convenient through the use of complex numbers and an important relationship between complex numbers and the trigonometric functions. The set of complex numbers are a generalization of the real numbers and the imaginary numbers—that is, numbers that are the square roots of negative numbers. A complex number c can be represented as a sum of a real number, a, and the product of the unit imaginary number ($i = \sqrt{-1}$) and a second real number, b:

$$c = a + ib$$

The number a is referred to as the "real part" of the complex number, and b is the "imaginary part." For some purposes, it is also convenient to represent complex numbers in the following exponential form:

$$c = me^{i\theta}$$

where m is called the *modulus* of the number and θ is called the *argument* or *phase*. The modulus can be thought of as the total magnitude of the complex number, roughly analogous to the absolute value of a real number, while the argument represents the relative contributions of the real and imaginary parts. The relationship between the two representations is defined by Euler's formula:

$$e^{i\theta} = \cos\theta + i \sin\theta$$

This relationship arises frequently in physics and applied mathematics, particularly when cyclic processes are considered. The properties of complex numbers and their relationship to the trigonometric functions will be discussed further in later chapters in the context of quantum mechanics, and a brief summary of these relationships is provided in Appendix B.

6.2.3   Fourier Transformation of the Time-domain Signal

The functions with which we are concerned are the time-domain signal from the spectrometer and the frequency-domain spectrum, which we will call $s(t)$ and $S(f)$, respectively. The Fourier series representing the time- and frequency-domain signals can be written as

$$s(t) \approx \sum_{k=0}^{N-1} c_k e^{i2\pi kt/T} = \sum_{k=0}^{N-1} c_k \left(\cos\left(2\pi k \frac{t}{T}\right) + i \sin\left(2\pi k \frac{t}{T}\right) \right)$$

and

$$S(f) \approx \sum_{j=0}^{N-1} C_j e^{-i2\pi jf/F} = \sum_{j=0}^{N-1} C_j \left(\cos\left(2\pi j \frac{f}{F}\right) - i \sin\left(2\pi j \frac{f}{F}\right) \right)$$

In these expressions, c_k and C_j are complex constants that represent the relative contributions of the terms with different frequencies, which are determined by the indices k and j in the exponential terms. The constants T and F represent the intervals of time and frequency, respectively, over which the series are valid. The number of terms in the series is given by N and is the same for the series representing

$s(t)$ and $S(f)$. For any given degree of accuracy we require, we can simply add more terms, at least in principle. As we add more terms, we are adding cosine and sine functions with higher frequencies, and the series becomes progressively better at approximating the changes in $s(t)$ and $S(f)$ that occur over short intervals of time or frequency. Note also that the signs of the exponents are different. This is arbitrary, but it makes what follows easier to write.

Describing the time-domain signal as a Fourier series may not seem so useful, because we already have the time signal stored in our computer. What we really want is the frequency-domain signal, and to know that, we have to know the values of the constants, C_j. Determining the values of these constants is possible because of a special relationship between the two Fourier series.

Before stating this relationship, we need to define the spacing between the points in the time and frequency domains. Suppose that we were to sample the time-domain signal at N time points, separated by Δt. For the jth time point, the value of t will be

$$t = j\Delta t$$

and the total time will be

$$T = N\Delta t$$

The ratio t/T then becomes

$$t/T = j/N$$

and we can write the Fourier series for the time-domain signal as

$$s(j\Delta t) \approx \sum_{k=0}^{N-1} c_k e^{i2\pi kj/N}$$

Notice that we have used the same number, N, for the number of time points and the number of terms in the series.

For the frequency spectrum we will define the total frequency range, F, as $1/\Delta t$, and we will consider points separated by $1/(N\Delta t)$. Because we are working with the heterodyne signal, the frequencies can be either positive or negative. We will assume, as is usually the case, that N is even, and we will identify the frequency points in the spectrum with the index k. The individual frequencies are given by $k/(N\Delta t)$, and k can take on the values

$$-\frac{N}{2}, -\frac{N}{2}+1, -\frac{N}{2}+2, \ldots, -1, 0, 1, \ldots, \frac{N}{2}-2, \frac{N}{2}-1$$

The frequencies then range from $-1/(2\Delta t)$ to $(N/2 - 1)/(N\Delta t)$, and the intensity of the spectrum at these frequency values is given by

$$S\big(k/(N\Delta t)\big) \approx \sum_{j=0}^{N-1} C_j e^{-i2\pi jk/N}$$

The remarkable relationship between the two series is this: The coefficients for the frequency-domain series are the sampled values of the time-domain signal

$$C_j = s(j\Delta t)$$

As a result, we can write the Fourier series for the frequency-domain spectrum as

$$S\big(k/(N\Delta t)\big) \approx \sum_{j=0}^{N-1} s(j\Delta t) e^{-i2\pi jk/N} \tag{6.2}$$

This provides just the relationship we want: We can now calculate the spectrum from the time-domain signal (i.e., from the FID).

This relationship also works in the opposite direction. In this case, the coefficients for the time-domain signal are given by

$$c_k = \frac{1}{N} S\big(k/(N\Delta t)\big)$$

and the Fourier series for the FID can then be written as

$$s(j\Delta t) \approx \frac{1}{N} \sum_{k=0}^{N-1} S\big(k/(N\Delta t)\big) e^{i2\pi kj/N}$$

So, in principle we can "back-calculate" the FID from the spectrum.

A very important feature of these relationships is that every point in the time domain contributes to every point in the frequency domain, and vise versa. Thus, we can't point to a part of the FID and say that it represents a particular resonance frequency.

But, what about the fact that we defined the coefficients to be complex numbers? How do we get a complex number from our real spectrometer? Remember that we actually collected two sets of data, one of which was $\pi/2$ rad out of phase with the other. To get a complex number for the Fourier transform, we take one of these signals, multiply it by the unit imaginary number, i, and add it to the other signal. We now have a complex number. The result of the transformation will also be a set of complex values. Generally only the real part is needed at the end of the data processing. However, the imaginary part is important for some of the intermediate steps, as discussed below.

Although the series shown above are attractive for their (relative) simplicity, they are not at all attractive from a computational point of view. If we want to calculate a spectrum made up of N points, we have to calculate a total of N^2 terms involving trigonometric functions. Because one of the major motivations for the pulse method is to record high-resolution spectra, the computational load can be quite heavy. Fortunately, there are fast Fourier transform (FFT) algorithms that are much more efficient than the equations shown above. These algorithms are based on certain symmetry relationships among the terms in the Fourier series that make it unnecessary to calculate all of the terms individually. Instead of the number of calculations being proportional to N^2, the computational load increases in proportion to $N \log N$. This actually makes a huge difference in the time required to transform large data sets.

6.3   The Shape of NMR Peaks: The Absorptive and Dispersive Lorentzian curves

We can see what individual peaks in the spectrum will look like by considering the Fourier transform of the signal from a single frequency contribution. As discussed earlier, the FID from a single frequency is described as a damped sine function:

$$e^{-rt} \sin(2\pi \nu t)$$

where r is the rate constant for the decay and ν is the Larmor frequency. Although computational methods are needed to Fourier transform an FID made up of multiple signals with unknown frequencies, a single damped sine function with known frequency can be transformed analytically to give a continuous function of f, the frequency values in the spectrum.

First, we need to consider what happens to the signal during quadrature detection. As discussed earlier, the signal is divided in two and then mixed with signals with the same reference frequency but different phases. One of the reference signals can be described by the function $\sin(2\pi \nu_0 t)$. After mixing and filtering the high-frequency component, the heterodyne signal has the following form:

$$e^{-rt} \cos\big(2\pi (\nu - \nu_0)t\big)$$

The other reference signal has the form $\cos(2\pi \nu_0 t)$, and mixing with this signal gives a heterodyne signal:

$$e^{-rt} \sin\big(2\pi (\nu - \nu_0)t\big)$$

To simplify the following expressions while also keeping in mind that we are dealing with a frequency difference, we will use the symbol ν' to represent $\nu - \nu_0$. These signals are used as input into the Fourier transform, writing them in the complex form as

$$s(t) = e^{-rt}\big(\cos(2\pi \nu't) + i\,\sin(2\pi \nu't)\big)$$

$$= e^{-rt}e^{i2\pi \nu't}$$

$$= e^{(i2\pi \nu't - rt)}$$

In a subsequent section, we will discuss why mixing the two heterodyned signals in this way gives rise to a spectrum with the sign of ν' correctly indicated. For now, though, we will just take this for granted.

For the Fourier transform of a continuous function, the summation in Eq. 6.2 (p. 127) is replaced by an integral. If the time-domain function is designated $s(t)$, then the frequency-domain function, $S(f)$, is given by

$$S(f) = \int_{-\infty}^{\infty} s(t)e^{-i2\pi ft}dt \qquad (6.3)$$

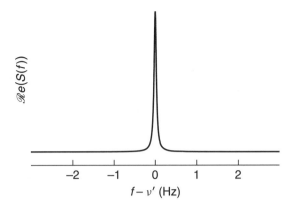

Figure 6.10
The real component of the Fourier transform of the quadrature-detected FID from a single frequency component.

The result of Fourier transforming the quadrature-detected signal is another complex function, called a Lorentzian function:

$$S(f) = \frac{r}{r^2 + 4\pi^2(f - \nu')^2} + i\frac{2\pi(\nu' - f)}{r^2 + 4\pi^2(f - \nu')^2} \tag{6.4}$$

First, let's look at the real part of the expression:

$$\mathcal{R}e\big(S(f)\big) = \frac{r}{r^2 + 4\pi^2(f - \nu')^2}$$

Inspection of this equation shows that $\mathcal{R}e\big(S(f)\big)$ is maximal when $f = \nu'$, which is what should happen in a spectrum! This can be seen in Fig. 6.10, which is a plot of the real part of the spectrum versus $f - \nu'$.

This function is called an *absorptive Lorentzian* function, and it has the qualitative shape one would expect for an absorption spectrum, though its mathematical form is different from that describing a UV or visible absorbance peak, which is usually approximated by a Gaussian (or "bell curve") function. We will discuss below how the shape of the absorptive Lorentzian is affected by the shape of the FID from which it is derived. First, though, it is also worthwhile to look at the imaginary part of the Fourier transformed signal, given by:

$$\mathcal{I}m\big(S(f)\big) = \frac{2\pi(\nu' - f)}{r^2 + 4\pi^2(f - \nu')^2}$$

This function has a somewhat more complicated shape, as shown in Fig. 6.11. This curve is called a *dispersive Lorentzian*. Notice that the function goes to zero when $\nu' = f$, corresponding to the peak in the absorptive Lorentzian. The two curves contain exactly the same information, and either can be used to represent an NMR spectrum. The absorptive curve is narrower, however, and is the form almost always used to present NMR spectra, where the goal is usually to resolve closely spaced peaks.

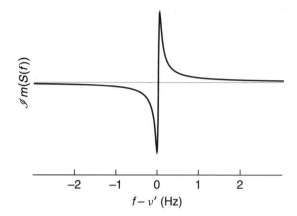

Figure 6.11
The imaginary component of the Fourier transform of the quadrature-detected FID from a single frequency component.

6.4   Relationships Between the FID and Spectrum

The best way to see how the shape of the absorptive peak is related to the FID is to look at some time-domain, frequency-domain pairs. Fig. 6.12 shows a damped sine curve and its corresponding absorptive peak in the frequency domain. For convenience, the damped sine function was calculated with a frequency of 1 Hz, and the rate constant for decay, r, is 0.2 s^{-1}. Curves that would be more realistic for NMR experiments would simply be scaled appropriately. One of the most important features of the Lorentzian function is that the width of the peak is proportional to the decay rate in the corresponding damped sine function. This is shown in Fig. 6.13, where r has been increased to 1 s^{-1}.

The width, w, of the peak is defined by its width at the position where the height is one-half of the maximum, which is given by

$$w = r/\pi$$

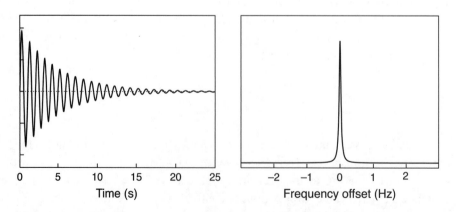

Figure 6.12   FID and spectrum from a single spin population, with a frequency of 1 Hz and decay rate of 0.2 s^{-1}.

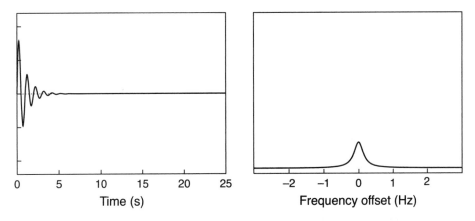

Figure 6.13   FID and spectrum from a single spin population, with a frequency of 1 Hz and decay rate of 1 s^{-1}.

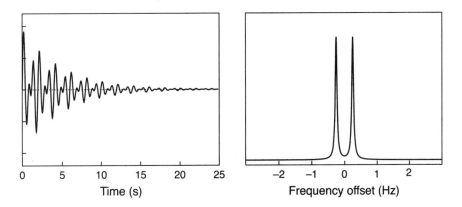

Figure 6.14   FID and spectrum from two spin populations, with frequencies of 1 and 1.5 Hz and decay rates of 0.2 s^{-1}.

In the example in Fig. 6.13, the half-height width is $1/\pi$ Hz. An increased decay rate also leads to broadening of the dispersive line shape.

The width of the peak represents the uncertainty in the frequency of the time-domain signal that gave rise to it. The faster the signal decays, the fewer cycles we are able to detect, thus increasing our uncertainty. The same thing would happen if, for some other reason, we had to limit the total time interval for data collection. This relationship is closely related to the one we discussed in Chapter 5, in the context of the frequency spectrum of the irradiating radiation. In that case, the pulse is deliberately made short in order to generate a broad range of frequencies. When we are detecting a signal, we generally want to collect data for as long as possible in order to get the sharpest peaks.

The peak width becomes especially important when we are trying to resolve the signals from two nuclei with similar Larmor frequencies. Fig. 6.14 shows the FID and spectrum arising when there are two signals, with frequencies of 1 and 1.5 Hz. For

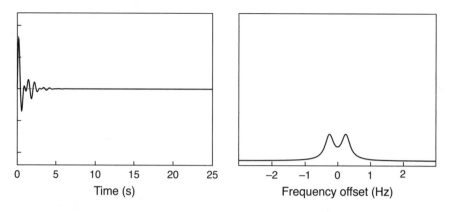

Figure 6.15   FID and spectrum from two spin populations, with frequencies of 1 and 1.5 Hz and decay rates of 1 s^{-1}.

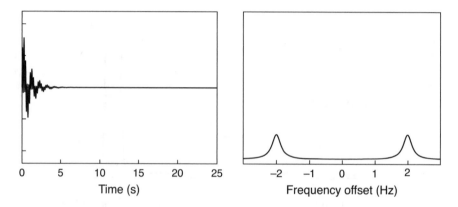

Figure 6.16   FID and spectrum from a two spin populations, with frequencies separated by 4 Hz and decay rates of 1 s^{-1}.

convenience, the frequency offset has been shifted to the midpoint between the two peaks. For both signals, the decay rate is 0.2 s^{-1}, and the two peaks are well resolved. If, however, the signals decay more rapidly, with $r = 1$ s^{-1}, then the resolution is much poorer, as shown in Fig. 6.15. If we compare the FIDs in Figs. 6.14 and 6.15, we can begin to see why the peaks are difficult to resolve when the signal decays more rapidly. When the FID remains strong for several seconds, there is enough time for the magnetization from the two spin populations to diverge from one another, which results in first destructive and then constructive interference between the two components. However, when the signal disappears quickly, the interference is barely detectable in the FID, and the resulting peaks are only partially separated.

If the frequency difference is larger, then the peaks can be resolved even with the shorter decay time, as shown in Fig. 6.16, where the frequency difference is now 4 Hz. Now, we don't need as much resolution. Notice that the features of the FID are now much more closely spaced, allowing the interference effects to be detected over a shorter time period. This is a general feature of Fourier transforms: When features are moved closer together in one domain, the corresponding features are separated in the other domain.

6.5   Sampling Rate and Spectral Width

The example shown in Fig. 6.16 illustrates another important point. In order to cover a wider range of frequencies, the FID must be sampled more finely. This was actually implicit in the way that we set the intervals for computing the Fourier transform. In order to cover a given frequency range, F, the sampling time interval is set so that

$$\Delta t = \frac{1}{F}$$

Recall also that the frequency range is set so that it is centered (or very nearly centered) at zero, so that the range of frequencies covered is from $-1/(2\Delta t)$ to $(N/2 - 1)/(N\Delta t)$. If, after setting the sampling interval, the signal actually contains frequencies that lie at the edges or outside of this range, some strange things start to happen!

To understand what happens when the signal frequency lies outside of the range specified above, it is helpful to look at a specific example. The curve in the leftmost panel of Fig. 6.17 shows a continuous sine function with a frequency of 1800 Hz. A full cycle of this function takes ≈ 0.56 ms. Suppose that we were to sample this signal at intervals of 0.5 ms, which might, at first glance, seem adequate. The sampled values are shown as filled circles in the graph. In the center panel, the sampled values are plotted without the curve, to emphasize an important point: After the data are sampled, all of the information about what happened between the sampled values is lost. Without the intermediate values, we can see that the points also lie on a sine wave with a lower frequency, as shown in the rightmost panel.

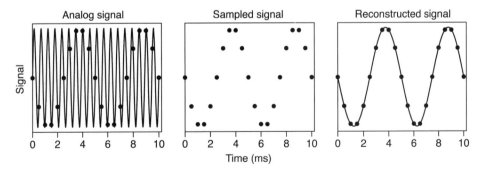

Figure 6.17   The effect of the sampling interval on the reconstruction of a signal. The original signal in the left-hand panel has a frequency of 1800 Hz and is sampled at 0.5-ms intervals. The sampled points also pass through a sine wave with a frequency of 200 Hz, as shown in the right-hand panel.

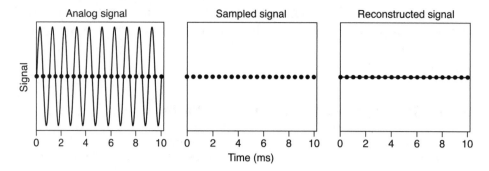

Figure 6.18   Sampling a sine wave for which the frequency (1000 Hz) matches $1/(2\Delta t)$. The sampled data are consistent with a frequency of zero.

For a given set of regularly spaced data sampled from a sine wave, there will be an infinite number of other sine waves with higher frequencies that will also pass through those points. But, there will be at most two frequencies, opposite in sign, that satisfy the condition $-1/(2\Delta t) \leq f < 1/(2\Delta t)$. Faced with this ambiguity, the Fourier transform algorithm implicitly assumes that the signal is composed of periodic functions with frequencies within the range defined by Δt. (We will return to the sign ambiguity in Section 6.7.) If the signal contains a frequency component that lies outside of the range determined by Δt, the transform algorithm will, in effect, assign it to frequencies within the range. In this case, the allowed range is -1000 Hz to 900 Hz, and the Fourier transform will return peaks at -200 and 200 Hz.

We can also see what happens if the frequency is exactly $1/(2\Delta t)$, as shown in Fig. 6.18, with a frequency of 1000 Hz and the sampling interval again set at 0.5 ms. In this case, each sampled value is 0, and the data are consistent with a frequency of 0, or any integer multiple of $1/(2\Delta t)$.

As a final example, Fig. 6.19 illustrates a case where the signal frequency (900 Hz) does satisfy the condition $-1/(2\Delta t) \leq f < 1/(2\Delta t)$. Although the pattern of sampled points may not be immediately recognizable as a single sine wave, they do define the original signal. In this case the average number of samples per cycle is ≈ 2.2, while in the previous example it was exactly 2. In general, an average sampling rate greater than 2 per cycle is required to define the frequency.

The way to avoid peaks in the Fourier transform with incorrect frequencies, called *aliases*, is to be sure to make the sampling interval small enough for the frequency range. This is usually easy enough to do in one-dimensional experiments, though it can be problematic in multidimensional experiments, where the constraints on the number of sampled points can become significant.

The relationship between sampling interval and frequency range is called the *sampling theorem* or *Nyquist theorem*, and the limiting frequency defined by a given sampling interval, $f_c = 1/(2\Delta t)$, is referred to as the *critical frequency*, or *Nyquist frequency*. This theorem plays an important role in virtually all applications of digital signal processing.

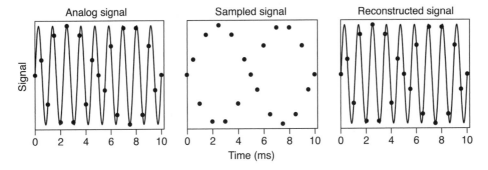

Figure 6.19 Sampling a sine wave for which the frequency (900 Hz) is less than $1/(2\Delta t)$. In this case the Nyquist condition is satisfied, and the reconstructed sine wave matches the original signal.

6.6 "Phasing" the Spectrum

Having learned how it all works, if you now go to the laboratory and record a pulse NMR spectrum, you are likely to have a nasty surprise. Instead of the nice absorptive Lorentzian peak discussed in the previous sections, or even the elegant and symmetric dispersive shape, at least some of the peaks in your spectrum will probably look something like the one shown in Fig. 6.20.

It appears as though the absorptive and dispersive components have somehow been mixed together, and this is exactly what has happened. Aside from the aesthetic shortcomings of this peak, notice that neither the major peak nor the zero cross-over point matches exactly the frequency corresponding to zero offset, making it difficult to accurately determine the frequency corresponding to the peak in a real spectrum. In order to generate the nice absorptive peaks that are usually shown in books and published papers, it is necessary to un-mix the two components, a processes that is called *phasing*, or *re-phasing*, the spectrum.

Figure 6.20
A mis-phased NMR peak, composed of a mixture of the absorptive and dispersive peak shapes.

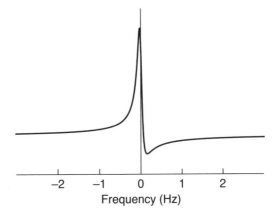

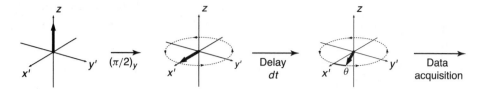

Figure 6.21   Displacement of the magnetization vector during the delay, dt, between a pulse and the beginning of data acquisition.

First, let's consider how things got mixed up in the first place. Although other artifacts can contribute, the most common source of distorted peak shapes is a lack of synchronization between the pulse and the data acquisition processes. Ideally, the FID is recorded immediately after the pulse. If the pulse is along the y'-axis, then at time zero in the data acquisition period, the initial y'-magnetization will be zero, and the x'-projection will represent the total initial magnetization. After quadrature detection, the x'- and y'-magnetization projections will correspond to real and imaginary components of the complex signal, respectively, with the mathematical forms $\cos(2\pi v't)$ and $\sin(2\pi v't)$, as discussed earlier. But, in reality there is always some delay between the pulse and data collection, largely because the same coils in the probe, as well as other electronic components, are used for both, and some recovery time is needed between transmitting a very strong signal and detecting a very weak one. If the Larmor frequency of interest does not match the rotating-frame frequency, then the magnetization vector will precess some distance during this delay, as shown Fig. 6.21.

The effect can be expressed as either the time delay, dt, or the resulting angle, θ, formed between the magnetization vector and the x'-axis. For now, it is easier to work with the angle. If the data acquisition begins at $t = 0$ with the vector at position θ, then the complex signal that is generated by quadrature detection and fed into the Fourier transform is given by

$$s(t) = e^{-rt}\left(\cos(2\pi v't + \theta) + i\,\sin(2\pi v't + \theta)\right) = e^{(i(2\pi v't+\theta)-rt)}$$

(We are ignoring the small decrease in amplitude due to decay of the signal during the delay.)

After Fourier transformation, we obtain a frequency domain spectrum with real and imaginary parts. The real part is given by

$$\mathcal{R}e\big(S(f)\big) = \cos(\theta)\frac{r}{r^2 + 4\pi^2(f - v')^2} - \sin(\theta)\frac{2\pi(v' - f)}{r^2 + 4\pi^2(f - v')^2}$$

Just as suggested by Fig. 6.20, this represents the sum of absorptive and dispersive Lorentzian functions, with the relative contributions determined by the displacement angle, θ. If we use the symbols $A(f)$ and $D(f)$ to represent the absorptive and dispersive functions, respectively, we can rewrite the real part of the spectrum as

$$\mathcal{R}e\big(S(f)\big) = \cos\theta\, A(f) - \sin\theta\, D(f)$$

If $\theta = 0$, the spectrum has a purely absorptive shape. If $\theta = \pi/2$, the shape is purely dispersive, but of opposite sign from what we saw earlier. For most other values, it will be some combination.

With a real spectrum, we can't simply throw away the dispersive component with some algebra. What we have to work with is a string of numbers stored in a computer that, under ideal circumstances, are described by the function above; but we don't have an algebraic representation. What we do have, though, is a second set of numbers that represent the imaginary part of the Fourier transform. For an ideal FID with a single component, the imaginary part is given by

$$\mathcal{I}m\big(S(f)\big) = \sin\theta A(f) + \cos\theta D(f) \tag{6.5}$$

Notice here that if $\theta = 0$, the imaginary part is purely dispersive, while if $\theta = \pi/2$, the shape is purely absorptive. Because the dispersive components of the real and imaginary parts of $S(f)$ are of opposite sign, we should be able to eliminate the dispersive component by adding the real and imaginary parts together in the correct proportions. We will call the resulting spectrum $S'(f)$, and the recipe for calculating the real part of the spectrum is given by

$$\mathcal{R}e\big(S'(f)\big) = \cos\theta\,\mathcal{R}e\big(S(f)\big) + \sin\theta\,\mathcal{I}m\big(S(f)\big)$$

That this gives the desired result, a purely absorptive Lorentzian, can be shown as follows:

$$\mathcal{R}e\big(S'(f)\big) = \cos\theta\,\mathcal{R}e\big(S(f)\big) + \sin\theta\,\mathcal{I}m\big(S(f)\big)$$
$$= \cos\theta\big(\cos\theta A(f) - \sin\theta D(f)\big) + \sin\theta\big(\sin\theta A(f) + \cos\theta D(f)\big)$$
$$= \cos^2\theta A(f) - \cos\theta\sin\theta D(f) + \sin^2\theta A(f) + \cos\theta\sin\theta D(f)$$
$$= (\cos^2\theta + \sin^2\theta)A(f)$$
$$= A(f)$$

In general, we don't actually know the value of θ. So, the usual practice is simply to try different values until the spectrum looks nice. This process is illustrated in Fig. 6.22, where each row of curves represents the summation of the real and imaginary parts of the original Fourier transform, weighted according to a different value of θ, which is also referred to as the phase-correction angle. In this case (which is, admittedly, contrived), combining equal parts of the real and imaginary parts, corresponding to $\theta = \pi/4$, yields a purely absorptive peak. This also indicates that the original offset angle, when the data were recorded, was $\pi/4$.

Correcting the phasing of a spectrum with multiple resonance peaks is, unfortunately, slightly more involved than this, because the phase errors depend, in general, on the resonance frequency. The primary origin of this frequency dependence can be seen by looking again at what happens during the inevitable delay between the pulse and the beginning of the data acquisition period. The drawings in Fig. 6.23 show the effect on the magnetization vectors representing nuclei with four different Larmor frequencies.

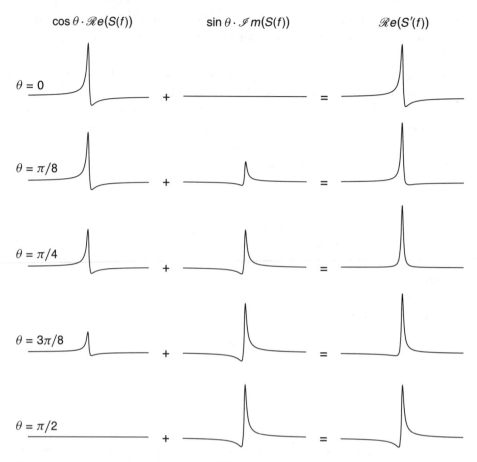

Figure 6.22 Simulated re-phasing of an NMR peak, in which the real and imaginary parts of the Fourier transform are combined in proportions defined by the phase-correction angle, θ.

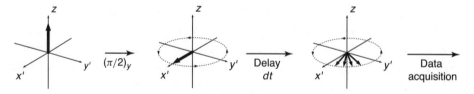

Figure 6.23 Displacement of magnetization vectors representing multiple spin populations during the delay, dt, between a pulse and the beginning of data acquisition.

For a given time delay, dt, the magnetization corresponding to each Larmor frequency will be displaced by a different angle θ. Following Fourier transformation, the uncorrected real part of the spectrum might look something like Fig. 6.24. We can use Eq. 6.5 with an appropriate value of θ to convert one of the peaks into a purely absorptive shape, but the others will, in general, still be distorted, as shown in Fig. 6.25.

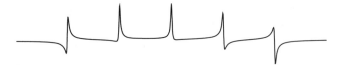

Figure 6.24 A simulated spectrum composed of five peaks, before re-phasing.

Figure 6.25 The simulated spectrum of Fig. 6.24, with a constant phase-correction angle, θ, applied to the full spectrum, so that the leftmost peak has an absorptive shape.

The solution to this problem is to make the phase correction angle, θ, in Eq. 6.5 dependent on the frequency plotted in the spectrum. Usually, this is done using a simple linear function of the form

$$\theta = \theta_0 + (f - f_\mathrm{p})\theta_1$$

where θ_0 and θ_1 are referred to as the zero- and first-order correction terms, respectively, and f_p is a "pivot" point in the spectrum, as explained below. The usual procedure is to first adjust θ_0 until a peak at one end of the spectrum has a pure absorptive shape, as in Fig. 6.25. The resonance frequency for this peak is defined as the pivot point, and then θ_1 is adjusted until the shapes of the other peaks are optimized. This process is simulated in Fig. 6.26, where the top row represents the spectrum after adjusting θ_0, and the rows below correspond to increasing values of θ_1.

Figure 6.26
The second step in the simulated re-phasing of an NMR spectrum. After adjusting θ_0 so that the peak at one end of the spectrum has an absorptive shape, θ_1 is adjusted to correct the phases of the other peaks.

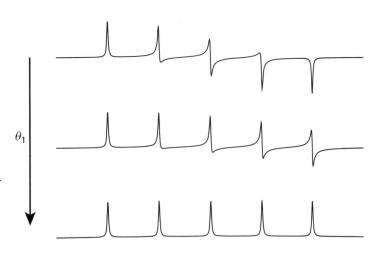

θ_1

In real spectra, phase errors can arise from a variety of different causes and may not be a simple linear function of frequency. As a consequence, it may not be possible to completely eliminate these errors for all frequencies, though the results are usually quite good.

With modern software and computers, phasing a one-dimensional spectrum is usually a very quick and easy procedure. Most data processing programs allow the user to adjust the correction parameters interactively as the spectrum is redrawn on the computer screen almost instantly. For multidimensional spectra, the phases cannot usually be adjusted interactively, but the underlying principles are the same.

6.7   More on Quadrature Detection

Now that we have spent some time considering the real and imaginary parts of Fourier transforms, it is worthwhile to go back to the subject of quadrature detection and see why the particular scheme outlined in the earlier sections actually works. Earlier we showed that mixing an FID that is described by a damped sine function with two reference signals that differ in phase by $\pi/2$ rad (and removing the high-frequency component) gives rise to two heterodyne signals described by the following functions, which, for convenience, we will call $COS(t)$ and $SIN(t)$:

$$COS(t) = e^{-rt} \cos\left(2\pi(\nu - \nu_0)t\right)$$
$$SIN(t) = e^{-rt} \sin\left(2\pi(\nu - \nu_0)t\right)$$

(6.6)

If the $COS(t)$ signal is treated as the real component of the input to a Fourier transform, without an imaginary component, the transformation results in a function with real and imaginary parts, as shown in Fig. 6.27.

Both the real and imaginary parts of the transform contain two peaks, positioned equal distances from zero. Importantly, exactly the same spectra will be generated whether $(\nu - \nu_0)$ is positive or negative, reflecting the mathematical property that the term $\cos(2\pi(\nu - \nu_0)t)$ has equal values for positive or negative values of $(\nu - \nu_0)$. In the absence of other information, there is no way to determine the sign of $(\nu - \nu_0)$.

In contrast, the Fourier transform of the $SIN(t)$ signal depends on the sign of $(\nu - \nu_0)$. For a positive value of $(\nu - \nu_0)$, the real and imaginary components of

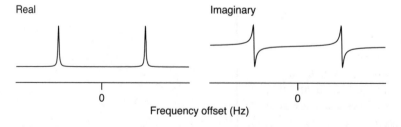

Real   Imaginary

0   0

Frequency offset (Hz)

Figure 6.27   Real and imaginary parts of the Fourier transform of the $COS(t)$ function defined in Eq. 6.6.

Real   Imaginary

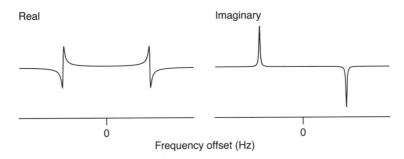

0   0

Frequency offset (Hz)

Figure 6.28   Real and imaginary parts of the Fourier transform of the SIN(t) function defined in Eq. 6.6, for ($v - v_0$) > 0.

Real   Imaginary

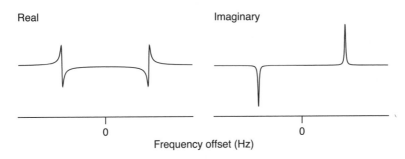

0   0

Frequency offset (Hz)

Figure 6.29   Real and imaginary parts of the Fourier transform of the SIN(t) function defined in Eq. 6.6, for ($v - v_0$) < 0.

the Fourier transform of this signal have the appearance shown in Fig. 6.28. Again, there are two peaks in both the real and imaginary components, but those in the real component have dispersive shapes and those in the imaginary component have absorptive shapes. Another important feature of the spectra in this figure is that the two peaks in each spectrum have opposite signs. If ($v - v_0$) is negative, the Fourier transformation will yield real and imaginary functions with the shapes shown in Fig. 6.29. Now, the signs of the peaks in both the real and imaginary parts have been reversed.

From these examples, it might appear that the solution to resolving the ambiguity of the sign of ($v - v_0$) is simply to use the SIN(t) signal, and then use the imaginary part of the Fourier transform as the spectrum. A positive value of ($v - v_0$) is indicated by a negative peak on the positive side of the spectrum, and a positive peak on this side of the spectrum indicates that ($v - v_0$) is negative. We could even multiply the spectrum by −1 to make it less confusing. There are, however, at least two problems with this solution. First, the sign of a peak can also be reversed by a phase error of π rad, and, as we discussed in the previous section, the magnitude of the appropriate phase correction is not usually known ahead of time. Second, if the sample contains nuclei with Larmor frequencies that are both greater and less than the reference frequency, both sides of the resulting spectrum will contain both positive and negative peaks, and closely spaced peaks will tend to cancel each other.

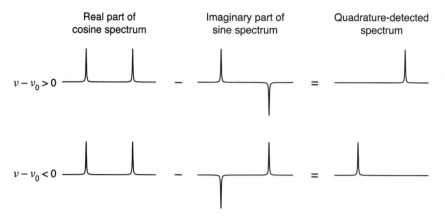

Figure 6.30   The construction of a quadrature-detected spectrum from the difference between the real part of the transform of the $\text{COS}(t)$ component and the imaginary part of the transformed $\text{SIN}(t)$ component.

What we really want is for each resonance frequency to generate a single positive peak on the side of the spectrum corresponding to the correct sign of $(\nu - \nu_0)$. This can be accomplished by taking the real part of the spectrum from one of the heterodyne signals and subtracting from it the imaginary part of the other spectrum. This is shown in Fig. 6.30 for the cases where $(\nu - \nu_0)$ is either positive or negative.

This manipulation can be performed either following the Fourier transform or, more commonly, by combining the $\text{COS}(t)$ and $\text{SIN}(t)$ functions into a complex function before the transformation, as described earlier. That the two procedures give rise to the same results reflects some important properties of the Fourier transform and, for this reason, is worth discussing further. In particular, the Fourier transform is said to be a linear operator, meaning that if we use the symbol $\mathcal{F}$ to represent the Fourier transformation of two functions, $s(t)$ and $r(t)$, and c is a constant, then the following two relationships hold:

$$\mathcal{F}\big(c \cdot s(t)\big) = c\mathcal{F}\big(s(t)\big)$$

$$\mathcal{F}\big(s(t) + r(t)\big) = \mathcal{F}\big(s(t)\big) + \mathcal{F}\big(r(t)\big)$$

Using this notation, we can write the Fourier transforms of $\text{COS}(t)$ and $\text{SIN}(t)$ as

$$\mathcal{F}\big(\text{COS}(t)\big) = \mathcal{R}e\Big(\mathcal{F}\big(\text{COS}(t)\big)\Big) + i\mathcal{I}m\Big(\mathcal{F}\big(\text{COS}(t)\big)\Big)$$

$$\mathcal{F}\big(\text{SIN}(t)\big) = \mathcal{R}e\Big(\mathcal{F}\big(\text{SIN}(t)\big)\Big) + i\mathcal{I}m\Big(\mathcal{F}\big(\text{SIN}(t)\big)\Big)$$

If we then multiply $\text{SIN}(t)$ by i, the transform of the product is

$$\mathcal{F}\big(i\,\text{SIN}(t)\big) = i\mathcal{F}\big(\text{SIN}(t)\big)$$

$$= i\Big[\mathcal{R}e\Big(\mathcal{F}\big(\text{SIN}(t)\big)\Big) + i\mathcal{I}m\Big(\mathcal{F}\big(\text{SIN}(t)\big)\Big)\Big]$$

$$= i\mathcal{R}e\Big(\mathcal{F}\big(\text{SIN}(t)\big)\Big) - \mathcal{I}m\Big(\mathcal{F}\big(\text{SIN}(t)\big)\Big)$$

Adding together COS(t) and i SIN(t) and then carrying out the Fourier transform gives

$$\mathcal{F}\big(\text{COS}(t) + i\ \text{SIN}(t)\big) = \mathcal{F}\big(\text{COS}(t)\big) + i\mathcal{F}\big(\text{SIN}(t)\big)$$

$$= \mathcal{R}e\Big(\mathcal{F}\big(\text{COS}(t)\big)\Big) + i\mathcal{I}m\Big(\mathcal{F}\big(\text{COS}(t)\big)\Big)$$

$$+ i\mathcal{R}e\Big(\mathcal{F}\big(\text{SIN}(t)\big)\Big) - \mathcal{I}m\Big(\mathcal{F}\big(\text{SIN}(t)\big)\Big)$$

The real part (i.e., the part that is *not* multiplied by i) is then given by

$$\mathcal{R}e\Big(\mathcal{F}\big(\text{COS}(t) + i\ \text{SIN}(t)\big)\Big) = \mathcal{R}e\Big(\mathcal{F}\big(\text{COS}(t)\big)\Big) - \mathcal{I}m\Big(\mathcal{F}\big(\text{SIN}(t)\big)\Big)$$

Thus, the real part of the transform of the combined signal represents just the difference that we showed earlier will give rise to a single absorptive peak with the sign of the frequency offset correctly indicated. Using the idealized spectra shown earlier, you should also be able to see that the imaginary part of this transform will give rise to a single dispersive peak. As noted earlier, the combined signal can be represented in either trigonometric or exponential forms:

$$\text{COS}(t) + i\ \text{SIN}(t) = e^{-rt}\cos\big(2\pi(v - v_0)t\big) + ie^{-rt}\sin\big(2\pi(v - v_0)t\big)$$

$$= e^{\big(i2\pi(v-v_0)t - rt\big)}$$

For the remainder of this book, we will generally ignore the issues of quadrature detection and the phasing of a spectrum. For purposes of explaining the design of most one- and two-dimensional experiments, it will be sufficient to simply assume that the original FID signal has a defined phase and that Fourier transformation of the signal from a spin population with a given Larmor frequency gives rise to a single absorptive peak. Once you begin working with real spectra, however, you will need to be aware of the issues raised in this and the previous section.

6.8   Other Data Manipulations

In addition to phase errors, which could arise even if we had continuous sampling of the FID for an indefinite period of time, there can be significant artifacts in the transformed spectra that reflect practical limitations on data acquisition. To illustrate these, and some of the methods used to deal with them, we will now look at some examples of discretely sampled data. As in the earlier examples, the spectra are based on idealized FIDs, but now sampled at intervals and subjected to a numerical Fourier transformation.

6.8.1   Zero-filling

One practical limitation in many real NMR experiments is the number of data points that can be recorded. While this is not usually an issue with simple one-dimensional experiments, it often is with multidimensional spectra, for which the demands on

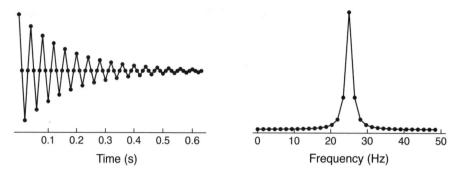

Figure 6.31   An example of a discrete Fourier transform of a simulated FID, with a frequency of 25 Hz and a relaxation rate of 6 s^{-1}. The FID is sampled for a total time of 0.64 s at 0.01-s intervals.

spectrometer time can be great. As discussed earlier, the number of data points recorded can limit both the resolution of the final spectrum and the frequency range it covers. First, we will consider the limitations on resolution.

The simulated FID and spectrum shown in Fig. 6.31 represent a single spin population with a Larmor frequency of 25 Hz (in the rotating frame) and a relaxation rate of 6 s^{-1}. The sampling interval is 0.01 s. The discrete sample points are represented by dots, and the line segments simply connect the dots to guide the eye. The sampling interval in the FID is 0.01 s, and a total of 64 points were recorded. (The fast Fourier transform algorithm is most efficient when the number of points is a power of two, though increases in computer speed have made this less of an issue.) Notice that the signal has almost completely decayed by the end of the acquisition period. Following the transformation, there are a total of 64 points in the frequency domain, but 32 of them represent negative values and are not shown in this and following spectra. Because the sampling time was 0.01 s, the total frequency range is $1/\Delta t - 1/(N \Delta t) \approx 98.4$ Hz, and the maximum positive frequency is ≈ 48.4 Hz. As expected, there is a peak in the spectrum at 25 Hz.

While this spectrum clearly shows the resonance frequency that generated the signal, there are only a few points that actually define the peak, and we might want to be able to draw a smoother curve. To generate more points in the spectrum, we need to have more points in the FID. In particular, we need data points representing longer time values in the FID in order to obtain more finely spaced points in the spectrum. While we could record the signal for a longer time, it has already decayed to almost zero by the last point already recorded. But, we can simply add zeros to the end of the recorded FID. Fig. 6.32 shows the result of doubling the FID data set with zeros.

Now, there are twice as many points in the spectrum, which covers the same frequency range as before, and the peak is defined by twice as many points. If this seems too good to be true, that is because it largely is. No new information has been added to the spectrum and the new points simply represent interpolations between the original ones. Nonetheless, "zero-filling" does lead to a smoother appearance. In addition, it is sometimes necessary to add zeros to an FID in order to make the

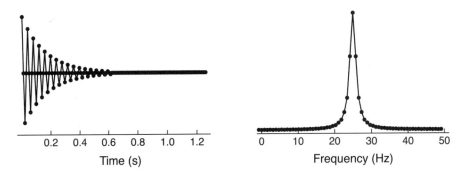

Figure 6.32 The effect of zero-filling the FID from Fig. 6.31. Adding points to the end of the FID results in more closely spaced points in the spectrum.

number of points a power of two, and certain other data processing procedures place further constraints on the number of points.

Beyond conserving valuable spectrometer time, under some circumstances, zero-filling is actually preferable to simply recording the FID for a longer time. This is because of the inevitable presence of noise in the FID and the effects of this noise on the spectrum. The FID–spectrum pair in Fig. 6.33 was simulated using the same parameters as in Fig. 6.31, but now a random number (with a maximum absolute value of 5% of the maximum signal amplitude) has been added to each of the recorded points. In spite of the noise in the FID, the peak is clearly detectable in the spectrum, but the baseline shows spurious peaks and valleys distributed over the full frequency range. This, in fact, is a defining feature of "random noise" (sometimes called "white noise"): Fourier transformation of the time-domain signal leads to random contributions throughout the frequency domain. Suppose that we were to record this noisy FID for twice as long, as illustrated in Fig. 6.34. As when we zero-filled the FID, the peak is now defined by more points, but in addition, the noise in the spectrum has increased in intensity, because there are more noisy points contributing to the FID. Since the total signal intensity has not increased significantly, while the noise has been doubled, the signal-to-noise ratio has been reduced nearly two-fold.

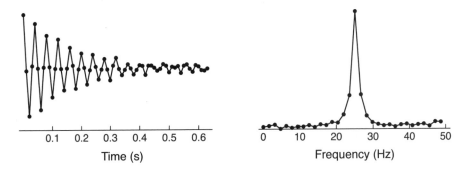

Figure 6.33 The FID from Fig. 6.31 with added noise, and the resulting spectrum.

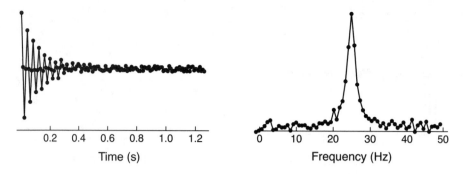

Figure 6.34 The effect of recording and transforming a noisy FID beyond the time the signal has decayed below the noise level.

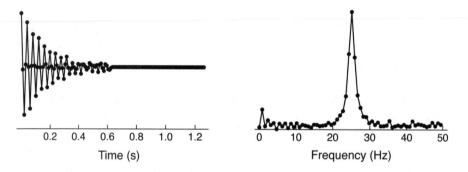

Figure 6.35 The effect of zero-filling the noisy FID in Fig. 6.33.

Suppose, now, that instead of collecting more points, with average values of zero, we just add zeros. The results are shown in Fig 6.35. Again, the peak is defined by more points, but the noise level is not as high as when the second half of the FID contained random noise. On the other hand, there is an artifactual peak near 0 Hz that arises from the abrupt elimination of the noise.

In the examples shown above, the primary effect of zero-filling is to improve the definition of the peaks in the plotted spectrum, without increasing the information content. For rather subtle reasons, however, when quadrature detection is used to create real and imaginary FID components, zero-filling to double the number of points does actually increase the signal-to-noise ratio of the spectrum.[1] Zero-filling beyond that, however, has only cosmetic effects.

6.8.2 Apodization

Under some circumstances, it may be advantageous to record more closely spaced data points to generate a spectrum covering a wider frequency range. But, if there

---

[1] See Bartholdi, E. & Ernst, R. R. (1973). *J. Magn. Reson.*, 11, 9–19. http://dx.doi.org/10.1016/0022-2364(73)90076-0

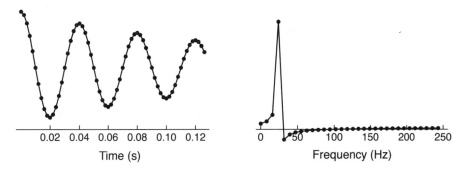

Figure 6.36   An FID truncated before the signal has decayed and the resulting spectrum. The total number of points is the same as in Fig. 6.31, but the sampling interval has been shortened from 0.01 s to 0.002 s in order to extend the frequency range.

are limitations on data collection, this may come at the expense of stopping the data collection before the signal has fully decayed. An example of a prematurely truncated FID (this time without noise) and the resulting spectrum is shown in Fig. 6.36. In this example, the time interval between sampled data points has been decreased by a factor of five (relative to the example in Fig. 6.31), leading to a five-fold increase in the frequency range. Because the damped sine function of the FID is now much more well defined, we might have expected that the peak in the spectrum would also be better defined. But, because the total time for data collection is reduced, there are actually fewer points to define the peak, and the peak also shows distortion that arises from the abrupt truncation of the signal. The only advantage gained by reducing the sampling interval is the ability to cover a wider frequency range, an advantage that isn't realized in this simple one-peak spectrum, but which can be critical in real experiments.

Suppose that we now try zero-filling this FID, as shown in Fig. 6.37. This is not so good either! While we do have more closely spaced points in the frequency-domain spectrum, as expected, the peak is now flanked by quite pronounced oscillations. These artifacts are a consequence of the abrupt transition between the still significant FID signal and the zeros that were added.

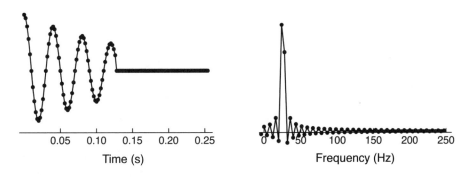

Figure 6.37   The effect of zero-filling the truncated FID shown in Fig. 6.36.

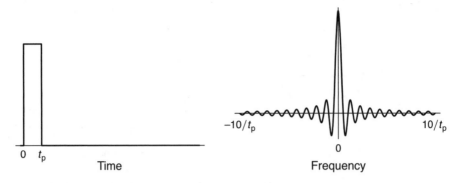

Figure 6.38   A square pulse and its Fourier transform, the sine cardinal (sinc) function.

Before discussing how to deal with this problem, which arises frequently in practical NMR, it is worthwhile to consider its origin and mathematical description. The pattern of decaying undulations that now flank the spectral peak are associated with the Fourier transform of a rectangular pulse, which can be defined as a function equal to 1 (or any other arbitrary value) between $t = 0$ and $t = t_p$, and 0 elsewhere. This function and the real part of its Fourier transform have the shapes shown in Fig. 6.38.

The relationship between the rectangular function and its Fourier transform may not be intuitively obvious, but notice, first of all, that the spectrum has a peak at 0 Hz, reflecting the fact that the signal does not *appear* to repeat itself with time. It is possible, however, that signal does repeat itself, but we just haven't waited long enough, resulting in the finite width of the main peak. The fluctuations on either side of 0 Hz reflect the fact that an accurate representation of the instantaneous change at $t = t_p$ requires an infinite series of sine and cosine functions with increasing frequency, but decreasing amplitude. The mathematical form of the real part of the transform is given by

$$\frac{\sin(\pi 2 f t_p)}{\pi 2 f t_p} \qquad (6.7)$$

The general form of this function, $\sin(\pi x)/(\pi x)$, is called the *sine cardinal* function, but more often goes by the shortened name "sinc." The fluctuations that lie on either side of the major peak are known colloquially as "sinc wiggles." Note that the sinc function, as defined above, does not have a value for $x = 0$, and a more proper definition includes the stipulation that sinc(0) = 1. To return to the case of the truncated FID, we can think of this signal as the result of multiplying the original FID by the rectangular function.

An important theorem of Fourier analysis states that the transform of a product of two signals, in this case the rectangle function and the damped sine function, is a special kind of merging, called a *convolution*, of the transforms of the individual components. The definition of convolution is a bit, well, convoluted and is discussed using illustrations in Appendix C. For now, it is sufficient to say that the Fourier transform of the truncated FID has a shape similar to that of the sinc function, but is centered at the frequency of the original signal, as seen in Fig. 6.37.

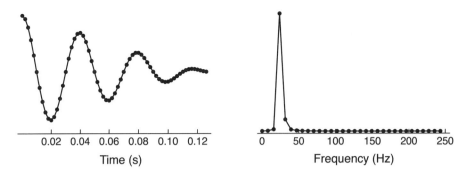

Figure 6.39 The effect of multiplying the truncated FID of Fig. 6.36 by the cosine apodization function.

In order to minimize the artifacts of truncation, the common practice is to modify the FID, so that its amplitude more smoothly approaches zero, by multiplying the signal by a function that goes to zero at the appropriate time value. This manipulation is called *apodization*, which translated literally from Greek means "cutting off the feet." One function that is often used for apodization is a portion of the simple cosine function. If the last recorded point represents $t = t_{max}$ then the cosine apodization function is given as

$$\cos\left(\pi t / (2t_{max})\right)$$

for $0 \leq t \leq t_{max}$. If we multiply the original (truncated) FID by this function before the Fourier transformation, we obtain the results shown in Fig. 6.39.

Now, the FID has an appearance quite similar to what we would expect if the FID had simply decayed to nearly zero by t_{max}. The resulting peak in the spectrum shows much less distortion than when the FID was abruptly truncated, but it is still rather poorly defined.

If we now add zeros to the FID, we obtain more closely spaced points in the spectrum, but without the distortions we saw when the truncated FID was zero-filled (Fig. 6.40). Now, at last, we have both a wide frequency range and a reasonably,

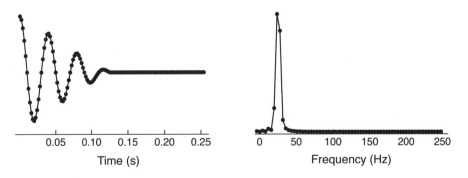

Figure 6.40 The truncated FID of Fig. 6.36 after apodization and zero-filling, and the resulting spectrum.

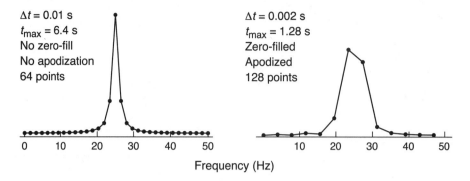

Figure 6.41   Close-up comparison of the simulated peak shown in Fig. 6.31 with that in Fig. 6.40. The peak on the left is from an FID recorded until the signal has decayed nearly completely, while the one on the right comes from a truncated FID, with apodization and zero-filling.

though not perfectly, shaped peak defined by multiple points. Again, this seems almost too good to be true, so what have we lost? The answer can be seen in Fig. 6.41, which compares this peak with the very first one we generated (Fig. 6.31), with the same scaling for the frequency axis. The frequency-domain peak has become much wider in the spectrum designed to cover a wider frequency range. This reflects two factors affecting the second spectrum: The initial data were recorded for a shorter time, and the later data points were smoothed to zero, leading to a wider peak just as if the relaxation rate were faster. Note also that even though the total number of points in the second spectrum is greater, there is still less dense coverage of a given frequency range, because the points are spread over a wider range. This could be changed by adding yet more zeros to the FID, but this would not increase the real resolution.

A variety of other apodization functions are sometimes used for NMR and are discussed in some of the references at the end of this chapter. For a given spectrum, careful choice of the apodization function may favor increased resolution, improved peak shapes or a greater signal-to-noise ratio.

To cover the wider frequency range *and* maintain the original resolution, we would have to record the FID at the shorter intervals *and* for the same total time as in the original example, as shown in Fig. 6.42. The cost here is that five times as many data points have to be recorded. Depending on the circumstances, the cost of recording additional data points may or may not be a significant consideration, and the relaxation rate of the signal will determine the value of additional data recorded at later times.

These examples provide an introduction to some of the practical methods most commonly used to process the data from pulse NMR experiments, as well as the inherent limitations of these methods. Data processing is, in itself, a large subject, and new methods are constantly being pursued, some of which involve alternatives to the Fourier transform. While none of these techniques can magically create new information from the finite data collected during an experiment, careful design of the experiment and appropriate use of data processing can make a huge difference in the quality of the final spectrum.

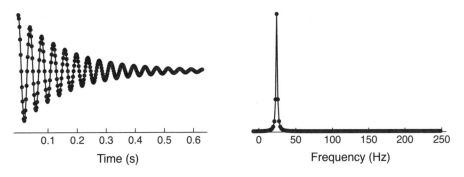

Figure 6.42 Combining closely spaced data points, for a wide frequency range, and a long total time, for narrow, well-defined spectral peaks.

6.9   Fourier Description of Pulses

The Fourier transform can also be used to gain additional insights into the distribution of frequencies generated during the rf pulses used to produce NMR spectra. In Section 5.2, we used a largely qualitative argument to conclude that the range of frequencies is inversely related to the pulse duration. For a more mathematical description, we can consider a conventional rectangular pulse as the product of a rectangle function and a cosine function describing the oscillating magnetic field, as illustrated in the left-hand panel of Fig. 6.43. The frequency distribution generated by the pulse is given by the Fourier transform of the product of functions and is shown in the right-hand side of Fig. 6.43. The transform is the convolution of those of the cosine

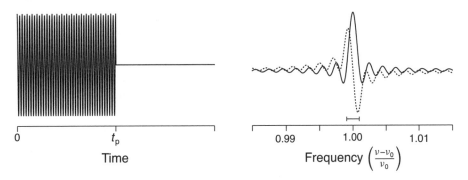

Figure 6.43 Representations of a rectangular pulse in the time (left) and frequency (right) domains. The pulse can described as the product of a cosine function with frequency v_0 and a rectangle function covering the range from $t = 0$ to t_p, which results in a frequency spectrum described by a sinc function centered at v_0. For purposes of illustration, the frequency suggested in the left-hand panel is much lower, relative to the pulse duration, than in a real experiment. In the right-hand panel, the frequency profile is calculated for a pulse with a frequency of 500 MHz and a duration of 1 μs, and the frequency axis is expressed as a relative offset from v_0. The real component of the frequency spectrum is shown as a solid curve and the imaginary component is shown as dashes. The bar below the curve indicates the range of frequencies separating the first zero-crossing points in the real component, which are found at $v_0 \pm 1/(2t_p)$, or 500 ± 0.5 MHz for this case.

function and that of the rectangle function and has both real and imaginary parts, which are shown as solid and dashed curves, respectively, in the right-hand panel of Fig. 6.43. The real part is

$$\mathcal{R}e\big(S(f)\big) = \text{sinc}\big(2(f - v_0)t_p\big) = \frac{\sin\big(2\pi(f - v_0)t_p\big)}{2\pi(f - v_0)t_p}$$

and the imaginary part is

$$\mathcal{I}m\big(S(f)\big) = -\sin\big(\pi(f - v_0)t_p\big)\text{sinc}\big((f - v_0)t_p\big)$$

The real and imaginary parts represent oscillations of the electric current (and therefore, in the resulting magnetic field) that differ in phase by $\pi/2$ rad. If the pulse is along the y'-axis, the real component will rotate nuclear magnetization toward the x'-axis, and the imaginary component will rotate magnetization toward the y'-axis. Also note that both the real and imaginary components are positive for some frequencies and negative for others, which will lead to rotation in opposite directions. Near the center of the frequency range, the real component is positive and much larger than the absolute value of the imaginary component, so nearly all of the rotation will be in the desired direction. Furthermore, the small amounts of magnetization rotated by the imaginary component will be in opposite directions, depending on the frequency, and will tend to cancel out. At frequencies further from v_0, however, the imaginary components become more significant, as does the negative real component.

As a measure of the width of the frequency spectrum, it is convenient to use the distance between the first frequencies on either side of v_0 where the real component crosses zero As illustrated in Fig. 6.43, the majority of the energy lies between these frequencies, which are

$$v_0 - \frac{1}{2t_p} \quad \text{and} \quad v_0 + \frac{1}{2t_p}$$

Thus, the total frequency range between the first zero crossings is simply $1/t_p$. The example illustrated in Fig. 6.43 corresponds to a pulse of 1 μs at 500 MHz, so the range is 1 MHz, or 2000 ppm, consistent with the qualitative argument presented in Section 5.2. This range is more than adequate for stimulating all of the nuclei of a given type in a sample.

For some purposes, it may be desirable to apply a more selective pulse by increasing its duration, while simultaneously decreasing the strength so as to maintain the same rotation angle. In Fig. 6.44, a series of frequency profiles are shown for increasing pulse durations, all assuming that $v_0 = 500$ MHz. Although the central region of the frequency profile narrows with increased pulse duration, as expected, the plots show that fluctuations of the sinc function also become prominent at frequencies closer to v_0 and extend quite far in both directions. Because these fluctuations, together with the imaginary components, cause rotations of magnetization in all possible directions, they can lead to a variety of artifacts in the final spectrum if the pulse duration is made too long. For these reasons, simply extending the duration of a rec-

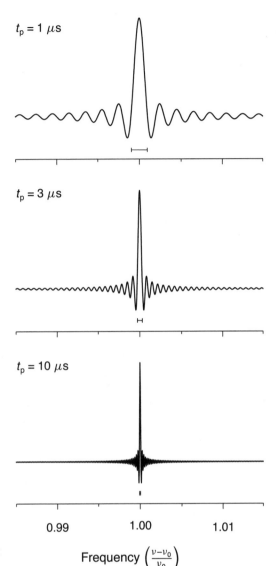

Figure 6.44

Frequency profiles for rectangular pulses of increasing duration, all assuming a frequency of 500 MHz. For clarity, only the real components of the Fourier transforms are shown, and the frequency axis is expressed as a relative offset from ν_0, The bars indicate the range of frequencies separating the first zero-crossing points in the real component, which are found at $\nu_0 \pm 1/(2t_p)$.

tangular pulse is usually not optimal for selectively stimulating a limited frequency range.

As noted earlier, in the context of a truncated FID, the sinc "wiggles" shown in Figs. 6.43 and 6.44 arise because of the sharp edges of the rectangular pulse, which generate a series of higher-frequency components. It might then be expected that the wiggles could be reduced by modulating the pulse shape so as to minimize sharp transitions, and a variety of pulse shapes have been developed along these lines, A particularly useful shape, at least as a starting point for pulse design, is a Gaussian function, the classic "bell curve," which can be written in the following form:

$$f(x) = e^{\pi x^2}$$

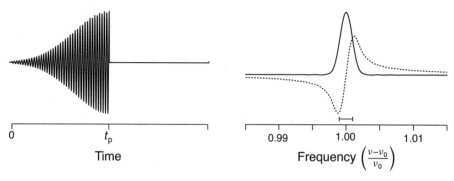

Figure 6.45 Representations of a "half-Gaussian" pulse in the time (left) and frequency (right) domains. The pulse can described as the product of a cosine function with frequency ν_0 and the first half of a Gaussian bell curve, starting at a point where the Gaussian function is a few percent of its maximum and ending at the maximum. The bar below the frequency spectrum indicates the range between $\nu_0 \pm 1/(2t_p)$. Other details of the plots are as in Fig. 6.43.

As written above, the function is centered at zero and has the interesting property of being equal to it's own Fourier transform. The Gaussian function never goes to zero, however, and an NMR pulse that extends in time from $-\infty$ to ∞ is not very practical. Still, a rather well behaved pulse frequency spectrum can be generated by modulating the shape so that the signal amplitude is proportional to the first half of a Gaussian function, beginning at the point where its value is a few percent of the maximum, as illustrated in Fig. 6.45. Although one might have expected that the abrupt termination of the pulse at the peak of the Gaussian function would result in sinc-like wiggles in the frequency spectrum, the symmetry of the Gaussian function (along with symmetry properties of the Fourier transform) lead to a close approximation of the full Gaussian function in the frequency domain. Notice, however, that the width of the frequency spectrum is significantly greater than the central portion of the one generated by a rectangular pulse of the same duration (Fig. 6.43), and there is still an imaginary component that will lead to rotation of nuclear magnetism toward the y'-axis as well as the x'-axis (assuming a y'-pulse). Nonetheless, the half-Gaussian represents a significant improvement over the rectangular shape for selective pulses.

It may also be advantageous, in some cases, to have relatively sharp boundaries to the range of frequencies generated in a pulse, something that the half-Gaussian pulse shape does not provide. One of the important properties of Fourier transforms is that they are (almost) reciprocal, so that if the Fourier transform of a function $g(t)$ in the time domain is $G(f)$ in the frequency domain, then the transform of $G(t)$ is $g(-f)$. Therefore, it should be possible to generate a frequency spectrum with a rectangular shape using a pulse with the shape of a sinc function. Again, there are practical limitations, because the sinc function has non-zero values over the range from $-\infty$ to ∞, but a rectangular frequency spectrum can be approximated using a pulse that begins at the maximum of the sinc function and extends to the second zero crossing, as shown in Fig. 6.46. The frequency spectrum is somewhat broader than that generated by the half-Gaussian pulse of the same duration, but the boundaries on each edge make up a smaller fraction of the total frequency range.

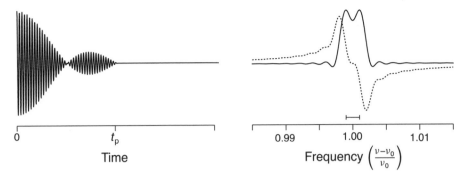

Figure 6.46 Representations of a shaped pulse based on the sinc function, in the time (left) and frequency (right) domains. The pulse can be described as the product of a cosine function with frequency ν_0 and the second half of a sinc function, starting at its maximum and ending at the second zero-crossing time. Other details of the plots are as in Fig. 6.43.

The examples above illustrate how Fourier analysis can be used to understand the relationships between the duration and shape of a pulse and the frequency components of the resulting radiation. However, the practice of designing shaped pulses for specific purposes is considerably more challenging than these examples may suggest. Significant complications arise because the change in magnetization, for instance from z- to x'-magnetization during a y-pulse, is a non-linear response to the intensity and duration of the radiation: When the rotation angle, a, is small, $\sin(a) \approx a$ and the change in x'-magnetization is nearly proportional to a, but as a approaches $\pi/2$, the projection on the x'-axis reaches a maximum and then decreases. Relatively small differences in the pulse strength for nuclei with different Larmor frequencies can lead to significant differences in final magnetization components as the rotation angle increases. Over the last few decades, the development of shaped pulses has become a quite advanced subject in itself and is generally based on both numerical simulations and extensive experimentation.

In closing this chapter, it is worth noting that the ideas and techniques discussed here also form the basis for other forms of digital data processing that are becoming increasingly important in our ever more digitized world. Examples include digitized music, telephone communications and images. In all of these, the goal is to maintain the highest fidelity to the original signal while using a limited amount of data, and concepts such as Fourier analysis and the Nyquist sampling theorem play central roles.

Summary Points for Chapter 6

- The signal recorded from a pulse NMR experiment is called a free induction decay and, for an idealized single spin population, has the form of an exponentially damped sine function.
- The major steps in converting the FID into a spectrum are the following:
  - Conversion of the high-frequency signal from the spectrometer coil to a lower frequency by mixing with a reference signal. Quadrature detection is

the means used to discriminate between frequencies greater and less than that of the reference signal.

- Sampling of the signal at discrete intervals and conversion to a digital form.
- Fourier transformation of the digitized time-domain signal into a frequency-domain spectrum.

- Peaks in the transformed spectrum are, for an ideal signal, Lorentzian functions with a real absorptive component and an imaginary dispersive component.
- Spectra usually require phase correction to produce purely absorptive peaks.
- The width of a spectral peak is inversely related to the rate at which the time-domain FID signal decays.
- The separation of spectral peaks is inversely related to the frequencies of fluctuations in the time-domain signal.
- The frequency range covered in a spectrum is determined by the intervals between sampled points in the FID, according to the sampling theorem.
- Common data processing steps used to improve spectra include the following:
  - Zero-filling—to better define peaks in the transformed spectra
  - Apodization—to eliminate "sinc-wiggle" artifacts due to truncation of the FID
- Fourier analysis is also used to analyze the effects of pulse duration and shape on the spectrum of frequencies stimulated during the preparation period of a pulse NMR experiment.

Exercises for Chapter 6

Some of the exercises for this chapter are intended to give you a feel for what happens when signals with different frequencies are combined in various ways. For these exercises, you will almost certainly want to use a computer program that can generate graphs of numerically evaluated functions. There are a variety of programs, both commercial and free, that can be used for this purpose, ranging from spreadsheets, like Microsoft Excel, to more sophisticated tools like Matlab, Mathematica, and Maxima.

6.1. This exercise illustrates the effects of adding signals with different frequencies, as in the FID from nuclei with different Larmor frequencies.

(a) Begin by creating a data set representing a series of time points, say from 0 to 25 in arbitrary units. You should probably use 200 to 500 points in this data set.

(b) Create and plot a data set representing a single frequency, v, calculated as

$$s(t) = \sin(2\pi v t)$$

where t represents time and the argument of the sine function is assumed to have units of radians. For this first signal, make $v = 1$, representing a frequency of 1 inverse time unit. (If the time unit is seconds, the frequency unit is Hz; if the time unit is ms, the frequency unit is kHz.)

(c) Create and plot a data set representing the sum of two frequency components:

$$s(t) = \sin(2\pi \nu_1 t) + \sin(2\pi \nu_2 t)$$

with $\nu_1 = 1$ and $\nu_2 = 1.1$.

(d) Repeat the step above with $\nu_2 = 1.2$ and 1.4. How does changing the difference between the two frequencies alter the signal?

(e) Try adding three signals with frequencies equal to 1, 1.1 and 1.2. Describe the resulting signal.

6.2. Now, generate some signals that decay with time.

(a) Create and plot a data set representing a single frequency, ν, and a decay rate, r:

$$s(t) = e^{-rt} \sin(2\pi \nu t)$$

with $\nu = 1$ and $r = 0.2$.

(b) Create and plot two additional data sets, each with the same frequency, $\nu = 1$, but with $r = 0.5$ and 1.

(c) Create and plot three data sets, each with two frequencies, 1.0 and 1.2, and with $r = 0.2, 0.5$ and 1. How does the decay rate affect the ability to distinguish the two frequencies?

6.3. This exercise simulates the generation of a heterodyne signal, by mixing the signal of interest with a reference signal.

(a) Create a data set representing the reference signal, with frequency $\nu_0 = 1$:

$$s_r(t) = \sin(2\pi \nu_0 t)$$

(b) Create and plot a data set representing the heterodyne signal generated by multiplying the reference signal and a signal with $\nu = 1.1$:

$$s_h = s_r(t) \sin(2\pi \nu t)$$

The resulting signal should be easily recognizable as being made up of two different frequencies, one of which is much smaller than that of either the original signal or the reference signal. According to Eq. 6.1, the high-frequency component of the heterodyne signal should have the following form:

$$-\frac{1}{2} \cos\big(2\pi (\nu + \nu_0)t\big)$$

Confirm this by subtracting this signal from the heterodyne signal. The result should be a single cosine function with the frequency $\nu - \nu_0$.

(c) Create two more heterodyne signals, using the same reference frequency and signal frequencies of 1.05 and 1.2. How does changing the input signal frequency change the heterodyne signal?

(d) Create a second reference signal, differing in phase from the first by $\pi/2$ rad:

$$s_r(t) = \sin(2\pi \nu_0 t + \pi/2) = \cos(2\pi \nu_r t)$$

Create a heterodyne signal with this reference signal and the sine functions with $\nu = 1$. Confirm that the high-frequency component has the form

$$\frac{1}{2} \sin\left(2\pi(\nu + \nu_0)t\right)$$

and that subtracting this component leaves a low-frequency component that differs in phase by $\pi/2$ from the heterodyne signal generated by mixing with the first reference signal.

6.4. Suppose that you want to collect a one-dimensional ^{1}H spectrum of a protein in a 600-MHz spectrometer. You would like to be able to cover a chemical shift range of 12 ppm, and you need to resolve peaks separated by as little as 0.01 ppm. Your criterion for resolution is that two peaks are resolved if their maxima are separated by more than the sum of their half-widths, measured at one-half of their respective heights. In the following, assume that all of the individual peaks in your spectrum will have equal intensities and line widths.

(a) Which of your requirements will determine how frequently you must sample and digitize the FID signal? What is the minimum sampling interval?

(b) If the transverse relaxation rate for your signals is 10 s^{-1}, would you be able to meet your requirements? Why or why not? If not, what relaxation rate would allow them to be met?

(c) Suppose that you are especially concerned about resolving the splittings of the signals from the amide protons due to scalar coupling with the α-protons? Are the criteria specified above likely to be adequate to do this for all, some or none of the amide protons?

Further Reading

Additional details on data acquisition and processing:

⋄ Cavanagh, J., Fairbrother, W. J., Palmer III, A. G., Rance, M. & Skelton, N. J. (2007). *Protein NMR Spectroscopy*, 2nd ed., pp. 124–165. Elsevier Academic Press, Amsterdam.

A book devoted to the details of data processing, with material on some more advanced techniques not discussed here:

⋄ Hoch, J. C. & Stern, A. S. (1996). *NMR Data Processing*. Wiley-Liss, New York.

More on shaped pulses:

⋄ Freeman, R. (1998). Shaped radiofrequency pulses in high resolution NMR. *Prog. Nucl. Mag. Res. Sp.*, 32, 59–106. http://dx.doi.org/10.1016/S0079-6565(97)00024-1

There are many mathematics textbooks that cover the theoretical and practical aspects of the Fourier transformation in great depth and rigor. A widely used text that provides a nice balance:

⋄ Bracewell, R. N. (2000). *The Fourier Transform and its Applications*, 3rd ed. McGraw-Hill, New York, NY.

A text with more of an emphasis on applications in spectroscopy:

◇ Marshall, A. G. & Verdun, F. R. (1990). *Fourier Transforms in NMR, Optical and Mass Spectrometry: A User's Handbook*. Elsevier, Amsterdam.

For a quite different approach, the following book presents the Fourier transform in a very descriptive and engaging form, assuming very little prior knowledge of mathematics:

◇ Transnational College of LEX (1995). *Who is Fourier? A Mathematical Adventure*. Translated by A. Gleason. Language Research Foundation, Boston.

CHAPTER 7

Relaxation

So far, we have stressed the importance of relaxation in two contexts: (1) The establishment of the initial population difference that is required to generate an observable NMR signal, and (2) the decay of the signal as the magnetization in the $x-y$ plane is lost in a pulse experiment. As we have seen, this decay is especially important in determining line widths and, therefore, the resolution of the spectrum. Here, we will focus on the contributions to the decay of the FID, though some of the same factors also play a role in establishing the original Boltzmann distribution.

7.1 Two Kinds of Relaxation

The decay of the signal following a pulse represents the combined effects of two quite different phenomena. The first of these is the conversion of the non-eigen superposition states back to the equilibrium distribution in which the number of nuclei with the $|\alpha\rangle$ wavefunction slightly exceeds that of $|\beta\rangle$ nuclei. In terms of the vector diagram, this is the conversion of magnetization in the $x-y$ plane back into net magnetization along the z-axis, as illustrated in Fig. 7.1 (in the rotating frame).

This process is usually referred to as *longitudinal relaxation*, since it represents the return of magnetization to the longitudinal direction. For simplicity, Fig. 7.1 shows the transverse magnetization as being stationary in the rotating frame, but magnetization precessing at frequencies faster or slower than the rotating frame will also return to the longitudinal orientation.

The second process that leads to signal decay is the loss of phase coherence among the magnetization vectors arising from different spin populations. Immediately after the pulse, the wavefunctions for the individual nuclei in the sample are correlated with one another (to a limited degree) so that they give rise to a net transverse magnetization. For a variety of reasons, however, the phase relationships for different nuclei

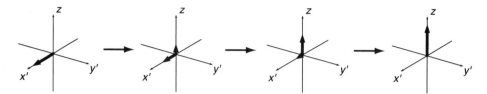

Figure 7.1 Vector diagram representation of the longitudinal relaxation of transverse magnetization. Longitudinal relaxation is measured as the return of z-magnetization to the equilibrium value.

may diverge with time, which is manifested as precession of the magnetization at different rates. In a vector diagram this can be represented in two ways, as shown in Fig. 7.2. In the upper row of diagrams, five magnetization vectors representing five subpopulations in the sample are shown precessing at different frequencies; some faster and some slower than the rotating frame. One way in which this can happen is if the external field is inhomogeneous, so that molecules located at different positions in the sample experience different field strengths. We will discuss this further below, along with more interesting mechanisms by which phase coherence is lost. The bottom row shows the net magnetization arising from the sum of the populations.

As shown by the diagrams, the loss of phase coherence leads to a decrease in the net magnetization in the x–y plane, without a corresponding increase of longitudinal magnetization. This might seem to violate at least one law of thermodynamics. What is happening thermodynamically, however, is that the total energy of the system is remaining constant, but the entropy is increasing. The total loss of magnetization in the x–y plane is referred to as *transverse relaxation*, and includes both the return to the equilibrium distribution of states and the loss of phase coherence among the nuclei in

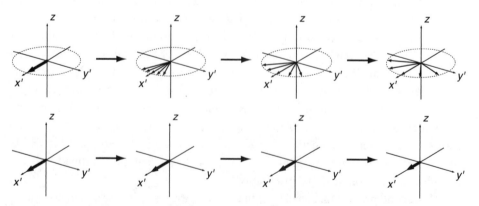

Figure 7.2 Disappearance of transverse magnetization through the loss of phase coherence. The figures in the upper row represent the separate precession of magnetization from subpopulations of the sample. The bottom row indicates the decrease in the net transverse magnetization.

non-eigenstates that give rise to transverse magnetization. As a consequence, the rate of transverse relaxation is always at least as large as the longitudinal relaxation rate.

The study of relaxation is important for two reasons. First, the rates at which the magnetization returns to its equilibrium state or loses its coherence can be critical constraints in the design and outcome of NMR experiments. Second, these rates can provide valuable information about the structure and dynamics of a molecule, including its size and shape, its rate of overall tumbling or the presence of internal structural fluctuations. In the following two sections, we will define the relaxation rate constants, R_1 and R_2, describe experiments by which each can be measured and consider the major molecular mechanisms that contribute to relaxation. In addition to their importance for studying molecular dynamics, the experiments for measuring relaxation rates also offer a good introduction to multi-pulse experiments.

7.2 Longitudinal Relaxation

In discussing relaxation, we are primarily concerned with the net observable magnetization that arises from a large population of nuclei, as opposed to the magnetization associated with individual spins. To emphasize this distinction, we will use symbols such as $\bar{I}_z$ to represent the component of the macroscopic magnetization vectors, in this case along the z-axis. With yet another symbol introduced, it may be helpful to summarize the various quantities related to magnetization and their meanings:

- I_z: The magnetization along the z-axis arising from a single spin.
- $\langle I_z \rangle$: The average value of I_z measured for a large population of nuclei, usually assumed to be in the same quantum state (i.e., described with the same wavefunction).
- $\bar{I}_z$: The net magnetization along the z-axis arising from a macroscopic sample of nuclei, which are generally not in the same quantum state.

In Chapter 11 we will introduce yet another symbol, $\hat{I}_z$, to represent the quantum mechanical "operator," or rule, that is used to calculate the average outcome of a measurement of I_z.

At equilibrium, the net magnetization along the z-axis of a sample is proportional to the difference between the number of nuclei in the $|\alpha\rangle$ and $|\beta\rangle$ state, N_α and N_β, respectively:

$$\bar{I}_z \propto N_\alpha - N_\beta$$

For nuclei with a positive gyromagnetic ratio, N_α is slightly larger than N_β at equilibrium, and $\bar{I}_z$ has a positive value. When the sample is irradiated, the number of nuclei in the two states are changed, and longitudinal relaxation represents the return of $\bar{I}_z$ to its equilibrium value.

7.2.1   The Longitudinal Relaxation Rate Constant, R_1

Longitudinal relaxation is almost always assumed to follow first-order kinetics, as described by the following differential equation:

$$\frac{d\bar{I}_z(t)}{dt} = -R_1 \left(\bar{I}_z(t) - \bar{I}_z^{\,0} \right)$$

where $\bar{I}_z(t)$ is the z-magnetization at time t, $\bar{I}_z^{\,0}$ is the equilibrium magnetization and R_1 is the longitudinal relaxation rate constant. Note that the instantaneous rate depends on the difference between the current z-magnetization and the equilibrium value. Just like the rate expression for a first-order chemical reaction, this equation can be rearranged and integrated to give an expression for $\bar{I}_z$ as a function of time:

$$\int_{\bar{I}_z(0)}^{\bar{I}_z(t)} \frac{d\bar{I}_z(t)}{\bar{I}_z(t) - \bar{I}_z^{\,0}} = \int_0^t -R_1 dt$$

$$\ln \left(\bar{I}_z(t) - \bar{I}_z(0) \right) \Big|_{\bar{I}_z(0)}^{\bar{I}_z(t)} = -R_1 t \Big|_0^t$$

$$\ln \left(\frac{\bar{I}_z(t) - \bar{I}_z^{\,0}}{\bar{I}_z(0) - \bar{I}_z^{\,0}} \right) = -R_1 t$$

$$\frac{\bar{I}_z(t) - \bar{I}_z^{\,0}}{\bar{I}_z(0) - \bar{I}_z^{\,0}} = e^{-R_1 t}$$

$$\bar{I}_z(t) = \bar{I}_z^{\,0} + \left(\bar{I}_z(0) - \bar{I}_z^{\,0} \right) e^{-R_1 t} \tag{7.1}$$

Note the distinction between $\bar{I}_z^{\,0}$, the equilibrium magnetization, and $\bar{I}_z(0)$, the magnetization at the beginning of the relaxation process. The rate constant for this process, R_1, is often expressed as a relaxation time, T_1:

$$T_1 = 1/R_1$$

The actual value of R_1 depends on a number of factors, including the structure of the molecule and the rates at which it undergoes different types of motion, as we will discuss later.

From the simple one-pulse experiment we have discussed so far, there is no easy way to determine the longitudinal relaxation rate, since the decay of the FID reflects the total loss of magnetization from the transverse plane. There is, however, a relatively simple two-pulse experiment that can be used to measure R_1 specifically. This measurement, called an *inversion-recovery* experiment, involves the following four steps:

1. A π-pulse.
2. A delay period.
3. A $\pi/2$-pulse.
4. Data acquisition, exactly as for a simple one-pulse experiment.

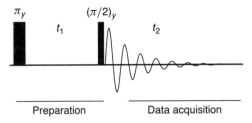

Figure 7.3

Pulse-sequence diagram for an inversion-recovery experiment to measure R_1, the longitudinal relaxation rate constant.

The experiment is usually repeated several times, with different delay periods.

The experiment is illustrated in Fig. 7.3, using the style of diagram introduced in Chapter 5. In addition to the two pulses, the diagram also defines two time periods; the first is the delay between the two pulses, t_1, and the second, t_2, represents the times at which data are recorded to generate the FID. (Do not confuse t_1 and t_2 with the relaxation times, T_1 and T_2.) The effects of these manipulations on the behavior of a simple, and idealized, system with just a single spin population are illustrated in Fig. 7.4.

The initial π-pulse inverts the equilibrium magnetization so that it now points along the $-z$-axis. In terms of the wavefunctions, the pulse converts all of the nuclei with $|\alpha\rangle$ wavefunctions into nuclei with $|\beta\rangle$ wavefunctions and vice versa. The small excess of $|\alpha\rangle$ nuclei is thus converted into an excess of $|\beta\rangle$ nuclei. If a signal were to be recorded from the sample at this point, nothing would be detected, because there is no magnetization in the transverse plane.

The first pulse is then followed by a delay period, t_1, and then a $(\pi/2)_y$-pulse. If the second pulse is immediately after the first, the full magnetization will now point along the $-x'$-axis, opposite the direction it would be pointing if we had just applied a simple $(\pi/2)_y$-pulse to the equilibrium magnetization. When the FID is now collected, it is 180° out of phase with the one that would be generated after a single $(\pi/2)_y$-pulse, and when this is Fourier transformed, the resulting peak in the spectrum is negative.

As the delay is made longer, longitudinal relaxation converts some of the excess $|\beta\rangle$ spins into $|\alpha\rangle$ spins, and the z-magnetization becomes less negative. When half of the excess has been converted, the population contains equal numbers of nuclei in the two eigenstates. At this point, there is no z-magnetization, and no signal is detected after the $(\pi/2)_y$-pulse. With still longer delays, the excess of $|\alpha\rangle$ nuclei is partially re-established, corresponding to positive z-magnetization. When this is converted to positive x'-magnetization by the second pulse, a positive peak is generated in the spectrum. With sufficiently long delays, the equilibrium magnetization is fully re-established, and the spectrum is exactly the same as would be generated by a single $(\pi/2)_y$-pulse.

Plotting the peak intensity (taking into account its sign) versus time gives a graph like that shown in Fig. 7.5. The data can then be fit to Eq. 7.1 to estimate R_1. In a real experiment, with nuclei with different resonance frequencies, the R_1 rate constants can be determined for each resonance that can be resolved in the spectrum. In practice, somewhat more complex versions of the inversion-recovery experiment are often

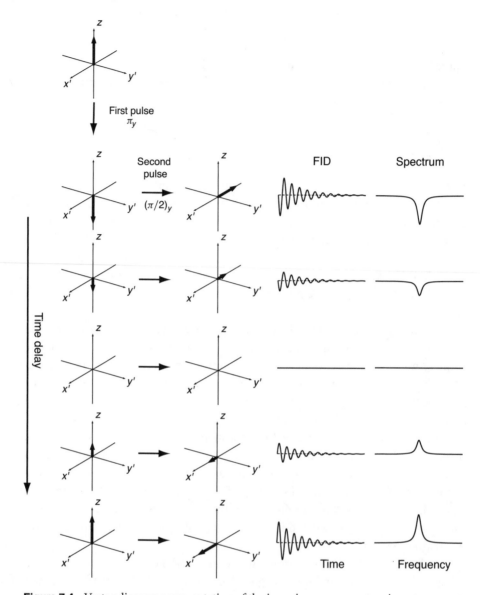

Figure 7.4   Vector diagram representation of the inversion-recovery experiment.

Figure 7.5

Result of the inversion-recovery experiment: a plot of peak intensity as a function of the delay time, t_1. Fitting the experimental data to Eq. 7.1 yields the longitudinal relaxation constant, R_1.

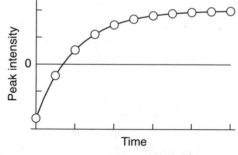

used to overcome technical limitations that we will not consider here. These experiments can also be extended to multidimensional spectra in order to obtain relaxation rate constants for individual nuclei in more complicated molecules.

7.2.2 Mechanisms of Longitudinal Relaxation in Liquids

In order to understand the physical basis for relaxation, we have to consider mechanisms by which the wavefunctions of the nuclei can return to their equilibrium distribution. As we have discussed previously, the net contributions of the $|\alpha\rangle$ and $|\beta\rangle$ eigenfunctions do not change spontaneously as the wavefunction evolves under the influence of a steady magnetic field. In order to restore the longitudinal magnetization, there must be some transfer of energy to or from the nuclei. As in the original conversion from the equilibrium distribution, this occurs through the absorption of radiation generated by fluctuating magnetic fields. During relaxation, however, it is the molecules in the sample that generate the fluctuating fields. As the molecules tumble in solution, the nuclei experience changes in their magnetic environments. If these fluctuations are near the Larmor frequency, then there will be a finite probability of changing the relative contributions of the $|\alpha\rangle$ and $|\beta\rangle$ eigenfunctions to the total wavefunction.

The two major factors that make the local field experienced by a nucleus sensitive to the orientation of the molecule are dipolar coupling and chemical shift anisotropy.

Dipolar coupling

This is the effect of the magnetic dipole of one nucleus on the local magnetic field experienced by another, as diagrammed schematically in Fig. 7.6. In the drawing, the molecule is represented as an ellipsoid with two spin-1/2 nuclei, A and B. As the molecule tumbles, the nuclei retain their orientation with respect to the external field, but the fields that they experience from one another fluctuate. To illustrate this, field lines are drawn to indicate the field created by nucleus A. When nucleus B is directly "above" A, its local field is slightly enhanced by the dipole of nucleus A.

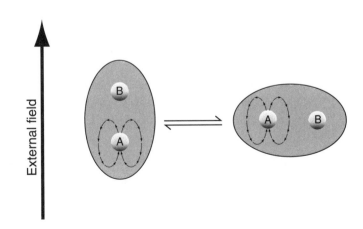

Figure 7.6

Fluctuation of the local magnetic field due to dipolar interactions in a tumbling molecule.

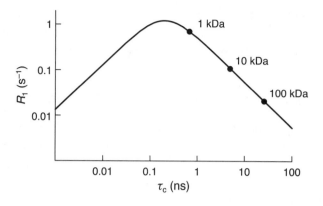

Figure 7.7
Predicted longitudinal relaxation rate constants due to dipolar interactions as a function of molecular correlation time for two ^{1}H nuclei separated by 2 Å in a field of 11.7 T.

When, however, B is to the side of A, the field from A is in the direction opposite that of the external field. As the molecule tumbles, the effect of dipolar coupling on the resonance frequency of each nucleus averages to zero, as discussed in Chapter 4, but the motion generates fluctuations in the local field. Under most circumstances that are relevant to solution NMR, the dipolar interaction is the major contributor to longitudinal relaxation.

The rate at which the wavefunction for a nucleus will change depends on both the strength of the fluctuating field and the rate of fluctuation. Relaxation is fastest when the fluctuation has a frequency close to the Larmor frequency. The tumbling of a molecule is a highly random process, and there are motions with a wide range of frequencies. As an approximation, however, the distribution of frequencies can be characterized by a single number, the *correlation time*, τ_c. A good working definition of τ_c is that it is the average time between changes in orientation of about 1 rad. For macromolecules in water near room temperature, τ_c is typically 1 to 100 ns. Although there are always slower and faster motions, the probability of a fluctuation with a given frequency, ν, is greatest when τ_c is approximately equal to $1/\nu$. Molecules with different sizes will have different tumbling rates and thus will display quite different relaxation rates at a given external field strength.

For simple systems, the relaxation rate can be predicted using theory developed in Chapter 8, as illustrated in Fig. 7.7 for the specific case of two ^{1}H nuclei separated by 2 Å in a field strength of 11.7 T (where the ^{1}H Larmor frequency is 500 MHz). The filled circles in the plot represent, very approximately, the expected correlation times and R_1 values for proteins of different molecular weights. The graph shows that R_1 is maximal when $1/\tau_c$ approximately matches the Larmor frequency of the proton. For smaller molecules, which tumble more rapidly, most of the local magnetic field fluctuations are too fast to stimulate the quantum transitions, and, for larger molecules, most of the fluctuations are too slow.

Chemical shift anisotropy

Remember that the shielding a nucleus experiences depends on both the effect of the external field on the electrons and the distribution of these electrons in the vicinity of the nucleus, relative to the external field. The distribution of electrons is generally asymmetric, so the shielding changes as a molecule tumbles in solution. This, then, generates a fluctuation in the local field that a particular nucleus feels. Like dipolar coupling, relaxation due to chemical shift anisotropy is generally most efficient when the tumbling rate is near the Larmor frequency. For liquid samples, the contribution from chemical shift anisotropy is usually much smaller than that from the dipolar interaction. Since the shielding effect reflects the influence of the external field on the electrons, its magnitude increases with field strength, and this contribution to relaxation becomes more significant at higher fields.

Other mechanisms

There are a variety of other mechanisms by which molecular motions can give rise to magnetic fluctuations that promote relaxation. For biomolecular samples, the most important additional contribution usually comes from unpaired electrons either within the same molecule or in other molecules. Like the ^{1}H, ^{13}C and ^{15}N nuclei, the electron has a spin of $1/2$ and has a much larger gyromagnetic ratio than any of the nuclei. As a consequence, its dipole is very strong and can promote relaxation over relatively long distances. Dissolved molecular oxygen, which is paramagnetic, in a sample can promote relaxation, as can unpaired electrons in organic compounds or in metal complexes. While these effects usually represent a nuisance to be avoided (because they tend to broaden the lines of a spectrum), they can also be exploited to gain specific information about a molecule. For instance, proteins and other molecules can be modified with *spin labels* containing unpaired electrons (typically nitroxides or metal chelates), and the effects of these spins on the relaxation of nuclei in the molecule are measured. These effects can then be used to estimate distances between the spin label and the nuclei. Another approach involves adding an extrinsic spin reagent to the solution and measuring the effects on relaxation. Here, the expectation is that the largest effects will be seen for the nuclei that are most accessible to the reagent. While these methods have provided valuable information about molecules that are often otherwise difficult to characterize, the interpretation is not always straightforward.

7.3   Transverse Relaxation

During the data acquisition period of a simple one-pulse experiment, the return of longitudinal magnetization is accompanied by a corresponding loss of magnetization in the transverse plane, and the transverse magnetization disappears at least as fast as the z-magnetization reappears. Often, the transverse magnetization disappears even faster. As discussed earlier, this can happen if the magnetization vectors associated with different subpopulations precess at slightly different frequencies, leading to the loss of phase coherence.

7.3.1   The Transverse Relaxation Rate Constant, R_2, and Peak Widths

Like longitudinal relaxation, transverse relaxation is usually a first-order process, at least approximately, and is characterized by a rate constant, R_2. If we look at the x'-magnetization for a case where the direction of the average magnetization is stationary in the rotating frame, then the change in the x'-magnetization with respect to time is given by

$$\frac{d\bar{I}_x(t)}{dt} = -R_2 \bar{I}_x(t)$$

Integrating this equation gives

$$\bar{I}_x(t) = \bar{I}_x(0)e^{-R_2 t} \tag{7.2}$$

Because the regain of magnetization in the z-direction always contributes to the loss of transverse magnetization, R_2 is always equal to or greater than R_1. The transverse relaxation rate is also often referred to by a relaxation time, T_2, defined as

$$T_2 = 1/R_2$$

There are two general ways in which R_2 can be measured. The first is to simply measure the width of the peaks in a spectrum, which reflects the rate of decay of transverse magnetization during the data acquisition period. The second method involves introducing a time delay, along with some special pulses, between the time at which the transverse magnetization is created and the beginning of the data acquisition period. As with the inversion-recovery experiment for measuring R_1, a series of spectra are recorded with increasing delay times, and the value of R_2 is determined by the rate of change in peak intensity. As discussed below, the two methods are not entirely equivalent and can yield different values of R_2, reflecting different mechanisms by which the phase coherence in the transverse plane is lost.

As discussed in Chapter 6, the width of an absorptive Lorentzian peak is proportional to the decay rate of the FID. The peak width, w, is defined as the distance between the two sides of the peak at the position where the intensity is half of its maximum, as illustrated in Fig. 7.8. The transverse relaxation rate constant can then be calculated according to

$$R_2 = \pi \cdot w$$

For some purposes, this may be an adequate method for measuring R_2, and, in fact, it is the peak width that is the practical concern with respect to spectral resolution. However, peak splitting by scalar coupling may make it difficult to measure peak widths accurately, and R_2 measured in this way is also sensitive to field inhomogeneity, as discussed further below.

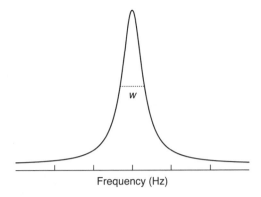

Figure 7.8
The Lorentzian peak width.

7.3.2   Spin Echoes and the Measurement of R_2

To understand why another method of measuring R_2 might be useful, it is necessary to consider two general mechanisms by which the coherence of the transverse
magnetization can be lost. First, there is the inevitable inhomogeneity of the external field, which causes equivalent nuclei within molecules located in different parts of
the sample to have slightly different precession frequencies. Provided that the translational diffusion of the molecules is relatively slow, the precession frequencies in
different locations will remain constant during the data acquisition period. Second,
there are processes that give rise to *random* fluctuations in precession frequency over
time. In solution NMR, the most pervasive source of random changes in Larmor
frequency is molecular tumbling. Because the effects of dipolar coupling and chemical shift anisotropy depend on the orientation of the molecule with respect to the
external field, the precession frequency of a nucleus will change randomly as the molecule tumbles. In addition to this effect, internal motions in the molecule may cause
random changes in the chemical shift of a nucleus. In terms of characterizing the
properties of a molecule, the rates of molecular tumbling and any internal motions
are generally much more interesting than the effects of magnetic field inhomogeneity.
This makes it desirable to have an experimental method that can distinguish between
these mechanisms.

   Consider, first, the loss of phase coherence due to field inhomogeneity. A useful
way to think about this effect is to imagine that the sample is divided up into pieces,
called *isochromats*, which are small enough that any differences in field strength
within them are negligible. We can then imagine that each isochromat gives rise to a
magnetization vector that precesses with a distinct frequency. After the initial $\pi/2$-
pulse, these vectors begin to spread out like a folding fan in the transverse plane, as
illustrated diagrammatically in Fig. 7.9.

   Fig. 7.10 shows a simulation of the FID generated by this effect. The simulated
FID on the left is similar to ones shown earlier and was calculated assuming a single
homogeneous spin population and a relaxation rate constant that is 0.04 times the
precession frequency. The FID on the right is based on the same assumptions except
that the sample is assumed to be made up of 10 isochromats, for which the resonance
frequencies are evenly spaced from 96% to 104% of the average precession frequency.

Figure 7.9

Vector diagram representing the precession of magnetization from different isochromats in an inhomogeneous magnetic field.

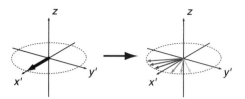

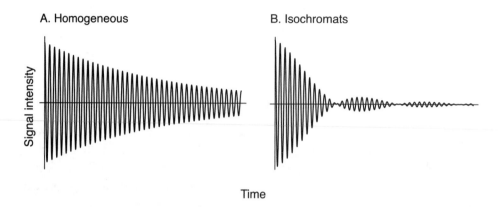

A. Homogeneous

B. Isochromats

Signal intensity

Time

Figure 7.10 Simulated FIDs for (A) a sample in a homogeneous magnetic field and (B) a sample made up of 10 isochromats with different precession frequencies.

Note that the inhomogeneity leads to a much faster decay of the FID, as well as a shape that is somewhat different from that of an exponentially damped sine function.

Notice, also, that the FID drawn in Fig. 7.10B initially disappears completely, and then seems to reappear with greatly reduced intensity. This reappearance is due to the tendency of some of the vectors to come back into alignment temporarily as they precess at different rates. If the individual precession frequencies remain constant, you might imagine that the vectors would eventually come back into phase completely. In practice, though, the FID is always observed to decay irreversibly. This is because it takes a very long time for the vectors representing all of the isochromats to come back into alignment, during which time the net magnetization returns to the z-direction via the longitudinal relaxation mechanisms discussed earlier. In general, the time required for two vectors precessing at different frequencies to become realigned is inversely proportional to the frequency difference. Since the sample is divided up into small volumes that differ only very slightly in effective field strength, the time required is far longer than the longitudinal relaxation time.

While spontaneous recovery of phase coherence cannot usually be observed, a simple trick can induce a nearly complete recovery of the signal intensity due to field inhomogeneity—a trick that plays an important role in measurements of R_2, as well as other NMR experiments. To illustrate this technique, consider the example shown in Fig. 7.11, where the average magnetization has precessed from the x'-axis to the y'-axis, but the vectors representing various isochromats have fanned out over nearly half of the transverse plane. Suppose, now, that we apply a π-pulse along the x'-axis,

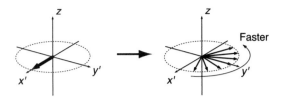

Figure 7.11   Vector diagram representing the precession of magnetization from different isochromats in an inhomogeneous magnetic field, as the average magnetization has moved to the y'-axis.

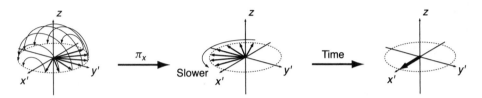

Figure 7.12   Refocusing of the fanned-out magnetization illustrated in Fig. 7.11 by a π-pulse followed by an equal delay time.

as shown in Fig. 7.12. The pulse flips each of the vectors to the other side of the x'-axis, where they continue to precess. The slowest vector is now closest to the original starting point, in the direction of precession, while the fastest is furthest away. If the vectors are allowed to precess for exactly the same length of time as before the π-pulse, they will all converge on the x'-axis again! This process is called *refocusing*. If we continue to record data, after the π-pulse, the signal will reverse its decay until the time after the π-pulse equals that preceding it, and then the decay will begin again. A simulation of the FID generated by this phenomenon is shown in Fig. 7.13. Notice that the full intensity of the FID is not recovered, because the signal is weakened by longitudinal relaxation and other processes discussed below.

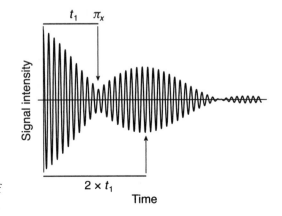

Figure 7.13
Simulated FID generated by application of a refocusing pulse to a sample composed of 10 isochromats, as in Fig. 7.10.

Figure 7.14

Pulse sequence for a refocusing experiment to measure the transverse relaxation rate constant, R_2.

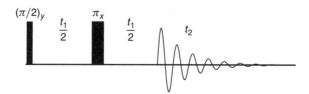

This rather spooky phenomenon was first observed in 1950 by Erwin Hahn, who called it a *spin echo*. Hahn's observation was totally unexpected and was actually made following a second $\pi/2$-pulse, rather than the π-pulse illustrated above. Rather than dismissing the echo as a technical artifact, Hahn systematically pursued its origin and derived its correct explanation. The resulting paper is considered one of the great classics of NMR, because it not only describes the spin echo and the first multi-pulse experiment, but is also one of the first papers to use vector diagrams to describe the effects of pulses and precession. (The use of three-dimensional vector diagrams to describe the spin-echo effect apparently was suggested by Purcell.) Indeed, this paper can be viewed as the predecessor of all pulse NMR experiments and was published nearly 15 years before Anderson and Ernst were able to implement Fourier transform NMR, thanks to the intervening advances in digital data acquisition and computing.

Hahn also demonstrated that the spin echo could be used as the basis for an experiment to measure transverse relaxation rates. This idea was subsequently refined by Carr and Purcell, who replaced the second $\pi/2$-pulse in Hahn's experiment with a π-pulse along the y'-axis, and then by Meiboom and Gill, who changed the orientation of the second pulse to make it a π_x-pulse, which helps compensate for imperfections in the pulses. The simplest form of such an experiment, as implemented using Fourier transform NMR, is based on the pulse sequence shown in Fig. 7.14. To measure R_2, a series of spectra are recorded using increasing values of t_1. For each spectrum, the π-pulse is made halfway through the delay period between the initial $\pi/2$-pulse and the beginning of the data acquisition period, t_2. In this way, any loss of phase coherence due to field inhomogeneity is reversed. The resulting FIDs are Fourier transformed as usual, and the peak intensities are measured. The data are plotted and fit to Eq. 7.2 (p. 170) to estimate R_2, as shown in Fig. 7.15.

A key assumption in the description of this experiment is that the rates of precession before and after the refocusing pulse remain the same for the individual sub-

Figure 7.15

Result of a refocusing experiment: a plot of peak intensity as a function of the delay time, t_1. Fitting the experimental data to Eq. 7.2 yields the transverse relaxation constant, R_2.

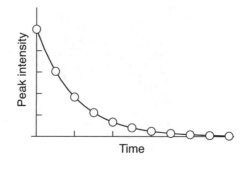

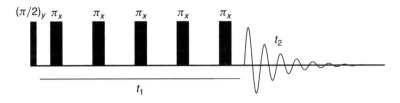

Figure 7.16 Multi-refocusing-pulse version of the CPMG experiment for measuring transverse relaxation rates.

populations. This will only be true if there is no significant diffusion of molecules between regions of the sample that experience different field strengths. In addition to introducing the π-refocusing pulse, Carr and Purcell implemented a further refinement in which the single pulse in the middle of the delay period is replaced with a series of such pulses, as diagrammed in Fig. 7.16.

Progressively longer total delay periods are generated by increasing the number of times the π_x-pulse is repeated, while keeping the time interval between successive pulses constant. Even if a molecule diffuses into a region with a different magnetic field strength during one of these shorter periods, the effect of the change in Larmor frequency will be limited to the difference in precession during this single period, rather than the much larger effect that would accumulate over a single long period. Recognizing the contributions of Carr, Purcell, Meiboom and Gill, the form of the experiment diagrammed in Fig. 7.16 is now called a CPMG pulse sequence.

The values of R_2 measured in the CPMG experiment will in general be smaller than those estimated from peak widths, because the latter are influenced by field inhomogeneities. Some authors distinguish the two measurements by using the symbol R_2^* (or its inverse T_2^*) to indicate a value derived from peak widths. If field inhomogeneity is the only factor contributing to the loss of phase coherence, then the value of R_2 measured by a CPMG experiment should be equal to R_1 measured by the inversion-recovery experiment. As we will discuss below, however, there are additional mechanisms, involving random variations of the precession frequency, that lead to loss of phase coherence that cannot be refocussed.

7.3.3 Random Fluctuations and Transverse Relaxation

As noted above, refocusing pulses are effective only when the precession frequencies of individual molecules in the sample remain constant for the period preceding and following the pulse. If the molecules undergo random and independent changes in Larmor frequency, these changes will lead to an irreversible loss in phase coherence and, therefore, signal intensity.

In liquid samples, the major source of random changes in precession frequency is the tumbling of the molecule. Because the orientation of a molecule with respect to the static external field influences both the shielding from electrons and the interaction of nearby dipoles, the Larmor frequency for each nucleus undergoes constant fluctuations as the molecule tumbles. In a simplified picture, we can imagine that each

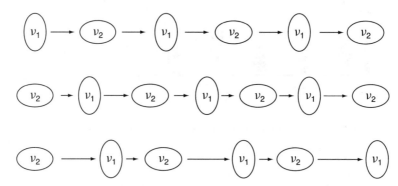

Figure 7.17   Schematic representation of the independent tumbling of molecules, alternating between two orientations giving rise to different Larmor frequencies, v_1 and v_2. Each row represents the trajectory of a separate molecule, with the lengths of the arrows indicating the time of interconversion between the two orientations.

molecule interconverts between two orientations that give rise to distinct precession frequencies, as illustrated schematically in Fig. 7.17 for just a few molecules.

The vector diagrams in Fig. 7.18 suggest how the magnetization from a single nucleus might fluctuate, assuming that it can take on two precession frequencies, with an average frequency of zero in the rotating frame. At first glance, you might imagine that the magnetization from each nucleus would remain quite close to the x'-axis, because it is equally likely to move in a clockwise or counterclockwise direction during each time interval. Over time, however, each vector will have an increasing probability of being displaced from the x'-axis, in one direction or the other.

The net effect of the fluctuations in precession frequency are illustrated by the simple simulation shown in Fig. 7.19. The graphs were generated assuming that a single molecule can exist in one of two orientations, which give rise to different precession frequencies (with an average frequency of zero), and that at every nanosecond interval the molecule randomly "chooses" either to remain in the same orientation or to change to the other. The top graph shows the instantaneous precession frequency, while the graph at the bottom represents the net angular displacement from the starting position.

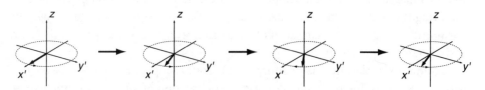

Figure 7.18   Precession of magnetization from a single nucleus as the molecule tumbles, as in Fig. 7.18, leading to an alternation between two precession frequencies centered on the rotating-frame frequency.

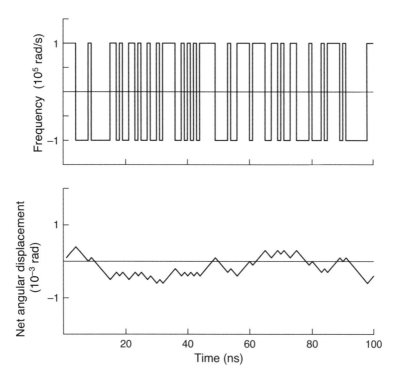

Figure 7.19   Simulation of a random alternation between two precession frequencies. The upper panel plots the instantaneous precession frequency as a function of time. In the lower panel, the total angular displacement from the initial position is plotted versus time.

Notice that when the molecule shifts between states frequently, the net displacement does not change much. But, when, as inevitably happens, there is an extended period in the same orientation, there is a substantial displacement. For this particular trajectory, the vector has shifted in the negative direction at the end of the 100-ns time period. As the time period increases, so does the likelihood of a significant net displacement.

The drawings in Fig. 7.20 illustrate the simulated effects of longer time periods on multiple molecules, under the same conditions as in Figure 7.19. Here, each dot in the figure represents the end of the magnetization vector corresponding to a single molecule, looking down on the transverse plane from the positive z-axis. Initially, all of the vectors are aligned with the x'-axis, but with time an increasing fraction of them have moved toward either the positive or negative y'-axis. While the average projection along the y'-axis remains close to zero, the net projection along the x'-axis decreases. Although this process is similar to the fanning out of vectors due to field inhomogeneity, it differs in an important respect; the process cannot be reversed by a refocusing pulse. This is because there is no correlation between the net precession of a single vector before and after the refocusing pulse. As a consequence, the random fluctuations *do* contribute to the value of R_2 measured by CPMG-type experiments.

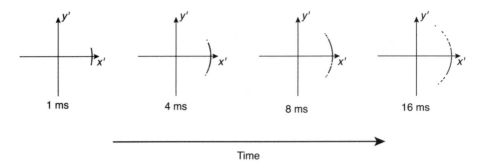

Figure 7.20 Simulation of the loss of phase coherence due to random changes in precession frequency. Each dot represents the end of the magnetization vector for a single molecule undergoing random alternation between two precession frequencies, as in Fig. 7.19. Initially, all of the vectors are aligned with the x'-axis.

7.3.4   The Relationship Between Tumbling Rate and Transverse Relaxation Rate

The process described above is an example of a *random walk*, in which we can think of each magnetization vector taking a series of small "steps" with a defined average length but random direction. Random walks arise in many aspects of physics, chemistry and biology, and the theory developed to analyze these problems can provide additional insights into the factors that determine transverse relaxation rates. Because it involves the fluctuation of only a single parameter, the angular displacement, we can consider the process a random walk in one dimension, albeit along a circular path.

One of the most important results from a mathematical analysis of a random walk is the relationship between the average displacement from the starting point and the number and length of steps in the walk. If we think of the displacement as being in either the positive or negative direction, the mean displacement for a large number of walks will be very close to zero. If, however, we calculate the square of each displacement and then take the average, the net displacement of each vector in the population then contributes to a positive value of the mean. The term "mean-square" is used to indicate this kind of average, which provides a measure of the average net displacement even when the simple average is zero. If the average length of a step is l and the number of steps is N, then the mean-square angular displacement, $\langle \theta^2 \rangle$, for our random walk is given by

$$\langle \theta^2 \rangle = l^2 N$$

This average is sometimes expressed as its square root, the root-mean-square (RMS) average, so as to give it the same dimensions as the original measurement. The RMS angular displacement is then given by

$$\mathrm{RMS}(\theta) = l\sqrt{N}$$

Thus, the RMS displacement is proportional to the length of the step, but only increases with the square root of the number of steps, reflecting the general inefficiency of a process in which the direction of motion changes randomly. The length of each step in the random walk in the transverse plane will be the time interval between frequency changes—that is, the correlation time (τ_c) for the molecular motion that causes the frequency change—times the angular frequency during that time period. For simplicity, we will assume that the average frequency is zero (in the rotating frame) and designate the alternate frequencies as $+\Delta\omega$ and $-\Delta\omega$. The absolute value of the step length, l, is then $\tau_c \Delta\omega$. The number of steps is the total time, t, divided by the average length of a step, τ_c, so that $N = t/\tau_c$. Substituting these expressions into the equation for the mean-square displacement gives:

$$\langle \theta^2 \rangle = \frac{t}{\tau_c}(\tau_c \Delta\omega)^2 = t\tau_c \Delta\omega^2$$

This simple equation has an important implication: For a given total time period, t, the mean-square displacement will be greater if the time is divided into a small number of long periods (large τ_c) than if it is divided into a larger number of short periods. This is a key distinction between a random walk and motion in a uniform direction, in which total displacement over a given time period is the same regardless of how many intervals this time is divided into. This point can be illustrated by an analogy with gambling: A single wager of a large sum is likely to cause a more substantial change in wealth (for better or worse) than the same sum gambled over many smaller bets. For transverse relaxation, the consequence is that slow interconversions among states with different precession frequencies are the most important in causing a loss of phase coherence among multiple molecules.

Further analysis of the random-walk problem leads to a probability distribution function, $p(\theta)$, that describes the relative probability of a net displacement as a function of the magnitude of that displacement, θ. This relationship is given by the following equation:

$$p(\theta) = \frac{1}{\sqrt{2\pi \langle \theta^2 \rangle}} e^{-\theta^2/(2\langle \theta^2 \rangle)} = \frac{1}{\sqrt{2\pi t\tau_c \Delta\omega^2}} e^{-\theta^2/(2t\tau_c \Delta\omega^2)}$$

which describes a *Gaussian* or "bell curve" function, illustrated in Fig. 7.21. The graph was calculated assuming a total time of 10 ms, a correlation time, τ_c, of 1 ns and a value of $\Delta\omega$ of 10^5 rad/s. The probability that θ will lie between two values, a and b, is calculated as the following integral:

$$p(a \leq \theta \leq b) = \int_a^b p(\theta)d\theta$$

In Fig. 7.22, probability distribution functions are plotted for increasing correlation times—that is, the average interval separating transitions between the alternating precession frequencies. These curves were again calculated assuming a total time of 10 ms and $\Delta\omega = 10^5$ rad/s, but with correlation times of 1, 10 and 100 ns. As suggested by the discussion of $\langle \theta^2 \rangle$, larger values of τ_c lead to increasing probabilities

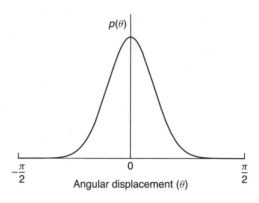

Figure 7.21
Probability distribution function for the total angular displacement of magnetization after 10 ms. The spin is assumed to interconvert randomly at 1-ns intervals between two precession frequencies differing by 2×10^5 rad/s.

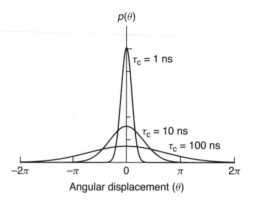

Figure 7.22
Probability distribution function for the total angular displacement of magnetization, as in Fig. 7.21, with different values of τ_c.

of larger angular displacements, which correspond to smaller net projections in the transverse plane.

We can use this probabilistic description to derive an expression for the decay of the net transverse magnetization as a function of time. If we assume that the average precession frequency matches the rotating-frame frequency, then the average net y'-magnetization will remain zero. For each spin, the expected x'-magnetization will be proportional to the cosine of the angular displacement, θ. The average x'-magnetization after time t can then be calculated by multiplying $\cos\theta$ by the probability of that value of θ and then integrating over all possible values of θ:

$$\bar{I}_x(t) = \int_{-\infty}^{\infty} p(\theta)\cos(\theta)d\theta = \int_{-\infty}^{\infty} \frac{1}{\sqrt{2\pi\langle\theta^2\rangle}}e^{-\theta^2/(2\langle\theta^2\rangle)}\cos(\theta)d\theta$$

Evaluation of this integral gives a remarkably simple result:

$$\bar{I}_x(t) = e^{-\langle\theta^2\rangle/2} = e^{-t\tau_c\Delta\omega^2/2}$$

Thus, the model predicts that the transverse magnetization will decrease exponentially, as is observed experimentally, and that the rate constant for this process will be

$$R_2 = \tau_c\Delta\omega^2/2 \tag{7.3}$$

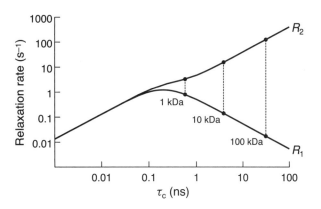

Figure 7.23
Predicted longitudinal and transverse relaxation rate constants due to dipolar interactions as a function of molecular correlation time for two ^{1}H nuclei separated by 2 Å in a field of 11.7 T.

Note that the rate constant is proportional to τ_c, so it increases with both molecular size and solvent viscosity. This is very different from the behavior of R_1, which displays a maximum value when τ_c is approximately equal to the reciprocal of the Larmor frequency.   It is important to note that the treatment above does not include the contribution of longitudinal relaxation to the loss of transverse magnetization. A more detailed analysis of both effects, as discussed in Chapter 8, can be used to predict the total transverse relaxation rate constant due to dipolar coupling as a function of the molecular correlation time, τ_c. In Fig. 7.23, R_1 and R_2 are both plotted as a function of τ_c, again for the case of two protons separated by 2 Å, in the field of a 500-MHz spectrometer.

The graph also indicates the approximate τ_c values and relaxation rate constants expected for protein molecules of different molecular masses in water at room temperature. For very small molecules (less than about 100 daltons), the values of R_1 and R_2 converge. In this regime, tumbling is so fast that there is no significant transverse relaxation due to the loss of phase coherence. For even small peptides, however, the values of τ_c are such that R_1 is less than its maximum value, and R_2 is greater than R_1, with the difference increasing monotonically for larger molecules. As shown in the figure, the R_2 rate constants for even moderately large proteins can be very high, leading to broad resonance peaks and poorly resolved spectra. Over the past two decades or so, a variety of clever methods have been developed to help deal with this problem, notably the TROSY method described in Chapter 8 (page 242), but large transverse relaxation rates still represent one of the major obstacles to using NMR on molecules larger than about 50 kDa.

7.3.5   Longitudinal Relaxation in the Rotating Frame: $R_{1\rho}$

This experimental relaxation measurement is quite closely related to R_2 and can be thought of as a limiting case of the CPMG experiment, with the delay between refocusing pulses eliminated. Under these conditions, the oscillating magnetic field from the spectrometer coils is applied constantly. As discussed in Chapter 5 (Fig. 5.19), an oscillating transverse field can be thought of as being composed of two components— one that is stationary in the rotating frame and one that rotates at twice the rotating-frame frequency and can be ignored. If we apply a $(\pi/2)_y$-pulse and then follow this

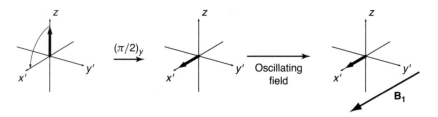

Figure 7.24 The establishment of a non-equilibrium state and a stationary magnetic field in the rotating frame.

with a continuous oscillating field such that the stationary component is aligned with the x'-axis, we can describe the situation as illustrated in Fig. 7.24.

The $R_{1\rho}$ rate constant is measured by recording and Fourier transforming an FID at the end of a period of applying the oscillating field. Typically, a series of measurements are made with increasing time periods, and the intensities of the resulting peaks are fit to an exponential decay function, just as in the CPMG experiment.

The $R_{1\rho}$ experiment can be interpreted most easily by recalling that the constant external field strength (B_0) can be considered to be zero in the rotating frame (for those nuclei with Larmor frequencies that match the rotating-frame frequency). Thus, the net field acting on these nuclei, as viewed from the rotating frame, will be the component of the oscillating field aligned with the x'-axis—$\mathbf{B}_1$ in Figure 7.24. Because the magnetization is aligned with the x'-axis, the situation resembles the equilibrium state as viewed from the laboratory frame, in which the magnetization is aligned with the stationary field along the z-axis. Importantly, however, the magnitude of the magnetization now aligned with the x'-axis was determined by the equilibrium established in the laboratory frame by the $\mathbf{B}_0$ field, which is much larger than the field generated by the pulse coils. Thus, if we consider the situation from the rotating frame, the system is quite far from equilibrium, just as it is in the laboratory frame. The natural tendency is for the x'-magnetization to decay toward the equilibrium value that would be established by the $\mathbf{B}_1$ field.

The decay of the x'-magnetization can be facilitated by exactly the same processes that normally promote longitudinal relaxation, especially the fluctuation of local fields caused by molecular tumbling. In the rotating frame, however, the most effective motions are those that match the Larmor frequency established by the $\mathbf{B}_1$ field. Because this field is generally weaker than the $\mathbf{B}_0$ field by several orders of magnitude, a plot of $R_{1\rho}$ versus molecular correlation time will display a maximum at proportionally larger values of τ_c. Typical values of the Larmor frequency in the rotating frame lie in the range of 1000 to 10,000 Hz, much smaller than $1/\tau_c$ for even very large macromolecules. As a consequence, $R_{1\rho}$ increases in proportion to τ_c in virtually all practical situations with liquid samples, much as R_2 does, consistent with the suggestion above that the $R_{1\rho}$ measurement is analogous to the CPMG experiment in the limit of a very short delay between the refocusing pulses.

Strictly speaking, this experiment is only valid when the Larmor frequency matches the rotating-frame frequency, which is also the frequency at which the $\mathbf{B}_1$ field oscillates. Consider, though, what happens if a particular nucleus is slightly "off

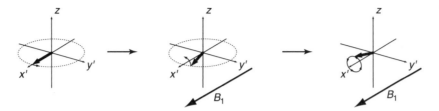

Figure 7.25   Spin-lock effect of a stationary magnetic field in the rotating frame. Magnetization from spins that do not match the rotating-frame frequency remains localized around the rotating-frame axis.

resonance." Immediately after the $(\pi/2)_y$-pulse, the average magnetization will begin to precess in the rotating frame. Once the magnetization has deviated from the x'-axis, however, the $\mathbf{B}_1$ field will cause it to begin precessing about the x'-axis, as diagrammed in Fig. 7.25.

The precession around the x'-axis tends to keep the average magnetization stationary in the rotating frame. This form of irradiation is often referred to as a *spin lock* and has a number of other applications, including the TOCSY (total correlation spectroscopy) experiment described in Chapter 15. The effectiveness of the spin lock depends on the relative rates of precession about the z-axis (in the rotating frame) and about the x'-axis. Provided that the $\mathbf{B}_1$ field is sufficiently strong, there will be relatively little loss of x'-magnetization due to precession about the axis, and the $R_{1\rho}$ measurement can be applied to spins with a modest range of chemical shifts.

For liquid samples, R_2, as measured by a CPMG experiment, and $R_{1\rho}$ provide roughly equivalent information about relatively slow molecular motions, and the choice of one over the other is determined by technical details that are beyond the scope of this text. In solids, however, $R_{1\rho}$ has properties that are quite different from R_2. This is because the net effective field experienced by a nucleus in the rotating frame is strongly influenced by the dipoles of other nearby stationary nuclei, effects that can exceed that of the $\mathbf{B}_1$ field. The relatively large contribution of other nuclei alters the mechanisms and time scales of rotating-frame relaxation in solids, generally leading to $R_{1\rho}$ values that are smaller than R_2 and may approach that of R_1.

7.4   Effects of Internal Motions on Relaxation

So far, we have only considered the effects of the overall tumbling motions of a molecule. Internal motions can also give rise to fluctuating magnetic fields and changes in precession frequency, and can thus also influence both longitudinal and transverse relaxation.

As indicated in Fig. 7.23, proteins and other macromolecules generally have tumbling rates such that R_1 is less than the maximum for relatively high spectrometer fields, and R_2 is usually greater than R_1. When nuclei within such a molecule undergo motions that are faster than overall tumbling, the effect, qualitatively, is to decrease the effective correlation time. As a consequence, such motions will always decrease

R_2, leading to sharper lines in the spectrum. In extreme cases, it is sometimes possible to detect very sharp lines for particularly flexible regions of a large molecule or complex, while the signals from the rest of the molecule decay too rapidly to even be detected.

The effect of internal motions on R_1 are somewhat harder to predict. If the motions are faster than overall tumbling, but not greater than the Larmor frequency, they will shift the effective correlation time toward the region where it approximates the reciprocal of the Larmor frequency, thereby enhancing longitudinal relaxation. On the other hand, if the frequency of the motions is greater than the Larmor frequency, then R_1 will decrease.

Motions slower than the overall tumbling rate generally do not affect longitudinal relaxation, because it is the fastest magnetic field fluctuations that determine the efficiencies of these mechanisms. However, slower motions can give rise to changes in resonance frequency that lead to the loss of phase coherence and thus increased transverse relaxation. For historical reasons, processes that give rise to changes in resonance frequency (i.e., chemical shift) are traditionally called *chemical exchange*, though there may not be any change in the covalent structure of the molecule. The term *conformational exchange* is sometimes used in reference to cases where transient changes in three-dimensional structure give rise to changes in precession frequency. These processes can influence both the spectral line widths and the value of R_2 measured in a CPMG experiment.

The effects of chemical or conformational exchange on the line width depends on several factors, including the rate of exchange, the relative concentrations of the alternate states and their difference in Larmor frequency. We consider first a relatively simple case, in which there are two equally populated states, A and B, that interconvert with a rate constant, k_{ex}, defined as the sum of the forward and reverse rate constants:

$$A \overset{k_{ex}}{\underset{}{\rightleftharpoons}} B$$

The plots in Fig. 7.26 show the predicted spectra for different exchange rates, assuming that the Larmor frequencies differ by 100 Hz ($\nu_A = -50$ Hz and $\nu_B = 50$ Hz).

The *difference* in Larmor frequencies, $\Delta\nu$, is a critical factor in determining the effects of exchange because it is this difference that determines the time period required for the magnetization vectors associated with the two frequencies to diverge from one another during the data acquisition period. The different exchange possibilities are usually classified as slow, fast, or intermediate:

1. *Slow exchange.* When the exchange rate constant is much smaller than the difference in Larmor frequencies (e.g., $k_{ex} = 2$ s^{-1} and $\Delta\nu = 100$ Hz), then there will not be any significant exchange during the data acquisition period. The result will be the same as if there were two populations of nuclei with different chemical shifts—that is, two peaks in the spectrum.
2. *Fast exchange.* At the other extreme, if the exchange rate constant is much greater than the frequency difference (e.g., $k_{ex} = 5000$ s^{-1} and $\Delta\nu = 100$ Hz),

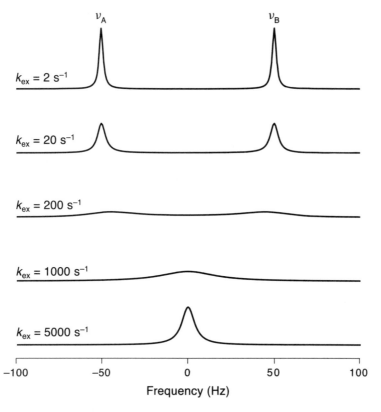

Figure 7.26   Simulated spectra illustrating the effects of exchange between two equally populated molecular states that give rise to different Larmor frequencies. For each simulation, the resonance frequencies for the two states differ by 100 Hz. k_{ex} represents the sum of the rate constants for the interconversion from A to B and back. (The simulated spectra in this figure and Fig. 7.27 were calculated using equations from Cavanagh, J., Fairbrother, W. J., Palmer III, A. G., Rance, M. & Skelton, N. J. (2007). *Protein NMR Spectroscopy*, 2nd ed., pp. 392–400. Elsevier Academic Press, Amsterdam.)

then the two states will interconvert many times during the time the FID is recorded. Because this happens so often, the two frequencies are effectively averaged, and the magnetization vectors associated with different populations never diverge very far from one another. This results in a single peak in the spectrum. Notice, however, that even for the example in Fig. 7.26 where k_{ex} is 50 times greater than $\Delta\nu$, the resulting peak is significantly broader than the peaks seen under conditions of very slow exchange. Even relatively fast exchange leads to significant broadening.

3. *Intermediate exchange.* When the exchange rate constant is closer to the frequency difference, then most of the molecules will undergo a small number of exchanges. Because these events occur randomly, the nuclei in different

molecules will change Larmor frequencies at different times during the data acquisition period. As we saw earlier these are the conditions that lead to the most pronounced loss of phase coherence. Depending on the exact rate of exchange, relative to the Larmor frequency difference, there may be either one or two broad peaks, as illustrated in Fig. 7.26. In extreme cases, the peaks may be so broadened as to almost disappear.

It is common to speak of a process as being either fast or slow on the "NMR time scale." Again, it is the *difference* in Larmor frequencies that determines this time scale. For ^1H nuclei, a typical frequency difference might be 1 ppm $\times$ 500 MHz = 500 Hz. Thus, a process that has a half-time of 10^{-4} s would be considered fast on this time scale, while one with a half-time of 0.1 s would be slow. The rate of exchange is usually quite sensitive to temperature, so it is sometimes possible for a process to shift from the slow or intermediate exchange regime to the fast regime as the temperature is increased.

In Fig. 7.27, simulated spectra are shown for two unequal populations undergoing exchange at different rates, leading to unequal equilibrium populations. For these simulations, 90% of the molecules were assumed to be in conformation A and 10% in conformation B, as indicated by the spectrum for $k_{ex} = 2$ s^{-1}. These plots demonstrate that exchange with even a quite minor population can lead to significant broadening of the major peak and disappearance of the minor peak. Under conditions of fast exchange, $k_{ex} \geq 5000$ s^{-1}, there is, as before, a single peak. But, now the frequency associated with this peak is the population weighted average, –40 Hz, of the two Larmor frequencies. Notice also that the spectra are very similar when $k_{ex} = 20$ or 5000 s^{-1}. In situations such as this, where only a single peak can be identified, it can be quite difficult to establish whether that peak represents an average of two conformations in relatively rapid exchange or a major component in slower exchange with a minor population.

Relatively slow internal motions will also influence R_2 rate constants determined in CPMG-type experiments. In this case, the length of the delay between the individual refocusing pulses also influences the apparent value of R_2. If the exchange rate is very slow relative to the refocusing delay, there will be no exchange during this delay, and there will be no loss of phase coherence or signal intensity due to exchange. On the other hand, if there are thousands of exchange events during the delay, the average net displacement in the transverse plane will be small. As before, it is the case where there are only a few random exchanges during the delay that leads to the largest loss of phase coherence. In typical experiments with modern spectrometers, the delays are usually limited to the range of 1 to 1000 μs, and internal motions on similar time scales will lead to pronounced increases in the measured R_2 rate constants, above those caused by the overall tumbling of the molecule. In favorable cases, the rates of exchange processes can be estimated by repeating the experiment with different delay times and comparing the observed changes in the measured R_2 values with those predicted by theory or simulations. These relationships are discussed in much more detail in Section 8.6.

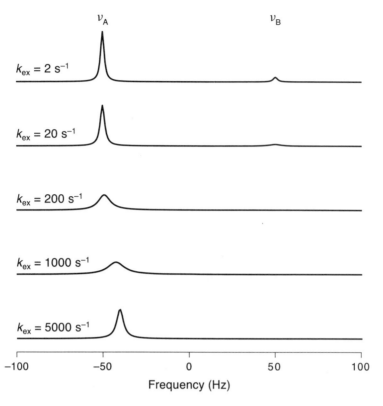

Figure 7.27   Simulated spectra illustrating the effects of exchange between two molecular states with unequal populations. For these simulations 90% of the molecules are assumed to be in conformation A and 10% in conformation B. The resonance frequencies for the two states differ by 100 Hz. k_{ex} represents the sum of the rate constants for interconversion from A to B and back.

Heteronuclear Relaxation Experiments to Study Molecular Motions

Although the rates of molecular motion are critical determinants of relaxation rates, it is not always easy to say much about the rates of motion from relaxation rates alone. A major reason for this is that the relaxation rates usually depend on other factors as well, especially the strength of the dipolar interactions that a particular nucleus feels, which depends, in turn, on the number, type and proximity of other spins (i.e., the three-dimensional structure). The situation is greatly simplified, however, for covalently linked heteronuclear pairs, such as ^1H–^{15}N or ^1H–^{13}C. In these cases, the relaxation rate of the ^{15}N or ^{13}C nucleus is determined almost entirely by interactions with the covalently attached ^1H. This property has been widely used to study the internal dynamics of proteins and, to a lesser extent, nucleic acids.

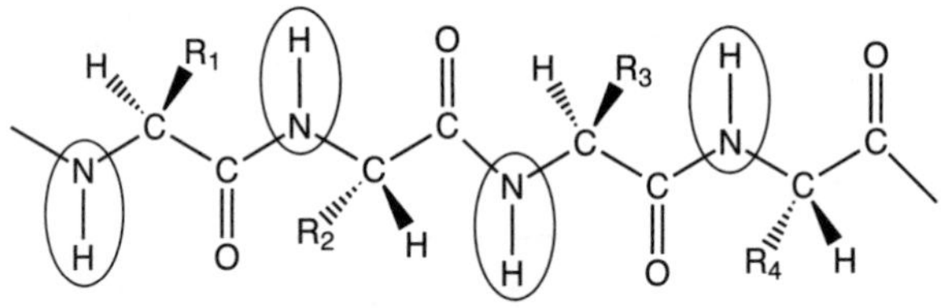

Figure 7.28
Chemical structure of
a polypeptide segment,
highlighting the H and N
atoms of the backbone amide
groups.

The use of ^{15}N relaxation measurements to study protein dynamics is particularly attractive, for several reasons. First, each amino-acid residue contains a nitrogen atom covalently bonded to a single hydrogen atom (except for the amino-terminal residue and any prolyl residues), as illustrated in Fig. 7.28. If the protein is uniformly labeled with ^{15}N, each amide group will give rise to a single peak in a two-dimensional ^{15}N– ^{1}H correlation spectrum, as is described in Chapter 16. The peak intensities in a series of such spectra can then be used to follow relaxation in either an inversion-recovery experiment for R_1, or a CPMG experiment for measuring R_2. Fig. 7.29 shows equivalent sections from a series of spectra collected as part of an R_2 experiment.

Because these spectra generally have excellent resolution and low background signals, the peak intensities can be measured quite accurately and then fit to an exponential decay function to determine the relaxation rate, as illustrated in Fig. 7.30. As predicted from our previous discussion, the transverse relaxation rate is considerably greater than the longitudinal rate. (Note the different scales on the time axes for the two plots.)

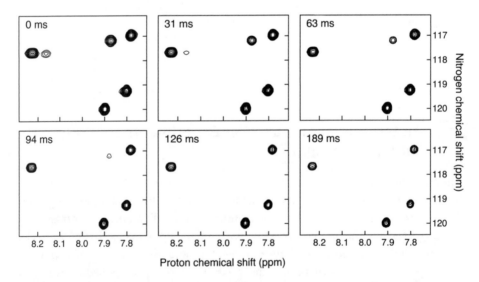

Figure 7.29   Contour plots from two-dimensional ^{1}H–^{15}N spectra of bovine pancreatic trypsin inhibtior (BPTI), used to measure R_2. The relaxation delay time in the CPMG experiment is indicated for each spectrum. (Adapted from Beeser, S. A. (1999). Ph.D. thesis, University of Utah.)

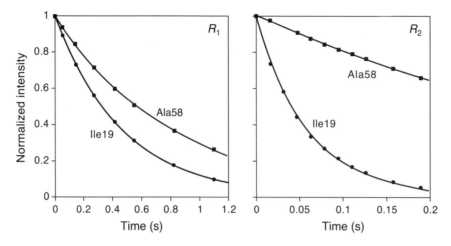

Figure 7.30   Decay of longitudinal and transverse magnetization of backbone amide ^{15}N nuclei in BPTI, from data like that shown in Fig. 7.29. Although the R_1 data represented in the left panel were derived from an inversion-recovery type experiment (page 165), the spectra were recorded in such a way as to show an exponential decay of the signal intensity. (Adapted from Beeser, S. A. (1999). Ph.D. thesis, University of Utah)

Because both the transverse and longitudinal relaxation mechanisms are determined almost entirely by interactions with the covalently attached ^{1}H nucleus, for which the geometry is nearly identical for each peptide unit, differences among ^{15}N relaxation rates within a protein are due predominately to differences in internal motions. As an example, ^{15}N R_1 rate constants measured for the amide nitrogens of BPTI are plotted in Fig. 7.31A. In this case, there are only relatively small variations among the rates for individual nuclei, except near the carboxyl terminus, where the rates are significantly lower. This pattern suggests that these amide groups undergo rapid internal motions that move the effective correlation time away from the optimum for longitudinal relaxation.

Panel B of Fig. 7.31 shows ^{15}N R_2 rate constants measured for the same protein. Here the striking observation is the large R_2 rate constants observed for just a few of the residues, centered at positions 14 and 38. These large rates arise from internal motions that are *slower* than tumbling, which leads to loss of phase coherence during the R_2 measurement.

There are other types of ^{15}N relaxation measurements that can be used to obtain information about internal motions on different time scales. In addition, mathematical methods have been developed for analyzing the data so that information about the frequencies and amplitudes of the motions can be extracted, in favorable cases at least. The quantitative analysis of heteronuclear relaxation rate constants is discussed in some depth in Chapter 8. Experiments of this type have been a major application of NMR in the study of proteins in the last two decades. The motion of side chains can also be studied by monitoring the relaxation of either ^{13}C or ^{2}H in isotopically labeled proteins.

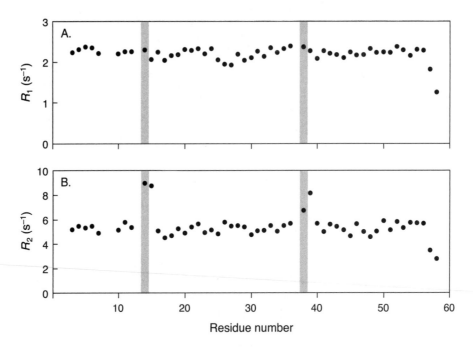

Figure 7.31 Longitudinal (A) and transverse (B) ^{15}N relaxation rate constants for the backbone amide groups of BPTI. Gray vertical bars indicate the position of Cys14 and Cys38, which are linked by a disulfide bond that undergoes conformational exchange on the μs time scale. (Adapted from Beeser, S. A., Goldenberg, D. P. & Oas, T. G. (1997). *J. Mol. Biol.*, 269, 154–164)

Summary Points for Chapter 7

- Relaxation is the restoration of equilibrium magnetization following a perturbation, such as a pulse.
- Relaxation depends on molecular motions, and measurements of relaxation rates can be used to characterize these motions.
- Relaxation is usually characterized by two rate constants, the longitudinal relaxation rate constant, R_1, and the transverse relaxation rate constant, R_2.
- Longitudinal relaxation:
  - R_1 describes the restoration of equilibrium z-magnetization and is measured using an inversion-recovery experiment.
  - Longitudinal relaxation requires fluctuation of local magnetic fields at frequencies close to the Larmor frequency.
  - In liquids, the major mechanism of longitudinal relaxation involves molecular tumbling and magnetic fluctuations due to dipolar interactions with other spins.
  - R_1 is greatest for molecules that tumble with a correlation time, τ_c, close to the inverse of the Larmor frequency.

- Transverse relaxation:
  - R_2 describes the disappearance of transverse magnetization and includes the effect of longitudinal relaxation as well as the loss of phase coherence among different subpopulations of the sample.
  - R_2 is always greater than or equal to R_1.
  - R_2 can be estimated from peak widths. Values measured in this way include the effects of magnetic field inhomogeneity and internal chemical or conformational exchange processes. Transverse relaxation rate constants estimated from line widths are often designated R_2^*.
  - Experiments based on refocusing pulses are used to measure R_2 without the contribution from static field inhomogeneity.
  - Loss of phase coherence in the transverse plane can be caused by random changes in the precession frequency, including those caused by molecular tumbling and dipolar coupling.
  - R_2 increases monotonically with increasing molecular weight and slower tumbling rates, leading to line broadening.
- Longitudinal relaxation in the rotating frame, with rate constant $R_{1\rho}$, is a complement to R_1 and R_2 measurements that is sensitive to slower molecular motions.
- Internal molecular motions that are faster than molecular tumbling can increase or decrease R_1, but always decrease R_2, leading to sharper peaks.
- Internal molecular motions slower than molecular tumbling do not affect R_1, but can contribute to increased R_2 and line broadening.
- Measurements of heteronuclear relaxation rates are particularly useful for characterizing the dynamics of macromolecules.

Exercises for Chapter 7

7.1. Consider the relaxation of a single ^{1}H spin population in a liquid sample, for which $R_1 = 0.1$ s^{-1} and $R_2 = 2$ s^{-1}.

(a) Assuming that the resonance frequency for this nucleus matches the rotating-frame frequency, draw a series of vector diagrams showing the *net* longitudinal and transverse magnetization components for this nucleus at the following times after a $(\pi/2)_y$-pulse:

$$0.2 \text{ s}, 1 \text{ s}, 4 \text{ s}, 10 \text{ s}, 20 \text{ s}.$$

These drawings should be based on calculations of the residual magnetization components, not just guesses!

(b) Some of your diagrams should show a situation in which the longitudinal magnetization has disappeared completely, but the longitudinal magnetization has not yet fully recovered. In your own words, briefly explain the physical processes by which this situation can arise.

7.2. The rates of both longitudinal and transverse relaxation are strongly dependent on the rotational correlation time, τ_c, which depends on the size and shape of the molecule of interest, as well as the viscosity of the surrounding fluid. For an idealized spherical particle in a homogeneous solvent, τ_c can be calculated from the Stokes–Einstein equation as:

$$\tau_c = \frac{\eta V}{kT}$$

where η is the solvent viscosity, V is the particle volume, k is Boltzmann's constant and T is the absolute temperature.

  (a) Before trying to use this equation, show that the units on the right-hand side cancel out to give a unit of time. As a starting point, the SI unit for viscosity is the Pa·s. The pascal (Pa), in turn, is the unit of pressure: $1\,\text{Pa} = 1\,\text{N}\cdot\text{m}^{-2}$.

  (b) Defining the appropriate volume, V, for a molecule is not entirely straight-forward, but a reasonable first-order approximation can be made by simply dividing the mass by an assumed density of 1 g/mL. From this assumption, derive an equation to calculate τ_c from the molar mass (M), the solvent viscosity (η) and the temperature (T).

  (c) The viscosity of water at 25°C is approximately 0.9×10^{-3} Pa·s. Calculate the expected rotational correlation times for molecules with molar masses of 100, 1000, 10,000 and 100,000.

  (d) For a molecule with molar mass of 10,000, calculate the expected correlation time in water at 10, 20, 30, 40 and 50°C. You should *not* assume that the viscosity of water is independent of temperature!

7.3. The treatment in Section 7.3.4 yields a simple equation for the contribution to R_2 from random tumbling:

$$R_2 = \tau_c \Delta\omega^2/2$$

where $\Delta\omega$ is one-half of the difference in precession frequency (in units of angular velocity, rad/s) of the nucleus when the molecule is in alternate orientations relative to the external magnetic field. For two ^1H nuclei separated by 2 Å, $\Delta\omega$ is about 2×10^5 rad/s. (See Fig. 4.12 on page 73.)

  (a) Assuming that R_2 is due entirely to this interaction, and using the estimates for τ_c from the previous problem, calculate the expected transverse relaxation rate constants for one of these nuclei in molecules with molar masses of 100, 1000, 10,000 and 100,000 in water at 25°C.

  (b) For a molecule of molar mass 10,000 in water at 25°C, calculate the expected line width of one of the ^1H nuclei described above. Calculate the line width in Hz and in ppm for spectrometers with ^1H frequencies of 400, 600 and 800 MHz.

7.4. For the same molecule, calculate the expected line width in water at 5, 20, 50 and 60°C, in a spectrometer with a ^1H frequency of 600 MHz.

Further Reading

The online resource, *eMagRes* (formerly the Encyclopedia of Nuclear Magnetic Resonance), has several excellent articles on relaxation mechanisms and rates, including the following:

⋄ Traficante, D. D. (2007). Relaxation: An introduction. *eMagRes*. http://dx.doi.org/10.1002/9780470034590.emrstm0452

⋄ Woessner, D. E. (2007). Relaxation effects of chemical exchange. *eMagRes*. http://dx.doi.org/10.1002/9780470034590.emrstm0454

The classic paper on spin echoes:

⋄ Hahn, E. L. (1950). Spin echoes. *Phys. Rev.*, 80, 580–594. http://dx.doi.org/10.1103/PhysRev.80.580

The treatment of transverse relaxation as a random walk is described in this early paper:

⋄ Pines, D. & Slichter, C. P. (1955). Relaxation times in magnetic resonance. *Phys. Rev.*, 100, 1014–1020. http://dx.doi.org/10.1103/PhysRev.100.1014

CHAPTER 8

A More Mathematical Look at Molecular Motion and NMR Relaxation

In Chapter 7 we considered, in a largely qualitative way, the relationships between the rates of molecular motions, especially global tumbling, and the observed rates of NMR relaxation. In the case of longitudinal relaxation, motions that match the Larmor frequency are expected to be most important, because this form of relaxation requires the stimulation of quantum transitions, whereas transverse relaxation is also promoted by slower motions that cause a loss of phase coherence. Up to now, we have described the rates of motion by a single parameter, the correlation time, but you might well wonder how an essentially random process can be characterized by one number. In this chapter, we will take a closer look at the nature of random motions and their relationships to relaxation rates, employing two mathematical tools-the autocorrelation function and the spectral density function. We also take a more thorough look at the effects of relatively slow internal motions on transverse relaxation. The more mathematical treatments presented in this chapter are not essential for the subsequent chapters in this text, and you may want to skim or even skip it if this is your first encounter with NMR relaxation. If, on the other hand, you want to use relaxation measurements to obtain quantitative information about molecular motions, an important application of NMR in the study of larger molecules, understanding the spectral density function and its interpretation is key.

8.1 Molecular Tumbling as Brownian Motion

The movements of molecules in solution, both translational and rotational, comes about because of random collisions with other molecules. Over time, the net force exerted by these collisions is zero, but at any instant, an imbalance between collisions from different directions may cause a molecule to move in one direction or another, or rotate in a particular direction. One of the first reports of this type of motion was made in 1827 by the botanist Robert Brown, who observed random movements of

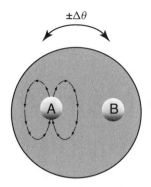

Figure 8.1

A molecule containing two spin-1/2 nuclei undergoing random tumbling.

tiny particles within plant pollen using a light microscope. Later, the nature of *Brownian motion* provided important evidence for the existence of molecules, which was still a subject of controversy in 1905, when Einstein published his classic treatment of the phenomenon.

Mathematically, Brownian motion can be described by a model in which the molecule takes random steps over very small time intervals, with the magnitudes of the steps determined by a Gaussian (bell-shaped) probability distribution function. The mathematical description of this type of model is called a *Wiener process*, for the twentieth-century mathematician Norbert Wiener, and is closely related to the random walk we discussed in Chapter 7 when considering the loss of phase coherence in the transverse plane. Now, we use a similar treatment to describe the motions of individual molecules, as opposed to the change in orientation of magnetization vectors.

As noted in chapter 7, the major mechanism of relaxation for most molecules in solution is the fluctuating magnetic fields due to dipolar interactions, and the cartoon in Fig. 8.1 is meant to represent an idealized molecule with two interacting nuclei. At any instant, the field experienced by one nucleus depends on the angle, θ, between the vector connecting the two nuclei and the stationary external field, as discussed in Section 4.2 (p. 70). A rigorous description of how θ changes with time requires some rather advanced mathematics in order to account for the fact that the molecule is tumbling in three dimensions. We can use a simplified model, however, in which angular motion is restricted to a plane, which captures the essential features of the full treatment. The simpler description leads to the correct mathematical form, within a constant of proportionality, of the full treatment. In this model, we will assume that after each small time interval the molecule undergoes a small random change in θ, determined by a Gaussian distribution. Fig. 8.2 shows the change in $\cos\theta$ with time, calculated from a simulation based on this model.

The curve was calculated by assuming that the time interval between motions is one picosecond (10^{-12} s) and that the distribution of rotation angles has a mean of zero and a standard deviation of 0.05 rad. Each of these motions causes only a small change in $\cos\theta$, but over time the changes accumulate so that eventually $\cos\theta$ changes from +1 to -1, representing a rotation of π rad.

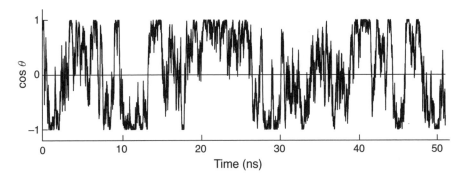

Figure 8.2   Simulation of the random reorientation of a molecule, expressed as the cosine of the total change in angle from the starting orientation. The time interval between motions is 1 ps, and the changes in angle are described by a Gaussian distribution with a standard deviation of 0.05 rad.

An important feature of Brownian motion is that the time required for a movement of a particular magnitude, such as the time for $\cos\theta$ to change from $+1$ to -1, can vary greatly. In the example shown in Fig. 8.2, the first such change takes place in about 1 ns, but it takes about 6 ns during the interval between 20 and 30 ns. At any instant, it is impossible to know how long the next large change will take, but it is clear from the graph that the sign of $\cos\theta$ will almost always change a few times over a 10-ns period. Although we cannot describe such motion with a single frequency, we can describe a *distribution* of frequencies using the autocorrelation function and the spectral density function discussed in the following sections.

8.2   The Autocorrelation Function

To describe random changes in orientation, such as illustrated in Fig. 8.2, it is useful to consider the product of $\cos\theta$ at a given time, t, and the value of $\cos\theta$ at a later time, $t + \Delta t$. For convenience, we will use $f(t)$ to represent $\cos\theta$ at time, t. The product we are interested in, then, is

$$f(t)f(t + \Delta t)$$

This product can take on values from -1 to $+1$. If Δt is very small, then we expect that $f(t)f(t + \Delta t) \approx f(t)^2$. For very large values of Δt, on the other hand, both $f(t)$ and $f(t + \Delta t)$ are equally likely to have any value between -1 and $+1$, and the *average* value of $f(t)f(t + \Delta t)$ is expected to be zero. The average rate at which this product decays to zero provides a measure of how fast the molecule changes orientation.

This idea is incorporated in the definition of the autocorrelation function, $g(\Delta t)$:

$$g(\Delta t) = \frac{\langle f(t)f(t + \Delta t)\rangle}{\langle f(t)f(t)\rangle} \tag{8.1}$$

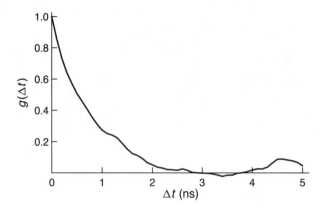

Figure 8.3

Autocorrelation function for random reorientation of a molecule, calculated from the simulation shown in Fig. 8.2.

The brackets indicate an average, either over time for a single molecule or over an ensemble of molecules, and the average value of $f(t)f(t + \Delta t)$ is divided by the average of $f(t)f(t)$ so as to account for different initial values of $f(t)$. Note that while $f(t)$ is a function describing the fluctuations of a single particle with time, $g(\Delta t)$ reflects the average loss of correlation with the original orientation. Although we have defined $f(t)$ as the value of $\cos \theta$ at time t, the autocorrelation function can be applied to many other types of functions and is used for a variety of applications, such as the analysis of noise in electronic circuits.

The graph in Fig. 8.3 shows the autocorrelation function calculated for the simulation of Brownian rotational motion of a single particle shown in Fig. 8.2. In this case, the average was calculated for the single particle in the simulation by taking each value of t as a new starting time. As expected from qualitative arguments, $g(\Delta t)$ decays to zero, and then randomly fluctuates a bit around this value. If the autocorrelation function is averaged over a large number of molecules, the fluctuations at later times will diminish. The half-time for the decay is about 0.5 ns.

The decay shown in the simulated example above appears to be exponential, and it is relatively easy to show that this is the expected result. The key to calculating the autocorrelation function for $f(t) = \cos \theta(t)$ is to determine the average net change in the angle θ after time interval Δt. This problem is analogous to the rotational random walk we considered in Chapter 7. If we think of the Brownian motion as a random walk with individual steps taking place over a very short time dt (1 ps in the example), then the number of steps after a total time Δt is $N = \Delta t/dt$. If the mean square length (in angular units) of each step is called σ^2 ($(0.05 \text{ rad})^2$ in the example), then the mean square total angular displacement over time Δt is given by

$$\langle \Delta \theta^2 \rangle = \sigma^2 \Delta t/dt$$

and the probability distribution function for $\Delta \theta$ is:

$$p(\Delta \theta) = \sqrt{1/(2\pi \langle \Delta \theta^2 \rangle)}e^{-\Delta \theta^2/(2\langle \Delta \theta^2 \rangle)} \tag{8.2}$$

The average in the numerator of Eq. 8.1 is then calculated by integrating over all possible values of θ and $\Delta \theta$. The integration over $\Delta \theta$ can be written as

$$\int_{-\infty}^{\infty} p(\Delta\theta) \cos\theta \cos(\theta + \Delta\theta) d\Delta\theta$$

$$= \int_{-\infty}^{\infty} \sqrt{1/(2\pi\langle\Delta\theta^2\rangle)} e^{-\Delta\theta^2/(2\langle\Delta\theta^2\rangle)} \cos\theta \cos(\theta + \Delta\theta) d\Delta\theta \qquad (8.3)$$

Here the integration is from $-\infty$ to ∞ because this is the range over which the Gaussian probability distribution function is defined. While rather formidable looking, the integral has the relatively simple form of

$$\int_{-\infty}^{\infty} p(\Delta\theta) \cos\theta \cos(\theta + \Delta\theta) d\Delta\theta = e^{-\langle\Delta\theta^2\rangle/2} \cos^2\theta$$

Integrating this expression over the range $\theta = 0$ to 2π and dividing by 2π (to give the average over this range) gives the numerator of Eq. 8.1:

$$\langle f(t)f(t + \Delta t)\rangle = \frac{1}{2\pi} \int_0^{2\pi} e^{-\langle\Delta\theta^2\rangle/2} \cos^2\theta d\theta = \frac{1}{2} e^{-\langle\Delta\theta^2\rangle/2}$$

Similarly, the average in the denominator of Eqn. 8.1 is determined by integrating $f(t)f(t)$ over θ from 0 to 2π and dividing by 2π:

$$\langle f(t)f(t)\rangle = \frac{1}{2\pi} \int_0^{2\pi} \cos^2\theta d\theta = \frac{1}{2}$$

Thus, the autocorrelation function for Brownian rotation is given by

$$g(\Delta t) = \frac{\langle f(t)f(t + \Delta t)\rangle}{\langle f(t)f(t)\rangle} = e^{-\langle\Delta\theta^2\rangle/2}$$

This can be rewritten in terms of the parameters that describe Brownian motion, dt and σ, and the time interval Δt:

$$g(\Delta t) = e^{-\Delta t\sigma^2/(2dt)}$$

As suggested by the simulation, $g(\Delta t)$ is an exponential decay function of Δt, and the reciprocal of the constant in the exponent is defined as the correlation time, τ_c:

$$\tau_c = 2dt/\sigma^2$$

For the simulation, dt was set to 1 ps, and σ was 0.05 rad, giving a correlation time of 0.8 ns, consistent with the graph shown in Fig. 8.3.

When the delay time, Δt, equals the correlation time, the mean-square angular displacement is given by

$$\langle\Delta\theta^2\rangle = \sigma^2\Delta t/dt = \sigma^2\tau_c/dt = \sigma^2(2dt/\sigma^2)/dt = 2$$

Thus, the correlation time is the period over which the mean-square angular displacement is 2 rad^2, and the root-mean-square angular displacement is $\sqrt{2}$ rad $\approx$ 1.4 rad; a result roughly equivalent to the more commonly cited definition of τ_c as the average time required for a rotation of 1 rad.

Although the numerical simulation required specific values for the short time interval, dt, and the associated mean angular deviation, σ, the derivation above assumes only that the angular deviation over longer time periods, Δt, is described by a Gaussian probability distribution function, and the nature of the microscopic events that cause this motion are not critical. The choice of dt is arbitrary and, in fact, the formal description of a Wiener process makes dt infinitesimally small. The important parameter in describing the motion of a specific particle under given conditions is the ratio of dt and σ^2, as given above by the expression for τ_c. It should also be noted, however, that the simulation and derivation both assume that the tumbling motion of a molecule can be described as a Wiener process with a single time constant. If a molecule is not spherically symmetrical, then angular displacements about the three orthogonal axes of the molecule may have different rates, requiring a more complex model of motion. Internal motions may also contribute to the average motions of a particular nucleus, and, as discussed in Chapter 7, the measurements of relaxation rates can be a powerful method for characterizing these motions. The usual approach to treating more complex motions is to assume that the autocorrelation function can be expressed as a product of two or more exponential decays, each with a characteristic time constant. We will discuss such a model in Section 8.5.2.

8.3   The Spectral Density Function

The autocorrelation function discussed in Section 8.2 is a description of molecular motion in the time domain. However, the probabilities of the quantum transitions that contribute to relaxation are most readily expressed in terms of the *frequencies* of magnetic fluctuations. For this reason, it is convenient to use the frequency domain to describe molecular motions. Just as the time and frequency domains represented by a free induction decay and NMR spectrum, respectively, are related by Fourier transformation, the autocorrelation function can be converted to the frequency domain by the same mathematical manipulation. The result is called the spectral density function and describes the probabilities of motions with different frequencies.

If the autocorrelation function is an exponential decay with time constant τ_c, as for a tumbling molecule with no internal motions,

$$g(\Delta t) = e^{-\Delta t/\tau_c}$$

then the spectral density function is defined as the real part of the Fourier transform of $g(\Delta t)$. The transform is calculated as the following integral:

$$\mathcal{F}(g(\Delta t)) = \int_{-\infty}^{\infty} g(\Delta t) e^{-i\Delta t f 2\pi} d\Delta t$$

where f is the frequency in Hz. Because $g(\Delta t)$ is defined only for non-negative values of Δt, the integral is evaluated as

$$\mathcal{F}(g(\Delta t)) = \int_{0}^{\infty} e^{-\Delta t/\tau_c} e^{-i\Delta t f 2\pi} d\Delta t = \frac{\tau_c}{1 + i2\pi f \tau_c}$$

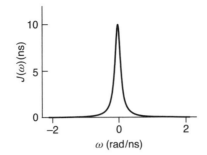

Figure 8.4

Spectral density function for a randomly tumbling molecule with $\tau_c = 10$ ns.

The real part of the transform is

$$\mathcal{R}e\left(\mathcal{F}\left(g(\Delta t)\right)\right) = \frac{\tau_c}{1 + (2\pi f \tau_c)^2}$$

As you may have noticed, $g(\Delta t)$ has the same form as an FID with zero frequency offset, and the Fourier transform has the same form as an NMR peak centered at 0 Hz. This can be rewritten in terms of an angular frequency, $\omega = 2\pi f$, to give the more commonly used form of the spectral density function:

$$J(\omega) = \frac{\tau_c}{1 + \omega^2 \tau_c^2} \tag{8.4}$$

The form shown above omits a constant of proportionality (usually 2 or 2/5) that is incorporated in the definitions found in some texts and articles. When these constants are included in the definition of the spectral density function, equations using $J(\omega)$ must be adjusted accordingly. The form used here has the convenient property that the value of $J(0)$, the significance of which is discussed below, is simply τ_c. The spectral density function has the shape of an absorptive Lorentzian peak centered at $\omega = 0$, as shown in Fig. 8.4 for $\tau_c = 10$ ns.

The spectral density function can be thought of as a continuous probability function, such that if $J(\omega)$ is integrated over a range of frequencies, then the value of the integral is proportional to the probability of fluctuations with frequencies lying within this range. In this context, positive and negative frequencies correspond to motions in opposite directions. Because the spectral density function is symmetrical, and motions in either direction are equally effective in stimulating relaxation, often only the positive frequency axis is shown.

The value of $J(\omega)$ at $\omega = 0$ is deserving of some special attention. By substituting $\omega = 0$ into Eq. 8.4, we obtain $J(0) = \tau_c$, as noted above. It should also be apparent that $J(0)$ is the maximum value of J, irrespective of τ_c. Thus, if we interpret $J(\omega)$ as the probability of motions of different frequencies, we reach the surprising conclusion that the most likely frequencies are those close to 0, irrespective of how fast the molecule tumbles! We can make some sense out of this by realizing that a frequency of 0 implies that the molecule has the same orientation at the end of a given time interval, Δt, as it did at the beginning. Because microscopic motions in any direction are equally likely, over time, the most likely outcome is, indeed, that the orientation of the molecule at the end of the interval is close to that at the beginning. As the value of τ_c is made smaller, corresponding to faster motions, the probability of this outcome decreases, but is still larger than the probability of any other outcome.

The value of $J(0)$ also reflects an important property of all pairs of functions related by Fourier transformation: The value of a function at 0 is proportional to the integral of its Fourier transform over all values for which it is defined. In this case, $J(0)$ is equal to the integral of the autocorrelation function, $g(\Delta t)$, from 0 to ∞:

$$\int_0^\infty g(\Delta t)d\Delta t = \int_0^\infty e^{-\Delta t/\tau_c}d\Delta t = -\tau_c e^{-\Delta t/\tau_c}\Big|_0^\infty = \tau_c$$

Because the integral can be thought of as the area under the curve defined by $g(\Delta t) = e^{-\Delta t/\tau_c}$, this result shows that the area becomes smaller as τ_c is made smaller, reflecting the faster decay of $g(\Delta t)$.

The inverse relationship also holds: The integral of $J(\omega)$ over all values of ω is proportional to the value of $g(0)$, which, from the results above, we know is simply 1, for all values of τ_c. Because $J(0)$ decreases for faster motions, these relationships require that J increase for other values of ω.

Additional insights into the relationships among the rate of tumbling, the auto-correlation function and spectral density function can be gained by looking at some simulated examples, as shown in Fig. 8.5. Each row in the figure shows, for a given value of τ_c, a simulated trajectory of $\cos\theta$ versus time for a single molecule, the autocorrelation function and the spectral density function. The top row represents the extreme condition where the molecule is frozen, so that $\tau_c \rightarrow \infty$. In this case, the autocorrelation function never decreases, and the spectral density function only has a non-zero value for $\omega = 0$. Thus, the only frequency possible for a fixed molecule is 0 Hz.

As τ_c takes on finite values, $\cos\theta$ fluctuates about 0, and the autocorrelation function decays with time constant τ_c. Correspondingly, the peak in the plot of $J(\omega)$ takes on a finite width, representing the increased likelihood of fluctuations with a finite frequency. As τ_c is made smaller, corresponding to faster motions, $g(\Delta t)$ decays more rapidly; the value of $J(0)$ decreases, and the value of $J(\omega)$ at large positive or negative values increases, reflecting the increased likelihood of higher frequencies.

Because the rates of relaxation can depend on motions of widely different frequencies, it is common to plot the spectral density function with a logarithmic axis for ω, as shown Fig. 8.6 for a few different values of τ_c. Consider, first, the example where $\tau_c = 3$ ns. For values of ω up to about 10^8 rad/s, $J(\omega)$ remains close to $J(0) = 3$ ns. In this regime, $\omega^2\tau_c^2 \ll 1$, so the denominator of Eq. 8.4 is approximately 1, and $J(\omega) \approx \tau_c$. As ω increases, the term $\omega^2\tau_c^2$ approaches and then exceeds 1. As this happens, the value of the denominator grows rapidly, and when $\omega^2\tau_c^2 \gg 1$, the denominator is approximately $\omega^2\tau_c^2$. In this regime, $J(\omega) \approx 1/(\tau_c\omega^2)$.

For smaller values of τ_c, as shown in Fig. 8.6 for $\tau_c = 0.3$ ns and 0.1 ns, the initial plateau has a correspondingly lower value, but also extends to higher frequencies before dropping toward zero. These effects compensate one another so that the integral

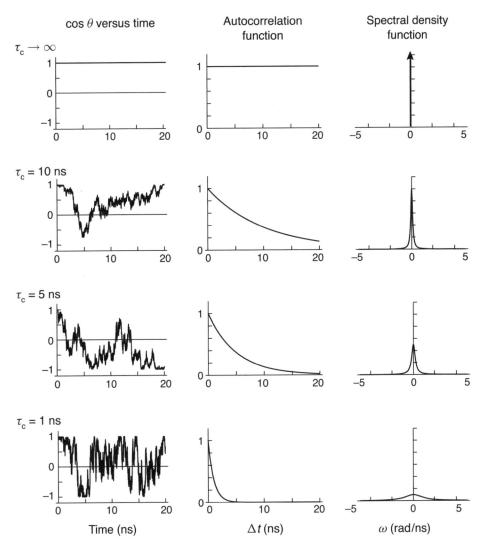

Figure 8.5 Simulated trajectories for random tumbling of molecules with different correlation times, and the corresponding autocorrelation and spectral density functions. The vertical arrow used to represent the spectral density function for the limiting case where $\tau_c = 0$ indicates a function with zero width and infinite height.

of $J(\omega)$ is the same for all values of τ_c. (The areas under the curves in Fig. 8.6 decrease with smaller values of τ_c, but this is due to the use of a logarithmic axis for ω.)

Next, consider what happens if we choose a particular value of ω and then ask how the spectral density function depends on the value of τ_c. In Fig. 8.6, arrows mark the value of ω corresponding to the resonance frequency of a ^1H nucleus in a magnetic field of 11.7 T, 500 MHz $\approx 3.1 \times 10^9$ rad/s. For values of τ_c greater than about 3 ns, the point corresponding to 500 MHz lies to the right of the plateau region, where $J(\omega) \approx 1/(\tau_c \omega^2)$. For $\tau_c = 3$ ns, $J \approx 0.035$ ns. On the other hand, if τ_c is less than

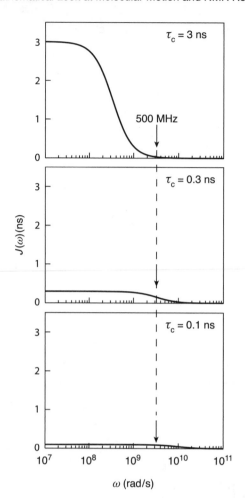

Figure 8.6

The spectral density function for random tumbling, on a logarithmic angular frequency scale, for different values of τ_c.

about 0.1 ns, then the 500 MHz point lies within the plateau region, where $J \approx \tau_c$. Importantly, the maximum value of J, for a given value of ω, occurs when $\tau_c = 1/\omega$. This situation is approximated by the center graph in Fig. 8.6, where $\tau_c = 0.3$ ns, and the 500 MHz point lies in the region where the plot of $J(\omega)$ decreases from approximately τ_c to $1/(\tau_c\omega^2)$. For $\tau_c = 0.3$ ns, $J \approx 0.16$ ns.

The relationship between J and τ_c at a specified angular frequency is shown in Fig. 8.7. As suggested by the qualitative arguments, there is a clear maximum in the value of J when $\tau_c = 1/\omega$. Because both axes have been drawn with logarithmic scales, the plot is approximately linear on both sides of the maximum: For small values of τ_c, $J = \tau_c$, while for large values, J is proportional to τ_c^{-1}.

The general shape of the graph in Fig. 8.7 should be familiar, since it is almost identical to the ones shown in Chapter 7 to illustrate the relationship between the longitudinal relaxation rate, R_1, and τ_c for the case of relaxation mediated by the dipolar interaction between two ^{1}H nuclei. The similarity between the graphs reflects the fact that relaxation is most efficient when the average rate of molecular tumbling generates the highest probability of motions with frequencies that match the resonance

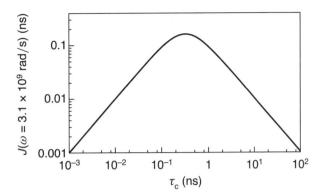

Figure 8.7
The spectral density function
for random tumbling
(evaluated at $\omega = 3.1 \times 10^9$ rad/s) as a function of τ_c.

frequency—that is, when the spectral density function evaluated at the resonance frequency is maximal. In the next sections, we will look at the relationships between relaxation rates and spectral density functions in more detail.

8.4   Calculating Relaxation Rates

Calculations of relaxation rates represent one of the most challenging areas of NMR theory, even for quite simple systems. Treatments can range from largely qualitative arguments, as presented in Chapter 7, to the use of quantum mechanics to describe both the nuclei of interest and the molecules with which they exchange energy. For many purposes, the most practical approach is one that uses a quantum mechanical description of the relaxing spins along with classical physics to describe the transfer of energy to the surrounding molecules. While even this type of treatment is only practical for systems involving a few interacting nuclei, the results provide important insights and the basis for interpreting relaxation rates in terms of molecular motions.

There are two general aspects to calculating relaxation rates by this approach. The first requires some careful thinking about how random transitions between quantum states will change the observable magnetization of a macroscopic sample, and the second involves establishing the relationships between the frequencies of molecular motions, as expressed in the spectral density function, and the rates of the quantum transitions. Of the two aspects, the first should be familiar to students of chemistry or biochemistry, since the analysis is similar to traditional treatments of chemical reaction kinetics. The calculation of the intrinsic rates of the transitions, on the other hand, requires some rather involved quantum mechanical and mathematical arguments that are beyond the scope of this text. Furthermore, the quantum mechanical treatment of transverse relaxation rates is considerably more difficult than that of longitudinal relaxation rates.

The following subsections are intended to provide a flavor of the theoretical treatment of relaxation rates, rather than a rigorous and complete derivation. We will first analyze the relationship between transition probabilities and longitudinal relaxation rates for an isolated spin and a dipolar coupled spin pair. Then, we will discuss the qualitative features of the equations for calculating the transition rates themselves and use these to write equations for longitudinal relaxation rates as a function of the correlation time for the case of a dipolar-coupled spin pair. The equations for transverse relaxation rates will also be discussed briefly, without derivation, and compared to the random-walk analysis presented in Chapter 7.

8.4.1   Longitudinal Relaxation Rates for an Isolated Nucleus

The simplest case to consider is that of an uncoupled spin-1/2 nucleus. At equilibrium, there is a small excess of molecules in the $|\alpha\rangle$ state, because it has a lower energy than the $|\beta\rangle$ state. As discussed in Chapter 7, the most common way of measuring the longitudinal relaxation rate involves applying a π-pulse, which converts the excess of $|\alpha\rangle$ spins into an excess of $|\beta\rangle$ spins, and then measuring the return of the net z-magnetization to the positive direction. The interconversion between the two states can be represented as

$$|\alpha\rangle \overset{W}{\longleftrightarrow} |\beta\rangle \qquad (8.5)$$

where W is defined as the probability of the transition per unit time. We will refer to parameters of this type as *transition rate constants*, which should be distinguished from the macroscopically observed rate constants, such as R_1. If the transition is stimulated by an external oscillating magnetic field, as during a pulse, the transition rates are the same in both directions. The differential equation describing the change in the number of nuclei in the $|\alpha\rangle$ state is given by

$$\frac{dN_\alpha}{dt} = WN_\beta - WN_\alpha$$

where N_α and N_β are the number of nuclei in the two states. If the system is irradiated for a sufficient time, it will reach an equilibrium in which the net rate of interconversion between the two states is zero. Under these conditions,

$$\frac{dN_\alpha}{dt} = WN_\beta - WN_\alpha = 0$$

and

$$\frac{N_\alpha}{N_\beta} = \frac{W}{W} = 1$$

Thus, irradiation for a sufficiently long time will equalize the two populations, a situation described as *saturation*, which we return to in Chapter 9 when discussing the steady-state nuclear Overhauser effect.

Like the transitions due to irradiation, relaxation processes are stimulated by fluctuating magnetic fields, but the situation is subtly different, leading to an equilibrium state in which the state with the lower energy, $|\alpha\rangle$ for a nucleus with positive gyromagnetic ratio, is present at a slightly higher concentration. This is because the energy for stimulating the transition comes from the thermal motions of the molecules, and the equilibrium between the $|\alpha\rangle$ and $|\beta\rangle$ states is coupled to the kinetic energy of the solution. At a microscopic level, thermal motions that lead to a transition from $|\alpha\rangle$ to $|\beta\rangle$

require more energy than those that convert $|\beta\rangle$ to $|\alpha\rangle$. As a consequence, the latter are slightly more probable, with the relative rates determined by the energy difference between the states and the temperature.

The usual practice for accounting for the different transition probabilities during relaxation is a bit artificial: The transition rate constant, W, is treated as if it were the same for both directions, and then an additional constant, K, is introduced to reflect the energy difference between the two states. For a simple single-spin transition, the modified differential equation is

$$\frac{dN_\alpha}{dt} = WN_\beta - WN_\alpha + K \tag{8.6}$$

We can also write

$$\frac{dN_\beta}{dt} = WN_\alpha - WN_\beta - K \tag{8.7}$$

At equilibrium, the net rates are both zero, and K can be expressed in terms of the equilibrium populations of $|\alpha\rangle$ and $|\beta\rangle$ nuclei, N_α^0 and N_β^0:

$$WN_\beta^0 - WN_\alpha^0 + K = 0$$

$$K = W(N_\alpha^0 - N_\beta^0)$$

In an NMR experiment, of course, we don't actually determine the number of nuclei in the two states, but, rather, the relative net magnetization in a particular direction. The bulk magnetization in the z-direction is proportional to the difference between N_α and N_β:

$$\bar{I}_z = C(N_\alpha - N_\beta)$$

where C is a constant of proportionality. Differentiating with respect to time gives the rate of change in the z-magnetization:

$$\frac{d\bar{I}_z}{dt} = C\left(\frac{dN_\alpha}{dt} - \frac{dN_\beta}{dt}\right)$$

which can be expressed in terms of the constants W and K as

$$\frac{d\bar{I}_z}{dt} = C\Big((WN_\beta - WN_\alpha + K) - (WN_\alpha - WN_\beta - K)\Big)$$

$$= C(2WN_\beta - 2WN_\alpha + 2K)$$

Substituting for K gives

$$\frac{d\bar{I}_z}{dt} = C(2WN_\beta - 2WN_\alpha + 2WN_\alpha^0 - 2WN_\beta^0)$$

$$= 2W\Big(C(N_\alpha^0 - N_\beta^0) - C(N_\alpha - N_\beta)\Big)$$

$$= 2W(\bar{I}_z^{\,0} - \bar{I}_z)$$

where $\bar{I}_z^{\,0}$ is the equilibrium z-magnetization. This result indicates that the rate of relaxation depends on the difference between the instantaneous magnetization and its equilibrium value. This is equivalent to Eq. 7.1 (p. 164), which was based on the assumption that the magnetization, itself, would behave somewhat like the concentration of a molecular species that decays with time. Here, the derivation is based on treating the observable magnetization as arising from the aggregate effects of numerous nuclei that interconvert between quantum states with a given probability per unit time. Comparison with the phenomenological equation shows that the observed rate constant, R_1, is related to the transition rate constant according to

$$R_1 = 2W$$

The factor of 2 reflects the fact that each transition from $|\beta\rangle$ to $|\alpha\rangle$ both increases N_α and decreases N_β.

8.4.2   Longitudinal Relaxation Rates for a Dipolar-Coupled Spin Pair

So far, we have treated the relaxation of a single nucleus as if it were isolated, so that its transitions do not affect those of other nuclei. Under most circumstances involving liquid samples, however, the major mechanism of relaxation involves the dipolar interaction among nuclei, which can induce coupled transitions of two or more spins. This phenomenon and its consequences can be illustrated by examining the case of two dipolar-coupled spin-1/2 nuclei. Because the two spins influence one another, a description of the eigenstates of the system includes four classes of molecules, as illustrated for the case of two spins of the same type (e.g., two ^{1}H nuclei) in the energy diagram of Fig. 8.8.

The pairs of kets in the diagram represent molecules in which each spin is in one or the other eigenstates. By convention, the two spins are usually designated I and S,

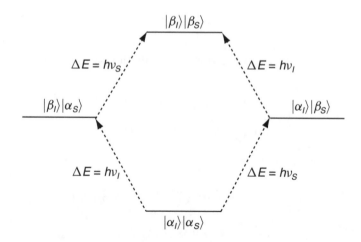

Figure 8.8
Energy diagram for two spin-1/2 nuclei of the same type, I and S, linked by dipolar interactions.

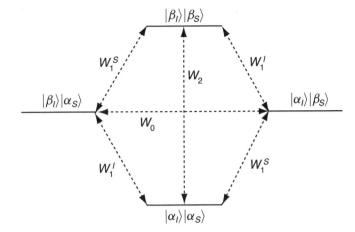

Figure 8.9
Energy diagram for a dipolar-coupled spin pair, indicating zero-, single-, and double-quantum transitions.

and their Larmor frequencies are ν_I and ν_S. At equilibrium, the most highly populated state will be $|\alpha_I\rangle|\alpha_S\rangle$, and the least populated will be $|\beta_I\rangle|\beta_S\rangle$. The two mixed states, $|\alpha_I\rangle|\beta_S\rangle$ and $|\beta_I\rangle|\alpha_S\rangle$ will be present at concentrations intermediate between those of the other two states. If the two nuclei are of the same type, such as two ^{1}H nuclei, the energy differences for the transitions will be very similar to one another, as suggested in Fig. 8.8. On the other hand, if the nuclei have different gyromagnetic ratios, the energy differences, and the corresponding equilibrium population differences, will be significantly different for the two nuclei.

The arrows in Fig. 8.8 represent the quantum transitions that can be stimulated by the absorption of radiation from an external source. Each of these transitions involves only one of the spins at a time and is described as a *single-quantum transition*. However, when two or more spins influence one another, they can also undergo concerted transitions during relaxation. In particular, $|\alpha_I\rangle|\alpha_S\rangle$ and $|\beta_I\rangle|\beta_S\rangle$ can interconvert directly, as can the $|\alpha_I\rangle|\beta_S\rangle$ and $|\beta_I\rangle|\alpha_S\rangle$ states. The various transitions that can occur as a consequence of dipolar coupling and molecular motion are shown in Fig. 8.9.

In this diagram, the symbols W_1^I and W_1^S represent the rate constants of the indicated single transitions involving the I- or S-spins, respectively. Similarly, W_2 is the rate constant for the *double-quantum transition* between $|\alpha_I\rangle|\alpha_S\rangle$ and $|\beta_I\rangle|\beta_S\rangle$. Transitions between $|\alpha_I\rangle|\beta_S\rangle$ and $|\beta_I\rangle|\alpha_S\rangle$ are referred to as *zero-quantum transitions* because the total quantum number is unchanged, and the rate constant for this type of transition is W_0. As before, each of the rate constants is assumed to be the same in both directions, with a small constant introduced into the differential equations to account for the thermal equilibrium.

After the system has been moved away from equilibrium, all of the pathways shown in Fig. 8.9 can contribute to relaxation. There are now four differential equations required to describe the changes in population with time:

$$\frac{dN_{\alpha\alpha}}{dt} = N_{\beta\alpha}W_1^I + N_{\alpha\beta}W_1^S + N_{\beta\beta}W_2 - N_{\alpha\alpha}(W_1^I + W_1^S + W_2) + K_1$$

$$\frac{dN_{\alpha\beta}}{dt} = N_{\alpha\alpha}W_1^S + N_{\beta\beta}W_1^I + N_{\beta\alpha}W_0 - N_{\alpha\beta}(W_1^I + W_1^S + W_0) + K_2$$

$$\frac{dN_{\beta\alpha}}{dt} = N_{\alpha\alpha}W_1^I + N_{\beta\beta}W_1^S + N_{\alpha\beta}W_0 - N_{\beta\alpha}(W_1^I + W_1^S + W_0) + K_3$$

$$\frac{dN_{\beta\beta}}{dt} = N_{\alpha\beta}W_1^I + N_{\beta\alpha}W_1^S + N_{\alpha\alpha}W_2 - N_{\beta\beta}(W_1^I + W_1^S + W_2) + K_4$$

By analogy to the constant, K, introduced for the case of the single spin, K_1, K_2, K_3 and K_4 depend on the equilibrium populations of the various states. The net z-magnetization arising from the I- and S- spins are identified as $\bar{I}_z$ and $\bar{S}_z$ and are related to the populations according to

$$\bar{I}_z = C_I(N_{\alpha\alpha} - N_{\beta\alpha} + N_{\alpha\beta} - N_{\beta\beta})$$

$$\bar{S}_z = C_S(N_{\alpha\alpha} - N_{\alpha\beta} + N_{\beta\alpha} - N_{\beta\beta})$$

where C_I and C_S are constants of proportionality specific to the indicated spins. Differentiating these equations with respect to time gives

$$\frac{d\bar{I}_z}{dt} = C_I \left(\frac{dN_{\alpha\alpha}}{dt} - \frac{dN_{\beta\alpha}}{dt} + \frac{dN_{\alpha\beta}}{dt} - \frac{dN_{\beta\beta}}{dt} \right)$$

$$\frac{d\bar{S}_z}{dt} = C_S \left(\frac{dN_{\alpha\alpha}}{dt} - \frac{dN_{\alpha\beta}}{dt} + \frac{dN_{\beta\alpha}}{dt} - \frac{dN_{\beta\beta}}{dt} \right)$$

Following the general approach detailed earlier for the isolated nucleus yields the following equations for the change in z-magnetization with time:

$$\frac{d\bar{I}_z}{dt} = -(\bar{I}_z - \bar{I}_z^{\,0})(W_0 + 2W_1^I + W_2) - (\bar{S}_z - \bar{S}_z^{\,0})(W_2 - W_0) \qquad (8.8)$$

$$\frac{d\bar{S}_z}{dt} = -(\bar{S}_z - \bar{S}_z^{\,0})(W_0 + 2W_1^S + W_2) - (\bar{I}_z - \bar{I}_z^{\,0})(W_2 - W_0) \qquad (8.9)$$

where $\bar{I}_z^{\,0}$ and $\bar{S}_z^{\,0}$ are the equilibrium magnetizations of the two spins. These very important equations were first published in 1955 by Ionel Solomon and are frequently identified by his name. The important thing that they reveal is that the rates of relaxation of dipolar-coupled spins are *not* independent of one another. The relaxation rate of each depends on the extent to which the magnetization of the other deviates from equilibrium.

If the two spins are of the same type, so that their gyromagnetic ratios are equal, and both are shifted from equilibrium to the same extent, then the situation simplifies considerably. For instance, consider an inversion-recovery experiment in which both spins are subjected to a π-pulse. Because both spins have the same gyromagnetic ratio, the equilibrium magnetization is the same for both:

$$\bar{I}_z^{\,0} = \bar{S}_z^{\,0}$$

Also, the probabilities of the single-quantum transitions, W_1^I and W_1^S, are equal:

$$W_1^I = W_1^S = W_1$$

Immediately after the pulse, the magnetization components will be

$$\bar{I}_z(0) = -\bar{I}_z^{\,0}$$

$$\bar{S}_z(0) = -\bar{S}_z^{\,0} = -\bar{I}_z^{\,0}$$

where $\bar{I}_z(0)$ and $\bar{S}_z(0)$ represent the magnetization at the beginning of the decay period. At this instant, the rate of decay for the I-spin is given by

$$\frac{d\bar{I}_z(0)}{dt} = -(\bar{I}_z(0) - \bar{I}_z^{\,0})(W_0 + 2W_1^I + W_2) - (\bar{S}_z(0) - \bar{S}_z^{\,0})(W_2 - W_0)$$

$$= -(-\bar{I}_z^{\,0} - \bar{I}_z^{\,0})(W_0 + 2W_1 + W_2) - (-\bar{I}_z^{\,0} - \bar{I}_z^{\,0})(W_2 - W_0)$$

$$= 2\bar{I}_z^{\,0}(W_0 + 2W_1 + W_2) + 2\bar{I}_z^{\,0}(W_2 - W_0)$$

$$= 4\bar{I}_z^{\,0}(W_1 + W_2)$$

Similarly, we can show that the decay of the S_z-magnetization immediately after the pulse is

$$\frac{d\bar{S}_z(0)}{dt} = 4\bar{S}_z^{\,0}(W_1 + W_2)$$

Because the equilibrium magnetization values, $\bar{I}_z^{\,0}$ and $\bar{S}_z^{\,0}$, are equal, we can conclude that the decay of $\bar{I}_z$ and $\bar{S}_z$ from the inverted state will begin at the same rate and proceed in parallel, so that at any time, $\bar{I}_z = \bar{S}_z$. Therefore, the decay of $\bar{I}_z$ is given by

$$\frac{d\bar{I}_z}{dt} = -(\bar{I}_z - \bar{I}_z^{\,0})(W_0 + 2W_1 + W_2) - (\bar{S}_z - \bar{S}_z^{\,0})(W_2 - W_0)$$

$$= -(\bar{I}_z - \bar{I}_z^{\,0})(W_0 + 2W_1 + W_2) - (\bar{I}_z - \bar{I}_z^{\,0})(W_2 - W_0)$$

$$= -(\bar{I}_z - \bar{I}_z^{\,0})2(W_1 + W_2)$$

Similarly

$$\frac{d\bar{S}_z}{dt} = -(\bar{S}_z - \bar{S}_z^{\,0})2(W_1 + W_2)$$

These equations will be valid for any situation in which the nuclei are of the same type and are subjected to the same perturbation from equilibrium to initiate the relaxation process. Under these conditions, the relaxation for each spin will be described by a single exponential decay, with the observed rate constant related to the transition rates according to

$$R_1^{dd} = 2(W_1 + W_2)$$

Here, the superscript dd has been introduced to emphasize that R_1^{dd} only reflects the contribution of the dipole-dipole interaction to the total observed relaxation rate. If the nuclei are of different types, such as ^1H and ^{15}N, then the rates of relaxation will be different and will not, in general, strictly follow a single exponential decay. Also, if one of the nuclei, but not the other, is dipolar coupled to a third nucleus, then the relaxation rates for the different spins may be different and the decay curve may display multiple phases.

In the example above, we assumed that the two spins, in addition to being of the same type, were both subjected to an inversion pulse at the beginning of the relaxation measurement. However, it is possible, and quite interesting, to apply selective pulses to the individual spins. Consider what would happen, for instance, if the I-magnetization were inverted but not the S. Examination of the Solomon equation for the I-magnetization (Eq. 8.8) shows that the relaxation rate will be different than when both spins are inverted, because the term $(S_z - S_z^0)$ is now zero. Depending on the relative values of W_0 and W_2, the rate may be either larger or smaller when the S-spin is inverted along with the I-spin. This is a very important result, because it shows that the signal from one nucleus can be altered by manipulation of another with which it interacts by dipolar coupling. This is the underlying principle of the nuclear Overhauser effect (NOE), which is the basis of most molecular structure determinations by NMR and is discussed further in Chapter 9.

The same general approach as described here can be used to analyze the relaxation kinetics of systems in which more than two nuclei are coupled by dipolar interactions. As you might guess, the resulting equations become much more complex as the number of interacting nuclei is increased, but the same general principles apply.

8.4.3   Transition Rates and Relaxation Rates for Dipolar-coupled Spins

Because the transitions between quantum states are most efficiently stimulated by magnetic fluctuations with frequencies corresponding to the energy difference, we might expect that the rates of these transitions, when stimulated by random tumbling of the molecule, would be proportional to the spectral density function evaluated at the Larmor frequency. This relationship is expressed as

$$W_{k,l} \propto J(\omega_{k,l}) = \frac{\tau_c}{1 + \omega_{k,l}^2 \tau_c^2}$$

where k and l identify the interconverting quantum states and $\omega_{k,l} = 2\pi(E_k - E_l)/h$. Confirming this relationship and determining the value of the constant of proportionality requires a good deal of mathematics and attention to details of the physical interactions that cause the magnetic fluctuations. Forgoing those details, the results for the case of two dipolar-coupled spins are given as

$$W_0 = \frac{1}{10} d^2 J(\omega_I - \omega_S) \tag{8.10}$$

$$W_1^I = \frac{3}{20} d^2 J(\omega_I) \tag{8.11}$$

$$W_1^S = \frac{3}{20} d^2 J(\omega_S) \tag{8.12}$$

$$W_2 = \frac{3}{5} d^2 J(\omega_I + \omega_S) \tag{8.13}$$

where d is the dipolar coupling constant, introduced in Chapter 4, which represents the strength of the magnetic interaction between the two nuclei. The numerical value of this constant is given by

$$d = \frac{\mu_0 \gamma_I \gamma_S h}{8\pi^2 r^3} \tag{8.14}$$

where μ_0 is the magnetic permeability of free space, and r is the distance between the two nuclei. Note that each of the transition rate constants depends on γ_I and γ_S, each to the second power, and the inverse sixth power of r. The factors 1/10, 3/20 and 3/5 in Eqns. 8.10–8.13 arise from some rather subtle geometrical arguments regarding the magnetic fields generated by each nucleus.

Using the results from the previous subsection, the equations for the transition rate constants can be used directly to calculate the relaxation rate constant for the specific case of two dipolar-coupled spins of the same type, where $\gamma_I = \gamma_S$ and $\omega_I \approx \omega_S$:

$$R_1^{dd} = 2(W_1 + W_2)$$

$$= \frac{3}{10} d^2 \big(J(\omega_I) + 4J(2\omega_I) \big)$$

$$= \frac{3}{10} d^2 \left(\frac{\tau_c}{1 + \omega_I^2 \tau_c^2} + 4 \frac{\tau_c}{1 + 4\omega_I^2 \tau_c^2} \right) \tag{8.15}$$

The solid curve in Fig 8.10 shows the calculated value of R_1 as a function of τ_c for the case of two ^{1}H nuclei separated by 2 Å, in an 11.7-T magnetic field. The dashed curves represent the contributions of the rate constants for the single- and double-quantum transitions to the overall relaxation rate constant.

Note that both the single- and double-quantum transitions make significant contributions to the observed relaxation rate. When τ_c is relatively small—that is, when the molecule tumbles rapidly compared to the Larmor frequency—the double-quantum transition makes the dominant contribution. For more slowly tumbling molecules, however, both transitions contribute equally.

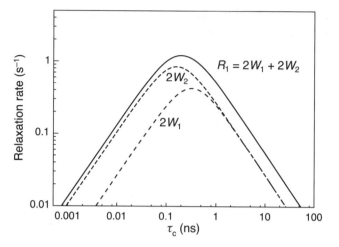

Figure 8.10

Calculated rate constant for longitudinal relaxation as a function of τ_c, for two ^1H nuclei separated by 2 Å in an 11.7-T field. The dashed lines show the contributions of the single- and double-quantum transitions.

8.4.4   Equations for Calculating R_2

So far in this chapter, we have focused on the longitudinal relaxation rate constant, R_1. The other important relaxation rate constant is R_2, the transverse relaxation rate, which, in addition to the processes that contribute to longitudinal relaxation, includes the loss of phase coherence among the individual nuclei in a sample. In Section 7.3.4, we used a simple random-walk model to show that the loss of phase coherence is expected to follow an exponential decay with a rate constant that is proportional to the average time between frequency changes. For relatively simple systems, such as a dipolar-coupled spin pair, the transverse relaxation rate can also be calculated from the spectral density function and the quantum mechanical transition rates.

In this approach, the loss of phase coherence is described in terms of quantum transitions among the superposition states that give rise to the transverse magnetization. These transitions include the single- and double-quantum transitions that we saw were important for longitudinal relaxation, and also zero-quantum transitions that leave the average z-magnetization unchanged but alter the average direction of magnetization in the transverse plane. As a consequence, the expression for R_2 in terms of the spectral density function includes $J(0)$ as well as $J(\omega)$ and $J(2\omega)$. The mathematics is rather involved, but for a pair of dipolar-coupled nuclei of the same type, the equation for R_2 is

$$R_2^{dd} = \frac{3}{20}d^2\big(3J(0) + 5J(\omega) + 2J(2\omega)\big)$$

$$= \frac{3}{20}d^2\left(3\tau_c + \frac{5\tau_c}{1 + \omega^2\tau_c^2} + \frac{2\tau_c}{1 + 4\omega^2\tau_c^2}\right) \tag{8.16}$$

To see how the various terms in this equation contribute to the overall relaxation rate, it is useful to consider the cases of extremely fast and extremely slow tumbling. If tumbling is fast, so that $\omega^2\tau_c^2 \ll 1$, then Eq. 8.16 simplifies to

$$R_2^{dd} = \frac{3}{20}d^2\left(3\tau_c + \frac{5\tau_c}{1+\omega^2\tau_c^2} + \frac{2\tau_c}{1+4\omega^2\tau_c^2}\right)$$

$$= \frac{3}{20}d^2(3\tau_c + 5\tau_c + 2\tau_c)$$

$$= \frac{3}{2}d^2\tau_c$$

Under the same conditions, the expression for R_1^{dd} (Eq. 8.15) simplifies to

$$R_1^{dd} = \frac{3}{10}d^2\left(\frac{\tau_c}{1+\omega_I^2\tau_c^2} + 4\frac{\tau_c}{1+4\omega_I^2\tau_c^2}\right)$$

$$= \frac{3}{10}d^2(\tau_c + 4\tau_c)$$

$$= \frac{3}{2}d^2\tau_c$$

Thus, R_1 and R_2 are equal under these conditions. This is because there is very little loss of phase coherence when the fluctuations of precession frequency occur very rapidly, and the overall rate of transverse relaxation is dominated by the contribution from longitudinal relaxation.

At the other extreme, if $\omega^2\tau_c^2 \gg 1$, the expression for R_2^{dd} simplifies to

$$R_2^{dd} = \frac{3}{20}d^2\left(3\tau_c + \frac{5\tau_c}{1+\omega^2\tau_c^2} + \frac{2\tau_c}{1+4\omega^2\tau_c^2}\right)$$

$$= \frac{3}{20}d^2\left(3\tau_c + \frac{5}{\omega^2\tau_c} + \frac{2}{4\omega^2\tau_c}\right)$$

If $\omega^2\tau_c^2 \gg 1$, then it is also true that $\tau_c \gg 1/(\omega^2\tau_c)$, so that only the first term contributes significantly, and the equation can be further simplified to

$$R_2^{dd} = \frac{9}{20}d^2\tau_c$$

In this regime, then, R_2^{dd} is again proportional to τ_c, but with a slightly different constant of proportionality. Under these conditions, R_1^{dd} approaches zero, and the observed rate of transverse relaxation is due almost entirely to the interconversion among the superposition states that give rise to the transverse magnetization.

Calculated values of R_1^{dd} and R_2^{dd} are plotted as a function of τ_c in Fig. 8.11, again for the case of two dipolar-coupled ^{1}H nuclei. As discussed earlier, when $\omega^2\tau_c^2 \ll 1$, the rates of longitudinal and transverse relaxation rates are nearly equal, and both are proportional to τ_c. When $\omega^2\tau_c^2 \gg 1$, R_1^{dd} is inversely proportional to τ_c and no longer makes a significant contribution to R_2^{dd}, which continues to increase with τ_c, but with a reduced proportionality constant. Note that when logarithmic scales are used for both axes in the plot, the change in proportionality constant in the two extreme τ_c

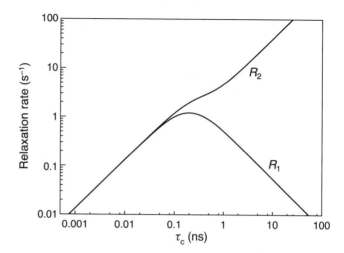

Figure 8.11

Calculated rate constants for longitudinal and transverse relaxation as a function of τ_c, for two ^1H nuclei separated by 2 Å in an 11.7-T field.

regimes appears as a relatively small displacement of the R_2^{dd} curve. This shift occurs when the contribution from R_1^{dd} plateaus and then begins to disappear.

We can also compare the results for transverse relaxation derived from the quantum mechanical treatment with that presented in Section 7.3.4 using the random-walk model. The random-walk model gave the following result:

$$R = \tau_c \Delta\omega^2/2$$

where the rate, R, refers just to the loss of phase coherence, and τ_c was defined rather loosely as the average time interval between tumbling motions that gave rise to a change in angular precession frequency of $\pm\Delta\omega$. For a dipolar-coupled spin pair, the change in precession frequency is due to a reorientation of the internuclear vector. The exact magnitude of $\Delta\omega$ depends on the orientation of this vector with respect to the external field, but is on the order of d, the dipolar coupling constant (see Eq. 4.2, p. 72). If we also assume that the average time between motions in this model is roughly equivalent to the correlation time derived from consideration of the autocorrelation function, then the expected relaxation rate due to the loss of phase coherence can be approximated as

$$R \approx d^2\tau_c$$

Thus, the two treatments give essentially identical results, but the quantum mechanical treatment is required in order to provide a precise relationship between τ_c and the observed relaxation rate constant.

8.5   Heteronuclear Relaxation and Internal Molecular Motions

Relaxation rates also depend on internal motions that generate fluctuating magnetic fields, and the measurement of R_1 and R_2 can be a powerful means of studying such motions. In practice, however, the rates of relaxation for any individual nucleus in a complex molecule are likely to depend on its interactions with several other nuclei, and the interpretation of these rates may be quite difficult. The situation is simplified considerably when two nuclei with magnetic dipoles are covalently attached to each other, because this leads to a particularly short internuclear distance and correspondingly strong dipolar coupling. Under these circumstances, the interactions between the two nuclei tend to dominate any effects involving more distant ones. For this reason, relaxation measurements involving covalently linked ^1H–^{15}N pairs or ^1H–^{13}C pairs are especially attractive. With the development of methods for biosynthetically enriching proteins and nucleic acids with the otherwise rare ^{15}N and ^{13}C isotopes, it has become possible to apply this approach to the study of internal motions in biological macromolecules, as introduced briefly in Section 7.5.

8.5.1   Rates of Heteronuclear Relaxation

For systems involving a heteronuclear pair, the rates of relaxation depend primarily on the dipolar interaction with the covalently attached nucleus and the chemical shift anisotropy due to the electronic structure of the covalent bond. The major difference between the effect of dipolar coupling in this situation from that involving a homonuclear pair is that the two mixed states, $|\beta_I\rangle|\alpha_S\rangle$ and $|\alpha_I\rangle|\beta_S\rangle$, have different energies, since the gyromagnetic ratios of the two nuclei are different. This gives rise to an energy diagram such as the one shown in Fig. 8.12.

With respect to relaxation, the important consequence is that the transition rates, W_1^I and W_1^S, depend on fluctuations on quite different time scales, reflecting the different Larmor frequencies for the two nuclei. In addition, the rates for the zero- and double-quantum transitions depend on different frequencies than they would for a homonuclear pair.

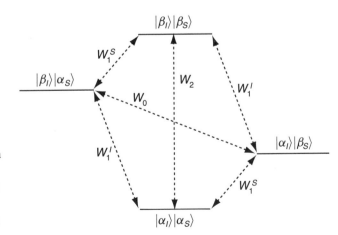

Figure 8.12
Energy diagram and relaxation pathways for a heteronuclear dipolar-coupled spin pair. The I-spin is assumed to have a larger gyromagnetic ratio than the S-spin.

The rates of longitudinal and transverse relaxation for the S-spin (usually ^{15}N or ^{13}C) due to dipolar coupling are given by

$$R_1^{\mathrm{dd}} = \frac{d^2}{10}\left(J(\omega_I - \omega_S) + 3J(\omega_S) + 6J(\omega_I + \omega_S)\right)$$

$$R_2^{\mathrm{dd}} = \frac{d^2}{20}\left(4J(0) + J(\omega_I - \omega_S) + 3J(\omega_S) + 6J(\omega_I) + 6J(\omega_I + \omega_S)\right)$$

where d is the dipolar coupling constant defined by Eq. 8.14.[1] Just as for the case of a homonuclear pair, the major difference between the expressions for R_1 and R_2 is the presence of the $J(0)$ term in the latter, which reflects the effectiveness of slow motions in promoting the loss of phase coherence among the nuclei in the sample.

A second major mechanism of relaxation for heteronuclear spin pairs is the effect of chemical shift anisotropy, which, like the dipolar interaction, causes local fluctuations of the magnetic field experienced by each nucleus. Unlike the dipolar interaction, however, there are no concerted changes in the wavefunctions of the two spins, and the contribution of chemical shift anisotropy to the longitudinal relaxation rate depends only on the probability of fluctuations with frequencies close to the Larmor frequency. For the S-spin, the rate of longitudinal relaxation due to chemical shift anisotropy is given by

$$R_1^{\mathrm{csa}} = \frac{2c^2}{5} J(\omega_S)$$

where c is a constant that reflects the magnitude of the anisotropy, defined as

$$c = \frac{\omega_S \Delta\sigma}{\sqrt{3}} \tag{8.17}$$

where $\Delta\sigma$ is the shielding anisotropy, as defined on page 70 of Chapter 4. Like the isotropically averaged chemical shift, $\Delta\sigma$ is independent of the strength of the magnetic field, but ω_S *does* depend on the external field. As a consequence, the contribution of chemical shift anisotropy to the observed relaxation rate increases with the second power of the field strength. Magnetic fluctuations associated with chemical shift anisotropy also contribute to the loss of phase coherence in the transverse plane, as reflected in the expression for the rate of transverse relaxation due to chemical shift anisotropy:

$$R_2^{\mathrm{csa}} = \frac{c^2}{15}\left(3J(\omega_S) + 4J(0)\right)$$

The total relaxation rates due to dipolar coupling and chemical shift anisotropy are given by

---

[1] In many papers on heteronuclear relaxation, the spectral density function is defined so as to include a factor of 2/5, relative to the definition used in the earlier sections of this chapter. For internal consistency, this factor has been incorporated into the equations for R_1 and R_2, but the reader is to be cautioned when comparing the equations presented here with those used elsewhere.

$$R_1 = \frac{d^2}{10}\left(J(\omega_I - \omega_S) + 3J(\omega_S) + 6J(\omega_I + \omega_S)\right) + \frac{2c^2}{5}J(\omega_S) \tag{8.18}$$

$$R_2 = \frac{d^2}{20}\left(4J(0) + J(\omega_I - \omega_S) + 3J(\omega_S) + 6J(\omega_I) + 6J(\omega_I + \omega_S)\right)$$

$$+ \frac{c^2}{15}\left(3J(\omega_S) + 4J(0)\right) \tag{8.19}$$

If a heteronuclear bond vector is rigidly constrained by the structure of the molecule, then the motions that give rise to relaxation are simply those associated with overall tumbling. In this case, the expected rates of relaxation can be calculated from the equations above and the form of the spectral density function given in Eq. 8.4 (p. 201). In general, however, covalent bonds are not rigidly constrained within molecules and, indeed, the major motivation for measuring heteronuclear relaxation rates is usually to detect internal motions.

When individual bonds have some degree of independent mobility, the form of $J(\omega)$ derived for overall molecular tumbling is no longer adequate, and we would like to use the relaxation rates to gain some description of the spectral density function for individual atom pairs. As indicated in Eqs. 8.18 and 8.19 above, the observed relaxation rates depend on the value of $J(\omega)$ evaluated at multiple frequencies, namely $J(0)$, $J(\omega_I)$, $J(\omega_S)$, $J(\omega_I - \omega_S)$ and $J(\omega_I + \omega_S)$. A fundamental challenge in the analysis is that the number of unknown parameters (five) exceeds the number of observables.

One approach to addressing this problem involves making additional types of relaxation measurements, designed to depend on the spectral density function in ways that are distinguishable from the expressions for R_1 and R_2. For instance, a very common measurement is the heteronuclear nuclear Overhauser effect (NOE), which is determined by first irradiating one nucleus of the pair (usually a proton) and then measuring the effect on the intensity of the signal from the other spin As predicted by the Solomon equations (Eqs. 8.8 and 8.9, p. 210), the relaxation of each nucleus, and therefore the intensity of the signal it generates, depends on the spin state of the other. The nuclear Overhauser effect is discussed in more detail in Chapter 9 (primarily in the context of homonuclear pairs), but for now, it is sufficient to define the measurement as the ratio of the intensities of the S-spin signal with and without pre-irradiation of the I-spin.[2] The heteronuclear NOE is related to the spectral density function according to

$$NOE \tag{8.20}$$

$$= 1 + \frac{(d^2/10)(\gamma_I/\gamma_S)\left(6J(\omega_I + \omega_S) - J(\omega_I - \omega_S)\right)}{(d^2/10)\left(J(\omega_I - \omega_S) + 3J(\omega_S) + 6J(\omega_I + \omega_S)\right) + (2c^2/5)J(\omega_S)}$$

---

[2] Here is yet another source of potential confusion: The definition of the NOE given above is the one commonly used in heteronuclear relaxation studies, but a different definition, given on page 265 in Chapter 9, is usually used in discussions of the homonuclear NOE. The two definitions are readily interconverted, but care must be exercised when comparing the equations or actual NOE values found in different texts or research articles.

Together with the R_1 and R_2 rates, we now have three observable parameters. There are other, more complex, types of relaxation measurements that can be added to these, thereby making it possible to solve a set of equations to estimate $J(\omega)$ for ω equal to 0, ω_I, ω_S, $(\omega_I - \omega_S)$ and $(\omega_I + \omega_S)$. The resulting plot of J versus ω is often referred to as a *spectral density map*.

Another approach to analyzing the relaxation data is to recognize that for heteronuclear pairs, such as ^1H–^{15}N or ^1H–^{13}C, the Larmor frequencies of the two nuclei are quite different, so that, for instance, ω_H, $(\omega_H - \omega_N)$ and $(\omega_H + \omega_N)$ have very similar values. By using a single value to approximate the spectral density function at all three of these frequencies, the number of unknown parameters in Eqs. 8.18, 8.19 and 8.20 is reduced to three, and the values of $J(\omega)$ at three frequencies can be determined from the three common relaxation measurements, R_1, R_2 and the heteronuclear NOE.

While the values of the spectral density function provide a quantitative description of the extent of motion at different frequencies, their meaning is rather abstract, and it is not so easy to derive from them a physical sense of the motions. An alternative approach to interpreting the data is to construct a simplified physical model of the internal motions and then derive a plausible mathematical form for the spectral density function with a minimum number of parameters. The values of these parameters can then be estimated by fitting to the observed relaxation parameters.

8.5.2   "Model-free" Analysis of Relaxation Rates

A specific form of the spectral density function that has been widely adopted for the analysis of heteronuclear relaxation data is one that was proposed by Giovanni Lipari and Atilla Szabo in the early 1980s. Lipari and Szabo derived their spectral density function making minimal assumptions about the nature of the physical motions and, therefore, described their method as being "model-free," though it is based on a quite specific mathematical model and an easily visualized physical description. In essence, the "model-free" model assumes that individual heteronuclear bonds vectors have varying degrees of freedom within the molecule, which is assumed to maintain the same overall structure, at least on the time scale of molecular tumbling events. This is depicted schematically in Fig. 8.13.

The overall tumbling motion is characterized by a correlation time, τ_c, and the internal motions are described by two parameters: an internal correlation time, τ_e, and an *order parameter* designated by the symbol S^2. The order parameter can take on values from 0 to 1, representing, respectively, completely unrestricted internal motions or a situation in which motions of the bond vector are constrained to those of the molecule as whole.

As before, when we considered only the effects of molecular tumbling, derivation of the spectral density function begins with an expression for the autocorrelation function. In this case, it is assumed that the autocorrelation function can be written as a product of two terms, one, g_0, representing overall tumbling and the other, g_I, the internal motions:

$$g(\Delta t) = g_0(\Delta t)g_I(\Delta t)$$

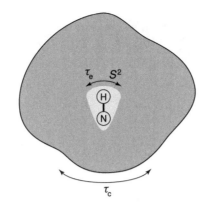

Figure 8.13

Schematic representation of the model-free description of internal motion of an H–N bond vector in a tumbling molecule.

Provided that the overall tumbling motions are isotropic, g_0 can be written as a simple exponential decay:

$$g_0(\Delta t) = e^{-\Delta t/\tau_c}$$

For the internal motions, the autocorrelation function must reflect both a time constant for these motions and their extent or amplitude.

To gain some insight into the general relationship between the extent of internal motions and the autocorrelation function, it is helpful to consider a specific case. Suppose that you are sitting on a molecule as it tumbles, so that it appears stationary in your frame of reference, and you are observing the motions of a particular N–H bond vector. After watching for a while, you conclude that the bond can randomly reorient over a space corresponding to half a sphere. That is, if, from your frame of reference, the average orientation is straight "up," the angle, θ, between the bond and the vertical axis can range from $-\pi/2$ to $\pi/2$. This motion can be simulated, as was done earlier for the overall tumbling motion, and the plot in Fig. 8.14 shows $\cos\theta$ as a function of time for a simulation in which the bond vector is assumed to undergo Brownian motion.

Because θ is now restricted, $\cos\theta$ is always positive. Recall that in calculating the autocorrelation function for a specific value of Δt, we take each value of $\cos\theta$ and multiply it by the value for Δt time units later. In this case, this product will always be positive, and we no longer expect the autocorrelation function to decay to zero. The autocorrelation function for the internal motions simulated in Fig. 8.14 is shown in Fig. 8.15.

In this case, $g_I(\Delta t)$ decays to a value of about 0.8. In general, the limiting value of the autocorrelation function will depend on the distribution of angles accessible to the bond vector, with more restricted distributions giving rise to larger values of $g_I(\infty)$. Without restricting ourselves to a particular model of the internal motions, we can define the order parameter, S^2, as $g_I(\infty)$.

The exact form of $g_I(\Delta t)$ will depend on the nature of the motions and the distribution of accessible angles, but a simple form that satisfies the general requirements of such a function is

$$g_I(\Delta t) = S^2 + (1 - S^2)e^{-\Delta t/\tau_e}$$

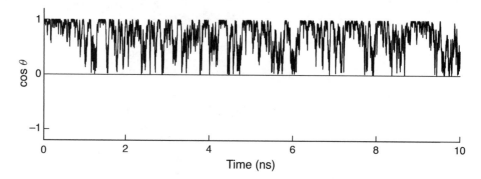

Figure 8.14 Simulated trajectory of internal motions of an H–N bond vector, with the direction of the vector constrained to lie within a half-sphere. θ is the angle between the bond vector and the axis defining the half-sphere. The motion is described as random changes in orientation at intervals of 1 ps. The standard deviation of the random motions is 0.1 rad.

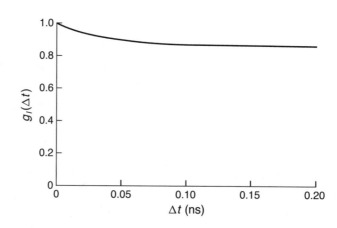

Figure 8.15

Autocorrelation function for simulated internal motions of an H–N bond vector, with the motions constrained to lie within a half-sphere.

Note that if $S^2 = 1$, $g_I(\Delta t)$ is constant for all values of Δt, implying no internal motion, while if $S^2 = 0$, $g_I(\Delta t)$ decays to zero, just as the autocorrelation function for molecular tumbling does. For any value of S^2, $g_I(\Delta t)$ approaches S^2 as Δt becomes much greater than τ_e.

The complete autocorrelation function, reflecting both tumbling and internal motions, is then given by

$$g(\Delta t) = g_0(\Delta t)g_I(\Delta t)$$

$$= e^{-\Delta t/\tau_c}\left(S^2 + (1 - S^2)e^{-\Delta t/\tau_e}\right)$$

$$= S^2 e^{-\Delta t/\tau_c} + (1 - S^2)e^{-\Delta t(1/\tau_c + 1/\tau_e)}$$

The graphs in Fig 8.16 show examples of the autocorrelation function for different values of S^2. In the top-most graph, $S^2 = 1$, so that the motions of the bond vector are restricted to those of the entire molecule, and the autocorrelation function simply

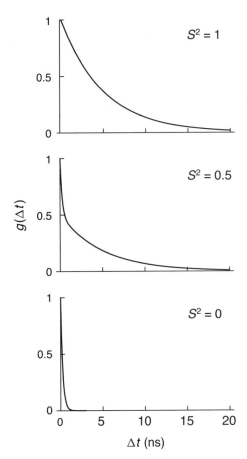

Figure 8.16
Autocorrelation functions calculated for the Lipari–Szabo treatment of overall tumbling and internal motions, with different values of S^2. The correlation time for overall tumbling, τ_c, is assumed to be 5 ns, and τ_e, the correlation time for internal motions, is 0.25 ns.

reflects the rate of overall tumbling. At the other extreme, $S^2 = 0$, implying that the internal motions of the bond vector are completely unrestricted. Because they are unrestricted and are much faster than overall tumbling, the internal motions are primarily responsible for the decay of the autocorrelation function. The center graph shows the case where $S^2 = 0.5$. Now, there is a quite rapid decay of the autocorrelation function to a value of about 0.5, and then a slower decay to zero. Although they are much faster than the tumbling motions, the internal motions are restricted spatially and only allow a portion of the initial correlation (or "memory" of the original orientation) to decay rapidly. The remainder decays at the rate reflecting overall tumbling. In general, the magnitude of the initial decay will reflect the value of S^2.

The shape of the decay curve for $g(\Delta t)$ also depends on the relative values of τ_c and τ_e. The graphs in Fig. 8.17 were each calculated assuming that $\tau_c = 5$ ns and $S^2 = 0.5$, with the indicated values of τ_e. When τ_e is much smaller than τ_c, the two phases of the decay curve are clearly distinguishable, as shown for the case in Fig. 8.17 for $\tau_e = 0.1$ ns. As the rate of the internal motions is made slower, however, the distinction becomes smaller, and the two parts of the curve merge when $\tau_e \approx \tau_c$. When the internal motions are even slower than overall tumbling, they become irrelevant to

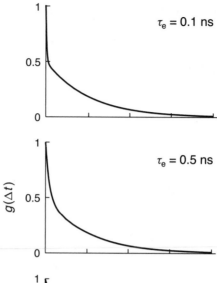

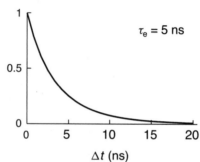

Figure 8.17
Autocorrelation functions
calculated for the Lipari–Szabo
treatment of overall tumbling
and internal motions, with
different values of τ_e. The
correlation time for overall
tumbling, τ_c, is assumed to
be 5 ns, and S^2, the order
parameter, is 0.5.

the autocorrelation function (though, as discussed further below, such motions can
contribute to the observed transverse relaxation rate).

The spectral density function, as before, is defined as the real part of the Fourier
transform of the autocorrelation function:

$$\mathcal{R}e\big(\mathcal{F}(g(\Delta t))\big) = \mathcal{R}e\left(\int_{-\infty}^{\infty} g(\Delta t)e^{-itf2\pi}\,d\Delta t\right)$$

$$= \mathcal{R}e\left(\int_{-\infty}^{\infty} \left(S^2 e^{-\Delta t/\tau_c} + (1-S^2)e^{-\Delta t(1/\tau_c + 1/\tau_e)}\right)e^{-itf2\pi}\,d\Delta t\right)$$

Evaluating this transform and expressing the results in terms of an angular frequency,
ω, gives the Lipari–Szabo model-free spectral density function:

$$J(\omega) = \frac{S^2\tau_c}{1+\omega^2\tau_c^2} + \frac{(1-S^2)/\left(1/\tau_c + 1/\tau_e\right)}{1+\omega^2/\left(1/\tau_c + 1/\tau_e\right)^2} \tag{8.21}$$

This is often written in a slightly simpler form by introducing a new correlation time,
τ, which is the reciprocal of the sum of reciprocals of τ_c and τ_e (the harmonic mean):

$$\tau = \frac{1}{1/\tau_c + 1/\tau_e}$$

so that:

$$J(\omega) = \frac{S^2 \tau_c}{1 + \omega^2 \tau_c^2} + \frac{(1 - S^2)\tau}{1 + \omega^2 \tau^2}$$

When written in this form, it is apparent that the model-free spectral density function is a sum of two Lorentzian terms, both centered at $\omega = 0$. The first term corresponds to the overall tumbling of the molecule, and its contribution is proportional to S^2, reflecting the extent to which the motions of the bond vector are correlated with those of the entire molecule. The second term reflects the internal motions, but, because these motions are not completely free, the correlation time associated with this term is influenced by τ_c as well as τ_e.

The relationship between the shape of the spectral density function and the order parameter S^2 is illustrated by the examples in Fig. 8.18. Each curve in the figure was calculated assuming that $\tau_c = 5$ ns and that $\tau_e = 0.25$ ns. When $S^2 = 1$, the spectral density function reduces to that for a rigidly tumbling molecule, with a single Lorentzian term, and $J(0) = \tau_c$. At the other extreme, when $S^2 = 0$, $J(\omega)$ is again a single Lorentzian term, but now $J(0) = \tau$. As we saw earlier, a smaller value of $J(0)$ is associated with larger values of the spectral density at higher frequencies.

At the intermediate value of the order parameter, $J(0)$ has an intermediate value, about 2.6 ns, while at high frequencies, the value of $J(\omega)$ is greater than for $S^2 = 1$. Qualitatively, this effect is not so different from what we would expect if the molecule were simply tumbling more rapidly. If, however, we look more closely at the shape of the spectral density function graph, we find that it is subtly different from that for a rigidly tumbling molecule. Fig. 8.19 compares the spectral density function for the case of $S^2 = 0.5$ with that for simple tumbling, but with the value of τ_c set so that $J(0)$ is the same for both.

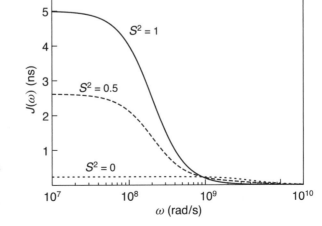

Figure 8.18
Spectral density function for the Lipari–Szabo model-free treatment, for different values of the order parameter, S^2. For each curve, $\tau_c = 5$ ns and $\tau_e = 0.25$ ns.

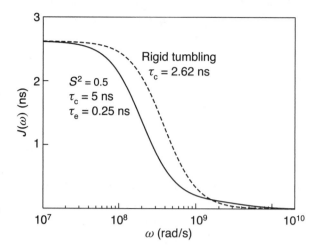

Figure 8.19

Lipari–Szabo spectral density function for $S^2 = 1$ (rigid tumbling) and $S^2 = 0.5$, with τ_c set so that $J(0)$ is the same for both cases.

While $J(0)$ is the same for both cases, the internal motions cause $J(\omega)$ to begin dropping at lower frequencies, but also to persist at somewhat higher frequencies. Thus, the general effect of the partially restricted internal motions is to increase the range of frequencies for which $J(\omega)$ takes on significant values.

The shape of the spectral density function also depends on the rates of the internal motions, as expressed by the internal correlation time, τ_e, and as illustrated in Fig. 8.20. Each of these curves was calculated with $S^2 = 0.5$ and $\tau_c = 5$ ns, and the indicated values of τ_e. If $\tau_e \ll \tau_c$, then $\tau \approx \tau_e$, and the spectral density function can approximated as

$$J(\omega) \approx \frac{S^2 \tau_c}{1 + \omega^2 \tau_c^2} + \frac{(1 - S^2)\tau_e}{1 + \omega^2 \tau_e^2}$$

In the limit of very small values of τ_e, the second term becomes much smaller than

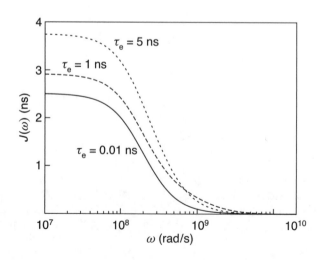

Figure 8.20

Lipari–Szabo spectral density function for different values of τ_e, the correlation time for internal motions. For each curve, $\tau_c = 5$ ns and $S^2 = 0.5$.

the first, leading to the following further approximation:

$$J(\omega) \approx \frac{S^2 \tau_c}{1 + \omega^2 \tau_c^2} \tag{8.22}$$

In this limit, the spectral density function is the same as for rigid tumbling, but scaled by the order parameter, S^2. $J(0)$, then, is approximately equal to $S^2 \tau_c$, as illustrated for the case where $\tau_e = 0.01$ ns in Fig 8.20. As τ_e increases, so does $J(0)$, but the values of $J(\omega)$ at high frequencies do not necessarily decrease, as shown on Fig 8.20 by the case where $\tau_e = 1$ ns.

As the examples above illustrate, the amplitudes and correlation times of internal motions determine the shape of the spectral density function and, therefore, the likelihood of relaxation processes associated with different frequencies. Given specific values for the parameters τ_c, τ_e and S^2, it is possible to predict the values of experimental relaxation parameters using Eqs 8.18, 8.19 and 8.20 (p. 219) for R_1, R_2 and the heteronuclear NOE, respectively. What we typically want to do, however, is to use experimental relaxation data to estimate the parameters that characterize the motions. The usual practice is to use an iterative fitting procedure to find values of τ_c, τ_e and S^2 that minimize the differences between the observed values of R_1, R_2 and the NOE and those predicted by the model. Usually, a single value of τ_c is used for all of the nuclei in a molecule, and individual values of τ_e and S^2 are estimated for each heteronuclear pair. Often, it is found that the experimental data can be well fit by assuming only that $\tau_e \ll \tau_c$, and a precise value of τ_e may be difficult to determine. The order parameter, S^2, on the other hand, is usually well defined by the experimental data and is a particularly useful description of internal motions. References to publications describing applications of this approach, as well as additional subtleties in the analysis, are provided at the end of this chapter.

The Lipari–Szabo treatment is but one approach to deriving a plausible functional form for the spectral density function accounting for the possibility of internal motions. Although other treatments have been developed, most of the resulting spectral density functions resemble that shown here in that they are composed of two or more Lorentzian terms. We have focused here on the Lipari–Szabo version both because it is widely used in the study of protein and nucleic acid dynamics and because it nicely illustrates the general features of a spectral density function while making a minimum of structural assumptions.

8.6   Slower Internal Motions: Chemical/Conformational Exchange

So far, we have only considered the effects of internal motions on relaxation mediated by dipolar interactions and chemical shift anisotropy. From our discussion of the autocorrelation and spectral density functions, it should be apparent that the effectiveness of these relaxation mechanisms depends on the rate at which the orientation of an internuclear vector loses its correlation with an initial orientation. When there are internal motions that are faster than the rate of overall tumbling, the decay of the autocorrelation function is more rapid than it would be for a rigid molecule, as discussed in the previous section. On the other hand, internal motions that are much slower than molecular tumbling do not affect the spectral density function, because the loss of correlation due to tumbling occurs before the internal motions can have any effect.

The above may seem a rather counterintuitive statement, because you might expect that slow motions should increase the value of the spectral density function at low frequencies. In resolving this apparent paradox, it may help to imagine yourself as a passenger rigidly fixed on a rapidly tumbling object with a clear view of a flag pole, which you can use to determine your orientation. By keeping track of your orientation with respect to the flag pole as a function of time, you could determine your own autocorrelation function. Now, suppose that your seat is undergoing its own random motion at a rate that is much slower than the overall tumbling motion. Would you be able to detect this motion? To do so, you would need to be able to determine whether your orientation has changed over a time period roughly equal to the correlation time for the slower motion. If the entire object you are riding has undergone multiple random changes of orientation between these times, you will not be able to tell whether or not the slower internal motion has occurred. For the same reason, the rate of overall tumbling will always obscure any slower changes in orientation of internal bond vectors with respect to the external magnetic field in an NMR experiment. As a consequence, these slower motions do not contribute to relaxation mediated by dipolar interactions and chemical shift anisotropy.

Nonetheless, slower internal motions can contribute to the observed rates of transverse relaxation by causing random changes in the precession frequencies of nuclei in individual molecules. When a molecule is tumbling rapidly, the effects of its orientation on the NMR frequencies of individual nuclei are effectively averaged, giving rise to the relatively sharp peaks characteristic of liquid samples. If, however, the molecule is also very slowly interconverting between two conformations that place a nucleus in alternate chemical environments, then we expect to see two separate peaks in the spectrum, as discussed in Section 7.4. At faster rates of conformational change, there will be multiple random changes in precession frequency during the course of recording a free induction decay, leading to a single peak, the width of which will depend on the relative rates of the conformational change and the difference in precession frequencies for the nucleus in the two states. Conformational changes of this type will also increase the value of R_2 determined from Carr–Purcell–Meiboom–Gill- (CPMG-) type experiments, and the measurement of transverse relaxation rates is a powerful means of characterizing such motions. The terms *chemical exchange* and *conformational exchange* are often used interchangeably to describe this relaxation mechanism, because they arise from changes in chemical shift, even if there is no change in covalent structure. Reversible association between molecules can also contribute to relaxation of this type.

The contribution to R_2 due to exchange processes, beyond that accounted for by molecular tumbling and faster internal motions, is referred to as R_{ex}, and the magnitude of this contribution depends on several related factors, including the refocusing delay time in the CPMG experiment. The outline of the experiment is shown again in Fig. 8.21.

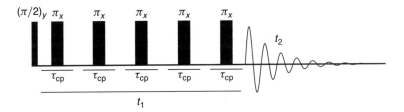

Figure 8.21 The Carr–Purcell–Meiboom–Gill (CPMG) experiment with multiple refocusing pulses for measuring transverse relaxation rates. The delay time for the refocusing pulses is τ_{cp}, and the total relaxation delay is $t_1 = n\tau_{cp}$, where n is the number of refocusing pulses.

The total relaxation delay time in the experiment, t_1, is divided into intervals of length τ_{cp}, each including a π_x refocusing pulse. In a typical experiment, peak intensities are measured in a series of spectra with increasing values of t_1 (corresponding to an increasing number of refocusing intervals of the same length), and the results are fit to an exponential decay function to estimate R_2. If the individual τ_{cp} delay periods can be made significantly shorter than the time associated with the conformational exchange, the contribution of R_{ex} can be minimized or eliminated, providing a means of estimating the rate of the conformational change. This is somewhat analogous to taking a series of photographs of a moving object using different shutter speeds; from the shutter speed that is just able to "freeze" the apparent motion of the object in the photograph (together with some geometrical considerations), the speed of the object can be deduced. However, the value of R_{ex} also depends on the difference in Larmor frequencies of the nucleus in the alternate conformations, and the inter-relationships of the various parameters is rather involved. We will begin by developing a physical picture of what goes on in some extreme cases, where the rate of exchange is either significantly smaller or greater than the *difference* in precession frequencies for a nucleus in the two alternate conformations. Then, the mathematical expression for calculating R_{ex} will be presented and used to compare the results under a wide range of conditions.

8.6.1 Slow Exchange

In all of the examples considered here, we will limit ourselves to the simplest case of a two-state conformational transition in which there are two interconverting species, A and B:

$$A \underset{k_{ba}}{\overset{k_{ab}}{\rightleftharpoons}} B$$

The equilibrium fractional populations of the A and B states are p_a and p_b, respectively, and the exchange rate constant, k_{ex}, is defined as the sum of the forward and reverse rate constants, k_{ab} and k_{ba}, respectively. The fractional populations and rate constants are related to one another as follows:

$$p_a + p_b = 1$$

$$\frac{p_a}{p_b} = \frac{k_{ba}}{k_{ab}}$$

$$k_{ab} + k_{ba} = k_{ex}$$

$$k_{ab} = p_b k_{ex}$$

$$k_{ba} = p_a k_{ex}$$

The angular Larmor frequencies of a given nucleus in the A and B conformations are ω_a and ω_b, respectively, and the absolute value of the difference between these frequencies is $\Delta\omega = |\omega_a - \omega_b|$.

As discussed in Section 7.4, three regimes for chemical or conformational exchange can be defined in terms of the relative rates of the exchange process, k_{ex}, and the divergence of the magnetization vectors associated with a nucleus in the two states ($\Delta\omega$):

- Slow exchange, where $k_{ex} < \Delta\omega$
- Intermediate exchange, where $k_{ex} \approx \Delta\omega$
- Fast exchange, where $k_{ex} > \Delta\omega$

To begin, we will assume that the two conformations are very unlikely to exchange during the total delay period, t_1. For purposes of illustrating the experiment, we will also assume that τ_{cp} is long enough so that the magnetization components representing the two conformations will diverge significantly during each of the repeated refocusing intervals. In Fig. 8.22, two vectors, displaced above and below the x'–y' plane, are used to represent the populations in the A and B conformations. At the end of the first half of the delay period, the vectors representing the two populations have diverged by $\tau_{cp}\Delta\omega/2$ rad, as indicated by the arc in the lower precession path. The π_x refocusing pulse and the second half of the delay return both vectors to their original positions, aligned with the x'-axis, as shown in Fig. 8.23.

If an FID is recorded after one or more refocusing periods and transformed into a spectrum, the result will be two peaks corresponding to the two Larmor frequencies. Although the intensities of these two peaks will reflect transverse relaxation during

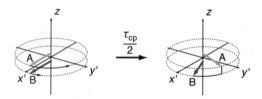

Figure 8.22   Vector diagrams representing the precession of nuclei in two populations of molecules in alternate conformations, A and B, during the first half of a CPMG delay period, in the absence of exchange during the delay period.

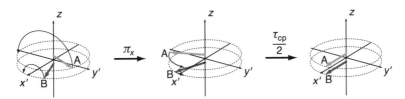

Figure 8.23 Vector diagrams representing the π_x refocusing pulse and precession during the second half of a CPMG delay period, for nuclei in two populations of molecules in alternate conformations, A and B, in the absence of exchange. At the end of this period, the average magnetization components for both populations are returned to their initial positions, aligned with the x'-axis.

the refocused delay period (due to the effects of dipolar coupling interactions and chemical shift anisotropy), there is no reduction in the peak intensity due to exchange processes.

Now, consider a case where the probability of an exchange event during a single delay period is significant (i.e., $k_{ex}\tau_{cp} \gtrsim 1$), but k_{ex} is less than $\Delta\omega$, so that we are in the slow-exchange regime. For this example, we will also assume that the delay period is long enough that the product $\tau_{cp}\Delta\omega$ is greater than 2π rad; the significance of this assumption will become apparent later.

As the vector representing the initial A conformation precesses, individual molecules will randomly convert to the B conformation, and the vectors representing these molecules leave a trail behind those that remain in the A state, as represented by the vectors shown above the $x'-y'$ plane in Fig. 8.24. Similarly, some of the molecules that begin in the B conformation randomly convert to the A state, and the vectors representing these molecules move ahead of those still in the B state, as shown in the vectors below the $x'-y'$ plane. In these and the next diagrams, the individual arrows are not meant to represent discrete populations, but rather describe, qualitatively, a continuous distribution of precession angles, with the thickness of the arrows suggesting the relative populations. The effects of the refocusing pulse and the second

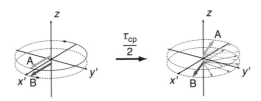

Figure 8.24 Vector diagrams representing the precession of nuclei in two populations of molecules that begin in alternate conformations, A and B, and undergo exchange, with $k_{ex} < \Delta\omega$ and τ_{cp} sufficiently long that $\tau_{cp}\Delta\omega \geq 2\pi$. The two sets of vectors, labeled A and B and drawn above and below the $x'-y'$ plane, represent the populations that begin in the two conformations.

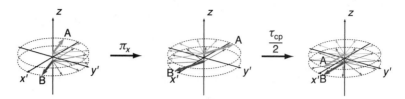

Figure 8.25 Vector diagrams representing the π_x refocusing pulse and precession during the second half of a CPMG delay period, for nuclei in two populations of molecules that begin the experiment in alternate conformations, A and B, for the case where $k_{ex} < \Delta\omega$.

half of the delay period are diagrammed in Fig. 8.25. Although the vectors representing the molecules that have *not* undergone exchange during either delay period are refocused, the vectors that have spread out are not, and the spreading increases during the second $\tau_{cp}/2$ delay period. This is because the exchange process is random, and the extent of precession for a given molecule after the refocusing pulse is uncorrelated with its precession before the pulse.

If the refocusing period is repeated multiple times to generate the total delay period, t_1, additional molecules undergo exchange, further decreasing the magnetization vector representing the unexchanged molecules. Because we specified, for this example, that $\tau_{cp}\Delta\omega > 2\pi$, the vectors representing the molecules that have undergone exchange are distributed about the z-axis after the first refocusing delay period. These vectors largely cancel one another, leaving only the magnetization of the molecules that never underwent an exchange to contribute to an FID and spectrum. The decay of this magnetization is determined only by the rate of conformational transitions and the total relaxation time, t_1. For the molecules that began in the A conformation, the fraction that remain in this state throughout the relaxation delay is given by

$$\frac{[A]}{[A_0]} = e^{-k_{ab}t_1} = e^{-p_b k_{ex}t_1}$$

The apparent exchange contribution to the total transverse relaxation rate for conformation A is then

$$R_{ex,A} = p_b k_{ex} \tag{8.23}$$

Conversely, the exchange contribution for conformation B is

$$R_{ex,B} = p_a k_{ex}$$

The conditions specified above, $k_{ex} < \Delta\omega$, define the "slow-exchange" regime. In this regime, R_{ex} depends only on the relative populations of the two states and their rate of interconversion, and is independent of $\Delta\omega$. Recall, however, that we also assumed that $\tau_{cp}\Delta\omega > 2\pi$, so that the vectors diverge by at least 2π radians during each CPMG delay period. If τ_{cp} is made shorter than this, then the apparent relaxation rate will decrease, as discussed further below.

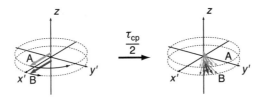

Figure 8.26 Vector diagrams representing the precession of nuclei during the first half of a CPMG delay period, in the case where $k_{ex}\Delta\omega > 1$ and $k_{ex}\tau_{cp} \gg 1$. The vectors drawn above and below the transverse plane represent the populations that begin in the A or B conformational states.

8.6.2 Fast Exchange

Now, consider what happens when $k_{ex} > \Delta\omega$, so that there are one or more transitions between the two conformations during the time required for the magnetization vectors to diverge. We will also assume for now that $k_{ex}\tau_{cp} \gg 1$, so that each molecule has time to undergo multiple exchange events between refocusing pulses. The effects on the magnetization components during the first half of the period are illustrated in the vector diagrams of Fig. 8.26. By the end of the first half of the τ_{cp} delay, the molecules have undergone multiple exchange events, and the average precession of the vectors representing the two initial populations are equal. But, the vectors representing both populations have spread out, because some molecules, by chance, spend more time in conformation A, while others spend more time in conformation B.

Fig. 8.27 illustrates the effects of the π_x refocusing pulse and the second half of the τ_{cp} delay period. Again, the refocusing pulse and delay do not restore the phase coherence of the magnetization vectors. During additional refocusing periods, there is further loss of coherence, but the center of the distribution of magnetization vectors is returned to the x'-axis.

The process leading to the spreading of magnetization vectors described above is formally similar to the random walk that was discussed in Section 7.3.4 in the context of transverse relaxation due to molecular tumbling. As in that case, the result here is a Gaussian distribution of precession angles, as indicated schematically in Fig. 8.27.

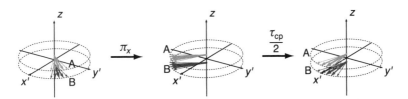

Figure 8.27 Vector diagrams representing the π_x refocusing pulse and precession during the second half of a CPMG delay period, for nuclei in two populations of molecules that begin the experiment in alternate conformations, A and B, for the case where $k_{ex}\Delta\omega > 1$ and $k_{ex}\tau_{cp} \gg 1$.

The width of this distribution, and the resulting loss of net transverse magnetization, depends on the average number of random changes in precession frequency and the average extent to which vectors representing the alternate conformations diverge between exchange events. Analysis of the situation here is a bit more complex than the one for molecular tumbling, because the probabilities of the alternate conformations, specified by p_a and p_b, are not necessarily equal.

We begin by defining the average precession frequency, $\bar{\omega}$, which is given by

$$\bar{\omega} = p_a\omega_a + p_b\omega_b$$

If an individual molecule is in the A conformation, the magnetization associated with the nucleus in that molecule will diverge from the mean position of the distribution at the following rate:

$$\omega_a - \bar{\omega} = \omega_a - (p_a\omega_a + p_b\omega_b) = (1 - p_a)\omega_a - p_b\omega_b = p_b(\omega_a - \omega_b)$$

The average time before a molecule in the A conformation undergoes exchange (to the B conformation) is $1/k_{ab} = 1/(p_b k_{ex})$. The step size, l, in the random walk is the absolute value of the average angular displacement from the mean position, which is calculated as the product of the average time and the absolute value of the rate of divergence:

$$l = \frac{1}{p_b k_{ex}} p_b |\omega_a - \bar{\omega}| = \frac{|\omega_a - \omega_b|}{k_{ex}} = \frac{\Delta\omega}{k_{ex}}$$

The same reasoning applied to molecules in the B conformation yields the same result, so that the average displacement from the mean is independent of the conformation and relative populations of the two states. However, the total number of exchange events in a time interval does depend on p_a and p_b. For an individual molecule, the average total number of exchanges from A to B, n_{ab}, in time, t_1, is proportional to the fraction of time that the molecule is in the A conformation and the rate of exchange from A to B:

$$n_{ab} = t_1 p_a k_{ab} = t_1 p_a p_b k_{ex}$$

Similarly, the number of exchanges from B to A is given by

$$n_{ba} = t_1 p_a p_b k_{ex}$$

and the total number of exchanges is

$$n_t = t_1 2 p_a p_b k_{ex}$$

If we now think of the process as a random walk away from the mean position of the magnetization vectors (which is repeatedly returned to the x'-axis by refocusing pulses), the mean-square angular displacement is calculated as

$$\langle \theta^2 \rangle = n_t l^2 = t_1 2 p_a p_b k_{ex} \left(\frac{\Delta \omega}{k_{ex}} \right)^2 = \frac{t_1 2 p_a p_b \Delta \omega^2}{k_{ex}}$$

Recall from the earlier discussion of random walks that a walk over a given total time period (t_1 in this case), leads to a broader distribution of end points when it is divided up into a smaller number of steps with larger average lengths. In this case, larger steps are implied by a smaller value of k_{ex}. The distribution is also influenced by the equilibrium population, with the widest distribution arising when $p_a = p_b = 0.5$. If one or the other conformation is more favored, the total number of exchanges decreases, but the average deviation in each step remains the same.

Applying the same analysis as presented in Section 7.3.4, we can use the distribution of angular displacements from the x'-axis to calculate the decay of I_x-magnetization with time:

$$\bar{I}_x(t_1) = e^{-\langle \theta^2 \rangle / 2} = e^{-t_1 p_a p_b \Delta \omega^2 / k_{ex}}$$

The rate constant for the exchange contribution to transverse relaxation is then given by

$$R_{ex} = \frac{p_a p_b \Delta \omega^2}{k_{ex}} \tag{8.24}$$

Note that this result, based on the assumption that $k_{ex} > \Delta \omega$, is very different than the one assuming slow exchange (Eq. 8.23, p. 232). For slow exchange, R_{ex} is independent of $\Delta \omega$ and increases with increasing k_{ex} (provided that the slow-exchange criterion, $k_{ex} < \Delta \omega$, remains satisfied). For fast exchange, R_{ex} increases in proportion to $\Delta \omega^2$ and decreases with increasing k_{ex}. In this treatment, however, we have also assumed that τ_{cp} is long enough with respect to k_{ex} that the molecule has an opportunity to undergo multiple exchange events between refocusing pulses. If τ_{cp} is made sufficiently short, the exchange contribution can be decreased.

8.6.3   Mathematical Expressions for the Full Range of Chemical Exchange Rates

A complete analysis of the effects of chemical exchange on transverse relaxation rates under the full range of exchange rates and values for the refocusing delay time is rather involved and will not be presented here, but the final expressions can be written in the following relatively compact, if not very simple, form:

$$R_{ex} = \frac{1}{2} \left[k_{ex} - \frac{1}{\tau_{cp}} \cosh^{-1} \left(D_+ \cosh(\eta_+) - D_- \cos(\eta_-) \right) \right] \tag{8.25}$$

Figure 8.28

Calculated values of R_{ex}, the exchange contribution to transverse relaxation, as a function of the exchange rate constant, k_{ex}, for different values of $\Delta\omega$, the angular Larmor frequency difference for a nucleus interconverting between two environments. Calculated from Eq. 8.25, with $\tau_{cp} = 0.1$ s and $p_a = 0.6$.

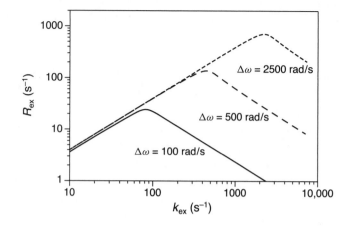

where cosh is the hyperbolic cosine function:[3] $\cosh^{-1}$ is the inverse of that function; and the parameters D_+, D_-, η_+ and η_- are defined as follows:

$$D_+ = \frac{1}{2}\left(\frac{\psi + 2\Delta\omega^2}{\sqrt{\psi^2 + \zeta^2}} + 1\right) \qquad D_- = \frac{1}{2}\left(\frac{\psi + 2\Delta\omega^2}{\sqrt{\psi^2 + \zeta^2}} - 1\right)$$

$$\eta_+ = \frac{\tau_{cp}}{\sqrt{2}}\sqrt{\sqrt{\psi^2 + \zeta^2} + \psi} \qquad \eta_- = \frac{\tau_{cp}}{\sqrt{2}}\sqrt{\sqrt{\psi^2 + \zeta^2} - \psi} \qquad (8.26)$$

$$\psi = k_{ex}^2 - \Delta\omega^2 \qquad \zeta = 2\Delta\omega k_{ex}(p_b - p_a)$$

The derivation of the expressions above assumes that $p_a > p_b$ and that the individual R_2 values for the two conformations, in the absence of exchange, are equal. Unfortunately, the complexity of these expressions tends to obscure the relationship between R_{ex} and the various parameters, but Eq. 8.23 can be evaluated numerically, and considerable insight can be gained from plots of R_{ex} as a function of k_{ex}, $\Delta\omega$ and τ_{cp}.

In Fig. 8.28 calculated values of R_{ex} are plotted as a function of the chemical exchange rate, k_{ex}, for different values of $\Delta\omega$, assuming a relatively long refocusing delay, with $\tau_{cp} = 0.1$ s. Consistent with the arguments presented in the previous pages, R_{ex} increases with k_{ex}, for a given value of $\Delta\omega$, until $k_{ex} \approx \Delta\omega$, and then decreases. This behavior corresponds to the effects of exchange on spectral line widths, discussed in Section 7.4, and defines the fast and slow time scales.

Although the time scale is defined by the relationship between k_{ex} and $\Delta\omega$, the influences of these parameters on R_{ex} are not fully reciprocal, as can be seen by plotting R_{ex} as a function of $\Delta\omega$ (Fig. 8.29). In this plot, the slow-exchange regime is represented on the right-hand side, where $k_{ex} < \Delta\omega$, and fast exchange is on the left. Notice that R_{ex} increases with $\Delta\omega$ in the fast-exchange regime, but plateaus once the

---

[3] The hyperbolic cosine (cosh) and hyperbolic sine (sinh) functions are analogs of the conventional trigonometric functions, but are based on the points that lie on a rectangular hyperbola for which $x^2 - y^2 = 1$ (and $x > 0$), rather than on a circle, for which $x^2 + y^2 = 1$. The cosh and sinh functions are related to the exponential function according to $\cosh(x) = \frac{1}{2}(e^x + e^{-x})$ and $\sinh(x) = \frac{1}{2}(e^x - e^{-x})$. The hyperbolic tangent is $\tanh(x) = \sinh(x)/\cosh(x)$.

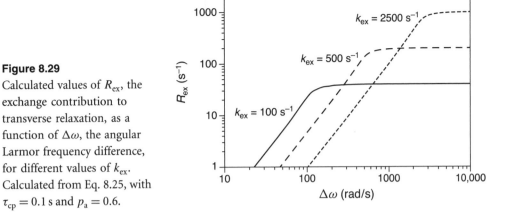

Figure 8.29

Calculated values of R_{ex}, the exchange contribution to transverse relaxation, as a function of $\Delta\omega$, the angular Larmor frequency difference, for different values of k_{ex}. Calculated from Eq. 8.25, with $\tau_{cp} = 0.1$ s and $p_a = 0.6$.

slow-exchange regime is entered, as predicted from Eqs. 8.23 and Eqs. 8.24 (pp. 232 and 235).

As shown by the plots in Figs. 8.28 and 8.29, R_{ex} depends on both k_{ex} and $\Delta\omega$ in the fast-exchange regime, where $k_{ex} > \Delta\omega$. Under these conditions, Eq. 8.25 is well approximated by a much simpler expression:

$$R_{ex} = \frac{p_a p_b \Delta\omega^2}{k_{ex}} \left(1 - \frac{2 \tanh(k_{ex}\tau_{cp}/2)}{k_{ex}\tau_{cp}} \right) \tag{8.27}$$

where tanh is the hyperbolic tangent function. This function can be written in terms of exponentials as

$$\tanh(x) = \frac{e^x - e^{-x}}{e^x + e^{-x}}$$

This expression shows that $\tanh(x)$ approaches 1 as x increases, and Eq. 8.24 approaches a limiting form,

$$R_{ex} = \frac{p_a p_b \Delta\omega^2}{k_{ex}}$$

as the product $k_{ex}\tau_{cp}$ increases beyond about 50. This result is the same as Eq. 8.24, which was derived under equivalent assumptions—that is, fast exchange relative to both $\Delta\omega$ and the CPMG delay time, τ_{cp}.

8.6.4   The Effects of Shorter Refocusing Delays

In the examples considered so far, the refocusing delay time was assumed to be significantly longer than the average time between exchange events. Now, we consider what happens when τ_{cp} is made shorter, so that the molecules may not, on average, have time to change conformation before the magnetization vectors are refocused.

In Fig. 8.30, calculated values of R_{ex} are plotted as a function of k_{ex}, assuming different values of τ_{cp} but the same value of $\Delta\omega$, 500 rad/s. As shown in the graph,

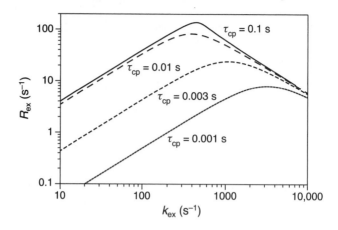

Figure 8.30

Calculated values of R_{ex}, the exchange contribution to transverse relaxation, as a function of the chemical exchange rate, k_{ex}, for different values of τ_{cp}. Calculated from Eq. 8.25, with $\Delta\omega = 500$ rad/s and $p_a = 0.6$.

the general shape of the curves remain the same with different values of τ_{cp}, with R_{ex} initially increasing with k_{ex} and then decreasing. As τ_{cp} decreases, however, the transition between the two parts of the curve becomes more gradual, and the position of the transition along the k_{ex}-axis moves to the right.

For the examples plotted in Fig. 8.30, where $\Delta\omega = 500$ rad/s, R_{ex} is largely unaffected by decreasing the refocusing delay time until τ_{cp} is made less than about 0.01 s, which corresponds to $\tau_{cp}\Delta\omega = 5$ rad—that is, just under the minimum value of 2π rad specified for the example illustrated in Figs. 8.24 and 8.25. The vector diagrams in Fig. 8.31 illustrate what happens when τ_{cp} is made short enough to limit the divergence of magnetization vectors representing the two conformations. In the case illustrated in Fig. 8.31, the total divergence at the end of the refocusing period is about $\pi/2$ radians and all of the vectors have significant projections onto the positive x'-axis. As a consequence, there is much more residual magnetization at the end of the delay than in the example shown in Figs. 8.24 and 8.25, where the vectors diverge by about 2π rad, resulting in a lower apparent relaxation rate, as plotted in Fig. 8.30.

A second consequence of shorter refocusing delay times, as illustrated in Fig. 8.30, is that R_{ex} continues to increase with k_{ex} into the region where $k_{ex} > \Delta\omega$, as if the

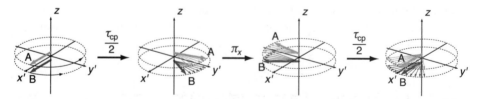

Figure 8.31 Vector diagrams representing the precession of nuclei in two populations of molecules that begin in alternate conformations, during a single CPMG refocusing cycle. The molecules undergo slow exchange, with $k_{ex} < \Delta\omega$, and τ_{cp} is relatively short, so that the divergence of the vectors is significantly less than 2π rad. The two sets of vectors, labeled A and B and drawn above and below the $x'-y'$ plane, represent the populations that begin in the two conformations.

slow-exchange regime were extended, or $\Delta\omega$ were made larger. In this "extended slow-exchange" region, $\tau_{cp}k_{ex} < 1$, so that the molecules undergo, on average, less than one exchange event during the delay time. As a consequence, the two conformations do not have time to equilibrate, and the magnetization vectors representing the molecules that do not undergo exchange are effectively refocused to the x'-axis, as in the examples of slow exchange illustrated in Figs. 8.24 and 8.25. As in those cases, R_{ex} is directly proportional to k_{ex}, until k_{ex} significantly exceeds $1/\tau_{cp}$, at which point R_{ex} begins to decrease, as expected in the fast-exchange regime.

8.6.5   Experimental Characterization of Exchange Processes

Upon carrying out a heteronuclear relaxation study, it is not at all uncommon to find that some nuclei display transverse relaxation rates that are markedly higher than others in similar functional groups in the same molecule or are higher than expected based on the size of the molecule. An example is illustrated in Fig. 7.31 (p. 190) for the backbone amide ^{15}N nuclei of a small protein. In such a case, the elevated R_2 values are very likely due to chemical exchange, and it is of interest to characterize the underlying motions. From the discussion above, however, it should be apparent that the relationship between the relaxation rate and the rate of the physical exchange process, k_{ex}, is not at all simple. In particular, the exchange contribution to R_2 depends on the difference between the Larmor frequencies of the nucleus in the alternate states, as well as the relative populations of the two (or more) states and the refocusing delay time. Disentangling these factors can be difficult, but, in favorable cases, it may be possible to place bounds on some of the parameters.

One of the first questions to be asked is whether the exchange process lies in the slow-, intermediate- or fast-exchange regime. In some cases, two peaks in a spectrum may be shown to arise from the same nucleus (usually by means of multidimensional spectra). Such an observation is consistent with slow exchange, but it may also arise from covalent heterogeneity, which can be difficult to rule out entirely. When only a single peak can be identified for a nucleus, but high R_2 values indicate chemical exchange, it is tempting to assume fast exchange, but it is also possible that a slow-exchange process is involved, as it may be difficult to detect the peak from an alternate conformation, especially if the populations are highly skewed (as illustrated in Fig. 7.27, p. 187) or there is significant overlap in the spectrum.

One important means of characterizing chemical or conformational exchange processes is to measure the transverse relaxation rate using spectrometers with different magnetic field strengths. As discussed earlier, the exchange contribution is, under some conditions, proportional to the square of $\Delta\omega$, which itself is proportional to the external field strength. The effect on R_{ex} is illustrated in Fig. 8.32 for a case involving an ^{15}N nucleus, assuming that the Larmor frequencies in the alternate conformations differ by 1 ppm and $\tau_{cp} = 0.1$ s. If, as assumed in Fig. 8.32, R_{ex} is measured at field strengths of 11.7 and 18.8 T, the measured value is predicted to be about 2.5-fold greater at the higher field than the lower, if the fast-exchange condition ($k_{ex} > \Delta\omega$) is satisfied. On the other hand, R_{ex} is insensitive to field strength

Figure 8.32

Calculated values of R_{ex} as a function of the chemical exchange rate, k_{ex}, for two different values of the external field strength, corresponding to ^{1}H Larmor frequencies of 500 and 800 MHz (11.7 and 18.8 T). Calculated from Eq. 8.25, with $\Delta\omega$ proportional to 1 ppm at each field strength, $\tau_{cp} = 0.1$ s and $p_a = 0.6$.

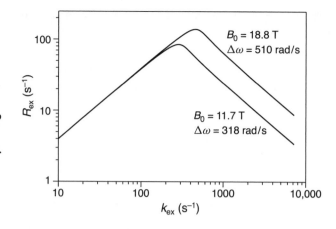

in the slow-exchange regime, *provided* that $\tau_{cp}\Delta\omega > 2\pi$ at both fields, as it is in this example (where $\tau_{cp} = 0.1\,\mathrm{s}^{-1}$).

An example of the possible interplay between the refocusing delay time and field strength is illustrated in Fig. 8.33, in which R_{ex} is plotted using the same parameters as used in the previous figures, except that τ_{cp} is assumed to be $0.01\,\mathrm{s}^{-1}$ for the dashed curves. As in the previous example, R_{ex} is decreased by about 2.5-fold in the fast-exchange regime when the field strength is lowered from 18.8 to 11.7 T, but with the shorter delay time, R_{ex} is also reduced in the slow-exchange regime. The reason for the latter effect can be found by examining the product $\tau_{cp}\Delta\omega$. When $\tau_{cp} = 0.01$ s, the product equals 5.1 rad at 18.8 T and 3.2 radians at 11.7 T. This difference corresponds, approximately, to the different situations depicted in Figs. 8.25 and 8.31. The shorter delay time in the latter situation limits the dispersion of the magnetization vectors and, thereby, leads to a smaller relaxation rate.

These examples show that an increase in R_{ex} with greater external field strength is not unambiguous evidence for fast exchange. Nonetheless, the magnitude of the difference is generally larger in the fast-exchange regime than for slow exchange, as

Figure 8.33

Calculated values of R_{ex} as a function of the chemical exchange rate, k_{ex}, for two different values of the external field strength. Calculated from Eq. 8.25, with $\Delta\omega$ proportional to 1 ppm at each field strength and $p_a = 0.6$. The solid curves represent the calculated exchange contribution to R_2 with $\tau_{cp} = 0.1$ s, whereas $\tau_{cp} = 0.01$ s for the dashed curves.

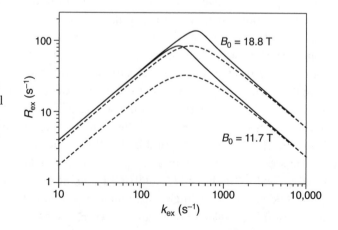

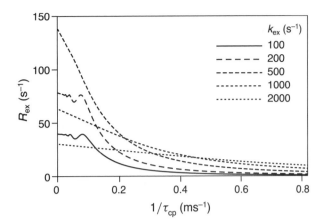

Figure 8.34

Simulated relaxation dispersion curves, in which R_{ex} is plotted as a function of $1/\tau_{cp}$. Calculated from Eq. 8.25, with $\Delta\omega = 500$ rad/s and $p_a = 0.6$, for different values of k_{ex}, as indicated by the key.

illustrated in the example shown in Fig. 8.33, and it can be shown that a change proportional to $\Delta\omega^2$ is, indeed, indicative of fast exchange.

Measuring R_{ex} using a wide range of refocusing delay times, especially when combined with measurements at two or more stationary field strengths, is particularly useful in characterizing chemical exchange processes. Such studies are commonly referred to as *relaxation dispersion* measurements, though the origins of this expression are somewhat obscure. More generally, the term *dispersion* has long been used in the context of NMR relaxation to describe the effect of magnetic field strength, or resonance frequency, on relaxation rates, and the refocusing delay time in the CPMG experiment can be interpreted as a frequency, $\nu_{cp} = 1/\tau_{cp}$. Indeed, the results of such experiments are typically shown as a plot of R_{ex} versus $1/\tau_{cp}$. Simulated dispersion curves are shown in Fig. 8.34 for different values of k_{ex}.

Each of the curves in Fig. 8.34 shows a decrease in R_{ex} as $1/\tau_{cp}$ is increased, but the magnitudes of the changes and the shapes of the curves depend on the value of k_{ex}. For a nucleus in slow exchange ($k_{ex} < \Delta\omega$), the dispersion curve is approximately flat until the product $\tau_{cp}\Delta\omega$ is less than about 2π rad ($1/\tau_{cp} \gtrsim 0.1$ ms in this example), at which point R_{ex} decreases sharply. This corresponds to the left-hand portion of the plot in Fig. 8.30. A curve with this shape thus identifies a nucleus in slow exchange and allows an estimate of $\Delta\omega$ to be made, usually by fitting the data to Eq. 8.25. If the peaks corresponding to both of the alternate states (assuming that there are only two) can be identified, the values of p_a, p_b and k_{ex} can also be estimated. If only one peak is identified, however, there is likely to be some ambiguity about the relative populations and exchange rate.

For intermediate or fast exchange ($k_{ex} \geq 500$ s^{-1} in Fig. 8.34), dispersion curves display a smooth decrease of R_{ex} with increased $1/\tau_{cp}$. The larger the value of k_{ex}, the higher $1/\tau_{cp}$ must be in order to cause a proportional decrease in R_{ex}, as also shown on the right-hand side of Fig. 8.30. For nuclei undergoing fast exchange, fitting dispersion data to Eq. 8.27, or related equations based on simplifying assumptions, can provide estimates of k_{ex} and the product, $p_a p_b \Delta\omega^2$. In practice, however, it may not be possible to make τ_{cp} short enough to observe a significant decrease in τ_{cp}, in which case only a lower limit can be placed on k_{ex}. Cryogenic probes, in particular, are susceptible to heat damage when the spacing between pulses is made much shorter than about 1 ms (circa 2015), placing a limit of about 1000 s^{-1} on the values of k_{ex} that can be determined in this way.

Even with the strategies described above, analysis of the exchange contribution to transverse relaxation often leaves some of the parameters undetermined, especially $\Delta\omega$ and the relative populations of interconverting states. Nonetheless, transverse relaxation rate measurements are currently one of the most effective means of characterizing conformational changes on the μs to ms time scale, a time scale that is of particular interest in the study of enzyme mechanisms and other macromolecular functions.

8.7   Relaxation Interference and TROSY

In previous sections of this chapter, the contributions of dipolar interactions and chemical shift anisotropy (CSA) to the relaxation of a heteronuclear pair were treated as though they were independent of one another. This implicit assumption is generally valid when the relaxation rates are averaged over an entire population, as is done in most NMR experiments. When we look at individual molecules, however, it becomes apparent that the two mechanisms are not necessarily independent, a fact that can be exploited to enhance the resolution of heteronuclear spectra.

Recall from Section 4.1.4 that the chemical shift for the ^{15}N nucleus in a peptide bond depends on the orientation of the bond with respect to the external magnetic field and is maximal when the N–H bond vector is nearly parallel with the external field, as shown in Fig. 8.35A. Thus, the local magnetic field experienced by the nucleus is greatest in this orientation. As the molecule tumbles in solution, the local magnetic field fluctuates and is minimal in the orientation shown in Fig. 8.35B.

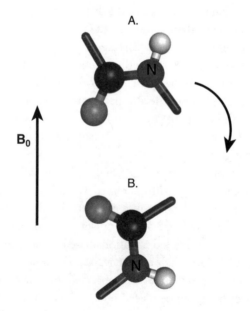

Figure 8.35
A peptide bond in the two orientations, relative to the external field, that give rise to the maximum (A) and minimum (B) chemical shifts.

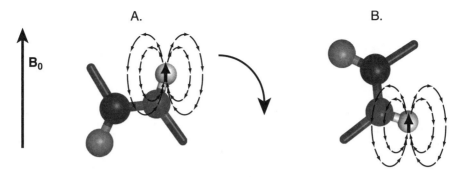

Figure 8.36 Influence of the ^{1}H dipole on the magnetic field experienced by the ^{15}N nucleus, with the ^{1}H nucleus in the $|\alpha\rangle$ (spin-up) state.

However, the local field is also influenced by the spin state of the ^{1}H nucleus. If, for instance, the ^{1}H nucleus is in the $|\alpha\rangle$ state, its influence on the ^{15}N nucleus will add to that of the CSA, as illustrated in Fig. 8.36 for the two molecular orientations. Importantly, the orientations of the nuclear dipoles remain fixed as the molecule rotates (until they are changed by irradiation or relaxation processes). In orientation A, the dipolar interaction, indicated by the field lines, felt by the ^{15}N nucleus adds to the external field, while in orientation B, the field from the ^{1}H nucleus is in the opposite direction. Note that changes in the CSA have the same sign as the molecule tumbles: In orientation A, the contribution from CSA is positive and maximal, while in orientation B this contribution is minimal. For this particular molecule, then, the two contributions add together in a way such that the total relaxation rate is greater than that expected for the average of the individual contributions.

Consider, now, a molecule in which the ^{1}H nucleus is in the $|\beta\rangle$ state, so that its dipole is aligned *against* the external field, as illustrated in Fig. 8.37. In this case, the dipolar interaction with the ^{1}H nucleus reduces the effective field when the molecule is in orientation A and increases the field in orientation B. The dipolar interaction and the CSA now act in opposition to one another, reducing the magnitude of the

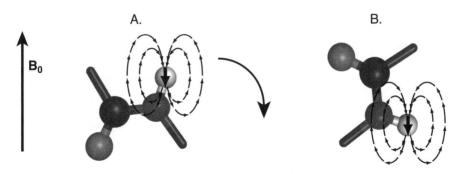

Figure 8.37 Influence of the ^{1}H dipole on the magnetic field experienced by the ^{15}N nucleus, with the ^{1}H nucleus in the $|\beta\rangle$ (spin-down) state.

Figure 8.38
An example of differential line widths due to interference between relaxation by dipolar coupling and chemical shift anisotropy. The up-field peak in each doublet is narrower than the down-field peak. (Adapted from Kaun, E., Rüterjans, H. & Hull, W. E. (1982). *FEBS Lett.*, 141, 217–221.)

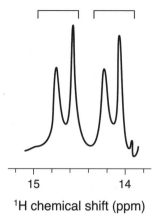

^{1}H chemical shift (ppm)

fluctuations in the local magnetic field experienced by the ^{15}N nucleus. As a consequence, the relaxation rate is smaller for this molecule than for the one in which the ^{1}H nucleus is in the $|\alpha\rangle$ state.

If the relaxation rates are measured collectively for all of the molecules in a sample, then the effects described above will be hidden, and the observed rates will represent an average for the two kinds of molecules. However, the state of the ^{1}H nucleus will, via scalar coupling, also influence the resonance frequency of the ^{15}N nucleus. If the two resulting peaks in the ^{15}N spectrum are resolved, they may have different line widths, depending on the relative contributions to the transverse relaxation rate. This effect, described as *relaxation interference*, was observed in the 1950s (in ESR experiments) and was rediscovered in different contexts over the next few decades.

The phenomenon remained rather obscure, however, until the early 1980s, when it was observed in ^{1}H spectra of ^{15}N-labeled RNA molecules. An example is shown in Fig. 8.38; a portion of an ^{1}H spectrum of ^{15}N-labeled tRNA recorded in a 500-MHz spectrometer. This region of the spectrum contains two doublets, with the components separated by the H-N scalar coupling constant, $\approx$90 Hz. Note that the downfield component in each doublet is significantly broader than the upfield component. Although a variety of explanations were initially proposed for this pattern, the correct one, relaxation interference, was suggested by Alfred Redfield, and a detailed theoretical analysis was developed by Maurice Goldman. As discussed further below, the particular circumstances required to see the effect were (1) a relatively large molecule with substantial line broadening, (2) labeling of the nucleic acid with ^{15}N and (3) a high magnetic field strength. Prior to the early 1980s, these conditions were unlikely to have been encountered.

A full treatment of relaxation interference is rather involved, but the magnitude of the effect can be seen by considering the relaxation interference rate constant, η, which represents the difference between the rates observed for the two components of the scalar-coupled doublet. Because of its importance to the apparent line widths, we will focus our attention on the transverse relaxation rate, R_2. The relaxation interference contribution to R_2, for the S-spin, is given by

$$\eta = -\frac{\sqrt{3}cd}{30}(3\cos^2\theta - 1)(4J(0) - 3J(\omega_S))$$

where d and c represent the magnitudes of the dipolar coupling and CSA, respectively, as defined by Eqs. 8.14 and 8.17 (pp. 213 and 218). The parameter θ is the angle between the internuclear vector and the principal axis of the CSA tensor. Notice that the term $(3\cos^2\theta - 1)$ is maximal when $\theta = 0$—that is, when the internuclear vector and the CSA axis are aligned, and goes to zero when $\theta = 54.7°$ (the "magic angle" discussed in Section 4.2.2). In the case of the amide ^{15}N nucleus, $\theta \approx 18°$, and $(3\cos^2\theta - 1) \approx 1.85$, close to its maximal possible value of 2.

An important feature of the interference effect is that it depends strongly on the strength of the external magnetic field, through its effect on the CSA term, c, which is proportional to the Larmor frequency of the S-spin (Eq. 8.17, p. 218). In Fig. 8.39, the value of η for the case of the amide ^{15}N nucleus is plotted as a function of the field strength, along with the individual contributions of dipolar coupling (R_2^{dd}) and chemical shift anisotropy (R_2^{csa}) to transverse relaxation. These curves were calculated assuming a molecular correlation time of 10 ns (as expected for a globular protein of about 20,000 Da) and neglecting any contribution from internal motions. Notice that the contribution from dipolar coupling actually decreases slightly at higher magnetic fields. This is because, as the Larmor frequencies of the nuclei increase, the values of the spectral density function at those frequencies decrease, thus making the longitudinal relaxation processes (which contribute to R_2) less efficient. In contrast, the contribution from CSA increases dramatically with field strength, as does the interference term, η.

When the two components of a doublet are examined individually, the interference term either adds to the relaxation rate or decreases it, as shown in the graphs of Fig. 8.40. For the component for which the interference term is positive, R_2 is nearly twice as large as the average value, even at relatively low magnetic fields. (10 T corresponds to a ^1H Larmor frequency of approximately 425 MHz.) On the other hand, the relaxation rate constant for the other component is less than half of the average at 10 T and decreases to a minimum value at a field strength of about 23 T (980 MHz).

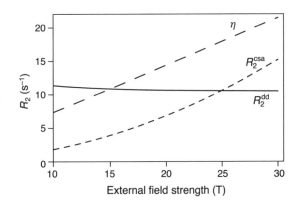

Figure 8.39

Contributions to the transverse relaxation rate constant of a peptide ^{15}N nucleus from dipolar coupling, chemical shift anisotropy and interference between these mechanisms, as a function of external field strength. The correlation time for tumbling is assumed to be 10 ns.

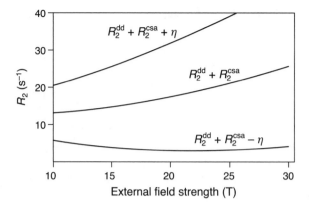

Figure 8.40   Transverse relaxation rate constants for a peptide ^{15}N nucleus. The top curve, $R_2^{dd} + R_2^{csa} + \eta$, represents the relaxation rate constant of the population with the ^{1}H nucleus in the spin-up state. The bottom curve, $R_2^{dd} + R_2^{csa} - \eta$, represents the rate constant for the population with the ^{1}H nucleus in the spin-down state. The middle curve is the average, as would be measured if the signals from the two populations were not resolved.

Although this phenomenon was well understood at least 10 years earlier, the practical significance was not fully grasped until about 1996, when Pervushin et al. described a method in which the interference effect was used to generate two-dimensional ^{1}H–^{15}N spectra with greatly reduced line widths. The authors gave the technique the name TROSY, for Transverse Relaxation Optimized SpectroscopY. In essence, the basic TROSY spectrum is a variation of the ^{1}H–^{15}N HSQC (Heteronuclear Single-Quantum Coherence) spectrum, discussed briefly in Chapter 1 and in more detail in Chapter 16, that is filtered in such a way that only the narrow components of the doublets appear in the final spectrum. This advance has been particularly important in the application of high-resolution NMR to larger proteins and nucleic acids. In addition, the underlying principle of TROSY has been incorporated into three-dimensional spectra and relaxation measurements, allowing these methods to be used with larger molecules.

In practice, the benefits of TROSY are not quite as dramatic as might be implied from the figures above. One important reason for this is that the transverse relaxation rate of each nucleus depends on the longitudinal relaxation rate of the other. As discussed above, cancellation of the dipolar and CSA contributions to the relaxation of the ^{15}N nucleus requires that the ^{1}H nucleus be in the $|\beta\rangle$ state (i.e., the high energy state). With time, a fraction of the ^{1}H spins that begin in this state will be converted to the $|\alpha\rangle$ state, leading to much faster transverse relaxation. Because the R_1 rate for the ^{1}H nucleus depends primarily on dipolar interactions with other protons, this effect can be minimized by replacing ^{1}H with ^{2}H in the molecule. A common strategy is to prepare protein or nucleic acid samples from bacteria or yeast grown in a medium containing ^{2}H and then allowing the amide deuterium atoms to exchange with ^{1}H in ordinary water. In this way, the amide protons can be detected in the ^{1}H NMR spectra but their relaxation rates are minimized.

Summary Points for Chapter 8

- The mathematical description of molecular tumbling:
  - Molecular tumbling can be described as a Wiener process, composed of small random changes in orientation.
  - The autocorrelation function describes the change in a parameter, such as the orientation of a molecule, over time.
  - The spectral density function is the real part of the Fourier transform of the autocorrelation function and describes the distribution of motions with different frequencies.
  - For random rotational tumbling, the autocorrelation function has the form of an exponential decay and the spectral density function is an absorptive Lorentzian function centered at 0 Hz.

- Longitudinal relaxation depends on quantum transitions that are stimulated by magnetic fluctuations with frequencies that are close to those corresponding to the energy differences between the quantum states.

- Spin pairs interacting via dipolar interactions can undergo both independent and concerted quantum transitions. These transitions are stimulated by molecular motions of the appropriate frequencies and are the major mechanism of relaxation in liquid samples.

- A secondary mechanism of relaxation in liquids is magnetic fluctuations due to tumbling and chemical shift anisotropy.

- For simple systems, the rates of longitudinal and transverse relaxation due to dipolar coupling and chemical shift anisotropy can be calculated from the spectral density function.

- Measurements of relaxation rates for covalently linked heteronuclear pairs, such as ^1H–^{13}C and ^1H–^{15}N, are particularly useful for characterizing internal motions of molecules.

- The "model-free" analysis of Lipari and Szabo is widely used to analyze heteronuclear relaxation rates in terms of global and internal motions. This treatment is based on an assumed form of the spectral density function with three parameters that are adjusted to fit the experimental data.

- Chemical or conformational transitions that occur on time scales longer than that of molecular tumbling can cause transient changes in precession frequencies and contribute to transverse relaxation. Measurements of the exchange contribution to R_2, R_{ex}, made at different external field strengths and different CPMG refocusing delays, can be used to characterize exchange processes and their rates.

- Under some conditions, the magnetic fluctuations due to dipolar coupling and chemical shift anisotropy can cancel one another, leading to the phenomenon of relaxation interference, which is exploited in TROSY experiments to reduce transverse relaxation rates.

Exercises for Chapter 8

8.1. Eq. 8.2 gives the probability distribution function for $\Delta\theta$, the change in orientation of a particle undergoing rotational Brownian motion, over a time period, in terms of the mean-squared value of $\Delta\theta$ over the given time period. (The equation is based on our simplifying assumption of rotation in only two dimensions, but the following exercises will give you a sense of the relative probabilities over time.)

(a) Using other relationships from Section 8.2, rewrite Eq. 8.2 in terms of the rotational correlation time, τ_c, and the length of the time period, Δt.

(b) Consider a molecule with a correlation time of 10 ns, a typical value for a medium-sized protein in water at room temperature. Draw a series of plots of the probability distribution function, $p(\Delta\theta)$, for time intervals of 1, 10 and 100 ns.

(c) To give some physical context to your plot, add to the horizontal axis drawings of a simple asymmetric object (such as an arrow) that has rotated by 0, 1, $\pi/2$ and π rad.

(d) Somewhat arbitrarily, we might say that a particle has retained its original orientation over a time interval if $\Delta\theta$ lies between –0.1 and 0.1 rad. Again assuming that $\tau_c = 10$ ns, calculate the probability that the molecule lies within this range of orientations after 1, 10 and 100 ns.

(e) Calculate the probability that the particle has rotated by more than 1 rad (approximately 57°) in either direction after 1, 10 and 100 ns.

(f) Calculate the probability that the particle has rotated by more than π rad in either direction after 1, 10 and 100 ns.

Note: Parts d, e and f require integrating the Gaussian probability distribution function over defined intervals. Sadly, these integrals do not have analytical solutions, and you will need to find a way to approximate them. A bit of Internet searching should lead you to a variety of ways to do this.

8.2. Suppose that you were told that the autocorrelation function for a particular motion (not necessarily that of an isolated molecule) had the following form:

$$g(\Delta t) = 0.5 + 0.5e^{-\Delta t/\tau_c}$$

where τ_c is a time constant for the motion.

(a) What can you conclude about the physical nature of the motion described by this autocorrelation function?

(b) It turns out that the spectral density function corresponding to this autocorrelation function has the same form as that for Brownian rotational motion:

$$J(\omega) = \frac{\tau_c}{1 + \omega^2\tau_c^2}$$

How can you explain the fact that the autocorrelation functions for the two types of motion are different, but the spectral density functions are the same?

8.3. In the treatment of relaxation due to dipolar interactions, much of the emphasis in this chapter was placed on the influence of the overall tumbling rate on the longitudinal and transverse relaxation rates, as highlighted in Fig. 8.11. Other parameters that appear in the expressions for calculating R_1 and R_2 are the gyromagnetic ratios of the nuclei (γ), the distance separating the nuclei (r) and the stationary field strength (B_0), which, together with the gyromagnetic ratios, determines the Larmor frequencies.

(a) Using relationships from this chapter, prepare plots of R_1 and R_2 as functions of the internuclear distance (over the range of 2 to 10 Å) between two ^{1}H nuclei. For this plot, assume that $\tau_c = 5$ ns and that the stationary external field strength is 12 T. What practical conclusion can you draw from this plot?

(b) Prepare log–log plots of R_1 and R_2 as functions of the correlation time for tumbling (τ_c), analogous to Fig. 8.11, but for homonuclear pairs of ^{13}C and ^{15}N nuclei, rather than a ^{1}H–^{1}H pair. In each case, assume that the internuclear distance is 2 Å. How does the gyromagnetic ratio affect the relaxation rates for molecules of different sizes?

(c) Prepare log–log plots of R_1 and R_2 as functions of τ_c, for a ^{1}H–^{1}H pair separated by 2 Å, as in Fig. 8.11, but for B_0 equal to 10, 15 and 20 T. What practical conclusions can you draw from these plots?

8.4. In anticipation of discussing the nuclear Overhauser effect (NOE) in Chapter 9, consider the behavior of a population of dipolar-coupled spin pairs when the magnetization of one spin has been inverted, but the other has not.

(a) Suppose that the two spins, I and S, are of the same type, with a positive gyromagnetic ratio (e.g., ^{1}H), and are both initially in the $|\alpha\rangle$ state, but the S-spin has been converted to the $|\beta\rangle$ state. Starting with the Solomon equations (Eqs. 8.8 and 8.9 on p. 210), write expressions for the rate of change in the z-magnetization of the two spins immediately after inversion of the S-spin. Write these expressions in terms of the equilibrium I_z-magnetization component ($\bar{I}_z^{\,0}$) and the transition probabilities (W_0, W_1 and W_2), which will be the same for both spins.

(b) What can you conclude about the direction in which the average S-magnetization will change immediately after its inversion?

(c) What can you conclude about how the average I-magnetization will change immediately after the selective inversion of the S-spin? Does this result surprise you? Why or why not?

(d) Write the expression for the initial rate of change in I_z-magnetization in terms of the angular Larmor frequency, ω_I, and the correlation time for tumbling, τ_c.

(e) From the expression derived in the previous part, determine the value of τ_c for which the I-magnetization will *not* change immediately following inversion of the S-spin.

8.5. As shown in Fig. 8.11, when τ_c is greater than the Larmor frequencies of a dipolar-coupled spin pair, the transverse relaxation rate, R_2, is greater than the longitudinal relaxation rate, R_1, and the two rates diverge further from one another as τ_c increases. The same qualitative pattern is seen for an ^{15}N nucleus in an amide bond, for which the relaxation rates are determined primarily by the dipolar inter-action with a covalently bonded hydrogen and by chemical shift anisotropy, as given by Eqs. 8.18 and 8.19 (p. 219). Under certain conditions, the ratio of R_2 and R_1 can be used to estimate the value of τ_c.

For this exercise, assume that the relaxation of a ^{15}N nucleus is determined only by dipolar interactions and chemical shift anisotropy. Assume further that the motions of the amide group are well described by the Lipari–Szabo spectral density function, with internal motions fast enough that the spectral density function is well approxi-mated by Eq. 8.22 (p. 227).

Eqs. 8.18 and 8.19 show that R_1 and R_2 are determined by $J(\omega)$ evaluated at the angular Larmor frequencies of the ^1H and ^{15}N nuclei (ω_I and ω_S, respectively) and at the frequencies 0, $\omega_I - \omega_S$ and $\omega_I + \omega_S$. For ^1H and ^{15}N, ω_I is approximately 10-fold greater than ω_S (in absolute value). Note also that $J(\omega)$ decreases sharply with ω, once the product $\omega\tau_c$ is greater than 1. Provided that S^2 is greater than about 0.5, and $\omega\tau_c$ is greater than 1, then $J(0)$ and $J(\omega_S)$ are much larger than $J(\omega)$ evaluated at ω_I, $\omega_I - \omega_S$ or $\omega_I + \omega_S$.

(a) Assuming that only the terms involving $J(0)$ and $J(\omega_S)$ are significant, write expressions for R_1 and R_2 (starting with Eqs. 8.18 and 8.19) in terms of only S^2, τ_c and ω_S.

(b) Write an expression for the ratio R_2/R_1 and rearrange this expression into an equation for calculating τ_c from the R_2/R_1 ratio.

8.6. The analysis of slow exchange in Section 8.6.1 suggests that the exchange contri-bution to R_2 depends only on the equilibrium populations of the two states (p_a and p_b) and the rate of exchange between the two, k_{ex}, provided that $k_{ex} < \Delta\omega$ and the product $\tau_{cp}\Delta\omega$ is greater than about 2π. Careful examination of the relaxation dis-persion curves in Fig. 8.34 shows, however, that the behavior predicted by Eq. 8.34 is somewhat more complicated: For the illustrated cases of slow exchange ($k_{ex} = 100$ or $200\ s^{-1}$), there is an oscillatory change in R_{ex} as $1/\tau_{cp}$ approaches the value of $0.1\ ms^{-1}$.

Suggest a physical explanation of the behavior described above. It may help to carefully reconsider the situation illustrated in Figs. 8.24 and 8.25 and the assump-tions leading to Eq. 8.23 (p. 232).

Further Reading

Nearly all of the important principles of NMR relaxation theory were first laid out in a classic paper published only a few years after the initial demonstrations of the NMR phenomenon in bulk phases:

◇ Bloembergen, N., Purcell, E. M. & Pound, R. V. (1948). Relaxation effects in nuclear magnetic resonance absorption. *Phys. Rev.*, 73, 679–712. http://dx.doi.org/10.1103/PhysRev .73.679

The Solomon equations were derived a few years later:

◇ Solomon, I. (1955). Relaxation processes in a system of two spins. *Phys. Rev.*, 99, 559–565. http://dx.doi.org/10.1103/PhysRev.99.559

A rigorous treatment of relaxation in liquid samples is presented in the classic text by Abragam:

◇ Abragam, A. (1961). *The Principles of Nuclear Magnetism*, pp. 264–353. Oxford University Press, Oxford.

A more concise discussion of much of the same material:

◇ Goldman, M. (1988). *Quantum Description of High-Resolution NMR in Liquids*, pp. 226–262. Oxford University Press, Oxford.

The theory of relaxation via dipolar coupling is very well treated in this more specialized text on the nuclear Overhauser effect:

◇ Neuhaus, D. & Williamson, M. P. (2000). *The Nuclear Overhauser Effect in Structural and Conformational Analysis*, 2nd. ed., pp. 23–61. Wiley-VCH, New York.

The "model-free" formalism widely used in the analysis of protein dynamics was first described in the following two papers:

◇ Lipari, G. & Szabo, A. (1982). Model-free approach to the interpretation of nuclear magnetic resonance relaxation in macromolecules. 1. Theory and range of validity. *J. Am. Chem. Soc.*, 104, 4546–4559. http://dx.doi.org/10.1021/ja00381a009

◇ Lipari, G. & Szabo, A. (1982). Model-free approach to the interpretation of nuclear magnetic resonance relaxation in macromolecules. 2. Analysis of experimental results. *J. Am. Chem. Soc.*, 104, 4559–4570. http://dx.doi.org/10.1021/ja00381a010

A useful graphical variation on the Lipari–Szabo treatment:

◇ Andrec, M., Montelione, G. T. & Levy, R. M. (2000). Lipari–Szabo mapping: A graphical approach to Lipari–Szabo analysis of NMR relaxation data using reduced spectral density mapping. *J. Biomol. NMR*, 18, 83–100. http://dx.doi.org/10.1023/A:1008302101116

The treatment of the effects of chemical exchange on R_2 as measured by CPMG experiments is described in the following paper, and references therein:

◇ Millet, O., Loria, J. P., Kroenke, C. D., Pons, M. & Palmer III, A. G. (2000). The static magnetic field dependence of chemical exchange linebroadening defines the NMR chemical shift time scale. *J. Am. Chem. Soc.*, 122, 2867–2877. http://dx.doi.org/10.1021/ja993511y

Two reviews on the use of relaxation measurements and other NMR techniques to study dynamics in proteins and nucleic acids:

◇ Palmer III, A. G. (2004). NMR characterization of the dynamics of biomacromolecules. *Chem. Rev.*, 104, 3623–3640. http://dx.doi.org/10.1021/cr030413t

◇ Kleckner, I. R. & Foster, M. P. (2011). An introduction to NMR-based approaches for measuring protein dynamics. *Biochim. Biophys. Acta*, 1814, 942–968. http://dx.doi.org/10.1016/j.bbapap.2010.10.012

A thorough treatment of interference between dipolar coupling and chemical shift anisotropy in relaxation:

◇ Goldman, M. (1984). Interference effects in the relaxation of a pair of unlike spin-1/2 nuclei. *J. Magn. Reson.*, 60, 437–452. http://dx.doi.org/10.1016/0022-2364(84)90055-6

The original description of TROSY:

◇ Pervushin, K., Riek, R., Wider, G. & Wüthrich, K. (1997). Attenuated T_2 relaxation by mutual cancellation of dipole-dipole coupling and chemical shift anisotropy indicates an avenue to NMR structures of very large biological macromolecules in solution. *Proc. Natl. Acad. Sci. USA*, 94, 12366–12371. http://dx.doi.org/10.1073/pnas.94.23.12366

CHAPTER 9

Cross-relaxation and the Nuclear Overhauser Effect

In Chapter 7 we implicitly assumed that the changes in wavefunction that occur during relaxation are essentially independent for individual spins. The more detailed analysis in Chapter 8, however, showed that when relaxation is promoted by dipolar interactions between spins, the two spins influence one another as each undergoes a change in wavefunction, and there can be concerted changes. These concerted changes give rise to the phenomenon of *cross-relaxation* and the very important nuclear Overhauser effect (NOE), by which the distances between nuclei in a molecule can be measured. In this chapter, we consider this phenomenon in some detail and discuss its use in structure determination.

9.1 Two Spins Linked by Dipolar Interactions

To review (and for those who may have chosen to skip the mathematical intricacies of Chapter 8), we consider a molecule containing two independent spins. The possible states of this molecule include four in which both spins are in eigenstates, as diagrammed in Fig. 9.1. All other states can be considered superpositions of these. By convention, the two spins are usually designated I and S, and their Larmor frequencies are ν_I and ν_S. In this chapter, we will restrict ourselves to cases where the two nuclei are of the same type, such as, two protons, within the same molecule. We will also assume that the gyromagnetic ratio is positive, so that the $|\alpha\rangle$ state has the lower energy, and the equilibrium magnetization lies along the positive z-axis. Although the nuclei have slightly different resonance frequencies, the energy differences associated with the transitions between their respective eigenstates, ΔE_I and ΔE_S, are nearly equal. In contrast to situations we will consider later, where the transitions for the two nuclei are linked by scalar (through-bond) coupling, the wavefunctions for the two nuclei are independent. The individual eigenstates are identified as $|\alpha_I\rangle$, $|\beta_I\rangle$, $|\alpha_S\rangle$ and $|\beta_S\rangle$.

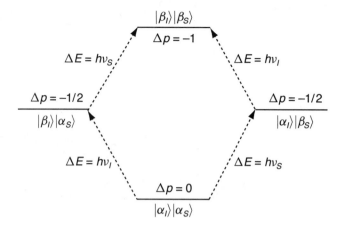

Figure 9.1
Energy diagram for two spin-1/2 nuclei of the same type, labeled I and S. Δp represents the relative population differences of the four states at equilibrium, expressed as a difference from the $|\alpha_I\rangle|\alpha_S\rangle$ population.

The key to understanding the nuclear Overhauser effect lies in a careful accounting of the populations of the various species in the scheme shown above, both at equilibrium and after perturbation by electromagnetic radiation. At equilibrium, the relative concentrations of molecules corresponding to the four states will be determined by the energy differences between them, as specified by the Boltzmann distribution.

For our purposes here, we are primarily concerned about the small *differences* in populations. For instance, we begin by calculating the difference between the number of molecules in the $|\alpha_I\rangle|\alpha_S\rangle$ and $|\alpha_I\rangle|\beta_S\rangle$ states, $N_{\alpha\alpha}$ and $N_{\alpha\beta}$, at equilibrium. The ratio of these numbers is given by the Boltzmann equation:

$$\frac{N_{\alpha\beta}}{N_{\alpha\alpha}} = e^{-\Delta E/kT}$$

where ΔE is the energy difference between the two states, k is Boltzmann's constant and T is the absolute temperature. Because the energy difference is much smaller than kT, the exponential term can be accurately approximated by

$$e^{-\Delta E/kT} \approx 1 - \Delta E/kT$$

The difference between $N_{\alpha\alpha}$ and $N_{\alpha\beta}$ is then calculated as follows:

$$\frac{N_{\alpha\beta}}{N_{\alpha\alpha}} = 1 - \Delta E/kT$$

$$N_{\alpha\beta} = N_{\alpha\alpha} - N_{\alpha\alpha}\Delta E/kT$$

$$N_{\alpha\alpha} - N_{\alpha\beta} = N_{\alpha\alpha}\Delta E/kT$$

Similarly, the difference between $N_{\alpha\alpha}$ and $N_{\beta\alpha}$ is calculated as

$$N_{\alpha\alpha} - N_{\beta\alpha} = N_{\alpha\alpha}\Delta E/kT$$

Since the nuclei have the same gyromagnetic ratio, ΔE is essentially the same for both transitions. On the other hand, the difference between $N_{\alpha\alpha}$ and $N_{\beta\beta}$ is determined by an energy difference that is twice ΔE:

$$N_{\alpha\alpha} - N_{\beta\beta} = N_{\alpha\alpha} 2\Delta E / kT$$

For simplicity in what follows, we can express all of the population differences in arbitrary units as follows:

$$N_{\alpha\alpha} - N_{\beta\beta} = 1$$

$$N_{\alpha\alpha} - N_{\alpha\beta} = N_{\alpha\alpha} - N_{\beta\alpha} = 1/2$$

$$N_{\alpha\beta} - N_{\beta\beta} = N_{\beta\alpha} - N_{\beta\beta} = 1/2$$

$$N_{\alpha\beta} - N_{\beta\alpha} = 0$$

These differences are indicated in Fig. 9.1 and subsequent diagrams by the value Δp, defined so that the differences are all expressed relative to the population of the $|\alpha_I\rangle|\alpha_S\rangle$ molecules. By absorbing radiation, the relative populations of the various states can change, but only via the four paths shown in the diagram: Transitions involving concerted changes in the wavefunctions are not stimulated by absorption of a single photon and are very improbable.

9.2   A Transient NOE Experiment

To introduce the concept of cross-relaxation, we will consider a case in which we start at equilibrium, and then *selectively* invert the population of S-spins. As discussed in Section 6.9, this can be done by applying a relatively long and appropriately shaped pulse of radiation, with a frequency that matches only the Larmor frequency of the S-spins. This is significantly different from the more common case, in which all of the nuclei of a given type are simultaneously stimulated, as in the experiments we have discussed previously. For our immediate purposes, we can ignore the technical issues associated with selective pulses, and consider just the outcome, as shown in vector diagrams in Fig. 9.2.

If the selective pulse is followed immediately by a *non-selective* $(\pi/2)_y$-pulse and an FID is recorded, we expect the FID and spectrum diagrammed in Fig. 9.3. After the second pulse, the magnetization vectors both lie in the transverse plane and then precess at their individual frequencies. The S-magnetization vector, however, initially points along the *negative x'-axis* and gives rise to a negative peak in the spectrum. If

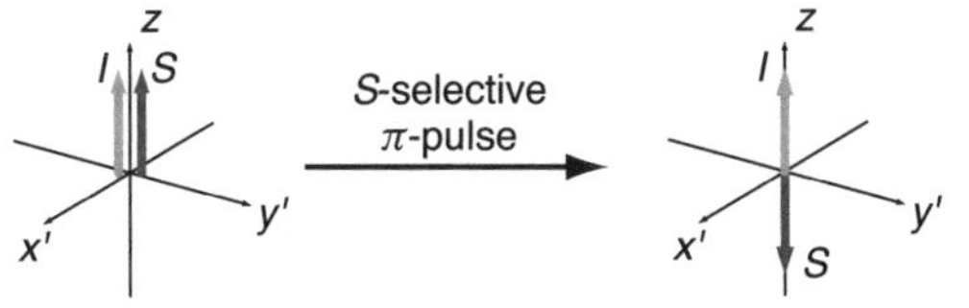

Figure 9.2

A selective π-pulse to invert the equilibrium magnetization of the S-spin.

Figure 9.3 A non-selective $\pi/2$-pulse following selective inversion of the S-spin, and the resulting FID and spectrum.

we introduce a short delay between the two pulses, the spins will begin to relax toward the equilibrium state, thus changing the intensities of the peaks in the final spectrum. The nuclear Overhauser effect is observed by comparing the spectra generated with and without the delay between the pulses.

To see how the NOE arises, we first consider the effect of the selective π-pulse, which interconverts the spins states within the various molecules as follows:

$$|\alpha_I\rangle|\alpha_S\rangle \rightarrow |\alpha_I\rangle|\beta_S\rangle$$
$$|\alpha_I\rangle|\beta_S\rangle \rightarrow |\alpha_I\rangle|\alpha_S\rangle$$
$$|\beta_I\rangle|\alpha_S\rangle \rightarrow |\beta_I\rangle|\beta_S\rangle$$
$$|\beta_I\rangle|\beta_S\rangle \rightarrow |\beta_I\rangle|\alpha_S\rangle$$

Note that in each case, only the S-spin is affected. By comparing the "new" spin states with the "old" ones from which they were generated, the population differences immediately after the selective pulse can be calculated:

$$\left(N_{\alpha\alpha} - N_{\beta\beta}\right)_{\text{new}} = \left(N_{\alpha\beta} - N_{\beta\alpha}\right)_{\text{old}} = 0$$
$$\left(N_{\alpha\alpha} - N_{\alpha\beta}\right)_{\text{new}} = \left(N_{\alpha\beta} - N_{\alpha\alpha}\right)_{\text{old}} = -1/2$$
$$\left(N_{\alpha\alpha} - N_{\beta\alpha}\right)_{\text{new}} = \left(N_{\alpha\beta} - N_{\beta\beta}\right)_{\text{old}} = 1/2$$
$$\left(N_{\alpha\beta} - N_{\beta\beta}\right)_{\text{new}} = \left(N_{\alpha\alpha} - N_{\beta\alpha}\right)_{\text{old}} = 1/2$$
$$\left(N_{\beta\alpha} - N_{\beta\beta}\right)_{\text{new}} = \left(N_{\beta\beta} - N_{\beta\alpha}\right)_{\text{old}} = -1/2$$
$$\left(N_{\alpha\beta} - N_{\beta\alpha}\right)_{\text{new}} = \left(N_{\alpha\alpha} - N_{\beta\beta}\right)_{\text{old}} = 1$$

The new population differences are indicated in the energy diagram of Fig. 9.4. As before, the population differences, Δp, are expressed relative to the $|\alpha_I\rangle|\alpha_S\rangle$ state. Note that the population differences across the two I-transitions are unaffected, while those corresponding to the S-transitions are inverted so that the higher-energy state is actually more populated than the low-energy state. This is clearly a system that is *not* in equilibrium.

Given enough time, the equilibrium population differences will be re-established via relaxation mechanisms involving the same single-quantum transitions as are stimulated by radiation. When the two spins can promote one another's relaxation via

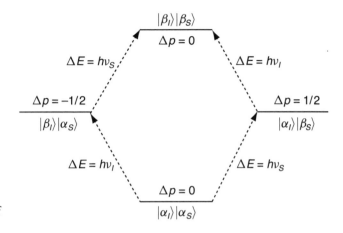

Figure 9.4
Energy diagram for the two-spin system, with the population differences, Δp, following selective inversion of the S-spin.

dipolar interactions, however, there are two additional transitions that can affect the *transient* population differences before the equilibrium is restored. These are the transitions

$$|\alpha_I\rangle|\alpha_S\rangle \leftrightarrow |\beta_I\rangle|\beta_S\rangle$$

and

$$|\alpha_I\rangle|\beta_S\rangle \leftrightarrow |\beta_I\rangle|\alpha_S\rangle$$

The first of these is called a *double-quantum* transition, and the second a *zero-quantum*, or *flip-flop* transition. Because they cannot be induced by the direct absorption or emission of radiation, these transitions are said to be *forbidden*. But, they add two new potential relaxation pathways to the energy diagram, as shown in Fig. 9.5.

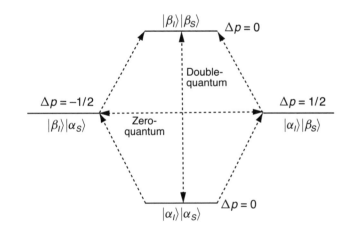

Figure 9.5
Energy diagram for the two-spin system following selective inversion of the S-spin, showing the potential relaxation pathways via zero- and double-quantum transitions.

Whether or not the double- and zero-quantum transitions will actually contribute to the relaxation process depends on two factors:

1. Whether or not there is a non-equilibrium population difference between the two states linked by the transition.
2. The presence of fluctuations in the dipole fields with frequencies that match the energy differences between the states.

For the case in which the S-spins have been selectively inverted, both of the population differences will be moved away from equilibrium. There will thus be a tendency for a net conversion of $|\alpha_I\rangle|\beta_S\rangle$ to $|\beta_I\rangle|\alpha_S\rangle$ and $|\beta_I\rangle|\beta_S\rangle$ to $|\alpha_I\rangle|\alpha_S\rangle$.

In order for the double-quantum transition to occur, there must be magnetic fluctuations of the appropriate frequencies, which requires molecular motions with approximately twice the Larmor frequency. On the other hand, the zero-quantum transition will be promoted by motions with frequencies corresponding to the *difference* between ν_I and ν_S. For ^{1}H nuclei in a typical spectrometer, this difference corresponds to only a few hundred Hz.

We can see how different types of motion can affect the NMR signal by considering two extreme situations, where tumbling is either very slow or very fast. To simplify matters further, we will also assume that there is no relaxation by other mechanisms, and we will consider the population changes that occur during only a short interval immediately after the selective inversion pulse.

First, consider the case where the molecule tumbles very rapidly, so that only the double-quantum transition is promoted. During the short time interval, there will be a small net conversion of $|\beta_I\rangle|\beta_S\rangle$ to $|\alpha_I\rangle|\alpha_S\rangle$, and we will call the relative number of spins thus converted δ. This change will affect nearly all of the population differences, as listed below:

$$\left(N_{\alpha\alpha} - N_{\beta\beta}\right) = 0 + 2\delta$$

$$\left(N_{\alpha\alpha} - N_{\alpha\beta}\right) = -1/2 + \delta$$

$$\left(N_{\alpha\alpha} - N_{\beta\alpha}\right) = 1/2 + \delta$$

$$\left(N_{\alpha\beta} - N_{\beta\beta}\right) = 1/2 + \delta$$

$$\left(N_{\beta\alpha} - N_{\beta\beta}\right) = -1/2 + \delta$$

$$\left(N_{\alpha\beta} - N_{\beta\alpha}\right) = 1$$

Now, we have to think about what happens after the $(\pi/2)_x$-pulse and the FID is converted into a spectrum. For each resonance, the intensity of the resulting peak depends upon the total population differences before the second pulse. For the S-spin, the intensity is proportional to

$$\Delta N_S = \left(N_{\alpha\alpha} - N_{\alpha\beta}\right) + \left(N_{\beta\alpha} - N_{\beta\beta}\right)$$

Before the relaxation period, this will be

$$\Delta N_S = -1/2 \ - \ 1/2 = -1$$

That the difference, as defined to indicate the excess of spins in the low-energy states, is negative reflects the inversion of the S-spin population. After the relaxation delay, the difference is

$$\Delta N_S = (-1/2 + \delta) + (-1/2 + \delta) = -1 + 2\delta$$

Thus, the absolute value of the signal from the S-spins is slightly decreased by the cross-relaxation via the double-quantum transition. This might seem like a reasonable result for a relaxation process. When we examine the I-spins, however, we find a more surprising result. For these spins, the population difference is given by

$$\Delta N_I = \left(N_{\alpha\alpha} - N_{\beta\alpha}\right) + \left(N_{\alpha\beta} - N_{\beta\beta}\right)$$

Before the short relaxation delay, this difference will be

$$\Delta N_I = 1/2 + 1/2 = 1$$

After the delay, the difference will be

$$\Delta N_I = (1/2 + \delta) + (1/2 + \delta) = 1 + 2\delta$$

Thus, the population difference for the I-spins is actually increased slightly from its equilibrium value. If this sounds suspicious, you would hardly be alone. When this effect was first predicted by Alfred Overhauser in 1953, he was greeted with great skepticism by the magnetic resonance community, including Bloch and Purcell. Only after careful consideration and experimental confirmation was it recognized that Overhauser's analysis was correct. In thinking about this, it may be helpful to consider a syphon, in which the potential energy difference between the liquid in two containers at different heights is used to transiently draw the liquid "uphill" through a tube. As in the case we are considering here, it is the accessibility of a particular pathway that allows this temporary deviation from equilibrium. Eventually, all of the populations will be restored to their equilibrium values.

If the relaxation delay is followed by a $(\pi/2)_y$-pulse and data acquisition, the I-signal will be a little more positive and the S-signal will be a little less negative than what they would be in the absence of the relaxation. What is most important about this phenomenon is that the intensity of the I-signal has been changed by a perturbation of the S-spins. This, then, is the famous *nuclear Overhauser effect* or NOE. The same initials can also stand for nuclear Overhauser *enhancement*, which is appropriate for the case of a small, rapidly tumbling molecule, but not for larger molecules as discussed below. The original phenomenon described by Overhauser involved changes in the NMR signal from nuclei in a metal caused by perturbing the *electron* spins, thus the designation of a *nuclear* Overhauser effect.

Now let's consider the case where relaxation occurs exclusively via the zero-quantum transition. This will occur for molecules that tumble slowly, so that there are significant magnetic fluctuations with frequencies that match the small energy difference between the $|\alpha_I\rangle|\beta_S\rangle$ and $|\beta_I\rangle|\alpha_S\rangle$ states. Because these states have nearly identical energies, they will have equal populations at equilibrium. After the selective inversion pulse, however, there is an excess of the $|\alpha_I\rangle|\beta_S\rangle$ state, and, during the relaxation period, the number of molecules in this state will decrease by δ and the number of $|\beta_I\rangle|\alpha_S\rangle$ molecules will increase by this amount. The population differences will then be

$$(N_{\alpha\alpha} - N_{\beta\beta}) = 0$$

$$(N_{\alpha\alpha} - N_{\alpha\beta}) = -1/2 + \delta$$

$$(N_{\alpha\alpha} - N_{\beta\alpha}) = 1/2 - \delta$$

$$(N_{\alpha\beta} - N_{\beta\beta}) = 1/2 - \delta$$

$$(N_{\beta\alpha} - N_{\beta\beta}) = -1/2 + \delta$$

$$(N_{\alpha\beta} - N_{\beta\alpha}) = 1 - 2\delta$$

Again, we calculate the total population differences for the two spins:

$$\Delta N_S = (N_{\alpha\alpha} - N_{\alpha\beta}) + (N_{\beta\alpha} - N_{\beta\beta})$$
$$= (-1/2 + \delta) + (-1/2 + \delta)$$
$$= -1 + 2\delta$$

and

$$\Delta N_I = (N_{\alpha\alpha} - N_{\beta\alpha}) + (N_{\alpha\beta} - N_{\beta\beta})$$
$$= (1/2 - \delta) + (1/2 - \delta)$$
$$= 1 - 2\delta$$

Thus, both population differences have been decreased slightly in absolute value. While the S-population has moved a little bit toward its equilibrium state, the I-population has again moved away from its equilibrium distribution. In this case, however, inversion of the S-spins has resulted in a *decrease* in the intensity of the signal from the I-spins. In general, cross-relaxation via zero- and double-quantum processes cause spectral intensity changes of opposite sign. Thus, the sign of the NOE reflects the tumbling rate and the size of the molecule. For molecules with intermediate tumbling rates, both cross-relaxation processes will contribute, and their effects can cancel one another, as discussed further in the next section.

In describing the transient NOE thus far, we have been a bit vague about the time delay between the inversion and read pulses, as well as the resulting change in the signals, which was expressed in terms of the population difference, δ. In practice, these two parameters are closely linked. If the delay is too short, there will be no significant cross-relaxation, but if it is too long, both spin populations will fully relax to their equilibrium levels. For a given pair of spins, the rate of build-up of the NOE, and its subsequent decay, will depend on the distance between the two nuclei, their interactions with other spins, the rate of molecular tumbling and, possibly, internal motions. As a consequence, it is often necessary to test different delay times to obtain optimum results.

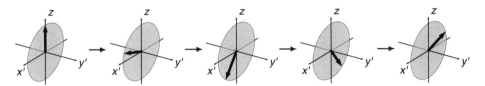

Figure 9.6 Concerted precession of magnetization during a relatively short pulse along the y'-axis.

Because the effectiveness of the dipolar interaction depends on the distance between the two nuclei, the NOE provides a method for determining intramolecular distances. For proton pairs, the effect is usually limited to nuclei that are within 5 Å of each other, and the magnitude of the effect increases dramatically at shorter distances. As a first approximation, the strength of an NOE depends on the negative sixth power of the internuclear distance. Although this provides a basis for quantitative distance measurements, there are a number of technical issues that often limit its use to providing only qualitative estimates. Nonetheless, the NOE forms the basis for determining the three-dimensional conformations of both small and large molecules.

9.3   Saturation and the Steady-state NOE

In the experiments described so far, we have restricted ourselves to short, intense radiation pulses that lead to rapid changes in the wavefunctions, and such pulses, in fact, are used in most modern NMR experiments. But, it wasn't always this way, and longer periods of weak irradiation can also have very interesting and useful effects. Indeed, the NOE was first described in the context of an experiment in which one spin is first exposed to a lengthy, but weak, irradiation.

As we have discussed previously, exposing a nucleus to electromagnetic radiation promotes a change in the wavefunction such that the relative contributions of the $|\alpha\rangle$ and $|\beta\rangle$ eigenfunctions are altered. In terms of the vector picture, this represents precession of the net magnetization about the axis of the pulse. For a y-pulse, the magnetization precesses in the $x'-z$ plane. If the pulse is relatively short, then we will have a concerted movement of the net magnetization vector, as illustrated in Fig. 9.6. But, if the radiation is weaker and the exposure correspondingly longer, then there will be the same sort of coherence loss that we have discussed for magnetization precessing in the $x'-y'$ plane. Subpopulations with different precession frequencies will give rise to vectors that become progressively more dispersed, as shown in Fig. 9.7. Eventually, there will be no net magnetization in *any* direction. A spin population in this state is said to be *saturated*.

It is important to stress the difference between saturation and the situation that is generated by a short and intense $\pi/2$-pulse. Although there is no net z-magnetization in either case, the two situations differ with respect to the coherence of the resulting wavefunctions. After a short $\pi/2$-pulse, all of the excess spins are in the same quantum state and their wavefunctions evolve together until transverse relaxation destroys

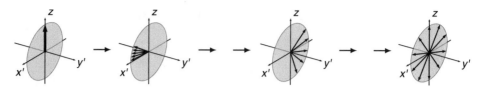

Figure 9.7 Saturation of magnetization by a period of long irradiation, typically with low power. Precession of different subpopulations at different frequencies leads to dispersion of the magnetization and the loss of coherence.

the coherence. If a signal is recorded immediately after the pulse, the usual FID is generated. On the other hand, after a saturating irradiation, there is no net magnetization in the transverse plane, and no signal can be generated in the coils. Subsequent pulses will generally not lead to signals, unless the spins have had time to return at least partially to their equilibrium state.

Remember also that short intense pulses give rise to radiation with a very broad frequency range. Conversely, a long, weak irradiation will have a much narrower frequency range. This makes it possible to selectively saturate nuclei with a narrow range of Larmor frequencies. If a saturation pulse is then followed by the standard $\pi/2$-pulse, the peaks corresponding to the saturated resonances will be selectively eliminated from the resulting spectrum. This, in fact, is a very useful method for eliminating the strong signal that would otherwise be generated from solvent molecules, such as water.

Just as inverting one-spin population can affect the signal from a nearby nucleus, saturating a spin can give rise to an NOE that alters the signal intensity of another nucleus. Although the various nuclei in the saturated sample will have many different wavefunctions, the practical consequences for relaxation can be described by treating the population as if it were composed of equal numbers of spins in the two eigenstates, $|\alpha\rangle$ and $|\beta\rangle$. Thus, there is no net z-magnetization and no net magnetization in any direction in the x–y plane.

The "classical" NOE experiment is carried out by first selectively saturating one spin using a weak but lengthy pulse and then applying the usual $\pi/2$-pulse to generate transverse magnetization and an FID. The spectrum from this experiment is then compared with that from an identical one but without the saturation period. The NOE is detected as a change in the intensity of the signal from the second nucleus.

To analyze the saturation NOE experiment, we start with the equilibrium populations, which are defined relative to that of the $|\alpha_I\rangle|\alpha_S\rangle$ state:

$$N_{\alpha\alpha} - N_{\beta\beta} = 1$$

$$N_{\alpha\alpha} - N_{\alpha\beta} = 1/2$$

$$N_{\alpha\alpha} - N_{\beta\alpha} = 1/2$$

The effect of selectively saturating the S-spin is to establish the following equalities:

$$N_{\alpha\alpha} = N_{\alpha\beta}$$

$$N_{\beta\alpha} = N_{\beta\beta}$$

In the absence of any relaxation, the population differences across the two I-transitions will be maintained:

$$N_{\alpha\alpha} - N_{\beta\alpha} = 1/2$$

$$N_{\alpha\beta} - N_{\beta\beta} = 1/2$$

The population differences that satisfy these conditions are shown in Fig. 9.8. The saturation of the S-spins is indicated by the double lines linking the $|\alpha_I\rangle|\alpha_S\rangle$ and $|\alpha_I\rangle|\beta_S\rangle$ states and the $|\beta_I\rangle|\alpha_S\rangle$ and $|\beta_I\rangle|\beta_S\rangle$ states, emphasizing the special relationships between the populations of these pairs. This diagram also uses four symbols introduced in Chapter 8:

W_0 : The transition rate constant for the zero-quantum transition

W_2 : The transition rate constant for the double-quantum transition

W_1^I : The transition rate constant for the single-quantum I-transition

W_1^S : The transition rate constant for the single-quantum S-transition

In the case of the transient NOE experiment, we first considered the perturbation of the inversion pulse and then the effects of cross-relaxation during a period before the observation pulse. Here, the situation is somewhat more complicated because relaxation occurs *during* the irradiation, and lots of things are going on at the same time.

First, we can consider what would happen in the absence of either of the cross-relaxation processes. The continued application of the radiation will ensure that there is no population difference across either of the S-transitions, so the two processes represented by W_1^S will not cause any further net population change. In addition, the population differences across the two I-transitions are already at equilibrium (with

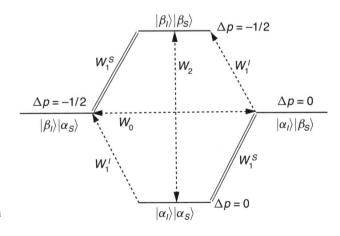

Figure 9.8

Energy diagram for the two-spin system showing the population differences established by saturation of the S-spin. The double line segments indicate the transitions for which the population difference has been eliminated.

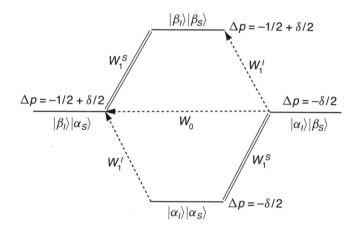

Figure 9.9
Energy diagram for the two-spin system with the S-spin saturated, showing the shifts in population due to the zero-quantum transition.

an excess of the species in the $|\alpha_I\rangle$ states), so the W_1^I processes will not have any effect, either. In short, saturating the S-spins will not affect the I-spin populations if there is no cross-relaxation.

Next, consider what would happen if the zero-quantum transition were to facilitate relaxation (Fig. 9.9). From before, we know that this is most likely to happen in a large, slowly tumbling molecule. Because the $|\alpha_I\rangle|\beta_S\rangle$ and $|\beta_I\rangle|\alpha_S\rangle$ states have nearly identical energies, the W_0 process will try to equalize the populations of these species with a net shift from $|\alpha_I\rangle|\beta_S\rangle$ to $|\beta_I\rangle|\alpha_S\rangle$. As before, we will call the change in population δ. But, as soon as this process begins, things get complicated. First of all, the S-spins are still being irradiated, so that any population differences across the S-transitions are quickly eliminated. Thus, the increase in the population of $|\beta_I\rangle|\alpha_S\rangle$ is "shared" with $|\beta_I\rangle|\beta_S\rangle$, and the decrease in the population of $|\alpha_I\rangle|\beta_S\rangle$ is split with $|\alpha_I\rangle|\alpha_S\rangle$. This results, temporarily, in the population differences indicated in Fig. 9.9. Things don't stop here, however, because now the population differences across the I-transitions have been moved away from equilibrium, with extra molecules in the higher-energy $|\beta_I\rangle$ states. This will be counteracted by relaxation via the two W_1^I processes.

Given sufficient time, the system will approach a steady state. Deriving the relative concentrations at steady state is a bit involved, but we can easily reach a qualitative conclusion. Provided that W_0 is much greater than W_1^I, then the population differences that give rise to the I-signal will be given by

$$\Delta N_I = \left(N_{\alpha\alpha} - N_{\beta\alpha}\right) + \left(N_{\alpha\beta} - N_{\beta\beta}\right)$$
$$= -\delta/2 - (-1/2 + \delta/2) + -\delta/2 - (-1/2 + \delta/2)$$
$$= 1 - 2\delta$$

Thus, we expect the signal from the I-spins to be somewhat weaker than normal if the S-spins are saturated. How much weaker will depend on the relative values of W_0 and W_1^I, as well as any other relaxation mechanisms that might be active during the saturation period.

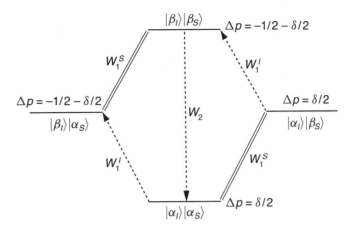

Figure 9.10
Energy diagram for the two-spin system with the S-spin saturated, showing the shifts in population due to the double-quantum transition.

In the same way, we can consider a case where only the double-quantum cross-relaxation process is significant. Under these conditions, the driving force for the available cross-relaxation pathway is the disequilibrium between the $|\beta_I\rangle|\beta_S\rangle$ and $|\alpha_I\rangle|\alpha_S\rangle$ states, with an excess of the former (relative to the equilibrium distribution). This will cause a net flux from $|\beta_I\rangle|\beta_S\rangle$ to $|\alpha_I\rangle|\alpha_S\rangle$, and these changes will be shared with $|\beta_I\rangle|\alpha_S\rangle$ and $|\alpha_I\rangle|\beta_S\rangle$. This will result in the distribution diagrammed in Fig. 9.10. As before, the changes from cross-relaxation will be counterbalanced by the W_1^I processes. But, if W_2 is greater than W_1^I, we can conclude qualitatively that the population differences across the I-transitions will be

$$\Delta N_I = \left(N_{\alpha\alpha} - N_{\beta\alpha}\right) + \left(N_{\alpha\beta} - N_{\beta\beta}\right)$$
$$= \delta/2 - (-1/2 - \delta/2) + \delta/2 - (-1/2 - \delta/2)$$
$$= 1 + 2\delta$$

In this case, the signal from the I-spins is enhanced by the saturation of the S-spins.

The magnitude of the steady-state NOE on spin I caused by saturation of spin S is generally defined as

$$\eta_I(S) = \frac{I - I_0}{I_0} \tag{9.1}$$

where I and I_0 are the observed intensities with and without presaturation. When the zero-quantum transition dominates, $I < I_0$ and $\eta_I(S)$ is negative. On the other hand when the double-quantum transition dominates, $I > I_0$ and $\eta_I(S)$ is positive.

A more complete treatment will show that the observed NOE depends on W_0, W_2 and W_1^I according to

$$\eta_I(S) = \frac{W_2 - W_0}{2W_1^I + W_2 + W_0} \tag{9.2}$$

Note that this relationship is consistent with our qualitative conclusions for the extreme cases where either W_0 or W_2 dominates.

As discussed in Chapter 8, the relative rates for the different transitions due to dipolar coupling will depend on their associated energy differences and the rate of tumbling, which can be expressed as the correlation time, τ_c. Although the derivations of the mathematical relationships between τ_c and the rate constants are rather lengthy, the final expressions are actually quite simple in form:

$$W_0 = \frac{1}{10}d^2 \frac{\tau_c}{1 + (\omega_I - \omega_S)^2 \tau_c^2}$$

$$W_1^I = \frac{3}{20}d^2 \frac{\tau_c}{1 + \omega_I^2 \tau_c^2}$$

$$W_2 = \frac{3}{5}d^2 \frac{\tau_c}{1 + (\omega_I + \omega_S)^2 \tau_c^2}$$

where ω_I and ω_S are the angular Larmor frequencies of the two nuclei (expressed in rad/s). The dipolar coupling constant, d, is

$$d = \frac{\mu_0 \gamma_I \gamma_S h}{8\pi^2 r^3}$$

where γ_I and γ_S are the gyromagnetic ratios of the nuclei, r is the distance between the nuclei and μ_0 is the permeability of free space, introduced previously.

Note that each of the expressions for the rate constants contains a term in the denominator that represents the product of a frequency (ω_I, $\omega_I - \omega_S$ or $\omega_I + \omega_S$) and τ_c. The correlation time also appears in the numerator. For small values of τ_c (i.e., fast tumbling), the product in the denominator is less than one, and the relaxation rates increase as τ_c increases. When the product becomes greater than one, however, the denominator increases more rapidly than the numerator, and the relaxation rate decreases with further increases in τ_c. As a consequence, each of the rates is maximal when $1/\tau_c$ is approximately equal to the corresponding frequency. This is illustrated for a specific case in Fig. 9.11.

The rates plotted in this graph are for two protons separated by 2 Å, in a 500-MHz spectrometer. The Larmor frequencies for the two nuclei are assumed to differ by 1 ppm. Note that W_1^I and W_2 have maxima when τ_c is approximately 10^{-9} s; that is, when $\tau_c \approx 1/\omega_I$. The rate of the zero-quantum transition, W_0, also shows a maximum, but only at very large correlation times.

The transition rate constants can then be used to predict the magnitude of the steady-state NOE, as defined previously and plotted in Fig. 9.12. As predicted from our qualitative arguments, the NOE is positive, with a maximum value of 0.5, for rapidly tumbling molecules, and is negative for more slowly tumbling ones, with a minimum value of −1.

Note also that there is range of τ_c values for which the predicted NOE is close to zero. The NOE is exactly zero when $\tau_c = 1.12/\omega_I$. For the field strengths that are typically used for solution NMR, this cross over between positive and negative NOEs generally occurs for molecules with molecular weights of about 1000–2000 Da, such

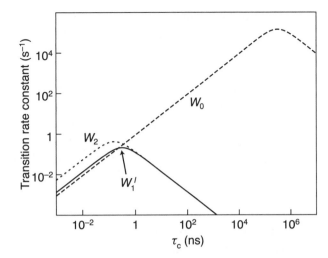

Figure 9.11

Transition rate constants as a function of the correlation time, τ_c, for two protons separated by 2 Å in a 500-MHz spectrometer.

as peptides of 10 to 20 residues. For this and other reasons, NMR studies of relatively small peptides can often be more challenging than studies of larger molecules.

An astute reader may have noticed something troublesome about the equations for calculating the steady-state NOE. In each of the expressions for the transition rate constants (W_0, W_1^I, and W_2), the internuclear distance r appears (as part of the constant d), in each case raised to the negative sixth power, so that the rate constants increase very dramatically at shorter distances. When these expressions are combined into the equation for the NOE (Eq. 9.2), however, the distance terms all cancel out. This predicts that the NOE does not depend on distance at all! Fortunately, this prediction is not borne out in the real world, though the reason is a bit subtle. The vertical axis in Figure 9.12 is labeled "Maximum NOE" because the values plotted represent the NOE that would be observed in the complete absence of relaxation mechanisms other than intramolecular dipolar coupling. Other mechanisms, such as those due

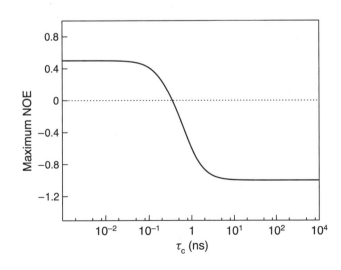

Figure 9.12

Predicted maximum NOE values as a function of correlation time, τ_c, for two protons in a 500-MHz spectrometer.

to chemical shift anisotropy or intermolecular dipolar coupling, will contribute to the apparent W_1^I term in a way that does not depend on any intramolecular distance and will decrease the absolute value of the NOE. Because all of the other terms *do* depend on distance, the extent to which the NOE approaches the maximum values plotted above shows a strong distance dependence. Thus, it is actually the existence of competing relaxation processes that makes NOE measurements useful in terms of determining intramolecular distances. It should also be noted that we have assumed that the saturation period is long enough to fully equalize the transitions of the S-spin. In practice, this may not be the case, thereby reducing the magnitude of the NOE.

In the discussion of both the transient and steady-state NOEs, we have only considered the homonuclear case, especially that of two ^1H nuclei. However, the NOE can also be observed between heteronuclei, such as ^1H–^{13}C and ^1H–^{15}N pairs. The heteronuclear NOE was briefly discussed in Chapter 8 in the context of studying internal molecular motions (page 219). Historically, heteronuclear NOEs were also used as a means of enhancing the signals from ^{13}C nuclei in small molecules. ^{13}C signals are much weaker than those from protons, both because the gyromagnetic ratio of ^{13}C is only about one-fourth that of ^1H and because its natural abundance is low (approximately 1% of total carbon). Because the equilibrium population difference for ^1H nuclei is also about four-fold greater than that of ^{13}C, the NOE generated by the saturation of a ^1H nucleus can have a large relative effect on the intensity of a ^{13}C signal. For the reasons discussed above, the NOE is a signal enhancement for a small molecule and can increase the ^{13}C signal intensity by as much as two-fold. (For large molecules, the NOE would only reduce the signal intensity.) This method was for a time widely used to enhance ^{13}C signals but has now been largely replaced by other heteronuclear methods, such as the INEPT pulse sequence described in Section 16.4, which works for both small molecules and large, slowly tumbling ones.

9.4   Transferred NOEs

As discussed in the previous section, the magnitude and sign of the NOE is especially sensitive to molecular size, and small peptides, in particular, often yield very small NOEs. While the weak NOEs from peptides of 10–20 amino acid residues can represent a limitation to the use of NMR to study them, there are situations in which this very phenomenon can be cleverly exploited to obtain unique information about the structures of peptides or similarly sized molecules when they are bound to much larger macromolecules.

In many cases, small peptides function by binding to larger structures, and the conformation of the bound form may well be different from that of the isolated peptide. Such a situation is diagrammed schematically in Fig. 9.13.

If the binding and release of the smaller molecule (usually referred to as a *ligand*) is relatively rapid, then a given molecule may spend significant amounts of time both bound and free during either the delay period of a transient NOE experiment or the saturation period of a steady-state experiment. In principle, cross-relaxation can occur when the ligand is either free or bound, and the signals from both forms can

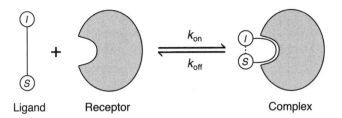

Figure 9.13 Reversible binding of a small-molecule ligand to a receptor. In a transferred NOE experiment, cross-relaxation between the I- and S-spins is maximal when the ligand is bound to the receptor and tumbles slowly, but the NMR signal is detected from the free ligand.

contribute to the FID during data acquisition. As indicated in Fig. 9.11, the zero-quantum transition for a relatively large molecule can be much more significant than either of the transitions for smaller ones. As a consequence, the cross-relaxation that occurs in the bound form can often dominate the NOE, even if the bound form represents only a small fraction of the total ligand concentration.

In a typical transfer NOE experiment, a rather large excess (10- to 100-fold) of ligand is mixed with the receptor protein. As a consequence, most of the signal in the spectrum arises from the free form, and the peaks are typically quite sharp, reflecting the free ligand's relatively long transverse relaxation time. On the other hand, the signals from the receptor and the ligand bound to it are much weaker, and often very broad. If, however, the cross-relaxation rate in the complex is much larger than that in the free ligand, the NOEs measured will reflect the conformation and tumbling rate of the ligand when it is bound.

In order for this to work, the rate of exchange between free and bound forms must be much larger than the transverse relaxation rate when the ligand is bound, otherwise the signal will essentially be lost to relaxation before it can be detected from the free form. For a moderately sized protein, this relaxation rate might be on the order of 10 to 100 s^{-1}, so the dissociation rate constant, k_{off}, must be at least this large. This also places restrictions on the thermodynamic stability of the complex. The equilibrium dissociation constant, K_d, is related to the rate constants for binding and dissociation according to

$$K_d = \frac{[L][R]}{[L \cdot R]} = \frac{k_{off}}{k_{on}}$$

where [L], [R] and [L · R] represent the concentrations of the free ligand, the receptor and the complex, respectively. The fastest intermolecular association rate constants, limited by the rate of diffusion, are about 10^8 s^{-1} M^{-1}. Thus, K_d is required to satisfy the condition that

$$K_d \geq \frac{10 \text{ s}^{-1}}{10^8 \text{ s}^{-1} \text{M}^{-1}}$$

$$K_d \geq 10^{-7} \text{ M} = 0.1 \,\mu\text{M}$$

This represents a relatively weak binding interaction. On the other hand, if K_d is too large, then it will be difficult to maintain the condition that a significant fraction of the ligand is bound to the receptor at any time, and the NOE signal will be correspondingly weakened. As a consequence, the transferred NOE experiment is really only practical for interactions for which K_d lies in the range of about 0.1 to 100 μM, and k_{off} is greater than about 10 s^{-1}.

In spite of these limitations, the transferred NOE has been extremely useful in studying a variety of biologically important interactions.

9.5   Spin Diffusion and Internal Motions— Some Cautionary Notes

In all of our discussion of the NOE we have only considered a highly idealized situation in which there are only two nuclei influencing one another via cross-relaxation, and the distance between these nuclei is constant. In virtually any real situation, things are likely to be more complicated, and the interpretation of NOE data is rarely as simple as implied by the previous sections.

One of the major complications arises from the fact that each nucleus of interest is likely to be within "NOE distance" of several others. Although the initial inversion or saturation may be very specific to one nucleus, as soon as cross-relaxation begins, the populations of nearby spins will move away from equilibrium, by definition. The disequilibrium at these sites will then begin to influence the distributions at their neighboring sites. As a consequence, perturbing one spin can sometimes lead to an apparent NOE at a site with which it does not interact directly. This phenomenon, often called *spin diffusion*, is expected to become more pronounced the longer the period allowed for cross-relaxation. As a consequence, it is usually possible to minimize spin diffusion by using shorter saturation or delay periods. In addition, it may be possible to distinguish direct NOEs from indirect ones by comparing the spectra obtained using different cross-relaxation periods.

Another difficulty arises when there are conformational changes that alter the distances between the nuclei undergoing cross-relaxation. Provided that these changes are relatively rapid, each conformation will contribute to the observed NOE. However, because the NOE depends so strongly on the internuclear distance, the conformations with short distances will generally dominate, even if these conformations only represent a small fraction of the population. As a consequence, there may be a tendency to underestimate the average distance between two sites in a flexible molecule and, thus, overestimate the populations of the conformations that give rise to the NOE.

These and other effects place some quite serious limitations on our ability to interpret observed NOEs quantitatively. Much more complete and rigorous treatments of the NOE have been developed, and these analyses, together with more refined experiments, can help resolve some of the ambiguities. In most practical situations, however, the best recourse is usually the measurement of NOEs for as many nuclei pairs as possible in a molecule. A large number of approximate distance measurements can often define a more reliable structural model than a few more accurate ones.

Summary Points for Chapter 9

- Cross-relaxation is the effect that spin populations exert on one another, via dipolar coupling, as they relax toward thermal equilibrium.
- The nuclear Overhauser effect (NOE) is the change in signal intensity from one spin population due to cross-relaxation with another spin population that has been selectively shifted away from equilibrium.
- In a transient NOE experiment, one spin population is inverted by a selective pulse, and a spectrum is recorded after a short relaxation delay, but before the system has re-equilibrated.
- In a saturation NOE experiment, one spin is subjected to a relatively long irradiation at low power, eliminating the net magnetization from that spin population, before recording the spectrum.
- Cross-relaxation and the NOE are mediated by zero- and double-quantum transitions of dipolar-coupled spin pairs.
- The sign and magnitude of the NOE depend on the rate of molecular tumbling and, therefore, molecular size:
    - Fast tumbling, with a correlation time less than the inverse of the Larmor frequency, leads to an enhancement of signal strength and a positive NOE.
    - Slow tumbling, relative to the Larmor frequency, leads to a reduction in signal strength and a negative NOE.
- The magnitude of the NOE is extremely sensitive to the distance between dipoles and is usually only detectable for ^1H pairs if the distance is less than about 5 Å. This provides the basis for determining the three-dimensional structures of molecules.
- In transferred-NOE experiments, cross-relaxation occurs while a small-molecule ligand is bound to a larger molecule and tumbles slowly, but the NMR signal is generated while the ligand is free and tumbling more rapidly.
- The NOE can be propagated from one spin to another in a molecule, through the process of spin diffusion, and can significantly complicate the interpretation of NOEs for structural analysis.

Exercises for Chapter 9

9.1. In Section 9.2, we considered the effects of cross-relaxation via dipolar coupling when the spin population of one nucleus is inverted and that of a nearby nucleus is left in its equilibrium state. Now, consider an experiment in which both spin populations are initially inverted with a non-selective pulse as shown in Fig. 9.14. Following a

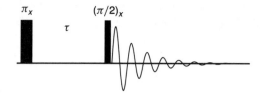

Figure 9.14
Pulse sequence for Exercise 9.1.

short delay, τ, a $(\pi/2)_x$-pulse is applied, and the FID is recorded. As in the example presented in Section 9.2, assume that we are dealing with a molecule in which there are two ^{1}H nuclei, I and S, that have different resonance frequencies and are within 5 Å of one another.

(a) Calculate the population differences among the four different kinds of molecules, defined by the spin states of I and S, immediately after the π-pulse. Draw an energy diagram representing the different states and their relative population differences after the initial pulse.

(b) Suppose that the molecule is a relatively large protein (with a molecular weight of about 20,000 Da). Do you expect any significant cross-relaxation to occur during the τ delay time? Explain your reasoning. If there is cross-relaxation, determine how this process will affect the two signals acquired after the second pulse.

(c) Suppose that the molecule is very small (with a molecular weight of about 100 Da). Do you expect any significant cross-relaxation to occur during the delay time? Explain your reasoning. If there is cross-relaxation, determine how this process will affect the two signals acquired after the second pulse.

9.2. The absolute magnitude of the nuclear Overhauser effect depends on a number of parameters, including the distance between the interacting nuclei, the tumbling rate of the molecule, the external field strength and the rates of competing relaxation processes. As a consequence, no attempt is usually made to interpret absolute NOEs. However, within a single molecule under defined conditions, the relative NOEs observed for different nuclei pairs can be used to estimate the corresponding distances. One approach is to assume that the NOE magnitude, η, is related to the internuclear distance, r, by a function of the following form:

$$\eta = Ar^{-6}$$

where A is a constant. The constant can be estimated from the NOE measured between nuclei separated by a known distance, such as two geminal ^{1}H nuclei bound to a common carbon atom, which are separated by approximately 1.7 Å.

(a) If the NOE observed between two geminal protons is defined to have a value of 100, calculate the distances that would correspond to NOE intensities of 10 and 1 on the same scale.

(b) Calculate the NOE intensity, on the scale defined above, for two protons separated by 5 Å.

(c) What factors might limit the use of a single calibration function to different inter-nuclear distances in the same molecule?

(d) As a practical matter, the use of NOEs to estimate interatomic distances is almost always limited to ^1H–^1H pairs. Suggest a reason that this approach is not easily applied to other types of nuclei, such as ^{15}N or ^{13}C.

Further Reading

The nuclear Overhauser effect is thoroughly covered in the folowing text:

◇ Neuhaus, D. & Williamson, M. P. (2000). *The Nuclear Overhauser Effect in Structural and Conformational Analysis*, 2nd ed. Wiley, New York.

A personal account of the discovery of the (non-nuclear) Overhauser effect and its initial reception:

◇ Overhauser, A. W. (2007). Dynamic nuclear polarization. *eMagRes*. http://dx.doi.org/ 10.1002/9780470034590.emrhp0135

A review on transferred NOE experiments and their applications:

◇ Post, C. B. (2003). Exchange-transferred NOE spectroscopy and bound ligand structure determination. *Curr. Op. Struct. Biol.*, 13, 581–588. http://dx.doi.org/10.1016/j.sbi.2003 .09.012

CHAPTER 10

Two-dimensional NMR Experiments

The invention of multidimensional experiments in the early 1970s represents one of the major advances in NMR spectroscopy and has been especially important for studies of macromolecules, as well as macroscopic imaging based on NMR. The ability to spread the signals from individual nuclei into two or more dimensions is essential for resolving the hundreds of signals from a protein or nucleic acid. Of equal importance, multidimensional experiments provide information about the structural relationships among the many nuclei. Depending on exactly how the experiment is performed, a cross peak linking two resonances can mean either that the two nuclei are close in space or are linked by a relatively small number of covalent bonds.

In this chapter, we will consider the general features of two-dimensional experiments and the details of one particular type of experiment, the NOESY, which is based on the nuclear Overhauser effect discussed in Chapter 9. This experiment has the important pedagogical virtue that it can be described quite well using a qualitative treatment of the interactions between the spins of interest. Experiments based on covalent connectivities among nuclei, on the other hand, can really only be described using a quantum mechanical treatment. For this reason, discussions of this second category of experiments is postponed for later chapters, after the necessary quantum mechanics has been developed. This chapter also includes a brief introduction to magnetic resonance imaging (MRI), which shares many features with the multidimensional experiments used to study molecular structure.

10.1 Constructing a Two-dimensional Experiment

Two-dimensional spectra are generated by assembling a series of simpler one-dimensional spectra into an array. The individual spectra are generated using a series of pulses, with some systematic difference introduced that distinguishes them. A generic description of one of the individual experiments can be represented with the diagram shown in Fig. 10.1.

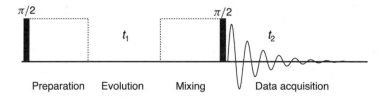

Figure 10.1   General form of the pulse diagram for one of the one-dimensional pulse experiments used to assemble a two-dimensional experiment.

In addition to the data acquisition phase that is common to all pulse experiments, the individual experiments are composed of three preceding elements:

1. A *preparation* period, which usually begins with a $\pi/2$-pulse and may include additional pulses and delay periods.
2. An *evolution* period, in which the quantum mechanical states of the nuclei are allowed to change over a defined period of time. The length of this time period, generally designated t_1, is what is systematically changed between the individual one-dimensional experiments.
3. A *mixing* period, which often, but not always, ends with a $\pi/2$-pulse and may include other pulses or delays.

At the end of the mixing period, the free induction decay is recorded and Fourier transformed in the same way we have described previously. The result is a series of one-dimensional spectra that are distinguished by the length of the evolution period, t_1. The times associated with data points in the FID are identified as t_2.

The easiest way to see how a two-dimensional spectrum is generated is to look at a highly simplified and idealized case. We will begin by considering a situation in which there are two nuclei with different resonance frequencies, but that do not influence one another at all. The example we will analyze consists of just two $\pi/2$-pulses separated by an evolution period. Thus, the preparation is just the initial $\pi/2$-pulse; the evolution is just a delay of time, t_1, and the mixing is a second $\pi/2$-pulse, which is sometimes referred to as a *read pulse*. We will define the two pulses so that they are both along the y'-axis, so that the experiment is described as shown in Fig. 10.2.

This simple pulse sequence is, in fact, the basis of a real two-dimensional experiment, the COSY experiment (for COrrelation SpectroscopY). In real systems, the COSY spectrum identifies pairs of nuclei that are linked by scalar coupling. Although

Figure 10.2
Pulse diagram for the simplest two-pulse two-dimensional experiment.

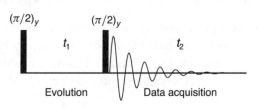

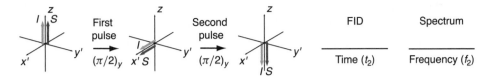

Figure 10.3   Vector diagrams representing the two-pulse experiment with $t_1 = 0$, and the resulting FID and spectrum.

the experiment is easy to describe and perform, its analysis for scalar-coupled pairs is actually quite involved, and we will consider here just the case for uncoupled spins.

First, consider what would happen when $t_1 = 0$, as shown in Fig. 10.3. The successive $(\pi/2)_y$-pulses are equivalent to a single π-pulse and simply invert the magnetization vectors for the two spins. Because there is no transverse magnetization, no FID signal is detected, and there are no peaks in the spectrum following Fourier transformation.

Next, consider what happens when there is a delay between the two pulses, as shown in Fig. 10.4 for three successive delays. For illustration, the precession frequency (in the rotating frame) of one of the spins (I) is set to twice that of the other (S). (Although the origin of the commonly used labels, I and S, is obscure, it may help in what follows to think of S as meaning "slow.") After the first delay shown above, the magnetization from spin I precesses from the x'-axis to the y'-axis, and the magnetization from spin S precesses half this far. The second $(\pi/2)_y$-pulse then has no effect on I, because its magnetization lies entirely along the direction of the pulse. For spin S, the magnetization component that projects along the y'-axis is similarly unaffected. However, the component of the S-magnetization that still projects

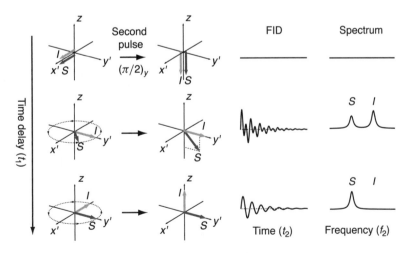

Figure 10.4   Vector diagrams representing the two-pulse experiment with incremented t_1 delay periods, and the resulting FIDs and spectra.

onto the x'-axis is rotated so that it now projects along the negative z-axis. Immediately after the second pulse, the transverse components of both magnetization vectors point along the y'-axis, and both make contributions to the FID, which are initially in-phase with one another.[1] However, the two spins precess at different frequencies, and Fourier transformation yields a spectrum with two peaks. The intensity of the peak from spin S is lower than that from I, because only part of the S-magnetization projects onto the transverse plane during the FID.

After a longer t_1 delay, as shown in the bottom row of drawings in Fig. 10.4, the I-magnetization points along the negative x'-axis, and the S-magnetization points along the positive y'-axis. In this case, the second $(\pi/2)_y$-pulse leaves the S-magnetization unaltered, but rotates the I-magnetization so that it lies completely along the z-axis. Thus, only S contributes to the FID and the resulting spectrum.

Following the same logic, we can draw out the magnetization vectors, FIDs and spectra that would arise for progressively longer delay times, as shown in Fig. 10.5. Note the following about each set of vector diagram, FID and spectrum:

- During the evolution period, the two vectors precess independently at their Larmor frequencies.
- After the second pulse, any magnetization that projected along the x'-axis is converted to a projection along the z-axis, while the y'-magnetization component is unaffected and can point in either the positive or negative direction.
- The greater the x'-component before the second pulse, the smaller the y'-component, both before and after the pulse.
- The sign and magnitude of the y'-magnetization at the end of the t_1 delay determines the sign and strength of the corresponding peak in the spectrum.

The net result of these manipulations is that the intensities and signs of the peaks in the individual spectra are determined by the Larmor frequencies, relative to the lengths of the individual evolution periods. In essence, differences in frequency are used to create differences in the amplitudes of the peaks in the final spectrum. In many respects, this is similar to the way FM radio works, in which amplitude differences are first converted into frequency differences in the transmitted signal and then, in the receiver, the frequency differences are converted back into amplitude differences.

---

[1] Note that the vector diagrams in these figures are drawn so that the precession frequency is positive in the rotating frame. For a positive gyromagnetic ratio, this corresponds to a Larmor frequency that is *less negative* than the rotating-frame frequency, as illustrated in Fig. 5.16 (page 100). The FIDs are drawn so that the signal represents the projection along the $-x'$-axis of the rotating frame. The spectra are drawn so that a positive y'-projection at the beginning of the data acquisition period gives rise to a positive absorptive peak in the spectrum. In the simulated spectra, the rotating-frame frequencies increase from left to right. In practice, the phase of the FID is determined by the phase of the heterodyne reference signal and can be changed by manipulating the Fourier transformation. The important thing right now is that all of the FIDs and spectra from the individual one-dimensional experiments are treated identically.

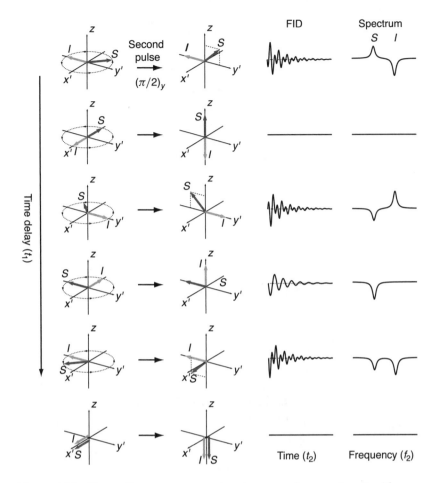

Figure 10.5 Vector diagrams representing the two-pulse experiment with progressively longer t_1 delay periods, and the resulting FIDs and spectra.

10.2   Generating a Two-dimensional Spectrum

After collecting many different FIDs using different t_1 delays and then Fourier transforming them, we have a set of one-dimensional spectra (Fig. 10.6). Because the frequency axis for these spectra comes from Fourier transforming the data collected in the second time interval, t_2, this axis is labeled f_2. (This convention for labeling the axis inevitably causes some confusion, since the f_2-axis is generated from the first of what will be two Fourier transformations.)

We can also think of this collection of spectra as a two-dimensional array of data, with one dimension representing a time domain, and the second representing a frequency domain. If we now plot out the amplitudes observed at a single frequency, say the one corresponding to the lower Larmor frequency, but detected at different t_1 delay times, we will see the oscillating signal shown in the top part of Fig. 10.7. The amplitudes corresponding to the individual spectra are shown as discrete points, and it can be seen that they follow a sinusoidal pattern with the same frequency as the

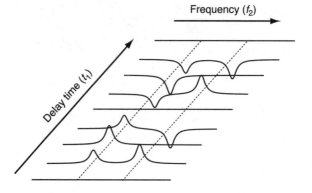

Figure 10.6
Result of transforming the
FIDs from the individual
one-dimensional experiments.

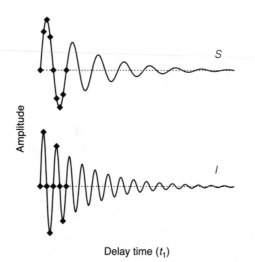

Figure 10.7
Fluctuations in the peak
amplitudes in the one-
dimensional spectra shown
in Fig. 10.6.

precession during the evolution period. If the precession is allowed to go for longer periods, the amplitude will decay due to relaxation, just as it does during an FID. Similarly, the points taken from the individual spectra at the frequency corresponding to the higher Larmor frequency define another damped sine wave, shown in the lower graph. Again, the frequency of this signal is the same as the precession frequency for the corresponding nucleus.

Each of the "slices" along the t_1 axis can be thought of as an indirect FID, generated by the incremented delay times, and can be Fourier transformed to generate a spectrum. The result is now a two-dimensional surface with each axis representing a frequency, as illustrated in Fig. 10.8.

The frequency dimension that results from transformation of the t_1 times is labeled f_1, and is often referred to as the *indirect dimension*, to distinguish it from the frequency dimension that is generated from the directly recorded FIDs. The indirectly recorded FID is also referred to as an *interferogram*, because of the constructive and destructive interference among the signals with different frequencies. The two-dimensional data set that is generated by transforming only the directly detected FIDs is said to be *half-transformed*.

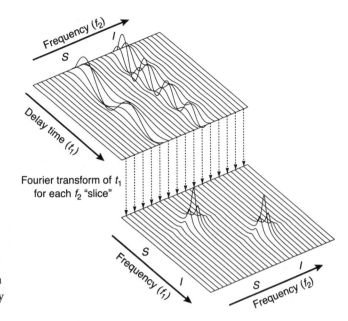

Figure 10.8
Fourier transformation of the second dimension of a two-dimensional spectrum. Each slice along the t_1 dimension of the half-transformed spectrum is transformed into a frequency slice in the final spectrum.

We now have a two-dimensional NMR spectrum, but not a very interesting one. The surface has only two peaks, each of which lies on a diagonal line where the frequencies on the two axes are equal. In order to be useful, a two-dimensional experiment must include some mechanism by which off-diagonal peaks are generated when there is some interaction between pairs of nuclei. Such a mechanism can involve either scalar coupling among covalently linked nuclei, or nuclear Overhauser effects between nuclei that are relatively close in space.

10.3   Magnetization Transfer Via the Nuclear Overhauser Effect: The NOESY Spectrum

One of the most important two-dimensional NMR experiments is the NOESY (for Nuclear Overhauser Effect SpectroscopY) experiment. As its name suggests, cross peaks in the resulting spectrum arise from the ability of the spin state of one nucleus to influence the spin state of another through cross-relaxation. As we have seen in our previous discussion of the nuclear Overhauser effect, this process is relatively easy to explain using vector diagrams, without relying on a detailed quantum mechanical treatment. The NOESY experiment thus provides a good starting point for discussing real two-dimensional experiments.

The pulse sequence for the NOESY experiment, shown in Fig. 10.9, is only slightly more complicated than the two-pulse sequence we have considered so far. There is now an additional $\pi/2$-pulse that follows the evolution period, t_1, which is then followed by a mixing time, τ_m, before the final read pulse. As before, the individual one-dimensional spectra that make up the experiment are distinguished by the length

Figure 10.9

Pulse sequence for a two-dimensional NOESY spectrum.

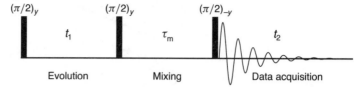

of the t_1 delay. The mixing time, however, remains constant through the experiment.[2] During this mixing time, cross-relaxation occurs, eventually leading to cross peaks in the spectrum.

To see how this happens, we again look at the magnetization vectors that are generated by different t_1 delay times, using an idealized pair of spins with different precession frequencies. An important feature of real NOESY experiments is that they are designed so that the only magnetization components that contribute to the final spectrum are those that were aligned with the z-axis after the second pulse. We will discuss how and why this is done later, but for now we will just take advantage of the fact that it simplifies the analysis considerably. Fig. 10.10 illustrates how the z-magnetization components depend on the t_1 delay and the precession frequencies of the two spins. Note that the z-components from both spin populations fluctuate with the delay time, according to their different precession frequencies.

Next comes the mixing time, τ_m, during which the non-equilibrium population differences have a chance to relax. Before considering this process, however, we will look at the FIDs that would be generated if the second pulse were immediately followed by the third, considering only the z-components present after the second pulse. In order to make what follows a little clearer, the third pulse is defined as being along the $-y$-direction, with the results shown in Fig. 10.11.

The FIDs arising from each of the t_1 delay times are plotted in Fig. 10.12, along with the individual one-dimensional spectra generated by Fourier transformation of each FID. The FIDs and spectra are drawn so that a magnetization projection along the positive x'-axis at the beginning of the FID gives rise to a positive peak in the spectrum. Note that the second and fourth spectra contain only contributions from spin S, while the third contains only the signal from spin I.

Now, we will consider the relaxation processes that can occur during a mixing time between the second and third pulses. This can involve both cross-relaxation and other relaxation mechanisms. For simplicity, we will consider just the case of cross-relaxation via the zero-quantum process, as would be observed for slowly tumbling molecules. For even moderately sized proteins, the single-quantum and double-quantum processes will make relatively small contributions.

The cross-relaxation effects will be essentially the same as those that were discussed in Chapter 9 in the case of the transient NOE experiment. In that case, we saw that cross-relaxation following inversion of one spin led to a decrease in the abso-

---

[2] To help distinguish between delay periods that are fixed within an experiment and those that are incremented, the Greek "τ" will be used to identify the former, and the Roman "t" for the latter.

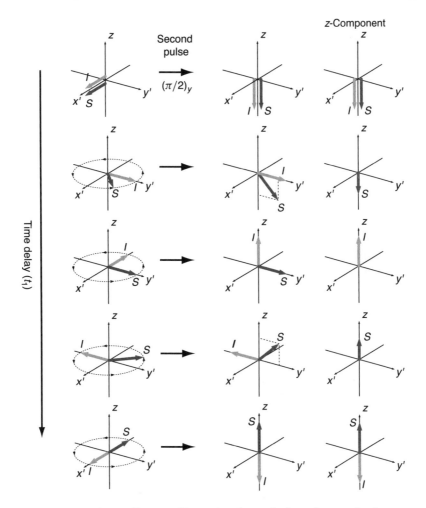

Figure 10.10 Vector diagrams illustrating the evolution of magnetization during the t_1 delay period of the NOESY experiment, and its conversion into z-magnetization components with the second pulse.

lute signal intensity from both spins (for slowly tumbling molecules). More generally, the zero-quantum transition will always tend to decrease the difference in the z-components of the two magnetization vectors. Furthermore, the extent of the change will depend on how large this difference is at the beginning of the mixing period. These changes are shown in Fig. 10.13, along with the vector diagrams showing the situation after the final $\pi/2$-pulse.

Note that when there is no difference in the two magnetization vectors (top row), there is no net cross-relaxation. On the other hand, when the difference is maximum (bottom row), both vectors are strongly affected. Because the relative magnitudes of the two vectors before cross-relaxation are determined by their independent precession during the t_1 delay period, the final magnetization vectors are each influenced by *both* precession frequencies. This process by which the magnetization of one spin

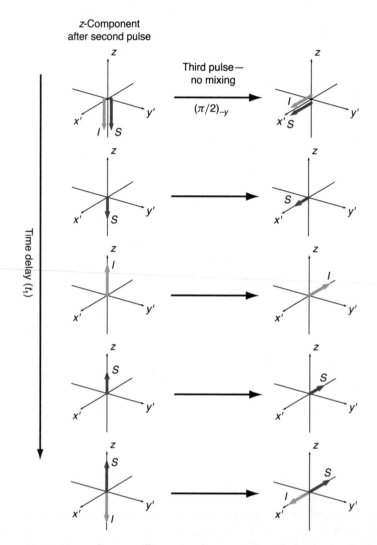

Figure 10.11   Vector diagrams illustrating the conversion of z-magnetization into x'-magnetization by the third pulse of the NOESY experiment, in the absence of the mixing period between the second and third pulses.

is influenced by the orientation of the other is referred to as *magnetization transfer*. The resulting FIDs and spectra are shown in Fig. 10.14. Carefully compare this set of FIDs and spectra with those in Fig. 10.12. In the absence of cross-relaxation, the second, third and fourth spectra contained only components from one or the other spin. Now, all three have contributions from both spins. The only spectrum that is not affected by cross-relaxation is the one corresponding to $t_1 = 0$, where there was no difference in the z-magnetization components to drive cross-relaxation.

Fig. 10.15 shows the result of transforming the t_1 dimension to generate the two-dimensional spectrum. Now there are cross peaks! These peaks are located at the off-diagonal positions where the frequency in one dimension is that of spin S, and

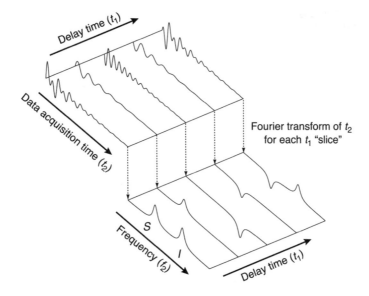

Figure 10.12   FIDs and Fourier transform to generate individual one-dimensional spectra in the NOESY experiment, in the absence of the mixing period between the second and third pulses.

the frequency in the other dimension is that of spin I. The presence of these peaks reflects the fact that the amplitudes of the signals in each one-dimensional spectrum are influenced by the precession of the other spin during the t_1 delay period.

The intensities of the cross peaks reflect the efficiency of the magnetization-transfer process. In the case of a NOESY spectrum, this depends on the tumbling rate and the distance between the nuclei. Generally, ^{1}H–^{1}H NOESY cross peaks are observed only when the two nuclei are within about 5 Å of one another, and the intensity depends strongly on this distance. Because of this distance effect, the NOESY spectrum is extremely useful for determining the three-dimensional structures of both small and large molecules. As a simple example, we can consider what would happen if there were three nuclei, A, B and C. A NOESY spectrum for these nuclei might look like Fig. 10.16. In this highly simplified example, there are two cross peaks above the diagonal and two below. One set of cross peaks appears at the positions where the frequencies for A and B intersect, and the other pair is located at the intersections between the frequencies for A and C. In principle, the spectrum should be symmetrical about the diagonal, and the information from the two sides is redundant, but often the signals on one side or the other may be stronger or clearer. Significantly, there is no cross peak between nuclei B and C. From this information we might reasonably conclude that B and C are more than 5 Å away from one another, but A lies between them.

For even moderately complex two-dimensional spectra, the surface representation used in the previous figures is too cluttered, and it is more common to use a

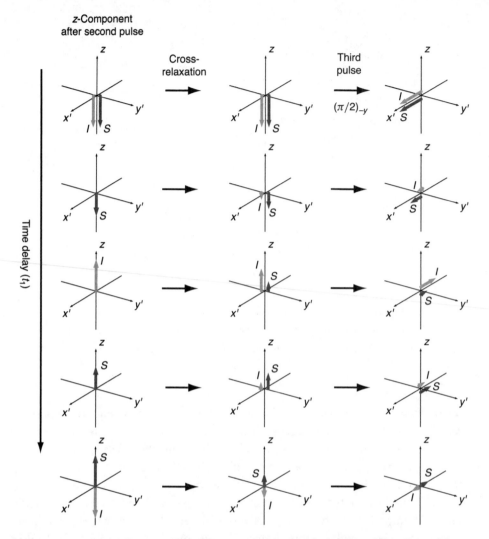

Figure 10.13 Vector diagrams illustrating cross-relaxation during the mixing time of the NOESY experiment and conversion of the z-magnetization into x' magnetization by the third pulse.

contour plot to represent the spectrum, as shown in Fig. 10.17 for the same spectrum. This representation makes it much easier to see which pairs of nuclei give rise to cross peaks, but it also makes it easier to miss (or deliberately hide) noise or weak signals below an arbitrary level used for the first contour.

There is another important difference between these simulated spectra and real ones. In real spectra, the diagonal peaks are much stronger than the cross peaks, reflecting the relatively small fraction of the magnetization of any one spin that is influenced by another during the mixing period. The individual cross peaks might typically be only 1% as strong as a diagonal peak. The contour plot shown in Fig. 10.18 is a real ^{1}H–^{1}H NOESY spectrum, collected for a small protein with a well-defined three-dimensional structure (a conotoxin).

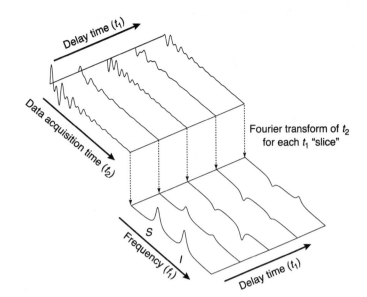

Figure 10.14
FIDs and Fourier transform to generate individual one-dimensional spectra in the NOESY experiment, with cross-relaxation during the mixing period between the second and third pulses.

Figure 10.15
Fourier transformation of the second dimension of the NOESY spectrum. Each slice along the t_1 dimension of the half-transformed spectrum is transformed into a frequency slice in the final spectrum.

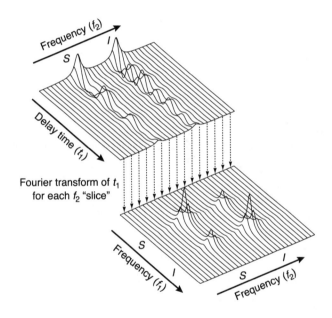

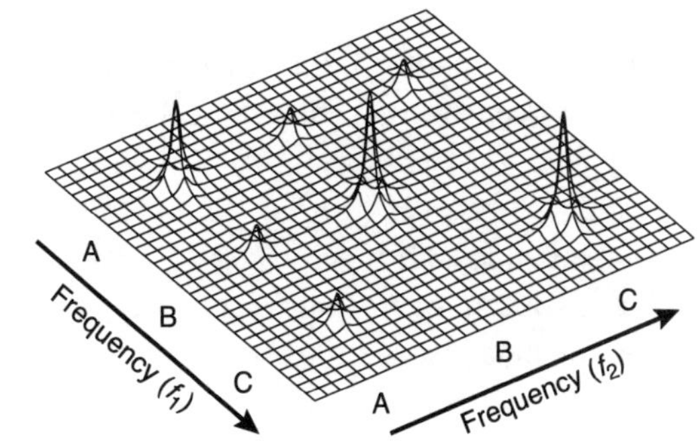

Figure 10.16
A simple example of a NOESY spectrum for three nuclei. Pairs A–B and A–C form cross peaks, but B–C does not.

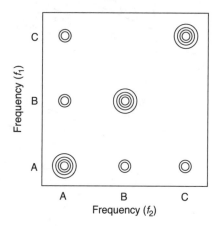

Figure 10.17

A contour plot representation of a simple NOESY spectrum for three nuclei.

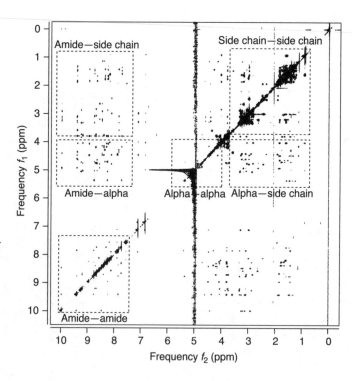

Figure 10.18

A ^{1}H–^{1}H NOESY spectrum of a small protein, ω-conotoxin MVIIA-Gly, composed of 27 amino acid residues. Regions of the spectrum usually associated with protons of different chemical classes are indicated. The strong vertical feature near the center of the spectrum is due to ^{1}H nuclei of the solvent water molecules.

As discussed in Chapter 4, the different classes of protons in a protein generally have resonance frequencies in characteristic regions of a one-dimensional spectrum. As a consequence, different regions of the two-dimensional spectrum generally represent interactions between protons from identifiable classes. The borders between these regions are shown *very approximately* by the dashed lines in the spectrum above. While these classifications are useful in orienting a spectrum, exceptions are common.

In order to obtain useful information from a NOESY spectrum, it is necessary to assign the individual peaks to specific proton pairs in the covalent structure. Except for very simple molecules, this requires additional multidimensional experiments that provide information about covalent connectivities. Experiments based on scalar coupling, including the COSY and TOCSY (TOtal Correlation SpectroscopY) experiments, are described in Chapters 15 and 16.

In addition to cross-relaxation, there is another mechanism by which cross peaks can appear in a NOESY spectrum—namely, chemical or conformational processes that alter the Larmor frequency of a nucleus. As discussed in Section 7.4, an exchange processes that is relatively slow, compared to the difference in Larmor frequencies associated with alternate conformations, will give rise to two peaks in a one-dimensional spectrum. If such an exchange event occurs during the mixing time of a NOESY experiment, the nucleus may have one precession frequency during the t_1 evolution period and an alternate frequency during the t_2 data-acquisition period, which will give rise to a cross peak in the resulting spectrum. Although this effect can be a source of confusion in the interpretation of NOESY spectra, it can also be exploited to characterize exchange processes, and variations of the NOESY experiment have been devised to eliminate the true NOE cross peaks, leaving only those arising from conformational or chemical exchange.

10.4   Eliminating Unwanted Magnetization: Phase Cycling

In describing the NOESY experiment, it was noted that the experiment is conducted in a way that eliminates all but the longitudinal component of the magnetization that is present at the beginning of the mixing period—that is, right after the second $\pi/2$-pulse. We will now return to this point and discuss why and how this is done.

First of all, consider what happens to the transverse magnetization that is present after the second pulse. The vector diagrams in Fig. 10.19 illustrate the changes in magnetization due to the pulses and the intervening delay time, t_1, for a particular precession frequency.

After the second pulse, there are projections of the magnetization along the z- and y'-axes, but not along the x'-axis. Our treatment of the NOESY experiment ignored the y'-component. Now, let's see if that was legitimate. For this purpose, it is useful to draw the two components of the net magnetization separately, as shown in the second diagram of the set in Fig. 10.20. During the mixing time, the z-magnetization is unaffected (except, of course, through relaxation and cross-relaxation), whereas the transverse component precesses, to a degree that depends on the Larmor frequency and the length of the delay time. Provided that the transverse component does not lie

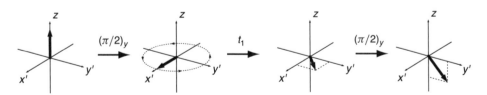

Figure 10.19 Evolution of transverse magnetization during the t_1 delay time in the NOESY experiment.

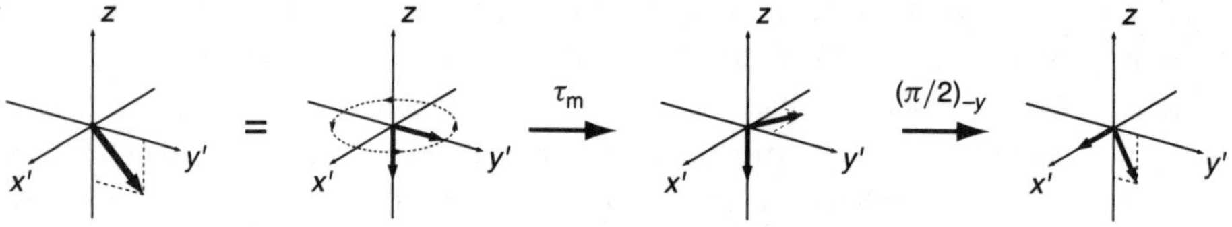

Figure 10.20 Evolution of transverse and longitudinal magnetization components during the mixing time in the NOESY experiment.

entirely along the x'-axis at the end of the mixing time, some of it will remain in the transverse plane after the final pulse, as shown in the rightmost diagram in Fig. 10.20. This component will contribute to the signal induced in the spectrometer coil during the FID, changing both its amplitude and phase. Clearly, the transverse component present after the second pulse does influence the spectrum. If, however, the evolution of the magnetization is as simple as indicated above, then the transverse components will not be influenced by other spins and will only affect the diagonal peaks of the final spectrum.

The situation is more complicated, however, because the transverse components, in fact, can be influenced by other spins. First, longitudinal relaxation during the mixing time, including cross-relaxation, will reduce the magnitude of the transverse components. Second, there are other mechanisms by which magnetization can be transferred among the transverse components arising from different spins. These effects will then also contribute to cross peaks, sometimes in rather complicated ways. In particular, scalar coupling between nuclei can give rise to magnetization transfer among the transverse components, and this is the basis for an entirely different class of multidimensional experiments, such as COSY. If, however, these effects are added to the nuclear Overhauser transfer, the spectrum would be almost impossible to interpret. This is the primary reason that the net effects of the transverse components must somehow be eliminated during the mixing period.

The most common way of eliminating the net contributions of the transverse components involves recording a series of individual FIDs with pulse sequences that are designed so that the contributions from the transverse components cancel out when the FIDs are added together. As an example, consider the same experiment as described above, but with the first two pulses modified so that they are directed along the *negative* y'-axis. In this case, the magnetization after the first two pulses and the t_1 delay is as shown in Fig. 10.21. Note that this is exactly the same as before, but now the transverse component points along the negative y'-axis.

Following the longitudinal and transverse magnetization components through the mixing period and the final pulse, we see the result shown in Fig. 10.22. Again, everything is the same, except for the orientation of the magnetization component that was transverse at the beginning of the mixing period. After the mixing time and the final pulse, a component of this magnetization now lies along the *negative* y'-axis, whereas this component pointed in the opposite direction at the end of the

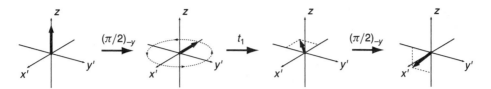

Figure 10.21 Evolution of transverse magnetization during the t_1 delay time in the NOESY experiment, as in Fig. 10.19, but with the directions of the first and second pulses reversed.

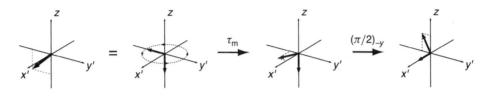

Figure 10.22 Evolution of transverse and longitudinal magnetization components during the mixing time in the NOESY experiment, as in Fig. 10.19, but with the directions of the first and second pulses reversed.

Table 10.1   A phase-cycling scheme for the NOESY experiment

Sequence	Pulse 1	Pulse 2	Pulse 3
A	$(\pi/2)_y$	$(\pi/2)_y$	$(\pi/2)_{-y}$
B	$(\pi/2)_{-y}$	$(\pi/2)_{-y}$	$(\pi/2)_{-y}$
C	$(\pi/2)_x$	$(\pi/2)_x$	$(\pi/2)_{-y}$
D	$(\pi/2)_{-x}$	$(\pi/2)_{-x}$	$(\pi/2)_{-y}$

original pulse sequence. If the FIDs from the two pulse sequences are combined, these contributions will cancel out, thus eliminating all but the component that was longitudinal during the mixing period. (Note that this was done by modifying the phases of two of the three pulses. What would happen if all three of the pulses were reversed?)

This procedure is called *phase cycling* and is an important part of all multidimensional experiments. In such experiments, each FID that is subjected to Fourier transformation is actually the sum of several acquired signals using different combinations of pulse phases. In real experiments, including the NOESY, the cycle is almost always more complex than indicated in the example above. This is because the magnetization transfer processes that can occur among the transverse components will generally generate multiple new magnetization components that must be eliminated. For instance, a phase cycle for the NOESY spectrum can be made up of the four pulse sequences outlined in Table 10.1.

In addition to eliminating unwanted magnetization components, this process also increases the signal-to-noise ratio in the spectrum, because the signals from the multiple FIDs combine linearly, whereas the noise accumulates in proportion to its square root. In fact, the phase cycle is often repeated several times for each t_1 delay in order to further increase the signal-to-noise ratio.

10.5   Quadrature Detection for the Indirect Dimension

In Chapter 6, the use of quadrature detection in one-dimensional Fourier transform NMR was discussed. Recall that the need for this method arises because the technique for recording the FID involves mixing the signal detected in the spectrometer coils with a reference signal that has the same frequency as the rotating frame. This mixing (followed by removal of a high-frequency component by electronic filters) reduces the frequency of the signal to a range that can be readily converted to a digital form. But, the cost of this convenience is that it is not possible to distinguish between frequencies that are less than or greater than the reference frequency by a given amount. Quadrature detection allows discrimination between positive and negative frequencies by incorporating the information from two signals that differ in phase by 90°. When two-dimensional spectra are recorded, quadrature detection is used for each of the directly detected FIDs that are transformed to generate the individual one-dimensional spectra.

An analogous problem also arises with respect to the indirect frequency dimension that is generated in the second Fourier transformation. Consider, for instance, the first steps in the NOESY experiment, as illustrated again in Fig. 10.23 for a specific t_1 interval. As discussed earlier, the z-projections following the second pulse reflect the rates at which the magnetization vectors precess, relative to the rotating frame, during the t_1 period. In the example shown in Fig. 10.23, the precession frequency is greater than the rotating-frame frequency. Now, let's look at what happens if the vectors precess at a rate less than that of the rotating frame, as in Fig. 10.24. These vectors have been drawn to represent frequencies (in the rotating frame) that are equal but opposite in sign from those shown in Fig. 10.23. Note that the resulting z-projections are exactly the same. Because the subsequent steps in the NOESY experiment reflect only the z-magnetization components present at this point, the two situations will give rise to the same FIDs and spectra following transformation.

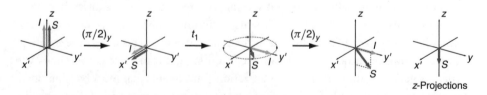

Figure 10.23 Evolution of transverse magnetization components during the t_1 delay time in the NOESY experiment.

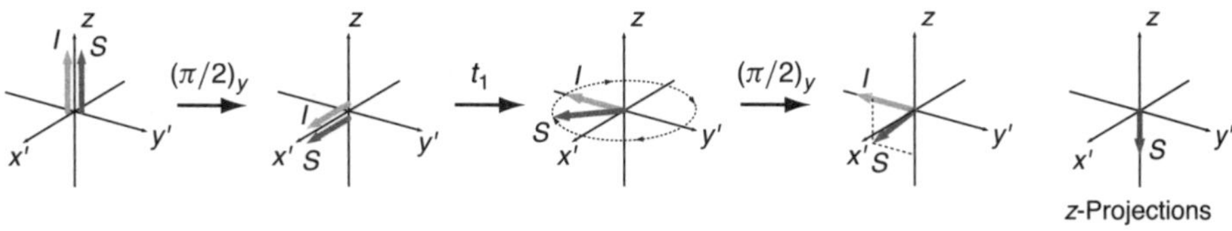

z-Projections

Figure 10.24 Evolution of transverse magnetization components during the t_1 delay time in the NOESY experiment, as in Fig. 10.23, but with precession frequencies less than the rotating frame frequency.

As we saw in the case of one-dimensional spectra, the key to distinguishing positive and negative frequencies is to use signals that differ in phase by 90° as the real and imaginary parts of a complex signal in the Fourier transformation. For directly detected FIDs, this is usually done by using separate reference signals with different phases to generate the two heterodyne signals simultaneously. For the indirect dimension of a two-dimensional signal, the phase-shifted signals are generated by sequentially carrying out two versions of the experiment for each t_1 interval.

To see how this works, we need to carefully consider how the amplitude of the directly detected FIDs change as t_1 is incremented. As defined in the example used in the earlier sections, the final pulse prior to data acquisition rotates the magnetization in such a way that negative z-magnetization becomes positive x'-magnetization. Ignoring for a moment the effects of the cross-relaxation during the mixing period, the amplitude of the resulting FID will be proportional to the magnitude of the negative z-magnetization after the second pulse. It should be apparent from the vector diagrams in Fig. 10.11 that the amplitudes will be maximum for $t_1 = 0$ and will then vary cyclically as t_1 is incremented. For each spin population, this pattern is described by a cosine function of t_1:

$$A(t_1) = A_0 \cos(t_1 \nu 2\pi)$$

where A_0 is the initial amplitude for $t_1 = 0$ and ν is the Larmor frequency of the spin of interest in the rotating frame. The amplitude is said to be *cosine modulated* with respect to t_1.

By analogy with quadrature detection for a one-dimensional spectrum, what we now need is a pulse sequence that will generate equivalent FIDs with amplitudes that are *sine modulated* with respect to t_1. That is, the amplitude should be zero for $t_1 = 0$ and then increase and decrease cyclically as the vectors precess for longer periods of time during the t_1 delay. This can be accomplished by changing the direction of the second pulse so that it is applied along the $-x'$-axis instead of y'. When $t_1 = 0$, the first two pulses lead to the results shown in Fig. 10.25.

The z-projections for both spins are zero and neither spin will contribute to the FID after the third pulse (provided that the transverse components present during the mixing time are eliminated by phase cycling, as described in the previous section). Fig. 10.26 shows the situation after a relatively short t_1 interval, for the case where the Larmor frequency is greater than the rotating-frame frequency. Now, the magnitudes

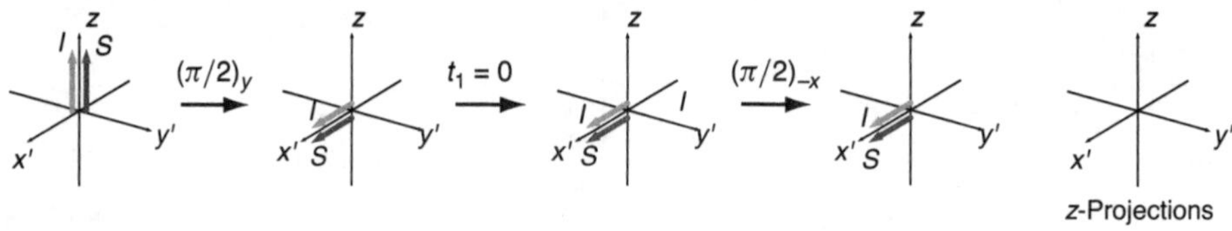

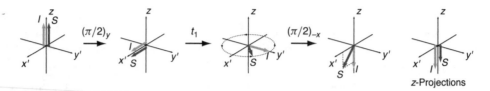

Figure 10.25 The first iteration of the NOESY experiment, $t_1 = 0$, with the direction of the second $\pi/2$-pulse set to $-x$, rather than y, as in the previous examples.

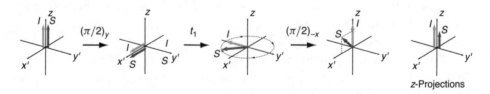

Figure 10.26 A subsequent iteration of the NOESY experiment, with the direction of the second $\pi/2$-pulse set to $-x$, as in Fig. 10.25.

of the z-components of both magnetization vectors have increased (in the negative direction). After the third pulse, these components will contribute to the initial amplitude of the FID. Note also that the z-magnetization component for the I-spin is larger than that for the S-spin, reflecting the fact that I-magnetization precessed further during the t_1 period. With longer delays, the z-components of both vectors will fluctuate cyclically. Importantly, the phase of this process is that of a sine function, as compared to the cosine modulation seen with the original form of the experiment, with the second pulse aligned with the y'-axis.

To illustrate how the addition of this second version of the pulse sequences allows discrimination between positive and negative frequencies, Fig. 10.27 shows what happens when the $-x$-pulse is used and the Larmor frequency is less than that of the rotating frame. Note that following the second pulse, the z-magnetization components have the same magnitude, but the opposite direction, as when the magnetization precesses in the opposite direction relative to the rotating frame. When the third pulse is applied after the mixing time, the sign of the FID will be reversed.

In two-dimensional experiments, the usual practice is to record separate FIDs using each of two pulse sequences that differ by the phase of one or more pulses so as

Figure 10.27 The same t_1 increment in the NOESY experiment as in Fig. 10.26, but for the case of negative precession frequencies (relative to the rotating frame).

to generate one set of signals that is cosine modulated with respect to t_1 and another set that is sine modulated. These are analogous to the two signals that are generated during one-dimensional quadrature detection by using either two perpendicular coils in the spectrometer or, more commonly, by mixing the signal from a single coil with two reference signals that differ in phase by 90°. Following the Fourier transformation of the first dimension, the real part of one of the two resulting one-dimensional spectra, for a given t_1 interval, is multiplied by i and added to the real part of the other to generate a spectrum made up of complex numbers. The full set of complex spectra are then subjected to Fourier transformation in the second dimension. The mathematics is completely analogous to the treatment outlined in Section 6.7 for the one-dimensional spectrum, and the final result is a spectrum in which each resonance gives rise only to positive peaks and with the correct sign of the frequency identified.

The alteration in pulse phases to facilitate the discrimination between positive and negative frequencies is distinct from that discussed in Section 10.4, where the goal was to eliminate unwanted magnetization components. In practice, both goals are critically important, and the cosine- and sine-modulated signals used for quadrature detection are each generated by combining the FIDs from a phase cycle designed to eliminate spurious magnetization. Thus, a quite large number of individual FIDs must usually be recorded for each value of t_1.

There are a few commonly used variations on the quadrature-detection scheme outlined here, and they go by a few different names. The technique described above is often called the *States method*, for the first author of the paper in which it was first described, or the *hypercomplex method*, a name that is suggestive of the idea that the real and imaginary parts of the data used for the second transform are, themselves, generated from quadrature detection in the first dimension. A slightly different approach is described as *time-proportional phase incrementation* or TPPI. The specific details are beyond the scope of this text, but they do become important in the design of multidimensional experiments and in processing the resulting data.

10.6   Field Gradients

Along with the phase-cycling method described in Section 10.4 for specifically eliminating magnetization components, a second, complementary method for achieving this goal has become widely used since approximately the 1980s. This method involves briefly introducing an inhomogeneity, described as a *pulsed field gradient* (PFG), in the external magnetic field, using special coils incorporated in the spectrometer probe. Although it may seem odd to deliberately create an inhomogeneous magnetic field, the use of field gradients dates back to the work of Carr and Purcell, who described them in their classic paper of 1954, in which they also introduced the use of π-pulses to measure transverse relaxation rates. Since then, field gradients have been used for a variety of purposes, some of which are introduced here.

The design of modern field-gradient coils is rather complex, but the general idea is illustrated schematically in Fig. 10.28, with a coil for generating a magnetic field gradient along the z-axis of an NMR sample. As shown in the figure, the axes of the

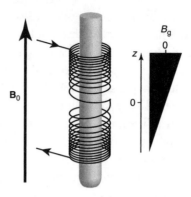

Figure 10.28   Schematic representation of a coil used to generate a field gradient along the z-axis of an NMR sample. The uniform magnetic field due to the superconducting magnet, $\mathbf{B}_0$, is supplemented by the much weaker field produced by the gradient coil, the strength of which (B_g) varies linearly along the z-axis, as indicated by the width of the triangle.

z-gradient coil and sample tube coincide, and the wire is wound so that the spacing between the coils varies along the z-axis. The direction of the windings is reversed at the center point, defined as $z = 0$. With the current direction as indicated in the figure, the top half of the coil adds positively to the field from the main magnet, $\mathbf{B}_0$, while the bottom half of the gradient coil generates a field with the opposite direction of $\mathbf{B}_0$. The windings are closest at the two ends of the coil, generating the strongest field, and become more widely spaced toward the center, where they change direction. As the center point is approached from the top, the strength of the gradient field, B_g, approaches zero, before changing sign and becoming more negative toward the bottom. If properly implemented, the gradient coil results in a field strength that is a linear function of z, given by

$$B_g = zG$$

where G is the gradient strength, specified in units of T/m or G/cm. Though somewhat more complex in their construction, coils to create gradients along the x- and y-directions (in the laboratory frame) are also frequently incorporated in modern NMR probes.

10.6.1   Elimination of Transverse Magnetization

During the time that the z-gradient coil is activated, chemically equivalent nuclei in molecules at different positions along the z-axis precess at different angular frequencies, $\omega(z)$:

$$\omega(z) = -\gamma(B_0 + B_g) = -\gamma B_0 - \gamma zG = \omega_0 - \gamma zG$$

where γ is the gyromagnetic ratio, and ω_0 is the angular frequency at the gradient center. If the rotating-frame frequency is taken to be ω_0, the angular displacement of the magnetization vector from its original position in the rotating frame, after a time

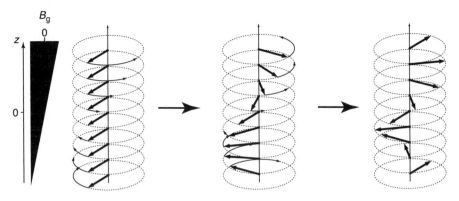

Figure 10.29 Vector diagrams illustrating the generation of position-dependent phase differences in an NMR sample by application of a field gradient. Vectors representing nuclei at different positions along the z-axis of the sample are similarly displaced in the vector diagram. The nuclei at different positions precess with different Larmor frequencies, determined by the magnetic field established by the gradient. After a sufficient time period, the phase angles of the magnetization vectors are uniformly distributed about the z-axis, thus eliminating the net transverse magnetization.

period, t, is

$$\phi(z) = -\gamma zGt$$

The effect on the precession of nuclei at different positions along the gradient is illustrated using vector diagrams in Fig. 10.29.

The variation in precession frequency results in a loss of phase coherence among the magnetization vectors from molecules in different parts of the sample, eliminating any net transverse magnetization. Any magnetization along the z-axis, however, is unaffected and can be recovered in the $x'-y'$ plane with a $\pi/2$-pulse. In the case of the NOESY experiment, the pulsed field gradient is applied during the mixing period.

Although the vector diagrams in Fig. 10.29 may suggest that the net transverse magnetization can be eliminated by application of a field gradient for a time sufficient to generate a phase difference of 2π rad along the total length of the sample, longer times are, in practice, required to effectively eliminate the magnetization. This is because the residual magnetization oscillates with time during application of the field gradient, as the magnetization vectors for different sample positions come in and out of phase, and small timing errors or other field inhomogeneities may leave a net transverse magnetization in some direction. The amplitude of this oscillation decays with time, however, making it possible to reduce the residual magnetization to an acceptable level.

The reduction in the total transverse magnetization during application of a field gradient can be calculated by considering the integral of the magnetization along the full length of the field gradient. If the initial magnetization lies along the x'-axis and is $I_x(0)$, then the I_x-magnetization at position z will, after time, t, be

$$I_x(z) = I_x(0) \cos\big(\phi(z)\big) = I_x(0)\cos(\gamma zGt)$$

If the total length of the sample enclosed by the field gradient is l, and the net average magnetization at time zero is $\bar{I}_x(0)$, then the average magnetization after time, t, is the integral of I_x along the z-axis, from $-l/2$ to $l/2$, divided by the total length:

$$\bar{I}_x(t) = \bar{I}_x(0)\frac{1}{l}\int_{-l/2}^{l/2}\cos(\gamma zGt)dz$$

$$= \bar{I}_x(0)\frac{\sin(\gamma zGt)}{\gamma Gt}\bigg|_{-l/2}^{l/2}$$

$$= \bar{I}_x(0)\frac{2\sin(\gamma lGt/2)}{\gamma lGt} \tag{10.1}$$

As suggested above, the residual magnetization displays an oscillatory behavior, with an amplitude that decreases with time. Once the denominator of Eq. 10.1 is greater than one, the upper envelope of the function, which represents the maximum value of the remaining magnetization, is given by

$$\bar{I}_x(t)_{\text{env}} = \frac{2\bar{I}_x(0)}{\gamma Glt} \tag{10.2}$$

In modern (circa 2015) spectrometers for liquid samples, a typical gradient strength lies in the range of 0.1 to 1 T/m, and the sample length might be about 1 cm. In Fig. 10.30, the predicted reduction in net transverse magnetization is plotted as a function of time, for a sample of ^{1}H nuclei, a gradient strength of 0.1 T/m and a gradient length of 1 cm. The dashed curve in the figure represents the upper envelope of the decay function, calculated according to Eq. 10.2. Eq. 10.2 can be used to estimate the duration of a PFG required to suppress transverse magnetization to a given level.

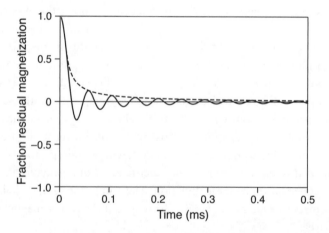

Figure 10.30 Predicted reduction of transverse magnetization for a sample of ^{1}H nuclei under the influence of a magnetic field gradient of 0.1 T/m and a gradient length of 1 cm. The solid curve, representing the oscillating residual magnetization, was calculated from Eq. 10.1. The dashed curve represents the envelope of the decay function and was calculated according to Eq. 10.2.

If, for instance, the transverse magnetization of a ^{1}H sample is to be reduced to 1% of its initial value using a gradient of 0.1 T/m and a sample length of 1 cm, a pulse of at least 0.75 ms would be used. Note that the residual magnetization is also inversely proportional to the gyromagnetic ratio, so that stronger gradients or longer times are required for nuclei with smaller magnetic dipoles, such as ^{13}C or ^{15}N.

10.6.2   Diffusion and Solvent Signal Suppression

The use of pulsed field gradients to eliminate transverse magnetization, often referred to as *spoiler* or *homospoil* gradients, is but one of several applications of such gradients. Other applications take advantage of the ability to reverse the effects of one gradient by application of a second with the opposite direction. If the direction of the current applied to the coil diagrammed in Fig. 10.28 is reversed, then the direction of the gradient is also reversed, with the largest field at the bottom of the sample, and the smallest at the top. Fig. 10.31 illustrates the effects of this reversed gradient on a dephased sample.

With the direction of the gradient reversed, so is the gradient of precession frequencies. Provided that the strength and duration of this gradient are the same as those for the first, and the molecules have not changed positions significantly, the phase coherence of the nuclei will be restored, as indicated in the vector diagrams. The same effect can be generated by introducing a π-refocusing pulse between two gradient pulses with the same direction.

If the molecules in a sample diffuse significantly during a delay between two gradient pulses of opposite direction (or two PFGs of the same direction but separated by a refocusing pulse), the phase coherence will not be restored completely, because some of the molecules will precess at different frequencies during the two periods.

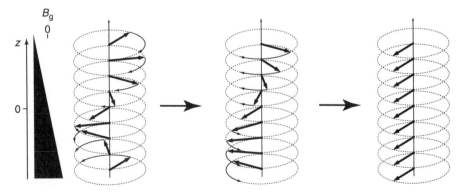

Figure 10.31   Vector diagrams illustrating the reversal of the position-dependent phase differences generated by application of a PFG, as in Fig. 10.29. The direction of the field gradient is reversed by reversing the direction of the electric current through the gradient coil, so now the strongest field is at the bottom of the sample. If the gradient is applied for exactly the same duration as the original gradient, the precession angles are brought back into phase.

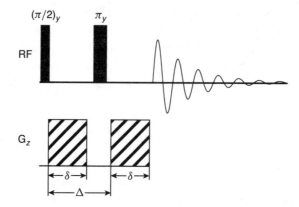

Figure 10.32 A pulse sequence for an experiment incorporating PFGs to measure molecular diffusion. The upper line represents standard radio-frequency pulses and the resulting FID, and the lower line shows two PFGs with the same direction. The π-pulse separating the two PFGs refocuses magnetization that has diverged because of chemical-shift differences and reverses the effective direction of the second PFG. δ is the duration of each PFG, whereas Δ is the delay between the beginning of the first PFG and the beginning of the second.

A simple pulse sequence designed to illustrate this effect is shown in Fig. 10.32 The duration of the gradient pulses is δ and the delay between the beginning of the first pulse and the beginning of the second is Δ. The π-pulse refocuses differences in precession frequencies due to chemical-shift differences and reverses the effect of the second PFG relative to the first. The reduction in the signal due to diffusion is described by an equation published in 1965 by E. O. Stejskal and J. E. Tanner:

$$\frac{S}{S_0} = e^{-D\gamma^2 G^2 \delta^2 (\Delta - \delta/3)} \tag{10.3}$$

where S and S_0 are the signal intensities observed with and without the gradient pulses, respectively, and D is the diffusion coefficient of the molecule. This relationship forms the basis for experiments to measure molecular diffusion coefficients. In practical experiments of this type, additional refinements are introduced to reduce a variety of artifacts, as described in references listed at the end of this chapter.

Another important application of PFGs is in the suppression of signals from solvent molecules. Particularly in the study of biological macromolecules, where the concentration of the macromolecular species is typically on the order of 1 mM and water is present at about 55 M, suppression of the solvent signal is critical. As indicated by Eq. 10.3, the rate of signal decay between paired PFGs is proportional to the diffusion coefficient, D, which is approximately 10-fold greater, or more, for water than for most proteins. (At room temperature, $D \approx 2 \times 10^{-9}$ m^2/s for water and 1×10^{-10} m^2/s or less for proteins.) With careful choice of the gradient strength and delay periods, it is possible, in principle, to reduce the solvent signal by a factor of 10^5 or more while leaving the protein signal nearly unaffected. In practice, however, the much greater transverse relaxation rates for the protein nuclei may limit the usefulness of this method, by itself, for solvent suppression. Nonetheless, pulsed field gradients are an important component, along with solvent-selective pulses, of more complex pulse sequences currently used for solvent suppression.

10.7   A Brief Introduction to Imaging by NMR

Perhaps the most far-reaching application of magnetic field gradients in NMR is their use to create images of macroscopic objects, widely referred to as *magnetic resonance imaging*, or MRI. Although MRI is a vast subject in itself, and well beyond to the focus of this book, the basic principles follow directly from our discussion of pulsed field gradients and two-dimensional spectroscopy, making a brief diversion into imaging appropriate here.

To introduce the basic idea of MRI, consider first what would happen if we were to apply a field gradient along the z-axis while recording the FID for a simple one-dimensional NMR spectrum of a homogenous sample, such as water. In the absence of the gradient, Fourier transformation of the FID would generate a spectrum with a single peak. But with the gradient, there will be a continuous range of precession frequencies, reflecting the different positions of nuclei along the direction of the gradient, and the resulting spectrum will, under ideal circumstances, have the shape of a rectangle, as illustrated in Fig. 10.33.

Note that the FID has the form of the plot shown earlier in Fig. 10.30, which represents the evolution of net magnetization along a transverse direction in a field gradient. In this and the other hypothetical examples shown in this section, the decay of the spectrometer signal is due entirely to the cancelation of magnetization arising from nuclei at different positions in the gradient; no relaxation by other mechanisms has been assumed. The mathematical representation of this FID is the

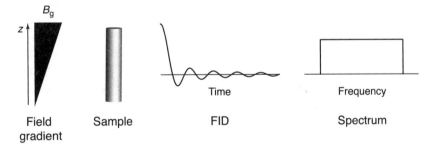

Figure 10.33 Schematic representation of a simple NMR imaging experiment applied to a uniform sample in an NMR tube. A field gradient is applied along the z-axis during the data acquisition period of a one-dimensional pulse NMR experiment. Nuclei at different positions along the z-axis have different Larmor frequencies and generate the rectangular spectrum. In the illustrations shown here, the region of the sample from which the FID is recorded is assumed to be somewhat smaller than the field gradients, and any effects due to the edges of the gradient are ignored.

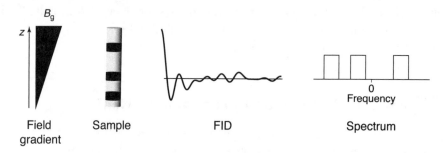

Figure 10.34 Schematic representation of a simple NMR imaging experiment applied to a sample consisting of three sub-volumes containing water (represented in black) embedded in a matrix devoid of ^{1}H nuclei (gray). The sub-volumes are exposed to different portions of the field gradient along the z-axis and generate corresponding rectangular peaks in the resulting NMR spectrum.

sine cardinal, or sinc, function: $\text{sinc}(x) = \sin(\pi x)/(\pi x)$, which was introduced in Chapter 6 (page 148). As discussed there in the context of a sharply truncated FID, the Fourier transform of the sinc function is a rectangular function, as seen here in the spectrum. Different frequencies in the spectrum correspond to different positions in the sample along the z-axis, and the spectral intensity indicates the total number of spins at a given position.

Consider, now, a sample containing three small volumes of water embedded in a solid containing no ^{1}H nuclei (such as fully deuterated wax) at different positions along the length of the sample tube, as illustrated in Fig. 10.34 In this case, only the volumes containing water contribute to the signal, and the signal intensity in the spectrum is limited to the corresponding frequencies. The spectrum thus represents the spatial distribution of the ^{1}H nuclei in the sample—that is, a one-dimensional image. Note also that the FID has a more complex shape than in the previous example, representing a sum of sinc functions.

Fig. 10.35 depicts the same hypothetical experiment applied to a cubic sample, in which the three sub-volumes containing ^{1}H nuclei are located at different positions along the three axes defining a coordinate system in the sample. As in the previous example, the precession frequencies are determined by the positions of the nuclei along the z-axis, and the intensity observed at any point in the spectrum indicates the density of nuclei in a plane perpendicular to the axis at that point. Thus, the spectrum corresponds to a projection of the object along the z-axis.

Now, suppose that the NMR probe contains a second gradient coil, so that we can generate field gradients along the y-axis of the sample. If we apply only this gradient during the data acquisition period, the precession frequencies will be determined by the positions of the nuclei along the y-axis, and the resulting spectrum represents a projection of the object along that axis, as shown in Fig. 10.36

Note that the spectral peaks corresponding to volumes B and C are reversed in position when projected along the z- and y-axes, although this could not be determined from examination of these spectra alone. Following the same logic, a gradient

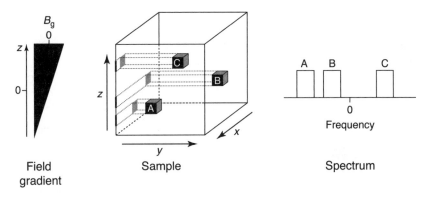

Figure 10.35 Schematic representation of a one-dimensional NMR imaging experiment applied to a sample consisting of three sub-volumes containing water (represented in black) embedded in a cubic volume. The representation of the sample volume shows projections of the water-containing sub-volumes onto the x–z plane and the z-axis. As in Figs. 10.27 and 10.28, a field gradient is applied along the z-axis during the data acquisition period, so the observed frequencies correspond to positions along the z-axis. The resulting spectrum represents a projection of the three-dimensional object along the z-axis. The labels on the spectral peaks identify the corresponding sub-volumes in the sample.

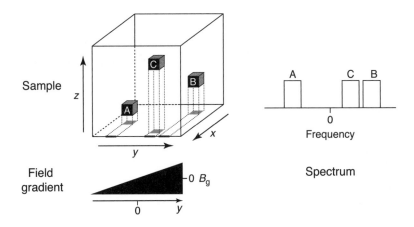

Figure 10.36 Schematic representation of a one-dimensional NMR imaging experiment applied to the same hypothetical example illustrated in Fig. 10.35, but with the field gradient applied along the y-axis during the data acquisition period.

coil along the x-axis of the laboratory frame can be used to generate a spectrum that represents a projection of the sample along that axis.

In general, the projections along the x-, y- and z-axes are insufficient to define the structure of a three-dimensional object, unless the object is known to have special symmetry properties. However, additional projections, along arbitrary axes, can be

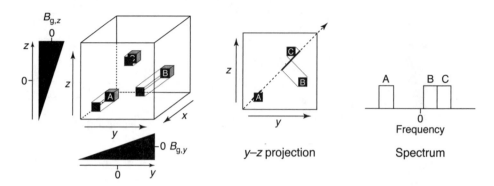

Figure 10.37 Schematic representation of a one-dimensional NMR imaging experiment applied to the same hypothetical example illustrated in Fig. 10.35, with field gradients of equal strength applied along the y- and z-axes during the data acquisition period. The center panel shows a projection of the water-containing volume elements onto the y-z plane. The Larmor frequencies of nuclei at different positions are determined by the sum of their y- and z- coordinates, as represented by the projections of these points onto the diagonal in the y-z plane. The resulting spectrum represents the projection of the three-dimensional object along this diagonal.

obtained by applying multiple gradients in combination during the data acquisition period, as illustrated in Fig. 10.37.

In this example, the y- and z-gradient coils are energized simultaneously. For an arbitrary point in the sample, the local field strength, and therefore the Larmor frequency, depends on the sum of the fields generated by the two gradients, G_y and G_z. The precession frequency is given by

$$v_{y,z} = \frac{\gamma}{2\pi}(yG_y + zG_z)$$

where y and z are the sample coordinates. If the two gradients are of equal strength, this simplifies to

$$v_{y,z} = \frac{\gamma G}{2\pi}(y + z) \tag{10.4}$$

Points in the sample that give rise to a given Larmor frequency will lie on a plane defined by Eq. 10.4. This plane is perpendicular to the diagonal of the y-z projection of the sample, as illustrated in the center panel of Fig. 10.37. The resulting spectrum thus represents a projection of the sample onto this diagonal axis. By adjusting the relative strengths of the y- and z-gradients, the angle of the iso-frequency planes relative to the y-axis can be adjusted, and spectra representing projections along any line in the y-z plane can be generated. By also using an x-gradient coil, a spectrum representing a projection along any direction in the three-dimensional space is possible.

The procedure described above for using field gradients to generate spectra that represent projections of an object along arbitrary axes was first described by Paul C. Lauterbur in 1973. In the same paper, Lauterbur demonstrated the reconstruction

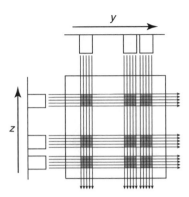

Figure 10.38   Back projection from one dimensional spectra to generate a two-dimensional image consistent with the projections along the z- and y-axes illustrated in Figs. 10.35 and 10.36. Rays from the spectral peaks are projected from the corresponding axes into the area representing the sample. Areas in which the rays from the two directions overlap are indicated by a darker gray and identify possible locations of the nuclei that gave rise to the spectra.

of a two-dimensional image (representing a projection along the perpendicular axis) from multiple one-dimensional projections, using methods analogous to those now widely used to reconstruct three-dimensional representations from two-dimensional x-ray images in computed tomography (CT).

The simplest form of reconstruction, referred to as "back projection," is illustrated schematically in Fig. 10.38, using the hypothetical spectra from Figs. 10.35 and 10.36. To construct the image, the spectra generated using the z- and y-gradients are placed along the corresponding axes, and rays from the spectral peaks are drawn through the area representing the sample. The points where rays from different directions intersect identify possible locations of nuclei that could give rise to the corresponding peaks in the spectra. In the simple example shown in Fig. 10.38, there are nine areas, indicated with gray patches, with overlapping rays, but comparison with Figs. 10.35 and 10.36 shows that the spectra can be accounted for by just three ^{1}H-containing volumes. The two spectra are thus insufficient, by themselves, to reconstruct the original object, even as a projection in the z–y plane.

The image can be improved by the addition of back projections from spectra generated using other gradient directions, as illustrated in Fig. 10.39, using the projection along the y–z diagonal. With the addition of the third spectrum, there are only three patches in the y–z plane where rays back-projected from all three spectra overlap, corresponding to the projections of the three sub-volumes in the original sample. In real imaging experiments, the number of spectra required to reconstruct an image will depend on the complexity of the object and the desired resolution.

A two-dimensional projection image can also be created using a two-dimensional Fourier transform experiment analogous to those described earlier in this chapter. The outline of such an experiment, to generate a projection image onto the x–y plane, is shown in Fig. 10.40. This experiment is very similar to the one described in

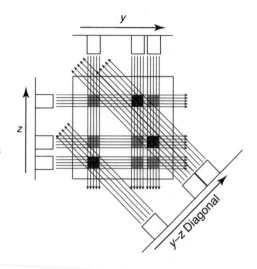

Figure 10.39
Back projection from one-dimensional spectra, as in Fig. 10.38, with the addition of information from the spectrum diagrammed in Fig. 10.37. The additional rays from the y–z diagonal projection help define the positions of the nuclei in the sample, as indicated by the darker gray patches.

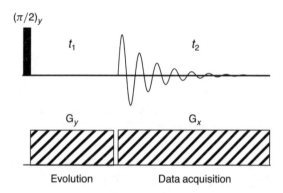

Figure 10.40   A pulse sequence for a two-dimensional NMR imaging experiment. The upper line represents standard radio-frequency pulses and the resulting FID, and the lower line indicates the application of field gradients along the y- and x-axes during the evolution and data acquisition periods, respectively. The experiment is conducted by recording individual FIDs with incremented delay times, t_1, or with incremented G_y gradient strengths.

Sections 10.1 and 10.2, with the addition of field gradients applied along the y- and x-gradients during the evolution (t_1) and data acquisition (t_2) periods, respectively. Note also that there is no pulse separating the evolution and data acquisition periods. As in the other two-dimensional experiments we have discussed, the experiment is conducted as a series of one-dimensional experiments with incremented values for the delay period, t_1, between the two pulses. During the evolution period, nuclei at different positions along the y-axis (in the laboratory frame) will precess at different frequencies, as illustrated in Fig. 10.41 using vectors to represent nuclei in three positions of the hypothetical three-dimensional sample diagrammed in Figs. 10.35–10.37.

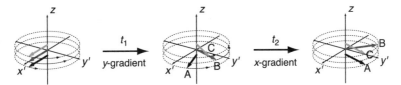

Figure 10.41   Vector diagrams representing the precession of magnetization during the evolution and data acquisition periods of the two-dimensional imaging experiment outlined in Fig. 10.40. The three vectors represent nuclei at the centers of the sub-volumes in the hypothetical three-dimensional sample diagrammed in Figs. 10.35–10.37, as indicated by the labels A, B and C. During the evolution period, the precession frequencies are determined by the positions of the nuclei along the y-axis of the sample, and these frequencies, along with the length of the evolution period (t_1), determine the initial positions of the vectors at the beginning of the data acquisition period. During data acquisition, the precession frequencies are determined by the positions of the nuclei along the x-axis of the sample.

After the t_1 period, the y-gradient is turned off, and the x-gradient coil is activated. The nuclei continue to process and generate the FID, with their precession frequencies now determined by their positions along the x-axis. The individual FID signals are digitized and Fourier transformed to generate a series of one-dimensional spectra, as illustrated in Fig. 10.42. When the t_1 delay time is zero, all of the magnetization vectors begin the data acquisition period aligned with the x'-axis, and each of the three spin populations represented in the figures gives rise to a simple absorptive peak in the one-dimensional spectrum. The frequencies of these peaks represent the positions of the nuclei along the x-axis of the sample. As the t_1 delay is incremented, and the spins are allowed to process under the influence of the y-gradient, the positions of the magnetization vectors diverge during the evolution period, so they begin the data acquisition period at different positions. These differences lead to shifts in the phases of the FID signals arising from nuclei located at different positions along the y-sample axis, very much like the phase shifts that were discussed in Section 6.6. As in the case of a mis-phased spectrum, the different FID phases lead to the mixing of absorptive and dispersive peak shapes in the spectra.

If the amplitudes of the spectra, at the individual frequencies indicated by the dashed vertical lines in Fig. 6.43, are followed as a function of t_1, the amplitudes are found to fluctuate cyclically, with frequencies that are determined by the locations of the nuclei along the y-axis of the sample, as shown in Fig. 10.43. Fourier transformation along t_1 slices of the array of f_2 spectra generates the f_1 frequency domain, with peaks corresponding to the precession frequencies determined by the G_y gradient.

Fig. 10.44 illustrates the first Fourier transformation step of simulated FIDs generated from the two-dimensional experiment, using the hypothetical three-dimensional sample illustrated in Figs. 10.35–10.37. The rectangular peaks in the f_2 spectra represent the projections of the ^{1}H-containing volumes along the x-axis. As the t_1 delay is

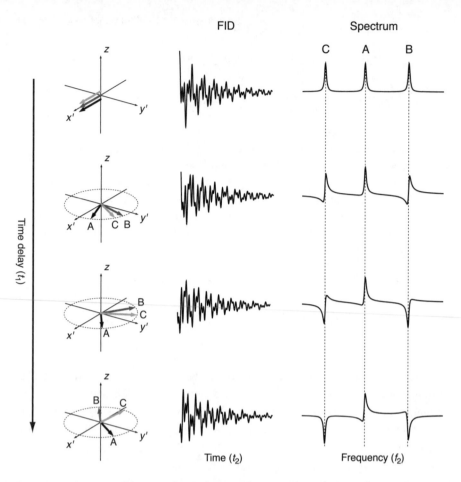

FID   Spectrum

Time (t₂)   Frequency (f₂)

Figure 10.42 Vector diagrams, free induction decays and spectra representing incremented t_1 delay periods in the two-dimensional imaging experiment diagrammed in Fig. 10.40. The phases of the FIDs are determined by the precession of the magnetization components during the evolution period. After Fourier transformation of the FIDs, the phase differences lead to different mixtures of the absorptive and dispersive peak shapes in the spectra.

Figure 10.43

Fluctuations in amplitudes of the one dimensional spectra generated by Fourier transformation of individual FIDs in the two-dimensional imaging experiment, as a function of the delay time, t_1. The filled symbols indicate the amplitudes of the individual spectra shown in Fig. 10.42, with the labels A, B and C indicating the corresponding vectors and frequency positions in that figure.

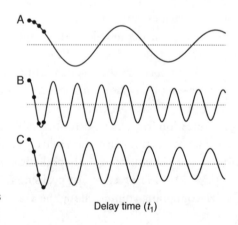

Delay time (t_1)

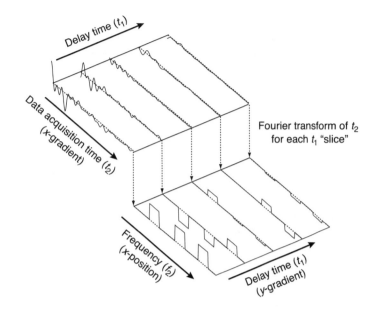

Figure 10.44

FIDs and Fourier transform to generate individual one-dimensional spectra in the two-dimensional imaging experiment diagrammed in Fig. 10.40. Frequencies in the f_2 dimension represent the positions of nuclei along the x-axis of the sample.

incremented, the phases of the FIDs are shifted, resulting in fluctuations of the amplitudes in these peaks, as illustrated in Fig. 10.43. (For clarity, the volume elements are assumed to be significantly larger than the peak widths, so that the fluctuations in peak shape are not visible.)

Slices perpendicular to the f_2 spectra are then transformed along the t_1 dimension, to generate a two-dimensional frequency spectrum, as shown in Fig. 10.45. Because the precession frequencies during the t_1 and t_2 periods are determined by

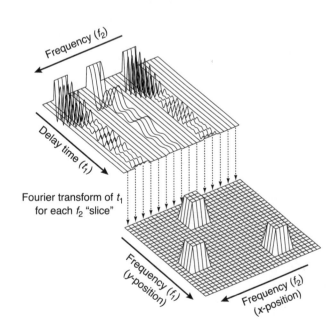

Figure 10.45

Fourier transformation of the second dimension of a two-dimensional imaging experiment. Each slice along the t_1 dimension of the half-transformed spectrum is transformed into a frequency slice in the final spectrum.

the position of the nuclei relative to the y- and x-field gradients, respectively, the two frequency axes in the final spectrum correspond to the y- and x-coordinates of the samples. Note, however, that the intensities shown in the simulated spectra do not represent the z-axis of the original object, but rather indicate the total density of nuclei along the axis perpendicular to the $x-y$ plane—that is, a projection of the density onto this plane. By application of other combinations of the x-, y- and z-gradients, projections onto any defined plane can be generated.

Although the experiment above was described as using increments in the length of the t_1 delay time, the same effect can be generated using a constant t_1 interval and incrementing the strength of the G_y gradient for the individual FIDs. Using either procedure, the effect is to increment the degree to which magnetization vectors representing different positions along the gradient diverge. As illustrated in Fig. 10.42, these differences in position lead to phase shifts in the FIDs and, finally, amplitude differences in the spectra. Because the effect of the evolution period is to establish a distance-dependent shift in phase angle, this process is described in the MRI literature as *phase encoding* of the position along the gradient. The position along the direction of the gradient activated during data acquisition (the x-axis in the example above) is said to be *frequency encoded*, but it should be apparent that differences in precession frequency are used to establish positional information in both directions.

When the individual one-dimensional spectra differ by the strength of the gradient applied, rather than delay time, it is not quite right to refer to the incremented dimension as being in the time domain. A more general term used in MRI to describe both time intervals and gradient strengths is *k-space*. Since the final image can be Fourier transformed back into k-space, all of the dimensions in this space formally have the units of inverse length.

Before going on, an important distinction should be noted between a two-dimensional imaging experiment and experiments based on molecular structure, such as the NOESY experiment described earlier in this chapter. In the NOESY experiment, individual nuclei have the same precession frequencies during the t_1 and t_2 periods, leading to peaks on the diagonal of the spectrum. Cross peaks are generated by the transfer of magnetization from one nucleus to another during the mixing period between the evolution and data-acquisition periods. In the imaging experiment, there is no magnetization transfer, but the nuclei have, in general, different Larmor frequencies during the two periods. Only the nuclei that happen to experience the same external field strength during the two periods give rise to a signal on the diagonal. A spectrum recorded with the same gradient applied during both t_1 and t_2 would contain density only along the diagonal and would not provide any information beyond that contained in a one-dimensional spectrum.

All of the imaging experiments described above give rise to projection images, along either a single axis or onto a plane. In real applications, the most useful images are usually ones representing either single cross-sections of an object or a three-dimensional stack of such cross-sections. One way of generating a three-dimensional image is to extend the two-dimensional Fourier transform experiment into a third dimension, with an additional delay interval during which a field gradient is ap-

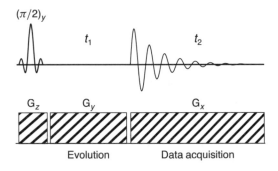

Figure 10.46 A pulse sequence for generating a two-dimensional image representing a cross-section along the z-axis of a sample. The initial pulse is applied while a field gradient along the z-axis is active, and the pulse is shaped to stimulate nuclei with a limited range of frequencies, corresponding to a slab within the sample.

plied along the third axis. This approach, however, suffers from serious practical limitations, because each delay interval for the third dimension requires recording a full two-dimensional spectrum. As discussed in Chapter 6, the frequency range of a Fourier transform spectrum is determined by the spacing between measured time points, and the resolution depends on the total time over which data are collected. For imaging experiments, the total frequency range is determined by the strengths of the field gradients and can be adjusted to allow convenient sampling, but higher resolution always requires a larger number of data points over a longer time period. When the method is extended into three dimensions, the total time required can become very long, especially for a human who happens to be serving as the sample! A related difficulty is that the individual planes in the data set do not individually contain readily interpreted information, so that a significant number of planes must be recorded before an image can be reconstructed, even with limited resolution.

For these reasons, most three-dimensional imaging uses an alternative strategy, which generates images representing individual slices of the object along a defined direction. This type of image can be produced using a variation of the two-dimensional Fourier transform experiment outlined in Fig. 10.40, with a frequency-selective pulse applied as the first step, as diagrammed in Fig. 10.46.

As suggested by the symbol used to represent the initial pulse in the diagram, the selective pulses used for imaging experiments often have the shape of a sinc function, or a related function. Because the Fourier transform of this function is a rectangle function, the pulse represents a band of frequencies with a width that is inversely related to the duration of the pulse. Importantly for this experiment, a gradient along one direction (z in the example) is activated during the selective pulse. As a result, the band of frequencies corresponds to a range of distances along the z-axis, and only the nuclei lying within this range are stimulated.

The rest of the experiment is the same as discussed earlier, with y- and x-gradients applied during the t_1 and t_2- periods. The data are processed as illustrated in

Figs. 10.44 and 10.45 to generate a two-dimensional spectrum. Now, however, only the nuclei that were stimulated by the initial selective pulse contribute to the spectrum, which represents a cross-sectional image of the sample. The thickness of the slab represented by the spectrum can be controlled by the duration of the selective pulse, and multiple cross-sections can be produced and assembled to create a three-dimensional image. As with a three-dimensional Fourier transform experiment, the total time required increases as additional two-dimensional spectra are recorded, but in this case the signals can be limited to a region of interest. The orientation of the cross-sections can be controlled by activating different combinations of the gradient coils during the different stages of the experiment. All of this is done without any mechanical manipulation of the instrument or sample. Particularly for diagnostic imaging of patients, this can be a significant practical advantage over, for instance, x-ray imaging.

Over the past four decades, there has been rapid progress in the development of magnetic resonance imaging techniques and their applications, and this brief introduction barely scratches the surface. The largest impact of MRI has been in biomedical applications, with millions of diagnostic studies performed each year. For this application, ^{1}H nuclei are used almost exclusively, and the resulting images reflect primarily the distribution of water in a subject. A variety of experiments have been devised to generate images that reflect specific properties of the detected nuclei, including their relaxation rates and rates of molecular diffusion. Indeed, much of the early motivation for the development of MRI came from observations by Raymond Damadian in 1971 that the relaxation rates for ^{1}H nuclei of water molecules were markedly lower in cancer cells than in normal cells. Though this difference has not proven to be a reliable diagnostic marker for cancer, methods incorporating inversion-recovery and spin-echo pulse sequences are routinely used to generate images in which contrast reflects differences in R_1 or R_2. More recently, imaging methods have been developed that are particularly sensitive to the presence of hemoglobin with and without oxygen bound, which have distinct magnetic properties. Deoxyhemoglobin, in particular, interacts with the external magnetic field and alters the Larmor frequencies of nearby ^{1}H nuclei, and this difference can be used to create contrast in an appropriately designed imaging experiment. This method is referred to as blood oxygen level dependent (BOLD) contrast imaging and, more commonly now, as functional magnetic resonance imaging (fMRI). fMRI has become widely used to monitor metabolic activity in the brain, often with the goal of identifying specific brain regions associated with different neurological functions.

For molecular spectroscopy, imaging now plays an important role in the very practical task of adjusting the shim coils in a spectrometer to produce a highly homogeneous field. In this application, gradient coils in the probe are used to generate an image of the sample, which is expected to be homogeneous. Any inhomogeneity revealed in the image is assumed to be due to distortions in the stationary field, and the shim coils are adjusted iteratively to minimize the inhomogeneity. This procedure can be readily automated, thus nearly eliminating one of the traditionally more tedious tasks for NMR spectroscopists.

Summary Points for Chapter 10

- Two-dimensional NMR spectra are created by recording the FIDs from a series of one-dimensional experiments in which a delay time (t_1) is systematically incremented.
- The data from a two-dimensional experiment undergo two Fourier transforms:
  - Each of the individual FIDs (with time axis, t_2) is transformed to generate a spectrum, with the frequency axis labeled f_2.
  - For each value of f_2, the spectral intensity as a function of t_1 is transformed to generate a frequency spectrum, with axis labeled f_1.
  - The result is a spectrum with intensity as a function of two frequencies.
- If two spins do not interact with one another to transfer magnetization, each will create a peak on the $f_1 = f_2$ diagonal, but there will not be a cross peak.
- The $^1H–^1H$ NOESY experiment includes a mixing time during which cross-relaxation between spins causes a transfer of z-magnetization.
- NOESY cross peaks usually arise when two ^1H nuclei are within about 5 Å of one another.
- Phase cycling is a process in which multiple FIDs are recorded for each t_1 delay time, with a cycle of changes in the phases of one or more of the pulses in a multidimensional experiment. The major function of phase cycling is to cancel out magnetization components that would otherwise complicate the spectrum.
- Quadrature detection for the indirect (f_1) dimension is implemented by recording two sets of phase-cycled FIDs for each t_1 delay, with another pulse-phase difference designed to alter the phase of the indirectly detected FID by 90°.
- Pulsed field gradients are another method for eliminating undesirable magnetization components and are also used to measure rates of molecular diffusion and to suppress solvent signals.
- Magnetic resonance imaging is based on the use of field gradients, which establish relationships between the locations of nuclei in a sample and their resonance frequencies. Two- and three-dimensional images are generated using field gradients with different orientations and are closely analogous to two- and three-dimensional molecular spectra.

Exercises for Chapter 10

10.1. In the example two-pulse experiment diagrammed in Fig. 10.2, both pulses are along the y'-axis. Suppose that this experiment were modified so that the first pulse was along the y'-axis but the second was along the $-x'$-axis. Draw out the vector diagrams that would represent the situation before and after the second pulse for the three t_1 delay times illustrated in Fig. 10.4. Also, sketch the expected one-dimensional spectra that would be generated for the three delay times. For this purpose, assume that a magnetization vector aligned with the positive x'-axis at the beginning of the data acquisition period will generate a positive peak in the frequency spectrum. How would this modification to the experiment alter the data recorded from the experiment? Would the final spectrum be affected?

10.2. The vector diagrams in Fig. 10.13 illustrate the effects of cross-relaxation during the mixing time of a NOESY experiment, for the case of a slowly tumbling molecule. Redraw the central column of vector diagrams in this figure for the case of a rapidly tumbling molecule. (It may helpful to review the discussion of the transient NOE experiment in Chapter 9.) How will the final NOESY spectrum differ from the one expected for a slowly tumbling molecule? Explain as clearly as possible why the spectra will differ.

10.3. Colchicine is a plant alkaloid that inhibits microtubule polymerization and has been used for millennia in treating a number of medical conditions, including gout and inflammatory conditions. The structure of colchicine and a predicted ^{1}H NMR spectrum are shown in Fig. 10.47.

Note that only ^{1}H nuclei directly bonded to carbon atoms are included in the predicted spectrum, as predictions of chemical shifts for protons bound to nitrogen are considerably less reliable. Also note that several of the peaks are split by scalar coupling, and the peaks due to methyl hydrogens are very strong. This latter effect is due to the fact that each ^{1}H nucleus of the methyl group displays the same chemical shift.

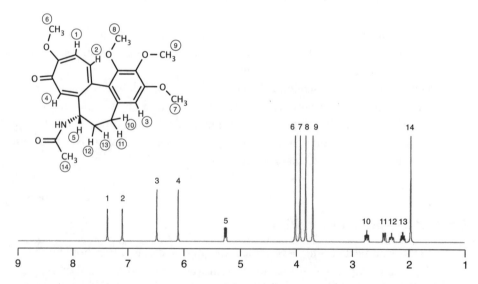

Figure 10.47   Chemical structure of colchicine and a predicted ^{1}H NMR spectrum. The predicted spectrum was generated using a free online service, http://www.nmrdb.org, which predicts ^{1}H chemical shifts using a neural network trained on an extensive database of experimental data, as described in Binev, Y. & Aires-de Sousa, J. (2004). *J. Chem. Inf. Comp. Sci.*, 44, 940–945. The reference data were all recorded from samples dissolved in deuterated CHCl$_3$. For Exercise 10.3.

From the one-dimensional spectrum, draw the two-dimensional ^{1}H–^{1}H NOESY spectrum that you would expect for colchicine. To simplify the drawing, you do not need to include the splitting due to scalar coupling. Also assume that a relatively short mixing time was used in recording the NOESY spectrum so that cross peaks are only generated for ^{1}H nuclei that are within 4 Å of each other.

To estimate distances between atoms, you may want to either build a physical model or use a molecular graphics program such as PyMOL (http://pymol.org) or Jmol (http://jmol.sourceforge.net). Before using one of these programs, however, you will need an atomic coordinate file for the molecule in a suitable format, such as the Protein Data Bank (pdb) format. One way to generate such a file for this or other small molecules is as follows:

(a) Look up colchicine on Wikipedia or a specialized molecular database such as PubChem (http://pubchem.ncbi.nlm.nih.gov) or ChEBI (http://www.ebi.ac .uk/chebi).

(b) Copy from the source the SMILES (simplified molecular-input line-entry system) string for the molecule. This string of characters specifies the covalent structure of a molecule.

(c) Use an online service for calculating a predicted three-dimensional structure of the molecule, such as http://cactus.nci.nih.gov/translate or http://www .molecular-networks.com/online_demos/corina_demo. As well as pdb files, these sites can output the atomic coordinates as MDL Molfile files, which can in turn be used as inputs for calculating ^{1}H NMR spectra at http://www .nmrdb.org.

10.4. Modern (circa 2015) MRI instruments for human diagnostics and research typically have a fixed magnetic field strength in the range of 0.3 to 3 T and can create field gradients up to about 0.1 T/m. Consider an instrument with a field strength of 1.5 T and a gradient strength of 0.05 T/m.

(a) What is the working ^{1}H resonance frequency in such an instrument?

(b) What is the range of ^{1}H resonance frequencies that would be observed over a spatial distance of 1 m along the field gradient, in both Hz and ppm? How does this range compare to the range of ^{1}H chemical shifts typically associated with protons in different chemical environments (in liquid samples)?

(c) Suppose that one wanted to resolve features in an object that were separated by 0.1 mm. Again assuming a gradient strength of 0.05 T/m, what range of frequencies does this thickness correspond to, in both Hz and ppm? What does this imply about the resolution requirements for imaging as compared to the molecular spectroscopy of liquid samples?

(d) Suppose that one wanted to selectively stimulate the nuclei in a cross-section of an object with a thickness of 0.1 mm. Using a sinc-shaped pulse as diagrammed in Fig. 6.46, what should the approximate duration of the pulse be to selectively irradiate this slice?

10.5. The two-dimensional imaging experiment diagrammed in Fig. 10.40 does not include a read pulse before the FID is recorded. However, an image could also be generated by an experiment that included a $\pi/2$-pulse at the beginning of the data acquisition period, more like the two-dimensional experiments described earlier in this chapter. Using vector diagrams, show how the magnetization would behave for nuclei in different regions of the sample in this version of the experiment and explain how the two-dimensional image would be generated. What practical reasons might favor including or omitting the second pulse?

Further Reading

A more advanced treatment of multidimensional experiments, based on the product-operator formalism:

◇ Cavanagh, J., Fairbrother, W. J., Palmer III, A. G., Rance, M. & Skelton, N. J. (2007). *Protein NMR Spectroscopy*, 2nd ed pp. 271–332. Elsevier Academic Press, Amsterdam.

Further discussion of the NOESY experiment, as well the general structure of two-dimensional experiments:

◇ Neuhaus, D. & Williamson, M. P. (2000). *The Nuclear Overhauser Effect in Structural and Conformational Analysis*, 2nd ed. pp. 282–330. Wiley-VCH, New York.

A general review on field gradients:

◇ Hurd, R. E. & John, B. K. (2011). Field gradients and their applications. *eMagRes*. http://dx.doi.org/10.1002/9780470034590.emrstm0164.pub2

A classic paper on the use of field gradients to measure diffusion constants:

◇ Stejskal, E. O. & Tanner, J. E. (1965). Spin diffusion measurements: Spin echoes in the presence of a time-dependent field gradient. *J. Chem. Phys.*, 42, 288–292. http://dx.doi.org/10.1063/1.1695690

A review on methods for solvent suppression:

◇ Zheng, G. & Price, W. S. (2010). Solvent signal suppression in NMR. *Prog. Nuci. Mag. Res. Sp.*, 56, 267–288. http://dx.doi.org/10.1016/j.pnmrs.2010.01.001

The paper that introduced the MRI method based on field gradients:

◇ Lauterbur, P. C. (1973). Image formation by induced local interactions: Examples employing nuclear magnetic resonance. *Nature*, 242, 190–191. http://dx.doi.org/10.1038/242190a0

An introductory overview of NMR and MRI:

◇ Kozlowski, P. (2009). *Magnetic resonance imaging*. In *Medical Imaging: Principles, Detectors and Electronics*, (Iniewski, K. Ed.), pp. 223–284. Wiley, Hobojen, NJ.

An excellent online resource for learning more about magnetic resonance imaging:

◇ Hornak, J. P. (2014). The basics of MRI. http://www.cis.rit.edu/htbooks/mri/

CHAPTER 11

The Mathematical Formalism of Quantum Mechanics

In Chapter 2, some of the fundamental ideas of quantum mechanics were presented in a very qualitative way. One of the most important of these ideas is that the properties of a quantum mechanical system are described by a rather abstract mathematical object, called the "wavefunction" and designated Ψ, that, in general, cannot be determined experimentally. If the wavefunction is known, however, it can be used to calculate the average outcome of a particular measurement. For an individual particle, the experimental result will have a discrete value, but usually it is only possible to calculate the average result. For a large population of particles with spin, the calculated average magnetization components can be described using vector diagrams. For populations of uncoupled spins, this classical description is entirely adequate for most purposes in NMR. When two or more spins interact closely, however, as when they are linked through covalent bonds, new and rather surprising features arise that are not at all easy to describe with simple vector diagrams. These interactions are critical to some of the most important modern NMR experiments, and an understanding of these experiments depends on a quantum mechanical treatment.

With this chapter, we begin to develop the mathematical formalism needed to treat a system composed of more than one spin, starting with a general discussion of the relationship between the wavefunction and measurements and moving on to the specific case of an uncoupled spin-1/2 particle. In Chapter 12, we will continue the treatment of an uncoupled spin and consider how the wavefunction changes with time. Chapters 13 and 14 extend this treatment to spins linked by scalar coupling, thereby laying the groundwork for the treatment of multidimensional experiments based on this kind of interaction.

At the outset, it is important to stress that quantum mechanics is a theory that is built upon a set of postulates—that is, rules that have not been proven from other principles, and that these postulates do not follow easily from our everyday experience, as the assumptions of classical mechanics do. Nonetheless, starting with these postulates, the theory has been developed with great mathematical rigor (though that may not be obvious from the rather ad hoc treatment presented here), and has passed essentially every experimental test that has been devised to challenge it. Although there continue to be debates about the ultimate physical (and philosophical) meaning of quantum mechanics, the predictive power of the theory is remarkable. Before delving into the mathematics, we will revisit the Stern–Gerlach experiment introduced in Chapter 2, with some extensions of that experiment that help illustrate some of the special properties of quantum mechanical systems that demand a non-classical theory.

11.1   Some Variations on the Stern–Gerlach Experiment

Recall from Chapter 2 that the basic Stern–Gerlach experiment involved generating a beam of spin-1/2 particles (originally silver atoms with a single unpaired electron) and directing the particles through a non-uniform magnetic field created by a special magnet, as shown in Fig. 11.1. Even though the magnetic dipoles in the initial population of spins are presumed to have no preferential orientation, the particles are separated into two beams, representing distinct values of the magnetization along the z-axis. Now, suppose that the original magnet is rotated 90°, so that the field is now aligned with the x-axis of the target, as in Fig. 11.2.

We will call this an *x-filter* and the original magnet a *z-filter*. The atoms are again separated into two beams, but are now deflected to different positions on the x-axis.

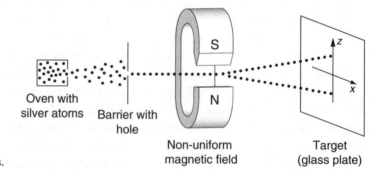

Figure 11.1
The Stern–Gerlach experiment oriented to measure magnetization along the z-axis.

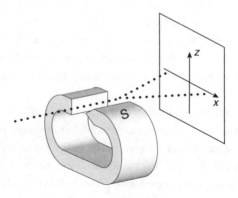

Figure 11.2
The Stern–Gerlach filter re-oriented to measure magnetization along the x-axis.

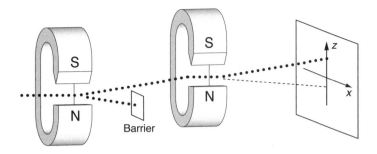

Figure 11.3
Two Stern–Gerlach z-filters
acting in sequence.

This isn't too surprising, since the original orientation of the magnet was arbitrary, but it does show that it is the magnet that is imposing an orientation on the magnetic dipoles.

Things become more interesting if we direct the beam from one filter into a second one. While the experiments described here probably haven't been performed in exactly this way, equivalent ones have been. The experiments are idealized here by assuming that the magnets have instantaneous effects on the particles and that the particles maintain their orientation as they move from one magnet to another.

In the first of these experiments, illustrated in Fig. 11.3, one of the beams leaving the first magnet is blocked, and the other is directed through a second, identically oriented z-filter. Only one beam is seen leaving the second magnet, and it is deflected in exactly the same way as the "up" beam from the first magnet. This seems reasonable if each magnet is acting as a filter. Once the atoms are separated by the first filter, they behave homogeneously and are deflected in the same way by the second filter. But, the observation is also a bit odd. In the first magnet, the entire population was somehow converted into two types of atoms, with roughly equal populations, but in the second magnet, there is no conversion of "ups" into "downs."

Next, let's pass the atoms through a z-filter and then take one of the beams and pass it through an x-filter, as shown in Fig. 11.4. Even after the atoms have been filtered along the z-axis, they can be filtered again along the x-axis. If we took both beams from the z-filter and then passed each through an x-filter, we would get four spots on the target. Maybe each atom can be described by two independent

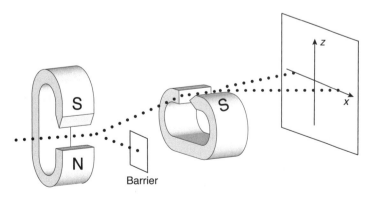

Figure 11.4
A z-filter followed by an x-filter.

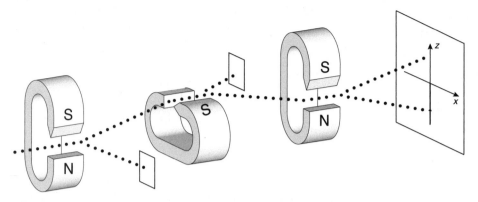

Figure 11.5   A z-filter followed by an x-filter, followed by another z-filter.

parameters, one that allows it to be oriented with respect to the z-axis and a second that defines its orientation with respect to the x-axis.

Now we will do one more experiment, with three filters—namely, passing the beam through a z-filter, an x-filter and then a second z-filter, as shown in Fig. 11.5. The result is exactly like the very first experiment, when the beam was passed through just one z-filter! Even though all of the "z-down" atoms were eliminated in the first filter, some of the remaining atoms have been converted into the "z-down" state in the third filter. Remember, this didn't happen when two z-filters were placed directly one after the other.

This last experiment is especially bewildering. It is as if information is being created and destroyed as the beam of atoms passes through the various magnetic fields. This is, in fact, what is happening, and it is a particularly vivid demonstration of some of the non-intuitive aspects of quantum mechanics. It is very difficult, if not impossible, to imagine an analogous process occurring with macroscopic objects, such as blocks of different shapes passing through holes with matching shapes. Just when we think we know what the shape is, it changes! To paraphrase the great Danish physicist Niels Bohr speaking in a similar context: If this doesn't seem very strange to you, you haven't thought about it enough!

11.2   A Demonstration with Polarized Light

While the Stern–Gerlach experiments with free atoms are actually quite difficult to execute, analogous effects can easily be demonstrated with light and polarizing filters. Some of what follows is likely to be familiar, but it is worth thinking about carefully with attention to the parallels with the Stern–Gerlach experiments.

The phenomenon of polarized light is usually described in terms of the orientation of the fluctuating electrical field vector as a light wave propagates. We say that the light is vertically polarized if the electrical vectors form a wave that lies in a plane defined by the vertical axis and the direction of propagation, the z- and y-axes as defined in the diagrams in Fig. 11.6. For horizontally polarized light, the wave lies in the x-

A. Vertical

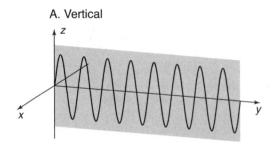

B. Horizontal

Figure 11.6
Vertically and horizontally
polarized light. The waves
represent the fluctuations of
the electrical vectors of the
propagating light.

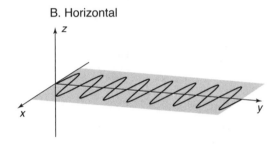

y plane. Determining the polarization requires an experimental measurement, which usually involves passing the light through a material that interacts differently with the radiation depending on its direction of polarization. Such materials include certain crystals, such as Iceland spar (a form of calcite), and the polarizing filters widely used in sunglasses. (Polaroid is the trade name for the first inexpensive polarizing filter, invented by Edwin Land in 1929.)

If light from an ordinary source is passed through a polarizing filter, only about one-half of the light intensity is detected on the other side, as diagrammed in Fig. 11.7 In this and the following diagrams, the intensity of light is indicated by the width of the arrows, and the orientation of the filter is indicated by the bars. By most criteria, the light that passes through the filter is indistinguishable from the light that entered it; there is just less of it, as if the light were directed through a piece of dark glass. If, however, we direct this light through a second filter, we can see that it has changed. If the second filter is oriented in the same way as the first, then nearly all of the light will pass through (assuming that the filters are not also tinted to further diminish the intensity), as shown in Fig. 11.8. If the second filter is rotated by 90° relative to the first, almost no light is transmitted (Fig. 11.9).

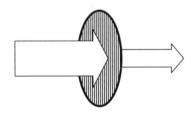

Figure 11.7
Unpolarized light passing
through a polarizing filter.

Figure 11.8
Unpolarized light passing though two vertically oriented polarizing filters.

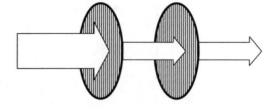

Figure 11.9
Unpolarized light passing through a vertical polarizer and then blocked by a horizontally oriented polarizer.

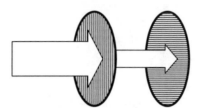

These phenomena are usually explained by saying that the light waves making up the original beam were oriented randomly around the axis of propagation and that only those that were vertically polarized can pass through the first filter. The vertically polarized light can then pass freely through a second vertical filter, but is blocked by the horizontal filter. This can be illustrated with a diagram like the one in Fig. 11.10, which represents a view down the axis of propagation, and the arrows represent the directions of the electrical vectors. This picture is appealing and is consistent with the representation of the polarizers as grates with microscopic bars that only allow waves with the correct orientation to pass. A bit of thought, however, suggests that maybe things aren't this simple. If the waves are initially oriented randomly, why can nearly half of the light pass through the filter, rather than the very small fraction for which the electrical vectors were initially vertical? As in the Stern–Gerlach filter it appears that the filter itself is imposing its character on the light that passes through it. The major differences between the two types of filter are that (1) the optical polarizer absorbs half of the light, rather than splitting the beam in half, and (2) orientations of optical polarization differ by 90° (vertical vs. horizontal) while those for the magnetic dipoles differ by 180° (up and down).

If the second filter is rotated so that its orientation is midway between vertical and horizontal, as in Fig. 11.11, about half of the light passes through. Again, this is rather

Figure 11.10
Unpolarized and verticallypolarized light, as represented by the orientation of the electrical vectors, viewed down the direction of propagation.

A. Unpolarized   B. Vertically polarized

Figure 11.11
Unpolarized light passing
through a vertical polarizer
and then partially blocked by
a polarizer at an intermediate
angle.

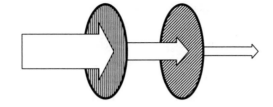

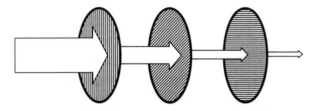

Figure 11.12 Unpolarized light passing through a series of three polarizers. The
first and last filters are oriented vertically and horizontally, respectively, and the
central filter is at an intermediate angle. Passing through the central filter allows a
fraction of the light to pass through the horizontal one.

difficult to reconcile with a picture in which the polarizers simply allow light with the
correct polarization to pass through, since none of the light passing through the first
filter should be oriented correctly to pass through the second. This arrangement is
analogous to the Stern–Gerlach experiment with two filters, one oriented along the
z-axis and the other along the x-axis.

In the experiment illustrated in Fig. 11.12, a third polarizer has been added,
aligned with the horizontal axis. Now, a fraction of the light makes it through the
horizontal filter as well as the original vertical filter! If the filter in the middle is
removed, all of the light will be blocked by the horizontal filter, as in Fig. 11.9.

This last experiment corresponds to the experiment with three Stern–Gerlach
filters, and it can only be understood if the filters are presumed able to alter the
properties of the light. In terms of the wave description of light illustrated in Fig. 11.6,
the effects can be accounted for by interactions between the electrical fields and the
polarizing material. But, in other experiments, light clearly displays the properties of
particles, and it is much more difficult to reconcile the effects of the polarizers in a
particle description, just as it is for the Stern–Gerlach experiments.

11.3 An Overview of the Mathematical Structure: Operators, Eigenfunctions and Eigenvalues

Although the behavior of the particles in the various Stern–Gerlach-type experiments
is very odd (from our experience in the macroscopic world at least), there are some
rather simple rules that can be deduced:

1. Whenever a beam of particles passes through a filter, the beam either leaves the filter undivided or is divided into two classes of particles that are deflected along the direction of the magnetic field.
2. If the beam has already passed through one filter oriented in a particular direction and is then immediately passed through a second filter with the same orientation, there is no further splitting.
3. If the immediately previous filter was not oriented in the same direction, the beam will be split.

In addition to these specific rules, the Stern–Gerlach experiments illustrate two much more general features of quantum mechanics: First, the results of experimental measurements take on discrete values and, second, the measurement itself can alter the state of the system. To describe these phenomena and predict the outcomes of experiments requires a mathematical formalism that is fundamentally different from the mathematics usually used to describe objects and their behaviors on the macroscopic scale.

As discussed briefly in Chapter 2, the state of a quantum mechanical system is defined by the wavefunction, usually designated Ψ. The form of the wavefunction depends on the nature of the system being described, and the form that is perhaps most widely known, especially to students and practitioners of chemistry, describes the spatial distribution of particles, such as electrons. Manipulations of this type of wavefunction usually require integration over the spatial coordinates, which can be very complicated mathematically, and only the simplest systems are amenable to exact analysis. On the other hand, the wavefunctions describing the property that we call spin are independent of the position of an isolated particle, and there are no spatial coordinates requiring integration, making the mathematics much simpler. Furthermore, even when the particle is part of an atom or even a molecule composed of thousands of atoms, the spin properties can, for most purposes, be treated independently of all of the other aspects of the system. This is because the energies and, therefore, the frequencies associated with the spin quantum states are much smaller than those involving the other interactions the nucleus is exposed to. Thus, motions of the atoms, that is, their changes in spatial coordinates, are usually much faster than changes in spin state and are averaged out on the time scales relevant to NMR. As a consequence, the influence of these other interactions on an individual spin can be treated by introducing a few simple parameters, such as the chemical shift and scalar coupling constant discussed in Chapter 4. This allows for a tremendous simplification of what would otherwise be a nearly intractable problem.

In addition to the wavefunction, the second essential element in the calculation is called an *operator*, which is, in essence, a rule that changes one function into another function. For instance, a very simple form of operator is one that adds a constant to a function or, alternatively, divides the function by a constant. Differentiation of a function is another, more sophisticated, kind of operator. Later, we will discuss the specific operators corresponding to measurements of the magnetic orientation of a particle with spin.

11.3.1 Eigenfunctions and Eigenvalues

To start with, we will treat the wavefunction and operators in a rather abstract way. We begin by stating that for any measurement there will be an operator called $\hat{A}$. We will use the circumflex, or "hat", over the symbols for operators to distinguish them from the actual outcome of the measurement, which would be A in this case. For any operator there are some special wavefunctions, Ψ_j, for which the result of $\hat{A}$ operating on Ψ_j will simply be Ψ_j multiplied by a constant called λ_j. The subscript j is simply an index to specify a particular function and its corresponding constant. This relationship is written as

$$\hat{A}\Psi_j = \lambda_j \Psi_j \qquad (11.1)$$

Note that when $\hat{A}$ is written in front of Ψ_j, it implies the action of the operator on the wavefunction and not, in general, simple multiplication by a constant, as is implied on the other side of the equation. When spatial coordinates are involved, the operation will include integration over these coordinates, but we will ignore this aspect, since it is not required for the wavefunctions and operators for spin systems.

Eq. 11.1 is called an *eigenvalue equation*, and λ_j is called the *eigenvalue* corresponding to the *eigenfunction*, Ψ_j. (Eigen is German for "characteristic.") In general, a given operator will only convert some of all possible functions into the same function multiplied by a constant. In the context of quantum mechanics, the values of the eigenvalues represent the possible outcomes of the measurement associated with the operator. This is how the discrete character of quantum mechanics emerges from the mathematical formalism. If the wavefunction describing the system happens to be one of the eigenfunctions for that measurement, then the result of the measurement (the constant λ_j) is absolutely predictable. However, Ψ doesn't have to be an eigenfunction of $\hat{A}$. If it isn't, the result of an individual measurement will still be one of the eigenvalues, but which one can't be predicted with certainty.

As noted in Chapter 2, the rules of quantum mechanics do *not* require that systems exist only in discrete states, but they do require that the results of experimental measurements are quantized. One important observable property of a system is its energy, for which the corresponding operator is called the Hamiltonian. The Hamiltonian is discussed further in Chapter 12. Like the other operators, it has discrete eigenfunctions, and the possible energies of a system are discrete.

Another very important thing happens during the measurement. Whatever the wavefunction was before the measurement, afterward it will be one of the eigenfunctions of the observational operator. If the wavefunction was already an eigenfunction, Ψ_j, then the result will be λ_j, and the wavefunction will be unaltered. If the wavefunction wasn't an eigenfunction, however, the result of the measurement will still be one of the eigenvalues, λ_j, and the wavefunction will be changed to the corresponding eigenfunction, Ψ_j. This rather mysterious process is referred to as the "collapse" of the wavefunction and ensures that the same result will be obtained if an identical measurement is made immediately after the first.

11.3.2   Calculating the Average Outcome

Although the result of a single measurement on a state described by a non-eigen-function cannot be predicted, the average outcome of many repetitions of the experiment, designated $\langle A \rangle$, can be predicted according to Eq. 11.2:

$$\langle A \rangle = \Psi^* \hat{A} \Psi \tag{11.2}$$

This equation represents the operation, $\hat{A}$, on Ψ, followed by multiplication of the resulting function by another function related to Ψ, called Ψ^* and described below. Note that the order of operations is backward from the usual left-to-right way of interpreting mathematical expressions. In addition, the multiplication is of a special kind, called an *inner product*, that reflects the nature of the wavefunction. We will return to this kind of multiplication when the specific wavefunctions for spin-1/2 particles are introduced.

The function Ψ^* introduced above is called the *complex conjugate* of Ψ. In general, the values of Ψ can be complex, with real and imaginary parts. If a complex number or function, c, is written as the sum of the real and imaginary parts,

$$c = a + bi$$

where a and b are real numbers (or functions), then the complex conjugate is defined as

$$c^* = a - bi$$

If c is real, then it and its complex conjugate are identical.

If a complex number is multiplied by its conjugate, the result is a positive real number:

$$cc^* = (a + bi)(a - bi) = a^2 + b^2$$

If c is real (i.e., $b = 0$), then cc^* is simply the square of the number. The product of a complex number and its conjugate can be thought of as a generalization of the square of a real number. Just as the square of either a negative or positive real number is always a positive number, the product of a complex number and its conjugate is always a positive real number. The square root of this product is called the *modulus* and is analogous to the absolute value of a real number. Accordingly, the modulus is also represented as $|c|$. The modulus is always positive and reflects the magnitude of the number, irrespective of whether it is composed of real or imaginary parts.

Similarly, the inner product of a wavefunction and its complex conjugate has a positive value. This property and those of the wavefunctions used to describe quantum mechanical systems guarantee that the predicted average outcome of the experiment will be a real number, which is good, since it should be a real experiment![1]

---

[1] Even though NMR signals and other kinds of experimental data are often represented as complex numbers, the data that are recorded directly are real, and their representation as complex

11.3.3   The Dirac Notation: "Bras" and "Kets"

The wavefunctions and their complex conjugates are often written using a special notation devised by Paul A. M. Dirac, one of the founders of quantum mechanics. In Dirac notation, the wavefunction is written between a vertical line and a right angle bracket,

$$\Psi \equiv |\Psi\rangle$$

while the complex conjugate is written as

$$\Psi^* \equiv \langle\Psi|$$

The symbol $|\Psi\rangle$ is called a *ket*, and $\langle\Psi|$ is called a *bra*. When they are written together, they are called a *bracket* and represent the inner product:

$$\langle\Psi|\Psi\rangle = \Psi^*\Psi \tag{11.3}$$

For each ket, there is a bra, and various bras and kets can be combined. For instance, operating on Ψ by the observational operator, $\hat{A}$, yields another function, or ket, written $\hat{A}|\Psi\rangle$, and the expected average result of the observation (Eq. 11.2) is now written as

$$\langle A\rangle = \langle\Psi|\hat{A}|\Psi\rangle \tag{11.4}$$

An advantage of the Dirac notation is that the bracket implicitly includes the appropriate integrations over spatial variables, making it clearer and more compact than a full representation when spatial coordinates are involved. In addition, it helps clearly identify the wavefunction and its conjugate among the many other symbols that arise in quantum mechanical calculations. Again, it is important to keep in mind that the operations are carried out from right to left.

This is only a very sketchy outline of the mathematical formalism of quantum mechanics, and some liberties have been taken. As noted above, we have ignored the integration over spatial coordinates that is an important aspect of quantum calculations that involve more than spin systems, such as the distributions of electrons in the orbitals that define atomic and molecular structures. Also, we have blurred a distinction, made in more rigorous treatments, between the wavefunction, Ψ, and a mathematical object called a *state vector*, which is more properly associated with the Dirac ket. Wavefunctions and state vectors represent the same information, and for our purposes the terms can be interchanged. From this brief, if abstract, outline, we can now go on to introduce the wavefunctions and operators for a simple spin system and show how the Stern–Gerlach experiments are described.

---

values is a matter of convenience, whereas the use of complex functions in quantum mechanics is absolutely fundamental.

11.4   The Spin-1/2 System

11.4.1   The $\hat{I}_z$ Operator

When a spin-1/2 particle passes through the inhomogeneous magnetic field of the Stern–Gerlach filter, its angular momentum along the z-axis is being measured and, as dictated by the postulates outlined above, it is converted into a state corresponding to one of the two eigenfunctions for this measurement. By convention, the wavefunctions for these two states are written as

$$|\alpha\rangle: \text{ magnetic dipole "up"}$$

$$|\beta\rangle: \text{ magnetic dipole "down"}$$

The corresponding eigenvalues are 1/2 and $-1/2$. No matter what starting state is analyzed by the filter, the only possible outcomes are 1/2 and $-1/2$, and, in the course of the experiment, the wavefunction will be converted to $|\alpha\rangle$ or $|\beta\rangle$. The operator corresponding to the measurement of the spin angular momentum along the z-axis is designated $\hat{I}_z$, and is defined by the following equations:

$$\hat{I}_z|\alpha\rangle = \frac{1}{2}|\alpha\rangle$$

$$\hat{I}_z|\beta\rangle = -\frac{1}{2}|\beta\rangle$$

The kets, $|\alpha\rangle$ and $|\beta\rangle$, have corresponding complex conjugates, or bras, which are written as $\langle\alpha|$ and $\langle\beta|$.

So what are these mysterious $|\alpha\rangle$ and $|\beta\rangle$? The most common way of writing and manipulating the wavefunctions and their operators is as vectors and matrices. The quantum mechanical behavior then emerges from the rules for vector and matrix multiplication. These rules and their application to spin-1/2 particles are presented in Appendices D and E. For our purposes here, however, we will just use $|\alpha\rangle$ and $|\beta\rangle$ as symbols and accept the rules for their manipulation and operators as givens.

Some of the most important rules for the manipulation of the wavefunctions are as follows:

Rule 1. The eigenfunctions have unit "length"

From the Stern–Gerlach experiments and its variants, we learned that if a particle is in the $|\alpha\rangle$ state and its z-magnetization is measured, the outcome will always be the same, and this outcome is defined as 1/2. Thus, the average outcome must also be 1/2. In Dirac notation,

$$\langle\alpha|\hat{I}_z|\alpha\rangle = \frac{1}{2}$$

Because

$$\hat{I}_z|\alpha\rangle = \frac{1}{2}|\alpha\rangle$$

and 1/2 is just a constant that can be factored out,

$$\langle\alpha|\frac{1}{2}|\alpha\rangle = \frac{1}{2}$$

$$\langle\alpha|\alpha\rangle = 1$$

This means that the inner product of $\langle\alpha|$ and its complex conjugate, $|\alpha\rangle$, must be 1. By the same argument, the relationship

$$\langle\beta|\beta\rangle = 1$$

must hold. If the bras and kets are thought of as vectors, these relationships are interpreted as meaning that the lengths of the vectors are each equal to 1. Vectors with length 1 are said to be *normalized*.

Rule 2. The eigenfunctions are orthogonal

This rule is not as obvious from the definitions given so far, but is expressed succinctly as

$$\langle\alpha|\beta\rangle = 0 \quad \text{and} \quad \langle\beta|\alpha\rangle = 0$$

If viewed as vectors, the two eigenfunctions are perpendicular, or orthogonal, to one another. Together, rules 1 and 2 are summarized by saying that the eigenfunctions are *orthonormal*.

Rule 3. Any wavefunction for a spin-1/2 particle can be written as a linear combination of $|\alpha\rangle$ and $|\beta\rangle$

Any $|\Psi\rangle$, including those describing the various states we generated in our hypothetical experiments, can be written as

$$|\Psi\rangle = c_\alpha|\alpha\rangle + c_\beta|\beta\rangle \tag{11.5}$$

where c_α and c_β are constants. The constants are, in general, complex numbers, with a real and an imaginary part. The corresponding bra for this wavefunction will be

$$\langle\Psi| = c_\alpha^*\langle\alpha| + c_\beta^*\langle\beta|$$

where c_α^* and c_β^* are the complex conjugates of c_α and c_β, respectively.

Having expressed $|\Psi\rangle$ and $\langle\Psi|$ as linear combinations, we now need to be more specific about the meaning of the inner product. With this representation, the inner product follows from the distributive property of ordinary algebra:

$$\langle\Psi|\Psi\rangle = \left(c_\alpha^*\langle\alpha| + c_\beta^*\langle\beta|\right)\left(c_\alpha|\alpha\rangle + c_\beta|\beta\rangle\right)$$

$$= c_\alpha c_\alpha^*\langle\alpha|\alpha\rangle + c_\beta c_\alpha^*\langle\alpha|\beta\rangle + c_\alpha c_\beta^*\langle\beta|\alpha\rangle + c_\beta c_\beta^*\langle\beta|\beta\rangle$$

From rules 1 and 2 above, we can simplify this to

$$\langle\Psi|\Psi\rangle = c_\alpha c_\alpha^* + c_\beta c_\beta^*$$

Other pairs of bras and kets are multiplied in the same way. For instance, if $\langle \Psi_1 |$ and $| \Psi_2 \rangle$ are defined as

$$\langle \Psi_1 | = c_{\alpha,1}^* \langle \alpha | + c_{\beta,1}^* \langle \beta |$$

$$| \Psi_2 \rangle = c_{\alpha,2} | \alpha \rangle + c_{\beta,2} | \beta \rangle$$

then the inner product, or bracket, is given by

$$\langle \Psi_1 | \Psi_2 \rangle = (c_{\alpha,1}^* \langle \alpha | + c_{\beta,1}^* \langle \beta |)(c_{\alpha,2} | \alpha \rangle + c_{\beta,2} | \beta \rangle)$$

$$= c_{\alpha,1}^* c_{\alpha,2} \langle \alpha | \alpha \rangle + c_{\alpha,1}^* c_{\beta,2} \langle \alpha | \beta \rangle + c_{\beta,1}^* c_{\alpha,2} \langle \beta | \alpha \rangle + c_{\beta,1}^* c_\beta^2 \langle \beta | \beta \rangle$$

$$= c_{\alpha,1}^* c_{\alpha,2} + c_{\beta,1}^* c_{\beta,2}$$

In order for the measurement operators to yield correct results, any wavefunction that they operate on must, like $| \alpha \rangle$ and $| \beta \rangle$, be normalized. Thus, the possible values of c_α and c_β are constrained so that

$$c_\alpha c_\alpha^* + c_\beta c_\beta^* = 1$$

The rules for manipulating the wavefunctions and operators also follow the distributive property. Thus, for any wavefunction written as a linear combination of $| \alpha \rangle$ and $| \beta \rangle$, the result of $\hat{I}_z$ operating on the wavefunction is

$$\hat{I}_z | \Psi \rangle = \hat{I}_z \left(c_\alpha | \alpha \rangle + c_\beta | \beta \rangle \right)$$

$$= c_\alpha \hat{I}_z | \alpha \rangle + c_\beta \hat{I}_z | \beta \rangle$$

$$= \frac{1}{2} c_\alpha | \alpha \rangle - \frac{1}{2} c_\beta | \beta \rangle \qquad (11.6)$$

Note that for most possible wavefunctions, $\hat{I}_z | \Psi \rangle$ is not simply $| \Psi \rangle$ multiplied by a constant, since the sign of one coefficient, but not the other, is changed. That is, most wavefunctions are not eigenfunctions of $\hat{I}_z$.

Be aware that not all of the properties that we take for granted in ordinary algebra are valid for quantum mechanical operators. In particular, the operators are not generally commutative. For instance, if $\hat{A}$ and $\hat{B}$ are two operators, the expression

$$\hat{A} \hat{B} | \Psi \rangle$$

represents the result of $\hat{A}$ acting on the result of $\hat{B}$ operating on $| \Psi \rangle$, whereas the expression

$$\hat{B} \hat{A} | \Psi \rangle$$

represents the opposite order of operation. In general, the kets generated by the two sets of operation are not equal to one another. Only if the measurements of the two properties, A and B, are fully independent will the operators be commutative. From the Stern–Gerlach experiments, we know that the measurements of magnetization along different directions are not independent, and, accordingly, the corresponding operators do not commute.

The average outcome of measuring I_z for a wavefunction expressed as a superposition of $|\alpha\rangle$ and $|\beta\rangle$ can be calculated as

$$\langle I_z \rangle = \langle \Psi | \hat{I}_z | \Psi \rangle = (c_\alpha^* \langle \alpha | + c_\beta^* \langle \beta |) \hat{I}_z \left(c_\alpha | \alpha \rangle + c_\beta | \beta \rangle \right)$$

$$= \left(c_\alpha^* \langle \alpha | + c_\beta^* \langle \beta | \right) \left(\frac{c_\alpha}{2} | \alpha \rangle - \frac{c_\beta}{2} | \beta \rangle \right)$$

$$= \frac{1}{2} \left(c_\alpha^* c_\alpha \langle \alpha | \alpha \rangle - c_\alpha^* c_\beta \langle \alpha | \beta \rangle + c_\beta^* c_\alpha \langle \beta | \alpha \rangle - c_\beta^* c_\beta \langle \beta | \beta \rangle \right)$$

$$= \frac{1}{2} \left(c_\alpha^* c_\alpha - c_\beta^* c_\beta \right)$$

This provides a simple way to calculate the average outcome of measuring the z-magnetization for particles with an arbitrary wavefunction defined by the coefficients c_α and c_β.

The ability to write an arbitrary wavefunction as a linear combination of eigenfunctions is a general feature of quantum mechanics. Any observational operator appropriate for a system will have a number of eigenfunctions that is just large enough so that any possible wavefunction can be written as a linear combination of these functions. This is called the *superposition principle*.

Rule 4. The probability equation

The probability of a measurement yielding a specific eigenvalue, λ_i, can be calculated from the corresponding eigenfunction, $|\Psi_i\rangle$, and the wavefunction describing the measured state, $|\Psi\rangle$, according to the equation

$$P_{\lambda_i} = |\langle \Psi_i | \Psi \rangle|^2 \tag{11.7}$$

For instance, we can calculate the probability that a measurement of I_z, applied to a general wavefunction written as a superposition of $|\alpha\rangle$ and $|\beta\rangle$, will produce the value $1/2$, for which the eigenfunction is $|\alpha\rangle$, as

$$P_{1/2} = |\langle \alpha | \Psi \rangle|^2$$

$$= |\langle \alpha | (c_\alpha | \alpha \rangle + c_\beta | \beta \rangle)|^2$$

$$= |c_\alpha \langle \alpha | \alpha \rangle + c_\beta \langle \alpha | \beta \rangle|^2$$

$$= |c_\alpha|^2$$

$$= c_\alpha^* c_\alpha$$

Similarly, the probability that the measurement will yield the value $-1/2$ is

$$P_{-1/2} = |\langle \beta | \Psi \rangle|^2 = c_\beta^* c_\beta$$

Note that the probabilities of the two possible outcomes sum to 1 (because the wave-function is normalized), as they should. Thus, the coefficients, c_α and c_β, have a simple physical interpretation in terms of the operator for which $|\alpha\rangle$ and $|\beta\rangle$ are eigenfunctions: The squares of their moduli represent the probabilities of the measurement giving rise to the corresponding eigenvalues.

Again, this rule is general, and the probability equation can be used to calculate the probabilities associated with other measurements. Brackets of the form $\langle\Psi_b|\Psi_a\rangle$ are referred to as *probability amplitudes* and are not, in general, real numbers. When multiplied by their complex conjugates, however, the product is real and represents the probability of the conversion of a state with wavefunction $|\Psi_a\rangle$ into one with wavefunction $|\Psi_b\rangle$, under appropriate physical conditions. Because functions and their complex conjugates are not, in general, equal to one another, $\langle\Psi_b|\Psi_a\rangle$ is not equivalent to $\langle\Psi_a|\Psi_b\rangle$. Furthermore, $\langle\Psi_b|\Psi_a\rangle$ is not equivalent to $|\Psi_a\rangle\langle\Psi_b|$. The latter expression represents another kind of multiplication, called an outer product, and can be used to represent an operator.

11.4.2   Other Angular Momentum Operators

As we saw in the multiple Stern–Gerlach experiments, we are not at all restricted to measuring the orientation of the magnetic dipole along the z-axis. By rotating the filter by 90°, we can measure the orientation along the x-axis. The corresponding operator is called $\hat{I}_x$ and can be defined by its action on an arbitrary wavefunction:

$$\hat{I}_x|\Psi\rangle = \hat{I}_x\left(c_\alpha|\alpha\rangle + c_\beta|\beta\rangle\right) = \frac{1}{2}c_\beta|\alpha\rangle + \frac{1}{2}c_\alpha|\beta\rangle$$

If this seems just like the relationship for $\hat{I}_z$ (Eq. 11.6), look more closely!

Importantly, $|\alpha\rangle$ and $|\beta\rangle$ are *not* eigenfunctions for $\hat{I}_x$. But, $\hat{I}_x$ has its own pair of eigenfunctions, which may be written as superpositions of $|\alpha\rangle$ and $|\beta\rangle$:

$$\frac{1}{\sqrt{2}}|\alpha\rangle + \frac{1}{\sqrt{2}}|\beta\rangle$$

and

$$-\frac{1}{\sqrt{2}}|\alpha\rangle + \frac{1}{\sqrt{2}}|\beta\rangle$$

The corresponding eigenvalues are $1/2$ and $-1/2$. Thus,

$$\hat{I}_x\left(\frac{1}{\sqrt{2}}|\alpha\rangle + \frac{1}{\sqrt{2}}|\beta\rangle\right) = \frac{1}{2}\left(\frac{1}{\sqrt{2}}|\alpha\rangle + \frac{1}{\sqrt{2}}|\beta\rangle\right)$$

and

$$\hat{I}_x\left(-\frac{1}{\sqrt{2}}|\alpha\rangle + \frac{1}{\sqrt{2}}|\beta\rangle\right) = -\frac{1}{2}\left(-\frac{1}{\sqrt{2}}|\alpha\rangle + \frac{1}{\sqrt{2}}|\beta\rangle\right)$$

These relationships follow directly from those given above, and you should be able to confirm them.

The eigenfunctions for $\hat{I}_x$ are also required to be orthonormal, just like $|\alpha\rangle$ and $|\beta\rangle$:

$$\left(\frac{1}{\sqrt{2}}\langle\alpha| + \frac{1}{\sqrt{2}}\langle\beta|\right)\left(\frac{1}{\sqrt{2}}|\alpha\rangle + \frac{1}{\sqrt{2}}|\beta\rangle\right) = 1$$

$$\left(-\frac{1}{\sqrt{2}}\langle\alpha| + \frac{1}{\sqrt{2}}\langle\beta|\right)\left(-\frac{1}{\sqrt{2}}|\alpha\rangle + \frac{1}{\sqrt{2}}|\beta\rangle\right) = 1$$

$$\left(\frac{1}{\sqrt{2}}\langle\alpha| + \frac{1}{\sqrt{2}}\langle\beta|\right)\left(-\frac{1}{\sqrt{2}}|\alpha\rangle + \frac{1}{\sqrt{2}}|\beta\rangle\right) = 0$$

$$\left(-\frac{1}{\sqrt{2}}\langle\alpha| + \frac{1}{\sqrt{2}}\langle\beta|\right)\left(\frac{1}{\sqrt{2}}|\alpha\rangle + \frac{1}{\sqrt{2}}|\beta\rangle\right) = 0$$

Again, from the relationships for $|\alpha\rangle$ and $|\beta\rangle$, and their bras, you should be able to confirm each of these relationships.

The orientation of the magnetic dipole can also be measured in a third direction, perpendicular to both the x- and z-axes. In our Stern–Gerlach experiments, the y-axis corresponds to the direction of the atomic beam, and the measurement would have to be made in a different way. Ignoring the experimental difficulties, the operator for measurement along this axis is called, not surprisingly, $\hat{I}_y$. Again, the operator can be defined in terms of its action on an arbitrary wavefunction:

$$\hat{I}_y|\Psi\rangle = \hat{I}_y\left(c_\alpha|\alpha\rangle + c_\beta|\beta\rangle\right) = -\frac{i}{2}c_\beta|\alpha\rangle + \frac{i}{2}c_\alpha|\beta\rangle \tag{11.8}$$

Like the other operators, $\hat{I}_y$ has two eigenfunctions:

$$\frac{1}{\sqrt{2}}|\alpha\rangle + \frac{i}{\sqrt{2}}|\beta\rangle \tag{11.9}$$

and

$$\frac{1}{\sqrt{2}}|\alpha\rangle - \frac{i}{\sqrt{2}}|\beta\rangle \tag{11.10}$$

and the corresponding eigenvalues are $1/2$ and $-1/2$.

The coefficients of these wavefunctions include imaginary components, which makes things a little more interesting. In particular, the bras are not simply equal to the kets with the angle brackets reversed, as they were before. Instead, the bras, or complex conjugates, are

$$\frac{1}{\sqrt{2}}\langle\alpha| - \frac{i}{\sqrt{2}}\langle\beta|$$

and

$$\frac{1}{\sqrt{2}}\langle\alpha| + \frac{i}{\sqrt{2}}\langle\beta|$$

However, the eigenvalues, 1/2 and -1/2, are real, and the average calculated outcomes of experiments are real numbers. For instance, let's calculate the average outcome if a particle with the $|\alpha\rangle$ wavefunction is tested in an experiment that measures magnetization along the y-axis.

$$\langle I_y \rangle = \langle \alpha | \hat{I}_y | \alpha \rangle = \langle \alpha | \frac{-i}{2} | \beta \rangle = \frac{-i}{2} \langle \alpha | \beta \rangle = 0$$

Because the two possible outcomes are 1/2 and $-1/2$, each must arise with equal probability. Similarly, if the measurement is made on particles represented by one of the eigenfunctions of the $\hat{I}_y$ operator, the result will be real:

$$\langle I_y \rangle = \left(\frac{1}{\sqrt{2}} \langle \alpha | - \frac{i}{\sqrt{2}} \langle \beta | \right) \hat{I}_y \left(\frac{1}{\sqrt{2}} | \alpha \rangle + \frac{i}{\sqrt{2}} | \beta \rangle \right)$$

$$= \left(\frac{1}{\sqrt{2}} \langle \alpha | - \frac{i}{\sqrt{2}} \langle \beta | \right) \frac{i}{2} \left(\frac{-i}{\sqrt{2}} | \alpha \rangle + \frac{1}{\sqrt{2}} | \beta \rangle \right)$$

$$= \frac{i}{2} \left(-\frac{i}{2} \langle \alpha | \alpha \rangle + \frac{1}{2} \langle \alpha | \beta \rangle - \frac{1}{2} \langle \beta | \alpha \rangle - \frac{i}{2} \langle \beta | \beta \rangle \right)$$

$$= \frac{i}{2} \left(-\frac{i}{2} - \frac{i}{2} \right)$$

$$= \frac{1}{2}$$

This makes sense, since the result of this measurement should always be 1/2.

In the previous subsection, we derived an expression to calculate the average outcome of measuring I_z for an arbitrary wavefunction. Similar expressions can also be derived for I_x and I_y, and all three expressions are given below for $|\Psi\rangle = c_\alpha | \alpha \rangle + c_\beta | \beta \rangle$:

$$\langle I_x \rangle = \langle \Psi | \hat{I}_x | \Psi \rangle = \frac{1}{2} \left(c_\alpha c_\beta^* + c_\beta c_\alpha^* \right)$$

$$\langle I_y \rangle = \langle \Psi | \hat{I}_y | \Psi \rangle = \frac{i}{2} \left(c_\alpha c_\beta^* - c_\beta c_\alpha^* \right)$$

$$\langle I_z \rangle = \langle \Psi | \hat{I}_z | \Psi \rangle = \frac{1}{2} \left(c_\alpha c_\alpha^* - c_\beta c_\beta^* \right)$$

These relationships will prove handy later on. In spite of the imaginary number in the expression for $\langle I_y \rangle$, the result will still be a real number for any valid wavefunction.

So far, we have considered measurements along three axes; x, y and z. For each of these measurements, there are two possible outcomes, 1/2 and $-1/2$. Also, for each measurement there are two eigenfunctions, each of which will yield one of the eigenvalues, all of the time. The same measurement applied to particles with any other wavefunction will also yield measurements of either 1/2 or $-1/2$, with a probability that can be calculated from Eq. 11.7.

We could, in principle, make measurements along any other direction, and this could go on forever! For each measurement, there would be the same two eigenvalues, 1/2 and $-1/2$, but a different pair of eigenfunctions. However, once the operators for $\hat{I}_x$, $\hat{I}_y$ and $\hat{I}_z$ are defined, the operators for measurements in any other direction can be calculated from a set of simple trigonometric rules. In fact, any set of three orthogonal axis can be used in this way.

11.5 A Possible Source of Confusion

Because of the language and mathematical formalism used to describe quantum mechanics, it is sometimes easy to confuse mathematical operations with physical processes. It is particularly important to distinguish between the following:

- The mathematical manipulation of an operator acting on a wavefunction.
- The collapse of a non-eigen wavefunction during the physical process of measurement.

The mathematical operation converts a wavefunction that represents a physical state into a function that generally does not represent a real state but, rather, serves as an intermediate in the mathematical manipulations that are used to predict the average outcome of the experiment. In contrast, the physical measurement changes the system so that it is now described by a new wavefunction, in particular one of the eigenfunctions for the operator associated with the measurement.

To make this distinction more concrete, consider the case of measuring I_z for particles with the following wavefunction:

$$|\Psi\rangle = \frac{1}{\sqrt{2}}|\alpha\rangle + \frac{1}{\sqrt{2}}|\beta\rangle$$

From our previous definitions, the result of the $\hat{I}_z$ operator acting on the wavefunction is

$$\hat{I}_z|\Psi\rangle = \frac{1}{2}c_\alpha|\alpha\rangle - \frac{1}{2}c_\beta|\beta\rangle = \frac{1}{2\sqrt{2}}|\alpha\rangle - \frac{1}{2\sqrt{2}}|\beta\rangle$$

This wavefunction is then multiplied by the complex conjugate of the original wavefunction in order to calculate the average outcome of the experiment (in this case, 0).

On the other hand, the physical effect of the measurement is to convert the wavefunction for each particle into one of the eigenfunctions for $\hat{I}_z$, $|\alpha\rangle$ or $|\beta\rangle$. In this particular case, both eigenfunctions are generated with equal probability.

If the wavefunction before the measurement happens to be one of the eigenfunctions, then the wavefunction of the particle will be unchanged, and the operator will simply multiply the wavefunction by a constant (the eigenfunction, 1/2 or $-1/2$). In this special case, the result of the mathematical operator is *almost* the same as the wavefunction of the particle following the measurement (which is the same as the wavefunction before the measurement.)

11.6   Application of the Spin Operators to the Stern–Gerlach Experiments

Having defined the magnetization operators and their eigenfunctions, we are now in a position to use this formalism to describe the outcome of the various permutations of the Stern–Gerlach experiment described earlier.

In the basic experiment (Fig. 11.1), we start with a population of spin-1/2 particles (silver atoms) that come from the oven. At this point, we really don't know anything about their spin states. When they pass through the z-filter, we are measuring the orientation along the z-axis. We know that there are only two possible outcomes of this measurement: 1/2 and $-1/2$, irrespective of what their starting spin states were. In addition, as soon as the measurement is made, the wavefunction is converted to either $|\alpha\rangle$ or $|\beta\rangle$. The atoms in the $|\alpha\rangle$ state are deflected upward along the z-axis, and those in the $|\beta\rangle$ state are deflected downward.

If we now take the atoms in the $|\alpha\rangle$ state and again measure their orientation along the z-axis (Fig. 11.3), we know that applying the $\hat{I}_z$ operator to $|\alpha\rangle$ will only result in the value 1/2. In this case, the result is absolutely predictable and there is no change in the wavefunction.

Suppose, however, that we now take the purified $|\alpha\rangle$ state and pass it through the Stern–Gerlach filter that has been rotated 90°, so that it measures orientation along the x-axis (Fig. 11.4). When the $\hat{I}_x$ operator is applied to $|\alpha\rangle$, we get

$$\hat{I}_x|\alpha\rangle = \frac{1}{2}|\beta\rangle$$

In this case, $|\alpha\rangle$ is not an eigenfunction of the operator, and there is not a unique outcome of the experiment. There are, however, still only two possible outcomes; the orientation along the x-axis can be either 1/2 or $-1/2$. We can also now calculate the average outcome of the experiment if it is repeated for many different particles with the $|\alpha\rangle$ wavefunction. From before, we calculate the average outcome as

$$\langle I_x\rangle = \langle\alpha|\hat{I}_x|\alpha\rangle = \langle\alpha|\left(\frac{1}{2}|\beta\rangle\right) = \frac{1}{2}\langle\alpha|\beta\rangle = 0$$

Thus, the average value of the x-magnetization would be 0. (The parentheses are drawn as a reminder that first $|\alpha\rangle$ is operated on by $\hat{I}_x$ and the resulting function is then multiplied by $\langle\alpha|$.) Because there are only two possible outcomes, 1/2 and $-1/2$, and the average outcome is 0, we can conclude that each outcome is observed for 50% of the particles that are tested. Once the measurement is made, the wavefunctions will be altered to be one or the other of the eigenfunctions for $\hat{I}_x$:

$$\frac{1}{\sqrt{2}}|\alpha\rangle + \frac{1}{\sqrt{2}}|\beta\rangle$$

or

$$-\frac{1}{\sqrt{2}}|\alpha\rangle + \frac{1}{\sqrt{2}}|\beta\rangle$$

Now, let's take one of these wavefunctions (the first one), and again measure the z-orientation (Fig. 11.5). Applying the operator to the wavefunction gives

$$\hat{I}_z\left(\frac{1}{\sqrt{2}}|\alpha\rangle + \frac{1}{\sqrt{2}}|\beta\rangle\right) = \frac{1}{2\sqrt{2}}|\alpha\rangle - \frac{1}{2\sqrt{2}}|\beta\rangle$$

This does not satisfy the eigen condition, so there is not a single predictable outcome. As before, though, we can calculate the average outcome:

$$\langle I_z\rangle = \left(\frac{1}{\sqrt{2}}\langle\alpha| + \frac{1}{\sqrt{2}}\langle\beta|\right)\hat{I}_z\left(\frac{1}{\sqrt{2}}|\alpha\rangle + \frac{1}{\sqrt{2}}|\beta\rangle\right)$$

$$= \left(\frac{1}{\sqrt{2}}\langle\alpha| + \frac{1}{\sqrt{2}}\langle\beta|\right)\left(\frac{1}{2\sqrt{2}}|\alpha\rangle - \frac{1}{2\sqrt{2}}|\beta\rangle\right)$$

$$= \frac{1}{4}\langle\alpha|\alpha\rangle + \frac{1}{4}\langle\beta|\alpha\rangle - \frac{1}{4}\langle\alpha|\beta\rangle - \frac{1}{4}\langle\beta|\beta\rangle$$

$$= \frac{1}{4}\cdot 1 + \frac{1}{4}\cdot 0 - \frac{1}{4}\cdot 0 - \frac{1}{4}\cdot 1$$

$$= 0$$

This is consistent with the observation that the z-filter, following the x-filter, again splits the beam into two populations.

11.7 Other Angles

At this point, you might begin to suspect that the average outcome is always either $1/2$, $-1/2$ or 0. The outcome will always be one of the eigenvalues if the starting state has a wavefunction that is an eigenfunction of the operator, and it looks as though the average outcome is always 0 otherwise. This, however, is not the case. So far, we have always made our measurements using successive pairs of filters that either have the same orientation or are turned $\pi/2$ rad (90°) from one another. For any pair of perpendicular filters, the average outcome of the second measurement will be 0. However, if the second filter is rotated by some angle other than $\pi/2$ rad (or an integer multiple thereof), the average outcome will be something other than -1/2, 0 or 1/2. As a final Stern–Gerlach experiment, illustrated in Fig. 11.13, consider a z-filter followed by a second filter that can be considered a z-filter that has been rotated by an angle, ϕ. (In the figure, the rotation angle is negative.)

We will call the operator for the second filter $\hat{I}_\phi$. As before, the operator can be defined by its actions on a general wavefunction written as a superposition of $|\alpha\rangle$ and $|\beta\rangle$:

$$\hat{I}_\phi(c_\alpha|\alpha\rangle + c_\beta|\beta\rangle) = \frac{1}{2}(c_\alpha\cos(\phi) + c_\beta\sin(\phi))|\alpha\rangle + \frac{1}{2}(c_\alpha\sin(\phi) - c_\beta\cos(\phi))|\beta\rangle$$

Although the operator is presented here with no justification, note that it does give the correct results if $\phi = 0$ ($\hat{I}_\phi = \hat{I}_z$) or if $\phi = \pi/2$ rad ($\hat{I}_\phi = \hat{I}_x$). As before, there will

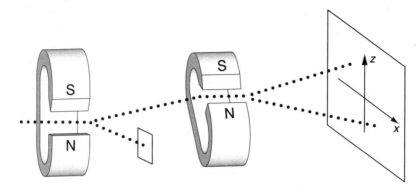

Figure 11.13 A Stern–Gerlach z-filter followed by a second filter rotated to a position between the x- and z-axes.

be two possible outcomes, 1/2 and $-1/2$, and two eigenfunctions, which are

$$|\Psi_{\phi 1}\rangle = \cos(\phi/2)|\alpha\rangle + \sin(\phi/2)|\beta\rangle$$

and

$$|\Psi_{\phi 2}\rangle = -\sin(\phi/2)|\alpha\rangle + \cos(\phi/2)|\beta\rangle$$

Again, these relationships are consistent with those shown previously for $\hat{I}_z$ and $\hat{I}_x$. Consider, now, the specific case of $\phi = -\pi/4$ rad ($-45°$), as shown in the drawing above. The eigenfunctions will be

$$|\Psi_{\phi 1}\rangle = \cos(-\pi/8)|\alpha\rangle + \sin(-\pi/8)|\beta\rangle \approx 0.924|\alpha\rangle - 0.383|\beta\rangle$$

and

$$|\Psi_{\phi 1}\rangle = -\sin(-\pi/8)|\alpha\rangle + \cos(-\pi/8)|\beta\rangle \approx 0.383|\alpha\rangle + 0.924|\beta\rangle$$

So, each of these has some $|\alpha\rangle$ and some $|\beta\rangle$ character, in unequal amounts. If the $-\pi/4$-filter is applied to particles in the $|\alpha\rangle$ state, the average outcome can be calculated as

$$\langle I_{(-\pi/4)}\rangle = \langle\alpha|\hat{I}_{(-\pi/4)}|\alpha\rangle$$

$$= \langle\alpha|\left(\frac{1}{2}\cos(-\pi/4)|\alpha\rangle + \frac{1}{2}\sin(-\pi/4)|\beta\rangle\right)$$

$$= \frac{1}{2}\cos(-\pi/4)\langle\alpha|\alpha\rangle - \frac{1}{2}\sin(-\pi/4)\langle\alpha|\beta\rangle$$

$$= \frac{1}{2}\cos(-\pi/4)$$

$$= \frac{1}{2\sqrt{2}} \approx 0.35$$

Since the two possible outcomes are 1/2 and -1/2, this result indicates that most, but not all, of the particles will leave the filter in the state with the eigenfunction corresponding to 1/2. The probabilities of each outcome can be calculated using Eq. 11.7. For the particles deflected upward, corresponding to the eigenvalue 1/2, the probability is calculated as

$$P_{1/2} = |\langle \Psi_{\phi 1}|\alpha\rangle|^2$$

$$= |\left(\cos(-\pi/8)\langle\alpha| + \sin(-\pi/8)\langle\beta|\right)|\alpha\rangle|^2$$

$$= \cos(-\pi/8)^2 \approx 0.854$$

For the eigenvalue $-1/2$, the probability is

$$P_{-1/2} = |\langle \Psi_{\phi 2}|\alpha\rangle|^2$$

$$= |\left(-\sin(-\pi/8)\langle\alpha| + \cos(-\pi/8)\langle\beta|\right)|\alpha\rangle|^2$$

$$= |-\sin(-\pi/8)|^2 \approx 0.146$$

Note that the sum of the two probabilities equals 1, as expected for two alternative outcomes.

In terms of the information represented by the wavefunctions, we can say that when a particle is first passed through one filter, to generate a particular state, and then passed through a perpendicular filter, all of the information from the first filter is lost. When the second filter is *not* perpendicular to the first, however, some of the original information is retained.

11.8   The Shift Operators, $\hat{I}^+$ and $\hat{I}^-$

There is another type of operator commonly used in the description of spin states, which are collectively referred to as *ladder operators* or *shift operators*. These operators are simply linear combinations of the $\hat{I}_x$- and $\hat{I}_y$-magnetization operators. The *raising operator*, $\hat{I}^+$, is defined as:

$$\hat{I}^+ = \hat{I}_x + i\hat{I}_y$$

while the *lowering operator*, $\hat{I}^-$, is

$$\hat{I}^- = \hat{I}_x - i\hat{I}_y$$

These operators can also be expressed in terms of their action on an arbitrary wavefunction, $|\Psi\rangle = c_\alpha|\alpha\rangle + c_\beta|\beta\rangle$:

$$\hat{I}^+|\Psi\rangle = c_\beta|\alpha\rangle$$

and

$$\hat{I}^-|\Psi\rangle = c_\alpha|\beta\rangle$$

Consider the results of applying these operators to $|\alpha\rangle$ and $|\beta\rangle$, the eigenfunctions for the I_z operator:

$$\hat{I}^+|\alpha\rangle = 0$$

$$\hat{I}^+|\beta\rangle = |\alpha\rangle$$

$$\hat{I}^-|\alpha\rangle = |\beta\rangle$$

$$\hat{I}^-|\beta\rangle = 0$$

Notice, first of all, that neither $|\alpha\rangle$ nor $|\beta\rangle$ is an eigenfunction for either of the shift operators, except in the trivial sense that $\hat{I}^+|\alpha\rangle$ and $\hat{I}^-|\beta\rangle$ both equal zero. In fact, there are *no* non-trivial eigenfunctions for either $\hat{I}^+$ or $\hat{I}^-$. This reflects the fact that the shift operators do not correspond to any physical measurement. Instead, these operators embody the rules for the interconversions between states and the corresponding changes in the spin quantum number.

By convention, the $|\alpha\rangle$ state of a spin-1/2 particle is said to have a spin quantum number of 1/2, and the $|\beta\rangle$ state has a quantum number of $-1/2$, corresponding to the $\hat{I}_z$ eigenvalues. The raising operator represents the conversion of a wavefunction to $|\alpha\rangle$, an increase in the quantum number, and the lowering operator represents a conversion to $|\beta\rangle$, a decrease in the quantum number. The raising operator changes the $|\beta\rangle$ state to $|\alpha\rangle$, but this operator returns the value of 0 when applied to $|\alpha\rangle$, because the quantum number of $|\alpha\rangle$ cannot be raised further. Conversely, $\hat{I}^-$ converts $|\alpha\rangle$ to $|\beta\rangle$, but returns 0 when applied to $|\beta\rangle$. More generally, $\hat{I}^+$ converts the $|\beta\rangle$ portion of a superposition state to $|\alpha\rangle$, and eliminates the original $|\alpha\rangle$ component, whereas $\hat{I}^-$ converts the $|\alpha\rangle$ portion to $|\beta\rangle$ and eliminates the original $|\beta\rangle$ component.

In the context of NMR, a major application of the shift operators is to calculate the probabilities of transitions specifically associated with upward and downward transitions when nuclei are influenced by oscillating magnetic fields of the proper frequency, a subject that we will return to in Chapters 12 and 13.

11.9   Final Comments

From the time it was first described, the quantum behavior of polarized particles has attracted a great deal of attention. Both real and conceptual experiments have been used to illustrate and probe some of the more mysterious aspects of quantum mechanics. There has also been a great deal of recent interest in possible applications of these phenomena in areas such as cryptography and computing. The fact that making a measurement on a quantum system potentially changes its state can, in principle, be exploited to devise communication systems in which any attempt to listen in can be detected. A related idea involves using quantum mechanical objects as the basis of computers. Because a single particle can exist in a superposition state in which the wavefunctions of multiple states are combined, manipulating particles in this superposition state is equivalent, in some respects, to manipulating multiple states in a parallel fashion. This has raised the possibility of quantum computers that can carry out multiple processes in parallel. Although various schemes for quantum computing have been proposed, one of the first implementations was the use of an NMR spectrometer to manipulate the spins in a liquid sample and then "read out" their state at the end of the computation.

Summary Points for Chapter 11

- Experiments with multiple Stern–Gerlach filters highlight some of the strangest features of quantum mechanics, including its probabilistic nature and the ability of observations to alter the state of a quantum system.
- Behavior similar to that seen with Stern–Gerlach filters can also be demonstrated with light and polarizing filters.
- In quantum mechanics, the state of a system at a given time is described by the wavefunction, designated Ψ.
- The wavefunction cannot, in general, be determined by experiments, but if Ψ is known, it can be used, together with the appropriate operator, to predict the average outcome of a measurement.
- For some wavefunctions, the outcome of a given measurement is always the same:
    - The wavefunctions for which the outcome is certain are the eigenfunctions of the measurement operator.
    - The result of applying the operator to one of its eigenfunctions is the eigenfunction multiplied by a real number, called the eigenvalue, which is the expected outcome of the measurement.
- The states described by the eigenfunctions are not the only possible states of a system, but any other state can be described by a linear superposition of the eigenfunctions of a valid measurement.
- When a measurement is made of a system with a wavefunction that is not one of the eigenfunctions for the measurement operator, the result will still be one of the eigenvalues.
- The specific outcome of a measurement applied to a state with a wavefunction that is not one of the eigenfunctions of the measurement operator cannot be predicted, but the probabilities of the possible outcomes and the average outcome can be calculated.
- A measurement applied to a non-eigenstate converts the wavefunction to one of the eigenfunctions of the corresponding operator.
- For a spin-1/2 particle, measurements of the magnetization component along the three coordinate axes are associated with three operators, $\hat{I}_x$, $\hat{I}_y$ and $\hat{I}_z$, each of which has two eigenvalues, 1/2 and $-1/2$.
- The shift operators for a spin-1/2 particle, $\hat{I}^+$ and $\hat{I}^-$, do not correspond to physical measurements, but define upward and downward transitions in terms of the spin quantum numbers 1/2 and $-1/2$.

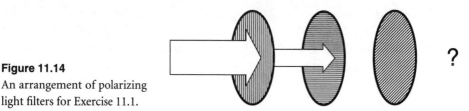

Figure 11.14
An arrangement of polarizing
light filters for Exercise 11.1.

Exercises for Chapter 11

The mathematics of quantum mechanics can quickly become quite awkward, even for relatively simple systems. Computer algebra systems (CAS), as implemented in programs such as Maple, Mathematica and Maxima, can greatly aid these calculations. A set of macros for calculating wavefunctions and operators for spin-1/2 particles (1spinLib.mac), along with some worksheets with examples, have been prepared for the open-source program Maxima and are available for download through http://uscibooks.com/goldenberg.htm. You are, however, strongly encouraged to try the simpler problems by hand before resorting to computer algebra!

11.1. Section 11.2 describes an experiment using a set of three polarizing filters arranged as shown in Fig. 11.12. Rather miraculously, introducing the middle filter, with an orientation midway between those of the other two, allows light to pass through the apparatus, while in the absence of this filter none would pass through. Suppose that we were to rearrange the filters as shown in Fig. 11.14.

Now, no light passes through the second filter, but what happens when the third filter, with its intermediate position is added? Will any light emerge from the apparatus? Carefully explain your reasoning and how this situation differs from that shown in Fig. 11.12.

11.2. The operator for measuring angular momentum in the y-direction, $\hat{I}_y$, is given by Eq. 11.8, and the eigenfunctions for this operator are given in Eqs. 11.9 and 11.10.

(a) Show that the two eigenfunctions are orthonormal.
(b) Show that the eigenfunctions are, in fact, eigenfunctions of $\hat{I}_y$. That is, show that when the $\hat{I}_y$ operator acts on the eigenkets, the results are the kets multiplied by a constant. What are the eigenvalues?

11.3. Suppose that you were to have a population of spin-1/2 particles with the following wavefunction:

$$|\psi\rangle = \frac{1}{\sqrt{5}}|\alpha\rangle + \frac{2}{\sqrt{5}}|\beta\rangle$$

(a) Confirm that this wavefunction is properly normalized.
(b) Calculate the average outcome of measuring I_x with this population. Note: If you don't get a real value between $-1/2$ and $1/2$, something has gone wrong!
(c) Calculate the probabilities that the I_x measurement of a single particle with this wavefunction will yield $1/2$ or $-1/2$.

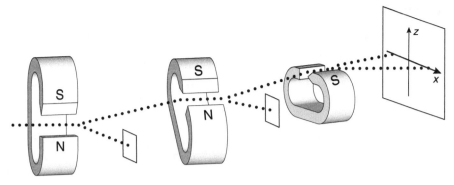

Figure 11.15 A Stern–Gerlach z-filter followed by a second filter rotated to a position $\phi = -\pi/4$ radians from the z-axis, followed by an x-filter. For Exercise 11.4.

(d) Calculate the average outcome of measuring I_y for the spin population with this wavefunction.

(e) Calculate the average outcome of measuring I_z for the spin population with this wavefunction.

(f) Draw a vector diagram on a three-dimensional coordinate system indicating the average magnetization in the three directions.

11.4. Suppose that the pair of Stern–Gerlach filters diagrammed in Fig. 11.13 were to be modified by the addition of a third filter, oriented along the x-axis and placed so that the upper beam from the filter oriented with $\phi = -\pi/4$ radians passes through the third filter, as illustrated in Fig. 11.15.

(a) Describe qualitatively how the third filter would affect the beam.

(b) Calculate the probabilities that a given particle would take each of the two paths from the third filter.

11.5. The probability equation, stated as rule 4 on page 331, is arguably more fundamental than the equation for calculating average outcomes (Eq. 11.2 and, in Dirac notation, Eq. 11.4). The average outcome, or expectation, of any random process with n discrete outcomes can be calculated from the probabilities of those outcomes as

$$\bar{X} = \sum_{j=1}^{n} P_j X_j$$

where X_j is the jth possible outcome and P_j is the probability of that outcome. Use this equation and the probability equation to derive the general form of the quantum expectation, Eq. 11.2. Do not assume that the number of possible outcomes is limited to two, as it is for a single spin-1/2.

Hint: Begin by expressing the arbitrary wavefunction upon which the measurement is to be made, $|\Psi\rangle$, as a linear combination of the eigenfunctions of the operator, $\hat{A}$, associated with the measurement.

Further Reading

For a more global introduction to quantum mechanics, its history, formalism and more mysterious aspects, the following books are highly recommended:

◇ Susskind, L. & Friedman, A. (2014). *Quantum Mechanics: The Theoretical Minimum*. Basic Books, New York.

This book is the second in a series based on lectures given by Leonard Susskind at Stanford University and is targeted at students with just enough mathematical background to follow a rigorous but accessible treatment. The majority of the book focuses on spin-1/2 particles and uses a treatment very close to the one presented here, though different symbols are often used. Stern–Gerlach filters aren't mentioned by name, but are the basis for the more abstract measurement devices described by the authors.

◇ Baggott, J. (1992). *The Meaning of Quantum Theory*. Oxford University Press, Oxford, UK.

Jim Baggott is a physical chemist and the author of several science books for general audiences. This early book of his was directed toward students of chemistry and physics and uses a level of mathematics roughly comparable to that in this book. It should be very accessible to readers of this book and offers a valuable perspective on the more philosophical aspects of quantum mechanics and some of its possible applications for cryptography and computing.

◇ Townsend, J. S. (2012). *A Modern Approach to Quantum Mechanics*, 2nd ed. University Science Books, Mill Valley, CA.

This book is intended primarily as a text for a course in quantum mechanics for advanced undergraduates or graduate students in physics. Unlike most other textbooks on this subject, this one begins by considering spin-1/2 particles and Stern–Gerlach filters, just as we have here. The mathematics is developed more rigorously than it is here, but in a very clear and accessible way.

◇ Feynman, R. P., Leighton, R. B. & Sands, M. (2013). *The Feynman Lectures on Physics*, Vol. III. Basic Books, New York. http://www.feynmanlectures.caltech.edu/III_toc.html

Feynman's approach to teaching quantum mechanics, beginning with the treatment of spin and the use of Stern–Gerlach filters, has inspired a growing number of other text books (including the ones listed above by Susskind and Friedman and by Townsend). But, Feynman begins with spin-1 particles, for which there are three energy levels, making things a bit more involved. His distinctive style, wit and clarity make this book a classic. The new edition has been completely re-typeset, with extensive corrections, and is available to read for free on the internet.

A more detailed discussion of the demonstration with polarizing filters and its formal relationship to the Stern–Gerlach experiment:

◇ Brom, J. M. & Rioux, F.(2002). Polarized light and quantum mechanics: An optical analog of the Stern-Gerlach experiment. *Chem. Educator*, 7, 200–204. http://dx.doi.org/10.1007/s00897020580a

CHAPTER 12

More Quantum Mechanics: Time and Energy

12.1 Time Dependence of the Wavefunction: The Hamiltonian Operator

So far, we have treated the wavefunctions as if they were static. Since the world obviously changes with time, the wavefunctions that describe it must, under some conditions at least, vary with time, too. The time dependence of the wavefunction describing a system is intimately linked to the energy of the system through a special operator, called the *Hamiltonian*, usually written as a rather stylized "$\mathcal{H}$." The change in the wavefunction is given by the Schrödinger equation, which, in the Dirac notation, can be written as

$$\frac{\partial}{\partial t}|\Psi(t)\rangle = \frac{-i}{\hbar}\mathcal{H}|\Psi(t)\rangle$$

where $|\Psi(t)\rangle$ represents the ket, or wavefunction, as a function of time. The symbol $\hbar$ represents Planck's constant, $h \approx 6.6262 \times 10^{-34}$ J·s, divided by 2π. This, actually, is one of two "Schrödinger equations" and is more specifically referred to as the time-dependent Schrödinger equation. Like the other operators we have considered, the Hamiltonian has eigenfunctions—that is, particular wavefunctions, $|\psi_j\rangle$, such that

$$\mathcal{H}|\psi_j\rangle = \lambda_j|\psi_j\rangle$$

where λ_j is a real constant. Because of their special properties and roles, the eigenfunctions of the Hamiltonian are usually designated with the lower case ψ, to distinguish it from the more general symbol for a wavefunction, Ψ. The Hamiltonian is related to the energy of the system because the value of λ_j is simply the energy of the corresponding eigenstate, E_j, so that

$$\mathcal{H}|\psi_j\rangle = E_j|\psi_j\rangle$$

This relationship is the time-independent Schrödinger equation for the special case where the Hamiltonian is independent of spatial coordinates.

For a system that starts out in a state represented by an eigenfunction of the Hamiltonian,

$$|\Psi(0)\rangle = |\psi_j\rangle$$

we can write

$$\frac{\partial}{\partial t}|\Psi(t)\rangle = -i\frac{E_j}{\hbar}|\Psi(t)\rangle$$

If the Hamiltonian is independent of time (which is the case for the situations we will be considering initially), then the time-dependent Schrödinger equation can be rearranged and integrated as follows:

$$\frac{\partial|\Psi(t)\rangle}{|\Psi(t)\rangle} = -i\frac{E_j}{\hbar}dt$$

$$\int_{|\Psi(0)\rangle}^{|\Psi(t)\rangle}\frac{\partial|\Psi(t)\rangle}{|\Psi(t)\rangle} = \int_{t=0}^{t}-i\frac{E_j}{\hbar}dt$$

$$\ln\left(|\Psi(t)\rangle\right) - \ln\left(|\Psi(0)\rangle\right) = -i\frac{E_j t}{\hbar}$$

$$\ln\left(\frac{|\Psi(t)\rangle}{|\Psi(0)\rangle}\right) = -i\frac{E_j t}{\hbar}$$

$$\frac{|\Psi(t)\rangle}{|\Psi(0)\rangle} = e^{-it(E_j/\hbar)}$$

$$|\Psi(t)\rangle = e^{-it(E_j/\hbar)}|\Psi(0)\rangle$$

$$|\Psi(t)\rangle = e^{-it(E_j/\hbar)}|\psi_j\rangle$$

That is, if the wavefunction at time 0 is an eigenfunction of the Hamiltonian, ψ_j, then after a time period, t, the wavefunction will have changed by the factor $e^{-it(E_j/\hbar)}$. The time-dependent wavefunction can also be written in terms of trigonometric functions:

$$|\Psi(t)\rangle = \cos(tE_j/\hbar)|\Psi(0)\rangle + i\sin(tE_j/\hbar)|\Psi(0)\rangle$$

In this sense, any quantum mechanical wavefunction is, indeed, wavy; its change with time is described by the cosine and sine functions. The complex conjugate of e^{ix} is e^{-ix}. From this relationship, the bra after time, t is

$$\langle\Psi(t)| = e^{it(E_j/\hbar)}\langle\psi_j|$$

Provided, as assumed above, that the starting function is an eigenfunction of the Hamiltonian, these time-dependent changes in the bra and ket have absolutely no measurable consequences! This can be shown by considering any possible physical measurement, with its corresponding operator, $\hat{A}$. The average outcome of the measurement is calculated as

$$\langle A(t) \rangle = \langle \Psi(t) | \hat{A} | \Psi(t) \rangle$$

If $|\Psi(0)\rangle$ is the eigenstate $|\psi_j\rangle$, then

$$\langle A(t) \rangle = e^{it(E_j/\hbar)} \langle \psi_j | \hat{A} e^{-it(E_j/\hbar)} | \psi_j \rangle$$

$$= e^{it(E_j/\hbar)} e^{-it(E_j/\hbar)} \langle \psi_j | \hat{A} | \psi_j \rangle$$

$$= \langle \psi_j | \hat{A} | \psi_j \rangle$$

Thus, the predicted average outcome is exactly the same as for the starting eigenfunction, ψ_j. Mathematically, there is no change in $\langle A(t) \rangle$ because the term $e^{-it(E_j/\hbar)}$ is a constant that can be factored out of the ket before it is operated upon by $\hat{A}$. When this factor is then multiplied by its complex conjugate, the result is just 1.

This, then, demonstrates the key property of the eigenfunctions of the Hamiltonian: They do not change in any meaningful way with time. For this reason, the physical states they represent are called the *stationary states* of the system, or the *eigenstates*.

Although other states of the system are not generally eigenstates, their wavefunctions can be written as a linear combination of the eigenfunctions for the Hamiltonian, just as we wrote, in Chapter 11, arbitrary wavefunctions as superpositions of the eigenfunctions of the $\hat{I}_z$ operator. Expressed as a superposition of the eigenfunctions, any wavefunction can be written as

$$|\Psi\rangle = \sum c_j |\psi_j\rangle$$

and the corresponding bra is given by

$$\langle\Psi| = \sum c_j^* \langle\psi_j|$$

The state is thus defined by the values of the constants, c_j. As the wavefunction changes with time, this change can be expressed in terms of changes in the individual terms in the sum, so that if the wavefunction starts out at time $t = 0$ as

$$|\Psi(0)\rangle = \sum c_j(0) |\psi_j\rangle$$

then, at time, t, it will be

$$|\Psi(t)\rangle = \sum e^{-it(E_j/\hbar)} c_j(0) |\psi_j\rangle$$

and

$$\langle\Psi(t)| = \sum e^{it(E_j/\hbar)} c_j^*(0) \langle\psi_j|$$

Now, you might be thinking that this means that *any* observable parameter will remain unchanged with time. But, this isn't true, as your everyday experience will tell you! The reason is that each of the terms in the previous two sums will have a different value of E_j, and the exponential terms will not, in general, factor out. This will become apparent in the next section, when we look at a specific case of importance for NMR spectroscopy.

12.2   Spin-1/2 Particles in a Magnetic Field

As implied in the previous section, there will only be interesting time-dependent changes in the wavefunction of a system if there are differences in the energies of the states making up that system. For particles with spin, the different spin states will have different energies only if they are placed in some kind of magnetic field. We will consider a uniform field oriented along the z-axis—that is, in the direction of the field used in our original Stern–Gerlach experiment. The eigenstates, or stationary states, under these conditions are those that are defined by our old friends $|\alpha\rangle$ and $|\beta\rangle$. This makes sense, because the original measurement involved a field along the z-axis, and this measurement forced any particle into either the $|\alpha\rangle$ or $|\beta\rangle$ state. If the particle was already in one of these states, it was unaffected by the measurement.

12.2.1   Time Dependence of the Wavefunction

Next, we need to know the Hamiltonian—that is, the operator that represents the energies of the eigenstates. Because the energy depends on the alignment of the magnetization of the spin with the direction of the external field, defined as the z-axis, we can write the Hamiltonian operator in terms of $\hat{I}_z$, the operator that gives the z-magnetization component of the spin. The energy also depends on the strength of the external field, B_0, and the gyromagnetic ratio, γ. For a spin with a positive gyromagnetic ratio, the $|\alpha\rangle$ state has a lower energy than the $|\beta\rangle$ state, whereas the $|\beta\rangle$ state has lower energy if the gyromagnetic ratio is negative. The following expression for the Hamiltonian incorporates all of these conditions:

$$\mathcal{H}|\Psi\rangle = -\hbar\gamma B_0 \hat{I}_z|\Psi\rangle$$

From Chapter 11, we know that $\hat{I}_z|\alpha\rangle = 1/2|\alpha\rangle$ and $\hat{I}_z|\beta\rangle = -1/2|\beta\rangle$. For these states we can then write

$$\mathcal{H}|\alpha\rangle = -\frac{\hbar}{2}\gamma B_0|\alpha\rangle$$

and

$$\mathcal{H}|\beta\rangle = \frac{\hbar}{2}\gamma B_0|\beta\rangle$$

Thus, $|\alpha\rangle$ and $|\beta\rangle$ are the eigenfunctions of the Hamiltonian operator, as well as for $\hat{I}_z$, and they represent the stationary states in a fixed magnetic field. The energies for the two eigenstates are then given by

$$E_\alpha = -\frac{\hbar}{2}\gamma B_0 \tag{12.1}$$

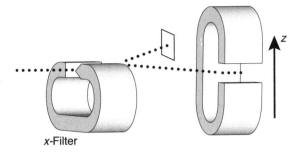

Figure 12.1
An imaginary experimental setup to place spin-1/2 particles with an eigenfunction of the $\hat{I}_x$ operator within a uniform magnetic field aligned with the z-axis.

x-Filter

and

$$E_\beta = \frac{\hbar}{2}\gamma B_0 \tag{12.2}$$

The energy difference, ΔE, is then

$$\Delta E = E_\beta - E_\alpha = \hbar\gamma B_0 = -h\nu \tag{12.3}$$

Maintaining the sign conventions introduced earlier, a positive gyromagnetic ratio is associated with a positive value for ΔE and a negative Larmor frequency. As discussed in Chapter 1, these energies are very small, on the order of 10^{-25} J/particle for ^{1}H, ^{15}N or ^{13}C in typical NMR spectrometers.

If a particle is in one of the eigenstates, it will have the energy indicated above, and it will remain in this state unless it is somehow perturbed. What happens, however, if a spin-1/2 particle is not in one of the eigenstates and it is placed in a magnetic field? We might imagine this happening if we were to first pass particles through an "x-filter" and then somehow trapped them in a uniform magnetic field oriented along the z-direction, as diagrammed in Fig. 12.1. For this gedanken experiment we use a uniform field so that we don't actually force a separation of particles. Because there is no measurement, the particles don't simply fall into one of the eigenstates. The wavefunction for a particle as it enters the uniform field will be

$$\frac{1}{\sqrt{2}}|\alpha\rangle + \frac{1}{\sqrt{2}}|\beta\rangle$$

At a given time after it enters the field, the wavefunction will be

$$|\Psi(t)\rangle = e^{-it(E_\alpha/\hbar)}\frac{1}{\sqrt{2}}|\alpha\rangle + e^{-it(E_\beta/\hbar)}\frac{1}{\sqrt{2}}|\beta\rangle$$

and its bra will be

$$\langle\Psi(t)| = e^{it(E_\alpha/\hbar)}\frac{1}{\sqrt{2}}\langle\alpha| + e^{it(E_\beta/\hbar)}\frac{1}{\sqrt{2}}\langle\beta|$$

As in the general case discussed above, each of the components of the wavefunction is modified by a complex exponential term that varies with time. Note that the rates at which the two components vary are different.

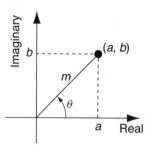

Figure 12.2

Representation of a complex
number, $a + ib$, as a point
on a plane defined by axes
representing the magnitudes of
the real and imaginary parts.

From the expressions for the bra and ket, we can predict the average outcome
of any experiment at a given time, provided that we know the appropriate operator.
Before trying some examples, however, it may be helpful to think some more about
what is happening mathematically to the wavefunctions with time. For this purpose,
it is useful to employ another representation of complex numbers (Fig. 12.2), in
which they are envisioned as points lying on a "complex plane," defined by a "real"
and an "imaginary" axis. The coordinates of a point correspond to the real (a) and
imaginary (b) parts of the complex number. A vector drawn from the origin to the
point has length m and forms an angle θ with the x-axis. The length corresponds to
the modulus of the complex number, as defined earlier:

$$m = |c| = \sqrt{cc^*} = \sqrt{a^2 + b^2}$$

The values of a and b are related to the modulus and the angle, θ, according to

$$a = m\cos(\theta)$$
$$b = m\sin(\theta)$$

and the complex number, c, can be written as

$$c = a + bi = m\,(\cos(\theta) + i\sin(\theta))$$

This can also be written in the exponential form

$$c = me^{i\theta}$$

The angle, θ, is sometimes called the *argument* or *phase* of the complex number and
represents the relative contributions of the real and imaginary parts to the overall
magnitude, or modulus, of the number. When $\theta = 0$, c is a pure real number, and
when $\theta = \pi/2$ rad, c is a pure imaginary number. (Rather confusingly, some authors
also refer to θ as the "amplitude" of the complex number, while others use this term
for the modulus.)

For the wavefunction we are interested in right now, the coefficient for $|\alpha\rangle$ be-
gins as

$$c_\alpha(0) = \frac{1}{\sqrt{2}}$$

Thus, the modulus is initially $1/\sqrt{2}$ and the phase is 0. With time, the coefficient changes to

$$c_\alpha(t) = \frac{1}{\sqrt{2}} e^{-it(E_\alpha/\hbar)}$$

By comparison with the general expression for a complex number, it should be apparent that the modulus does not change with time, while the phase changes with time according to:

$$\theta = -t(E_\alpha/\hbar)$$

Similarly, the coefficient for $|\beta\rangle$ at time, t, is

$$c_\beta(t) = \frac{1}{\sqrt{2}} e^{-it(E_\beta/\hbar)}$$

Again, the modulus of the complex coefficient remains constant. Thus, the relative magnitudes of the contributions from the $|\alpha\rangle$ and $|\beta\rangle$ states are unchanged with time. The phases, however, do change, and the rates of change for the two phases are different, because they depend on the energies of the respective eigenstates, E_α and E_β.

Although we have considered just a single starting state, where the contributions from $|\alpha\rangle$ and $|\beta\rangle$ are equal, the general results can be extended to other states composed of $|\alpha\rangle$ and $|\beta\rangle$: The moduli of the coefficients will remain constant with time, and the phases will change according to the energies of the corresponding eigenstates. In order to change the moduli of the coefficients (i.e., the relative contributions of $|\alpha\rangle$ and $|\beta\rangle$), the nuclei have to experience a *change* in the magnetic field, as when they pass through a Stern–Gerlach filter with a different orientation or when a fluctuating field is applied in an NMR experiment, as we will discuss later.

12.2.2   Time Dependence of the z-Magnetization

Let's start by calculating the magnetization along the z-axis, using the $\hat{I}_z$ operator. From Chapter 11, the action of $\hat{I}_z$ on an arbitrary wavefunction is

$$\hat{I}_z|\Psi\rangle = \hat{I}_z \left(c_\alpha|\alpha\rangle + c_\beta|\beta\rangle \right)$$

$$= c_\alpha \hat{I}_z|\alpha\rangle + c_\beta \hat{I}_z|\beta\rangle$$

$$= \frac{1}{2}c_\alpha|\alpha\rangle - \frac{1}{2}c_\beta|\beta\rangle$$

From this relationship and one for $\langle\Psi|$, we could calculate the average outcome of measuring I_z for any wavefunction. For convenience, we will use the expression we derived earlier,

$$\langle I_z \rangle = \langle \Psi|\hat{I}_z|\Psi\rangle = \frac{1}{2} \left(c_\alpha c_\alpha^* - c_\beta c_\beta^* \right)$$

which allows us to calculate the average outcome from the coefficients of the bra and ket. For the example introduced in the previous subsection, the coefficients are initially as follows:

$$c_\alpha = \frac{1}{\sqrt{2}} \qquad c_\alpha^* = \frac{1}{\sqrt{2}}$$

$$c_\beta = \frac{1}{\sqrt{2}} \qquad c_\beta^* = \frac{1}{\sqrt{2}}$$

and the average outcome is calculated as

$$\langle I_z \rangle = \frac{1}{2} \left(c_\alpha c_\alpha^* - c_\beta c_\beta^* \right)$$

$$= \frac{1}{2} \left(\frac{1}{\sqrt{2}} \frac{1}{\sqrt{2}} - \frac{1}{\sqrt{2}} \frac{1}{\sqrt{2}} \right)$$

$$= 0$$

Since the average outcome is 0, the possible outcomes must occur with equal probability.

After time, t, the coefficients are

$$c_\alpha(t) = \frac{1}{\sqrt{2}} e^{-it(E_\alpha/\hbar)} \qquad c_\alpha^*(t) = \frac{1}{\sqrt{2}} e^{it(E_\alpha/\hbar)}$$

$$c_\beta(t) = \frac{1}{\sqrt{2}} e^{-it(E_\beta/\hbar)} \qquad c_\beta^*(t) = \frac{1}{\sqrt{2}} e^{it(E_\beta/\hbar)}$$

and the average outcome is

$$\langle I_z(t) \rangle = \frac{1}{2} \left(c_\alpha(t) c_\alpha^*(t) - c_\beta(t) c_\beta^*(t) \right)$$

$$= \frac{1}{2} \left(\frac{1}{\sqrt{2}} e^{-it(E_\alpha/\hbar)} \frac{1}{\sqrt{2}} e^{it(E_\alpha/\hbar)} - \frac{1}{\sqrt{2}} e^{-it(E_\beta/\hbar)} \frac{1}{\sqrt{2}} e^{it(E_\beta/\hbar)} \right)$$

$$= \frac{1}{2} \left(\frac{1}{2} - \frac{1}{2} \right)$$

$$= 0$$

So, no matter how long the particle is left in the field, the average result of measuring the z-magnetization remains zero! (Keep in mind that this represents a highly idealized situation in which there are no external influences, including relaxation, that can promote changes among the spin states.)

We can also consider a more general case, where the initial coefficients are

$$c_\alpha(0), \quad c_\alpha^*(0), \quad c_\beta(0), \quad \text{and} \quad c_\beta^*(0)$$

and the initial average z-magnetization is

$$\langle I_z(0)\rangle = \frac{1}{2}\left(c_\alpha(0)c_\alpha^*(0) - c_\beta(0)c_\beta^*(0)\right)$$

The time-dependent coefficients are

$$c_\alpha(t) = c_\alpha(0)e^{-it(E_\alpha/\hbar)} \qquad c_\alpha^*(t) = c_\alpha^*(0)e^{it(E_\alpha/\hbar)}$$

$$c_\beta(t) = c_\beta(0)e^{-it(E_\beta/\hbar)} \qquad c_\beta^*(t) = c_\beta^*(0)e^{it(E_\beta/\hbar)}$$

The average outcome is calculated in the same way as before:

$$\langle I_z(t)\rangle = \frac{1}{2}\left(c_\alpha(t)c_\alpha^*(t) - c_\beta(t)c_\beta^*(t)\right)$$

$$= \frac{1}{2}\left(c_\alpha(0)e^{-it(E_\alpha/\hbar)}c_\alpha^*(0)e^{it(E_\alpha/\hbar)} - c_\beta(0)e^{-it(E_\beta/\hbar)}c_\beta^*(0)e^{it(E_\beta/\hbar)}\right)$$

$$= \frac{1}{2}\left(c_\alpha(0)c_\alpha^*(0) - c_\beta(0)c_\beta^*(0)\right)$$

Thus, the average z-magnetization remains constant for particles with any wavefunction!

The reason for this constancy can be found in the general expression for calculating the average z-magnetization:

$$\langle I_z\rangle = \langle \Psi|\hat{I}_z|\Psi\rangle = \frac{1}{2}\left(c_\alpha c_\alpha^* - c_\beta c_\beta^*\right)$$

Note that the terms in this expression are simply the squares of the moduli of the coefficients for the two eigenfunctions. Thus, the z-magnetization represents the difference in the relative contributions of the two eigenfunctions. The magnetization is 1/2 for the pure $|\alpha\rangle$ state, $-1/2$ for pure $|\beta\rangle$ and 0 when the two make equal contributions. As we saw earlier, these contributions remain constant in an unchanging field, and the z-magnetization remains constant.

12.2.3   Time Dependence of the x- and y-Magnetization Components

What about the magnetization along the x-axis? This is calculated by the $\hat{I}_x$ operator, defined for a general wavefunction as

$$\hat{I}_x|\Psi\rangle = \frac{1}{2}c_\beta|\alpha\rangle + \frac{1}{2}c_\alpha|\beta\rangle$$

The expression for the average value of I_x is

$$\langle I_x\rangle = \frac{1}{2}\left(c_\alpha c_\beta^* + c_\beta c_\alpha^*\right)$$

so the time-dependent x-magnetization is

$$\langle I_x(t) \rangle = \frac{1}{2} \left(c_\alpha(t) c_\beta^*(t) + c_\beta(t) c_\alpha^*(t) \right)$$

$$= \frac{1}{2} \left(c_\alpha(0) e^{-it(E_\alpha/\hbar)} c_\beta^*(0) e^{it(E_\beta/\hbar)} + c_\beta(0) e^{-it(E_\beta/\hbar)} c_\alpha^*(0) e^{it(E_\alpha/\hbar)} \right)$$

$$= \frac{1}{2} \left(c_\alpha(0) c_\beta^*(0) e^{it(E_\beta - E_\alpha)/\hbar} + c_\beta(0) c_\alpha^*(0) e^{-it(E_\beta - E_\alpha)/\hbar} \right)$$

For the particular case we introduced earlier, where all of the coefficients are $1/\sqrt{2}$, the initial magnetization is

$$\langle I_x(0) \rangle = \frac{1}{2} \left(c_\alpha(0) c_\beta^*(0) + c_\beta(0) c_\alpha^*(0) \right)$$

$$= \frac{1}{2} \left(\frac{1}{\sqrt{2}} \frac{1}{\sqrt{2}} + \frac{1}{\sqrt{2}} \frac{1}{\sqrt{2}} \right)$$

$$= \frac{1}{2} \left(\frac{1}{2} + \frac{1}{2} \right)$$

$$= \frac{1}{2}$$

This is consistent with the fact that we prepared the system so that the initial wavefunction was an eigenfunction of the I_x operator. The average outcome will then change with time according to

$$\langle I_x(t) \rangle = \frac{1}{2} \left(c_\alpha(0) c_\beta^*(0) e^{it(E_\beta - E_\alpha)/\hbar} + c_\beta(0) c_\alpha^*(0) e^{-it(E_\beta - E_\alpha)/\hbar} \right)$$

$$= \frac{1}{2} \left(\frac{1}{\sqrt{2}} \frac{1}{\sqrt{2}} e^{it(E_\beta - E_\alpha)/\hbar} + \frac{1}{\sqrt{2}} \frac{1}{\sqrt{2}} e^{-it(E_\beta - E_\alpha)/\hbar} \right)$$

$$= \frac{1}{4} \left(e^{it(E_\beta - E_\alpha)/\hbar} + e^{-it(E_\beta - E_\alpha)/\hbar} \right)$$

This time, not everything seems to be canceling out. At this point, it is convenient to use the following relationship:

$$e^{ix} + e^{-ix} = 2 \cos(x)$$

so that

$$\langle I_x(t) \rangle = \frac{1}{2} \cos \left(t(E_\beta - E_\alpha)/\hbar \right)$$

Finally, something is changing as a function of time! At $t = 0$, the average outcome will be 1/2, as expected, and the average will fluctuate from this value to -1/2 and back to 1/2, over and over again. Because the outcome of any single measurement must be either 1/2 or -1/2, our result indicates that the *probabilities* of the two outcomes change cyclically.

We can do a similar calculation for the magnetization along the y-axis, using the analogous relationship for the $\hat{I}_y$ operator:

$$\langle I_y \rangle = \langle \Psi | \hat{I}_y | \Psi \rangle = \frac{i}{2} \left(c_\alpha c_\beta^* - c_\beta c_\alpha^* \right)$$

The initial y-magnetization for our starting wavefunction is

$$\langle I_y(0) \rangle = \frac{i}{2} \left(c_\alpha(0) c_\beta^*(0) - c_\beta(0) c_\alpha^*(0) \right)$$

$$= \frac{i}{2} \left(\frac{1}{\sqrt{2}} \frac{1}{\sqrt{2}} - \frac{1}{\sqrt{2}} \frac{1}{\sqrt{2}} \right)$$

$$= 0$$

and the general time-dependent expression is

$$\langle I_y(t) \rangle = \frac{i}{2} \left(c_\alpha(t) c_\beta^*(t) - c_\beta(t) c_\alpha^*(t) \right)$$

$$= \frac{i}{2} \left(c_\alpha(0) e^{-it(E_\alpha/\hbar)} c_\beta^*(0) e^{it(E_\beta/\hbar)} - c_\beta(0) e^{-it(E_\beta/\hbar)} c_\alpha^*(0) e^{it(E_\alpha/\hbar)} \right)$$

$$= \frac{i}{2} \left(c_\alpha(0) c_\beta^*(0) e^{it(E_\beta - E_\alpha)/\hbar} - c_\beta(0) c_\alpha^*(0) e^{-it(E_\beta - E_\alpha)/\hbar} \right)$$

For the particular starting coefficients we are considering

$$\langle I_y(t) \rangle = \frac{i}{2} \left(\frac{1}{\sqrt{2}} \frac{1}{\sqrt{2}} e^{it(E_\beta - E_\alpha)/\hbar} - \frac{1}{\sqrt{2}} \frac{1}{\sqrt{2}} e^{-it(E_\beta - E_\alpha)/\hbar} \right)$$

$$= \frac{i}{4} \left(e^{it(E_\beta - E_\alpha)/\hbar} - e^{-it(E_\beta - E_\alpha)/\hbar} \right)$$

This may look worrisome, since we seem to have an imaginary result for the average outcome of an experiment! This apparent problem disappears, however, when we simplify the expression using another relationship between trigonometric functions and the exponentials of imaginary numbers:

$$e^{ix} - e^{-ix} = i2 \sin(x) \tag{12.4}$$

so that

$$\langle I_y(t) \rangle = -\frac{1}{2} \sin \left(t(E_\beta - E_\alpha)/\hbar \right) \tag{12.5}$$

This is exactly the same as the result for the magnetization along the x-axis, except that it is a sine function of t rather than the cosine. From Eq. 12.3 (p. 349), we can express the energy difference, $E_\beta - E_\alpha$, as $-h\nu$ to give

$$\langle I_x(t) \rangle = \frac{1}{2} \cos \left(-2\pi \nu t \right)$$

$$\langle I_y(t) \rangle = -\frac{1}{2} \sin \left(-2\pi \nu t \right)$$

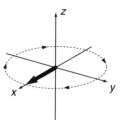

Figure 12.3 Rotation of a magnetic moment about z-axis, as described by the average x- and y-magnetization components calculated for a spin-1/2 particle with a positive gyromagnetic ratio (and a negative Larmor frequency), in a uniform magnetic field.

Since $\cos(-x) = \cos(x)$, and $\sin(-x) = -\sin(x)$, the expressions simplify further to

$$\langle I_x(t) \rangle = \frac{1}{2} \cos(2\pi \nu t)$$

$$\langle I_y(t) \rangle = \frac{1}{2} \sin(2\pi \nu t)$$

As illustrated in Fig. 12.3, these equations describe the motion of a magnetic moment about the z-axis. Thus, we have found the quantum mechanical analog of the precession of a spinning top! In the quantum mechanical interpretation, what changes at the Larmor frequency is the average magnetization expected for a measurement along any given direction in the x–y plane. This fluctuation arises because of the changes in the phase angles of the complex coefficients of $|\alpha\rangle$ and $|\beta\rangle$ in the wavefunction. The two phases change at different rates, reflecting the energies of the two eigenstates, and the difference in these rates corresponds to the frequency of precession of the average magnetization. Note that a positive gyromagnetic ratio gives rise to a negative frequency and rotation in the negative sense in the right-handed coordinate system, as represented in Fig. 12.3.

12.3 When the Hamiltonian Changes with Time: The Effect of Radiation on the Wavefunction

Thus far, we have carefully restricted ourselves to cases where the Hamiltonian operator is independent of time, as exemplified by the case of a spin-1/2 particle placed in a stationary magnetic field. We have seen that under these conditions the phases of the complex coefficients that define the wavefunction change with time, but the moduli of these coefficients remain constant. While assuming a constant Hamiltonian is valid under some conditions, and has allowed us to derive the quantum mechanical precession of a particle with spin, there are important cases where this assumption is not valid at all. In the context of NMR spectroscopy, the two most important situations in which the Hamiltonian changes with time are when the sample is irradiated by a fluctuating external field and when molecular motions promote a return of the system to equilibrium, the process of relaxation. Both of these processes are associated with changes in the moduli of the coefficients of $|\alpha\rangle$ and $|\beta\rangle$.

When the Hamiltonian changes with time, the mathematics of solving the time-dependent Schrödinger equation becomes much more difficult, and often it is not possible to find an exact solution. In some cases, however, either approximations or a clever transformation of coordinates can be used to make the problem manageable. For the specific case of the oscillating magnetic field in an NMR spectrometer, the problem can be solved exactly by using the rotating-frame transformation, in analogy with the vector description of a pulse used in Chapter 5. As discussed in that chapter, the transformation to the rotating frame can be viewed as a reduction of the stationary magnetic field strength to zero for a nucleus with a precession frequency that exactly matches the rotating-frame frequency. If we imagine ourselves on the rotating frame equipped with an instrument to measure I_x, I_y and I_z (all in the rotating frame), the average values of all of the magnetization components will remain constant with time, in the absence of any other field. This special set of conditions can only be met if the energies of the $|\alpha\rangle$ and $|\beta\rangle$ states are equal. Thus, in the rotating frame $E_\alpha = E_\beta$.

12.3.1   An x-Pulse

As we saw in Chapter 5, a magnetic field oscillating in the transverse plane of the laboratory frame can be treated as a stationary field in the rotating frame (together with a field that rotates at twice the rotating-frame frequency, which we can ignore). For an x-pulse, the stationary component lies along the x'-axis. We can then treat this field just as we did the stationary z-field in the laboratory frame. Under these conditions, the Hamiltonian operator, in the rotating frame, is proportional to $\hat{I}_x$, just as the Hamiltonian in a stationary field along the z-axis is proportional to $\hat{I}_z$ in the laboratory frame. Provided that the measurement is made from the rotating frame of reference, then the operator for measuring I_x is the same as the one that was introduced in Chapter 11 for the laboratory frame, and the Hamiltonian can be written as

$$\mathcal{H}_x|\Psi\rangle = -\gamma B_1 \hat{I}_x|\Psi\rangle = -\gamma B_1 \hat{I}_x(c_\alpha|\alpha\rangle + c_\beta|\beta\rangle) = -\frac{\gamma B_1}{2}(c_\beta|\alpha\rangle + c_\alpha|\beta\rangle)$$

Now, however, the $|\alpha\rangle$ and $|\beta\rangle$ states are no longer stationary. But, we can continue to use $|\alpha\rangle$ and $|\beta\rangle$ as a basis set for writing any other wavefunction. In particular, the eigenfunctions of $\hat{I}_x$ (and, therefore, for the Hamiltonian) can be written as superpositions of $|\alpha\rangle$ and $|\beta\rangle$:

$$|A\rangle = \frac{1}{\sqrt{2}}|\alpha\rangle + \frac{1}{\sqrt{2}}|\beta\rangle$$

and

$$|B\rangle = -\frac{1}{\sqrt{2}}|\alpha\rangle + \frac{1}{\sqrt{2}}|\beta\rangle$$

where we use the symbols $|A\rangle$ and $|B\rangle$ to represent the stationary states in the rotating frame during an x-pulse.

Just as an arbitrary wavefunction can be written as a superposition of $|\alpha\rangle$ and $|\beta\rangle$,

$$|\Psi\rangle = c_\alpha |\alpha\rangle + c_\beta |\beta\rangle$$

it can also be written as a superposition of $|A\rangle$ and $|B\rangle$,

$$|\Psi\rangle = c_A |A\rangle + c_B |B\rangle$$

where c_A and c_B are complex coefficients. However, the two representations have to correspond to the same wavefunction, and this condition is satisfied by requiring that the two sets of coefficients are related according to

$$c_A = \frac{c_\alpha}{\sqrt{2}} + \frac{c_\beta}{\sqrt{2}}$$

and

$$c_B = -\frac{c_\alpha}{\sqrt{2}} + \frac{c_\beta}{\sqrt{2}}$$

You can check these relationships by calculating the values of c_A and c_B for the two eigenfunctions of $\hat{I}_x$ given earlier; you should find that the eigenfunctions are simply $|A\rangle$ and $|B\rangle$ when expressed in this new basis set.

Once a wavefunction is expressed as a superposition of the stationary states $|A\rangle$ and $|B\rangle$, the time dependence of the wavefunction in the oscillating field can be calculated in the same way as before:

$$|\Psi(t)\rangle = e^{-itE_A/\hbar} c_A(0)|A\rangle + e^{-itE_B/\hbar} c_B(0)|B\rangle$$

where $c_A(0)$ and $c_B(0)$ are the coefficients of $|A\rangle$ and $|B\rangle$ at time zero, and E_A and E_B are the energies of these states. Because the strengths of the oscillating fields used for irradiation are typically much smaller than the stationary field strength, the difference between E_B and E_A is smaller than the difference between the energies of $|\alpha\rangle$ and $|\beta\rangle$ (as originally defined in the laboratory frame).

This expression for the time dependence of the wavefunction is written in terms of $|A\rangle$ and $|B\rangle$, but it would be more useful to have an expression in terms of $|\alpha\rangle$ and $|\beta\rangle$, since that is the basis set that we have used for everything else. This can be done by first substituting in the expressions for $|A\rangle$ and $|B\rangle$ in terms of $|\alpha\rangle$ and $|\beta\rangle$:

$$|\Psi(t)\rangle = e^{-itE_A/\hbar} c_A(0) \left(\frac{1}{\sqrt{2}}|\alpha\rangle + \frac{1}{\sqrt{2}}|\beta\rangle \right) + e^{-itE_B/\hbar} c_B(0) \left(-\frac{1}{\sqrt{2}}|\alpha\rangle + \frac{1}{\sqrt{2}}|\beta\rangle \right)$$

Next, we replace the starting coefficients $c_A(0)$ and $c_B(0)$ with expressions in terms of the initial coefficients for $|\alpha\rangle$ and $|\beta\rangle$, $c_\alpha(0)$ and $c_\beta(0)$:

$$|\Psi(t)\rangle = e^{-itE_A/\hbar} \left(\frac{c_\alpha(0)}{\sqrt{2}} + \frac{c_\beta(0)}{\sqrt{2}} \right) \left(\frac{1}{\sqrt{2}}|\alpha\rangle + \frac{1}{\sqrt{2}}|\beta\rangle \right)$$

$$+ e^{-itE_B/\hbar} \left(-\frac{c_\alpha(0)}{\sqrt{2}} + \frac{c_\beta(0)}{\sqrt{2}} \right) \left(-\frac{1}{\sqrt{2}}|\alpha\rangle + \frac{1}{\sqrt{2}}|\beta\rangle \right)$$

Grouping together the coefficients for $|\alpha\rangle$ and $|\beta\rangle$ gives

$$|\Psi(t)\rangle = \frac{1}{2}\left(e^{-itE_A/\hbar}(c_\alpha(0) + c_\beta(0)) + e^{-itE_B/\hbar}(c_\alpha(0) - c_\beta(0))\right)|\alpha\rangle$$

$$+ \frac{1}{2}\left(e^{-itE_A/\hbar}(c_\alpha(0) + c_\beta(0)) + e^{-itE_B/\hbar}(-c_\alpha(0) + c_\beta(0))\right)|\beta\rangle$$

By analogy to the case of $|\alpha\rangle$ and $|\beta\rangle$ in a stationary magnetic field aligned with the z-axis, the energies of $|A\rangle$ and $|B\rangle$ in the oscillating field are given by

$$E_A = -\frac{\hbar}{2}\gamma B_1$$

and

$$E_B = \frac{\hbar}{2}\gamma B_1$$

where B_1 is the strength of the oscillating field. Substituting into the expression above for the wavefunction gives

$$|\Psi(t)\rangle = \frac{1}{2}\left(e^{it\gamma B_1/2}(c_\alpha(0) + c_\beta(0)) + e^{-it\gamma B_1/2}(c_\alpha(0) - c_\beta(0))\right)|\alpha\rangle$$

$$+ \frac{1}{2}\left(e^{it\gamma B_1/2}(c_\alpha(0) + c_\beta(0)) + e^{-it\gamma B_1/2}(-c_\alpha(0) + c_\beta(0))\right)|\beta\rangle$$

This can be rearranged to group together the coefficients $c_\alpha(0)$ and $c_\beta(0)$ on each line:

$$|\Psi(t)\rangle = \frac{1}{2}\left(c_\alpha(0)\left(e^{it\gamma B_1/2} + e^{-it\gamma B_1/2}\right) + c_\beta(0)\left(e^{it\gamma B_1/2} - e^{-it\gamma B_1/2}\right)\right)|\alpha\rangle$$

$$+ \frac{1}{2}\left(c_\alpha(0)\left(e^{it\gamma B_1/2} - e^{-it\gamma B_1/2}\right) + c_\beta(0)\left(e^{it\gamma B_1/2} + e^{-it\gamma B_1/2}\right)\right)|\beta\rangle$$

We can again use the trigonometric expressions for $e^{ix} + e^{-ix}$ and for $e^{ix} - e^{-ix}$ to write the wavefunction as

$$|\Psi(t)\rangle = \left(c_\alpha(0)\cos(t\gamma B_1/2) + ic_\beta(0)\sin(t\gamma B_1/2)\right)|\alpha\rangle$$

$$+ \left(ic_\alpha(0)\sin(t\gamma B_1/2) + c_\beta(0)\cos(t\gamma B_1/2)\right)|\beta\rangle$$

From the vector description of the effect of a pulse, we expect a precessional motion of the average magnetization about the x'-axis, with an angular velocity, ω, related to the gyromagnetic ratio and the strength of the oscillating field according to

$$\omega = -\gamma B_1$$

The product of the time of irradiation, t, and the angular velocity, ω, then represents the angle of precessional motion, a, and the wavefunction can be expressed as

$$|\Psi(a)\rangle = \left(c_\alpha(0)\cos(a/2) - ic_\beta(0)\sin(a/2)\right)|\alpha\rangle$$

$$+ \left(-ic_\alpha(0)\sin(a/2) + c_\beta(0)\cos(a/2)\right)|\beta\rangle \tag{12.6}$$

Recall that in a stationary magnetic field along the z-axis, only the phases of the coefficients of $|\alpha\rangle$ and $|\beta\rangle$ change with time. During a pulse, however, the moduli generally change, altering the relative contributions of $|\alpha\rangle$ and $|\beta\rangle$ to the wavefunction. Note also that the sine terms in Eq. 12.6 have a negative sign, reflecting the convention that a positive gyromagnetic ratio gives rise to a negative rotation angle. To see that this change in wavefunction really does correspond to a precessional motion in the rotating frame, consider the case of a pulse applied to a population of spins in the $|\alpha\rangle$ state, so that $c_\alpha(0) = 1$ and $c_\beta(0) = 0$. The wavefunction at the end of the pulse will be

$$|\Psi(a)\rangle = \cos(a/2)|\alpha\rangle - i \sin(a/2)|\beta\rangle$$

If $a = \pi/2$, the new wavefunction is

$$|\Psi(\pi/2)\rangle = \cos(\pi/4)|\alpha\rangle - i \sin(\pi/4)|\beta\rangle = \frac{1}{\sqrt{2}}|\alpha\rangle - \frac{i}{\sqrt{2}}|\beta\rangle$$

Thus, the irradiation has converted the pure $|\alpha\rangle$ state to a superposition composed of equal parts $|\alpha\rangle$ and $|\beta\rangle$.

Using the expressions derived in Chapter 11, we can calculate the average magnetization in each of the three perpendicular directions of the coordinate system:

$$\langle I_x \rangle = \langle \Psi | \hat{I}_x | \Psi \rangle = \frac{1}{2}\left(c_\alpha c_\beta^* + c_\beta c_\alpha^* \right) = \frac{1}{2}\left(\frac{1}{\sqrt{2}}\frac{i}{\sqrt{2}} - \frac{i}{\sqrt{2}}\frac{1}{\sqrt{2}} \right) = 0$$

$$\langle I_y \rangle = \langle \Psi | \hat{I}_y | \Psi \rangle = \frac{i}{2}\left(c_\alpha c_\beta^* - c_\beta c_\alpha^* \right) = \frac{i}{2}\left(\frac{1}{\sqrt{2}}\frac{i}{\sqrt{2}} - \frac{-i}{\sqrt{2}}\frac{1}{\sqrt{2}} \right) = -\frac{1}{2}$$

$$\langle I_z \rangle = \langle \Psi | \hat{I}_z | \Psi \rangle = \frac{1}{2}\left(c_\alpha c_\alpha^* - c_\beta c_\beta^* \right) = \frac{1}{2}\left(\frac{1}{\sqrt{2}}\frac{1}{\sqrt{2}} - \frac{-i}{\sqrt{2}}\frac{i}{\sqrt{2}} \right) = 0$$

Just as expected, the pulse rotates the magnetization from the z-direction to the $-y'$-axis. At this point, it is very important to keep in mind that the derivation of the expression for the wavefunction was based on a transformation to the rotating frame, and we are now committed to interpreting the values of $\langle I_x \rangle$ and $\langle I_y \rangle$ in terms of the rotating-frame coordinates. As noted in Chapter 5, the definition of the direction of an initial pulse is arbitrary, but, once specified, it defines the orientation of the average magnetization with respect to the rotating frame. All future pulses will retain this definition.

12.3.2   y-Pulses

The same strategy can be used to derive an expression for the change in the wavefunction due to a pulse along the y'-axis. In this case, the stationary states in the rotating frame (during the pulse) are the eigenfunctions of the $\hat{I}_y$ operator, and the resulting equation is

$$|\Psi(a)\rangle = \big(c_\alpha(0) \cos(a/2) - c_\beta(0) \sin(a/2)\big)\, |\alpha\rangle$$
$$+ \big(c_\beta(0) \cos(a/2) + c_\alpha(0) \sin(a/2)\big)\, |\beta\rangle \tag{12.7}$$

If we again begin with the $|\alpha\rangle$ state and apply a $\pi/2$-pulse, but now along the y'-axis, the resulting wavefunction will be

$$|\Psi(\pi/2)\rangle = \cos(\pi/4)|\alpha\rangle + \sin(\pi/4)|\beta\rangle = \frac{1}{\sqrt{2}}|\alpha\rangle + \frac{1}{\sqrt{2}}|\beta\rangle$$

Again, the resulting wavefunction is a superposition of equal parts $|\alpha\rangle$ and $|\beta\rangle$, but now with a different phase relationship between the two coefficients. Applying the rules for calculating the average magnetization projections gives

$$\langle I_x\rangle = \frac{1}{2}$$

$$\langle I_y\rangle = 0$$

$$\langle I_z\rangle = 0$$

So, just as expected from the vector description, the pulse along the y'-direction rotates the magnetization from the positive z-axis to the x'-axis.

The vector picture also predicts that if the magnetization points along either the x'- or y'-axis, a pulse along that axis will leave the magnetization unchanged. We can examine the quantum mechanical prediction by considering the wavefunction generated by the $(\pi/2)_x$-pulse and the effects of a subsequent $(\pi/2)_y$-pulse. After the first pulse, the coefficients of $|\alpha\rangle$ and $|\beta\rangle$ are

$$c_\alpha = \frac{1}{\sqrt{2}}$$

$$c_\beta = -\frac{i}{\sqrt{2}}$$

These coefficients are then used as $c_\alpha(0)$ and $c_\beta(0)$ in the y'-pulse expression. For the $(\pi/2)_y$-pulse, the resulting wavefunction is calculated as

$$|\Psi(\pi/2)\rangle = \big(c_\alpha(0)\cos(\pi/4) - c_\beta(0)\sin(\pi/4)\big)|\alpha\rangle$$

$$+ \big(c_\beta(0)\cos(\pi/4) + c_\alpha(0)\sin(\pi/4)\big)|\beta\rangle$$

$$= \left(\frac{1}{\sqrt{2}}\cos(\pi/4) + \frac{i}{\sqrt{2}}\sin(\pi/4)\right)|\alpha\rangle$$

$$+ \left(-\frac{i}{\sqrt{2}}\cos(\pi/4) + \frac{1}{\sqrt{2}}\sin(\pi/4)\right)|\beta\rangle$$

$$= \left(\frac{1}{\sqrt{2}}\frac{1}{\sqrt{2}} + \frac{i}{\sqrt{2}}\frac{1}{\sqrt{2}}\right)|\alpha\rangle$$

$$+ \left(-\frac{i}{\sqrt{2}}\frac{1}{\sqrt{2}} + \frac{1}{\sqrt{2}}\frac{1}{\sqrt{2}}\right)|\beta\rangle$$

$$= \left(\frac{1}{2} + \frac{i}{2}\right)|\alpha\rangle + \left(-\frac{i}{2} + \frac{1}{2}\right)|\beta\rangle$$

At first glance, it appears as if this result contradicts our expectation, since the wavefunction has changed. If, however, we go ahead and calculate the magnetization components from this new wavefunction, we obtain the following:

$$\langle I_x \rangle = 0$$

$$\langle I_y \rangle = -\frac{1}{2}$$

$$\langle I_z \rangle = 0$$

Thus, the magnetization is still aligned with the y'-axis, just as we expected. This apparent paradox can be resolved by taking a closer look at the form of the wavefunction generated by the pulse. If we take each of the new complex coefficients and factor out the corresponding coefficient in the previous wavefunction ($1/\sqrt{2}$ and $-i/\sqrt{2}$, respectively), the new wavefunction can be written as

$$|\Psi(\pi/2)\rangle = \frac{1}{\sqrt{2}}\left(\frac{1}{\sqrt{2}} + \frac{i}{\sqrt{2}}\right)|\alpha\rangle - \frac{i}{\sqrt{2}}\left(\frac{1}{\sqrt{2}} + \frac{i}{\sqrt{2}}\right)|\beta\rangle$$

$$= \left(\frac{1}{\sqrt{2}} + \frac{i}{\sqrt{2}}\right)\left(\frac{1}{\sqrt{2}}|\alpha\rangle - \frac{i}{\sqrt{2}}|\beta\rangle\right)$$

Thus, the original wavefunction has been multiplied by a complex constant, $1/\sqrt{2} + i/\sqrt{2}$. This number can also be written in an exponential form:

$$\frac{1}{\sqrt{2}} + \frac{i}{\sqrt{2}} = e^{i\pi/4}$$

Recall from earlier in this chapter (page 347) that multiplying any wavefunction by a constant of the form $e^{i\theta}$ (i.e., a complex number with modulus 1) has no effect on any observable parameter. In this case, there is no change in the observable magnetization because the starting wavefunction is an eigenfunction of the Hamiltonian that is acting during the pulse (as viewed in the rotating frame), just as $|\alpha\rangle$ and $|\beta\rangle$ are eigenfunctions for the Hamiltonian corresponding to the static magnetic field along the z-axis, as viewed in the laboratory frame.

12.3.3   A 2π-Pulse

There is one more curiosity that is worth mentioning before moving on. You might reasonably suppose that the effect of a pulse corresponding to a rotation of 2π rad would return the wavefunction to its original value. But, consider what happens if we were to start with the $|\alpha\rangle$ state and apply a 2π-pulse along the x'-axis. In this case, $c_\alpha(0) = 1$ and $c_\beta(0) = 0$. Using the equation derived earlier

$$|\Psi(2\pi)\rangle = c_\alpha(0)\cos(\pi)|\alpha\rangle + c_\alpha(0)\sin(\pi)|\beta\rangle = -|\alpha\rangle$$

Thus, the wavefunction has been multiplied by −1, rather than being returned to its original value. This is another example of a wavefunction being multiplied by a number of the form $e^{i\theta}$, in this case $e^{i\pi}$, and all of the observable properties are the same as if the wavefunction were $|\alpha\rangle$. Still, it is rather odd that a full rotation hasn't returned the wavefunction to its original form. You should be able to show that a rotation of 4π rad (i.e., two complete rotations) *does* return the wavefunction to its original form.

This result is quite general: A rotation of the magnetization by a full cycle in any direction is associated with a change in sign of the wavefunction, while a return of the wavefunction to its original value requires two complete rotations of the magnetization.

Mathematical objects, in this case the wavefunction, that have this peculiar property are called *spinors*, and particles (including spin-1/2 nuclei) that are described with such wavefunctions are called *fermions*. In general, fermions have half-integral spin quantum numbers (1/2, 3/2, 5/2, etc.), while particles with integral spin numbers (including zero), such as photons, are called *bosons*. The differences between the two classes of particles are fundamental and have consequences well beyond the realm of NMR. In most situations, in fact, the spinor property of the wavefunction for a fermion cannot usually be observed by NMR, because the magnetic properties of a particle are unaffected by the sign of the wavefunction. When two such particles interact strongly, however, clever experiments can be devised to reveal the sign of the wavefunction of one of them and the spinor property can be experimentally demonstrated.

12.3.4   Precession Following a Pulse

So far, we have assumed that the Larmor frequency of the nucleus of interest exactly matches that of the rotating frame and the oscillating field. After a $\pi/2$-pulse, the magnetization will precess at the Larmor frequency in the laboratory frame, but will remain stationary in the rotating frame. If we were only interested in one Larmor frequency at a time, then there wouldn't be much more to add. For practical NMR, though, we need to be able to describe the behavior of spins with different Larmor frequencies. As discussed briefly in Section 5.5 (page 111), the magnetization of nuclei with Larmor frequencies slightly different from the frequency of the radiation will not move quite as far during a pulse as when there is an exact match. In addition, the magnetization will slightly precede or lag behind the rotating frame. While these effects can sometimes have practical consequences that must be addressed, for the purposes of describing idealized experiments, we can safely assume that the equations derived for an on-resonance pulse will be a very good approximation for the change in wavefunction even when there are nuclei with slightly different Larmor frequencies. We can then consider what happens to the wavefunction following the pulse, when the Larmor frequency differs from the rotating-frame frequency.

As we saw before, after a $(\pi/2)_y$-pulse is applied to the $|\alpha\rangle$ state, the wavefunction will be

$$|\Psi\rangle = \frac{1}{\sqrt{2}}|\alpha\rangle + \frac{1}{\sqrt{2}}|\beta\rangle$$

Subsequent change in the wavefunction will depend on the energy difference between the $|\alpha\rangle$ and $|\beta\rangle$ states:

$$|\Psi(t)\rangle = e^{-it(E_\alpha/\hbar)} \frac{1}{\sqrt{2}} |\alpha\rangle + e^{-it(E_\beta/\hbar)} \frac{i}{\sqrt{2}} |\beta\rangle$$

But, because we derived the form of the new wavefunction from the reference point of the rotating frame, we must continue to express these energies as they would appear in this frame. If the nucleus is exactly on resonance, then $E_\alpha = E_\beta$, and the exponential terms can be factored out. As expected, the magnetization remains stationary in the rotating frame. More generally, however, if the rotating-frame frequency is ν_0, and a nucleus has a Larmor frequency of ν, the apparent frequency in the rotating frame will be $\nu - \nu_0$, and the apparent energy difference between the $|\alpha\rangle$ and $|\beta\rangle$ states will be $\Delta E = -h(\nu - \nu_0)$. The energies of the individual states are then

$$E_\alpha = \frac{h(\nu - \nu_0)}{2}$$

and

$$E_\beta = -\frac{h(\nu - \nu_0)}{2}$$

The time evolution of the wavefunction, in the rotating frame, can then be written as

$$|\Psi(t)\rangle = e^{-it\pi(\nu - \nu_0)} \frac{1}{\sqrt{2}} |\alpha\rangle + e^{it\pi(\nu - \nu_0)} \frac{i}{\sqrt{2}} |\beta\rangle$$

If the wavefunction is then used to predict the movement of magnetization in the rotating frame, we will obtain the expected result that the magnetization precesses with frequency $\nu - \nu_0$. Thus, everything is consistent with the picture derived from the analogy with a spinning ball, including the sign relationships for the rotating frame, as illustrated in Figs. 5.16 and 5.17.

12.3.5 Transition Probabilities and the Absorption of Energy

The treatment developed in the previous sections describes the continuous conversion of an initial wavefunction when a spin is exposed to an oscillating magnetic field (i.e., electromagnetic radiation), leading to rotation of the average magnetization. The interaction of the particles with radiation is also associated with the absorption or emission of energy, depending on the direction of the change in wavefunction. For a spin with positive gyromagnetic ratio, a transition that increases the contribution of the $|\beta\rangle$ component leads to the absorption of energy, whereas an increase in the $|\alpha\rangle$ component is associated with emission. Starting with a sample at thermal equilibrium, both processes will occur and will tend to equalize the contributions of the $|\alpha\rangle$ and $|\beta\rangle$ in the population as a whole, leading to a net absorption of energy. At the same time, the system will attempt to return to equilibrium, transferring the absorbed energy to thermal motions. With the application of particularly strong irradiation for extended times, the temperature of the sample can increase significantly.

The net absorption of energy was the basis for detection of the signal in the original Purcell NMR experiment (Chapter 3), and the signal intensities from all NMR experiments are proportional to the net probabilities of stimulated transitions between the I_z eigenstates, though this relationship may not be so obvious when the signal is detected by the induction of an electric current, as in the Bloch experiment or pulse experiments. Derivation of the expressions for calculating the probabilities of these transitions is somewhat involved, requiring the methods of perturbation theory, but the key results can be stated rather succinctly.

For a transition that increases the $|\beta\rangle$ character of a wavefunction (a decrease in the spin quantum number), the probability of a transition is calculated using the lowering operator, $\hat{I}^-$. If the starting and ending states are $|\Psi_a\rangle$ and $|\Psi_b\rangle$, respectively, the probability, $P_{a,b}$, is proportional to the quantity,

$$|\langle\Psi_b|\hat{I}^-|\Psi_a\rangle|^2$$

(Remember that Dirac brackets are evaluated from right to left, and the outer vertical bars represent the modulus of a complex number.) Thus, for the downward (in the quantum-number sense) transition from $|\alpha\rangle$ to $|\beta\rangle$

$$P_{\alpha,\beta}^- \propto |\langle\beta|\hat{I}^-|\alpha\rangle|^2$$

$$\propto |\langle\beta|\beta\rangle|^2$$

$$\propto 1$$

On the other hand, if a spin is already in the $|\beta\rangle$ state, the relative downward probability is

$$P_{\beta,\beta}^- \propto |\langle\beta|\hat{I}^-|\beta\rangle|^2$$

$$\propto |\langle\beta|\alpha\rangle|^2$$

$$= 0$$

because the quantum number cannot be reduced further. Conversely, the probabilities of upward transitions are calculated using the operator $\hat{I}^+$ and follow the proportionality

$$P_{a,b}^+ \propto |\langle\Psi_b|\hat{I}^+|\Psi_a\rangle|^2$$

For instance, for the transition from an arbitrary wavefunction, $|\Psi\rangle = c_\alpha|\alpha\rangle + c_\beta|\beta\rangle$, to $|\alpha\rangle$), the relative probability is calculated as

$$P_{\Psi,\beta}^+ \propto |\langle\beta|\hat{I}^+|\Psi\rangle|^2$$

$$\propto |\langle\beta|\hat{I}^+(c_\alpha|\alpha\rangle + c_\beta|\beta\rangle)|^2$$

$$\propto c_\beta c_\beta^*$$

Thus, the probability of an upward transition is determined by the contribution of $|\beta\rangle$ to the starting wavefunction.

The transition probabilities take on more interesting forms when two or more spins are coupled to one another, as discussed in Chapter 13.

Summary Points for Chapter 12

- The time-dependent Schrödinger equation describes the change of a wavefunction with time, which is determined by the Hamiltonian operator, $\mathcal{H}$.
    - The eigenfunctions of the Hamiltonian do not change with time and describe the stationary states of the system.
    - The eigenvalues of the Hamiltonian are the discrete energy levels of the system.
    - Any wavefunction for the system can be expressed as a linear superposition of the eigenfunctions of the Hamiltonian.
- For a spin-1/2 particle in a static uniform magnetic field aligned with the z-axis, $|\alpha\rangle$ and $|\beta\rangle$ are the eigenfunctions of the Hamiltonian and represent the stationary states.
- For a spin-1/2 particle with any wavefunction in a static z-magnetic field, the moduli of the complex coefficients of $|\alpha\rangle$ and $|\beta\rangle$ in the wavefunction remain constant with time, and the average z-magnetization remains constant.
- For a spin-1/2 particle in a static z-magnetic field, the phases of the coefficients of $|\alpha\rangle$ and $|\beta\rangle$ change with time, and the average x- and y-magnetization components oscillate, as described by the rotation of a magnetic dipole about the z-axis.
- An oscillating magnetic field transverse to the static z-magnetic field, with a frequency matching the energy difference between the $|\alpha\rangle$ and $|\beta\rangle$ states, causes a change in the moduli of the complex coefficients describing the wavefunction, corresponding to a rotation about the axis of the oscillation in the rotating reference frame.

Exercises for Chapter 12

The Maxima macros for calculating wavefunctions and operators for spin-1/2 particles will simplify many of the calculations required for these problems. But, you are again encouraged to try the simpler problems by hand.

12.1. Suppose that you have a population of ^{1}H nuclei with the wavefunction

$$|\psi\rangle = \frac{i}{2}|\alpha\rangle + \frac{\sqrt{3}}{2}|\beta\rangle$$

and they have been placed in a uniform magnetic field of 12 T (with the field oriented in the z-direction).

(a) Derive a relationship to calculate the average z-magnetization for this population as a function of time.

(b) Derive a relationship to calculate the average x-magnetization as a function of time.

(c) Derive a relationship to calculate the average y-magnetization as a function of time.

(d) Make a graph in which $\langle I_x \rangle$, $\langle I_y \rangle$ and $\langle I_z \rangle$ are plotted as a function of time over the period from $t = 0$ to $t = 2 \times 10^{-9}$ s.

(e) Draw a series of four vector diagrams that show how the average magnetization changes as a function of time.

12.2. Following the approach used in Section 12.3.1, derive the expression given in Eq. 12.7 (p. 360) for the effect of a pulse of rotation angle a along the y'-axis on an initial wavefunction, $|\Psi(0)\rangle = c_\alpha(0)|\alpha\rangle + c_\beta(0)|\beta\rangle$.

12.3. Suppose that you could manipulate a pure population of spin-1/2 particles using a combination of rf pulses and evolution periods in a stationary magnetic field along the z-axis, and that you wanted to convert a population of spins in the $|\alpha\rangle$ state to a population with the average magnetization components illustrated by the vector diagram in Fig. 12.4. The average magnetization components (in the rotating frame) are

$$\langle I_x \rangle = 1/4$$

$$\langle I_y \rangle = -1/4$$

$$\langle I_z \rangle = 1/(2\sqrt{2})$$

(a) Describe a suitable combination of pulse(s) and evolution period(s) to convert the starting $|\alpha\rangle$ state to a wavefunction with the average magnetization components specified above. When considering evolution period(s), assume that the precession frequency in the rotating frame, $\nu_{rot} = \nu - \nu_0$, is positive. Specify the rotation angles of any pulses and the duration of any evolution periods. Specify the evolution time in terms of the rotating-frame precession frequency, ν_{rot}.

(b) Calculate the wavefunction that will describe the state after these manipulations and confirm that this wavefunction will give rise to the specified magnetization components.

Figure 12.4
Vector diagram representing the average magnetization components for a hypothetical pure spin population, for Exercise 12.3.

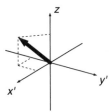

(c) If the state described above were irradiated with an oscillating electromagnetic field of the appropriate frequency to stimulate transitions, what would be the relative probabilities of an upward transition to the $|\alpha\rangle$ state and a downward transition to the $|\beta\rangle$ state?

Further Reading

More complete mathematical treatments of the time dependence of quantum states can be found in the following books originally cited in Chapter 11:

◇ Townsend, J. S. (2012). *A Modern Approach to Quantum Mechanics*, 2nd ed., pp. 111–140. University Science Books, Mill Valley, CA.

◇ Susskind, L. & Friedman, A. (2014). *Quantum Mechanics: The Theoretical Minimum*, pp. 93–127. Basic Books, New York.

The derivation of transition probabilities using perturbation theory:

◇ Harris, R. K. (1986). *Nuclear Magnetic Resonance Spectroscopy: A Physicochemical View*, pp. 240–244. Longman Scientific and Technical, Essex, UK.

CHAPTER 13

Quantum Description of a Scalar-Coupled Spin Pair

Having laid out the mathematical formalism of quantum mechanics and used this formalism to describe the behavior of a single spin, we are now ready to consider two nuclei that interact via scalar coupling. The concept of scalar coupling was briefly introduced in Chapter 4, in a very qualitative way. If we think of a nucleus as being able to take on two eigenstates, the magnetic field experienced by each nucleus in a coupled pair is influenced by the state of the other. As a consequence, nucleus I, for instance, in different molecules in a sample will have two Larmor frequencies, determined by the state of a coupled nucleus, S. The difference between these frequencies is defined as the coupling constant, J, and depends on the gyromagnetic ratios of the nuclei and the covalent structure by which they are linked. This situation leads to the familiar splitting of the NMR peaks arising from the two nuclei, as illustrated schematically in Fig. 13.1. In most cases, scalar coupling is only detected when nuclei are separated by one, two or three covalent bonds. In the case of two nuclei separated by three bonds, the strength of the coupling is strongly influenced by the dihedral angle of the central bond, as described by the Karplus equation (Eq. 4.4, page 80).

While this view provides an explanation for the splitting seen in simple spectra, it is a great oversimplification, since we know, for instance, that each nucleus can take on innumerable states that represent superpositions of the eigenstates. In order to better understand the phenomenon of scalar coupling and see how it can be exploited in multidimensional experiments, we need to examine in some detail the quantum mechanics of a two-spin system. One result of this analysis will be to show that the spectrum in Fig. 13.1 only arises under certain conditions, specifically when the magnitude of the coupling constant, J, is substantially smaller than the difference in the unperturbed Larmor frequencies of the two spins. The quantum mechanical analysis also reveals the existence of correlations between the magnetization components of coupled spins, which form the basis of the multidimensional experiments discussed in the subsequent chapters.

Figure 13.1

Simulated spectrum for two
weakly coupled spins, I and
S. ν_I and ν_S are the Larmor
frequencies of the two spins
in the absence of the coupling
interaction, and J is the scalar
coupling constant.

13.1   Interacting Spins in the Absence of an External Field

In the context of NMR experiments, the scalar coupling interaction between two
nuclei is much weaker than that between each nucleus and the external magnetic field.
In considering the quantum effects of the scalar interactions, however, it is helpful to
begin with the spins in the absence of an external field, so that they are influenced
only by their mutual interaction.

13.1.1   Magnetization Operators

We start by considering an idealized molecule with two nuclei, as diagrammed in
Fig. 13.2. In the absence of an external field, there is no reason to believe that the
magnetic dipoles will be oriented in any particular direction with respect to an arbitrary coordinate system. In addition, the dipoles might be pointed in any direction
with respect to each other, though we suspect that the interaction between them will
make some arrangements more likely than others. We can, however, do an experiment to measure the average orientation of the nuclei in a given direction. The most
direct way to do this would be with a Stern–Gerlach apparatus such as we discussed
earlier, but now with the two nuclei passing through the filter as a pair. As before, we
define the axis of the Stern–Gerlach filter as the z-axis in a coordinate system.

If the two nuclei have the same gyromagnetic ratio, the result illustrated in
Fig. 13.3 is expected. As we saw previously with a single spin, the measurement gives
rise to quantized results, this time with three values. Interestingly, the beam that corresponds to no net z-magnetization is twice as strong as the other two. We can account
for this result by saying that there are four eigenstates for this measurement, which
we will represent as

$$|\alpha\alpha\rangle, \qquad |\alpha\beta\rangle, \qquad |\beta\alpha\rangle \quad \text{and} \quad |\beta\beta\rangle$$

Figure 13.2

Schematic representation of
a molecule containing two
spin-1/2 nuclei linked by scalar
coupling.

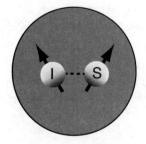

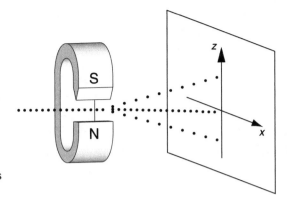

Figure 13.3
An idealized Stern–Gerlach experiment with a molecule composed of two coupled spins of the same type.

The eigenstates are named so as to suggest that the I- and S-spins are independently aligned with (α) or against (β) the direction defined by the measuring apparatus. The first and second symbols in the kets indicate the states of the I- and S-spins, respectively. We can attribute the total z-magnetization detected in the experiment as arising from the sum of the z-magnetization from the two nuclei. Thus, the z-magnetization values associated with the four eigenstates of the z-magnetization measurement are

$$|\alpha\alpha\rangle: 1/2 + 1/2 = 1$$
$$|\alpha\beta\rangle: 1/2 - 1/2 = 0$$
$$|\beta\alpha\rangle: -1/2 + 1/2 = 0$$
$$|\beta\beta\rangle: -1/2 - 1/2 = -1$$

If we use a molecule in which the nuclei have different gyromagnetic ratios, then the $|\alpha\beta\rangle$ and $|\beta\alpha\rangle$ molecules will be deflected differently by the magnet, as shown in Fig. 13.4 for the case where γ_I is about twice as large as γ_S. In this case, we are able to simultaneously determine the z-magnetization components of each of the spins. The magnetization components in any direction can, in principle, also be determined independently for the two nuclei. In an NMR experiment, the average x- and y-components for a population of molecules are distinguished by their different precession frequencies.

The four eigenfunctions for the z-magnetization measurement serve as a convenient basis set with which to express any arbitrary wavefunction for the two-spin

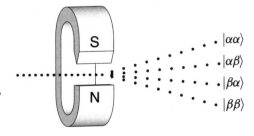

Figure 13.4
An idealized Stern–Gerlach experiment with a molecule composed of two coupled spins, with different gyromagnetic ratios.

system. Thus, the wavefunction, $|\Psi\rangle$, can be written as

$$|\Psi\rangle = c_{\alpha\alpha}|\alpha\alpha\rangle + c_{\alpha\beta}|\alpha\beta\rangle + c_{\beta\alpha}|\beta\alpha\rangle + c_{\beta\beta}|\beta\beta\rangle$$

where $c_{\alpha\alpha}$, $c_{\alpha\beta}$, $c_{\beta\alpha}$ and $c_{\beta\beta}$ are complex coefficients that represent the relative contributions of the basis functions to the superposition state. Just as for the single spin, there are operators that allow us to calculate the average outcomes of measuring the magnetization components for each of the spins in any given direction. The average magnetization from the I-spin along the x-axis, for instance, is calculated according to

$$\langle I_x\rangle = \langle\Psi|\hat{I}_x|\Psi\rangle$$

where $\hat{I}_x$ now represents the mathematical operator corresponding to the measurement of magnetization along the x-axis. As before, we will not try to derive the operators, but will accept them as givens. In this case, the operator is defined so that

$$\hat{I}_x|\Psi\rangle = \frac{1}{2}\left(c_{\beta\alpha}|\alpha\alpha\rangle + c_{\beta\beta}|\alpha\beta\rangle + c_{\alpha\alpha}|\beta\alpha\rangle + c_{\alpha\beta}|\beta\beta\rangle\right)$$

The resulting wavefuncion is then multiplied by the bra for the wavefunction, which is

$$\langle\Psi| = c_{\alpha\alpha}^*\langle\alpha\alpha| + c_{\alpha\beta}^*\langle\alpha\beta| + c_{\beta\alpha}^*\langle\beta\alpha| + c_{\beta\beta}^*\langle\beta\beta|$$

Like the eigenfunctions for the one-spin system, those for the two-spin system are orthonormal, which means that

$$\langle\alpha\alpha|\alpha\alpha\rangle = 1$$
$$\langle\alpha\beta|\alpha\beta\rangle = 1$$
$$\langle\beta\alpha|\beta\alpha\rangle = 1$$
$$\langle\beta\beta|\beta\beta\rangle = 1$$

while all of the other possible products have values of zero. This greatly simplifies calculating the average outcome, and it can be readily shown that

$$\langle I_x\rangle = \langle\Psi|\hat{I}_x|\Psi\rangle = \frac{1}{2}\left(c_{\beta\alpha}c_{\alpha\alpha}^* + c_{\beta\beta}c_{\alpha\beta}^* + c_{\alpha\alpha}c_{\beta\alpha}^* + c_{\alpha\beta}c_{\beta\beta}^*\right) \qquad (13.1)$$

For magnetization from the S-spin along the x-axis, the operator is defined by

$$\hat{S}_x|\Psi\rangle = \frac{1}{2}\left(c_{\alpha\beta}|\alpha\alpha\rangle + c_{\alpha\alpha}|\alpha\beta\rangle + c_{\beta\beta}|\beta\alpha\rangle + c_{\beta\alpha}|\beta\beta\rangle\right)$$

and the average magnetization is calculated according to

$$\langle S_x\rangle = \langle\Psi|\hat{S}_x|\Psi\rangle = \frac{1}{2}\left(c_{\alpha\beta}c_{\alpha\alpha}^* + c_{\alpha\alpha}c_{\alpha\beta}^* + c_{\beta\beta}c_{\beta\alpha}^* + c_{\beta\alpha}c_{\beta\beta}^*\right) \qquad (13.2)$$

In total, there are three operators for measuring the magnetization in the three orthogonal directions for each of the two spins. These six operators are given in Appendix F, along with the six equations for calculating each of the average magnetization quantities.

13.1.2 Shift Operators

There are also shift operators for each of the two spins. For the I-spin, the raising and lowering operators are

$$\hat{I}^+ = \hat{I}_x + i\hat{I}_y$$

$$\hat{I}^- = \hat{I}_x - i\hat{I}_y$$

For the S-spin, the corresponding operators are

$$\hat{S}^+ = \hat{S}_x + i\hat{S}_y$$

$$\hat{S}^- = \hat{S}_x - i\hat{S}_y$$

The effects of these operators on an arbitrary wavefunction, $|\Psi\rangle$, represented as a superposition of $|\alpha\alpha\rangle$, $|\alpha\beta\rangle$, $|\beta\alpha\rangle$ and $|\beta\beta\rangle$, are given by the following relationships:

$$\hat{I}^+|\Psi\rangle = c_{\beta\alpha}|\alpha\alpha\rangle + c_{\beta\beta}|\alpha\beta\rangle$$

$$\hat{I}^-|\Psi\rangle = c_{\alpha\alpha}|\beta\alpha\rangle + c_{\alpha\beta}|\beta\beta\rangle$$

$$\hat{S}^+|\Psi\rangle = c_{\alpha\beta}|\alpha\alpha\rangle + c_{\beta\beta}|\beta\alpha\rangle$$

$$\hat{S}^-|\Psi\rangle = c_{\alpha\alpha}|\alpha\beta\rangle + c_{\beta\alpha}|\beta\beta\rangle$$

As discussed in Chapter 12 for the case a single spin-1/2 particle, we can use the shift operators to calculate the relative probabilities of transitions between states, assuming the presence of a fluctuating magnetic field with the appropriate frequency. For instance, the probability of an upward transition of the I-spin from $|\beta\alpha\rangle$ to $|\alpha\alpha\rangle$ is calculated as

$$P^+ \propto |\langle\alpha\alpha|\hat{I}^+|\beta\alpha\rangle|^2$$

$$\propto |\langle\alpha\alpha|\alpha\alpha\rangle|^2$$

$$\propto 1$$

Keep in mind that the designation "upward" or "downward" indicates the change in the spin quantum number (1/2 or −1/2), rather than the energy of the state. For a nucleus with a positive gyromagnetic ratio, an "upward" transition corresponds, in fact, to a *decrease* in energy. Also remember that the probabilities are calculated in a right-to-left fashion, beginning with the shift operator acting on the starting wavefunction, followed by multiplication by the bra for the ending wavefunction.

The probability of an upward transition of the I-spin from $|\alpha\beta\rangle$ to $|\alpha\alpha\rangle$ is

$$P^+ \propto |\langle\alpha\alpha|\hat{I}^+|\alpha\beta\rangle|^2$$

$$\propto |\langle\alpha\alpha|0\rangle|^2$$

$$\propto 0$$

This result simply reflects the fact that the I-spin is already in the "up" state in the $|\alpha\beta\rangle$ state and cannot be raised further. As an example of a transition of the S-spin, consider the probability of a downward conversion from $|\beta\alpha\rangle$ to $|\beta\beta\rangle$:

$$P^- \propto |\langle\beta\beta|\hat{S}^-|\beta\alpha\rangle|^2$$

$$\propto |\langle\beta\beta|\beta\beta\rangle|^2$$

$$\propto 1$$

The shift operators for the I- and S-spins can be combined to represent the total probability of an upward or downward transition. The combined operators are

$$\hat{F}^+ = \hat{I}^+ + \hat{S}^+$$

$$\hat{F}^- = \hat{I}^- + \hat{S}^-$$

The action of these operators on an arbitrary wavefunction are given by

$$\hat{F}^+|\Psi\rangle = (c_{\beta\alpha} + c_{\alpha\beta})|\alpha\alpha\rangle + c_{\beta\beta}|\alpha\beta\rangle + c_{\beta\beta}|\beta\alpha\rangle$$

$$\hat{F}^-|\Psi\rangle = c_{\alpha\alpha}|\alpha\beta\rangle + c_{\alpha\alpha}|\beta\alpha\rangle + (c_{\beta\alpha} + c_{\alpha\beta})|\beta\beta\rangle$$

These operators are used to calculate the total probability of a stimulated transition between two arbitrary wavefunctions, irrespective of the spin affected. For instance, the probability of an upward transition from $|\beta\beta\rangle$ to $|\beta\alpha\rangle$ is

$$P^+ \propto |\langle\beta\alpha|\hat{F}^+|\beta\beta\rangle|^2$$

$$\propto \left|\langle\beta\alpha|(|\alpha\beta\rangle + |\beta\alpha\rangle)\right|^2$$

$$\propto (0 + 1)^2$$

$$\propto 1$$

which is the same as the probability of an upward transition from $|\beta\beta\rangle$ to $|\alpha\beta\rangle$:

$$P^+ \propto \left|\langle\alpha\beta|\hat{F}^+|\beta\beta\rangle\right|^2$$

$$\propto \left|\langle\alpha\beta|(|\alpha\beta\rangle + |\beta\alpha\rangle)\right|^2$$

$$\propto (1 + 0)^2$$

$$\propto 1$$

The probability of an upward transition from $|\alpha\beta\rangle$ to $|\alpha\alpha\rangle$ has the same value:

$$P^+ \propto \left| \langle \alpha\alpha | \hat{F}^+ | \alpha\beta \rangle \right|^2$$

$$\propto \left| \langle \alpha\alpha | \alpha\alpha \rangle \right|^2$$

$$\propto 1$$

We obtain a result that is, perhaps, less obvious when we consider the probability of the downward transition from $|\alpha\alpha\rangle$ to $|\beta\beta\rangle$:

$$P^- \propto \left| \langle \beta\beta | \hat{F}^- | \alpha\alpha \rangle \right|^2$$

$$\propto \left| \langle \beta\beta | (|\alpha\beta\rangle + |\beta\alpha\rangle) \right|^2$$

$$\propto 0$$

This result corresponds to the statement that double-quantum transitions are forbidden, in the sense that they do not occur by the absorption or emission of a single photon. The $|\alpha\alpha\rangle$ state can, though, be converted to $|\beta\beta\rangle$ by the sequential absorption of two photons with energies matching the individual transitions. Concerted transitions can also occur via cross-relaxation, as discussed in Chapters 8 and 9. Application of the shift operators also shows that zero-quantum transitions between $|\alpha\beta\rangle$ and $|\beta\alpha\rangle$ are forbidden. The probability of an upward transition from $|\alpha\beta\rangle$ to $|\beta\alpha\rangle$, for instance, is given by

$$P^+ \propto | \langle \beta\alpha | \hat{F}^+ | \alpha\beta \rangle |^2$$

$$\propto | \langle \beta\alpha | \alpha\alpha \rangle |^2$$

$$\propto 0$$

These applications of the raising and lowering operators show that the following transitions all have equal probabilities, provided that they are stimulated by radiation with the appropriate frequencies:

$$|\alpha\alpha\rangle \leftrightarrow |\alpha\beta\rangle$$
$$|\alpha\alpha\rangle \leftrightarrow |\beta\alpha\rangle$$
$$|\alpha\beta\rangle \leftrightarrow |\beta\beta\rangle$$
$$|\beta\alpha\rangle \leftrightarrow |\beta\beta\rangle$$

On the other hand, the forbidden transitions are

$$|\alpha\alpha\rangle \leftrightarrow |\beta\beta\rangle$$
$$|\alpha\beta\rangle \leftrightarrow |\beta\alpha\rangle$$

In these cases, the calculated probabilities are either 1 or 0. The shift operators can also be applied to states representing superpositions of $|\alpha\alpha\rangle$, $|\alpha\beta\rangle$, $|\beta\alpha\rangle$ and $|\beta\beta\rangle$, and, in these cases, the relative probabilities often have fractional values. As we will see in a subsequent section, these fractional probabilities can correspond to sets of NMR peaks with unequal intensities.

13.1.3   The Hamiltonian Operator

In order to determine the energy of a given state, we need to know the Hamiltonian operator, the special operator that is also used to calculate how the wavefunction changes with time. The energy of a given state depends on the relative orientation of the magnetic moments of the two nuclei. It may seem obvious that the states in which the z-magnetization components point in the same direction, $|\alpha\alpha\rangle$ and $|\beta\beta\rangle$, would have different average energies than those in which the z-components point in opposite directions, $|\alpha\beta\rangle$ and $|\beta\alpha\rangle$. The situation is more complicated than that, however, because the $|\alpha\beta\rangle$ and $|\beta\alpha\rangle$ states are not eigenstates of the Hamiltonian.

Although the form of the Hamiltonian is not easily derived from first principles, we can gain some intuition from the algebra used to describe the relative orientations of two vectors in three-dimensional space, such as those representing classical magnetic dipoles. While vectors are often thought of graphically as "arrows" with a defined length and direction, a more general mathematical description of a vector is simply an ordered list of numbers or functions. For a vector in three-dimensional space, the list corresponds to three coordinates, usually expressed in a Cartesian coordinate system. Thus, we can describe a vector $\mathbf{A}$:

$$\mathbf{A} = [\, A_x \quad A_y \quad A_z \,]$$

The three elements of the vector can be thought of as the coordinates of the "head" of the arrow if the "tail" were placed on the origin of the Cartesian coordinate system. The elements also represent the projections onto the three axes of the coordinate system.

The extent to which two vectors, say $\mathbf{A}$ and $\mathbf{B}$, are oriented in the same direction is expressed by the *dot product* (also called the *scalar product* or *inner product*, though the latter term is more general), which is calculated from the vector elements as

$$\mathbf{A} \cdot \mathbf{B} = A_x B_x + A_y B_y + A_z B_z$$

As indicated by its alternative name, scalar product, the dot product is a simple number, as opposed to a vector, and it represents the projection of one of the vectors onto the other. The length of a vector is given by the square root of the dot product of the vector with itself:

$$l = \sqrt{\mathbf{A} \cdot \mathbf{A}} = \sqrt{A_x A_x + A_y A_y + A_z A_z}$$

If two vectors are parallel to one another and pointed in the same direction, the dot product is the positive product of their lengths. If they are parallel but pointed in opposite directions, the dot product is negative. On the other hand, if the vectors are perpendicular, then the dot product is zero. If this form of vector multiplication is not familiar to you, try a few simple examples to convince yourself of these relationships.

We can also form vectors out of functions or operators, such as the magnetization operators for a single spin. For the I-spin, for instance,

$$\hat{\mathbf{I}} = [\, \hat{I}_x \quad \hat{I}_y \quad \hat{I}_z \,]$$

This, in itself, doesn't really add anything to the information provided by the individual magnetization operators, but we can also form a product between the operator vectors for two spins:

$$\hat{\mathbf{I}} \cdot \hat{\mathbf{S}} = \hat{I}_x \hat{S}_x + \hat{I}_y \hat{S}_y + \hat{I}_z \hat{S}_z \tag{13.3}$$

Each of the products in this expression ($\hat{I}_x \hat{S}_x$, $\hat{I}_y \hat{S}_y$ and $\hat{I}_z \hat{S}_z$) is an operator that can act on a wavefunction. For instance, the expression $\hat{I}_z \hat{S}_z | \Psi \rangle$ represents the result of $\hat{I}_z$ acting on the result of $\hat{S}_z$ acting on $| \Psi \rangle$. If we write the wavefunction, $| \Psi \rangle$, as a linear combination of $|\alpha\alpha\rangle$, $|\alpha\beta\rangle$, $|\beta\alpha\rangle$ and $|\beta\beta\rangle$, then, from the definitions of $\hat{I}_z$ and $\hat{S}_z$ given in Appendix F, $\hat{I}_z \hat{S}_z | \Psi \rangle$ is given by

$$\hat{I}_z \hat{S}_z | \Psi \rangle = \hat{I}_z \left(\frac{1}{2} \left(c_{\alpha\alpha} |\alpha\alpha\rangle - c_{\alpha\beta} |\alpha\beta\rangle + c_{\beta\alpha} |\beta\alpha\rangle - c_{\beta\beta} |\beta\beta\rangle \right) \right)$$

$$= \frac{1}{4} \left(c_{\alpha\alpha} |\alpha\alpha\rangle - c_{\alpha\beta} |\alpha\beta\rangle - c_{\beta\alpha} |\beta\alpha\rangle + c_{\beta\beta} |\beta\beta\rangle \right)$$

Consider what happens if we apply this operator to the eigenfunctions for the z-magnetization measurement:

$$\hat{I}_z \hat{S}_z |\alpha\alpha\rangle = \frac{1}{4} |\alpha\alpha\rangle$$

$$\hat{I}_z \hat{S}_z |\alpha\beta\rangle = -\frac{1}{4} |\alpha\beta\rangle$$

$$\hat{I}_z \hat{S}_z |\beta\alpha\rangle = -\frac{1}{4} |\beta\alpha\rangle$$

$$\hat{I}_z \hat{S}_z |\beta\beta\rangle = \frac{1}{4} |\beta\beta\rangle$$

Thus, in addition to being eigenfunctions for the individual operators, $\hat{I}_z$ and $\hat{S}_z$, these wavefunctions are also eigenfunctions for the new operator, $\hat{I}_z \hat{S}_z$. The physical meaning of this operator should be apparent; it is the result of measuring I_z and S_z independently and multiplying the result. For $|\alpha\alpha\rangle$ and $|\beta\beta\rangle$ the product is 1/4, and for $|\alpha\beta\rangle$ and $|\beta\alpha\rangle$ the product is $-1/4$. $\hat{I}_z \hat{S}_z$, along with $\hat{I}_x \hat{S}_x$ and $\hat{I}_y \hat{S}_y$, are thus referred to as *product operators*.

At first glance, operators such as $\hat{I}_z \hat{S}_z$ may not look very interesting, since they seem to convey the same information as the individual operators from which they are formed. We will see later, however, that the product operators can reveal important, and sometimes very surprising, relationships between coupled spins. For now, we will return our focus to the dot product, $\hat{\mathbf{I}} \cdot \hat{\mathbf{S}}$. Because this is just a sum of operators (each of which is a product operator, Eq. 13.3), it, too, is an operator. Like the dot product of two ordinary vectors, this operator reflects the extent to which the magnetization components of the two nuclei are aligned, but now in a quantum mechanical sense.

The importance of $\hat{\mathbf{I}} \cdot \hat{\mathbf{S}}$ is that it is proportional to the Hamiltonian operator for two scalar-coupled spins in the absence of an external magnetic field:

$$\mathcal{H}_{sc} = h J \hat{\mathbf{I}} \cdot \hat{\mathbf{S}}$$

where J is the scalar coupling constant introduced earlier and has units of Hz. The sign of J can be either positive or negative. If $J > 0$, alignment of the two spins in the same direction increases their energies, while a negative coupling constant indicates a lower energy when the magnetization components are aligned. In many situations, however, the sign of J does not have any visible consequence.

With a bit of algebra, we can express the Hamiltonian for a general wavefunction, $|\Psi\rangle$, as

$$\mathcal{H}_{sc}|\Psi\rangle = \frac{hJ}{4}\left(c_{\alpha\alpha}|\alpha\alpha\rangle + (2c_{\beta\alpha} - c_{\alpha\beta})|\alpha\beta\rangle + (2c_{\alpha\beta} - c_{\beta\alpha})|\beta\alpha\rangle + c_{\beta\beta}|\beta\beta\rangle \right) \quad (13.4)$$

Notice that for the wavefunctions, $|\alpha\alpha\rangle$ and $|\beta\beta\rangle$, the results of operating with the Hamiltonian are

$$\mathcal{H}_{sc}|\alpha\alpha\rangle = \frac{hJ}{4}|\alpha\alpha\rangle$$

and

$$\mathcal{H}_{sc}|\beta\beta\rangle = \frac{hJ}{4}|\beta\beta\rangle$$

Thus, $|\alpha\alpha\rangle$ and $|\beta\beta\rangle$ are eigenfunctions of the Hamiltonian: The states they represent remain unchanged with time, and they have defined energies of $hJ/4$.

Importantly, the other two eigenfunctions for the z-magnetization operators, $|\alpha\beta\rangle$ and $|\beta\alpha\rangle$, are *not* eigenfunctions of the Hamiltonian. For instance,

$$\mathcal{H}_{sc}|\alpha\beta\rangle = \frac{hJ}{4}(-|\alpha\beta\rangle + 2|\beta\alpha\rangle)$$

As a consequence, both $|\alpha\beta\rangle$ and $|\beta\alpha\rangle$ will change with time. However, the Hamiltonian does have two other eigenfunctions in addition to $|\alpha\alpha\rangle$ and $|\beta\beta\rangle$, each of which is a superposition of $|\alpha\beta\rangle$ and $|\beta\alpha\rangle$. These we will call $|\Psi_0\rangle$ and $|\Psi_1\rangle$ and are given by

$$|\Psi_0\rangle = \frac{1}{\sqrt{2}}(|\alpha\beta\rangle - |\beta\alpha\rangle)$$

and

$$|\Psi_1\rangle = \frac{1}{\sqrt{2}}(|\alpha\beta\rangle + |\beta\alpha\rangle)$$

You should be able to show that the results of the Hamiltonian acting on these wavefunctions are

$$\mathcal{H}_{sc}|\Psi_0\rangle = \mathcal{H}_{sc}\frac{1}{\sqrt{2}}(|\alpha\beta\rangle - |\beta\alpha\rangle) = \frac{-3hJ}{4}|\Psi_0\rangle$$

and

$$\mathcal{H}_{sc}|\Psi_1\rangle = \mathcal{H}_{sc}\frac{1}{\sqrt{2}}(|\alpha\beta\rangle + |\beta\alpha\rangle) = \frac{hJ}{4}|\Psi_1\rangle$$

Thus, three of the eigenfunctions of the Hamiltonian, $|\alpha\alpha\rangle$, $|\beta\beta\rangle$ and $|\Psi_1\rangle$, have the same energy, $hJ/4$, and one, $|\Psi_0\rangle$, has an energy of $-3hJ/4$. The energy difference between $|\Psi_0\rangle$ and any of the three other eigenstates is hJ. We therefore expect that a transition between $|\Psi_0\rangle$ and one of the other eigenstates will be stimulated by radiation with a frequency of J Hz.

The eigenfunctions of the Hamiltonian form an orthonormal set, and we can write any arbitrary wavefunction as a superposition of these four:

$$|\Psi\rangle = c_{\alpha\alpha}|\alpha\alpha\rangle + c_0|\Psi_0\rangle + c_1|\Psi_1\rangle + c_{\beta\beta}|\beta\beta\rangle$$

The choice of using this basis set or the eigenfunctions for the z-magnetization measurements ($|\alpha\alpha\rangle$, $|\alpha\beta\rangle$, $|\beta\alpha\rangle$ and $|\beta\beta\rangle$) is arbitrary. In fact, these are just two of an infinite number of basis sets that can be used. For different purposes, however, one or another basis set may be more convenient. The set, $|\alpha\alpha\rangle$, $|\alpha\beta\rangle$, $|\beta\alpha\rangle$ and $|\beta\beta\rangle$, is useful for calculating the magnetization components in the x-, y- and z-directions, while the eigenfunctions of the Hamiltonian are most convenient for calculating the time-dependent change in a wavefunction.

To calculate the change in a wavefunction with time, we begin with the general expressions discussed in Chapter 12. A state that begins in an eigenfunction of the Hamiltonian, $|\psi_j\rangle$, will, after time, t, have the wavefunction,

$$|\Psi(t)\rangle = e^{-it(E_j/\hbar)}|\psi_j\rangle$$

And, a state that begins in an arbitrary superposition state of the Hamiltonian eigenfunctions given by

$$|\Psi(0)\rangle = \sum c_j(0)|\psi_j\rangle$$

will, at time, t, be described by the wavefunction,

$$|\Psi(t)\rangle = \sum e^{-it(E_j/\hbar)}c_j(0)|\psi_j\rangle$$

For the case we are considering here, two scalar-coupled spin-1/2 nuclei in the absence of an external field, the time-dependent wavefunction is

$$|\Psi(t)\rangle = e^{-it(J\pi/2)}c_{\alpha\alpha}(0)|\alpha\alpha\rangle + e^{it(3J\pi/2)}c_0(0)|\Psi_0\rangle$$
$$+ e^{-it(J\pi/2)}c_1(0)|\Psi_1\rangle + e^{-it(J\pi/2)}c_{\beta\beta}(0)|\beta\beta\rangle$$

If we have a wavefunction that is expressed as a superposition of $|\alpha\alpha\rangle$, $|\alpha\beta\rangle$, $|\beta\alpha\rangle$ and $|\beta\beta\rangle$, we can rewrite it as a superposition of the stationary states by using the following relationships:

$$c_0 = \frac{1}{\sqrt{2}}(c_{\alpha\beta} - c_{\beta\alpha})$$

$$c_1 = \frac{1}{\sqrt{2}}(c_{\alpha\beta} + c_{\beta\alpha})$$

The coefficients, $c_{\alpha\alpha}$ and $c_{\beta\beta}$, are equivalent for the two representations, because $|\alpha\alpha\rangle$ and $|\beta\beta\rangle$ are the same in both. This conversion then makes it easy to calculate the time dependence of the wavefunction. Having done this, we might want to convert back to the basis set for the z-magnetization components in order to calculate the various magnetization components. This transformation is made with the following equations:

$$c_{\alpha\beta} = \frac{1}{\sqrt{2}}(c_0 + c_1)$$

$$c_{\beta\alpha} = \frac{1}{\sqrt{2}}(c_1 - c_0)$$

In order to describe practical NMR experiments, we also have to consider the effect of the external field, which usually greatly exceeds those of the scalar coupling interaction. Before doing so, however, we will look at some other features of a scalar-coupled pair, features that are very mysterious and play a central role in multidimensional NMR experiments based on scalar coupling.

13.1.4   Spin Correlations and the Product Operators

The two wavefunctions, $|\Psi_0\rangle$ and $|\Psi_1\rangle$, have some remarkable properties. To begin with, if we use the magnetization operators to calculate the average magnetization components of the two spins along the x-, y- and z-axes, we find that all six of the averages are zero for both of these wavefunctions. Thus, if we have a population of molecules in one of these states, measurements of either of the spins along any direction will give values of 1/2 and $-1/2$ with equal probabilities. Although this may seem similar to situations we have discussed earlier, where a loss of phase coherence leads to a net average magnetization of zero in all directions, it is important to keep in mind that we are now talking about a homogeneous population, where each molecule has the same wavefunction. In contrast, a homogeneous population of single spin-1/2 particles will always give rise to an average magnetization of 1/2 in some specific direction, as represented by the vector diagrams we have used up to now.

The second interesting feature of $|\Psi_0\rangle$ and $|\Psi_1\rangle$ becomes apparent if we apply the product operators introduced on page 377. For $|\Psi_0\rangle$, the following results are obtained:

$$\hat{I}_x\hat{S}_x|\Psi_0\rangle = -1/4|\Psi_0\rangle$$

$$\hat{I}_y\hat{S}_y|\Psi_0\rangle = -1/4|\Psi_0\rangle$$

$$\hat{I}_z\hat{S}_z|\Psi_0\rangle = -1/4|\Psi_0\rangle$$

Consider what these results imply. First, $|\Psi_0\rangle$ is an eigenfunction of each of the product operators $\hat{I}_x\hat{S}_x$, $\hat{I}_y\hat{S}_y$ and $\hat{I}_z\hat{S}_z$, so that if we independently measure, for instance, $\hat{I}_z$ and $\hat{S}_z$ for individual molecules, the product of the measurements will always be $-1/4$. But, we also know that the average values of I_z and S_z are zero, so the two possible outcomes of the individual measurements, 1/2 and -1/2, have equal

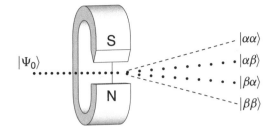

Figure 13.5

A hypothetical Stern–Gerlach experiment applied to a two-spin molecule described by the wavefunction, $|\Psi_0\rangle$.

probabilities. The product operator result implies, then, that whenever the result of measuring I_z for a molecule is 1/2, the value of S_z must be –1/2, and whenever I_z is –1/2, S_z must be 1/2. We will never see an outcome where $I_z = 1/2$ and $S_z = 1/2$ or where $I_z = -1/2$ and $S_z = -1/2$.

It may help in thinking about this to imagine the outcome of a hypothetical Stern–Gerlach experiment, as illustrated in Fig. 13.5, applied to a homogeneous population of molecules with the wavefunction, $|\Psi_0\rangle$: In this case, we see only two outcomes of the simultaneous measurement of I_z and S_z, and the measurement converts the molecules to either the $|\alpha\beta\rangle$ or $|\beta\alpha\rangle$ state. The outer two beams, corresponding to $|\alpha\alpha\rangle$ and $|\beta\beta\rangle$, are now absent. The same result would be obtained for measurements of I_x and S_x or I_y and S_y. Pairs of spins that are related in this way are said to be *entangled*, and the defining character of such states is that the individual measurements are correlated, even when the outcome of either measurement is unpredictable.

Applying the product operators to the wavefunction, $|\Psi_1\rangle$, gives slightly different results:

$$\hat{I}_x \hat{S}_x |\Psi_1\rangle = 1/4 |\Psi_1\rangle$$

$$\hat{I}_y \hat{S}_y |\Psi_1\rangle = 1/4 |\Psi_1\rangle$$

$$\hat{I}_z \hat{S}_z |\Psi_1\rangle = -1/4 |\Psi_1\rangle$$

For this wavefunction, the measurement of I_z and S_z are negatively correlated, as described above for $|\Psi_0\rangle$, but the measurement pairs, $I_x S_x$ and $I_y S_y$, are positively correlated. Thus, if we measure I_x and S_x, whenever one result is 1/2, the other will also be 1/2, and whenever the result of one measurement is –1/2, the other will be –1/2. What is surprising here is the different results obtained with $\hat{I}_z \hat{S}_z$ and $\hat{I}_x \hat{S}_x$. If we perform a Stern–Gerlach experiment with the magnet aligned with the z-axis, we will again see only the inner two beams, corresponding to $|\alpha\beta\rangle$ and $|\beta\alpha\rangle$. But, if we rotate the filter by 90°, making it an x-filter (Fig. 13.6), we expect to see only the outer two beams, corresponding to the eigenfunctions for the x-magnetization measurements with $I_x = S_x = 1/2$ and $I_x = S_x = -1/2$. Although, it may seem that there must now be something special about the x'-axis, there really isn't, except with respect to the particular wavefunction, $|\Psi_1\rangle$, that we happen to be looking at. There is another wavefunction for which the z-filter would generate only the outer two beams, and the x-filter would generate the inner two. (Can you find it?)

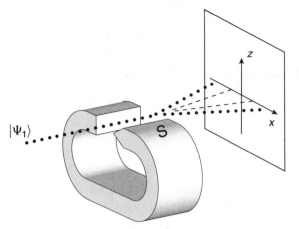

Figure 13.6

A hypothetical Stern–Gerlach experiment to measure x-magnetization, applied to a two-spin molecule described by the wavefunction, $|\Psi_1\rangle$.

We can also form product operators from other pairs of I- and S-magnetization operators. For instance, the product $\hat{I}_z\hat{S}_x$ corresponds to the product of independent measurements of I_z and S_x, though it's not obvious how we would actually make the measurements. Although there are a total of nine product operators for the I–S pair, the wavefunctions, $|\Psi_0\rangle$ and $|\Psi_1\rangle$, are eigenfunctions only for $\hat{I}_x\hat{S}_x$, $\hat{I}_y\hat{S}_y$ and $\hat{I}_z\hat{S}_z$. We can, however, use the other six product operators to calculate the *average* values of the other possible products if the simultaneous measurements are made with a large number of molecules.

It is important to emphasize that the average value of the product is not necessarily equal to the product of the average values of the individual measurements. For instance, we have seen already that for $|\Psi_0\rangle$ and $|\Psi_1\rangle$, the average values of I_x, I_y, I_z, S_x, S_y and S_z are all zero, but the average values of the products I_xS_x, I_yS_y and I_zS_z are either 1/4 or −1/4, reflecting the correlations between the measurements. For the other six product operators, however, the average values predicted for $|\Psi_0\rangle$ and $|\Psi_1\rangle$ are, in fact, zero. Thus, if we were to measure I_z and S_x independently we would find not only that the average values of the independent measurements are zero, but the possible values of the product I_zS_x, 1/4 and −1/4, arise with equal probabilities. In this case, there is no correlation between the measurements. There are other wavefunctions, however, for which these particular measurements are correlated, examples of which will emerge when we consider some specific NMR experiments.

You might also be wondering about products such as $\hat{S}_z\hat{I}_x$, as compared to $\hat{I}_x\hat{S}_z$, since multiplication of operators is not, in general, commutative. In this case, however, commutation between the I- and S-magnetization operators gives equivalent results. This reflects the fact that any one of the magnetization components of the I-spin can be measured without preventing measurement of one of the S-magnetization components. On the other hand, as we have seen earlier, we cannot independently measure two orthogonal magnetization components for a single spin. Mathematically, this is reflected by the result, for instance, that $\hat{I}_z\hat{I}_x|\Psi\rangle$ is not equal to $\hat{I}_x\hat{I}_z|\Psi\rangle$. You should be able to demonstrate this, and the result that $\hat{S}_z\hat{I}_x|\Psi\rangle$ does equal $\hat{I}_x\hat{S}_z|\Psi\rangle$, from the definitions of the individual operators.

13.2   Scalar-Coupled Spins in the Presence of an External Field

When a scalar-coupled spin pair is placed in an external magnetic field, the energy of the system is influenced by the interaction between each spin and the magnetic field, as well as the interaction between the spins. For a molecule that is tumbling isotropically, so that the effect of the external field is averaged, the Hamiltonian operator, $\mathcal{H}_T$, is made up of three terms,

$$\mathcal{H}_T = -\hbar B_0 \gamma_I \hat{I}_z - \hbar B_0 \gamma_S \hat{S}_z + h J \hat{\mathbf{I}} \cdot \hat{\mathbf{S}}$$

where B_0 is the magnetic field strength along the z-axis, and γ_I and γ_S are the gyromagnetic ratios of the two spins. The Hamiltonian can also be expressed in terms of the Larmor frequencies of the two uncoupled spins, v_I and v_S:

$$\mathcal{H}_T = h v_I \hat{I}_z + h v_S \hat{S}_z + h J \hat{\mathbf{I}} \cdot \hat{\mathbf{S}}$$

If the wavefunction, $|\Psi\rangle$, is written as a linear combination of $|\alpha\alpha\rangle$, $|\alpha\beta\rangle$, $|\beta\alpha\rangle$ and $|\beta\beta\rangle$, then the result of the Hamilton acting on $|\Psi\rangle$ is

$$\begin{aligned}
\mathcal{H}_T|\Psi\rangle = {} & h(c_{\alpha\alpha}/2)\left(J/2 + v_I + v_S\right)|\alpha\alpha\rangle \\
& + h\big((c_{\beta\alpha}/2)J + (c_{\alpha\beta}/2)(-J/2 + v_I - v_S)\big)|\alpha\beta\rangle \\
& + h\big((c_{\alpha\beta}/2)J + (c_{\beta\alpha}/2)(-J/2 - v_I + v_S)\big)|\beta\alpha\rangle \\
& + h(c_{\beta\beta}/2)\left(J/2 - v_I - v_S\right)|\beta\beta\rangle
\end{aligned}$$

Notice that $|\alpha\alpha\rangle$ and $|\beta\beta\rangle$ are eigenfunctions of $\mathcal{H}_T$, just as they were for the Hamiltonian in the absence of the external field, because

$$\mathcal{H}_T|\alpha\alpha\rangle = (h/2)\left(J/2 + v_I + v_S\right)|\alpha\alpha\rangle$$

and

$$\mathcal{H}_T|\beta\beta\rangle = (h/2)\left(J/2 - v_I - v_S\right)|\beta\beta\rangle$$

It should also be apparent that $|\alpha\beta\rangle$ and $|\beta\alpha\rangle$ are *not* eigenfunctions, except in the special case where $J = 0$—that is, when there is no interaction between the spins. For the general case, the other two eigenfunctions will be superpositions of $|\alpha\beta\rangle$ and $|\beta\alpha\rangle$. The relative contributions of $|\alpha\beta\rangle$ and $|\beta\alpha\rangle$ to the eigenfunctions can be expressed by introducing a parameter θ defined so that

$$\cos(2\theta) = (v_I - v_S)/D \tag{13.5}$$

where

$$D = \sqrt{J^2 + (v_I - v_S)^2} \tag{13.6}$$

Neither θ nor D has any real physical significance, but rather these parameters simply represent the relative values of ν_I, ν_S and J, as will become apparent when we consider some specific examples. With these definitions, one of the eigenfunctions, which we will call $|\Psi_A\rangle$, can be written as

$$|\Psi_A\rangle = -\sin\theta|\alpha\beta\rangle + \cos\theta|\beta\alpha\rangle$$

and the other is given by

$$|\Psi_B\rangle = \cos\theta|\alpha\beta\rangle + \sin\theta|\beta\alpha\rangle$$

The energies of the four eigenstates are

$$E_{\alpha\alpha} = h(\nu_I + \nu_S)/2 + hJ/4$$

$$E_A = -h\sqrt{J^2 + (\nu_I - \nu_S)^2}/2 - hJ/4$$

$$E_B = h\sqrt{J^2 + (\nu_I - \nu_S)^2}/2 - hJ/4$$

$$E_{\beta\beta} = -h(\nu_I + \nu_S)/2 + hJ/4$$

The relationships among the eigenstates is illustrated in the energy diagram shown in Fig. 13.7. This diagram indicates that there are four allowed transitions. The previous discussion of the shift operators showed that the potential transition between $|\alpha\alpha\rangle$ and $|\beta\beta\rangle$ is forbidden, and a similar analysis shows that the transition between $|\Psi_A\rangle$ and $|\Psi_B\rangle$ is also disallowed. Using the definitions given earlier, the

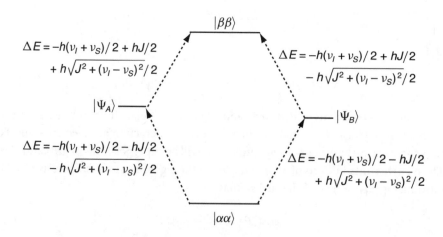

Figure 13.7   Energy diagram for two scalar-coupled spin-1/2 nuclei in the presence of a static external magnetic field.

relative probability of the upward transition is calculated as

$$P^+ \propto |\langle \Psi_A | \hat{F}^+ | \Psi_B \rangle|^2$$

$$\propto \left| (\sin \theta \langle \alpha\beta | - \cos \theta \langle \beta\alpha |) \hat{F}^+ (\cos \theta | \alpha\beta \rangle + \sin \theta | \beta\alpha \rangle) \right|^2$$

$$\propto \left| (\sin \theta \langle \alpha\beta | - \cos \theta \langle \beta\alpha |)(\cos \theta + \sin \theta) | \alpha\alpha \rangle \right|^2$$

$$\propto | \sin \theta (\cos \theta + \sin \theta) \langle \alpha\beta | \alpha\alpha \rangle - \cos \theta (\cos \theta + \sin \theta) \langle \beta\alpha | \alpha\alpha \rangle |^2$$

$$= 0$$

The probability of the transition in the opposite direction, calculated with $\hat{F}^-$, is also zero.

An important feature of the scalar-coupled spin system is that the allowed transitions will, in general, have different probabilities that depend on the value of θ (i.e., the relative contributions of $|\alpha\beta\rangle$ and $|\beta\alpha\rangle$ to the eigenstates), resulting in different intensities of the corresponding peaks in an NMR spectrum. For instance, the relative intensity of a peak corresponding to the transition from $|\alpha\alpha\rangle$ to $|\Psi_A\rangle$ is calculated as

$$P^- \propto |\langle \Psi_A | \hat{F}^- | \alpha\alpha \rangle|^2$$

$$\propto \left| (\sin \theta \langle \alpha\beta | - \cos \theta \langle \beta\alpha |) \hat{F}^- | \alpha\alpha \rangle \right|^2$$

$$\propto \left| (\sin \theta \langle \alpha\beta | - \cos \theta \langle \beta\alpha |)(| \alpha\beta \rangle + | \beta\alpha \rangle) \right|^2$$

$$\propto (\sin \theta - \cos \theta)^2$$

$$\propto 1 - 2 \cos \theta \sin \theta$$

$$\propto 1 - \sin(2\theta)$$

The expressions for the relative transition probabilities, along with the frequencies of the radiation required to stimulate these transitions (calculated from the eigenstate energies), are summarized in Table 13.1.

Fig. 13.8 shows the general form of a one-dimensional NMR spectrum that would be generated by two scalar-coupled spins of the same type (e.g., two ^{1}H nuclei), with

Table 13.1   Probabilities for transitions among the four eigenstates of a scalar-coupled spin pair in an external field, and the frequencies of radiation required to stimulate the transitions, as illustrated in Fig. 13.7.

Transition	Relative probability	Frequency
$|\alpha\alpha\rangle \rightarrow |\Psi_A\rangle$	$1 - \sin(2\theta)$	$\left(\nu_I + \nu_S + \sqrt{J^2 + (\nu_I - \nu_S)^2} + J \right) / 2$
$|\alpha\alpha\rangle \rightarrow |\Psi_B\rangle$	$1 + \sin(2\theta)$	$\left(\nu_I + \nu_S - \sqrt{J^2 + (\nu_I - \nu_S)^2} + J \right) / 2$
$|\Psi_A\rangle \rightarrow |\beta\beta\rangle$	$1 - \sin(2\theta)$	$\left(\nu_I + \nu_S - \sqrt{J^2 - (\nu_I - \nu_S)^2} - J \right) / 2$
$|\Psi_B\rangle \rightarrow |\beta\beta\rangle$	$1 + \sin(2\theta)$	$\left(\nu_I + \nu_S + \sqrt{J^2 + (\nu_I - \nu_S)^2} - J \right) / 2$

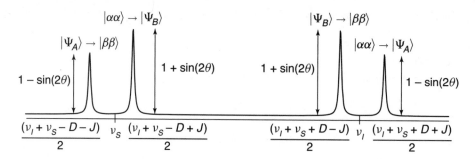

Figure 13.8 General form of the spectrum for two scalar-coupled spins of the same type. The diagram indicates the frequencies and the relative intensities of the four peaks. θ and D are the parameters defined by Eqs. 13.5 and 13.6.

the assumptions that $J > 0$ and $v_I > v_S$. In general, the two spins will generate four peaks in the spectrum with two intensities. The intensities of the two outer peaks will be equal to one another, as will those of the two inner peaks.

We can gain some insight into the relationships among the various parameters and the features of the spectrum by considering three extreme cases:

Case 1: $J = 0$

If there is no scalar coupling interaction between the nuclei, then $D = \sqrt{(v_I - v_S)^2}$, and, assuming that $v_I > v_S$, $\cos(2\theta) = 1$. This condition also implies that $\sin(2\theta) = 0$ and is satisfied if $\theta = 0$. The eigenfunctions of the Hamiltonian, in this case, reduce to $|\alpha\alpha\rangle$, $|\alpha\beta\rangle$, $|\beta\alpha\rangle$ and $|\beta\beta\rangle$—that is, the eigenfunctions for the z-magnetization operators. The corresponding energy diagram is shown in Fig. 13.9. In the absence of an interaction between the two spins, the transition from $|\alpha\alpha\rangle$ to $|\beta\alpha\rangle$ is energetically equivalent to that between $|\alpha\beta\rangle$ and $|\beta\beta\rangle$. Similarly, the transitions $|\alpha\alpha\rangle \rightarrow |\alpha\beta\rangle$ and $|\beta\alpha\rangle \rightarrow |\beta\beta\rangle$ have the same energy. The probabilities of the four transitions are all

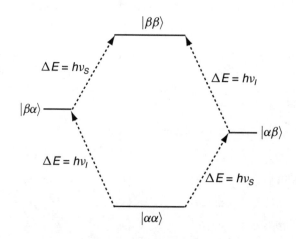

Figure 13.9
Energy diagram for two uncoupled spin-1/2 nuclei. For the case of two spins of the same type, the energy difference between the $|\alpha\beta\rangle$ and $|\beta\alpha\rangle$ states is exaggerated.

Figure 13.10
Simulated spectrum for two spins, in the absence of coupling (i.e., $J = 0$).

equal. As a consequence, only two peaks appear in the spectrum, with equal intensities and with frequencies, ν_I and ν_S, as shown in Fig. 13.10. Since these are, by definition, the frequencies observed in the absence of coupling, this result is hardly surprising, but it shows that the mathematical relationships derived for the general case are at least valid in this limit.

Case 2: $J^2 \gg (\nu_I - \nu_S)^2$

This condition, usually referred to as "strong" coupling, represents the alternative extreme, where the coupling constant is large enough to dominate the difference in resonance frequencies of the two spins. The parameter D is now approximately equal to $\sqrt{J^2}$, and $\cos(2\theta)$ is approximately zero. These conditions correspond to a value of θ equal to $\pi/4$. The four eigenfunctions, are the following:

$$|\alpha\alpha\rangle$$

$$|\Psi_A\rangle = -\sin(\pi/4)|\alpha\beta\rangle + \cos(\pi/4)|\beta\alpha\rangle = -\frac{1}{\sqrt{2}}|\alpha\beta\rangle + \frac{1}{\sqrt{2}}|\beta\alpha\rangle$$

$$|\Psi_B\rangle = \cos(\pi/4)|\alpha\beta\rangle + \sin(\pi/4)|\beta\alpha\rangle = \frac{1}{\sqrt{2}}|\alpha\beta\rangle + \frac{1}{\sqrt{2}}|\beta\alpha\rangle$$

$$|\beta\beta\rangle$$

These are the same as the eigenstates when there is no external field, with $|\Psi_A\rangle$ and $|\Psi_B\rangle$ corresponding to $|\Psi_0\rangle$ (multiplied by the constant -1) and $|\Psi_1\rangle$, respectively. The external field, however, does influence the energies of the eigenstates, which, for this case, are given by

$$E_{\alpha\alpha} = h(\nu_I + \nu_S)/2 + hJ/4$$

$$E_A = -h\sqrt{J^2}/2 - hJ/4$$

$$E_B = h\sqrt{J^2}/2 - hJ/4$$

$$E_{\beta\beta} = -h(\nu_I + \nu_S)/2 + hJ/4$$

Assuming that $\nu_I > \nu_S$ and $J > 0$, the energy diagram can be drawn as shown in Fig. 13.11. Note that two of the transitions, $|\alpha\alpha\rangle \rightarrow |\Psi_B\rangle$ and $|\Psi_B\rangle \rightarrow |\beta\beta\rangle$, have the same energy difference, $h(\nu_I + \nu_S)/2$, while the other two transitions have energies

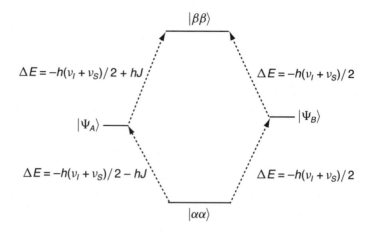

Figure 13.11

Energy diagram for two strongly coupled spin-1/2 nuclei, for which $J^2 \gg (v_I - v_S)^2$.

Figure 13.12

Simulated spectrum for two spins, in the strong-coupling limit, where $J^2 \gg (v_I - v_S)^2$. Only a single peak is observed, with a frequency equal to the average of the two Larmor frequencies for the uncoupled spins.

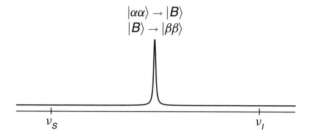

that differ from this value by either hJ or $-hJ$. This might lead to the expectation that the spectrum will contain three peaks. However, the probabilities of the transitions are quite different under these conditions. Indeed, in the limiting case where $\theta = \pi/4$, the probabilities of the two transitions involving the $|\Psi_A\rangle$ state approach zero. As a consequence, the spectrum contains only a single peak, with a position corresponding to the average of v_I and v_S, as shown in Fig. 13.12. In this case, the two inner peaks shown in Fig. 13.8 have merged to form a single peak, and the two outer peaks have disappeared.

A common example of the situation described here is a methyl group with its three ^{1}H nuclei. Because the methyl group can rotate very rapidly, the average environment of the three nuclei are the same, leading to identical average chemical shifts. Although a superficial description of scalar coupling, in which each of the spins can exist in only an "up" or "down" state, would suggest that a multiplet should be seen, methyl groups, in fact, give rise to only a single ^{1}H peak, bearing out the prediction made here.

Case 3: $J^2 \ll (v_I - v_S)^2$

Here we consider the case where scalar coupling is dominated by the frequency difference between the two spins, but J^2 is still greater than zero. This is referred to as the "weak-coupling" limit. As with the case where $J = 0$, we can assume that

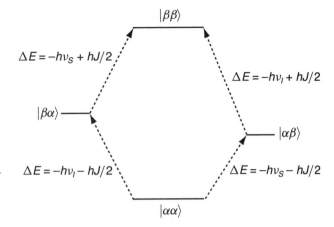

Figure 13.13
Energy diagram for two weakly coupled spin-1/2 nuclei, for which $J^2 \ll (v_I - v_S)^2$, but $J^2 > 0$.

$D \approx \sqrt{(v_I - v_S)^2}$, and $\theta \approx 0$, but J makes a finite contribution to the energies of the eigenstates. The eigenstates themselves can be approximated as $|\alpha\alpha\rangle$, $|\alpha\beta\rangle$, $|\beta\alpha\rangle$ and $|\beta\beta\rangle$, with energies (assuming $v_I > v_S$) given as

$$E_{\alpha\alpha} = h(v_I + v_S)/2 + hJ/4$$

$$E_{\alpha\beta} = h(v_I - v_S)/2 - hJ/4$$

$$E_{\beta\alpha} = -h(v_I - v_S)/2 - hJ/4$$

$$E_{\beta\beta} = -h(v_I + v_S)/2 + hJ/4$$

Note that, in the limit of weak coupling (and a positive coupling constant), the coupling interaction increases the energies of the eigenstates in which the spins are aligned ($|\alpha\alpha\rangle$ and $|\beta\beta\rangle$) and decreases the energies of the states in which they are anti-parallel ($|\alpha\beta\rangle$ and $|\beta\alpha\rangle$), consistent with a simple picture of interacting magnetic dipoles. The energy diagram describing the weak-coupling case is shown in Fig. 13.13. This leads to a spectrum in which the signal from each spin is "split" into two peaks separated by J Hz, as shown at the beginning of this chapter in Fig. 13.1 (p. 370). The intensities of the four peaks are equal. Under most circumstances, especially at the high magnetic fields commonly used now, the weak-coupling limit is at least approximately satisfied by homonuclear spin pairs, and it is always satisfied for heteronuclear pairs. Nonetheless, it is important to remember that for homonuclear spins the weak-coupling assumption may not be fully satisfied. Such deviations will give rise to slightly different intensities for the two components of the doublet, and the average frequency of the two components will not match exactly the frequency for the uncoupled spin.

Chapter 14 focuses on systems that can be described by the weak-coupling approximation, discussing how this system responds to pulses and time evolution periods. While the approximation greatly simplifies the analysis, we will see that even weak coupling leads to some quite surprising behavior, behavior that is exploited in many multidimensional NMR experiments.

13.3   Final Comments

The phenomenon of entanglement is arguably one of the most counterintuitive and fundamental results of quantum mechanics. Entanglement was first brought to prominence in a 1935 paper by Einstein, Podolsky and Rosen (EPR), who devised a thought experiment involving a system composed of two entangled particles that were then physically separated. Application of momentum and position operators predicted that measurement of a property of one of the particles could instantaneously influence the other. This counterintuitive result was presented by Einstein, Podolsky and Rosen as evidence that the fundamentals of quantum mechanics, as then and currently understood, could not possibly be correct, or at least could not be a complete description of reality. Subsequently, David Bohm formulated a more easily visualized variation on the EPR experiment, based on an entangled spin pair, and the nature of the EPR paradox was further refined by John Stewart Bell in 1964, in what is now known as Bell's inequality. Although the EPR experiment was initially formulated as a thought experiment, with little expectation that it would actually be performed, experimentalists rose to the challenge, beginning in the 1970s, and have implemented variations of the experiment with increasing sophistication and greater physical separation. The results have been exactly as predicted by the quantum theory, no matter how odd they might seem. NMR experiments involving scalar coupling are also direct manifestations of entanglement, but the more mysterious aspects of the phenomenon are not so prominent, because NMR measurements are applied to mixed, rather than pure, populations of quantum states.

Summary Points for Chapter 13

- The quantum treatment of a coupled spin pair (or larger group of spins) requires that both spins be described by a single wavefunction.
- The wavefunction for a coupled spin pair can be written as a superposition of the eigenfunctions for the measurement of the z-magnetization components of the two spins (I_z and S_z): $|\alpha\alpha\rangle$, $|\alpha\beta\rangle$, $|\beta\alpha\rangle$ and $|\beta\beta\rangle$.
- The eigenfunctions for the measurement of the z-magnetization components of the two spins are *not*, in general, the stationary states of the system (i.e., the eigenfunctions of the Hamiltonian operator).
- The Hamiltonian operator for a coupled spin pair depends on the interaction between the spins, expressed as the scalar coupling constant (J), as well as the interaction of each spin with an external field.
- Scalar-coupled spins represent an entangled system, meaning that measurements of the magnetization components of the individual spins are not always fully independent:
    - Correlations between magnetization components of entangled spins are reflections of the wavefunction, just as the individual magnetization components are.

- The average correlations for a coupled spin pair are calculated from the wavefunction using the product operators, such as $\hat{I}_z\hat{S}_z$, $\hat{I}_z\hat{S}_x$, etc., and represent the average of the products of the two measurements for individual spin pairs, such as $\langle I_z S_z \rangle$ and $\langle I_z S_x \rangle$.
- The behavior of scalar-coupled spins depends on the relative magnitudes of the interaction between the spins, expressed as J, and the difference in their independent Larmor frequencies: $\nu_I - \nu_S$.
- If $J = 0$, the spins are uncoupled and behave fully independently.
- If $J^2 \gg (\nu_I - \nu_S)^2$, the spins are said to be strongly coupled, and the NMR spectrum will contain only a single peak for the two spins, at the position corresponding to their average Larmor frequency.
- If $J^2 \ll (\nu_I - \nu_S)^2$, but J is non-zero, the spins are said to be weakly coupled.
  - In the limit of weak coupling, the stationary states are well approximated by the eigenfunctions for the measurement of the z-magnetization components of the two spins: $|\alpha\alpha\rangle$, $|\alpha\beta\rangle$, $|\beta\alpha\rangle$ and $|\beta\beta\rangle$).
  - Each spin of a weakly coupled pair will give rise to two peaks of equal intensity, separated by J Hz.

Exercises for Chapter 13

A set of macros for calculating wavefunctions and operators for a coupled pair of spin-1/2 particles (2spinLib.mac), along with some worksheets with examples, are available for download through http://uscibooks.com/goldenberg.htm.

13.1. The operators for measuring the z-magnetization of the I- and S-spins of a scalar-coupled spin pair are

$$\hat{I}_z|\Psi\rangle = \frac{1}{2}\left(c_{\alpha\alpha}|\alpha\alpha\rangle + c_{\alpha\beta}|\alpha\beta\rangle - c_{\beta\alpha}|\beta\alpha\rangle - c_{\beta\beta}|\beta\beta\rangle\right)$$

and

$$\hat{S}_z|\Psi\rangle = \frac{1}{2}\left(c_{\alpha\alpha}|\alpha\alpha\rangle - c_{\alpha\beta}|\alpha\beta\rangle + c_{\beta\alpha}|\beta\alpha\rangle - c_{\beta\beta}|\beta\beta\rangle\right)$$

(a) Derive equations for calculating the average I_z- and S_z-magnetization components for a general wavefunction, analogous to those for I_x and S_x given in Eqs. 13.1 and 13.2 (p. 372).

(b) Show that $|\alpha\alpha\rangle$, $|\alpha\beta\rangle$, $|\beta\alpha\rangle$ and $|\beta\beta\rangle$ are all eigenfunctions for both $\hat{I}_z$ and $\hat{S}_z$.

13.2. Show that the four eigenfunctions of the Hamiltonian (Eq 13.4, p. 378) for a coupled spin pair in the absence of an external field ($|\alpha\alpha\rangle$, $|\beta\beta\rangle$, $|\Psi_0\rangle$ and $|\Psi_1\rangle$) form an orthonormal set. Take as a given that $|\alpha\alpha\rangle$, $|\alpha\beta\rangle$, $|\beta\alpha\rangle$ and $|\beta\beta\rangle$ are all orthonormal.

13.3. Confirm that $|\Psi_0\rangle$ and $|\Psi_1\rangle$ are eigenfunctions of the Hamiltonian for a coupled spin pair in the absence of an external field.

13.4. Suppose that you were to have a population of coupled spins in a state described by the wavefunction, $|\Psi_0\rangle$, and you were somehow able to simultaneously measure I_x and S_y. Calculate the average values expected for the individual measurements I_x and S_y. Calculate the average product of I_x and S_y. What do you conclude about any correlation between these two measurements?

13.5. Consider the following two wavefunctions

$$|\Psi_a\rangle = \frac{1}{\sqrt{2}}|\alpha\alpha\rangle + \frac{1}{\sqrt{2}}|\beta\beta\rangle$$

and

$$|\Psi_b\rangle = \frac{1}{\sqrt{2}}|\alpha\alpha\rangle - \frac{1}{\sqrt{2}}|\beta\beta\rangle$$

(a) Calculate all six of the average Cartesian magnetization components for pure populations of spin pairs with these two wavefunctions, as well as the nine average products. What results would you expect if spin pairs with these wavefunctions were analyzed using Stern–Gerlach filters aligned with the z- and x-axes (as in Figs. 13.5 and 13.6)?

(b) Are the wavefunctions, $|\Psi_a\rangle$ and $|\Psi_b\rangle$, eigenfunctions of the Hamiltonian for a coupled spin pair in the absence of an external field?

(c) In what ways are $|\Psi_a\rangle$ and $|\Psi_b\rangle$ similar to and different from $|\Psi_0\rangle$ and $|\Psi_1\rangle$? In what ways are they similar to and different from $|\alpha\alpha\rangle$ and $|\beta\beta\rangle$?

13.6. In the text, we considered the extremes of strong and weak scalar coupling. Now, consider what might be called intermediate coupling, where $J = \nu_I - \nu_S$. Following the approach used to treat the other cases, calculate the parameters D and θ, and use these parameters to determine the frequencies and relative intensities of the four possible NMR peaks. Make a sketch of the expected spectrum.

Further Reading

For more detailed treatments of the quantum mechanics of coupled spin pairs, see the following books originally cited in Chapter 11:

◇ Townsend, J. S. (2012). *A Modern Approach to Quantum Mechanics*, 2nd ed. pp. 141–190. University Science Books, Mill Valley, CA.

◇ Susskind, L. (2014). *Quantum Mechanics: The Theoretical Minimum*, pp. 149–181. Basic Books, New York.

A clear discussion of the Einstein, Podolsky and Rosen experiment and Bell's inequality:

◇ Baggott, J. (1992). *The Meaning of Quantum Theory*. Oxford University Press, Oxford.

NMR Spectroscopy of a Weakly Coupled Spin Pair

In Chapter 13 we found that the weak-coupling condition leads to a situation in which the four eigenstates for a spin pair are well approximated by the set $|\alpha\alpha\rangle$, $|\alpha\beta\rangle$, $|\beta\alpha\rangle$ and $|\beta\beta\rangle$, the eigenfunctions for the z-magnetization operators, $\hat{I}_z$ and $\hat{S}_z$. This results in four allowed transitions and two pairs of split peaks. In order to analyze further the behavior of the system in an NMR experiment, we need tools to do the following:

1. Calculate the time-dependent change in the wavefunction for the coupled spins in a uniform magnetic field.
2. Calculate the change in the wavefunction for the coupled spins arising from pulses along the x'- and y'-axes.
3. Predict the average magnetization observed for a spin population with a given wavefunction along the x'-, y'- and z-axes.

Because the eigenfunctions for the Hamiltonian are the same as those for the z-magnetization operators, it is most convenient to use this set for all of the calculations. The operators for calculating the magnetization components were presented in Chapter 13, and are listed in Appendix F. In this chapter we consider the mathematics for calculating the time evolution of the wavefunction and the effects of pulses, and then apply these rules to some idealized NMR experiments. In addition, a special form of vector diagram is introduced to follow the behavior of the quantum correlations, in addition to the observable magnetization, as the spin pair is manipulated.

14.1 Time Evolution of the Wavefunction

As discussed in Chapter 12, the change in a wavefunction with time is determined by the energies of the eigenstates and is relatively simple to calculate, provided that the operator that defines this change, the Hamiltonian, does not itself change with time. For spin systems, the Hamiltonian is constant as long as the external field does not change with time—that is, if the sample is not being irradiated. If the starting wavefunction for the two-spin system is written as

$$|\Psi(0)\rangle = c_{\alpha\alpha}|\alpha\alpha\rangle + c_{\alpha\beta}|\alpha\beta\rangle + c_{\beta\alpha}|\beta\alpha\rangle + c_{\beta\beta}|\beta\beta\rangle$$

where $c_{\alpha\alpha}$, $c_{\alpha\beta}$, $c_{\beta\alpha}$ and $c_{\beta\beta}$ are the coefficients at time 0, then at time t, the wavefunction will be

$$|\Psi(t)\rangle = e^{-itE_{\alpha\alpha}/\hbar}c_{\alpha\alpha}|\alpha\alpha\rangle + e^{-itE_{\alpha\beta}/\hbar}c_{\alpha\beta}|\alpha\beta\rangle$$
$$+ e^{-itE_{\beta\alpha}/\hbar}c_{\beta\alpha}|\beta\alpha\rangle + e^{-itE_{\beta\beta}/\hbar}c_{\beta\alpha}|\beta\beta\rangle$$

where $E_{\alpha\alpha}$, $E_{\alpha\beta}$, $E_{\beta\alpha}$ and $E_{\beta\beta}$ are the energies of the eigenstates. The evolution of the wavefunction can also be expressed in terms of the Larmor frequencies of the individual spins, ν_I and ν_S, and the coupling constant, J:

$$|\Psi(t)\rangle = e^{-i\pi t(\nu_I+\nu_S+J/2)}c_{\alpha\alpha}|\alpha\alpha\rangle + e^{-i\pi t(\nu_I-\nu_S-J/2)}c_{\alpha\beta}|\alpha\beta\rangle$$
$$+ e^{i\pi t(\nu_I-\nu_S+J/2)}c_{\beta\alpha}|\beta\alpha\rangle + e^{i\pi t(\nu_I+\nu_S-J/2)}c_{\beta\alpha}|\beta\beta\rangle$$

Just as we discussed earlier in the case of the single spin, only the phases of the complex coefficients change with time. The moduli, which represent the relative contributions of the different eigenstates, remain constant (neglecting any relaxation processes). In terms of the observable magnetization, the longitudinal magnetization components ($\langle I_z\rangle$ and $\langle S_z\rangle$) will remain constant with time, while the transverse magnetization components will evolve cyclically, as we will see below.

14.2   Pulses

In contrast to the evolution of the wavefunction in a constant magnetic field, electromagnetic radiation alters the wavefunction so as to change the relative contributions of the eigenstates. For a heteronuclear pair, the resonance frequencies of the two nuclei are widely separated, and radiation can be applied selectively to one or the other nucleus. As an example, if $c_{\alpha\alpha}$, $c_{\alpha\beta}$, $c_{\beta\alpha}$ and $c_{\beta\beta}$ are the initial coefficients in the wavefunction, a selective pulse at the I-spin's resonance frequency along the y'-axis will result in the following wavefunction:

$$|\Psi(a)\rangle = \left(c_{\alpha\alpha}\cos(a/2) - c_{\beta\alpha}\sin(a/2)\right)|\alpha\alpha\rangle$$
$$+ \left(c_{\alpha\beta}\cos(a/2) - c_{\beta\beta}\sin(a/2)\right)|\alpha\beta\rangle$$
$$+ \left(c_{\beta\alpha}\cos(a/2) + c_{\alpha\alpha}\sin(a/2)\right)|\beta\alpha\rangle$$
$$+ \left(c_{\beta\beta}\cos(a/2) + c_{\alpha\beta}\sin(a/2)\right)|\beta\beta\rangle$$

where a is the angle of rotation of the magnetization, which depends on the intensity and duration of the irradiation. Appendix F includes a set of four such relationships, corresponding to pulses applied to either the I- or S-spin, along either the x'- or y'-axis.

In the case of a homonuclear pair, a pulse will influence both nuclei simultaneously. The effects on the wavefunction are calculated by first applying the pulse
operator for one spin and then applying to the resulting wavefunction the pulse operator for the other spin.

Equipped with the mathematical relationships outlined above, we can, in principle, calculate the results of any combination of pulses and time delays. Once we see
how the magnetization components change with pulses and time under different circumstances, we can use that information to understand a variety of multidimensional
experiments based on scalar coupling.

As before, a pulse experiment begins with a sample thermally equilibrated in
a uniform magnetic field. The spin state for each individual molecule will be one
of the four eigenstates or, much more likely, one of the innumerable superposition
states. These states will be populated according to their energies and the Boltzmann
distribution law. We can, however, introduce a major simplification in the analysis by
treating the sample *as if* it were composed only of molecules in the eigenstates, each
populated according to its energy. In considering a single spin, it was sufficient to
analyze only the magnetization arising from the initial small population excess in the
low-energy eigenstate ($|\alpha\rangle$ for nuclei with positive gyromagnetic ratios). This works
because the magnetization from the spins that begin in the other eigenstate, $|\beta\rangle$, is
always exactly balanced by spins that began as $|\alpha\rangle$. In the case of the two-spin system,
things do not always cancel out in such a simple fashion, and it will be necessary to
follow the spin populations that derive from all four of the eigenstates. We will begin,
however, by considering only the lowest energy state, $|\alpha\alpha\rangle$.

If we use the six magnetization operators to calculate the average magnetization
components along the three orthogonal axes arising from the two spins in the $|\alpha\alpha\rangle$
state, we obtain the following:

$$\langle I_x \rangle = 0 \qquad \langle I_y \rangle = 0 \qquad \langle I_z \rangle = 1/2$$

$$\langle S_x \rangle = 0 \qquad \langle S_y \rangle = 0 \qquad \langle S_z \rangle = 1/2$$

As we might have expected, the magnetization from each of the spins is fully aligned
with the z-axis, as represented in the vector diagram of Fig. 14.1. We can then ask
what happens if we apply a pulse of radiation that matches the resonance frequencies
of one of the spins, say the I-spin. If the pulse is applied along the y'-axis and the angle
of rotation is $\pi/2$, we designate the pulse as $(\pi/2)_{y,I}$, and the resulting wavefunction
will be

$$|\Psi\rangle = \frac{1}{\sqrt{2}}|\alpha\alpha\rangle + \frac{1}{\sqrt{2}}|\beta\alpha\rangle \tag{14.1}$$

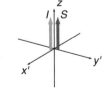

Figure 14.1
Vector diagram representing
the $|\alpha\alpha\rangle$ state.

Thus, the eigenstate is converted into a superposition state composed of equal parts $|\alpha\alpha\rangle$ and $|\beta\alpha\rangle$. Note that in both kets, the S-spin is still designated α. Although the two spins are part of a single-quantum system, we can continue to think of them as having some independent character.

Applying a $\pi/2$-pulse along the y'-axis to the S-spin of the $|\alpha\alpha\rangle$ eigenstate results in the following wavefunction:

$$|\Psi\rangle = \frac{1}{\sqrt{2}}|\alpha\alpha\rangle + \frac{1}{\sqrt{2}}|\alpha\beta\rangle$$

Now, the superimposed states are $|\alpha\alpha\rangle$ and $|\alpha\beta\rangle$, in both of which the I-spin is still in the α state.

Applying $\pi/2$-pulses along the y'-axis to both spins (simultaneously or in either order, provided there is no significant delay between the pulses), yields the following wavefunction:

$$|\Psi\rangle = \frac{1}{2}|\alpha\alpha\rangle + \frac{1}{2}|\alpha\beta\rangle + \frac{1}{2}|\beta\alpha\rangle + \frac{1}{2}|\beta\beta\rangle$$

with all four of the eigenstates contributing to the superposition state.

From the wavefunctions, we can calculate the average magnetization components for the two spins following the various pulses. After the $(\pi/2)_{y,I}$-pulse, the magnetization components are

$$\langle I_x\rangle = 1/2 \qquad \langle I_y\rangle = 0 \qquad \langle I_z\rangle = 0$$

$$\langle S_x\rangle = 0 \qquad \langle S_y\rangle = 0 \qquad \langle S_z\rangle = 1/2$$

As we would expect from the simple geometrical picture developed earlier, the pulse rotates the I-magnetization from the z-axis to the x'-axis, while leaving the S-magnetization unchanged, as shown in Fig. 14.2. If this is followed immediately by a $(\pi/2)_{y,S}$-pulse, the result is

$$\langle I_x\rangle = 1/2 \qquad \langle I_y\rangle = 0 \qquad \langle I_z\rangle = 0$$

$$\langle S_x\rangle = 1/2 \qquad \langle S_y\rangle = 0 \qquad \langle S_z\rangle = 0$$

Vector diagrams representing the effects of the pulse are shown in Fig. 14.3.

Pulses along the x'-axis have analogous effects. For instance, applying a $\pi/2$-pulse at the I-frequency converts the $|\alpha\alpha\rangle$ eigenstate to

$$|\Psi\rangle = \frac{1}{\sqrt{2}}|\alpha\alpha\rangle - \frac{i}{\sqrt{2}}|\beta\alpha\rangle$$

Figure 14.2

Vector diagrams representing a selective $(\pi/2)_{y,I}$-pulse applied to the $|\alpha\alpha\rangle$ state.

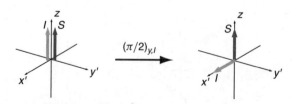

Figure 14.3

Vector diagrams representing a selective $(\pi/2)_{y,S}$-pulse, following the $(\pi/2)_{y,I}$-pulse applied to the $|\alpha\alpha\rangle$ state.

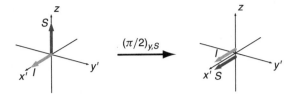

Figure 14.4

Vector diagrams representing a selective $(\pi/2)_{x,I}$-pulse applied to the $|\alpha\alpha\rangle$ state.

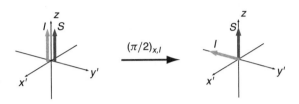

Note that this wavefunction is a superposition of $|\alpha\alpha\rangle$ and $|\beta\alpha\rangle$, just as it was when the $(\pi/2)_{y,I}$-pulse was applied to $|\alpha\alpha\rangle$, but the phase relationship between the two coefficients is different. This results in a different orientation of the I-magnetization:

$$\langle I_x\rangle = 0 \quad \langle I_y\rangle = -1/2 \quad \langle I_z\rangle = 0$$

$$\langle S_x\rangle = 0 \quad \langle S_y\rangle = 0 \quad \langle S_z\rangle = 1/2$$

Fig. 14.4 contains vector diagrams illustrating the effects of this pulse. As described earlier for the single-spin system, rotation about the x'-axis leads to magnetization along the $-y$-direction, in accordance with the "right-hand rule" introduced earlier.

These examples should enable you to predict the outcome from any pulse applied to any of the eigenstates.

14.3 Time Evolution Following a Selective Pulse

Just as in the case of the single spin, the rates of change in the observable magnetization components depend on the differences in energies among the eigenstates. In the following, the frequencies associated with the energy differences are identified as

$$\nu_{I,1} = \nu_I + J/2 \quad \text{and} \quad \nu_{I,2} = \nu_I - J/2$$

$$\nu_{S,1} = \nu_S + J/2 \quad \text{and} \quad \nu_{S,2} = \nu_S - J/2$$

Suppose that we begin again with the $|\alpha\alpha\rangle$ state and apply a selective $\pi/2$-pulse to the I-spin along the y'-axis, as illustrated in Fig. 14.2. The resulting wavefunction is

$$|\Psi\rangle = \frac{1}{\sqrt{2}}|\alpha\alpha\rangle + \frac{1}{\sqrt{2}}|\beta\alpha\rangle$$

If this wavefunction is allowed to evolve for time, t, it becomes

$$|\Psi(t)\rangle = \frac{1}{\sqrt{2}}e^{(-it E_{\alpha\alpha}/\hbar)}|\alpha\alpha\rangle + \frac{1}{\sqrt{2}}e^{(-it E_{\beta\alpha}/\hbar)}|\beta\alpha\rangle$$

Figure 14.5

Vector diagrams representing a selective $(\pi/2)_{y,I}$-pulse applied to the $|\alpha\alpha\rangle$ state, followed by precession.

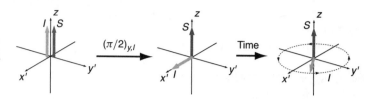

If we use the appropriate operators to calculate the average S-magnetization in the three directions, we will find that they do not change with time and remain $\langle S_x \rangle = 0$, $\langle S_y \rangle = 0$ and $\langle S_z \rangle = 1/2$. In addition, the mean I_z-magnetization remains constant, just as we saw with a single spin. But, the magnetization of the I-spin along the x'- and y'-axes changes with time according to

$$\langle I_x(t) \rangle = (1/2)\cos\left(t(E_{\beta\alpha} - E_{\alpha\alpha})/\hbar\right)$$

$$\langle I_y(t) \rangle = (1/2)\sin\left(t(E_{\beta\alpha} - E_{\alpha\alpha})/\hbar\right)$$

The term $(E_{\beta\alpha} - E_{\alpha\alpha})/\hbar$ in each of these equations can be replaced with $2\pi\nu_{I,1}$ to give

$$\langle I_x(t) \rangle = (1/2)\cos(2\pi\nu_{I,1}t)$$

$$\langle I_y(t) \rangle = (1/2)\sin(2\pi\nu_{I,1}t)$$

As we saw with the single-spin case, functions of this form describe the precession of the I-magnetization in the x–y plane, as shown in Fig. 14.5. Note that this wavefunction gives rise to a single precession frequency, $\nu_{I,1}$, corresponding to the α–β transition of the I-spin with the S-spin in the α state.

To predict the outcome of a real experiment, we also need to consider the three other eigenstates: $|\alpha\beta\rangle$, $|\beta\alpha\rangle$ and $|\beta\beta\rangle$. A $\pi/2$ pulse along the y'-axis applied to the I-spin of the $|\alpha\beta\rangle$ state gives rise to the wavefunction,

$$|\Psi\rangle = \frac{1}{\sqrt{2}}|\alpha\beta\rangle + \frac{1}{\sqrt{2}}|\beta\beta\rangle$$

The resulting magnetization components can be represented as shown in Fig. 14.6. As before, the I_z-, S_x-, S_y- and S_z-magnetization components remain constant with time, but the I_x- and I_y-components change according to

$$\langle I_x(t) \rangle = (1/2)\cos\left(t(E_{\beta\beta} - E_{\alpha\beta})/\hbar\right) = (1/2)\cos(2\pi\nu_{I,2}t)$$

$$\langle I_y(t) \rangle = (1/2)\sin\left(t(E_{\beta\beta} - E_{\alpha\beta})/\hbar\right) = (1/2)\sin(2\pi\nu_{I,2}t)$$

This is *almost* the same as the result we obtained starting with the $|\alpha\alpha\rangle$ state, but the precession frequency is now determined by the energy difference $E_{\beta\beta} - E_{\alpha\beta}$, corresponding to,

$$\nu_{I,2} = \nu_I - J/2$$

The two frequencies can be thought of as arising from the two different orientations of the S-spin.

Figure 14.6

Vector diagrams representing a selective $(\pi/2)_{y,I}$-pulse applied to the $|\alpha\beta\rangle$ state.

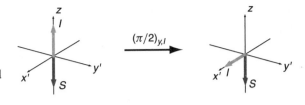

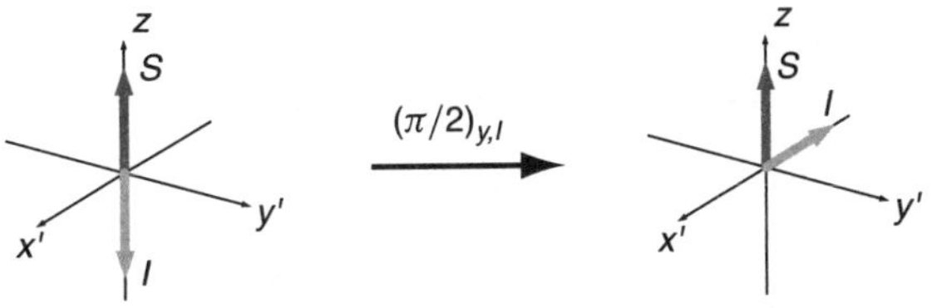

Figure 14.7

Vector diagrams representing a selective $(\pi/2)_{y,I}$-pulse applied to the $|\beta\alpha\rangle$ state.

Applying a $(\pi/2)_{y,I}$-pulse to the $|\beta\alpha\rangle$ state gives rise to the magnetization picture shown in Fig. 14.7. With time, the I_x- and I_y-magnetization components change according to

$$\langle I_x(t)\rangle = -(1/2)\cos\big(t(E_{\beta\alpha} - E_{\alpha\alpha})/\hbar\big) = -(1/2)\cos(2\pi\, v_{I,1}t)$$

$$\langle I_y(t)\rangle = -(1/2)\sin\big(t(E_{\beta\alpha} - E_{\alpha\alpha})/\hbar\big) = -(1/2)\sin(2\pi\, v_{I,1}t)$$

These expressions are exactly the same as those obtained when the pulse was applied to the $|\alpha\alpha\rangle$ state, *except* that they have opposite signs. This results in precession with the same frequency and direction, but with the vector initially pointed in the opposite direction, as indicated in the drawing. If we were to start out with equal populations of the $|\alpha\alpha\rangle$ and $|\beta\alpha\rangle$ states, the signals would cancel exactly. Because, however, $|\alpha\alpha\rangle$ has a lower energy, its equilibrium population is slightly higher than that of $|\beta\alpha\rangle$, and there will be a net signal with the same sign and frequency as predicted when we considered only the $|\alpha\alpha\rangle$ state, but with greatly reduced amplitude.

For completeness, we must also look at the effect of applying the $(\pi/2)_{y,I}$-pulse to the $|\beta\beta\rangle$ state, as diagrammed in Fig. 14.8. Following the pulse, the I_x- and I_y-magnetization components change according to

$$\langle I_x(t)\rangle = -(1/2)\cos\big(t(E_{\beta\beta} - E_{\alpha\beta})/\hbar\big) = -(1/2)\cos(2\pi\, v_{I,2}t)$$

$$\langle I_y(t)\rangle = -(1/2)\sin\big(t(E_{\beta\beta} - E_{\alpha\beta})/\hbar\big) = -(1/2)\sin(2\pi\, v_{I,2}t)$$

Now, the frequency is the same as that observed when we started with the $|\alpha\beta\rangle$ state, but the sign is opposite. The I-magnetization arising from the original $|\alpha\beta\rangle$ and $|\beta\beta\rangle$

Figure 14.8

Vector diagrams representing a selective $(\pi/2)_{y,I}$-pulse applied to the $|\beta\beta\rangle$ state.

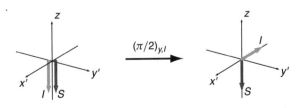

states will cancel one another, except for the small excess from the population that began as $|\alpha\beta\rangle$.

We can summarize the precession of magnetization following application of a $\pi/2$-pulse to the I-spin of the four eigenstates as follows:

1. The state of the S-spin determines the precession frequency of the I-magnetization after it has been moved to the x–y plane. Starting with $|\alpha\alpha\rangle$ or $|\beta\alpha\rangle$ gives rise to a frequency of $\nu_{I,1} = \nu_I + J/2$. The populations that begin in the $|\alpha\beta\rangle$ and $|\beta\beta\rangle$ states give rise to frequency, $\nu_{I,2} = \nu_I - J/2$.

2. The initial state of the I-spin determines the orientation of the I-magnetization immediately after the pulse and, therefore, the sign of the resulting signal. In the example shown here, the $|\alpha\alpha\rangle$ and $|\alpha\beta\rangle$ states give rise to I-magnetization with a positive orientation along the x'-axis, while $|\beta\alpha\rangle$ and $|\beta\beta\rangle$ give rise to negative I_x-magnetization.

3. The direction of precession (in the laboratory frame) is the same in all four cases.

The net result, assuming that the system is at equilibrium when the pulse is applied, is net magnetization along the positive y'-axis, which then precesses with two frequencies, $\nu_I \pm J/2$, giving rise to two peaks in the spectrum after the signal is Fourier transformed. If we were to consider the effects of a selective pulse to the S-spin of the four eigenstates, we would obtain analogous results for the evolution of the S-magnetization.

This picture is still quite consistent with the simple picture of scalar coupling introduced earlier. When both spins are irradiated, however, the results are subtly different, as discussed next.

14.4 Time Evolution Following a $\pi/2$-Pulse to Both Spins

We begin again with the $|\alpha\alpha\rangle$ eigenstate, but this time irradiate both spins simultaneously with a $\pi/2$-pulse along the y'-axis. The resulting wavefunction is

$$|\Psi\rangle = \frac{1}{2}|\alpha\alpha\rangle + \frac{1}{2}|\alpha\beta\rangle + \frac{1}{2}|\beta\alpha\rangle + \frac{1}{2}|\beta\beta\rangle$$

As we saw earlier, this places both the I- and S-magnetization vectors along the positive x'-axis. If the system is allowed to evolve with time, the phases of all four coefficients change:

$$|\Psi(t)\rangle = \frac{1}{2}e^{(-it E_{\alpha\alpha}/\hbar)}|\alpha\alpha\rangle + \frac{1}{2}e^{(-it E_{\alpha\beta}/\hbar)}|\alpha\beta\rangle$$

$$+ \frac{1}{2}e^{(-it E_{\beta\alpha}/\hbar)}|\beta\alpha\rangle + \frac{1}{2}e^{(-it E_{\beta\beta}/\hbar)}|\beta\beta\rangle$$

We now expect both the I- and S-magnetization components to change with time.

Figure 14.9 Vector representation of frequency components from the I-spin following a $(\pi/2)_y$-pulse to both spins of the $|\alpha\alpha\rangle$ state. The two frequencies come from a single population.

Looking first at the I-magnetization:

$$\langle I_x(t)\rangle = (1/4)\Big(\cos\big(t(E_{\beta\alpha} - E_{\alpha\alpha})/\hbar\big) + \cos\big(t(E_{\beta\beta} - E_{\alpha\beta})/\hbar\big)\Big)$$

$$= (1/4)\big(\cos(2\pi v_{I,1}t) + \cos(2\pi v_{I,2}t)\big)$$

$$\langle I_y(t)\rangle = (1/4)\Big(\sin\big(t(E_{\beta\alpha} - E_{\alpha\alpha})/\hbar\big) + \sin\big(t(E_{\beta\beta} - E_{\alpha\beta})/\hbar\big)\Big)$$

$$= (1/4)\big(\sin(2\pi v_{I,1}t) + \sin(2\pi v_{I,2}t)\big)$$

Now, there are two frequency terms in the expressions for $\langle I_x(t)\rangle$ and $\langle I_y(t)\rangle$. Together, the two expressions can be thought of as describing the precession of two magnetization vectors in the transverse plane, as shown in Fig. 14.9. For clarity, Fig. 14.9 shows only the magnetization components from the I-spin. The drawing is also a little misleading because it suggests that there are two populations of molecules for which the I-magnetization vectors have different precession frequencies. This is not the case: The magnetization is coming from a single population of spins, all with the same wavefunction. The experimentally determined average magnetization, however, evolves as if there were two small magnets revolving about the z-axis with different frequencies. When detected by the coil of the spectrometer and Fourier transformed, the signal will give rise to two peaks. These will represent the same two frequencies as arose from the I-spin in the separate populations we examined after a selective I-pulse.

The generation of two peaks for the I-spin from a population with a single wavefunction is a quite surprising result that is not at all obvious from a classical picture. Because there is no net magnetization from the S-spin along the z-axis, we might have expected that the I-magnetization would precess at a single frequency, as if the S-spin were not there at all. But, the S-spin is still there, though in a superposition state. If we were to somehow measure the z-magnetization of single S-spins from this population, we would obtain the results $1/2$ and $-1/2$ with equal probabilities. When we ask what the precession frequency of the coupled I-spin is, we also are asking, indirectly, about the z-magnetization of the S-spin, and as with other quantum measurements, we obtain the two quantized answers.

Figure 14.10   Vector representation of the frequency components from each of the two spins, following a $(\pi/2)_y$-pulse to both spins of the $|\alpha\alpha\rangle$ state. All four frequencies come from a single population.

We can also calculate the average magnetization from the S-spins as a function of time:

$$\langle S_x(t)\rangle = (1/4)\Big(\cos\big(t(E_{\alpha\beta} - E_{\alpha\alpha})/\hbar\big) + \cos\big(t(E_{\beta\beta} - E_{\beta\alpha})/\hbar\big)\Big)$$

$$= (1/4)\Big(\cos\big(2\pi\,\nu_{S,1}t\big) + \cos(2\pi\,\nu_{S,2}t)\Big)$$

$$\langle S_y(t)\rangle = (1/4)\Big(\sin\big(t(E_{\alpha\beta} - E_{\alpha\alpha})/\hbar\big) + \sin\big(t(E_{\beta\beta} - E_{\beta\alpha})/\hbar\big)\Big)$$

$$= (1/4)\Big(\sin(2\pi\,\nu_{S,1}t) + \sin(2\pi\,\nu_{S,2}t)\Big)$$

These equations have exactly the same form as those for the I_x- and I_y-magnetization components, except that the frequencies of the sine and cosine terms are determined by the energy differences for the transitions of the S-spin. Thus, the single population gives rise to two signals for the S-spin, and two for the I-spin. A vector picture that represents the signals generated by both spins is shown in Fig. 14.10. Again, it is important to stress that the four signals are coming from a single population of molecules containing two spins. The other three populations, originating from the eigenstates $|\alpha\beta\rangle$, $|\beta\alpha\rangle$ and $|\beta\beta\rangle$, will also each give rise to magnetization components precessing with all four frequencies, but the initial signs of the magnetization components will differ, leading to partial cancellation of the signals.

14.5   Some More Exotic Examples

In the previous two sections, we saw that for a pure population composed of molecules with two coupled spins, all with the same initial wavefunction, the evolution of the magnetization from one spin is highly dependent on the state of the other. If the S-magnetization is aligned with the positive z-axis, the I-magnetization precesses with the frequency $\nu_{I,1} = \nu_I + J/2$, but if the S-magnetization is aligned with the negative z-axis, the frequency observed for the I-spin is $\nu_{I,2} = \nu_I - J/2$. Rather remarkably, both frequencies are observed from a single population if the S-spin is in a superposition state so that its average z-magnetization is 0.

We might now ask, what happens if the average S-magnetization lies neither along the z-axis nor within the transverse plane? One way to generate such a state from

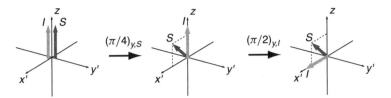

Figure 14.11 Vector diagrams representing a selective $(\pi/4)_{y,S}$-pulse applied to the $|\alpha\alpha\rangle$ state, followed by a $(\pi/2)_{y,I}$-pulse.

the $|\alpha\alpha\rangle$ eigenstate is to apply an S-pulse along the y'-axis for a period of time that rotates the magnetization only partway toward the x'-axis, say $\pi/4$ rad. If this pulse is followed by a $\pi/2$-pulse to the I-spin along the y'-axis, the state represented in Fig. 14.11 is generated. The wavefunction for the resulting state is

$$|\Psi\rangle = \frac{1}{\sqrt{2}}\Big(\cos(\pi/8)|\alpha\alpha\rangle + \sin(\pi/8)|\alpha\beta\rangle + \cos(\pi/8)|\beta\alpha\rangle + \sin(\pi/8)|\beta\beta\rangle\Big)$$

A quick sketch (or a few calculator keystrokes) will show that $\cos(\pi/8) > \sin(\pi/8)$, so that the net contributions of the eigenstates representing the S-spin in the α state ($|\alpha\alpha\rangle$ and $|\beta\alpha\rangle$) are greater than the contributions of the β state of the S-spin (from $|\alpha\beta\rangle$ and $|\beta\beta\rangle$). This corresponds to the partial alignment of the S-magnetization along the positive z-axis. The full set of magnetization components is

$$\langle I_x\rangle = 1/2 \qquad \langle I_y\rangle = 0 \qquad \langle I_z\rangle = 0$$

$$\langle S_x\rangle = 1/(2\sqrt{2}) \qquad \langle S_y\rangle = 0 \qquad \langle S_z\rangle = 1/(2\sqrt{2})$$

As before, we then calculate the time dependence of the wavefunction and the magnetization components at time, t. For the magnetization along the x-axis arising from the I-spin,

$$\langle I_x(t)\rangle = (1/2)\Big(\cos^2(\pi/8)\cos(2\pi\nu_{I,1}t) + \sin^2(\pi/8)\cos(2\pi\nu_{I,2}t)\Big)$$

The mean I_y-magnetization is described by an equivalent expression with the cosine terms replaced with sine terms. As in the case considered in the previous section, where the S-magnetization was placed entirely in the transverse plane, we find that there are two frequency components for the time evolution of the I-magnetization, with the same frequencies as before. This time, however, the relative amplitudes of the two components are different, determined by the terms $\cos^2(\pi/8)$ and $\sin^2(\pi/8)$. Thus, the signal corresponding to the $|\alpha\alpha\rangle$–$|\beta\alpha\rangle$ transition, with frequency $\nu_{I,1} = \nu_I + J/2$, is stronger than the signal corresponding to the $|\alpha\beta\rangle$–$|\beta\beta\rangle$ transition. A vector diagram representing this situation is shown in Fig. 14.12. The lengths of the two arrows are drawn to suggest the relative strengths of the signals that would be generated in the spectrometer coil. In fact, for the case of an initial $\pi/4$-pulse to the S-spin, the difference in magnitude of the two I-frequency components is considerably larger than indicated in the drawing.

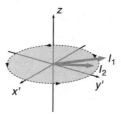

Figure 14.12   Vector representation of the frequency components from the I-spin, following the pulses shown in Fig. 14.11. The lengths of the vectors are drawn to suggest the relative intensities of the two signals that would be generated in a spectrometer. The diameter of the circle in the $x'-y'$ plane indicates the magnitude of the vectors that would arise if the S-magnetization lay entirely in the transverse plane.

What about the S-spin? Applying the S_x operator to the time-dependent wavefunction gives the following results:

$$\langle S_x(t)\rangle = \frac{1}{4\sqrt{2}}\left(\cos\left(t(E_{\alpha\beta} - E_{\alpha\alpha})/\hbar\right) + \cos(t(E_{\beta\beta} - E_{\beta\alpha})/\hbar)\right)$$

$$= \frac{1}{4\sqrt{2}}\left(\cos\left(2\pi\nu_{S,1}t\right) + \cos(2\pi\nu_{S,2}t)\right)$$

We see the same two S-frequency components as before, and, in contrast to the I-components, they have equal amplitudes. The amplitudes are equal because the I-magnetization, which modulates the relative amplitudes of the S-components, lies entirely in the transverse plane. Note also that because there is a significant z-component to the S-magnetization, both amplitudes are lower, by the factor $1/(2\sqrt{2})$, than when the S-magnetization lies entirely in the $x-y$ plane. Combining the vector representations for the two spins leads to the rather complicated drawing shown in Fig. 14.13.

We can ask, more generally, how the orientation of the S-magnetization influences the two frequency components of the I-signal by carrying out the same set of calculations for the case when the initial S-pulse rotates the magnetization by an arbitrary angle of a rad. Following the same mathematical steps as before, the general

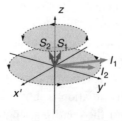

Figure 14.13   Vector representation of the frequency components from the I- and S-spins, following the pulses shown in Fig. 14.11. The lengths of the vectors are drawn to suggest the relative intensities of the two signals that would be generated in a spectrometer.

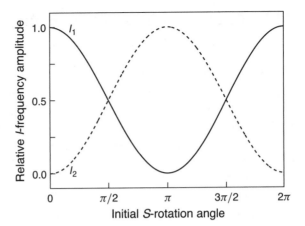

Figure 14.14

Relative amplitudes of the I-frequency components as a function of the rotation of the S-magnetization from the z-axis.

expression for the mean I_x-magnetization at time, t, is

$$\langle I_x(t)\rangle = (1/4)\big(1 + \cos(a)\big)\cos(2\pi\nu_{I,1}t) + \big(1 - \cos(a)\big)\cos(2\pi\nu_{I,2}t)$$

From this expression, it is apparent that the amplitudes of the two frequency components are determined by the terms $\big(1 + \cos(a)\big)$ and $\big(1 - \cos(a)\big)$, which vary sinusoidally from 0 to 2 as the initial angle of the S-magnetization vector is changed. The relative intensities of the two components are plotted as a function of the S-rotation angle in Fig. 14.14. The graph illustrates what we saw earlier for a few special cases: When the S-magnetization is aligned with the positive z-axis ($a = 0$), only the first I-frequency is observed. When the S-magnetization is inverted to point along the negative z-axis ($a = \pi$), only the second frequency contributes. When the S-magnetization is rotated by exactly $\pi/2$ or $3\pi/2$ rad, to place it along the positive or negative x'-axis, the two components have equal amplitudes. Notice that the amplitudes are always positive, however, so the sign of the signals are not affected by the position of S-magnetization. Note also that the combined amplitudes for the two I-frequencies remain constant.

From the above analysis, it appears that the critical factor in determining the relative amplitudes of the I-frequency components is the projection of the S-magnetization along the z-axis, since the same result is obtained whether the S-vector points in the positive or negative x'-direction. Also, note that the relative amplitudes from the I-spin do not change with time as the S-magnetization is also precessing. Just to be sure, we can predict the result of applying an initial pulse to the S-spin along the x'-axis, thereby placing the initial S-magnetization somewhere in the $y'-z$ plane at an angle a from the z-axis. Analyzing the wavefunction as described in the previous case leads to exactly the same expressions for the time evolution of the I_x- and I_y-magnetization, confirming that the relative amplitudes of the two I-frequency components are determined by the z-magnetization of the S-spin.

The important point here is that the state of one spin in a coupled pair can influence the signals from the other, providing a potential mechanism for generating cross peaks in multidimensional experiments.

14.6   Effects of Pulses and Evolution on Spin–Spin Correlations

Although the concepts of correlations and product operators introduced in Chapter 13 may have seemed rather esoteric, a full description of the behavior of a scalar-coupled spin pair requires that we consider not only changes in the magnetization components, but in addition, those of the correlations, which are also affected by pulses and evolution periods. Recall from Section 13.1.4 (page 380) that the correlation between two spin magnetization components are calculated using operators constructed by taking the products of the operators for the individual components. For instance, to calculate the average correlation between the z-components of the I- and S- spins in the wavefunction $|\Psi\rangle$, the product operator $\hat{I}_z\hat{S}_z$ is used as follows:

$$\langle I_zS_z\rangle = \langle\Psi|\hat{I}_z\hat{S}_z|\Psi\rangle$$

Applying all nine of the product operators to the $|\alpha\alpha\rangle$ wavefunction yields the following results:

$$\langle I_xS_x\rangle = 0 \qquad \langle I_yS_x\rangle = 0 \qquad \langle I_zS_x\rangle = 0$$
$$\langle I_xS_y\rangle = 0 \qquad \langle I_yS_y\rangle = 0 \qquad \langle I_zS_y\rangle = 0$$
$$\langle I_xS_z\rangle = 0 \qquad \langle I_yS_z\rangle = 0 \qquad \langle I_zS_z\rangle = 1/4$$

This result simply indicates that if the z-magnetization components of the two spins are each measured and the results multiplied together, the result is alway 1/4. This makes sense, since both I_z and S_z are always 1/2. None of the other magnetization components are correlated, however. If, for instance, I_y and S_x are measured, each will yield +1/2 and −1/2 with equal probability, and there will be no correlation between the values observed for individual spin pairs.

If the same procedure is applied to the wavefunction generated by applying a $(\pi/2)_{y,I}$-pulse to the $|\alpha\alpha\rangle$ state, the following results are obtained from the resulting wavefunction (Eq. 14.1, p. 395):

$$\langle I_xS_x\rangle = 0 \qquad \langle I_yS_x\rangle = 0 \qquad \langle I_zS_x\rangle = 0$$
$$\langle I_xS_y\rangle = 0 \qquad \langle I_yS_y\rangle = 0 \qquad \langle I_zS_y\rangle = 0 \qquad (14.2)$$
$$\langle I_xS_z\rangle = 1/4 \qquad \langle I_yS_z\rangle = 0 \qquad \langle I_zS_z\rangle = 0$$

Again, this makes sense, because the I- and S-magnetization components are aligned with the x'- and z-axes, respectively. As shown earlier (page 380), however, the results are not always so easily rationalized, as there are states for which there are no net magnetization components, but there are non-zero average correlations. In fact, it can be shown that there is at least one non-zero correlation for any wavefunction, reflecting the very fundamental nature of these correlations.

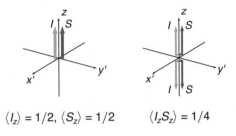

Figure 14.15
Vector diagrams to represent the observable magnetization components (left) and correlations (right) for the $|\alpha\alpha\rangle$ state.

$\langle I_z \rangle = 1/2, \langle S_z \rangle = 1/2$ $\langle I_z S_z \rangle = 1/4$

14.6.1   Correlation Vector Diagrams

To help visualize the behavior of the correlations, it is useful to introduce a new kind of vector diagram to represent them as distinct properties of the spin pair. As a first example, again consider the $|\alpha\alpha\rangle$ wavefunction. The left-hand drawing in Fig. 14.15 was introduced earlier to represent the observable magnetization components, whereas the drawing to the right is designed to represent the correlation between the I_z- and S_z-components. The new diagram contains two vector pairs, with each pair connected to show that a positive value of I_z is always correlated with a positive value of S_z, and negative values are always correlated with one another. It is important to emphasize that the two diagrams represent distinct features of the wavefunction. Note, for instance, that the same correlation diagram would be used for the $|\beta\beta\rangle$ wavefunction, even though the z-magnetization components are both negative for this wavefunction. Although the positive value of $I_z S_z$ could, in principle be represented by a single vector pair, as in the left-hand diagram, the use of two pairs helps to distinguish the two types of information. In addition, because the vectors in the correlation diagram cancel one another, the diagram helps to convey the important idea that the correlation, by itself, does not imply a net magnetization in any direction.

As another example, the wavefunction, $|\Psi_1\rangle$, was introduced in Chapter 13 and was defined as

$$|\Psi_1\rangle = \frac{1}{\sqrt{2}}(|\alpha\beta\rangle + |\beta\alpha\rangle)$$

For this wavefunction, all of the average magnetization components are zero, but the average correlations are

$$\langle I_x S_x \rangle = 1/4 \qquad \langle I_y S_x \rangle = 0 \qquad \langle I_z S_x \rangle = 0$$

$$\langle I_x S_y \rangle = 0 \qquad \langle I_y S_y \rangle = 1/4 \qquad \langle I_z S_y \rangle = 0$$

$$\langle I_x S_z \rangle = 0 \qquad \langle I_y S_z \rangle = 0 \qquad \langle I_z S_z \rangle = -1/4$$

These correlations are represented diagrammatically in Fig. 14.16. Notice that the arrows in the diagram representing the $I_z S_z$ correlation are connected by line segments that cross one another, indicating that a positive value of I_z is correlated with a negative value of S_z, and vice versa.

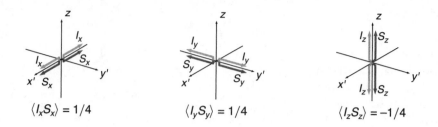

Figure 14.16   Correlation diagrams representing the $I_x S_x$, $I_y S_y$, and $I_z S_z$ correlations for the $|\Psi_1\rangle$ wavefunction. Adapted with permission from Goldenberg, D.P. (2010) *Concepts. Magn. Reson.* **36A**, 49-83. Copyright © 2010 by John Wiley and Sons.

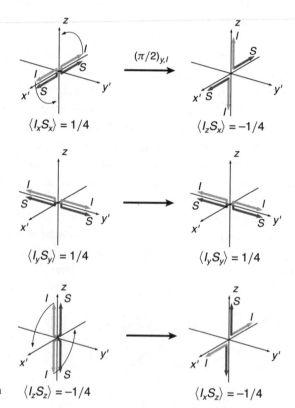

Figure 14.17
Effects of a $(\pi/2)_{y,I}$-pulse on the correlations associated with the $|\Psi_1\rangle$ wavefunction.

These diagrams can be used to illustrate and predict the changes in correlations caused by pulses, as illustrated in Fig. 14.17 for the case of a $(\pi/2)_{y,I}$-pulse applied to the $|\Psi_1\rangle$ wavefunction discussed above. For each of the I-vectors in the correlation diagrams, the pulse has the same effect as would be expected for a simple magnetization vector, converting $I_x S_x$ to $-I_z S_x$ and $-I_z S_z$ to $-I_x S_z$, while leaving $I_y S_y$ unchanged. Since there is no observable magnetization associated with $|\Psi_1\rangle$, there is none after the pulse, either.

The effects of any pulse on an existing correlation can be predicted by using the diagrams as illustrated by these examples. It is important to emphasize, again, that the correlations may or may not be associated with non-zero magnetization components, and the changes in the correlations and any net magnetization components must be treated separately.

14.6.2   Time Evolution of Correlations

Just as the net observable magnetization components evolve with time in the presence of a stationary external magnetic field, the correlations between coupled spins can evolve. If a $(\pi/2)_{y,I}$-pulse is applied to the $|\alpha\alpha\rangle$ state, the resulting transverse I-magnetization will evolve as described earlier. At the same time, the $I_x S_z$ correlation will evolve into an $I_y S_z$ correlation according to the following relationships:

$$\langle I_x S_z(t)\rangle = (1/2)\cos(2\pi \nu_{I,1}t)$$

$$\langle I_y S_z(t)\rangle = (1/2)\sin(2\pi \nu_{I,1}t)$$

The evolution of the correlations can be visualized as the precessional motion of the vectors representing the transverse I-components, as shown in Fig. 14.18. Initially, $\langle I_x\rangle = 1/2$ and $\langle S_z\rangle = 1/2$, so that, for each spin pair, the product $I_x S_z$ is always equal to 1/4. As the I-spin precesses, the probability that an individual measurement of I_x will yield a value of 1/2 decreases, so that the average value of $I_x S_x$ also decreases. When the I-magnetization has precessed to the y'-axis, the two possible outcomes of measuring the I_x-magnetization have equal probability, so that there is no correlation with the S_z-magnetization. But, the probability that $I_y = 1/2$ is now 1, so that the product $I_y S_z$ is now 1/4 for each spin pair. This cycle repeats as the I-spin continues to precess.

A somewhat more subtle relationship between the observable magnetization and correlations emerges from a situation in which both magnetization components lie in the transverse plane. This is illustrated by the case in which the starting positions of both the I- and S-magnetization vectors are aligned along the x'-axis—that is, the state following a non-selective $(\pi/2)_y$-pulse applied to the $|\alpha\alpha\rangle$ state. As discussed earlier, the magnetization components associated with the two spins each evolve with

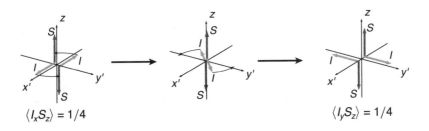

$\langle I_x S_z\rangle = 1/4$ $\langle I_y S_z\rangle = 1/4$

Figure 14.18   Interconversion of the $I_x S_z$ and $I_y S_z$ correlations during an evolution period.

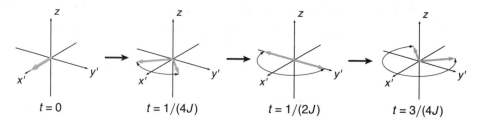

Figure 14.19   Vector representation of the evolution of the two I-frequency components in the rotating frame, with both the I- and S-magnetization components lying in the transverse plane, and the rotating-frame frequency set to ν_I.

two frequencies, even though there is only a single wavefunction. To simplify the resulting analysis and diagrams, it is convenient to set the average frequencies of both spins to zero. This is equivalent to using a separate rotating frame for each spin and setting the frequencies of these frames to those of the corresponding spins (in the absence of coupling). With this simplification, the evolution of the magnetization components becomes

$$\langle I_x(t) \rangle = (1/2) \cos(\pi J t)$$

$$\langle S_x(t) \rangle = (1/2) \cos(\pi J t)$$

With ν_I and ν_S set to zero, the evolution of the magnetization depends only on J, and the only net magnetization is along the x'-axis. This can be visualized as the precession of two vectors (for each spin) away from the x'-axis in opposite directions, as illustrated in Fig. 14.19 for the I-spin. The y'-components of the two vectors always cancel one another, and the x'-components are zero when $t = 1/(2J)$, so that the net magnetization disappears.

During this evolution period, the non-zero product averages are $\langle I_x S_z \rangle$, $\langle I_x S_x \rangle$ and $\langle I_z S_x \rangle$, which change with time according to

$$\langle I_x S_x(t) \rangle = 1/4$$

$$\langle I_y S_z(t) \rangle = (1/4) \sin(\pi J t) \qquad (14.3)$$

$$\langle I_z S_y(t) \rangle = (1/4) \sin(\pi J t)$$

Note first that the correlation that is present initially, $\langle I_x S_x \rangle = 1/4$, remains constant with time, while two new correlations appear and change cyclically. Also note that the cyclic patterns of $\langle I_y S_z \rangle$ and $\langle I_z S_y \rangle$ have the same frequency as the cycles for observable x'-magnetization for each spin, but are $\pi/2$ rad out of phase with the magnetization cycles.

The cycles for the observable magnetization and the correlations can be viewed as two manifestations of the same phenomenon, as illustrated in Fig. 14.20, using vector diagrams for the I_x-magnetization and the $I_y S_z$ correlation. In the leftmost part of the figure, a new variation on the vector diagrams is introduced to represent the initial I_x-magnetization. This diagram resembles those used to represent correlations, with two vector pairs showing a coupling between the I- and S-spins. Now, however, the two I-vectors point in the same direction, along the x'-axis, indicating the presence of

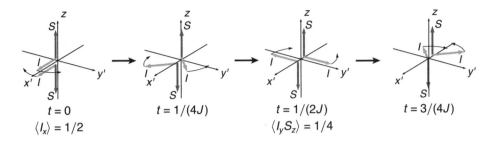

Figure 14.20   Vector representation of the interconversion of I_x-magnetization and the I_yS_z correlation. The rotating-frame frequency is assumed to be equal to ν_I.

a net I-magnetization, while the vectors for the S-spin point in opposite directions along the z-axis. Importantly, this diagram is designed to represent the *absence* of a correlation between I_x- and S_z-components, rather than its presence, since positive I_x-magnetization is associated equally with positive and negative S_z-magnetization. The two S_z-vectors in the drawing can be thought of as representing a *potential* correlation, which emerges as the two vectors associated with the I-magnetization precess in opposite directions.

As the two I-vectors precess away from the x'-axis, the y'-component of each increases and is correlated with an S_z-component of the same sign. When $t = 1/(2J)$, the I-magnetization disappears, and the correlation between I_y and S_z reaches a maximum. With further evolution, the I_yS_z correlation is converted back to a net I-magnetization. A similar set of diagrams can be drawn to represent the conversion of net S_x-magnetization into a correlation between S_y and I_z.

What about the I_xS_x correlation that was initially associated with the I_x- and S_x-magnetization components? It may seem surprising that Eq. 14.3 indicates that $\langle I_xS_x \rangle$ remains constant with time, but this reflects the choice to set the rotating-frame frequencies to the average frequencies of the I- and S-spins. If ν_I and ν_S are not constrained to be zero, I_xS_x is found to interconvert with three other correlations as follows:

$$\langle I_xS_x(t) \rangle = (1/4)\cos(2\pi \nu_I t)\cos(2\pi \nu_S t)$$

$$\langle I_xS_y(t) \rangle = (1/4)\cos(2\pi \nu_I t)\sin(2\pi \nu_S t)$$

$$\langle I_yS_x(t) \rangle = (1/4)\sin(2\pi \nu_I t)\cos(2\pi \nu_S t) \qquad (14.4)$$

$$\langle I_yS_y(t) \rangle = (1/4)\sin(2\pi \nu_I t)\sin(2\pi \nu_S t)$$

These equations can be visualized as arising from the precession of independent I- and S-components of the initial I_xS_x correlation, as illustrated in Fig. 14.21. Each expression in the equations above represents the product of the projections of the two precessing vectors onto the indicated axes. Initially, the I- and S-components are both aligned with the positive (or negative) x'-axis, and the I_xS_x correlation is maximal. As both vectors precess toward the y'-axis, the projections onto the x'-axis decrease, and the average I_yS_y correlation increases. The evolution of each correlation depends on

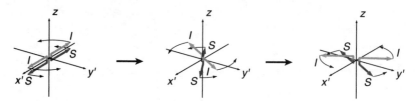

Figure 14.21   Vector representation of the interconversion of the I_xS_x, I_xS_y, I_yS_x and I_yS_y correlations, when the precession frequencies do not match the rotating-frame frequency. Adapted with permission from Goldenberg, D.P. (2010) *Concepts. Magn. Reson.* **36A**, 49-83. Copyright © 2010 by John Wiley and Sons.

the precession frequency of both spins. However, these expressions do not depend on the scalar coupling constant, J, because there is no z-component in any of the correlations. An important consequence is that these correlations do not interconvert with observable magnetization, because the individual I-components, for instance, always precess at the same rate and balance one another, never combining to generate a net magnetization component.

Although the I_xS_x, I_xS_y, I_yS_x and I_yS_y correlations do not give rise to observable magnetization during free precession, pulses can convert any of these correlations into one of the correlations that do interconvert with net magnetization (I_xS_z, I_yS_z, I_zS_x or I_zS_y). The magnitude of the resulting magnetization will depend on the relative orientations of the correlation vectors and the direction of the pulse. If the evolution period is incremented to generate a two-dimensional spectrum, the positions of peaks in the indirectly detected dimension will be determined by the frequencies associated with the evolution of the correlations, as discussed further below.

14.6.3   Multiple-Quantum Coherence and Some Nomenclature

The term *multiple-quantum coherence* is often used in the NMR literature to refer to states characterized by correlations between transverse components—that is, the I_xS_x, I_xS_y, I_yS_x and I_yS_y correlations. The origins of this term lie in the time evolution of the correlations, as expressed in Eq. 14.4 and illustrated in Fig. 14.21. Because the expressions in Eq. 14.4 are written as products of trigonometric functions, they do not reveal the actual frequencies for the evolution of the correlations. Using trigonometric identities, however, these equations can be rewritten as

$$\langle I_xS_x(t)\rangle = (1/8)\Big(\cos\big(2\pi(\nu_I + \nu_S)t\big) + \cos\big(2\pi(\nu_I - \nu_S)t\big)\Big)$$

$$\langle I_xS_y(t)\rangle = (1/8)\Big(\sin\big(2\pi(\nu_I + \nu_S)t\big) - \sin\big(2\pi(\nu_I - \nu_S)t\big)\Big)$$

$$\langle I_yS_x(t)\rangle = (1/8)\Big(\sin\big(2\pi(\nu_I + \nu_S)t\big) + \sin\big(2\pi(\nu_I - \nu_S)t\big)\Big) \tag{14.5}$$

$$\langle I_yS_y(t)\rangle = -(1/8)\Big(\cos\big(2\pi(\nu_I + \nu_S)t\big) - \cos\big(2\pi(\nu_I - \nu_S)t\big)\Big)$$

These equations show that the evolution of each correlation is determined by two frequencies, which represent the sum and difference of v_I and v_S. These frequencies also correspond to the energies associated with the transitions between the $|\alpha\alpha\rangle$ and $|\beta\beta\rangle$ states ($h(v_I + v_S)$) and between $|\alpha\beta\rangle$ and $|\beta\alpha\rangle$ ($h(v_I - v_S)$)—that is, multiple-quantum transitions. Although these transitions cannot be stimulated by absorption of single photons with the corresponding energies, spectral peaks at the frequencies $(v_I + v_S)$ and $(v_I - v_S)$ can be generated in the indirect dimension of an appropriately designed two-dimensional experiment. This, then leads to the practice of referring to such states as representing multiple-quantum coherence, with the last word referring to the fact that a detectable signal requires a population of molecules for which the wavefunctions are at least partially in phase.

Though widely used, the term multiple-quantum coherence does not provide a great deal of insight into the underlying physical phenomena, which concerns the evolution of correlations between transverse components. As an alternative, the following terminology is suggested to describe different classes of correlations:

1. A unique longitudinal correlation, involving only z-components: $I_z S_z$. Like longitudinal magnetization, this correlation does not change with time in a static magnetic field.

2. Transverse correlations, which involve only x'- or y'-components: $I_x S_x$, $I_x S_y$, $I_y S_x$ and $I_y S_y$. These correlations interconvert among themselves with time in a stationary field, but, in the absence of pulses (or relaxation), do not interconvert either with other classes of correlations or with observable magnetization. The evolution of these correlations is determined by the sum and difference of v_I and v_S, but is independent of the coupling constant, J.

3. Longitudinal-transverse correlations, or "mixed correlations," each of which involves a z-component and an x'- or y'-component: $I_x S_z$, $I_y S_z$, $I_z S_x$ and $I_z S_y$. In a static magnetic field, each of these correlations interconverts cyclically with observable magnetization components. The frequency of this cycle is determined by the coupling constant, J. The correlations $I_x S_z$ and $I_y S_z$ also interconvert with one another, as do $I_z S_x$ and $I_z S_y$, as determined by the precession rates of the transverse components.

It is also common practice to describe the states associated with the mixed correlations as possessing "anti-phase magnetization," both because the vectors representing the transverse components point in opposite directions and because these states can lead to anti-phase peaks in the resulting spectrum, as in some of the experiments discussed in the following section. The most important feature of these states, however, is the correlation between the transverse and longitudinal magnetization components.

Finally, with respect to nomenclature, the terms *coherence* and *correlation* are used almost interchangeably in many NMR papers and textbooks, a rather unfortunate tendency. As discussed above, a correlation is an intrinsic quantum property of an individual spin pair (or a larger group of coupled spins), while coherence describes the extent to which the wavefunctions for the spins (or coupled spins) in a population share a common phase relationship. For individual spin pairs, there will be at least one non-zero correlation between magnetization components, but the correlations may cancel out in a population, just as the net magnetization components for a population may be zero. These issues will be addressed further in Chapters 17 and 18, in which the density matrix and product-operator formalisms are introduced.

14.7   Some Demonstration Experiments

The mathematical manipulations and vector-diagram gymnastics presented in the earlier sections of this chapter may all seem to be very abstract and with no obvious relevance to practical NMR spectroscopy. Though there is no denying the abstract nature of the treatment, the very real consequences can be demonstrated in quite simple experiments.

14.7.1   Experiment 1

Consider, first, an experiment involving a single selective pulse to one spin of a coupled pair, as outlined in Fig. 14.22. The leftmost panel shows the pulse sequence. A single pulse is applied to the S-spin, and the FID is recorded from this spin either immediately (delay time $\tau = 0$) or after a delay $\tau = 1/(2J)$. For convenience, the rotating-frame frequency is set to match the average precession frequency of the S-spin.

The middle panel of Fig. 14.22 shows a spectrum from an actual experiment with $\tau = 0$, along with a vector diagram representing the expected magnetization components of the population of spin pairs that began in the $|\alpha\alpha\rangle$ state. When the FID is recorded immediately after the selective pulse, the spectrum is exactly as expected from our earlier analysis. The two peaks can be accounted for by the excess populations that began in the $|\alpha\alpha\rangle$ and $|\beta\alpha\rangle$ states, with the I-magnetization aligned with the positive or negative z-axis.

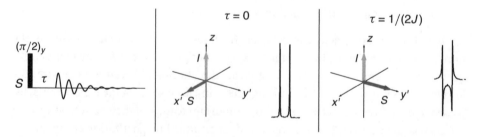

Figure 14.22   A one-pulse experiment with a selective $(\pi/2)_{y,S}$-pulse, with a delay time, τ, between the pulse and the beginning of data acquisition. The vector diagrams represent the magnetization components from the populations of spin pairs initially present in the $|\alpha\alpha\rangle$ state at equilibrium. The experimental spectra were recorded using chloroform ($CHCl_3$), with S representing the ^{13}C nucleus (at natural abundance) and I representing the single ^1H nucleus. Adapted with permission from Goldenberg, D.P. (2010) *Concepts. Magn. Reson.* **36A**, 49-83. Copyright © 2010 by John Wiley and Sons.

The rightmost panel of Fig. 14.22 shows the expected magnetization component from the original $|\alpha\alpha\rangle$ population following a delay of $1/(2J)$ and the spectrum generated when the FID is recorded at the end of this delay. During the delay, the S-magnetization components from the different populations precess at different frequencies according to the orientation of the I-spin. Because the rotating-frame frequency has been set to match the average S-frequency, the two components precess away from the x'-axis at equal rates, and are aligned with the positive and negative y'-axes at the end of the delay. If the FID is recorded and processed exactly as in the first experiment, the result is two dispersive peaks that are out of phase with one another. All of this is consistent with a simple picture based on magnetization vectors.

14.7.2   Experiment 2

Next, consider a modification of the experiments, in which a selective $(\pi/2)_{y,I}$-pulse is applied immediately before the FID is recorded, as illustrated in Fig. 14.23. When there is no delay between the pulses ($\tau = 0$, middle panel), both magnetization vectors are aligned with the x'-axis at the beginning of the data acquisition period. As discussed earlier, each of the populations that began in one of the four eigenstates gives rise to two S-frequency components, even though each of the I-spins is in a superposition state with no net z-magnetization. The spectrum contains the two expected peaks.

A more surprising result is obtained when a delay of $1/(2J)$ is introduced between the first and second pulses, as illustrated in the rightmost panel of Fig. 14.23. In this case, the double peak from the S-spin is almost completely eliminated by a pulse to the I-spin. Considering only the observable magnetization components arising from the initial $|\alpha\alpha\rangle$ state, there is no obvious way to explain this result.

In order to understand this result, we need to carefully examine the effects of the pulses and evolution time on each of the four eigenstates making up the equilibrium population. The expected effect on the $|\alpha\alpha\rangle$ state is shown by the vector diagram

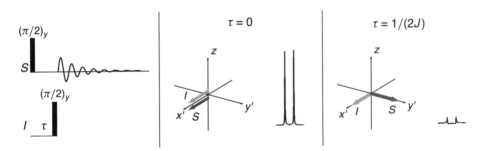

Figure 14.23   Modification of the experiment illustrated in Fig. 14.22, with a $(\pi/2)_{y,I}$-pulse immediately before the data acquisition period. The vector diagrams represent the magnetization components from the populations of spin pairs initially present in the $|\alpha\alpha\rangle$ state at equilibrium. Adapted with permission from Goldenberg, D.P. (2010) *Concepts. Magn. Reson.* **36A**, 49-83. Copyright © 2010 by John Wiley and Sons.

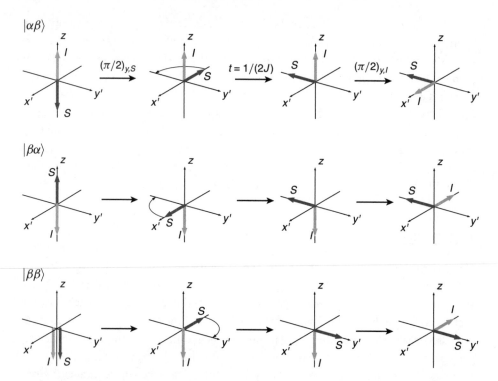

Figure 14.24   Vector diagrams describing the effects of the experiment illustrated in Fig. 14.23 on the $|\alpha\beta\rangle$, $|\beta\alpha\rangle$ and $|\beta\beta\rangle$ states, with $\tau = 1/(2J)$.

in the rightmost panel of Fig. 14.23. After the second pulse, the S-magnetization is aligned with the y'-axis, and the I-magnetization is in the transverse plane, aligned with the x'-axis. From our previous analysis, we know that the S-magnetization will now evolve with two frequencies, with each component making an equal contribution.

The effects on the other three starting states are illustrated in Fig. 14.24. In each case, the final position of the S-magnetization is aligned with the positive or negative y'-axis, and the I-magnetization is aligned with the positive or negative x'-axis. Since only the signal from the S-spin is recorded, the influence of the I-spin depends only on its magnetization projection onto the z-axis. For each of the four populations, the z-projection of the I-magnetization is zero, so that the two S-spin frequencies are expected to make equal contributions to the signal.

The total amplitude of the signal will depend on the relative contributions of the four populations, which depend in turn on the equilibrium populations of the starting eigenstates. For this purpose, it is convenient to express the equilibrium population differences for each of the two spins as

$$\Delta P_I = \frac{N_{\alpha\alpha} - N_{\beta\alpha}}{N_{\alpha\alpha} + N_{\beta\alpha}} = \frac{N_{\alpha\beta} - N_{\beta\beta}}{N_{\alpha\beta} + N_{\beta\beta}} \tag{14.6}$$

and

$$\Delta P_S = \frac{N_{\alpha\alpha} - N_{\alpha\beta}}{N_{\alpha\alpha} + N_{\alpha\beta}} = \frac{N_{\beta\alpha} - N_{\beta\beta}}{N_{\beta\alpha} + N_{\beta\beta}} \qquad (14.7)$$

where $N_{\alpha\alpha}$, $N_{\alpha\beta}$, $N_{\beta\alpha}$ and $N_{\beta\beta}$ are the number of spins in the four states at equilibrium. If the total population is N_T, the fractional populations of the four eigenstates at equilibrium can then be expressed as

$$f_{\alpha\alpha} = \frac{N_{\alpha\alpha}}{N_T} = \frac{1}{4}\left(1 + \Delta P_I + \Delta P_S\right)$$

$$f_{\alpha\beta} = \frac{N_{\alpha\beta}}{N_T} = \frac{1}{4}\left(1 + \Delta P_I - \Delta P_S\right)$$

$$f_{\beta\alpha} = \frac{N_{\beta\alpha}}{N_T} = \frac{1}{4}\left(1 - \Delta P_I + \Delta P_S\right)$$

$$f_{\beta\beta} = \frac{N_{\beta\beta}}{N_T} = \frac{1}{4}\left(1 - \Delta P_I - \Delta P_S\right) \qquad (14.8)$$

The total S_y-magnetization at the beginning of the data acquisition period is

$$\bar{S}_y = f_{\alpha\alpha}\langle S_y\rangle_{\alpha\alpha} + f_{\beta\alpha}\langle S_y\rangle_{\alpha\beta} + f_{\beta\alpha}\langle S_y\rangle_{\beta\alpha} + f_{\beta\beta}\langle S_y\rangle_{\beta\beta}$$

The subscripts are used to indicate the average S_y-magnetization components of the populations that begin in the indicated eigenstates, and the symbol $\bar{S}_y$ represents the average over the population of spins with different wavefunctions, as distinct from the quantum mechanical average for a pure population, $\langle S_y\rangle$. From the vector diagrams, the molecules that begin in the $|\alpha\alpha\rangle$ and $|\beta\beta\rangle$ states give rise to a positive S_y-component at the beginning of the data acquisition period, whereas the populations that begin in the $|\alpha\beta\rangle$ and $|\beta\alpha\rangle$ states give rise to a negative S_y-magnetization component. The final signal intensity is then calculated as

$$\bar{S}_y = f_{\alpha\alpha}\left(\frac{1}{2}\right) + f_{\alpha\beta}\left(\frac{-1}{2}\right) + f_{\beta\alpha}\left(\frac{-1}{2}\right) + f_{\beta\beta}\left(\frac{1}{2}\right)$$

$$= \frac{1}{8}\left[(1 + \Delta P_I + \Delta P_S) - (1 + \Delta P_I - \Delta P_S)\right.$$

$$\left. - (1 - \Delta P_I + \Delta P_S) + (1 - \Delta P_I - \Delta P_S)\right]$$

$$= 0$$

Thus, the final signal is predicted to cancel out. The residual signals observed in the real experiment are due to small timing errors that prevent exact cancellation of the signals. This experiment provides an example of how the signal from one spin can be manipulated by pulses to its scalar-coupled partner. The other important feature of the experiment is that the final outcome depends on the relative populations of the starting states, so that analysis of the experiment requires carefully accounting for the behavior of the populations that begin in all four eigenstates, not just the excess $|\alpha\alpha\rangle$ spin pairs.

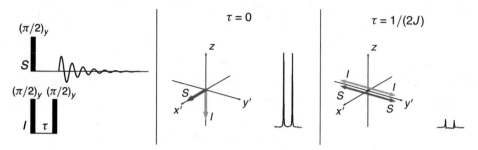

Figure 14.25 Further modification of the experiments illustrated in Figs. 14.22 and 14.23, with the original $(\pi/2)_{y,S}$-pulse at the beginning replaced with a non-selective $(\pi/2)_{y}$-pulse. The vector diagrams represent the magnetization components from the populations of spin pairs initially present in the $|\alpha\alpha\rangle$ state at equilibrium. Adapted with permission from Goldenberg, D.P. (2010) *Concepts. Magn. Reson.* **36A**, 49-83. Copyright © 2010 by John Wiley and Sons.

14.7.3   Experiment 3

Finally, we consider another variation, with an even more surprising result. In this case, the initial $\pi/2$-pulse is applied to both spins, as illustrated in Fig. 14.25. After the delay period, a second $\pi/2$-pulse is applied to the I-spin, and the FID is again recorded from the S-spin. If the delay time is zero, the second $(\pi/2)_{y,I}$-pulse simply rotates the I-magnetization to the negative or positive z-axis, depending on the initial state. Each of the four populations generates a signal with one of the two frequencies, and the expected doublet spectrum is observed. However, if a delay of $1/(2J)$ is introduced between the two pulses, the magnetization from both spins is expected to split during this period, resulting in no net magnetization, as illustrated by the vector diagrams in the third panel. Because the I-magnetization vectors are now aligned with the y'-axis, it would seem that a $(\pi/2)_{y,I}$-pulse would have no effect, and the magnetization components would reappear during the data acquisition period to generate an anti-phase signal, as in the first experiment. But, the pulse is found to eliminate all of the signal from the S-spin, as observed in Experiment 2. How can this be?

In this case, the key to understanding the result lies in the interconversion between the observable S_x-magnetization and the I_zS_y correlation during the delay period. For the molecules that begin in the $|\alpha\alpha\rangle$ state, the first pulse converts the S_z-magnetization to S_x-magnetization. Because the I_z-magnetization is also rotated to the transverse plane by this pulse, the S_x-magnetization is no longer correlated with I_z and evolves with both frequencies. Using the correlation diagrams introduced earlier, the evolution of the S_x-magnetization can be represented as shown in Fig. 14.26. When the delay period is $1/(2J)$, the net magnetization is entirely converted to a correlation. If the FID were recorded at this point, without the final pulse, the correlation would be cyclically converted to S_x-magnetization and back, generating a signal. If, however, a $(\pi/2)_y$-pulse is applied to the I-spin, the I_zS_y correlation is converted to an I_xS_y correlation, as illustrated in Fig. 14.27. Importantly, the I_xS_y correlation does

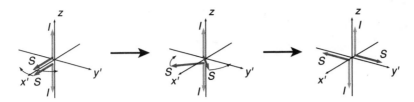

Figure 14.26 Correlation diagrams illustrating the conversion of S_x-magnetization to an $I_z S_y$ correlation during the delay period of the experiment shown in Fig. 14.25, with $\tau = 1/(2J)$.

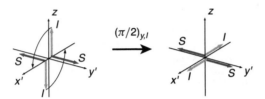

Figure 14.27 Correlation diagrams illustrating the conversion of the $I_z S_y$ correlation to an $I_x S_y$ correlation by the $(\pi/2)_{y,I}$-pulse of the experiment shown in Fig. 14.25, with $\tau = 1/(2J)$.

not interconvert with any net magnetization component. As a consequence, the population that begins in the $|\alpha\alpha\rangle$ state does not make any contribution to a signal from the S-spin. Similar analysis of the other three starting states will show that the initial S_x magnetization components from these states also do not contribute to a net S-magnetization after the final pulse.

There is a bit more to this, however, since there are other components that evolve between the first and second pulse. From the $|\alpha\alpha\rangle$ state, the first pulse also generates I_x-magnetization and an $I_x S_x$ correlation. After this pulse, the I_x-magnetization evolves into an $I_y S_z$ correlation, as shown in Fig. 14.28. Because the I-components are aligned with the y'-axis, the final $(\pi/2)_{y,I}$-pulse has no effect on this correlation, and it will continue to interconvert with observable I_x-magnetization. If the I-signal were to be recorded, this component would give rise to an FID.

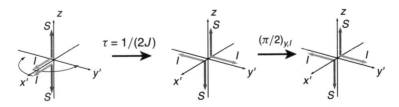

Figure 14.28 Conversion of I_x-magnetization into an $I_y S_z$ correlation in the experiment shown in Fig. 14.25, with $\tau = 1/(2J)$.

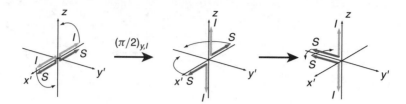

Figure 14.29   Conversion of $I_x S_x$ correlation into an $I_z S_x$ correlation by the second $(\pi/2)_{y,I}$-pulse in the experiment shown in Fig. 14.25, with $\tau = 1/(2J)$. The $I_z S_x$ correlation then interconverts with S_y-magnetization during the data-acquisition period.

Finally, we must consider the $I_x S_x$ correlation that was generated in the first pulse. Because both ν_I and ν_S are zero in the rotating frame, the $I_x S_x$ correlation does not change during the delay period. However, the $(\pi/2)_{y,I}$-pulse will convert this into an $I_z S_x$ correlation, which will interconvert with S_y-magnetization during the data acquisition period, as illustrated in Fig. 14.29. The vector diagrams in Fig. 14.29 would suggest that there should be an observable S-signal in the experiment. However, a full analysis of the contributions from all of the starting states will show that the S-magnetization components derived from the four eigenstates will cancel out, much as we saw in the second experiment. Thus, there is no signal predicted from this experiment.

The important lesson from this example is that the observable magnetization components, as detected in the spectrum, can be influenced by pulses that affect only the correlations. In this sense, the correlations are very "real" and can be exploited in NMR experiments, as we will see in the next chapters.

14.8   Final Comments

In this chapter, we have covered a great deal of material, some of which is likely to be quite counterintuitive. For some simple examples, we found that the NMR signal from a coupled spin pair is relatively easy to account for using traditional vector diagrams. In other cases, however, the behavior is more subtle. In particular, when the magnetization of both spins lies partially or completely in the transverse plane, each gives rise to two precession frequencies, and the evolution of the observable magnetization is associated with the evolution of quantum correlations between the spins.

Keeping track of the effects of multiple pulses and evolution periods using the wavefunctions and magnetization operators can quickly become very difficult, especially because a complete description requires treating all four of the starting eigenstates. The special vector diagrams introduced here to represent the correlations can help in visualizing the steps in an experiment, but it is important to remember that all of the magnetization components and correlations must be followed through each of the steps.

More powerful mathematical methods have been developed for dealing with these problems, and these are presented in Chapters 17 and 18. These treatments allow the behavior of the entire population to be treated with one set of calculations. The disadvantage of these methods, however, is that they tend to blur the distinction between those effects that are due to purely quantum mechanical phenomena, such as the correlations emphasized here, and those that arise from population phenomena (i.e., statistical mechanics). For this reason, the descriptions of several NMR experiments in Chapters 15 and 16 are based on the approach outlined here, in which the wavefunctions and magnetization operators are used to analyze the various steps in the experiments, and vector diagrams are used to illustrate the magnetization components and correlations as they change. Understanding these examples should help make the more abstract treatments more accessible.

Summary Points for Chapter 14

- In most cases, the behavior of scalar-coupled spins in liquid samples can be treated under the assumption of weak coupling, such that the eigenstates (stationary states) are the eigenfunctions of the z-magnetization operators: $|\alpha\alpha\rangle$, $|\alpha\beta\rangle$, $|\beta\alpha\rangle$ and $|\beta\beta\rangle$, for the case of two spins.
- Pulses applied to either or both spins in a coupled pair can be described using vector diagrams, with a separate vector for each spin.
- The evolution of the average transverse magnetization components of one spin (I) in a homogeneous population of coupled spin pairs depends on the z-magnetization of the other spin (S):
  - If the S-magnetization is aligned with the positive or negative z-axis, the transverse I-magnetization will evolve with a single frequency, either $v_I + J/2$ or $v_I - J/2$.
  - If the S-magnetization lies entirely in the transverse plane, the I-magnetization will evolve with two frequencies, $v_I + J/2$ and $v_I - J/2$, with equal amplitudes.
  - If the S-magnetization lies between the z-axis and the transverse plane, the I-magnetization will evolve with the same two frequencies as above, but with relative amplitudes that depend on the magnitude of the S_z-magnetization.
- Pulses applied to the spins of a coupled pair can change the correlations between magnetization components, as well as the magnetization components themselves, and these effects can be represented using special vector diagrams for the correlations.
- Correlations also evolve with time in a static magnetic field, depending on the nature of the correlation:
  - The correlation between longitudinal components, $I_z S_z$, does not change with time.

- The transverse correlations, $I_x S_x$, $I_x S_y$, $I_y S_x$ and $I_y S_y$, interconvert among themselves, but do not interconvert with other correlations or observable magnetization components.
- The mixed longitudinal-transverse correlations, $I_z S_x$, $I_z S_y$, $I_x S_z$ and $I_y S_z$, interconvert with each other *and* with observable transverse magnetization components.

- Predicting the results of NMR experiments with scalar-coupled spins requires tracking both the observable magnetization components and the correlations, starting with each of the equilibrium eigenstates.

Exercises for Chapter 14

14.1. Using equations found in Appendix F, calculate the wavefunctions that would be generated by applying the following pulses to spin pairs initially in the $|\beta\beta\rangle$ state.

   (a) A selective $\pi/4$-pulse along the x'-axis to the I-spin.

   (b) A selective $\pi/4$-pulse along the x'-axis to the I-spin and, immediately afterwards, a $(\pi/2)_x$-pulse to the S-spin.

   (c) A non-selective $\pi/4$-pulse along the x'-axis to both spins.

14.2. For each of the wavefunctions calculated in Exercise 14.1, calculate the average I_x-, I_y-, I_z-, S_x-, S_y- and S_z-magnetization components and draw vector diagrams representing the magnetization components. Do the pulses have the expected effects on the magnetization?

14.3. For each of the wavefunctions calculated in Exercise 14.1, describe qualitatively how the magnetization components would evolve with time after the pulses. Assuming that you could somehow record signals arising from pure populations of these wavefunctions, describe the resulting spectra.

14.4. Consider the gedanken experiment illustrated in Fig. 14.11. The analysis showed that the pulses applied to spin pairs starting in the $|\alpha\alpha\rangle$ state would give rise to two signals from the I-spin, with different amplitudes, and two signals from the S-spin with equal amplitudes. Show what the results would be if this experiment were applied to an equilibrium population of a coupled spin pair. This will require an analysis similar to the one for the experiment discussed in Section 14.7.2.

14.5. For each of the three experiments described in Section 14.7, predict the outcome expected if the delay time, τ, were set to $1/J$, where J is the scalar coupling constant. You will need to analyze the behavior of each of the populations that begin in one of the four eigenstates.

14.6. The terms introduced in Eqs. 14.6, 14.7 and 14.8 (the population differences, ΔP_I and ΔP_S, and the fractional populations, $f_{\alpha\alpha}$, $f_{\alpha\beta}$, $f_{\beta\alpha}$ and $f_{\beta\beta}$) will come in handy later, especially in Chapter 17, where the density matrix is introduced, so it is worthwhile to examine them a bit further.

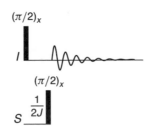

Figure 14.30
Pulse sequence for Exer-
cise 14.7.

(a) Using the definitions of ΔP_I and ΔP_S in Eqs. 14.6 and 14.7, confirm the expressions in Eq. 14.8.
(b) Using the Boltzmann distribution and appropriate approximations, derive expressions for ΔP_I and ΔP_S in terms of the gyromagnetic ratios of the I- and S-spins and the external field strength, B_0.

14.7. Consider the two-pulse experiment diagrammed in Fig. 14.30. For the spin-pair population that begins in the $|\alpha\alpha\rangle$ state, draw vector diagrams that represent the magnetization components and correlations that are present at each of the following points in the experiment:

(a) After the initial $(\pi/2)_{x,I}$-pulse.
(b) After the delay period.
(c) After the second $(\pi/2)_{x,I}$-pulse.
(d) A time during the data acquisition period, chosen to illustrate the evolution of the various components.

In drawing these diagrams, assume that the uncoupled Larmor frequency of the I-spin matches the rotating-frame frequency for that nucleus, but that the frequency of the S-spin is positive in the rotating frame.

Further Reading

Most quantum mechanical treatments of the manifestations of scalar coupling in NMR spectroscopy are based on the product-operator formalism, which is discussed in Chapter 18. Suggestions for further reading on that approach can be found on page 554.

CHAPTER 15

Two-Dimensional Spectra Based on Scalar Coupling: COSY and TOCSY Experiments

The very first Fourier transform two-dimensional NMR experiment was proposed by Jean Jeener in 1971 at a workshop in Yugoslavia.[1] The experiment was not actually executed until a few years later, however, when Richard Ernst and his colleagues worked out the practical details and demonstrated the feasibility of multidimensional spectroscopy. This first, and simplest possible, experiment was composed of just two $\pi/2$-pulses applied to a system composed of scalar-coupled spins of the same type, as diagrammed in Fig. 15.1. As in the NOESY experiment described in Chapter 10, the two-dimensional spectrum is generated by collecting a series of one-dimensional spectra, with increasing time delays between the two pulses. The individual free induction decay signals are Fourier transformed, and the resulting spectra are combined into a two-dimensional data set, which is then Fourier transformed in the second dimension.

We considered this pulse sequence earlier when we first introduced the concept of a two-dimensional spectrum. In Chapter 10, though, it was assumed that there was no mechanism by which magnetization could be transferred between the two spins of interest, and we saw that the resulting spectrum only contained peaks on the diagonal. Now, we will use the quantum mechanical principles outlined in the previous chapters to see how scalar coupling between the spins can give rise to cross peaks. This type of spectrum is usually referred to as a COSY spectrum, for COrrelation SpectroscopY. Although the most commonly used form of the COSY experiment is a ^{1}H homonuclear experiment, we will begin with a simplified heteronuclear version so as to make the analysis somewhat clearer. In this chapter we will also briefly consider another class of experiments by which scalar-coupled spins give rise to cross peaks, exemplified by the TOCSY (TOtal Correlation SpectroscopY) experiment.

---

[1] See page 456 for a reference to this lecture.

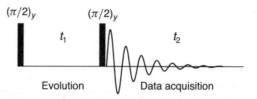

Figure 15.1

Pulse diagram for the homonuclear COSY experiment.

15.1   A Simplified Heteronuclear COSY Experiment

As before, we will identify the two spins in our system as I and S, which could correspond, for instance, to a scalar-coupled ^1H–^{13}C pair. In a simple one-dimensional
experiment, each spin will give rise to two signals with frequencies $\nu_{I,1}$, $\nu_{I,2}$, $\nu_{S,1}$ and
$\nu_{S,2}$, with the numbers 1 and 2 in the subscript indicating the higher- and lower-
frequency components, respectively, of each doublet. The frequency difference between the two components of the doublet is the scalar coupling constant, J, which is
usually expressed in Hz and is independent of external field strength.

Because the resonance frequencies for the two spins are very different, pulses can
be applied to them individually and their signals can be recorded independently. Our
model experiment, diagrammed in Fig. 15.2, consists of a $\pi/2$-pulse applied only to
the S-spin, a t_1 evolution period, a second $\pi/2$-pulse, now applied to both spins, and
finally a data acquisition period. Only the signal from the I-spin will be recorded.

15.1.1   Starting with the $|\alpha\alpha\rangle$ State

Using the results of the analysis presented in Chapter 14, we will follow the magnetiza-
tion arising from each of the four eigenstates individually, starting with $|\alpha\alpha\rangle$. The first
one-dimensional experiment is carried out with $t_1 = 0$, generating the results shown
in Fig. 15.3. Because there is no time allowed for the S-magnetization to evolve, the
second pulse places it along the negative z-axis, while the I-magnetization is rotated
to the positive x'-axis. Because the S-spin is now in the β state, the I-magnetization
precesses with the lower of its two frequencies, $\nu_{I,2}$. This gives rise to an FID with

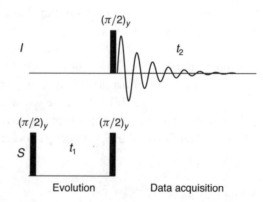

Figure 15.2

Pulse diagram for a heteronuclear COSY experiment.

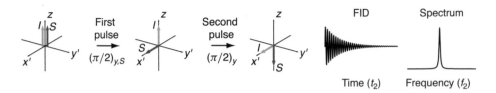

Figure 15.3 Vector diagrams representing the effects of the heteronuclear COSY experiment diagrammed in Fig. 15.2 applied to the $|\alpha\alpha\rangle$ state, with $t_1 = 0$, and the resulting FID and spectrum.

a single frequency component, which, when Fourier transformed, gives a spectrum with a single peak.

Successive one-dimensional spectra are recorded using progressively longer delays between the two pulses, allowing the S-magnetization to precess with frequency $\nu_{S,1}$, as illustrated in Fig. 15.4. In the second sequence shown in Fig 15.4, the S-magnetization precesses to a position halfway between the x'- and y'-axes during the t_1 delay period. When the second pulse is applied, the y'-component of the S-magnetization is unchanged, while the component still projecting along the x'-axis is rotated so that it now projects along the negative z-axis. Because the S-magnetization now has a transverse component, we expect the I-magnetization to evolve with two frequency components, $\nu_{I,1}$ and $\nu_{I,2}$, giving rise to two peaks in the spectrum. But, because the negative z-component of the S-magnetization is greater than the positive z-component, the signal corresponding to the lower frequency is still stronger than the higher-frequency component of the doublet. (Although the S-magnetization now

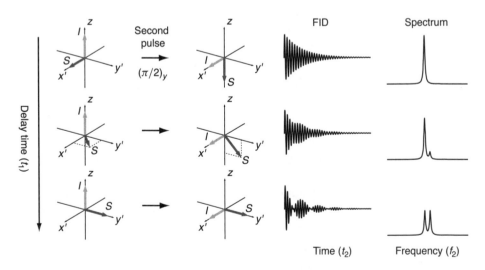

Figure 15.4 Vector diagrams representing the effects of the heteronuclear COSY experiment diagrammed in Fig. 15.2 applied to the $|\alpha\alpha\rangle$ state, for incremented t_1 intervals, and the resulting FIDs and spectra.

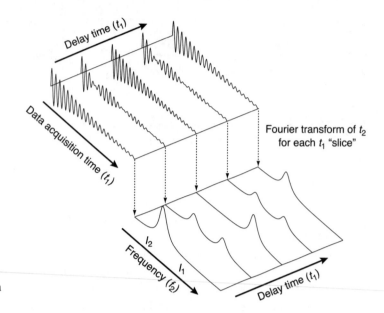

Figure 15.5

FIDs and Fourier transform
to generate individual one-
dimensional spectra in
the heteronuclear COSY
experiment, for the population
starting in the $|\alpha\alpha\rangle$ state.

also has a transverse component, it does not affect the spectrum, since we are only recording the signal from the I-spins.)

After a longer delay between the pulses, as in the third sequence in Fig. 15.4, the S-magnetization precesses to the y'-axis and, in this case, is unaffected by the second pulse. With the positive and negative z-components of the S-magnetization exactly balanced, the two frequency components of the I-magnetization's precession have equal amplitudes, generating equal peaks in the spectrum.

As the t_1 delay is made progressively longer, the S-magnetization precesses further, resulting in a continuous change in the S_z-magnetization component following the second pulse and a continuous change in the relative amplitudes of the two frequency components of the I-magnetization's precession during the data acquisition period. When the S-magnetization has precessed exactly one-half cycle during the t_1 delay, only the signal with the higher frequency, $\nu_{I,1}$, is generated. When the S-magnetization returns all the way back to the positive x'-axis, the I-signal will again have only the lower frequency, $\nu_{I,2}$.

Fig. 15.5 shows some of the resulting FID signals and corresponding spectra generated during a single cycle of S-magnetization precession. Notice that the first, third and fifth FIDs look nearly identical, because they represent single frequencies, but the frequency differences among them become clearly apparent after the Fourier transform. The amplitudes of the two I-frequency components vary sinusoidally as the t_1 delay is increased, but the cycles are out of phase by π rad. Both amplitudes remain positive and their sum remains constant.

As in the examples of two-dimensional spectra illustrated in Chapter 10, the data generated by the Fourier transforms of the FIDs are then transformed in the other direction, corresponding to the increasing t_1 delay times, as illustrated in Fig. 15.6. When plotted as shown in Fig. 15.6, the fluctuating curves along the t_1 dimension in the half-transformed data set look somewhat different from those we looked at for the

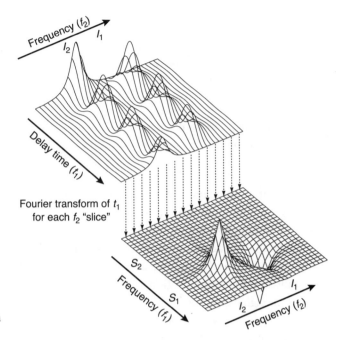

Figure 15.6

Fourier transformation of the t_1 time-domain dimension in the heteronuclear COSY experiment, for the population starting in the $|\alpha\alpha\rangle$ state.

two-pulse experiment in the absence of scalar coupling (see Fig. 10.8). In the previous examples, the damped sine waves fluctuated between positive and negative values, with an average value of zero. Now, the function is always positive, representing an offset in the average value. Along the t_1 direction in each of the f_2 slices, this offset is constant and does not represent a frequency component. As a consequence, the offset does not affect the spectrum generated after Fourier transforming the second dimension.

The second transform generates two peaks, corresponding to frequencies $\nu_{I,1}$ and $\nu_{I,2}$ in the f_2 dimension. In the f_1 dimension, both peaks are located at frequency $\nu_{S,1}$, reflecting the frequency of precession of the S-magnetization during the t_1 delay. The two peaks have opposite signs, however, reflecting the fact that the rise and fall of the amplitudes for the two I-frequencies are out of phase by π rad.

Although this might not appear to be an especially useful spectrum, since we still just have the two peaks for the I-spin that would be generated in a one-dimensional spectrum, the important point is that the position of these peaks in the two-dimensional plane is associated with one of the frequencies for the S-spin. For a molecule with multiple I–S pairs (e.g., covalently bonded ^1H–^{13}C pairs), each pair would give rise to peaks with characteristic I- and S-frequencies. Not only does this spread the peaks out into a second dimension, but it provides direct evidence of scalar coupling (i.e., covalent connectivity) between the nuclei.

15.1.2   Starting with the $|\beta\alpha\rangle$ State

At this point, we have considered only one of the four eigenstates. A complete description of the spectrum for a real sample, in which all four eigenstates are initially

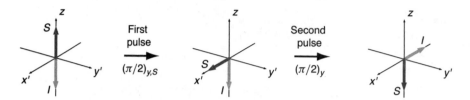

Figure 15.7 Vector diagrams representing the effects of the heteronuclear COSY pulse sequence on the $|\beta\alpha\rangle$ state with $t_1 = 0$.

present at populations determined by the Boltzmann distribution, requires that we carry out a similar analysis for each starting state.

We look next at the population that begins in the $|\beta\alpha\rangle$ state. With the spins in this state, the S-magnetization points along the positive z-axis, and the I-magnetization is aligned with the negative z-axis. Fig. 15.7 shows the effects of the first and second pulses on the magnetization vectors, with no delay between the pulses. The frequency of the signal generated by the I-spin is the same as for when we started with the $|\alpha\alpha\rangle$ population, $\nu_{I,2}$, since for both the S-magnetization is aligned with the negative z-axis during the data acquisition period. Now, however, the phase of the I-signal is π rad out of phase with that generated from the $|\alpha\alpha\rangle$ state.

As the t_1 delay time is increased, the S-magnetization precesses as before. For the population that began in the $|\beta\alpha\rangle$ state, however, the rate of this precession is lower than for the molecules that began in the $|\alpha\alpha\rangle$ state, because the I-magnetization is now aligned with the *negative* z-axis. The resulting FIDs and spectra for some of the initial iterations are shown in Fig. 15.8. Because the phase of the FID differs by π rad from that shown before, the sign of the spectral peaks is reversed. With zero delay, only the lower I-frequency is observed, but, as transverse S-magnetization

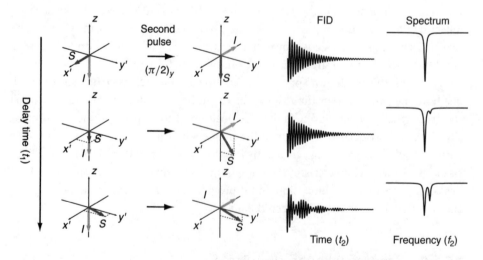

Figure 15.8 Vector diagrams representing the effects of the heteronuclear COSY experiment applied to the $|\beta\alpha\rangle$ state, for incremented t_1 intervals, and the resulting FIDs and spectra.

A. Starting state $|\alpha\alpha\rangle$

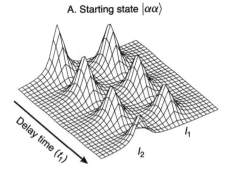

B. Starting state $|\beta\alpha\rangle$

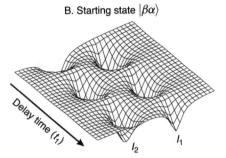

Figure 15.9

Comparison of the predicted half-transformed signals from the heteronuclear COSY experiment, from the populations beginning in the $|\alpha\alpha\rangle$ (panel A) and $|\beta\alpha\rangle$ (panel B) states, In panel C, the signal arising from the original $|\beta\alpha\rangle$ population has been inverted by multiplying by -1.

C. Starting state $|\beta\alpha\rangle$
(signal multiplied by -1)

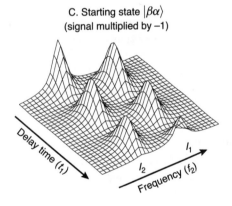

is generated by the delay and second pulse, the higher I-frequency also becomes detectable. As seen before, the amplitudes of the two peaks vary sinusoidally, and their sum remains constant.

In the actual experiment, the signals from this population will combine with those from the other three to generate the FID. For the purposes of predicting the expected spectrum, however, the expected contributions from the different populations can be Fourier transformed individually and then combined. The two-dimensional data set generated after transformation of the directly detected dimension is shown as a surface plot in panel B of Fig. 15.9, where it is compared with that generated from the molecules that began in the $|\alpha\alpha\rangle$ state (panel A).

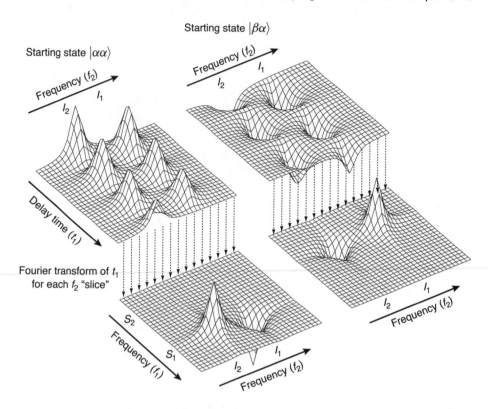

Figure 15.10   Fourier transformation of the t_1 time-domain dimension in the heteronuclear COSY experiment, for the population starting in the $|\alpha\alpha\rangle$ and $|\beta\alpha\rangle$ states.

As represented in panel B, it may be difficult to see the relationship between the spectra generated by the two starting states. If, however, the data generated from the $|\beta\alpha\rangle$ state are multiplied by -1, as shown in panel C of Fig. 15.9, it is apparent that they are quite similar. Indeed, the only difference (after inverting the $|\beta\alpha\rangle$ data) is that the sinusoidal fluctuation of the amplitudes along the t_1 dimension is slower when the starting state is $|\beta\alpha\rangle$. For the limited range of t_1 values plotted in the drawing above, the amplitudes from the $|\alpha\alpha\rangle$ state have undergone exactly three cycles, while those from the $|\beta\alpha\rangle$ state have undergone about two and a half cycles.

When the data arising from the $|\beta\alpha\rangle$ state are subjected to Fourier transformation along the t_1 dimension, the resulting two-dimensional spectrum differs from that generated by the $|\alpha\alpha\rangle$ state in two respects: The sign of the spectral peaks are reversed, reflecting the inversion of the peaks after the first transformation, and the position of the peaks along the f_1 dimension corresponds to $\nu_{S,2}$ rather than $\nu_{S,1}$. The two spectra are compared in Fig. 15.10.

15.1.3   Starting with the $|\alpha\beta\rangle$ or $|\beta\beta\rangle$ State

Next, we consider the signal arising from the population that begins in the $|\alpha\beta\rangle$ state, where the S-magnetization is aligned with the negative z-axis, and the I-magnetization is aligned with the positive z-axis. Fig. 15.11 diagrams the effects

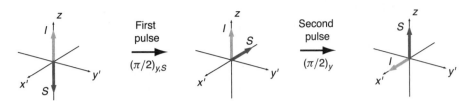

Figure 15.11   Vector diagrams representing the effects of the heteronuclear COSY pulse sequence on the $|\alpha\beta\rangle$ state with $t_1 = 0$.

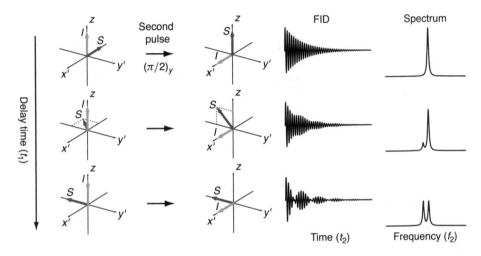

Figure 15.12   Vector diagrams representing the effects of the heteronuclear COSY experiment applied to the $|\alpha\beta\rangle$ state, for incremented t_1 intervals, and the resulting FIDs and spectra.

of the two sequential pulses, with no delay between them. During the data acquisition period, the I-spin generates a signal with a single frequency. Because the S-magnetization is now aligned with the *positive* z-axis, the frequency is the higher of the two, $\nu_{I,1}$. For the population that began in the $|\alpha\alpha\rangle$ state, the first spectrum contained only a peak corresponding to the lower frequency, $\nu_{I,2}$.

With a delay, t_1, introduced between the two pulses, the S-magnetization is allowed to precess. The frequency of this precession is the same as for the population that began in the $|\alpha\alpha\rangle$ state, since the I-magnetization is again aligned with the positive z-axis. As before, the second pulse converts the x'-component of the S-magnetization into a z-component, thereby changing the relative amplitudes of the two frequency components in the FID signal generated by the I-spin. A few examples are shown in Fig. 15.12. The only difference between these spectra and those generated from the $|\alpha\alpha\rangle$ state is that now the higher-frequency peak appears first and then is replaced by the lower-frequency peak.

After the second Fourier transformation, there will again be two peaks. Both will be located at frequency $\nu_{S,1}$ in the f_1 dimension, and at either frequency $\nu_{I,1}$ or $\nu_{I,2}$ in the f_2 dimension, just as seen for the spectrum generated by the $|\alpha\alpha\rangle$ population.

Table 15.1 Frequencies and signs associated with the peaks generated from the four eigenstates in the heteronuclear COSY experiment

Initial state	Frequency coordinates			
	$\nu_{I,1}, \nu_{S,1}$	$\nu_{I,2}, \nu_{S,1}$	$\nu_{I,1}, \nu_{S,2}$	$\nu_{I,2}, \nu_{S,2}$
$|\alpha\alpha\rangle$	−	+		
$|\alpha\beta\rangle$	+	−		
$|\beta\alpha\rangle$			+	−
$|\beta\beta\rangle$			−	+

The corresponding peaks in the spectra from the two populations will have opposite signs, however.

Finally, we consider the molecules that begin in the $|\beta\beta\rangle$ state. Using arguments like those outlined above for the other three states, you should be able to predict the general features of the signals generated by this population, before and after each of the Fourier transformations. This analysis will reveal that the spectrum generated from the $|\beta\beta\rangle$ state will contain two peaks, both at positions on the f_1 dimension corresponding to $\nu_{S,2}$, like the spectrum from the $|\beta\alpha\rangle$ state. However, the signs of the peaks in the $|\beta\beta\rangle$ spectrum will be opposite those from the $|\beta\alpha\rangle$ state.

The contributions to the spectrum from the populations of molecules that begin in the four eigenstates are summarized in Table 15.1, where the entries represent the signs of the spectral peaks at the indicated frequency coordinates. If the molecules in the four eigenstates were initially present in equal concentrations, their contributions would cancel exactly, leaving a blank spectrum. At equilibrium, however, the $|\alpha\alpha\rangle$ state will be present at higher concentration than the $|\alpha\beta\rangle$ state. This leads to a net positive peak at the position corresponding to $\nu_{I,2}, \nu_{S,1}$ and a net negative peak at position $\nu_{I,1}, \nu_{S,1}$. Similarly, the concentration of the $|\beta\alpha\rangle$ state will be higher than that of the $|\beta\beta\rangle$ state, leading to a positive peak at $\nu_{I,1}, \nu_{S,2}$ and a negative peak at $\nu_{I,2}, \nu_{S,2}$. Both of these initial population differences reflect the difference in energies between the α and β states of the S-spin.[2] Thus, the intensities of the four peaks will be identical, as illustrated in Fig. 15.13.

The group of four peaks is described as a *doubly anti-phase* peak and is an important characteristic of COSY-type spectra. For a molecule containing multiple heteronuclear pairs, each would give rise to an anti-phase group of four peaks. Notice

---

[2] If, as suggested earlier, the I-spin is an ^{1}H nucleus and S is ^{13}C, the fact that the important initial population difference reflects the S-spin energy difference identifies an inherent weakness of this experiment, because the I-spin is associated with a larger population difference. The roles of the I- and S-spins in this experiment can easily be reversed. But, if this is done, the smaller gyromagnetic ratio will be responsible for actually generating the FID, more than compensating for the greater population difference due to the I-spin. What we would really like is an experiment in which both the exploited population difference and the FID derive from the spin with the larger gyromagnetic ratio. Two such heteronuclear experiments, the HSQC and HMQC experiments, are described in Chapter 16.

Figure 15.13

Resultant spectrum from the heteronuclear COSY experiment after addition of the contributions from the four eigenstates, with the contributions weighted according to the equilibrium Boltzmann distribution.

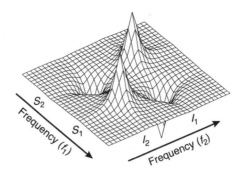

that the total integral of the peaks (i.e., the volume under the peaks) is zero. This reflects the fact that there is no true transfer of net magnetization from the S-spin to the I-spin, as there was in the NOESY experiment. Instead, the S-spin influences the signal from the I-spin only by changing the relative contributions of the two frequencies in the doublet.

Another difference between the spectrum generated by this heteronuclear COSY experiment and that from the homonuclear NOESY is the absence of diagonal peaks. This is because the effects of the precession frequencies of the two nuclei are separated from one another. During the t_1 evolution period, only the S-spin undergoes precession, and during the t_2 data acquisition period, only the I-spin contributes to the spectrometer signal. Although the S-spin precesses during the acquisition period, the spectrometer is tuned so as not to detect the resulting magnetic oscillation. It is, in principle, possible to carry out an experiment with heteronuclei in which both spin types are stimulated by each pulse and the free induction decays of both are detected. Although this would generate a spectrum with a diagonal, the two dimensions would have to cover an impractically wide range of resonance frequencies. In Section 15.2, we will consider the much more common homonuclear COSY experiment, in which both diagonal and cross peaks are generated.

15.2   The Homonuclear COSY Experiment

As described earlier and illustrated in Fig. 15.1, the pulse sequence for the homonuclear COSY experiment consists of two non-selective $\pi/2$-pulses separated by the variable t_1 period. The vector diagrams of Fig. 15.14 illustrate the effects of the first pulse and the evolution period, starting with the $|\alpha\alpha\rangle$ population. After the first pulse, the evolution of the magnetization components for the two spins is characterized by a total of four frequencies; $\nu_{I,1}$, $\nu_{I,2}$, $\nu_{S,1}$ and $\nu_{S,2}$, as represented by the split vectors. After the delay period, t_1, the second pulse is applied along the y'-axis. This pulse leaves the y'-magnetization unchanged, but rotates the x'-magnetization of each spin into z-magnetization, as illustrated in the vector diagrams in Fig. 15.15.

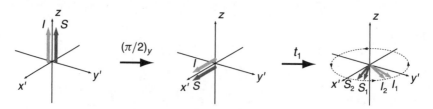

Figure 15.14   Vector diagrams illustrating the first pulse and evolution period for the homonuclear COSY experiment, applied to the $|\alpha\alpha\rangle$ population.

Figure 15.15
Vector diagrams illustrating the effects of the second pulse in the homonuclear COSY experiment, applied to the population beginning in the $|\alpha\alpha\rangle$ state.

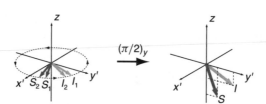

Immediately after the second pulse, single vectors are used to represent the I- and S-magnetization components. Although the current position of each vector was determined by two frequency components during the evolution period, that period is now history. A single vector for each spin represents the current state of the system, with components along the y'- and z-axes. The new values of the average magnetization components for the I-spin will be

$$\langle I_x \rangle = 0$$

$$\langle I_y \rangle = (1/4)\big(\sin(2\pi \nu_{I,1} t_1) + \sin(2\pi \nu_{I,2} t_1) \big) \tag{15.1}$$

$$\langle I_z \rangle = -(1/4)\big(\cos(2\pi \nu_{I,1} t_1) + \cos(2\pi \nu_{I,2} t_1) \big)$$

Similar expressions describe the S-magnetization components.

At this point, we might guess how the magnetization will precess further: Each of the spins, it would seem, should give rise to two frequency components, with the relative amplitude of each component depending on the z-magnetization of the other spin. In fact, however, the situation is more complicated than suggested by the vectors in Fig. 15.15. This is because additional magnetization components arise from spin–spin correlations that evolved during the delay period.

Using the wavefunctions and operators described in Chapters 13 and 14, expressions for the I- and S-magnetization components, as functions of both t_1 and t_2, can be derived. For the I-magnetization along the y'-axis, the expression is

$$\langle I_y(t_1, t_2) \rangle$$

$$= (1/8)\Big(\cos(2\pi v_{I,1}t_2)\big(\sin(2\pi v_{I,1}t_1) + \sin(2\pi v_{I,2}t_1)\big) \tag{15.2a}$$

$$+ \cos(2\pi v_{I,2}t_2)\big(\sin(2\pi v_{I,1}t_1) + \sin(2\pi v_{I,2}t_1)\big) \tag{15.2b}$$

$$+ \sin(2\pi v_{I,1}t_2)\big(\cos(2\pi v_{S,2}t_1) - \cos(2\pi v_{S,1}t_1)\big) \tag{15.2c}$$

$$- \sin(2\pi v_{I,2}t_2)\big(\cos(2\pi v_{S,2}t_1) - \cos(2\pi v_{S,1}t_1)\big) \tag{15.2d}$$

$$- \cos(2\pi v_{I,1}t_2)\big(\sin\big(2\pi (v_I - v_S)t_1\big) + \sin\big(2\pi(v_I + v_S)t_1\big)\big) \tag{15.2e}$$

$$+ \cos(2\pi v_{I,2}t_2)\big(\sin\big(2\pi (v_I - v_S)t_1\big) + \sin\big(2\pi(v_I + v_S)t_1\big)\big)\Big) \tag{15.2f}$$

Things have now become very complicated, indeed! The expression above has been written so that each line contains a single term representing precession of the I-magnetization during the t_2 period, on the left, multiplied by a term that determines the amplitude of the t_2 term according to what happened during the t_1 period.

Lines (a) and (b) in Eq. 15.2 are relatively easy to interpret. They represent precession of the I-magnetization immediately following the second pulse, with the two frequencies, $v_{I,1}$ and $v_{I,2}$. The amplitudes of these terms are the same and represent the projection of the I-magnetization along the y'-axis immediately after the second pulse, which, in turn, depends on the precession during the t_1 period. In the discussion that follows, these magnetization components will be referred to as the "in-phase" components. Note that they do not reflect in any way the S-frequency terms.

Lines (c) and (d) also reflect precession of I-magnetization during the t_2 period, but the amplitudes of these terms are modulated by the evolution of S-magnetization during the t_1 period, with frequencies $v_{S,1}$ and $v_{S,2}$. The evolution of these components during t_2 is described by the sine, rather than cosine, function. This means that the magnetization components represented by these terms evolve in a way that is $\pi/2$ rad out of phase with the contributions represented by the first two lines. These will be referred to as "anti-phase" terms.

15.2.1   Origin of the Anti-Phase COSY Components

The key to understanding the COSY experiment lies in how evolution of S-components during the t_1 delay gives rise to I-magnetization components during the t_2 data acquisition period, as expressed in lines (c) and (d) of Eq. 15.2. As discussed in Chapter 14, the evolution of transverse magnetization components in a coupled spin pair is associated with interconversion of observable magnetization with correlations, which cannot, themselves, be directly detected.

Using the correlation diagrams introduced earlier, the evolution of the S_x-magnetization during the t_1 period can be represented as shown in Fig. 15.16. Remember that diagrams like the one on the left side of Fig. 15.16 are used to represent an observable transverse magnetization and the *potential* correlation that can be generated

Figure 15.16

Vector diagrams illustrating the conversion of S_x-magnetization into a mixture of observable magnetization and correlations.

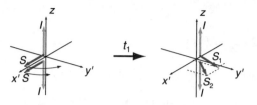

Figure 15.17

Resolution of the final vector diagram from Fig. 15.16 into two components.

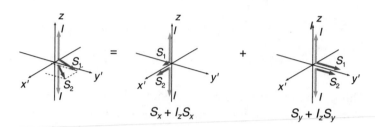

as the vectors representing S-magnetization precess with the two frequencies, $\nu_{S,1}$ and $\nu_{S,2}$.[3] After the first evolution period, the initial S_x-magnetization will have been converted into a mixture of S_x- and S_y-magnetization components *and* the correlations I_zS_x and I_zS_y. In the example shown in Fig. 15.16, after the t_1 period, there is a net projection of the S-vectors on both the x'- and y'-axes. But, there are also correlations: The transverse S-vector associated with positive I_z has a greater projection along the y'-axis than does the vector associated with negative I_z, which has a greater projection along the x'-axis. Thus, there is a net positive I_zS_y correlation and a negative I_zS_x correlation.

The magnetization components and correlations present at the end of the t_1 period can be more easily visualized by resolving the transverse vectors into x'- and y'-components, to generate the two separate diagrams shown in Fig. 15.17. Mathematically, the evolution of the S_x-magnetization and the components it generates during the t_1 period are described as follows:

$$\langle S_x(t_1) \rangle = (1/4)\left(\cos(2\pi \nu_{S,1}t_1) + \cos(2\pi \nu_{S,2}t_1) \right)$$

$$\langle S_y(t_1) \rangle = (1/4)\left(\sin(2\pi \nu_{S,1}t_1) + \sin(2\pi \nu_{S,2}t_1) \right)$$

$$\langle I_zS_x(t_1) \rangle = (1/8)\left(\cos(2\pi \nu_{S,1}t_1) - \cos(2\pi \nu_{S,2}t_1) \right)$$

$$\langle I_zS_y(t_1) \rangle = (1/8)\left(\sin(2\pi \nu_{S,1}t_1) - \sin(2\pi \nu_{S,2}t_1) \right)$$

$$(15.3)$$

Notice that the observable magnetization components represent the sum of the two S-frequency components, while the correlations evolve from the difference between

---

[3] In Chapter 14, most of the examples were simplified by assuming that the average S-frequency was zero (relative to the rotating frame), so that the two vectors precess at equal, put opposite, frequencies away from their starting position, and only the effects of scalar coupling are observed. In general, however, the average frequency will be non-zero, and we must consider the effects of both scalar coupling and differences in chemical shift.

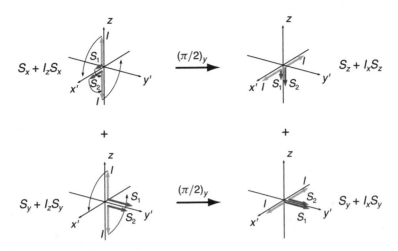

Figure 15.18   Vector diagrams representing the effects of the second pulse in the homonuclear COSY experiment on the magnetization components and correlation components derived from the S_x-magnetization generated from the first pulse. The relative magnitudes and signs of the various components will depend on the precession frequencies and the duration of t_1.

them. This is reflected in the drawings in Fig. 15.17 by the difference in the lengths of, for instance, S_x-vectors associated with positive and negative I_z-vectors.

The next step of the homonuclear COSY experiment is the second $(\pi/2)_y$-pulse, followed by the data acquisition period. Continuing to consider the components that arise from the S_x-magnetization, the effects of the second pulse are diagrammed in Fig. 15.18. After the pulse, there are two net magnetization components, S_z and S_y, and two correlations, I_xS_z and I_xS_y. (Here, we are ignoring the signs, which will depend on the precession frequencies and t_1.) The net S-magnetization components shown in this diagram are the same as those shown in Fig. 15.15. The S_y-component will precess during the data acquisition period, contributing to signals with frequencies $\nu_{S,1}$ and $\nu_{S,2}$. Neither the S_z-magnetization nor the I_xS_y correlation will contribute to any observable magnetization components during the data acquisition period. However, the I_xS_z correlation will evolve into a mixture of I_x- and I_y-magnetization components, as well as an I_yS_z correlation. To visualize this process, we can resolve the vector representation of the S_z-magnetization and I_xS_z correlation into representations of the individual components, as shown in Fig. 15.19. The evolution of the I_xS_z correlation into observable magnetization components is then visualized as shown in Fig. 15.20.

The net result, then, is that the S_x-magnetization that was present immediately after the first pulse has been partially converted into observable I-magnetization that evolves during the data acquisition period and contributes to the signal with I-frequencies. The magnitude of this signal depends on the evolution of that original S_x-magnetization during the t_1 period. This is the origin of the components in lines (c) and (d) of Eqn. 15.2. These components, as we will see, generate the cross peaks in the COSY spectrum.

Figure 15.19

Vector diagrams representing the S_z-magnetization and the $I_x S_z$ correlation components present after the second pulse of the homonuclear COSY experiment.

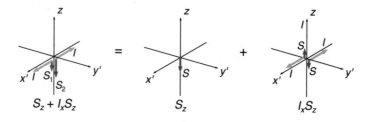

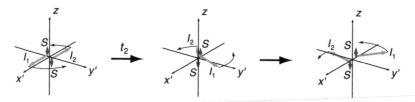

Figure 15.20   Vector diagrams representing the evolution of the $I_x S_z$ correlation into observable I_x- and I_y-magnetization components during the data acquisition period of the homonuclear COSY experiment.

To recap, the steps involved in this process, considering only the components that begin as S_z-magnetization and eventually contribute to the I-signal, are

$$S_z \xrightarrow{(\pi/2)_y} S_x \xrightarrow{t_1} I_z S_x \xrightarrow{(\pi/2)_y} I_x S_z \xrightarrow{t_2} I_x + I_y$$

Some important points to note are as follows:

- The magnitude of the final I-magnetization components that arise from the S-spin depends on the extent to which S_x-magnetization is converted into an $I_z S_x$ correlation during the t_1 delay period, as described in Eq. 15.3.
- At the beginning of the t_2 period, the I-components of the $I_x S_z$ correlation cancel one another. The signal they induce begins at zero, and increases according to the sine functions shown in lines (c) and (d) of Eq. 15.2. This is the origin of the anti-phase character of these signals.

The magnetization components represented by lines (e) and (f) of Eq. 15.2 are, in some regards, even stranger than the anti-phase components discussed above. These lines each represent a contribution to the precession of I-magnetization during t_2 with one of the frequencies, $\nu_{I,1}$ or $\nu_{I,2}$. But, the frequencies associated with the evolution during t_1 are $\nu_I + \nu_S$ and $\nu_I - \nu_S$, which represent the energy differences, $E_{\beta\beta} - E_{\alpha\alpha}$ and $E_{\beta\alpha} - E_{\alpha\beta}$, respectively, as shown in the energy diagram for a weakly coupled spin pair on page 389. As noted earlier, these "multiple-quantum" transitions cannot be stimulated in the usual way by the absorptive of single photons, though they can occur by cross-relaxation if the two nuclei are linked by dipolar coupling. In the COSY experiment, terms with these frequencies arise from the evolution during the t_1 delay of the $I_x S_x$ correlation that is present immediately after the first pulse. The evolution of this type of correlation was illustrated and discussed briefly on page 412.

As it happens, the multiple-quantum terms from the four starting eigenstates cancel out exactly, so they are not detectable in the final spectrum, and we will not divert further attention to them. There are, however, ways to manipulate the spins so as to make these contributions visible in other multidimensional experiments.

15.2.2   Generation of the COSY Spectrum

In order to describe the signal that would arise from a real sample, we must again consider the signals predicted for the four eigenstates and then sum the signals according to the relative populations of those states. If the four states were present at equal concentrations, the signals they generate would cancel exactly, as we saw above for the heteronuclear example. The actual signal depends on the relative populations of the four starting states, as shown below for the case of the I_y-magnetization, using the symbols introduced in Section 14.7.2:

$$\bar{I}_y = f_{\alpha\alpha}\langle I_y\rangle_{\alpha\alpha} + f_{\alpha\beta}\langle I_y\rangle_{\alpha\beta} + f_{\beta\alpha}\langle I_y\rangle_{\beta\alpha} + f_{\beta\beta}\langle I_y\rangle_{\beta\beta}$$

For the homonuclear case, the equilibrium population differences, ΔP_I and ΔP_S, are equal to one another, and the fractional populations are given by

$$f_{\alpha\alpha} = \left(1 + 2\Delta P_I\right)/4$$
$$f_{\alpha\beta} = 1/4$$
$$f_{\beta\alpha} = 1/4$$
$$f_{\beta\beta} = \left(1 - 2\Delta P_I\right)/4$$

When the predicted signals from the I-magnetization are summed according to these population weights, the net signal, $\bar{I}_y$, is proportional to the one predicted for the $|\alpha\alpha\rangle$ state, *except* that the zero- and double-quantum terms (lines (e) and (f) of Eq. 15.2) cancel out. Analysis of the signal from the S-spin results in an analogous expressions, as shown below for the S_y-component:

$$\bar{S}_y(t_1, t_2) = (1/8)\Big(\cos(2\pi v_{S,1}t_2)\big(\sin(2\pi v_{S,1}t_1) + \sin(2\pi v_{S,2}t_1)\big) \quad\quad\text{(15.4a)}$$

$$+ \cos(2\pi v_{S,2}t_2)\big(\sin(2\pi v_{S,1}t_1) + \sin(2\pi v_{S,2}t_1)\big) \quad\quad\text{(15.4b)}$$

$$+ \sin(2\pi v_{S,1}t_2)\big(\cos(2\pi v_{I,2}t_1) - \cos(2\pi v_{I,1}t_1)\big) \quad\quad\text{(15.4c)}$$

$$- \sin(2\pi v_{S,2}t_2)\big(\cos(2\pi v_{I,2}t_1) - \cos(2\pi v_{I,1}t_1)\big)\Big) \quad\quad\text{(15.4d)}$$

Just as for the I-magnetization, there are both in-phase components, in this case with amplitudes determined by the evolution of the S-spin during t_1, and anti-phase components, with amplitudes determined by the evolution of the I-spin during t_1. During the acquisition period, t_2, only the frequencies associated with the S-spin influence the observed signal.

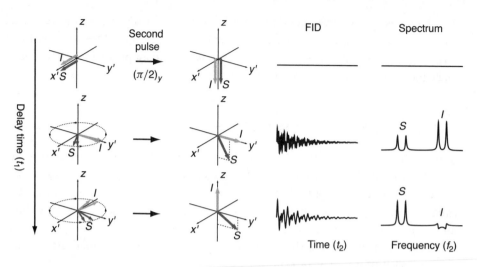

Figure 15.21 Vector diagrams representing precession of the in-phase magnetization components during the t_1 period and the effects of the second pulse in the heteronuclear COSY experiment, for incremented t_1 intervals, and the resulting FIDs and spectra. The vector diagrams represent the net magnetization components derived from the four initial eigenstates but do not include the correlations that evolve during t_1. The final contributions of the correlations *are* included in the FIDs and spectra.

Because the FID is generated by mixtures of signals with different phases, the peaks generated following Fourier transformation have rather complex shapes with both absorptive and dispersive components. The relative contributions of the two shapes will depend on the precession of the two spins during the t_1 period and, as discussed below, these differences in peak shape lead to the cross peaks in the two-dimensional spectrum following the second Fourier transformation.

The relationships for $\bar{I}_y$ and $\bar{S}_y$ allow us to predict how the observed signal changes as t_1 is incremented during the COSY experiment. Fig. 15.21 illustrates, for a few initial t_1 increments, how the in-phase magnetization components evolve during the t_1 period and the second pulse. When there is no delay between the first and second pulse, as in the top row of Fig. 15.21, the net magnetization from the I- and S-spins is simply rotated to the negative z-axis, and no signal is generated. With longer delays, the vectors precess in the transverse plane, with each spin displaying two frequency components. The second $(\pi/2)_y$-pulse converts the x'-magnetization to z-magnetization and leaves the y'-magnetization unchanged. These components then give rise to the in-phase signals defined by lines (a) and (b) of Eqs. 15.2 and 15.4. As in other examples we have looked at, the different Larmor frequencies of the I- and S-spins are apparent in both the relative peak positions in the individual spectra and in the different rates at which the amplitudes fluctuate as a function of t_1.

The vector diagrams in Fig 15.21 do *not* include any representation of the anti-phase magnetization components, since these components only make a net contribution to any observable magnetization once the t_2 period has begun. The drawings of the FIDs and spectra, however, do include these contributions. Whereas the in-phase components lead to spectral peaks with an absorptive shape, the anti-phase components lead to a dispersive shape. This is most apparent in the third row of drawings in Fig.15.21. At this value of t_1, the in-phase I-magnetization has precessed to the negative x'-axis, and the second pulse converts this to pure z-magnetization. In the absence of anti-phase components there would be no spectral peaks from the I-spin. Because of the scalar coupling between the spins, however, a detectable anti-phase component has emerged and gives rise to peaks with the dispersive shape. The rather oddly shaped peak structure actually represents two dispersive peaks with opposite signs. In general, the peaks in the spectra generated by the first Fourier transform will have some combination of absorptive and dispersive shape. The amplitude of the absorptive components for the I-spin will reflect the evolution of the I-spin during the t_1 period, whereas the amplitudes of the dispersive components will reflect the evolution of the S-spin during t_1. Conversely, the absorptive and dispersive components of the S-signal depend on the evolution of the S- and I-spins, respectively, during t_1.

The drawings in Fig. 15.22 illustrate how this process continues with still longer t_1 delays. The absorptive and dispersive components can have either negative or positive signs. However, the two absorptive peaks in either of the doublets always have the same sign, while the dispersive peaks in a doublet have opposite signs. Note also that the FIDs generated for the COSY spectrum seem more complex and irregular than those we have looked at earlier. This reflects both the contributions of more frequency components and the combination of signals with different phases.

Just as before, the individual FIDs are Fourier transformed to generate a two-dimensional data set, in which one dimension represents f_2 and the other t_1, and this data set is then transformed along the t_1 dimension to generate a two-dimensional

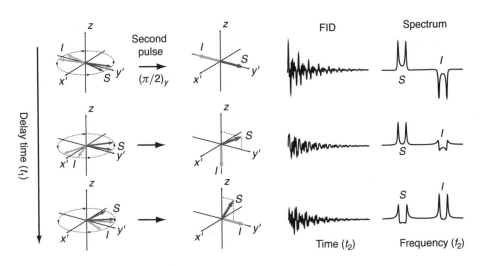

Figure 15.22 Vector diagrams representing precession of the in-phase magnetization components during the t_1 period, and the effects of the second pulse in the heteronuclear COSY experiment, for incremented t_1 intervals, and the resulting FIDs and spectra, as in Fig. 15.21.

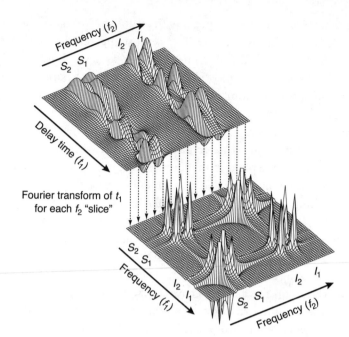

Fourier transform of t_1
for each f_2 "slice"

Figure 15.23
Fourier transformation of the
t_1 time-domain dimension
in the homonuclear COSY
experiment.

frequency spectrum, as illustrated in Fig. 15.23. For the simple case of just two coupled spins, the spectrum contains four sets of multiple peaks. Two of these multiplets
lie along the diagonal and contain essentially the same information as is present in
a simple one-dimensional spectrum. As in a one-dimensional spectrum, these appear as simple absorptive multiplets. The important new information is contained in
the cross peaks, which arise from the anti-phase magnetization components. Because
these components evolve out of phase with those that give rise to the diagonal peaks,
the cross peaks have a dispersive shape. Since the cross peaks are of greatest interest,
the phases of the final spectrum are usually changed to give these peaks an absorptive
shape that is much more easily detected and resolved from other peaks. This gives rise
to a spectrum with the appearance shown in Fig. 15.24. Now, the cross peaks have
the same doubly anti-phase absorptive shape that we saw earlier in the heteronuclear
COSY spectrum, while the diagonal multiplets have a dispersive shape.

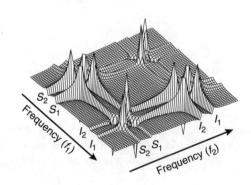

Figure 15.24
The simulated COSY spectrum
of Fig. 15.23 re-phased so that
the cross peaks are absorptive
(but anti-phase).

No matter how the data are manipulated, the diagonal and cross peaks will always have phases that differ by $\pi/2$ rad. While the phasing shown above makes the cross peaks more easily detected, the broad anti-phase peaks on the diagonal can still obscure nearby cross peaks. This represents a major disadvantage of the classic COSY experiment. A slightly more complicated pulse sequence, called a double-quantum filtered COSY, alleviates this problem by generating diagonal and cross peaks with the same phase. In spite of a loss of sensitivity, this form of the COSY experiment is the one that is now most commonly used.

This description of the COSY experiment has only considered a simple case of two nuclei linked by scalar coupling. In practice, individual nuclei may be coupled to two or more others. These additional interactions complicate the spectrum in two ways. First, each pair of coupled nuclei will create an additional cross peak multiplet. Second, each of these multiplets will contain more components, just as the peaks in a one-dimensional spectrum are split into additional components. The analysis of such a system becomes quite complex, and we will not pursue it further in this text, though the cross-peak fine structure can be a valuable source of structural information.

15.3   Application of the ¹H–¹H COSY Experiment to Proteins

One of the most common applications of the COSY experiment is its use in assigning the resonance frequencies of individual ¹H nuclei in proteins. Its particular usefulness in this application derives in part from the nature of the polypeptide backbone, as illustrated in Fig. 15.25 for a valine–asparagine pair within a polypeptide. Starting with the valine residue, note that the amide hydrogen (attached to the nitrogen) is linked by three covalent bonds to the hydrogen attached to the α-carbon. The α-hydrogen, in turn, is linked by three bonds to the β-hydrogen, which is similarly linked to each of the hydrogens attached to each γ-carbon. Thus, each of the hydrogen nuclei within this residue participates in scalar coupling interactions with at least one other hydrogen nucleus. Similarly, the ¹H nuclei of the asparagine residue form a network of

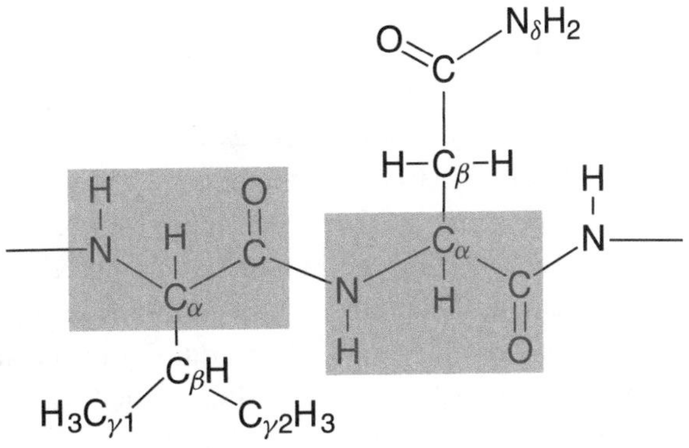

Figure 15.25

Chemical structure of a valine–asparagine dipeptide segment within a longer polypeptide chain. The shaded boxes enclose the backbone atoms of the individual residues, defining the residue boundaries.

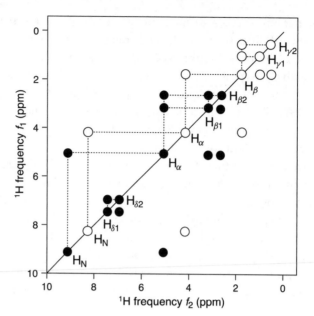

Figure 15.26

Idealized representation of a COSY spectrum for the valine–asparagine dipeptide shown in Fig. 15.25. The peaks are shown as circles, without the multiplet structures. Open and filled circles are used to represent the peaks from the valine and asparagine residues, respectively.

scalar coupling interactions, except for the two side-chain amide hydrogens which are coupled only to each other.

Importantly, the shortest covalent ^{1}H–^{1}H linkage *between* residues is between the α-hydrogen of one residue and the amide hydrogen of the next, a distance of four bonds. As a consequence, the ^{1}H nuclei in the individual amino acid residues are always isolated from one another with respect to scalar coupling, and a cross peak in a COSY spectrum must *always* reflect an interaction between two nuclei in the same residue.

Fig. 15.26 shows a schematic representation of the peaks in a COSY spectrum expected for a valine–asparagine dipeptide. In this drawing, the diagonal and cross peaks are shown as simple circles, ignoring the multiplet structures of a real spectrum. As in the NOESY spectrum, two sets of cross peaks are arranged symmetrically about the diagonal. For clarity, the circles representing the peaks from the asparagine residue are filled, while those from the valine residue are open. In real spectra, of course, the peaks are not automatically color-coded, but the clear patterns of connectivity among them often makes it quite easy to determine whether or not two cross peaks arise from the same or different amino acid residues. The analysis is greatly aided by the very high resolution and precision of solution NMR spectra, such that peaks separated by less than 0.1 ppm can often be resolved, and resonance frequencies can be determined with precisions of 0.01 ppm or finer.

Different amino acid residue types give rise to characteristic patterns of COSY cross peaks, reflecting the number of ^{1}H nuclei and their covalent connectivities. In the example shown above, the pattern from the asparagine residue reflects the fact that this residue contains, in addition to the amide and α-hydrogens, two β-hydrogens that are likely to be positioned in distinct chemical environments. Each of the β-hydrogens forms a cross peak with the α-hydrogen, as well as with each other. Several other residue types, including aspartic acid, serine and cysteine, give rise to similar patterns. The valine residue, on the other hand, yields a quite different pattern. In this case, the α-hydrogen forms only a single cross peak with the one β-hydrogen, which, in turn, forms cross peaks with the γ-hydrogens. Although there are six γ-hydrogens, rapid rotation of the methyl groups ensures that the three hydrogen nuclei attached to each γ-carbon atom experience the same average chemical environment, so they are not distinguished in the spectrum. On the other hand, each of the methyl groups is likely to be in a distinct environment, so the two sets of hydrogens give rise to peaks at two positions. The peaks arising from the methyl hydrogens will also be about three times as intense as, for instance, the peaks from the amide or α-hydrogens. Note also that there is no cross peak between the two sets of γ-methyl groups, since these hydrogens are separated by four covalent bonds.

Although not every residue type gives rise to a unique pattern in the COSY spectrum, the observed patterns provide a great deal of information about possible assignments between spectral peaks and specific residues in the amino acid sequence. In the classic "sequential" method for assigning protein NMR spectra, developed by Kurt Wüthrich, Gerhard Wagner and their colleagues, the NOESY spectrum provides another crucial piece of the puzzle. Because of the covalent constraints between them, the amide hydrogen of one residue and the α-hydrogen of the residue that precedes it in the sequence are always within 5 Å of one another and almost always give rise to a NOESY cross peak. Although there will inevitably be other cross peaks between amide and α-hydrogen resonances, these cross peaks identify possible sequential neighbors. Provided that the amino acid sequence of the protein is known, the NOESY cross peaks and the distinct COSY cross-peak patterns for different residue types allow one to begin identifying groups of spin systems that represent short fragments of the sequence. Once a few sets of cross peaks have been associated with specific residues, the investigator knows what additional neighbors to look for. This approach, using only ^{1}H–^{1}H two-dimensional spectra, is quite practical for relatively small proteins (less than 40 or 50 residues). For larger molecules, however, additional heteronuclear and multidimensional experiments are almost essential for making resonance assignments.

Of the cross peaks in a ^{1}H–^{1}H COSY spectrum of a protein, those between amide and α-resonances are usually the best resolved and provide the most unambiguous information. This part of the spectrum is sometimes referred to as a "fingerprint" region, and examination of this region can be a quick means of comparing modified or closely related proteins to determine whether or not they likely have similar three-dimensional structures. On the other hand, the region of the spectrum containing cross peaks from the side-chain hydrogens, especially those of methyl groups, is often quite congested and may be partially obscured by the diagonal peaks. As a consequence, it may be difficult or impossible to fully assign the side-chain resonances from a COSY spectrum.

15.4   The TOCSY Experiment

Although the COSY experiment played a critical role in the early development of protein NMR spectroscopy, it has to a degree been replaced by an alternative experiment,

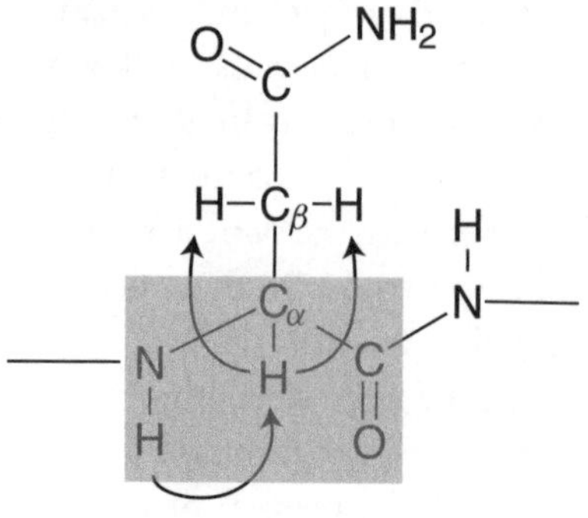

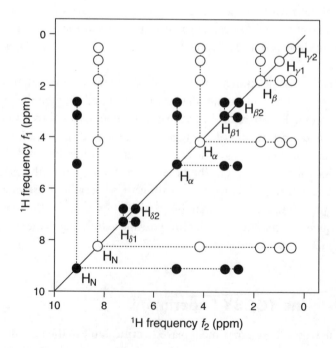

Figure 15.27

Pattern of sequential scalar coupling interactions within an asparagine residue. The shaded box encloses the backbone atoms.

the TOCSY (for TOtal Correlation SpectroscopY). As discussed above, cross peaks appear in a COSY spectrum only if a pair of nuclei are linked by three or fewer covalent bonds. In the TOCSY experiment, this requirement is relaxed somewhat, so that a cross peak appears if two nuclei are linked by a series of connectivities, each of which is still limited to three bonds or less. For example, the backbone amide hydrogen of an asparagine residue will generate cross peaks with each of the two β-hydrogens, as well as the α-hydrogen, following the linkages illustrated in Fig. 15.27. On the other hand, the two side-chain amide hydrogens remain isolated from the rest of the residue, because they are separated by four covalent bonds from the β-hydrogens. Similarly, the hydrogens of adjacent residues are unable to form cross peaks.

Fig. 15.28 shows an idealized TOCSY spectrum for the valine–asparagine dipeptide. As in Fig. 15.26, the peaks from the valine residue are shown in Fig 15.28 as open circles and those from the asparagine are shown as filled circles. Starting from the

Figure 15.28

Idealized representation of a TOCSY spectrum for the valine–asparagine dipeptide shown in Fig. 15.25. The peaks are shown as circles, without the multiplet structures. Open and filled circles are used to represent the peaks from the valine and asparagine residues, respectively.

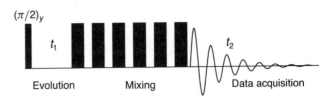

Figure 15.29

The basic TOCSY pulse sequence.

backbone amide hydrogen of each residue, scalar-coupled protons give rise to a series of peaks that form lines either vertically above the diagonal or horizontally below it. Similar, though shorter, lines originate from the other diagonal peaks.

Though the TOCSY spectrum contains more peaks than the corresponding COSY spectrum, the linear arrangement of cross peaks makes it very easy to identify the resonances arising from a single residue. In addition, the cross peaks that lie on the lines that originate from the amide ^{1}H resonances are generally well separated from the crowded diagonal. A third advantage is that the TOCSY diagonal and cross peaks are simple absorptive peaks, rather than the doubly anti-phase peaks characteristic of the COSY spectrum. On the other hand, the interpretation of the TOCSY spectrum may be more ambiguous. For instance, the resonance from the α-proton is readily identified in the COSY spectrum, where it forms the only cross peak with the backbone amide proton, but it may be confused with side-chain resonances in the TOCSY spectrum. Also, the relative intensities of the various cross peaks in a TOCSY spectrum can be quite sensitive to experimental details, and some may be difficult to detect. For these reasons, it is often desirable to record both COSY and TOCSY spectra.

Although the information content of the COSY and TOCSY spectra are similar, the pulse sequences and underlying principles of the two experiments are quite different. In fact, the quantum mechanical treatment we have developed so far is not really adequate for a thorough treatment of the TOCSY experiment. We can, however, gain some insights into the physical mechanism from qualitative considerations. The simplest version of the TOCSY pulse sequence is diagrammed in Fig. 15.29 and consists of a $\pi/2$-pulse, an incremented t_1 evolution interval and a mixing period. During the mixing period, a series of closely spaced π-pulses are applied.

Just as in the other two-dimensional experiments we have considered, the initial $\pi/2$-pulse followed by the evolution period leads to dispersion of the magnetization vectors in the transverse plane. This is illustrated in Fig. 15.30 for a simple system composed of three spins, which are labeled a, b and c and are assumed to form a scalar-coupled system. To minimize the complexity of the drawing in Fig. 15.30, the precession of each of the spins during the t_1 interval is represented by a single vector, but the precession of each will be determined by two or more frequencies, depending on how many direct scalar coupling interactions are involved.

During the mixing period, the sample is subjected to a series of π-pulses or, in more sophisticated versions of the experiment, more complex composite pulses. If these pulses are applied along the y'-axis, they have the effect of repeatedly rotating the magnetization about this axis, as illustrated Fig. 15.31. One consequence of this

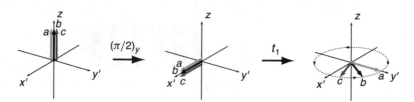

Figure 15.30   Vector diagrams representing the initial $(\pi/2)_y$-pulse and the t_1 evolution period in the TOCSY experiment.

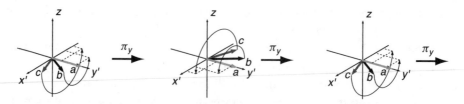

Figure 15.31   Vector diagrams representing the effects of the refocusing pulses during the mixing period of the TOCSY experiment.

irradiation is that the average magnetization along the y'-axis remains constant (neglecting relaxation), leading to the expression *spin lock* to describe the effect. (See the earlier discussion of refocusing pulses in experiments designed to measure transverse relaxation rates and the rotating-frame relaxation rate, $R_{1\rho}$, in Chapter 7.) During the spin-lock period, the z- and y'-axes reverse, in a sense, their usual roles. During normal precession, the magnetization projection along the z-axis remains constant, and the y-magnetization fluctuates sinusoidally. Now, the z-magnetization fluctuates, and we can think of the magnetization as precessing about the y'-axis. (Viewed from the laboratory frame of reference, things are more complicated, with the projections along all three axes changing during the spin-lock period.)

Importantly, the frequency of precession about the y'-axis during the spin lock is much lower than the normal precession about the z-axis, because the strength of the fluctuating field is lower than that of the static field. As a consequence, the differences in precession frequency due to scalar coupling, which are independent of the magnetic field strength, are quite large relative to the chemical-shift differences. Thus, the assumption of weak coupling, upon which all of our previous analysis was based, is no longer valid. Instead, the system falls into the realm of strong coupling.

A rigorous quantum mechanical treatment of strongly coupled spins is well beyond the scope of this book, but a qualitative description may be helpful. For a system of three spin-1/2 particles, the wavefunction representing a state can be expressed as a linear combination of eight kets defined in terms of the α and β states of the individual spins:

$$|\Psi\rangle = c_{\alpha\alpha\alpha}|\alpha\alpha\alpha\rangle + c_{\beta\alpha\alpha}|\beta\alpha\alpha\rangle + c_{\alpha\beta\alpha}|\alpha\beta\alpha\rangle + c_{\alpha\alpha\beta}|\alpha\alpha\beta\rangle$$

$$+ c_{\beta\beta\alpha}|\beta\beta\alpha\rangle + c_{\beta\alpha\beta}|\beta\alpha\beta\rangle + c_{\alpha\beta\beta}|\alpha\beta\beta\rangle + c_{\beta\beta\beta}|\beta\beta\beta\rangle$$

Under conditions of a constant magnetic field and *weak* scalar coupling, the set of kets shown in Eq. 15.8 represent the eigenstates of the system. Under these conditions, the phases of the complex coefficients change with time, as we saw before, but their relative magnitudes remain constant. This is reflected in a constant magnetization along the z-axis.

Under the conditions of the spin lock, the mixed kets, such as $|\beta\alpha\alpha\rangle$, $|\alpha\beta\alpha\rangle$, $|\beta\beta\alpha\rangle$, etc., no longer represent eigenstates. That is, if a spin in the $|\beta\alpha\alpha\rangle$ state, for instance is placed in the fluctuating magnetic field, its wavefunction will not remain constant, as it would in the steady field aligned with the z-axis. Instead, it will be converted into other wavefunctions, including linear combinations of the other kets. (Even though they do not all represent eigenstates, the eight kets can still serve as a basis set to represent any possible wavefunction.)

As a consequence of the strong coupling among the individual spins in a molecule, changes in their states are highly correlated. For instance, a transition such as $|\alpha\beta\alpha\rangle \rightarrow |\beta\alpha\alpha\rangle$ is more probable than one in which only a single spin is affected, such as $|\alpha\alpha\alpha\rangle \rightarrow |\beta\alpha\alpha\rangle$. These correlations among the spin states lead to correlations among the magnetization components along the x'-, y'- and z-axes. In particular, although the magnetization from any one spin in a given direction can change with time, the total magnetization in each direction from the set of coupled spins remains constant.

To visualize the magnetization transfer processes during the TOCSY mixing time, it is helpful to draw the net magnetization vectors as sums of x'- and y'-components. Using the example illustrated in the previous drawings, the state at the beginning of the mixing time can be drawn as shown in Fig. 15.32. During the mixing time, the relative magnitudes of the individual magnetization components along the y'-axis shift, while the total magnetization along this axis remains constant, as suggested in Fig. 15.33. In the example shown here, y'-magnetization from the b-spin is first transferred to the c-spin, and then magnetization is transferred from the a-spin to both b and c. As the process continues, the total initial y'-magnetization becomes more evenly distributed among the coupled spins.

In the simple form of the TOCSY experiment illustrated above, the mixing time is directly followed by the data acquisition period, t_2. During this period, the magnetization that was "locked" along the y'-axis precesses normally and induces a signal in the spectrometer coils. In principle, the magnetization components that were aligned

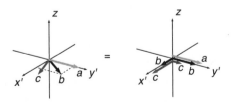

Figure 15.32 Vector diagrams representing the magnetization components present at the end of the t_1 evolution period in the TOCSY experiment, resolved into separate x'- and y'-components.

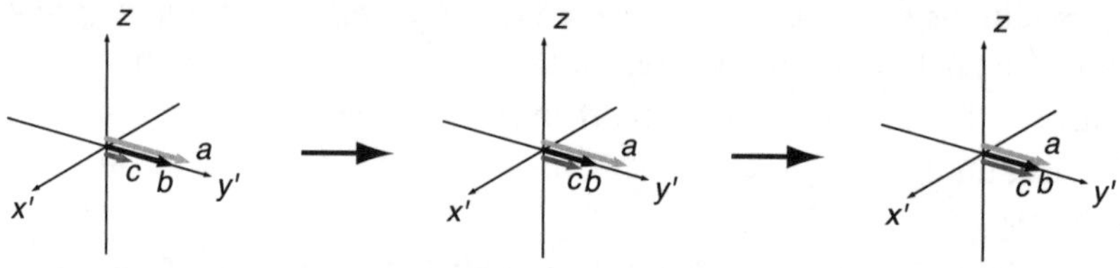

Figure 15.33   Vector diagrams representing the transfer of transverse magnetization during the mixing period of the TOCSY experiment.

with the x'-axis at the beginning of the mixing period will also contribute to the signal. In practice, however, these components lose most of their phase coherence during the mixing time and usually do not make significant contributions, particularly for large molecules that give rise to relatively broad peaks. These contributions can be further reduced using slightly more complex pulse sequences, phase cycling, or pulsed field gradients.

The FIDs generated by the individual experiments with different t_1 delays are Fourier transformed as usual and then transformed in the second dimension. Cross peaks appear in the resulting spectrum because the magnetization components present at the end of the mixing time reflect the starting magnetizations of all of the nuclei in the spin system.

The kinetics of the TOCSY transfer process can be quite complex and depend on the initial magnetization components and the strengths of the couplings among the various spins. As a consequence, the relative intensities of different cross peaks in the final two-dimensional spectrum can depend on the length of the mixing time. Short mixing times tend to emphasize couplings between nuclei separated by fewer covalent bonds, while longer times may be necessary to obtain efficient transfer among several sequentially linked nuclei, such as in a longer amino-acid side chain. For protein experiments, mixing times are typically on the order of 20–100 ms. In some cases, it may be helpful to carry out two or more experiments with different mixing times in order to obtain clear signals for all of the coupled nuclei.

Summary Points for Chapter 15

- COSY-type experiments can be applied to either heteronuclear (typically ^{1}H–^{15}N or ^{1}H–^{13}C) or homonuclear (usually ^{1}H) systems.
- Features of the simple heteronuclear COSY experiment described here include the following:
    - Evolution of S-spin magnetization during the t_1 evolution period determines the S_z-magnetization after the second pulse.
    - The S_z-magnetization determines the relative contributions of the two I-spin frequency components during the t_2 data acquisition period.
    - The final spectrum contains doubly anti-phase cross peaks that identify the scalar-coupled I–S spin pairs.

- Even though the homonuclear COSY experiment is based on the simplest possible two-pulse sequence, the description of the experiment is surprisingly complicated and requires the analysis of spin correlations as well as the observable magnetization components.
  - Both spins precess in the transverse plane during the t_1 evolution period.
  - During the evolution period, the initial transverse magnetization components interconvert with both other magnetization components and correlations with the z-component of the other spin.
  - The second pulse reverses the correlations between I_z and transverse S-components, and between S_z and transverse I-components.
  - During the data acquisition period, mixed correlations, such as $I_z S_x$ or $I_x S_z$, are interconverted with observable magnetization components, which give rise to cross peaks in the spectrum.
  - The diagonal peaks and cross peaks always have opposite phases.
- When applied to peptides or proteins, the ^{1}H–^{1}H COSY experiment gives rise to characteristic cross peaks that reflect the covalent linkages within amino acid residues, but not between them.
- In the traditional sequential-assignment strategy for proteins, NOESY spectra are used to establish the connectivity of the polypeptide chain, linking the amide ^{1}H nucleus of one reside with the α-^{1}H of the preceding residue.
- In the TOCSY experiment, sequential scalar coupling interactions (involving up to three covalent bonds each) give rise to cross peaks.
- During the mixing time of the TOCSY experiment, the transverse magnetization components are "spin-locked" along one direction, so that the scalar coupling interaction dominates the effect of the external field, leading to the strong-coupling regime and allowing magnetization transfer among multiple spins.

Exercises for Chapter 15

15.1. For the heteronuclear COSY experiment diagrammed in Fig. 15.2, draw a set of vector diagrams describing the behavior during the preparation period of spin pairs that begin the experiment in the $|\beta\beta\rangle$ state, like the drawings shown for the other three eigenstates in Figs. 15.4, 15.8 and 15.12. Also sketch the expected one-dimensional spectra for the corresponding t_1 intervals.

15.2. For simplicity, the vector diagrams in Figs. 15.21 and 15.22 represent only the observable magnetization components that evolve during the preparation period of the homonuclear COSY experiment, and not the correlations that actually give rise to the cross peaks in the final spectrum. For the specific example shown in the bottom row of Fig. 15.21, use correlation diagrams to illustrate how the anti-phase dispersive peaks at frequencies $\nu_{I,1}$ and $\nu_{I,2}$ appear in that particular one-dimensional spectrum. Your diagrams should begin with representations of the initial I_x- and S_x-magnetization components that are generated with the first $(\pi/2)_y$-pulse, and follow the evolution of these components during the t_1 period and the second pulse. As necessary, you should resolve the diagrams into simpler ones to show how the final signal is generated, as in the examples shown in Figs. 15.17 and 15.19.

15.3. This problem is intended to give you a sense for how the information from different types of two-dimensional spectra is utilized to make resonance assignments and derive structural information for proteins. Here, we will assume that the idealized spectra shown below represent the peaks that would arise from just three amino acid residues in a protein: an alanine (Ala), a glutamine (Gln) and a serine (Ser). Further, we will assume that the three residues form a tripeptide segment, but in an unknown order. In a more realistic situation, we would probably know the sequence, but in this problem, you will need to deduce the sequence from the spectra. Electronic versions of the figures for this exercise, suitable for printing at a larger scale, are available for download via the web site, http://uscibooks.com/goldenberg.htm.

(a) The drawing in Fig. 15.34 represents an idealized ^{1}H–^{1}H COSY spectrum that might be generated from the three residues. The drawing does *not* include the anti-phase fine structure of the peaks in a real COSY spectrum. We expect peaks from the following protons:
- Ala: amide-^{1}H (HN), ^{1}H$_\alpha$ (HA), ^{1}H$_\beta$ (HB)
- Gln: HN, HA, ^{1}H$_{\beta 1}$ (HB1), ^{1}H$_{\beta 2}$ (HB2), ^{1}H$_{\gamma 1}$ (HG1), ^{1}H$_{\gamma 2}$ (HG2), ^{1}H$_{\epsilon 1}$ (HE1), ^{1}H$_{\epsilon 2}$ (HE2)
- Ser: HN, HA, HB1, HB2 (the γ-hydrogen is unlikely to be detected because of exchange with solvent water.)

On the drawing, add line segments that represent the scalar coupling interactions among the hydrogen atoms. Then, make a table showing the hydrogen atoms listed above and their corresponding chemical shifts.

(b) The drawing in Fig. 15.35 shows only the diagonal peaks from the COSY spectrum. Using the information you deduced from the COSY, label each of the diagonal peaks and then add the peaks you would expect in a ^{1}H–^{1}H TOCSY spectrum. You only need to draw the portion of the spectrum that lies above the diagonal.

(c) The final spectrum, in Fig. 15.36, is a NOESY.

On this drawing, circle each of the peaks that represents an *inter-residue* NOE interaction. Next, make a list of all of the inter-residue NOEs. What is the maximum number of inter-residue NOEs you would expect from these three amino acid residues?

From the information you have assembled, deduce the sequence of the three residues. Explain how you reached your answer. If you believe that the information is insufficient to fully determine the sequence, explain the ambiguity and suggest how it might be resolved experimentally.

For an extra challenge, try building a model of the tripeptide (either with physical models or with a computer program), and see if you can deduce anything about the three-dimensional conformation of the segment. In particular, see if the NOE data are sufficient to at least rule out some of the possible conformations.

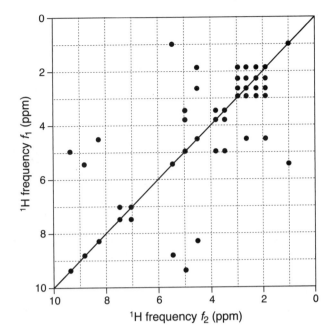

Figure 15.34
Idealized COSY spectrum for
Exercise 15.3.

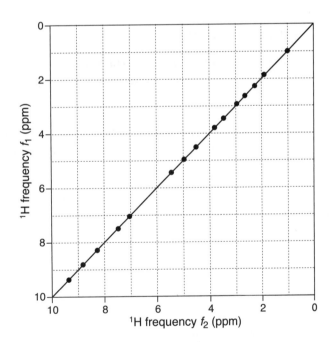

Figure 15.35
Spectral axes for Exercise 15.3,
part b. The peaks on the
diagonal are the same as in the
COSY spectrum diagrammed
in Fig. 15.34. Draw in the
peaks expected for a TOCSY
spectrum from the same
peptide segment.

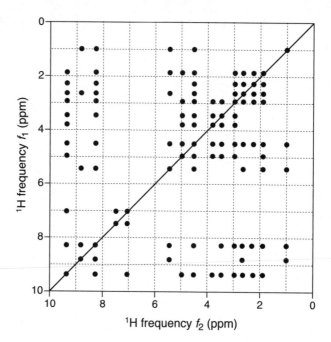

Figure 15.36
Idealized NOESY spectrum for
Exercise 15.3.

Further Reading

The notes from the lecture in which Jean Jeener first proposed a two-dimensional NMR experiment were eventually published in a festschrift volume for Anatole Abragam:

⬦ Jeener, J. (1994). The unpublished Basko Polje (1971) lecture notes about two-dimensional NMR spectroscopy. In *NMR and More: In Honour of Anatole Abragam* (Goldman, M. & Pourens, M., Eds.), pp. 265–278. Les Clis, France.

The heteronuclear COSY experiment discussed in this chapter was one of the first heteronuclear two-dimensional experiments performed:

⬦ Maudsley, A. A. & Ernst, R. R. (1977). Indirect detection of magnetic resonance by heteronuclear two-dimensional spectroscopy. *Chem. Phys. Lett.*, 50, 368–372. http://dx .doi.org/10.1016/0009-2614(77)80345-X

Very shortly after the introduction of the heteronuclear experiment described above, a slightly more complicated version was developed (commonly referred to as the HETCOR experiment), in which the anti-phase doublets are replaced with simple absorptive peaks:

⬦ Maudsley, A. A., Müller, L. & Ernst, R. R. (1977). Cross-correlation of spin-decoupled NMR spectra by heteronuclear two-dimensional spectroscopy. *J. Magn. Reson.*, 28, 463–469. http://dx.doi.org/10.1016/0022-2364(77)90288-8

Detailed discussions of the homonuclear COSY and TOCSY experiments, along with other homonuclear experiments:

◇ Cavanagh, J., Fairbrother, W. J., Palmer III, A. G., Rance, M. & Skelton, N. J. (2007). *Protein NMR Spectroscopy*, 2nd ed., pp. 405–532. Elsevier Academic Press, Amsterdam.

The first description of the sequential resonance assignment method for proteins, based on homonuclear COSY and NOESY spectra:

◇ Wagner, G., Kumar, A. & Wüthrich, K. (1981). Systematic application of two-dimensional ^{1}H nuclear-magnetic-resonance techniques for studies of proteins: 2. Combined use of correlated spectroscopy and nuclear Overhauser spectroscopy for sequential assignments of backbone resonances and elucidation of polypeptide secondary structures. *Eur. J. Biochem.*, 114, 375–384. http://dx.doi.org/10.1111/j.1432-1033.1981.tb05157.x

CHAPTER 16

Heteronuclear NMR Techniques

During the 1980s and 90s, techniques utilizing ^{13}C and ^{15}N together with ^{1}H became increasingly important for biomolecular NMR, largely as a consequence of the use of genetic engineering methods to produce proteins and nucleic acids enriched with the rarer isotopes. These developments, together with other technical advances, have made it possible to use NMR to study molecules of increasing size and complexity. In this chapter we consider some of the special techniques that can be applied to molecules containing two or more spin-1/2 isotopes, beginning with relatively simple manipulations originally developed for small organic molecules and moving on to some multidimensional experiments. Two key elements that are common to almost all heteronuclear experiments are (1) the use of selective pulses to manipulate the different spin types independently and (2) the exploitation of scalar coupling interactions between nuclei of different types. As you will see, the more complex experiments are generally built up from combinations of simpler ones. This modular structure aids greatly in designing and understanding these experiments.

16.1   Heteronuclear Coupling and Decoupling

The scalar coupling interaction between nuclei of different types, such as ^{1}H and ^{13}C, forms the basis for all heteronuclear experiments. At the same time, however, this coupling can lead to unwanted complexity in the resulting spectra, and methods have been devised to reduce this complexity. Previously, we discussed some simple cases involving just two nuclei, such as an ^{1}H nucleus covalently bonded to a ^{13}C nucleus, which causes a "splitting" of the peak in a ^{13}C spectrum, as in Fig. 16.1. This is the spectrum that would be generated in a simple one-dimensional pulse experiment in which the pulse is applied selectively to the ^{13}C nuclei and the FID generated by these nuclei is recorded. As discussed in Chapter 14, the two components of the doublet can be thought of as arising from the separate components of the original population in which the ^{1}H nuclei are initially in either the spin-up (α) or spin-down (β) state.

Figure 16.1

Splitting of a peak in a ^{13}C spectrum by a covalently attached ^{1}H nucleus.

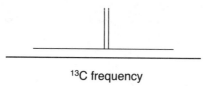

^{13}C frequency

More complicated multiplet structures can arise when additional spins are coupled to one another. For instance, if a single ^{13}C nucleus is bonded to two ^{1}H nuclei, the eigenstates of the system can be written as $|\alpha, \alpha, \alpha\rangle$, $|\beta, \alpha, \alpha\rangle$, $|\alpha, \beta, \alpha\rangle$, $|\alpha, \alpha, \beta\rangle$, etc., where the first two symbols in the ket represent the ^{1}H nuclei and the third represents the ^{13}C. The observed resonance frequencies associated with the ^{13}C nucleus are determined by the transitions from each of the eigenstates in which the ^{13}C is in the α state to the corresponding eigenstate with the ^{13}C in the β state, as listed below:

$$|\alpha, \alpha, \alpha\rangle \longrightarrow |\alpha, \alpha, \beta\rangle$$
$$|\beta, \alpha, \alpha\rangle \longrightarrow |\beta, \alpha, \beta\rangle$$
$$|\alpha, \beta, \alpha\rangle \longrightarrow |\alpha, \beta, \beta\rangle$$
$$|\beta, \beta, \alpha\rangle \longrightarrow |\beta, \beta, \beta\rangle$$

The energies of these states depend upon the states of the individual nuclei as well as the coupling between them. Just as we saw earlier with the simpler two-spin system, the energy of a state is increased when coupled spins are both α or both β and is decreased when they are unmatched (for positive coupling constants).

The eigenstate energies for any system of n weakly coupled spins can be calculated using a set of relatively simple rules. For a system of n spins, each of the eigenstates can be defined by a set of n numbers, m_1 through m_n, where m_k is 1/2 if nucleus k is in the α state and $-1/2$ if it is in the β state. First, we consider the energy contribution from the spins if they were uncoupled. For an uncoupled spin-1/2 nucleus, the energies of the α and β states are given by $\frac{1}{2}h\nu$ and $-\frac{1}{2}h\nu$, where ν is the (signed) resonance frequency, determined by the gyromagnetic ratio, the external field strength and shielding (i.e., chemical shift) experienced by the nucleus. In the absence of scalar coupling, the energies of each of the eigenstates is calculated by summing over the energies of the individual nuclei in that state:

$$E_{\text{uncoupled}} = h \sum_{k=1}^{n} \nu_k m_k$$

where ν_k is the Larmor frequency associated with the kth spin. The contribution from scalar coupling is given by

$$E_{\text{coupling}} = h \sum_{k<l}^{n} \delta_{k,l} m_k m_l J_{k,l}$$

Here, the summation is over each pair of spins, with indices k and l, and $\delta_{k,l}$ is 0 if the spins are degenerate (i.e., have identical Larmor frequencies) and 1 if they are non-degenerate. The symbols m_k and m_l have the same meaning as in Eq. 16.2, and $J_{k,l}$ is

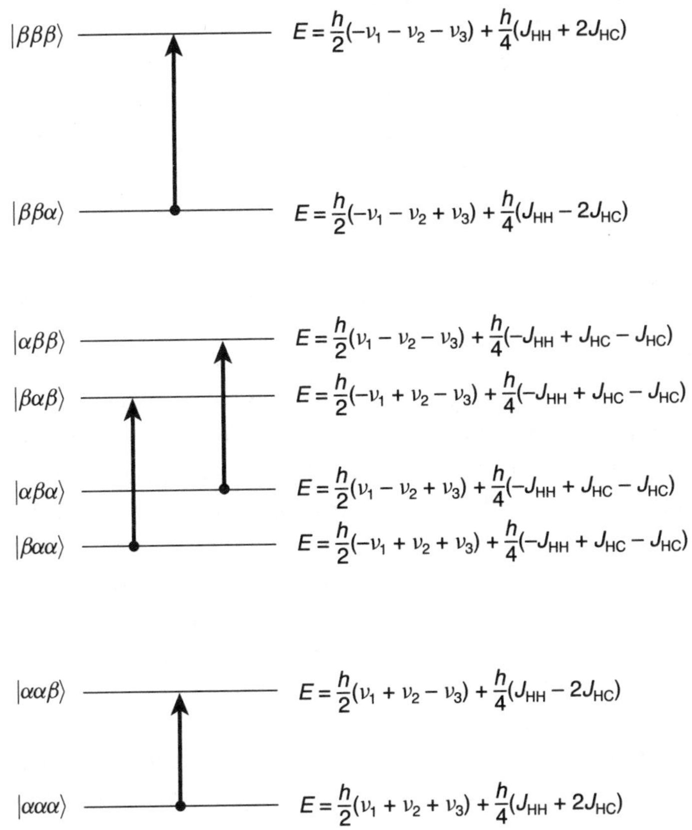

$|\beta\beta\beta\rangle$ —————— $E = \dfrac{h}{2}(-\nu_1 - \nu_2 - \nu_3) + \dfrac{h}{4}(J_{HH} + 2J_{HC})$

$|\beta\beta\alpha\rangle$ —————— $E = \dfrac{h}{2}(-\nu_1 - \nu_2 + \nu_3) + \dfrac{h}{4}(J_{HH} - 2J_{HC})$

$|\alpha\beta\beta\rangle$ —————— $E = \dfrac{h}{2}(\nu_1 - \nu_2 - \nu_3) + \dfrac{h}{4}(-J_{HH} + J_{HC} - J_{HC})$

$|\beta\alpha\beta\rangle$ —————— $E = \dfrac{h}{2}(-\nu_1 + \nu_2 - \nu_3) + \dfrac{h}{4}(-J_{HH} + J_{HC} - J_{HC})$

$|\alpha\beta\alpha\rangle$ —————— $E = \dfrac{h}{2}(\nu_1 - \nu_2 + \nu_3) + \dfrac{h}{4}(-J_{HH} + J_{HC} - J_{HC})$

$|\beta\alpha\alpha\rangle$ —————— $E = \dfrac{h}{2}(-\nu_1 + \nu_2 + \nu_3) + \dfrac{h}{4}(-J_{HH} + J_{HC} - J_{HC})$

$|\alpha\alpha\beta\rangle$ —————— $E = \dfrac{h}{2}(\nu_1 + \nu_2 - \nu_3) + \dfrac{h}{4}(J_{HH} - 2J_{HC})$

$|\alpha\alpha\alpha\rangle$ —————— $E = \dfrac{h}{2}(\nu_1 + \nu_2 + \nu_3) + \dfrac{h}{4}(J_{HH} + 2J_{HC})$

Figure 16.2 Energy levels of a spin system composed of one ^{13}C and two ^1H nuclei. Both kinds of nuclei have positive gyromagnetic ratios and *negative* Larmor frequencies. The vertical arrows indicate the single-quantum transitions of the ^{13}C nucleus.

the coupling constant for the pair, expressed in Hz. The total energy of each eigenstate is given by the sum of these two expressions.

The diagram in Fig. 16.2 illustrates schematically the energies of the eight eigenstates in the H_2C system. The energies of the various states are expressed in terms of the Larmor frequencies of the uncoupled spins (ν_1, ν_2 and ν_3, corresponding to the two ^1H nuclei and the ^{13}C nucleus, respectively) and the homonuclear (J_{HH}) and heteronuclear (J_{HC}) coupling constants. The vertical arrows represent the four single-quantum transitions associated with the ^{13}C nuclei. Although the drawing is intended to represent the relative energies of the different states, the differences associated with the scalar coupling constants and the differences in chemical shift of the two ^1H nuclei are greatly exaggerated.

Of the four transitions involving the ^{13}C nucleus, two are associated with identical energy differences, $|\beta, \alpha, \alpha\rangle \longrightarrow |\beta, \alpha, \beta\rangle$ and $|\alpha, \beta, \alpha\rangle \longrightarrow |\alpha, \beta, \beta\rangle$. As a consequence, three resonances are observed in a ^{13}C spectrum, and the one with the intermediate frequency has twice the intensity of the other two, as illustrated in Fig. 16.3. Each of the ^1H nuclei will also give rise to a multiplet of resonances, if a ^1H spectrum

Figure 16.3

Splitting of a peak in a ^{13}C spectrum by two covalently attached ^1H nuclei.

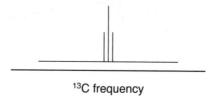

^{13}C frequency

is recorded. Systems composed of more spins generate increasingly complex multiplet structures.

For simple molecules or parts of molecules, the multiplets can provide important clues for interpreting the spectra. As the complexity of the spectrum increases, however, it can become very difficult to sort out the contributions from different nuclei, especially as the multiplets begin to overlap. In such situations, it is helpful to be able to suppress the effects of scalar coupling so that the components of a multiplet collapse into a single peak. This procedure is referred to as *heteronuclear decoupling*, although this term is rather misleading, since the physical interaction between the nuclei remains. Only the manifestation of the coupling in the spectrum is affected.

The idea upon which decoupling is based can be illustrated by considering a simple system composed of two nuclei of different types, designated, as usual, I and S. The I-spin might be an ^1H nucleus and the S-spin a ^{13}C nucleus. Suppose that we wish to record a spectrum from the S-nucleus. We apply a $\pi/2$-pulse to the S-spin and then record and Fourier transform the signal from its precession. The net signal will reflect the excess populations of the starting eigenstates, $|\alpha\alpha\rangle$ and $|\beta\alpha\rangle$. Vector diagrams representing the effects of the pulse and precession on these two states are shown in Fig. 16.4. As the spins in the two populations precess, they generate two frequency components in the FID, with the population that started out in the $|\alpha\alpha\rangle$ state generating the higher frequency. Consider what happens, however, if a π-pulse is applied to the I-spin after a short time, as diagrammed in Fig. 16.5. For the

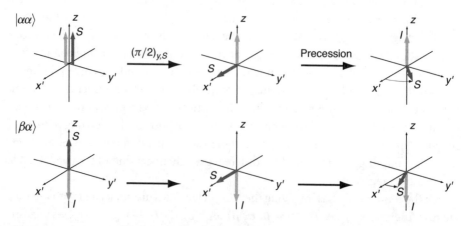

Figure 16.4 Vector diagrams representing the effects of a selective $(\pi/2)_{y,S}$-pulse to the excess $|\alpha\alpha\rangle$ and $|\beta\alpha\rangle$ populations of a scalar-coupled two-spin system, followed by precession.

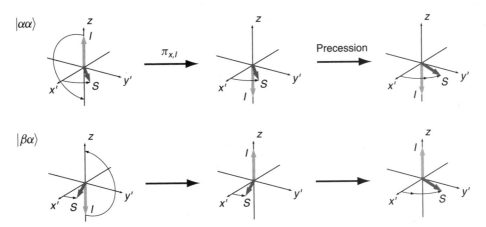

Figure 16.5   Vector diagrams representing the effects of a selective $\pi_{x,I}$-pulse on the precession of the S-magnetization arising from the excess $|\alpha\alpha\rangle$ and $|\beta\alpha\rangle$ populations of a scalar-coupled two-spin system.

population that began in the $|\alpha\alpha\rangle$ state, the I-spin is converted to the β state, and the precession frequency of the S-spin decreases. At the same time, the I-spin in the population that began in the $|\beta\alpha\rangle$ state is converted to the α state, and the precession frequency of the coupled S-spin increases. If the time intervals before and after the π-pulse are equal, the two vectors representing the magnetization from the S-spins will be in the same position after the second interval.

In its simplest form, heteronuclear decoupling involves applying a series of closely spaced π-pulses to the I-spin during the data acquisition period, when the signal from the S-spin is recorded. This ensures that vectors corresponding to the spin populations never diverge very far from one another as the signal is recorded, and the doublet is collapsed into a single peak. In practice, this simple scheme has important limitations, primarily because it is only effective if the decoupling pulses match the frequency of the I-spin very closely. In order to eliminate the splitting throughout the spectrum, more elaborate schemes involving composite pulses are usually used. These schemes are closely related, or even identical, to those used for the "spin lock" in the TOCSY experiment described in Chapter 15.

In Chapter 14, we saw that splitting can also arise from a single population when the magnetization components from the I- and S-spins are both placed in the transverse plane and allowed to precess. We represented this situation as shown in Fig. 16.6

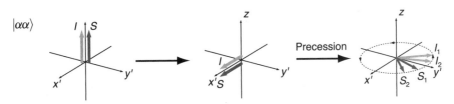

Figure 16.6   Vector diagrams representing the two frequency components arising from each of the two spins, when the magnetization of both lie in the transverse plane.

Figure 16.7

Vector diagram representing the effects of a selective $\pi_{x,I}$-pulse, when the magnetization of both spins lie in the transverse plane.

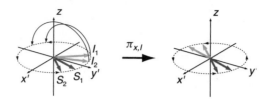

for a case starting with the $|\alpha\alpha\rangle$ state, to which is applied a non-selective $(\pi/2)_y$-pulse, followed by a period of free precession. The split vectors represent the two frequency components for each of the two spins. The mathematical representation of this situation, derived from the quantum mechanical treatment, is given by the expressions for the average transverse magnetization components of the two spins as a function of an evolution time, τ_1. For the average y'-magnetization components, the expressions are

$$\langle I_y(\tau_1)\rangle = (1/4)\big(\sin(2\pi v_{I,1}\tau_1) + \sin(2\pi v_{I,2}\tau_1)\big)$$

$$\langle S_y(\tau_1)\rangle = (1/4)\big(\sin(2\pi v_{S,1}\tau_1) + \sin(2\pi v_{S,2}\tau_1)\big)$$

where $v_{I,1}$, $v_{I,2}$, $v_{S,1}$ and $v_{S,2}$ are the precession frequencies associated with frequency components of the I- and S-spins. If this evolution period is followed by a selective π_x-pulse to the I-spin, the average y-magnetization components are then given by

$$\langle I_y(\tau_1)\rangle = -(1/4)\big(\sin(2\pi v_{I,1}\tau_1) + \sin(2\pi v_{I,2}\tau_1)\big)$$

$$\langle S_y(\tau_1)\rangle = (1/4)\big(\sin(2\pi v_{S,1}\tau_1) + \sin(2\pi v_{S,2}\tau_1)\big)$$

As expected, the pulse changes the sign of the mean I_y-magnetization component, but not the S_y-magnetization, as diagrammed in Fig. 16.7. From the vector picture, it's not really clear what will happen following the $\pi_{x,I}$-pulse. The split vectors represent the two frequency components that can be detected during the initial evolution period, but, at any given time, the average magnetization from each spin points in only a single direction. Thus, we might expect that the splitting process would simply start over after the π-pulse. Rather remarkably, however, analysis of the wavefunctions shows that the magnetization components retain a "memory" of what happened during the first evolution period. If the system is allowed to evolve for a second time period, τ_2, the average y-components of the magnetization from the two spins are given by

$$\langle I_y(\tau_1, \tau_2)\rangle = -(1/4)\big(\sin\big(2\pi v_{I,1}(\tau_1 - \tau_2)\big) + \sin\big(2\pi v_{I,2}(\tau_1 - \tau_2)\big)\big)$$

$$\langle S_y(\tau_1, \tau_2)\rangle = (1/4)\big(\sin\big(2\pi (v_{S,1}\tau_1 + v_{S,2}\tau_2)\big) + \sin\big(2\pi (v_{S,2}\tau_1 + v_{S,1}\tau_2)\big)\big)$$

These relationships indicate that the two spins are affected somewhat differently. For the two components of the I-magnetization, the precession frequency is the same before and after the π-pulse. However, the final position of the I-magnetization is determined by the *difference* between the two time intervals, reflecting the reversal of position caused by the π-pulse following the first interval. For the S-magnetization,

Figure 16.8

Vector diagram representing the evolution of S_x-magnetization into a combination of observable magnetization components and correlations.

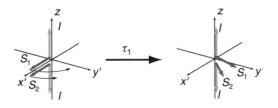

on the other hand, the final position is determined by the sum of the effects in the two intervals, but the precession frequencies are interchanged by the pulse.

This is yet another effect that is not easily explained by simple vector diagrams that represent only the observable magnetization components. However, the correlation diagrams introduced in Chapter 14 (pages 407–409) can help in visualizing the quantum mechanical phenomena. If we consider first the S-components, the x-magnetization present immediately after the first pulse can be represented as shown in the left-hand diagram in Fig. 16.8.

As the two S-components precess during the first delay period, τ_1, they diverge and the magnetization components are partially converted into correlations. The $\pi_{x,I}$-pulse then acts on the I-components of the correlations, as shown in Fig. 16.9. The S-component that was initially correlated with the positive I_z-component (S_1), and therefore had the higher precession frequency, is now correlated with the negative I_z-component and has the lower frequency. Conversely, the S-component (S_2) that initially had the lower frequency now has the higher frequency. These changes are indicated by the labels S_1 and S_2 in Fig. 16.9. If precession is allowed to continue for an equal period of time, τ_2, then the two S-components end up in the same position, as shown in Fig. 16.10. Note, also, that there is now no net correlation between the spins: S_x and S_y are both equally associated with positive or negative I_z-magnetization.

Similarly, we can follow the initial I_x-magnetization using correlation diagrams. The drawings in Fig. 16.11 show the effects of the first evolution period and the $\pi_{x,I}$-pulse. In this case, it is the transverse components that are affected by the pulse, and the two I-components, I_1 and I_2, remain associated with the positive and negative S_z-components, respectively. Thus, the precession frequencies remain the same, but the relative positions are switched. Precession for the same period of time then returns

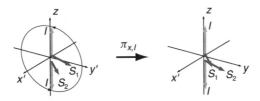

Figure 16.9 Vector diagram representing the effects of a selective $\pi_{x,I}$-pulse on the magnetization components and correlations shown in Fig. 16.8. Note that the positions of the labels S_1 and S_2 have been exchanged with respect to the two S-vectors, reflecting the change in precession frequencies after the pulse.

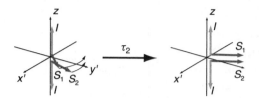

Figure 16.10   Vector diagram representing precession of the S-magnetization components following the selective $\pi_{x,I}$-pulse diagrammed in Fig. 16.9. If the time delays before and after the pulse are equal, the two S-vectors have the same final positions.

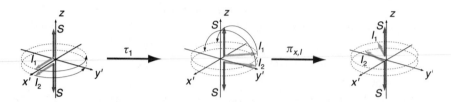

Figure 16.11   Vector diagram representing the evolution of I_x-magnetization into a combination of observable magnetization components and correlations, followed by a selective $\pi_{x,I}$-pulse.

both transverse I-components to the initial position, as shown in Fig. 16.12. The net effect of the refocusing sequence is summarized in Fig. 16.13. As shown, the effect on the S-spin is exactly the same as when the splitting arises from two separate populations, each displaying a single S-magnetization precession frequency. If the $\pi_{x,I}$-pulse is repeated continuously during the data acquisition period, only a single peak appears in the spectrum.

Decoupling is an important component in the analysis of both simple and complex molecules. When analyzing relatively simple one-dimensional spectra, a useful strategy is to compare the spectra generated with and without decoupling. This often makes it possible to distinguish the primary resonance frequencies from the peaks arising from scalar coupling. Once identified, the splitting patterns can then provide valuable information about the covalent structure. In multidimensional heteronu-

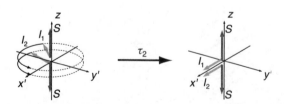

Figure 16.12   Vector diagram representing precession of the I-magnetization components following the selective $\pi_{x,I}$-pulse. If the time delays before and after the pulse are equal, the two I-vectors will return to their initial positions.

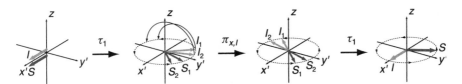

Figure 16.13   Summary of the effect of refocusing sequence on the S-spin of a single population, with both the I- and S-magnetization components in the transverse plane.

clear experiments with more complex molecules, decoupling is almost always incorporated in the data acquisition period in order to simplify the resulting spectra. In these cases, it is usually the ^{1}H signal that is directly recorded and decoupling pulses are applied to another nucleus type, such as ^{15}N or ^{13}C. Decoupling pulses are also frequently incorporated in the preparation periods of heteronuclear multidimensional experiments, where they suppress the effects of coupling on the evolution of indirectly detected nuclei.

16.2   Suppressing Chemical-Shift Differences While Detecting Scalar Coupling

The decoupling strategy described in Section 16.1 has the effect of simplifying a spectrum so that differences in the chemical shifts of different nuclei can be detected without the confusion generated by scalar coupling effects. In other situations, it is desirable to do the opposite—that is, make the splitting due to scalar coupling observable, while hiding differences in Larmor frequencies of nuclei in different chemical environments. In general, this is not very useful during the data acquisition period, but it can be an important component of the preparation period of multi-pulse experiments, as illustrated later in this chapter.

To illustrate the pulse sequence used to suppress chemical-shift differences, we again consider the $|\alpha\alpha\rangle$ and $|\beta\alpha\rangle$ states of the I–S-spin system, and begin by applying a selective $(\pi/2)_y$-pulse to the S-spins, as shown in Fig. 16.14. In the drawings, the precession time and the relative frequencies of the S-spins in the two populations have been chosen, for convenience of illustration, so that the spin that began in the $|\alpha\alpha\rangle$ state precesses $\pi/2$ rad, and the spin that began in the $|\beta\alpha\rangle$ state precesses $3\pi/8$ rad.

At the end of this precession period, a *non-selective* π-pulse is applied to the I- and S-spins, and precession is allowed to continue for exactly the same time period as before this pulse, as shown in Fig. 16.15. Following the π-pulse, the S-magnetization from the population that began in the $|\alpha\alpha\rangle$ state is now aligned with the negative y'-axis. Because the I-spin in this population is now oriented along the negative z-axis, the S-magnetization precesses at the lower frequency and moves by $3\pi/8$ rad during the second period. This places the S-magnetization $\pi/8$ rad before the x'-axis in the direction of precession.

For the population that began in the $|\beta\alpha\rangle$ state, the π-pulse places the I-magnetization along the positive z-axis, and the S-magnetization is rotated to a position

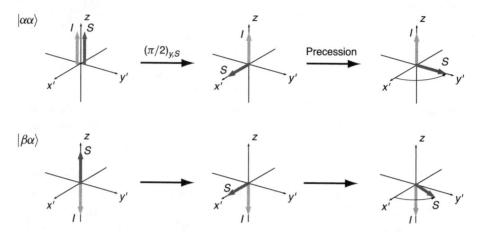

Figure 16.14 Vector diagrams representing the effects of a selective $(\pi/2)_{y,S}$-pulse to the excess $|\alpha\alpha\rangle$ and $|\beta\alpha\rangle$ populations of a scalar-coupled two-spin system, followed by precession.

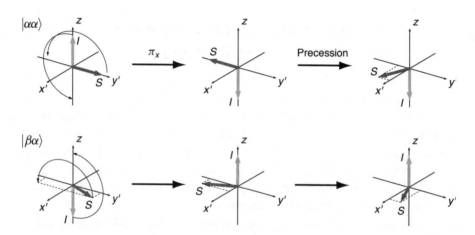

Figure 16.15 Vector diagrams representing the effects of a non-selective π_x-pulse to the magnetization components arising from the excess $|\alpha\alpha\rangle$ and $|\beta\alpha\rangle$ populations, as shown in Fig. 16.14, followed by precession for the same time period.

$\pi/8$ rad ahead (in the direction of precession) of the negative y'-axis. The S-magnetization now precesses at the higher frequency, and is located, at the end of the second period, $\pi/8$ rad ahead of the positive x'-axis.

The net result of these manipulations is that the S-magnetization vectors for the populations lie on opposite sides of the x'-axis. Their distances from the axis are equal and reflect the difference in their precession frequencies—that is, the coupling constant, J. It is as if the average precession frequency happened to match the rotating-frame frequency. Importantly, the final positions of the S-vectors do not depend on the chemical shift of the S-nucleus; the same result is obtained irrespective of the average precession frequency for the S-spins in the two populations. Note, however, that

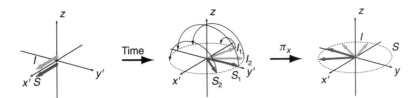

Figure 16.16   Vector diagrams representing the effects of an evolution beginning with I_x- and S_x-magnetization components arising from a single spin population, followed by a non-selective π_x-pulse.

the refocusing pulse has also reversed the direction of the I_z-magnetization in the two populations, a change that will influence further precession.

The final positions of the two S-magnetization vectors, expressed as rotations from the positive x-axis, are $\tau J 2\pi$ and $-\tau J 2\pi$, where τ is the time interval before and after the π_x-pulse. If τ is set to $1/(4J)$ a particularly useful result is obtained—namely, the two vectors point along the positive and negative y'-axes, thus canceling one another. We will see how this situation is utilized in more complex experiments later in this chapter.

This manipulation can also be used to suppress chemical-shift differences in a system in which a single population gives rise to split signals from each spin. The vector diagrams in Fig. 16.16 show the effects of the first evolution period and the π_x-pulse, starting with a state in which the I- and S-magnetization components are both initially aligned with the x'-axis. Once again, the simple vector diagrams do not make a clear prediction of how the frequencies of the magnetization components present after the pulse are related to those present before. But, the behavior can be predicted using correlation diagrams. As diagrammed in Fig. 16.17, the initial S_x-magnetization evolves into a mixture of observable magnetization components and correlations during the first delay period. The π_x-pulse then reverses the signs of the I_z-components, as well as the signs of the S_y-components, as diagrammed in Fig. 16.18. The frequencies of the two S-components are now reversed, as are their relative positions. The second evolution period places the S-components on opposite sides of the x'-axis, as shown in Fig. 16.19. The same effect is also observed for the I-magnetization. The final positions do not depend on the chemical shift of either spin, and the spacings are the same for the I- and S-spins, because the same coupling constant applies to both.

Figure 16.17
Evolution of S_x-magnetization into a combination of observable S_x- and S_y-magnetization components and $I_z S_x$ and $I_z S_y$ correlations.

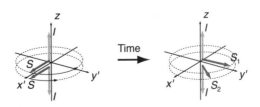

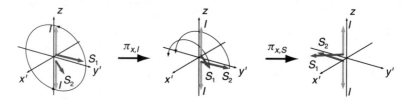

Figure 16.18 Vector diagrams representing the effects of a non-selective π_x-pulse, shown as sequential pulses to the I- and S-spins, applied to the magnetization and correlation components derived from the evolution of S_x-magnetization, as diagrammed in Fig. 16.17. The labels S_1 and S_2 indicate the frequencies of the associated magnetization components, which are reversed by the non-selective pulse.

Figure 16.19

Vector diagrams representing evolution of the magnetization and correlations following the non-selective refocusing pulse shown in Fig. 16.18.

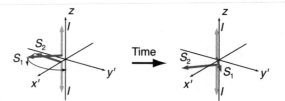

Note also that the initial S_x-magnetization is partially converted to an I_zS_y correlation. Simultaneously, the I_x-magnetization is partially converted into an I_yS_z correlation. If the time intervals before and after the π_x-pulse are set to $1/(4J)$, then the initial I_x and S_x-magnetization components are completely converted to correlations, as shown in Fig. 16.20. Though invisible, the I_zS_y and I_yS_z correlations can be manipulated by subsequent pulses and evolution periods into observable magnetization components. As in the earlier examples, the net effect of the refocusing pulse is to suppress the effects of chemical-shift differences so as to create the same result as would be seen if the average precession frequencies happened to match the rotating-frame frequencies for the I- and S-spins.

In Section 16.3, we will see one way in which a refocusing pulse that suppresses the effects of chemical-shift differences can be incorporated in a useful pulse sequence.

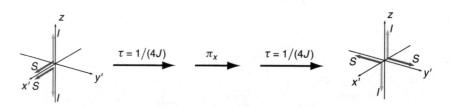

Figure 16.20 Summary of the conversion of the S_x-magnetization component into an I_zS_y correlation via a refocusing pulse sequence.

16.3   Polarization Transfer and the INEPT Experiment

Although they can provide invaluable information, spectra based on ^{13}C and ^{15}N suffer a major disadvantage: The signals generated by these nuclei are much weaker than those from ^1H. The weaker signals reflect the smaller gyromagnetic ratio, which has three effects, all of them bad:

1. The equilibrium population difference between the eigenstates is proportional to the gyromagnetic ratio, and, as we have seen, the net signal depends on this population difference.
2. The gyromagnetic ratio represents the strength of each nuclear magnet, and the strength of the electrical signal is proportional to this strength.
3. The strength of the signal is also proportional to the rate at which the field from each nuclear magnet passes across the coils in the spectrometer (i.e., the precession frequency), and this rate is also proportional to the gyromagnetic ratio.

Since the gyromagnetic ratios of ^{13}C and ^{15}N are approximately 26% and 10%, respectively, that of ^1H, these effects result in much reduced sensitivity. The sensitivity is decreased much further if the ^{13}C or ^{15}N is present only in natural abundance. Problems 1 and 3 listed above can be addressed, to some degree, by increasing the static field strength, though this is very expensive.

There is an another method for increasing the sensitivity of the signal from ^{13}C or ^{15}N that, surprisingly, involves almost no cost. By a clever manipulation of the spin populations, the population difference due to the ^1H nuclei can be "transferred" to a heteronucleus to which it is covalently attached. This process is called *polarization transfer* and is a central component of nearly all multidimensional heteronuclear experiments. To illustrate this concept, we begin with a simplified experiment that would work only for the special case where the average resonance frequency for the ^1H nucleus exactly matches the rotating-frame frequency. This idealized experiment can be extended to a more practical one by incorporating an appropriate refocusing pulse.

The pulse sequence for the simplified one-dimensional experiment involves an initial selective $\pi/2$-pulse to the ^1H nuclei (labeled I), followed by a delay of a defined interval and pulses to both nuclei, as illustrated in Fig. 16.21. Before we analyze this pulse sequence, consider the equilibrium populations of the eigenstates and how they would contribute to a simple one-dimensional spectrum from the less sensitive S-nucleus. The relevant population differences can be written as

$$\delta_I = N_{\alpha\alpha} - N_{\beta\alpha} = N_{\alpha\beta} - N_{\beta\beta}$$
$$\delta_S = N_{\alpha\alpha} - N_{\alpha\beta} = N_{\beta\alpha} - N_{\beta\beta}$$

(16.1)

These differences can be calculated to a very good approximation as

$$\delta_I = N_{\alpha\alpha}\hbar\gamma_I B/(kT)$$
$$\delta_S = N_{\alpha\alpha}\hbar\gamma_S B/(kT)$$

(16.2)

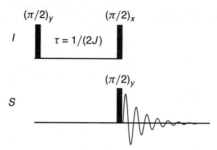

Figure 16.21

Pulse sequence for a simplified polarization transfer experiment.

where γ_I and γ_S are the gyromagnetic ratios of the two nuclei, B is the static field strength and k is Boltzmann's constant.

If S represents, for instance, a ^{13}C nucleus, and we apply a simple $\pi/2$-pulse to this spin and record the FID, we expect to see two frequency components. The signal with the lower of these frequencies will depend on the initial population difference between the $|\beta\alpha\rangle$ and $|\beta\beta\rangle$ states, and the higher-frequency signal will depend on the difference between the $|\alpha\alpha\rangle$ and $|\alpha\beta\rangle$ populations. Thus, the strengths of both signals reflect δ_S and the gyromagnetic ratio for ^{13}C.

Now, we consider the effects of the polarization-transfer pulse sequence on each of the four eigenstates. In Fig. 16.22, the first two steps are illustrated with magnetization vector diagrams for the molecules that begin in the $|\alpha\alpha\rangle$ state. For these molecules, the I-magnetization precesses in the positive direction with the frequency, $J/2$ Hz, relative to the rotating-frame frequency. Because the time interval is set to $1/(2J)$, the magnetization precesses 1/4 revolution, or $\pi/2$ rad, to the y'-axis. For directly bonded ^1H–^{13}C nuclei, the coupling constant is typically about 140 Hz, so the delay time would be set to about 3.5 ms. For ^1H–^{15}N pairs, $J \approx 90$ Hz, so the delay is about 5.5 ms. Following this delay, a $\pi/2$-pulse is applied to the I-spin along the x'-axis, and a $\pi/2$-pulse is applied to the S-spin along the y'-axis, as shown in Fig. 16.23. From its effects on the molecules in the $|\alpha\alpha\rangle$ state, this exercise seems rather pointless: Exactly the same thing could be accomplished by simply applying a $\pi/2$-pulse to the S-spin along the y'-axis. To understand the rationale for this pulse sequence, we have to examine its effects on the other three eigenstates.

Beginning with the $|\alpha\beta\rangle$ state, Fig. 16.24 shows the effects of the initial pulse and delay. Because the S-spin is in the β state now, the precession of the I-spin has a negative frequency relative to the rotating frame, and the I-magnetization moves to

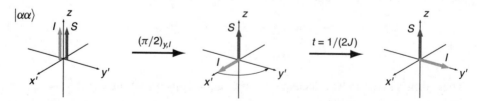

Figure 16.22   Vector diagrams representing the effects of the first two steps of the simplified polarization-transfer experiment on the population of molecules that begin in the $|\alpha\alpha\rangle$ state.

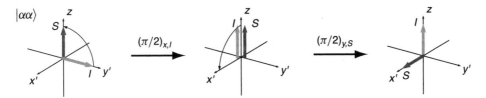

Figure 16.23 Vector diagrams representing the effects of the final two pulses of the simplified polarization-transfer experiment on the population of molecules that begin in the $|\alpha\alpha\rangle$ state.

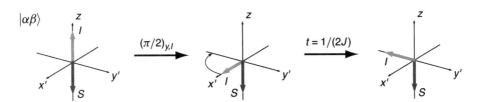

Figure 16.24 Vector diagrams representing the effects of the first two steps of the simplified polarization-transfer experiment on the population of molecules that begin in the $|\alpha\beta\rangle$ state.

the negative y'-axis during the delay period. The final pulses have the effects shown in Fig. 16.25. Note that the net result here is *not* the same as would be obtained by applying a single $(\pi/2)_{y,S}$-pulse to the $|\alpha\beta\rangle$ state. If we had done that, the S-magnetization would point along the negative x'-axis, as it does here, but the I-magnetization would still lie along the positive z-axis. Because the I-magnetization lies along the negative z-axis, the S-spin will now give rise to a signal that has the lower of its two frequencies. If the single $(\pi/2)_{y,S}$-pulse had been used, the molecules that began in the $|\alpha\alpha\rangle$ and $|\alpha\beta\rangle$ states would have given rise to signals with the same frequency (the higher one), but opposite signs.

The same manipulations applied to the $|\beta\alpha\rangle$ state give the results shown in Fig. 16.26. In this case, the effect is the same as that of a single $(\pi/2)_{y,S}$-pulse, just as

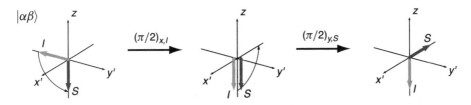

Figure 16.25 Vector diagrams representing the effects of the final two pulses of the simplified polarization-transfer experiment on the population of molecules that begin in the $|\alpha\beta\rangle$ state.

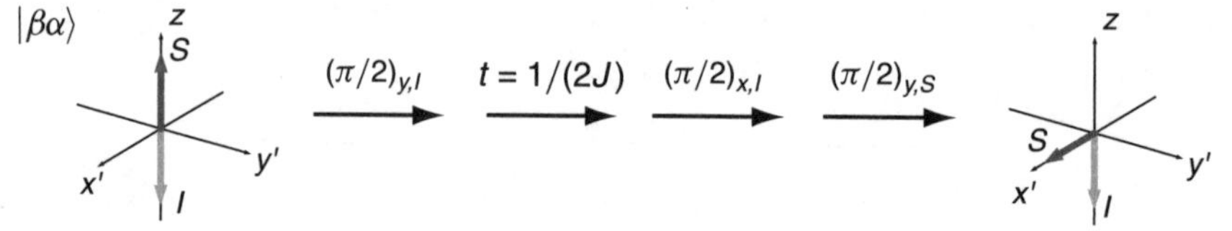

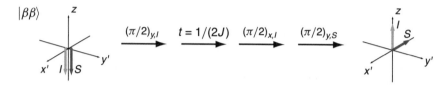

Figure 16.26 Vector diagrams representing the effects of the simplified polarization-transfer experiment on the population of molecules that begin in the $|\beta\alpha\rangle$ state.

Figure 16.27 Vector diagrams representing the effects of the simplified polarization-transfer experiment on the population of molecules that begin in the $|\beta\beta\rangle$ state.

it was for the $|\alpha\alpha\rangle$ state. This population will give rise to a signal with the lower S-frequency and of opposite sign from that generated by the molecules that begin in the $|\alpha\beta\rangle$ state.

Finally, Fig. 16.27 shows the result starting with the $|\beta\beta\rangle$ state. As seen for the $|\alpha\beta\rangle$ state, the combination of pulses is not equivalent to a single $(\pi/2)_{y,S}$-pulse. In fact, the end result for the $|\beta\beta\rangle$ state is the same as if a $(\pi/2)_{y,S}$-pulse had been applied to the $|\alpha\beta\rangle$ state. The converse is also true; the effect on the $|\alpha\beta\rangle$ state is the same as if a single $(\pi/2)_{y,S}$-pulse had been applied to the $|\beta\beta\rangle$ state. In essence, the combination of pulses has reversed the roles of the molecules that begin in the $|\alpha\beta\rangle$ and $|\beta\beta\rangle$ states. If the two kinds of molecules were initially present at equal concentrations, there would be no effect. But, because their energies are different, their equilibrium populations differ, and the effects on the spectrum are quite pronounced.

To predict the resulting spectrum, we need to consider how each of the populations will contribute to the signal and the relative strengths of those contributions. The signal with the higher S-frequency will be generated by the molecules that begin in the $|\alpha\alpha\rangle$ and $|\beta\beta\rangle$ states, and the intensity of the signal will be proportional to the equilibrium population difference between these two states. From Eq. 16.1, it is easy to show that this difference is given by

$$N_{\alpha\alpha} - N_{\beta\beta} = \delta_I + \delta_S$$

Remember that for a simple spectrum based on a direct pulse to the S-spin, the lower-frequency signal will be proportional to $N_{\alpha\alpha} - N_{\alpha\beta}$, which we defined earlier as δ_S. Remember, also, that each of the original population differences, δ_I and δ_S, is proportional to the gyromagnetic ratio for the corresponding nucleus. Thus, the signal strength for the higher S-frequency is changed from a value proportional to γ_S to one proportional to $\gamma_S + \gamma_I$. For a ^{13}C nucleus bonded to an ^1H nucleus, this is an increase in signal intensity of approximately five-fold, and about 11-fold for a ^{15}N nucleus bonded to an ^1H nucleus.

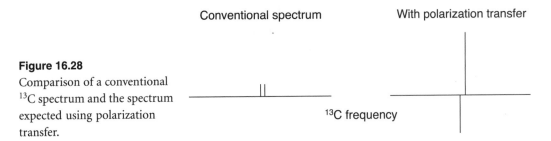

Figure 16.28

Comparison of a conventional ^{13}C spectrum and the spectrum expected using polarization transfer.

When we look at the lower S-frequency, the situation is a little bit different. For this frequency, the signal is proportional to the equilibrium population difference between the $|\beta\alpha\rangle$ state, which makes a positive contribution as we have defined it here, and the $|\alpha\beta\rangle$ state, which makes a negative contribution. Thus, the signal intensity will be proportional to the difference:

$$N_{\beta\alpha} - N_{\alpha\beta} = \delta_S - \delta_I$$

Since δ_I is greater than δ_S, this signal will have a sign opposite to that of the other component of the doublet, and the intensity of the signal will be slightly lower. Still, this represents a quite significant increase in sensitivity. Fig. 16.28 compares the expected spectrum from a conventional ^{13}C experiment and that predicted when the polarization-transfer pulse sequence is used.

The version of this procedure we have discussed so far is not very practical, since it works only if the average frequency of the I-spin doublet matches the rotating-frame frequency. The experiment can readily be made more general, however, by incorporating a refocusing pulse halfway through the delay period. As shown in the previous section, a non-selective π-pulse to the S- and I-spins refocusses the magnetization from both spins so that the effects of chemical-shift differences are hidden, while those due to scalar coupling remain visible, just as if the average precession frequency of the I-spin matched the rotating-frame frequency. The complete pulse sequence is shown in Fig. 16.29. This pulse sequence was first introduced in 1979 by Morris and Freeman, who gave it the rather ironic name "INEPT," for Insensitive Nuclei Enhanced by Polarization Transfer. The essence of the technique, which may be lost in the details, is that the (relatively) large population difference, or polarization, attributable to the I-spin is used to enhance the signal from the S-spin. It is not only

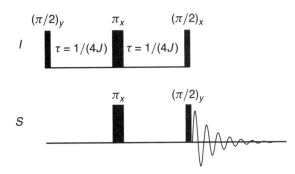

Figure 16.29

Pulse sequence for the INEPT (Insensitive Nuclei Enhanced by Polarization Transfer) experiment.

used for simple experiments with the less sensitive nuclei, but it is also incorporated as an important component of more complex multidimensional experiments, as illustrated in a subsequent section.

16.4   Reverse INEPT

Following the initial $\pi/2$-pulse to the I-spin, the INEPT sequence can be summarized as a refocussed evolution period, with total time $1/(2J)$, followed by $\pi/2$-pulses applied to the I- and S-spins. The reverse of this pulse sequence, illustrated in Fig. 16.30, also turns out to be very useful. Here, the goal is to detect the state of the S-spin through its effect on the magnetization of the I-spin, as observed in an FID recorded from the I-spin. The starting point for this manipulation usually has the S-magnetization in the transverse plane, and the final I-signal is determined by the initial direction of the S-magnetization.

The way this pulse sequence works can be illustrated by considering two cases of pure spin populations, one in which the S-magnetization points along the x'-axis (A) and one in which it points along the y'-axis (B), as shown in Fig. 16.31. In each case, the I-magnetization is initially aligned with the z-axis and is rotated to the x'-axis by the first pulse. A simultaneous $(\pi/2)_y$-pulse is applied to the S-spin. For case A, this pulse rotates the S-magnetization to the negative z-axis, but the pulse does not affect the S-magnetization in case B, because this vector is parallel to the direction of the pulse. The different positions of the S-magnetization then determine the behavior of the I-magnetization during the precession period, as shown in Fig. 16.32. For case A, the I-magnetization precesses with a single frequency, because the S-magnetization has no transverse component. During the first half of the evolution period, the frequency is $\nu_{I,2}$ (the lower of the two frequencies), while in the second half it is $\nu_{I,1}$. Precession for the total time $1/(2J)$ moves the I-magnetization to the positive y'-axis, irrespective of the value of ν_I.

For case B, the first two pulses place both the I- and S-magnetization components in the transverse plane, and both precess with two frequency components, represented in Fig. 16.32 as split vectors. The π-pulse to the two spins in the middle of the evolution period changes both the orientation of these vectors (relative to the x'-axis) and their precession frequencies, as indicated by the subscripts. (See Fig. 16.18.) At the end of the total time, the two frequency components for the I-spin are oriented

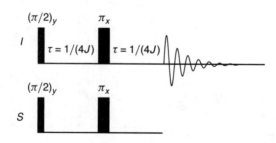

Figure 16.30

The reverse INEPT pulse sequence.

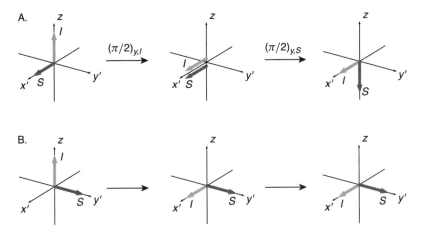

Figure 16.31   Vector diagrams illustrating the effects of the initial pulses of the reverse INEPT sequence, applied to pure populations beginning with different transverse S-magnetization components.

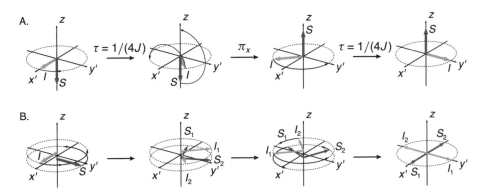

Figure 16.32   Vector diagrams illustrating precession during the refocused evolution period of the reverse INEPT sequence, for populations beginning with different transverse S-magnetization components, as in Fig. 16.31.

in opposite directions along the y'-axis, and the two components for the S-spin are pointed in opposite directions along the x'-axis.

If the FID signal is now recorded for the I-spin, the two cases will give rise to very different spectra. For case A, where the S-magnetization was initially aligned with the x'-axis, there is single peak, while for the case where the S-magnetization began along the y'-axis, there are two peaks with opposite signs (an antiphase doublet) as shown in Fig. 16.33. The difference between the two can be made even clearer by applying a decoupling signal to the S-spin during the data acquisition period. For case A, this will simply shift the observed frequency from $\nu_{I,1}$ to the average I-frequency, $\nu_{I,1} - J/2$. For case B, however, decoupling averages the two signals, and they cancel one another. Thus, a signal is only observed for the case where the initial S-magnetization is aligned with the x'-axis.

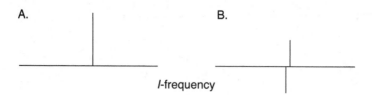

Figure 16.33   Expected spectra from application of the reverse INEPT pulse sequence to starting states with the S-magnetization aligned with the x'-axis (A) or y'-axis (B).

The applications of this pulse sequence would be rather limited if it could only distinguish between the extreme cases illustrated above. In fact, however, it works for any orientation of the S-magnetization in the transverse plane, converting this information into the amplitude of a signal from the I-spin. If the initial position of the S-magnetization is described by the angle, a, it forms with the x'-axis, the effects of the reverse INEPT sequence can be summarized by the vector diagrams in Fig. 16.34. The $(\pi/2)_{y,I}$-pulse rotates the I-magnetization to the x'-axis, as before, and the $(\pi/2)_{y,S}$-pulse rotates the S-magnetization from the transverse plane to a position between the z- and y'-axes, determined by the value of a. During the refocused evolution period, each spin gives rise to two frequency components. The S-frequency components have equal intensity, since the I-magnetization lies entirely in the transverse plane. But, the relative amplitudes of the two I-components are determined by the net z-magnetization of the S-spin (see pages 402–405). For the case illustrated in Fig. 16.34, the more intense I-component has the lower frequency during the first half of the evolution period, but then is changed to the higher frequency by the refocusing pulse. At the end of the refocusing period, the two I-magnetization components are pointing in opposite directions along the y'-axis and have different magnitudes. As a result, the higher-frequency component in the FID has a greater intensity than the lower frequency, as well as the opposite sign.

Using the wavefunction and appropriate operators, a general mathematical description of the resulting signal (measured along the y'-axis) can be derived:

$$\langle I_y(a, t)\rangle = \left(1/4\right)(1 + \cos(a))\cos(2\pi v_{I,1}t) - (1/4)\left(1 - \cos(a)\right)\cos(2\pi v_{I,2}t)$$

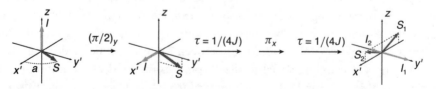

Figure 16.34   Vector diagrams illustrating the effects of the reverse INEPT sequence on a pure population, with the S-magnetization at an arbitrary position in the transverse plane, defined by the angle a.

where t is the time during the acquisition period. The equation shows that the two frequency components will always have opposite signs and that their relative intensities are determined by the cosine of angle a—that is, the initial position of the S-magnetization. If decoupling is applied during the acquisition period, the two frequency components are combined, as described by

$$\langle I_y(a, t) \rangle = (1/2) \cos(a) \cos \left(\pi (\nu_{I,1} + \nu_{I,2}) t \right)$$

Thus, the pulse sequence neatly converts the position of the S-magnetization into the amplitude of a signal from the I-spin. Note that the amplitude is determined by the cosine of a, which is equivalent to the initial projection of the S-magnetization along the x'-axis. This manipulation is especially useful in a multidimensional experiment, where the position of the S-magnetization is determined by an evolution period and then this information is effectively transferred to the I-spin.

In addition to having a structure that is the opposite of the INEPT pulse sequence, this sequence is in a very practical sense complementary to the INEPT. (Rather surprisingly, no one seems to have forced the word "adept" into a suitable acronym.) While the INEPT sequence converts the net z-magnetization of the I-spin to an enhanced transverse S-magnetization, the reverse INEPT converts the x-magnetization of the S-spin into transverse I-magnetization. It is worth noting, however, that the mechanisms by which these transfers take place are rather different. In the case of INEPT, the initial I_z-magnetization reflects a population difference, and the enhancement of S_x-magnetization, beyond what would be observed from a simple $(\pi/2)_{y,S}$-pulse, is due to a manipulation of the spin populations. The effect cannot be accounted for by considering only a single starting wavefunction. For the reverse INEPT, on the other hand, the transfer is based on the effect of one spin, S, on the frequency splitting of the other, I, during an evolution period. This quantum mechanical effect can be described for a single wavefunction, though any real experiment will, of course, reflect the net contributions of all of the spins in the sample. In Section 16.5, we will see how the INEPT and reverse INEPT sequences can be combined into a very effective two-dimensional heteronuclear experiment.

16.5 The Heteronuclear Single-Quantum Coherence (HSQC) Experiment

In Chapter 15, we considered a two-dimensional heteronuclear COSY experiment as an introduction to the more widely used homonuclear experiment. This experiment generates a spectrum in which each dimension represents the resonance frequency for one nucleus type, typically ^1H for the directly detected dimension and ^{13}C or ^{15}N for the indirectly detected dimension. Each heteronuclear pair linked by scalar coupling gives rise to a set of doubly antiphase peaks at a position reflecting the resonance frequencies of the two nuclei. This general type of spectrum is extremely useful because it reveals the covalent connectivity between nuclei that are otherwise identified only by their resonance frequencies. As a practical experiment, however, the

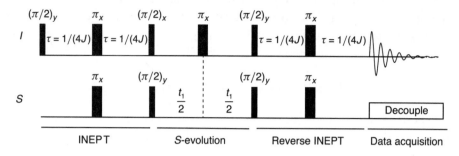

Figure 16.35   The pulse sequence for the HSQC experiment.

heteronuclear COSY is limited by its relative insensitivity, which reflects the smaller gyromagnetic ratio of the non-^{1}H spin. Although variations of this experiment were used extensively in the past, they have been largely replaced by two other experiments, which, while based on more complex pulse sequences, give rise to greater sensitivity.

These newer methods are the heteronuclear single-quantum coherence (HSQC) and heteronuclear multiple-quantum coherence (HMQC) experiments. As indicated by their names, the two are based on somewhat different quantum mechanical phenomena. Although it involves a considerably more complex pulse sequence than the HMQC, we will discuss the HSQC first because it can be readily explained using the elements described in the previous sections. We will return to the HMQC in Section 16.6.

The pulse sequence for the HSQC is illustrated in Fig. 16.35. This is a formidable sequence, but it should be recognizable as a composite of smaller ones we have discussed before, as indicated by the labels at the bottom. The major new element is a period of variable duration, with a selective π_x-pulse to the I-spin in the middle, which separates INEPT and reverse INEPT pulse sequences. To generate a two-dimensional spectrum, a series of one-dimensional ^{1}H spectra are recorded with the duration of this middle period increased incrementally. As before, the resulting data are converted to the frequency domain by a two-dimensional Fourier transform, and, as we will see below, the second dimension reflects the precession frequency of the S-spin.

As illustrated earlier, the net effect of the INEPT sequence on each of the starting eigenstates is to generate a state in which the I-magnetization is aligned with the positive or negative z-axis, and the S-magnetization is aligned with the positive or negative x'-axis. For now, we will consider only the population that begins in the $|\alpha\alpha\rangle$ state. The net effect of the INEPT sequence on this state is to simply rotate the S-magnetization to the x'-axis. During the two $t_1/2$ evolution periods, the S-magnetization precesses as shown in Fig. 16.36. The selective $\pi_{x,I}$-pulse in the middle of the evolution period has exactly the same role as a decoupling signal during a data acquisition period—namely, it ensures that the final position of the S-magnetization

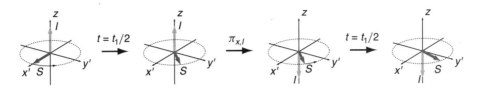

Figure 16.36   Vector diagrams representing precession of the S-magnetization during the evolution period of the HSQC experiment.

reflects the *average* Larmor frequency of the spin. As a consequence, all of the molecules in the sample contribute to a single frequency component, irrespective of the starting state of the I-spin.

The position of the S-magnetization in the transverse plane is then converted into the intensity of a transverse I-magnetization component through the reverse INEPT sequence, exactly as illustrated in Section 16.4. As the duration of the t_1 period is incremented, the intensity of the signal from the I-spin is modulated according to the Larmor frequency of the S-spin to which it is scalar coupled.

From the examples of two-dimensional spectra discussed in detail earlier, it should be apparent that Fourier transformation of the resulting signals generates a spectrum in which one dimension represents the I-spin frequency and the other the S-spin frequency. One of the great virtues of the HSQC and related experiments is that they utilize both the initial population difference associated with the ^1H nucleus, through the incorporation of the INEPT sequence, and the stronger signal that this nucleus generates during the data acquisition period. The S-spin, with its generally lower gyromagnetic ratio, participates only in the evolution period that is detected indirectly through the ^1H magnetization.

What about all of the other I-spins (i.e., ^1H nuclei) that are likely to be present in a molecule but are not scalar coupled to a heteronuclear S-spin? Since they have also been stimulated by the initial $(\pi/2)_{y,I}$-pulse, they will inevitably contribute to the signal induced in the spectrometer coil. If these other nuclei are completely isolated from other spins, then their magnetization components should be refocused by the three π-pulses, and their contribution to the NMR signal will be independent of the length of the t_1 delay period. When subjected to the second dimension of the Fourier transformation, these signals should contribute only a zero-frequency component. In practice, however, the length of the t_1 evolution period does affect the net magnetization from nuclei that are not coupled to heteronuclei, because their final magnetization components are influenced by other scalar coupling effects and relaxation processes (including the nuclear Overhauser effect) that act during this period. In order to obtain a spectrum containing only the desired cross peaks, the magnetization components must be eliminated either by combining the signals from individual iterations with systematically altered phases for selected pulses (phase cycling) or by using pulsed field gradients. These procedures effectively filter the

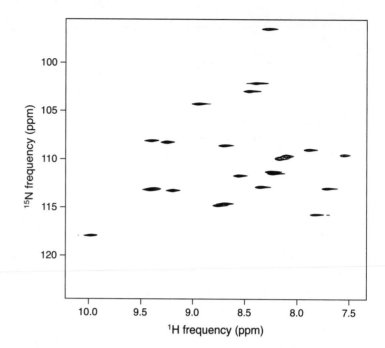

Figure 16.37

HSQC spectrum of a small protein, ω-conotoxin MVIIA-Gly.

signal so that only the protons directly coupled to the heteronucleus contribute to the spectrum.

The HSQC experiment is widely used in studies of proteins, most frequently with samples enriched with ^{15}N. Because each backbone amide group contains a nitrogen atom covalently bonded to a hydrogen atom, nearly every amino acid residue in a protein gives rise to a cross peak in the HSQC spectrum, the exceptions being proline residues and the amino-terminal residue (which has three hydrogen atoms that exchange rapidly with the solvent water). The side chains containing N–H bonds can also contribute to the HSQC spectrum. Fig. 16.37 shows a relatively simple HSQC spectrum, for an ^{15}N-enriched 27-residue peptide toxin, ω-conotoxin MVIIA-Gly. This spectrum shows excellent dispersion in both the ^{1}H and ^{15}N dimensions, indicating that the peptide has a well-defined three-dimensional structure in spite of its small size. Although the HSQC spectrum does not contain enough information on its own to allow resonance assignments to be made, the ability to resolve signals from most or all of the backbone amide groups is usually taken as a promising sign that more advanced experiments will be feasible with a particular sample. Thus, recording a high-quality HSQC spectrum often represents an important milestone in a protein NMR project.

Once the backbone ^{1}H and ^{15}N resonances have been assigned to specific atoms in the molecule, the basic HSQC spectrum can be used for a variety of biochemical experiments. For instance, interactions between an ^{15}N-labeled protein and another molecule can often be detected by changes in the chemical shifts of the amide ^{15}N or ^{1}H nuclei. The largest effects are usually limited to nuclei close to the interaction surface, making it possible to identify the residues that are most likely to interact with the other molecule.

Another application of the HSQC experiment is the measurement of the rates at which amide hydrogen atoms exchange with hydrogen atoms in the solvent. In a typical experiment, an ^{15}N-labeled protein dissolved in H_2O is rapidly subjected to a chromatographic process that replaces the H_2O solvent with D_2O. A series of HSQC spectra are then recorded at different times after the exchange. As ^2H from the solvent exchanges with the amide ^1H nuclei of the protein, the intensities of the peaks in the HSQC spectrum decrease. The rates of exchange depend greatly on the position of the amide groups in the protein as well as the solution conditions. In general, the nuclei that exchange most slowly are those that are buried in the folded structure and participate in intramolecular hydrogen bonds. This technique has been widely used to study protein stability, dynamics and folding mechanisms. Although other methods can be used to monitor hydrogen exchange, the ability of NMR to monitor individual amide groups makes it a particularly powerful approach.

The HSQC pulse sequence also serves as a building block for yet more complicated experiments. For instance, it can be used to measure relaxation rates by incorporating delay times during which inverted longitudinal magnetization (for R_1) or transverse magnetization (for R_2) are allowed to decay before the signal is recorded. A variety of three-dimensional experiments have been devised by combining the HSQC with pulse sequences for other two-dimensional experiments, such as a homonuclear NOESY or TOCSY, as we will discuss briefly later in this chapter.

16.6   The Heteronuclear Multiple-Quantum Coherence (HMQC) Experiment

A two-dimensional heteronuclear spectrum can also be created using the much simpler HMQC pulse sequence outlined in Fig. 16.38. Perhaps the first thing to notice about this pulse sequence is that the I-spin is subjected only to an initial $(\pi/2)_y$-pulse, followed by a delay period with a π_x-refocusing pulse in the middle. We might expect,

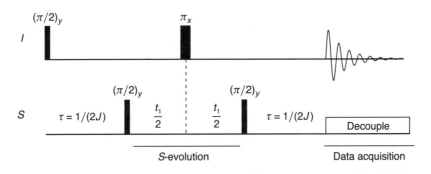

Figure 16.38   The pulse sequence for the HMQC experiment.

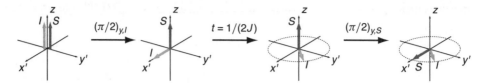

Figure 16.39 Vector diagrams representing the initial steps of the HMQC experiment, applied to the population of molecules initially in the $|\alpha\alpha\rangle$ state.

then, that any evolution of the I-magnetization before the data acquisition period would simply be refocussed to the x'-axis. As we will see, all of the I-magnetization does, in fact, return to the x'-axis, but the presence of transverse S-magnetization during part of the preparation period leads to splitting of the I-magnetization into two components, the relative intensities of which are influenced by the evolution of the S-magnetization.

The first stages of the HMQC sequence are illustrated in Fig. 16.39 with vector diagrams representing the effects on a population that begins in the $|\alpha\alpha\rangle$ state. During the first delay period, only the I-magnetization lies in the transverse plane, and the precession of this component reflects both the average chemical shift of the I-spin and the state of the S-spin. For the population starting in the $|\alpha\alpha\rangle$ state, the precession frequency is $\nu_{I,1} = \nu_I + J/2$.

After the initial delay, which is set to $1/(2J)$ for each of the one-dimensional spectra that make up the experiment, a $(\pi/2)_y$-pulse is applied to the S-spin. Because the magnetization components for both spins now lie in the transverse plane, the precession of each is described by two frequency components, as represented by the split arrows in the vector diagrams of Fig. 16.40. The I- and S-magnetization components are allowed to precess for a time period, $t_1/2$, where t_1, as usual, is the total time period that is incremented in the two-dimensional experiment. A π_x-pulse is then applied to the I-spin at a point that represents the mid-point of both the total preparation period and the incremented interval, t_1. This pulse flips the positions of the two I-magnetization components with respect to the x'-axis and also interchanges the frequencies for the two S-components (pages 463–466). The four magnetization components then precess for a second interval of length $t_1/2$. During this interval, the magnetization components from each spin re-converge. The position of the net S-magnetization reflects the precession during the t_1 period, while that of the I-magnetization reflects both the t_1 period and the initial period of $1/(2J)$.

The last pulse of the sequence is a selective $(\pi/2)_y$-pulse to the S-spin. As shown in Fig. 16.41, this pulse converts any x'-magnetization due to the S-spin to z-magnetization. Both spins are then allowed to precess for another period of $1/(2J)$. Since the I-magnetization lies entirely in the transverse plane, the S-magnetization precesses with two frequency components of equal amplitude. The I-magnetization will, in general, also split into two components, but the relative amplitudes of these components will depend on the net z-magnetization of the S-spin. The final evolution period, of length $1/(2J)$, brings the two I-magnetization components back to the x'-axis, pointing in opposite directions from one another. The transverse com-

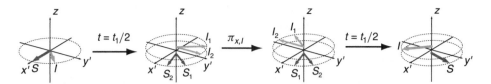

Figure 16.40 Vector diagrams representing the evolution period of the HMQC experiment, following the steps shown in Fig. 16.39 applied to the population of molecules initially in the $|\alpha\alpha\rangle$ state. The vectors representing the I-spin are drawn slightly above the x'–y' plane, and those for the S-spin are drawn below the plane.

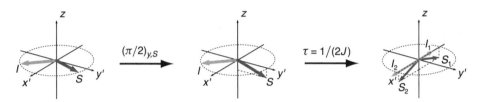

Figure 16.41 Vector diagrams representing the final steps of the HMQC experiment, following the steps shown in Fig. 16.40 applied to the population of molecules initially in the $|\alpha\alpha\rangle$ state.

ponents of the S-magnetization also end up pointing in opposite directions, but not necessarily along either of the coordinate axes. This final stage of the HMQC pulse sequence is very similar to the reverse INEPT sequence, and like that sequence, it converts the position of the S-magnetization (immediately prior to the final $(\pi/2)_{y,S}$-pulse) into the relative magnitudes of the I-magnetization components, which point in opposite directions at the end of the final delay.

During the data acquisition period, the two I-components continue to evolve, generating signals with the two frequencies, $\nu_{I,1}$ and $\nu_{I,2}$. If decoupling pulses are applied to the S-spin during data acquisition, the two I-signals are combined, resulting in a single frequency component with an amplitude that reflects the difference in the projections of the two I-magnetization components along the positive and negative x'-axes at the end of the preparation period. As we saw above, the relative intensities of these components are determined by the position of the S-magnetization vector at the end of the t_1 period. Thus, the net intensity of the I-signal is modulated by the S-spin frequency, and two-dimensional Fourier transformation of the data results in a two-dimensional spectrum very similar to that generated by the HSQC experiment.

The HMQC and HSQC experiments are, in many respects, interchangeable. The more complicated HSQC experiment does, however, have practical advantages that are not easily illustrated by the relatively simple treatment used here. The major disadvantages of the HMQC are that (1) there is faster relaxation of the signal during the S-evolution period, resulting in broader peaks, and (2) the peaks undergo splitting by scalar coupling interactions between the I-spin and other ^{1}H nuclei to which it is indirectly coupled (e.g., the α-hydrogen in an amino acid residue). As a consequence, the HSQC has become the more widely used experiment.

Finally, the reader might ask, what is the fundamental difference between the HMQC and HSQC and what do these names really mean? What is the difference between "multiple-quantum coherence" and "single-quantum coherence"? The key distinction between the two experiments is found in the t_1 evolution period, when the S-magnetization precesses to a position that will ultimately determine the intensity of the signal from the I-spin. In the HSQC experiment, this evolution takes place under conditions where the α- and β-character of the S-spin have been mixed in a superposition state, but the I-spin is in either the α or β state, depending on which of the four eigenstates the particular molecule was in at the beginning of the experiment. As a consequence, the S-spin for a given molecule evolves with a single frequency, corresponding to the single-quantum transition between the α and β states of the S-spin. In contrast, in the HMQC experiment, the state that undergoes evolution during t_1 represents a superposition of the α- and β-character of *both* spins and involves the transverse correlations, $I_x S_x$, $I_x S_y$, $I_y S_x$ and $I_y S_y$. As discussed on pages 412–414, these correlations evolve with frequencies that represent the sum and difference between ν_I and ν_S, which are also the frequencies associated with the double- and zero-quantum transitions. Thus, the designation "multiple-quantum coherence." It should be noted, however, that the evolution of the transverse correlations does not lead to any spectral peaks with frequencies corresponding to the multiple-quantum transitions.

16.7   Three-Dimensional Heteronuclear NMR Experiments

The principle of two-dimensional spectroscopy can be extended to a third or even higher dimension to provide additional resolution and structural information. Although three-dimensional experiments can, in principle, be designed to use only nuclei of one type, such as ^{1}H, the most practical experiments for biomolecular NMR have proven to be those based on combinations of ^{13}C, ^{15}N and ^{1}H.

16.7.1   HMQC-NOESY

To illustrate the principle of experiments in higher dimensions, we will consider one of the simplest examples, which combines two experiments that we have already discussed, the HMQC and the two-dimensional homonuclear NOESY. As a reminder, the NOESY experiment is diagrammed in Fig. 16.42. During the t_1 period, which is incremented in each iteration of the pulse sequence, each of the ^{1}H nuclei precesses in the transverse plane, and the second pulse of the experiment converts the remaining x'-component of this magnetization into longitudinal magnetization. When pairs of nuclei are within about 5 Å of one another, the nuclear Overhauser effect leads to cross relaxation during the mixing period, so that the z-magnetization of each spin is influenced by its neighbors. Following the mixing period, the z-magnetization is rotated to the transverse plane and the FID is recorded.

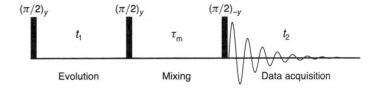

Figure 16.42
Pulse sequence for a two-dimensional homonuclear NOESY experiment.

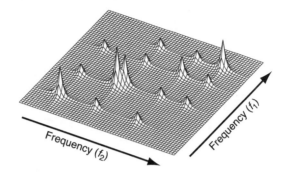

Figure 16.43
A simulated two-dimensional NOESY spectrum.

A simulated two-dimensional NOESY spectrum for a simple system composed of only five ^{1}H nuclei is shown in Fig.16.43. For this hypothetical case, there are five peaks on the diagonal, representing the five protons, but two of them have very similar chemical shifts, potentially making it difficult to determine which is associated with each of the three cross peaks that are associated with this resonance frequency. Furthermore, cross peaks arising from interactions between the two nuclei with similar chemical shifts are likely to be hidden by the much stronger diagonal peaks.

The pulse sequence for a three-dimensional experiment that can help resolve these ambiguities, based on the HMQC and two-dimensional NOESY sequences, is shown in Fig. 16.44. As indicated in the drawing, the preparation period can be divided into portions that are essentially the same as those for the separate HMQC and NOESY experiments. The preparation period now includes *two* delay periods that

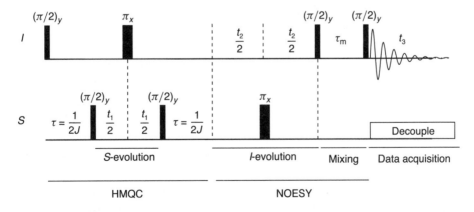

Figure 16.44   Pulse sequence for a three-dimensional HMQC-NOESY experiment.

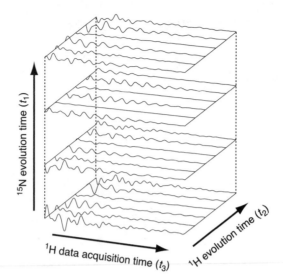

Figure 16.45

Arrayed FIDs for a simulated three-dimensional ^1H–^{15}N HMQC-NOESY experiment.

are systematically, and independently, incremented as successive FIDs are recorded. Each of these is divided by a refocusing pulse, but it is the total time of the periods, labeled t_1 and t_2, that are important. The data acquisition time is now labeled t_3.

The first part of the preparation period is the same as used in the HMQC experiment, and, as before, it results in I-magnetization aligned along the x'-axis, with the net magnitude and sign reflecting the precession of the S-spin during the t_1 period. The HMQC period, in effect, replaces the initial $\pi/2$-pulse in the NOESY experiment, which also has the function of aligning the I-magnetization with the x'-axis. Now, however, the intensity of that magnetization component and the resulting FID are determined, in part, by the precession of the S-spin during t_1. The rest of the pulse sequence is the same as for the two-dimensional NOESY, except that a decoupling pulse is included during the evolution period, and a decoupling signal is applied during the data acquisition period. These eliminate the splitting that would otherwise be observed for any I-spins that are coupled to S-spins.

The experiment is executed by first setting t_1 and collecting a series of FIDs with incremented values of t_2. The value of t_1 is then incremented, and another set of FIDs is recorded with the full range of t_2 values. This process is repeated to generate a set of FIDs that can be thought of as forming a stack of two-dimensional data sets, as illustrated in Fig. 16.45. For this example, the I-spins are assumed to be ^1H nuclei and the S-spins are assumed to be ^{15}N. Each FID is associated with a unique combination of t_1 and t_2 delay times, and the signal recorded in each reflects the evolution of magnetization during these two periods. The initial form of the data is a three-dimensional array, with each dimension representing the time domain.

The steps in processing the three-dimensional data set are illustrated in Fig. 16.46, using simulated data based on the same idealized case with five ^1H nuclei. In the first step, each of the FIDs is Fourier transformed to generate a one-dimensional spectrum, with the frequency axis labeled f_3. Note that the spectrum at the front of the bottom layer (corresponding to $t_1 = 0$ and $t_2 = 0$) contains five peaks, all

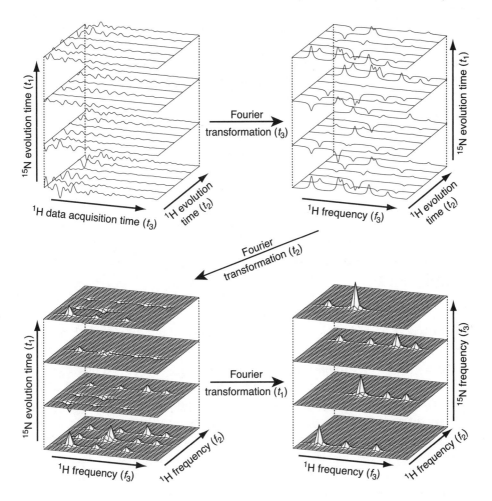

Figure 16.46 Fourier transformation of a simulated three-dimensional HMQC-NOESY experiment.

positive in sign. The peaks differ in intensity because they have been influenced differently by magnetization transfer during the NOESY mixing period. The relative intensities of the peaks in the other one-dimensional spectra are further influenced by the precession of the S- and I-spins during the t_1 and t_2 periods, respectively.

The second stage of data processing is a Fourier transformation along the t_2 dimension. For each layer (representing a single t_1 value), the intensities at a given frequency value along the f_3-axis are treated as an FID and transformed to generate a spectrum, with dimension labeled f_2. Now, we have a stack of two-dimensional spectra. Note that the bottom spectrum, corresponding to $t_1 = 0$, is quite similar to the original two-dimensional spectrum shown earlier. In the other planes, the same set of peaks are present, but with varying intensity and sign, reflecting the different lengths of the t_1 delay.

Finally, the data are subjected to a third set of Fourier transforms, this time along the t_1 dimension. For each f_2, f_3 pair, the points from the individual layers are combined into a data set and transformed to generate the f_1 dimension in the final spectrum. Since t_1 corresponds to the S-evolution period in the HMQC portion of the experiment, this third axis represents the S-frequency. All of the peaks are positive now, and each of the layers contains a subset of those that were observed in the simple two-dimensional homonuclear NOESY.

Peaks in the individual layers arise when one of two conditions are satisfied. If an I-nucleus is scalar coupled to an S-nucleus with a resonance frequency associated with the layer, a strong peak will appear on the diagonal of that layer, with a position in both the f_2 and f_3 dimensions corresponding to the I-spin frequency. In addition, any I-spin that interacts via the nuclear Overhauser effect with the I-spin of the I-S pair will give rise to a cross peak. Importantly, this second I-spin does not have to be scalar coupled to an S-spin. The position of the cross peaks along the f_2-axis corresponds to the resonance frequency of the I-spin that is scalar coupled to the S-spin, while the position along the f_3-axis corresponds to the I-spin that interacts via the NOE.

Like the shift from one dimension to two, adding a third dimension to a spectrum provides both new structural information and enhanced resolution. The appearance of a cross peak on an individual layer demonstrates that two I-spins, with resonance frequencies f_2 and f_3, are within about 5 Å of one another and that the spin with frequency f_2 is scalar coupled to an S-spin with frequency f_1. Because NOE cross peaks appear in a given layer only if one of the I-spins is scalar coupled to an S-spin of the appropriate frequency, each layer represents a simplified two-dimensional NOESY spectrum in which the chances of overlapping peaks are greatly reduced. In the hypothetical example illustrated above, the signals from the two I-spins with very similar resonance frequencies are coupled to S-spins that resolved into separate layers in the f_1 dimension. As a consequence, it is very easy to tell which of the NOE cross peaks is associated with which nucleus.

The HMQC-NOESY experiment shown here is but one of many different three-dimensional experiments that have been devised over the past three decades. In other variations, the HSQC has been combined with the homonuclear NOESY or TOCSY experiments. In some situations, the HSQC offers subtle advantages over the HMQC version in these applications, as does reversing the order of the homonuclear and heteronuclear components of the preparation period, to give experiments identified as the NOESY-HSQC and TOCSY-HSQC. These details are beyond the scope of this book, but become very important in the design of practical experiments. Other three-dimensional experiments have been designed to establish unambiguous scalar coupling relationships among multiple nuclei. Such experiments are particularly important for making resonance assignments in proteins and other complex molecules, as discussed in the following section.

16.7.2 Three-Dimensional Experiments for Resonance Assignments in Proteins: HNCA and HN(CO)CA

The "classic" strategy for making protein resonance assignments, described in Chapter 15 (page 447), depends on detecting an NOE interaction between the amide hy-

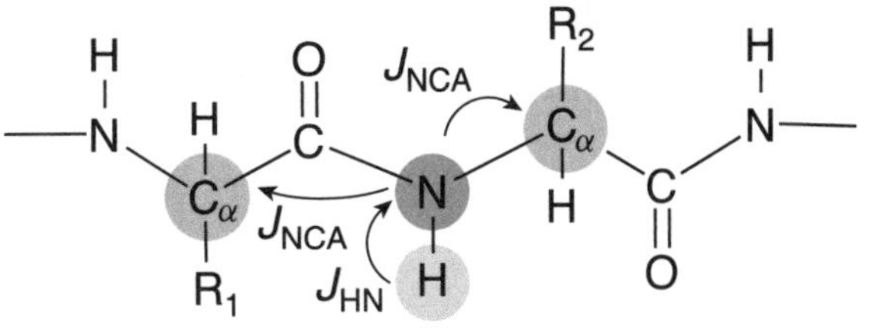

Figure 16.47
Covalent structure of a
dipeptide unit, indicating the
scalar coupling interactions
utilized in the HNCA
experiment.

drogen of one residue and the α-hydrogen of the preceding residue in the sequence. Although this interaction can usually be detected in a NOESY spectrum, its strength can differ among residue pairs, and it is not always easy to distinguish the sequential NOESY peak from others involving the same amide hydrogen. These problems can become particularly severe for larger proteins and for highly flexible molecules, which often display limited ^{1}H chemical-shift dispersion. In the past two decades, the NOE-based assignment strategy has been largely superseded by heteronuclear methods that detect scalar coupling interactions between nuclei in sequential residues.

Experiments of this type are often identified using names that indicate the path of magnetization transfer. For instance, an experiment widely used for backbone assignments in proteins is designated HNCA, indicating that it is based on transfers between the amide hydrogen (H), the amide nitrogen (N) and the α-carbon (CA) nuclei. The scalar coupling interactions are diagrammed in Fig. 16.47. The specificity of the transfers in these experiments is determined by the duration of the delay periods used to interconvert the scalar-coupled magnetization components. As discussed on page 470, these conversions are maximal when the evolution period is set to $t = 1/(2J)$. Notice that the drawing in Fig. 16.47 shows two N–CA coupling interactions, one between the amide nitrogen and the α-carbon of the same residue and one between the nitrogen and the α-carbon of the preceding residue. As it happens, the coupling constants for these two types of interaction are not greatly different, and the HNCA experiment does not directly distinguish between them.

The pulse sequence for one version of the HNCA experiment is shown in Fig. 16.48. Though this sequence is quite complex, and we will not analyze it in complete detail, it can be largely understood by recognizing the important units from which it is assembled. One point to note at the outset is that four different pulse frequencies are used. In addition to the ^{1}H and ^{15}N pulses, there are two kinds of ^{13}C pulses, one for α-carbons (CA) and one for carbonyl carbons (CO). The chemical shifts for these two types of carbon atoms are sufficiently different (about 50 ppm for CA and 175 ppm for CO) that they can be irradiated selectively. Although the CO nucleus does not play a direct role in this experiment, decoupling pulses are applied in some of the evolution periods.

The first segment of the HNCA experiment is an INEPT sequence, which, as described earlier, has the net effect of converting the equilibrium z-magnetization of the ^{1}H nucleus to transverse ^{15}N magnetization. The INEPT segment is then followed by an incremented evolution period, t_1, during which the transverse ^{15}N magnetization evolves. When the data are processed, the incremented t_1 times will be Fourier transformed into an ^{15}N frequency dimension. Refocusing pulses are applied to the ^{1}H and

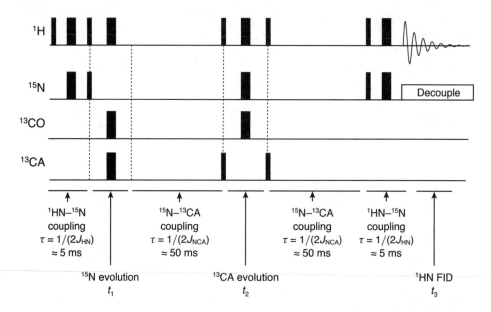

Figure 16.48 Pulse sequence for the basic HNCA experiment. Thick and thin solid vertical bars represent π- and $\pi/2$-pulses, respectively. The dashed vertical lines bracket the incremented t_1 and t_2 delay times.

^{13}C nuclei during the t_1 period to eliminate the effects of scalar coupling between these nuclei and the ^{15}N nucleus.

Following the t_1 ^{15}N evolution period, the ^{15}N spins are allowed to evolve for an additional period defined by the N–CA coupling constant, $\tau = 1/(2J_{NCA})$. During this period, the net ^{15}N magnetization in the transverse plane is converted into a correlation with the z-magnetization of the ^{13}CA nucleus. At the end of the period, $\pi/2$-pulses are applied to both the ^{1}H and ^{13}CA spins, resulting in transverse components for all three spins. All of these components then evolve for an incremented period, t_2, but refocusing pulses to the ^{1}H and ^{15}N nuclei eliminate any chemical-shift effects on these nuclei, and a refocusing pulse to the carbonyl carbon (CO) eliminates the effect of scalar coupling between this nucleus and the CA spin. As a consequence, transformation of the t_2 increments gives rise to a ^{13}C frequency dimension that reflects only the chemical-shift evolution of the CA nuclei.

The segments following the t_2 evolution period are essentially a mirror image of those that preceded it, except that the t_1 ^{15}N evolution period is not included. These segments have the effect of converting the ^{13}CA magnetization into ^{1}H magnetization, which is detected during the data acquisition period, t_3.

Fourier transformation of the data results in a three-dimensional spectrum, as illustrated schematically in Fig. 16.49. In this representation, each plane represents an ^{15}N frequency, with axes corresponding to ^{1}H and ^{13}C frequencies. Cross peaks in these planes correspond to α-carbons that are scalar coupled to amide nitrogens with the chemical shift associated with the plane. The ^{1}H frequencies associated with the cross peaks are those of the amide hydrogens. As noted earlier, the ^{1}J coupling

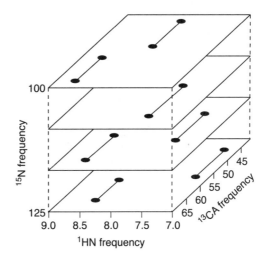

Figure 16.49

Schematic representation of an HNCA three-dimensional spectrum.

constant for an amide nitrogen and the α-carbon of the same residue is not greatly different from that of the ^2J constant for the interaction with the α-carbon of the preceding residue. As a consequence, the HNCA spectrum contains correlations between each amide nitrogen and two α-carbons (except for the N-terminal residue, where the amino hydrogen atoms exchange rapidly with the solvent).

In order to determine which of the two CA cross peaks associated with a given amide group arises from the intra-residue α-carbon, the HNCA experiment can be supplemented with a closely related one, the HN(CO)CA experiment, which is specific for interactions between an amide nitrogen and the α-carbon of the previous residue. This specificity is acquired by transferring magnetization from the ^{15}N nucleus in two steps, via the intervening carbonyl carbon (CO), as illustrated in Fig. 16.50. The use of parentheses in the name of this experiment indicates that magnetization is transferred through the carbonyl carbon, but that this nucleus is not frequency labeled to generate a new dimension in the spectrum. The pulse sequence, which is similar to that for the HNCA experiment, is shown in Fig. 16.51. Comparison with the HNCA experiment shows that the N–CA transfer of the former has been replaced with two transfer steps, N to CO and CO to CA. The coupling constants for these two linkages are approximately 15 and 55 Hz, respectively, significantly different from the direct N–CA coupling constants, which are 4–11 Hz.

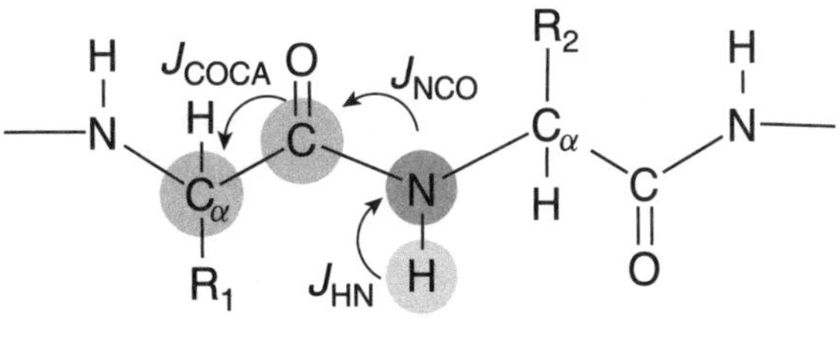

Figure 16.50

Covalent structure of a dipeptide unit, indicating the scalar coupling interactions utilized in the HN(CO)CA experiment.

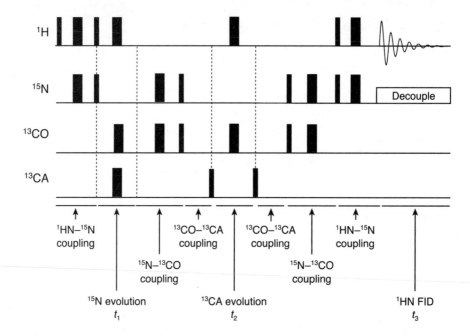

Figure 16.51   Pulse sequence for the basic HN(CO)CA experiment. Thick and thin solid vertical bars represent π- and $\pi/2$-pulses, respectively. The dashed vertical lines bracket the incremented t_1 and t_2 delay times.

Like the HNCA experiment, the HN(CO)CA experiment generates a three-dimensional spectrum with cross peaks that establish the connection between HN, N and CA nuclei. In the HN(CO)CA spectrum, however, each amide group is correlated with only a single CA nucleus, that of the preceding residue. The spectra generated by two experiments are often superimposed and displayed as strips extracted from the individual HN–CA planes, as illustrated schematically in Fig. 16.52. In this simplified example, the filled circles represent cross peaks that are present in the HNCA

Figure 16.52

Schematic representation of superimposed HNCA and HN(CO)CA spectra, displayed as strips from individual ^{1}H N–^{13}CA planes representing different ^{15}N frequencies. Filled circles represent cross peaks present in only the HNCA spectrum, and the open circles represent cross peaks present in both the HNCA and HN(CO)CA spectra.

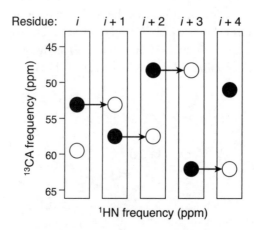

spectrum, but not in the HN(CO)CA spectrum, while the open circles are peaks that appear in both. (In displaying real data, the contours representing the two spectra are typically drawn with different colors.) Thus, in a given strip, the filled circle identifies the CA nucleus that is part of the same residue as the correlated HN and N nuclei—that is, the "self" CA nucleus. The open circle represents the CA nucleus from the preceding residue. The arrows in the drawing above show the sequential connectivity from the CA nucleus of one residue to the next.

The HNCA and HN(CO)CA experiments are usually supplemented with additional backbone coupling experiments and similar ones designed for side-chain assignments. Because of the very direct information these experiments provide, as well as the dispersion of the resonances in multiple dimensions, the interpretation is usually relatively straightforward and efficient. In recent years, computer programs have been developed to automate the assignment process based on these experiments, though some human intervention is still usually required. These advances have reduced much of the drudgery and potential errors in resonance assignments, as well as extended the techniques to larger proteins.

16.7.3   General Considerations for Three- (and Higher-) Dimensional Experiments

The major drawback to three-dimensional experiments is the greatly increased time required to record the spectra, since each layer in the third dimension represents another complete two-dimensional spectrum. To minimize the total time required, the number of t_1 time intervals for which data are recorded is usually much lower than the number of individual FIDs recorded as part of a two-dimensional spectrum. In addition, each of the two-dimensional planes is usually composed of a smaller number of t_2 intervals than is typically used for a two-dimensional experiment. As a consequence, the resolution in the resulting f_1 and f_2 dimensions is usually relatively low.

Recently, radically different, and much more rapid, methods for recording multidimensional spectra have been introduced. These methods take advantage of the fact that much of the data recorded by completely sampling the indirect time dimensions is highly redundant. With suitable experimental design and data processing, sparse data sets can be recorded much more rapidly and can yield spectra with almost the same quality as those generated by the traditional method. These techniques are currently an active area of research and promise to greatly enhance the efficiency of much NMR spectroscopy, especially for large biological molecules.

Another potential difficulty of three- and four-dimensional experiments comes from the longer preparation periods that they typically require. Significant relaxation, both transverse and longitudinal, may occur during these periods, resulting in lowered signal intensities and peak broadening. In spite of these limitations, however, the benefits of three-dimensional experiments very often outweigh their disadvantages and these methods have become indispensable for many projects, especially the study of all but the smallest proteins and nucleic acids.

Summary Points for Chapter 16

- Heteronuclear NMR experiments take advantage of scalar coupling between nuclei of different types, along with the ability to apply selective pulses to the nuclei and selectively record the signals they generate.
- A π-refocusing pulse, separating two evolution periods of equal length, applied to a heteronuclear spin pair can have different effects, depending on the nuclei targeted by the pulse:
  - If one spin is precessing in the transverse plane, with two frequencies due to scalar coupling, application of a selective π-pulse to the other spin will suppress the divergence of the vector components representing the two frequencies. This process is referred to as decoupling, though it does not eliminate the scalar coupling interaction.
  - If the refocusing pulse is applied to both spins of a coupled pair, the average precession of the two transverse vector components (relative to the rotating frame) is suppressed, but the splitting due to scalar coupling is still observed.
- The INEPT pulse sequence increases the signal intensity from a nucleus with a lower gyromagnetic ratio by transferring the equilibrium population difference associated with a coupled spin with a larger gyromagnetic ratio.
- The reverse INEPT pulse sequence converts information about the transverse magnetization of one spin into the amplitude of the signal recorded from a scalar-coupled spin.
- The HSQC (heteronuclear single-quantum coherence) experiment combines the INEPT and reverse INEPT pulse sequences, separated by an incremented evolution period, to generate a two-dimensional spectrum in which cross peaks arise from scalar-coupled spin pairs.
- The HMQC (heteronuclear multiple-quantum coherence) experiment produces spectra similar to those from the HSQC experiment, but uses a simpler pulse sequence and is based on a somewhat different quantum mechanical mechanism.
- The HMQC and HSQC heteronuclear experiments can be combined with homonuclear two-dimensional experiments, such as NOESY, COSY or TOCSY, to generate three-dimensional experiments in which the homonuclear cross peaks are separated into planes representing the frequency of a scalar-coupled heteronucleus.
- Triple-resonance experiments, such as HNCA and HN(CO)CA, are used to identify scalar coupling interactions among three or more nuclei of different types and have become very important for resonance assignments in proteins and other molecules.

Exercises for Chapter 16

16.1. Using magnetization vector diagrams, analyze the INEPT experiment (with refocusing pulse) diagrammed in Fig. 16.29. Following the logic used in the analysis of the simplified experiment, draw a set of diagrams representing each of the steps in the pulse sequence, starting with the $|\alpha\alpha\rangle$ and $|\alpha\beta\rangle$ states. (For this purpose, the contributions of the initial $|\beta\alpha\rangle$ and $|\beta\beta\rangle$ states can be ignored, since they will simply have the opposite signs as those from the $|\alpha\beta\rangle$ and $|\alpha\alpha\rangle$ states, respectively.) Your drawings should *not* assume that the Larmor frequency of the uncoupled I-spin matches the rotating-frame frequency, and the analysis should confirm that the results do not depend on the I-spin frequency.

16.2. Using correlation diagrams, analyze the reverse INEPT experiment, as applied to the situation illustrated in Fig. 16.34, where the initial position of the S-magnetization is at an arbitrary position in the transverse plane, forming an angle, a, with the positive x'-axis. Your analysis should describe the intermediate steps in the experiment and confirm the final magnetization components shown in the rightmost diagram in the figure. As appropriate, resolve the various diagrams into orthogonal components to simplify the drawings, as in the examples shown in Figs. 15.17 and 15.19.

16.3. Fig. 16.53 shows the covalent structure of 4-allylanisole, also called estragole, a natural compound found in a variety of trees and plants. It is widely used in perfumes and as flavoring in foods. One-dimensional ^1H and ^{13}C spectra of 4-allylanisole, kindly provided by C. Austin Service and Peter Flynn at the University of Utah, are shown in Fig. 16.54. The samples used for these spectra and those in Fig. 16.55 were not enriched in ^{13}C, so the ^{13}C signals are due to the natural abundance of this isotope. The three peaks in the ^1H spectrum labeled A, B and C are multiplets. Peak A comes from a single proton, but is split because of scalar coupling interactions. Peak B comes from two protons in nearly equivalent environments and is also split by scalar coupling interactions. Peak C also comes from two protons in nearly equivalent environments. The ^{13}C spectrum was recorded using decoupling in the data acquisition period to suppress splitting by scalar-coupled ^1H nuclei.

Two-dimensional ^1H–^1H COSY and ^1H–^{13}C HMQC spectra of the same compound are shown in Fig. 16.55. Because the concentration of the sample was very high, and the COSY spectrum was recorded with very high sensitivity (using an 800-MHz spectrometer), scalar couplings between ^1H nuclei separated by up to four covalent bonds are detectable in this spectrum.

(a) Before attempting to interpret the spectra, examine the structure of 4-allylanisole and identify the ^1H and ^{13}C nuclei that are likely to be in chemically equivalent, or nearly equivalent, environments. How do you expect these equivalencies to influence the spectra?

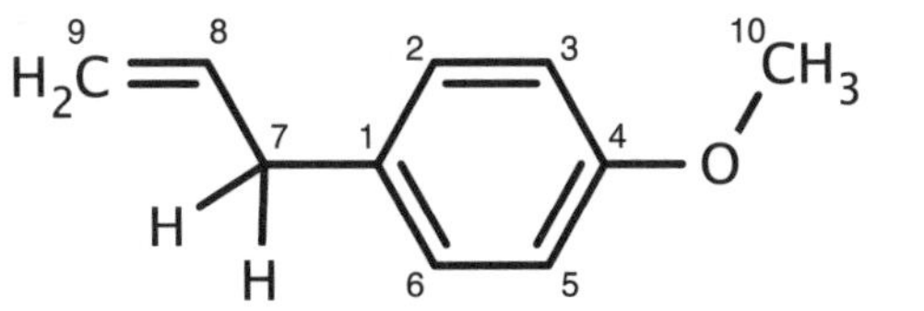

Figure 16.53

The structure of 4-allylanisole, for Exercise 16.3.

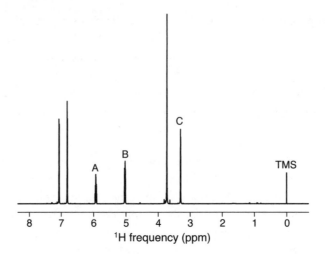

Figure 16.54
One-dimensional ^1H and ^{13}C
NMR spectra of 4-allylanisole,
for Exercise 16.3.

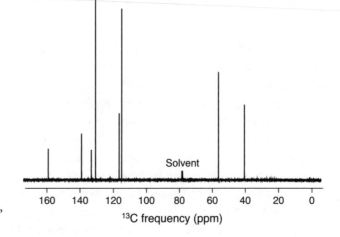

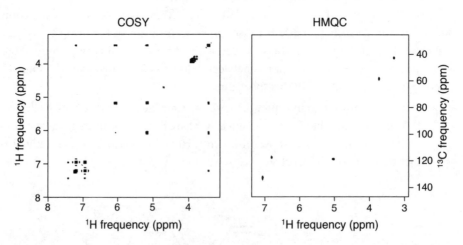

Figure 16.55 Two-dimensional COSY and HMQC spectra of 4-allylanisole, for Exercise 16.3.

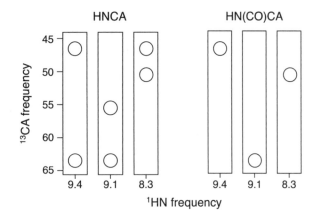

Figure 16.56
Strips from idealized HNCA and HN(CO)CA spectra for Exercise 16.4.

(b) Which of the ^{1}H nuclei do you expect to be influenced by scalar coupling? How do you expect these interactions to influence the spectra?

(c) Incorporating your answers to the questions above, assign resonance frequencies to each of the ^{1}H and ^{13}C nuclei in 4-allylanisole, using the numbers in Fig. 16.53 to identify the carbon and hydrogen atoms.

16.4. The drawings in Fig. 16.56 represent strips from idealized HNCA and HN(CO)CA spectra for a three-residue polypeptide segment containing an asparagine (Asn), a glycine (Gly) and a threonine (Thr) residue. Hint: Information in Table 4.1 will help in solving this problem!

(a) What is the sequence of the tripeptide?

(b) What are the chemical shifts of the amide hydrogens of the Asn, Gly and Thr residues?

(c) What are the chemical shifts of the α-carbons of the Asn, Gly and Thr residues?

Further Reading

More details on decoupling pulse sequences:

⋄ Shaka, A. J. & Keeler, J. (1987). Broadband spin decoupling in isotropic-liquids. *Prog. Nuci. Mag. Res. Sp.*, 19, 47–129. http://dx.doi.org/10.1016/0079-6565(87)80008-0

⋄ Freeman, R. (1998). *Spin Choreography: Basic Steps in High Resolution NMR*, pp. 223–249. Oxford University Press, Oxford.

The original description of the INEPT experiment for polarization transfer:

⋄ Morris, G. A. & Freeman, R. (1979). Enhancement of nuclear magnetic resonance signals by polarization transfer. *J. Am. Chem. Soc.*, 101, 760–762. http://dx.doi.org/10.1021/ja00497a058

Detailed descriptions, using the product operator formalism, of the HSQC, HMQC, HNCA and HN(CO)CA experiments, along with numerous of other two- and three-dimensional heteronuclear NMR experiments:

◇ Cavanagh, J., Fairbrother, W. J., Palmer III, A. G., Rance, M. & Skelton, N. J. (2007). *Protein NMR Spectroscopy*, 2nd ed., pp. 533–678. Elsevier Academic Press, Amsterdam.

A review on recent advances in non-traditional methods of recording and processing multidimensional NMR data:

◇ Hyberts, S. G., Arthanari, H. & Wagner, G. (2012). Applications of non-uniform sampling and processing. In *Novel Sampling Approaches in Higher Dimensional NMR* (Billeter, M. & Overkhov, V., Eds.), Vol. 316 of *Topics in Current Chemistry*, pp. 125–148. Springer Berlin. http://dx.doi.org/10.1007%2F128_2011_187, http://www.ncbi.nlm.nih.gov/pmc/articles/PMC3292636/

CHAPTER 17

Introduction to the Density Matrix

Throughout our analysis of various NMR experiments, we have followed a relatively straightforward, if sometimes tedious, approach: For each of the eigenstates for the system of interest (two for a single spin and four for a pair of scalar-coupled spins), we determined the effects of the various pulses and delays on the wavefunction and then used the appropriate operators to calculate the average magnetization components arising from the populations that began in each eigenstate. To calculate the observed time-dependent magnetization (i.e., the FID), which gives rise to the spectrum, the contributions from the populations were added together, weighted by the expected Boltzmann distribution of the starting eigenstates. These manipulations capture the essential aspects of any NMR experiment—namely, the quantum mechanics, which describes the average behavior of pure states defined by wavefunctions, and the statistical mechanics, which describes the distribution of states with different energies at thermal equilibrium. A conscious effort has been made to distinguish the quantum and statistical aspects, in the belief that this approach may provide the clearest understanding of what really goes on in an NMR experiment, as well as allowing for a more graphical description. This approach is somewhat unconventional, however, and Chapters 17 and 18 are intended to provide a link to the more sophisticated mathematical techniques that are used in most other texts and the primary literature. These methods, based on the density matrix and product operators, can make the analysis of even the most complex NMR experiments relatively straightforward. As you will see, they involve additional levels of mathematical abstraction, but they are based on the same principles that were used in the earlier chapters.

This chapter focuses on the density matrix, a mathematical representation that allows us to treat a population made up of molecules with different wavefunctions, while making minimal assumptions about the nature of the distribution of those molecules. The density matrix was first introduced by John von Neumann in 1927 as part of his work in establishing the formal foundations of quantum mechanics and extending its realm of applications to mixed populations of quantum states. The density matrix treatment also offers a justification for focusing on only the eigenstates, as was done in the earlier chapters. Following this approach requires a basic knowledge of matrix algebra. For those readers who are unfamiliar with this subject or who feel a little rusty, a brief introduction is included in Appendix D. As in the earlier chapters, the emphasis is on illustrating how the mathematics works and how it is related to the physical phenomena, often relying on examples at the expense of mathematical rigor. The references at the end of this chapter provide more rigorous treatments that do justice to the mathematical elegance that has been applied to the more subtle aspects of NMR spectroscopy.

17.1   The Density Matrix for a Population of Spin-1/2 Nuclei

17.1.1   Average Magnetization Components from a Mixed Population

The purpose of the density matrix and its associated mathematics is to provide a means of predicting the outcome of an experimental measurement when applied to a mixture of molecules in different quantum mechanical states. When we first considered the quantum mechanics of a spin-1/2 particle, we derived expressions that made it quite easy to calculate the average magnetization in the x-, y-, and z-directions. If the wavefunction for a single homogeneous population is given by

$$|\Psi\rangle = c_\alpha|\alpha\rangle + c_\beta|\beta\rangle$$

where c_α and c_β are the complex coefficients of the eigenfunctions, $|\alpha\rangle$ and $|\beta\rangle$, then the average magnetization components are given by

$$\langle I_x \rangle = \langle \Psi | I_x | \Psi \rangle = \frac{1}{2}\left(c_\alpha c_\beta^* + c_\beta c_\alpha^*\right) \tag{17.1a}$$

$$\langle I_y \rangle = \langle \Psi | I_y | \Psi \rangle = \frac{i}{2}\left(c_\alpha c_\beta^* - c_\beta c_\alpha^*\right) \tag{17.1b}$$

$$\langle I_z \rangle = \langle \Psi | I_z | \Psi \rangle = \frac{1}{2}\left(c_\alpha c_\alpha^* - c_\beta c_\beta^*\right) \tag{17.1c}$$

where c_α^* and c_β^* are the complex conjugates of c_α and c_β, respectively.

Suppose that instead of a simple homogeneous population, we have a mixture of subpopulations, each of which is described by a wavefunction, $|\Psi_j\rangle$, given by

$$|\Psi_j\rangle = c_{\alpha,j}|\alpha\rangle + c_{\beta,j}|\beta\rangle$$

If we wanted to predict, for instance, the mean x-magnetization, we would apply Eq. 17.1a to the coefficients for each of the wavefunctions and then sum the results according to the number of molecules in each of the subpopulations. Thus, if the number of molecules in subpopulation j is n_j, and the total number of molecules is

N_T, then the mean x-magnetization is given by

$$\bar{I}_x = \sum_j \frac{n_j}{N_T} \langle I_x \rangle_j = \sum_j \frac{n_j}{N_T} \frac{1}{2} \left(c_{\alpha,j} c_{\beta,j}^* + c_{\beta,j} c_{\alpha,j}^* \right)$$

Here, $\langle I_x \rangle_j$ represents the quantum mechanical average over the subpopulation, j, and we use the symbol $\bar{I}_x$ to distinguish those averages from the overall population average. This can be written in a more concise form if we introduce the following special symbols to represent the average values of the products, $c_{\alpha,j} c_{\beta,j}^*$ and $c_{\beta,j} c_{\alpha,j}^*$:

$$\overline{c_\alpha c_\beta^*} = \sum_j \frac{n_j}{N_T} c_{\alpha,j} c_{\beta,j}^*$$

$$\overline{c_\beta c_\alpha^*} = \sum_j \frac{n_j}{N_T} c_{\beta,j} c_{\alpha,j}^*$$

The bars over the symbols indicate that the values are averaged. Importantly, these symbols should always be interpreted as the averages of the products, *not* the product of the average values of c_α and c_β^*, for instance. The expression for $\bar{I}_x$ can then be written as

$$\bar{I}_x = \frac{1}{2} \left(\overline{c_\alpha c_\beta^*} + \overline{c_\beta c_\alpha^*} \right) \tag{17.2}$$

Similarly, the expressions for $\bar{I}_y$ and $\bar{I}_z$ can be written as:

$$\bar{I}_y = \frac{i}{2} \left(\overline{c_\alpha c_\beta^*} - \overline{c_\beta c_\alpha^*} \right) \tag{17.3}$$

$$\bar{I}_y = \frac{1}{2} \left(\overline{c_\alpha c_\alpha^*} - \overline{c_\beta c_\beta^*} \right) \tag{17.4}$$

Superficially, all that distinguishes these equations from those for a homogeneous population are the bars over the products. Conceptually, however, things have changed greatly, because we are now considering a heterogeneous population, with information about both the wavefunctions and the distribution of subpopulations contained within the four terms, $\overline{c_\alpha c_\alpha^*}$, $\overline{c_\beta c_\beta^*}$, $\overline{c_\alpha c_\beta^*}$ and $\overline{c_\beta c_\alpha^*}$. Keep in mind that each of the coefficients in the individual wavefunctions is a complex number, and the averages of the products can also have both real and imaginary parts. Note also that nothing has been said so far about either the wavefunctions or the number of molecules within each subpopulation. These expressions can, therefore, be applied to any situation, provided that we can somehow determine the appropriate averages.

Although it is not absolutely necessary, matrix algebra provides an elegant and efficient way of carrying out the mathematics necessary to calculate and utilize the averages defined above. The density matrix is defined as a square matrix whose elements are the averages of the products of the complex coefficients. For a population composed of identical uncoupled spin-1/2 nuclei, the density matrix is given by

$$\rho = \begin{bmatrix} \overline{c_\alpha c_\alpha^*} & \overline{c_\alpha c_\beta^*} \\ \overline{c_\beta c_\alpha^*} & \overline{c_\beta c_\beta^*} \end{bmatrix}$$

More generally, for a system with wavefunctions written as superpositions of n eigen functions, the density matrix is an $n \times n$ square matrix with elements $\rho_{k,l}$ given by

$$\rho_{k,l} = \overline{c_k c_l^*}$$

where k and l represent the row and column numbers, respectively, of the element, and c_k and c_l are the complex coefficients of the corresponding eigenfunctions. To use the density matrix, we need to be able to calculate how it changes with various pulses and time-evolution periods and then calculate the average magnetization components, just as we have done before using the wavefunction for a homogeneous population.

Before moving on to the rules for using the density matrix, the terms lying on the principle diagonal of the matrix deserve special attention. For the ensemble of single spins, the diagonal elements are

$$\overline{c_\alpha c_\alpha^*} = \sum_j \frac{n_j}{N_T} c_{\alpha,j} c_{\alpha,j}^*$$

$$\overline{c_\beta c_\beta^*} = \sum_j \frac{n_j}{N_T} c_{\beta,j} c_{\beta,j}^*$$

(The other diagonal, from bottom left to top right, is called the skew diagonal.) The sum of the principle diagonal terms of a square matrix is called the trace, and for the density matrix of a one-spin system is

$$\mathrm{Tr}(\rho) = \overline{c_\alpha c_\alpha^*} + \overline{c_\beta c_\beta^*}$$

$$= \sum_j \frac{n_j}{N_T} c_{\alpha,j} c_{\alpha,j}^* + \sum_j \frac{n_j}{N_T} c_{\beta,j} c_{\beta,j}^*$$

$$= \sum_j \frac{n_j}{N_T} (c_{\alpha,j} c_{\alpha,j}^* + c_{\beta,j} c_{\beta,j}^*)$$

For each subpopulation, j, normalization of the wavefunction makes $c_{\alpha,j} c_{\alpha,j}^* + c_{\beta,j} c_{\beta,j}^* = 1$, and the trace of the density matrix is 1. This is a general property of the density matrix for any system.

We begin by considering how to calculate the average magnetization components. In essence, this simply requires taking the various elements from the matrix and combining them as specified by Eqs. 17.2–17.4. Rather than using those equations, however, we will utilize some matrix manipulations that give the same results. In this formalism, the operators for the magnetization components are written as the following matrices:

$$\mathbf{I}_x = \frac{1}{2}\begin{bmatrix} 0 & 1 \\ 1 & 0 \end{bmatrix}$$

$$\mathbf{I}_y = \frac{i}{2}\begin{bmatrix} 0 & -1 \\ 1 & 0 \end{bmatrix}$$

$$\mathbf{I}_z = \frac{1}{2}\begin{bmatrix} 1 & 0 \\ 0 & -1 \end{bmatrix}$$

This set of matrices is known as the Pauli spin matrices, for Wolfgang Pauli who introduced them in 1925, when he deduced the quantum mechanical rules for spin-1/2 particles. Note that these, and most of the other matrices we will use, are identified with bold upright fonts, to distinguish them, for instance, from the actual magnetization components identified as I_x, I_y and I_z. These matrices are used to calculate the average values of the magnetization components by the following two-step procedure:

1. Multiply the density matrix by the operator matrix to generate a new matrix.
2. Calculate the sum of the principal diagonal elements (the trace) of the resulting matrix to determine the average outcome of the experimental observation.

As discussed further in Appendix 1, the rules of matrix multiplication are more involved than you might initially guess. In the case of two 2×2 matrices, the product is calculated as

$$\begin{bmatrix} a_{1,1} & a_{1,2} \\ a_{2,1} & a_{2,2} \end{bmatrix}\begin{bmatrix} b_{1,1} & b_{1,2} \\ b_{2,1} & b_{2,2} \end{bmatrix} = \begin{bmatrix} a_{1,1}b_{1,1}+a_{1,2}b_{2,1} & a_{1,1}b_{1,2}+a_{1,2}b_{2,2} \\ a_{2,1}b_{1,1}+a_{2,2}b_{2,1} & a_{2,1}b_{1,2}+a_{2,2}b_{2,2} \end{bmatrix}$$

More generally, the elements of matrix $\mathbf{C} = \mathbf{AB}$ are given by

$$c_{k,l} = \sum_{m=1}^{n} a_{k,m}b_{m,l}$$

where n is the number of columns in $\mathbf{A}$ and the number of rows in $\mathbf{B}$, which must be equal in order for the multiplication to be defined. Importantly, matrix multiplication is not, in general, commutative, so that if $\mathbf{A}$ and $\mathbf{B}$ are both matrices,

$$\mathbf{AB} \neq \mathbf{BA}$$

However, multiplication of a matrix by a single number, real or complex, simply involves the multiplication of each element by the number. This operation is commutative, and any numerical factors that arise may be factored out of an expression.

For the case of magnetization along the x-axis, the matrix multiplication step is

$$\mathbf{I}_x\boldsymbol{\rho} = \frac{1}{2}\begin{bmatrix} 0 & 1 \\ 1 & 0 \end{bmatrix}\begin{bmatrix} \overline{c_\alpha c_\alpha^*} & \overline{c_\alpha c_\beta^*} \\ \overline{c_\beta c_\alpha^*} & \overline{c_\beta c_\beta^*} \end{bmatrix} = \frac{1}{2}\begin{bmatrix} \overline{c_\beta c_\alpha^*} & \overline{c_\beta c_\beta^*} \\ \overline{c_\alpha c_\alpha^*} & \overline{c_\alpha c_\beta^*} \end{bmatrix}$$

The average magnetization is then determined by calculating the trace:

$$\bar{I}_x = \mathrm{Tr}(\mathbf{I}_x\boldsymbol{\rho}) = \mathrm{Tr}\left(\frac{1}{2}\begin{bmatrix} \overline{c_\beta c_\alpha^*} & \overline{c_\beta c_\beta^*} \\ \overline{c_\alpha c_\alpha^*} & \overline{c_\alpha c_\beta^*} \end{bmatrix}\right) = \frac{1}{2}(\overline{c_\beta c_\alpha^*} + \overline{c_\alpha c_\beta^*})$$

This is the same result as we have seen before, and nothing has really been added by using the matrices. The multiplication step simply moves the appropriate terms from the density matrix to the diagonal of a new matrix, and these terms are added when we calculate the trace. Although the matrix manipulations may not have any obvious advantage for this simple example, they do make dealing with more complicated situations much more manageable. The same procedure is used to calculate $\bar{I}_y$ and $\bar{I}_z$.

17.1.2   Effects of Pulses on the Density Matrix

Next, we consider how pulses are treated in the density matrix formulation. On pages 357–360, we derived an equation to calculate the wavefunction for a single spin following a pulse of angle, a, along the x'-axis:

$$|\Psi(a)\rangle = \big(c_\alpha \cos(a/2) - ic_\beta \sin(a/2)\big)|\alpha\rangle + \big(c_\beta \cos(a/2) - ic_\alpha \sin(a/2)\big)|\beta\rangle$$

where c_α and c_β are the complex coefficients that define the wavefunction before the pulse. The complex coefficients after the pulse, $c_\alpha(a)$ and $c_\beta(a)$, are then

$$c_\alpha(a) = c_\alpha \cos(a/2) - ic_\beta \sin(a/2)$$

$$c_\beta(a) = c_\beta \cos(a/2) - ic_\alpha \sin(a/2)$$

and their complex conjugates are

$$c_\alpha^*(a) = c_\alpha^* \cos(a/2) + ic_\beta^* \sin(a/2)$$

$$c_\beta^*(a) = c_\beta^* \cos(a/2) + ic_\alpha^* \sin(a/2)$$

To calculate the average magnetization components for the new state, using Eqs. 17.2(a)–17.2(c), the new coefficients and their complex conjugates must be multiplied. For instance, the product of the new c_α and its conjugate is written as

$$c_\alpha(a)c_\alpha^*(a) = \big(c_\alpha \cos(a/2) - ic_\beta \sin(a/2)\big)\big(c_\alpha^* \cos(a/2) + ic_\beta^* \sin(a/2)\big)$$

$$= \cos^2(a/2)c_\alpha c_\alpha^* - i \sin(a/2) \cos(a/2)c_\beta c_\alpha^*$$

$$+ i \sin(a/2) \cos(a/2)c_\alpha c_\beta^* + \sin^2(a/2)c_\beta c_\beta^* \tag{17.5}$$

Note that this expression is a linear combination of the four possible products of the starting coefficients and their complex conjugates. Each of the original products is multiplied by a simple constant (determined by the pulse angle, a), and the four terms are added together.

The next step in constructing the new density matrix is to average the products over all of the subpopulations. The linear form of the preceding equations makes this very easy. For example, using the same notation as before to indicate subpopulations, we can write the following:

$$\overline{c_\alpha(a)c_\alpha^*(a)} = \sum_j \frac{n_j}{N_T}\Big(\cos^2(a/2)c_{\alpha,j}c_{\alpha,j}^* - i\sin(a/2)\cos(a/2)c_{\beta,j}c_{\alpha,j}^*$$

$$+ i\sin(a/2)\cos(a/2)c_{\alpha,j}c_{\beta,j}^* + \sin^2(a/2)c_{\beta,j}c_{\beta,j}^*\Big)$$

$$= \cos^2(a/2)\sum_j \frac{n_j}{N_T}c_{\alpha,j}c_{\alpha,j}^* - i\sin(a/2)\cos(a/2)\sum_j \frac{n_j}{N_T}c_{\beta,j}c_{\alpha,j}^*$$

$$+ i\sin(a/2)\cos(a/2)\sum_j \frac{n_j}{N_T}c_{\alpha,j}c_{\beta,j}^* + \sin^2(a/2)\sum_j \frac{n_j}{N_T}c_{\beta,j}c_{\beta,j}^*$$

$$= \cos^2(a/2)\overline{c_\alpha c_\alpha^*} - i\sin(a/2)\cos(a/2)\overline{c_\beta c_\alpha^*}$$

$$+ i\sin(a/2)\cos(a/2)\overline{c_\alpha c_\beta^*} + \sin^2(a/2)\overline{c_\beta c_\beta^*} \qquad (17.6)$$

At first glance, the difference between this and Eq. 17.5 may, again, appear to be a trivial change in notation, but it has a very important implication that may not be so obvious. Namely, Eq. 17.6 shows that once the density matrix elements of one state are known, the density matrix for the state following the pulse can be calculated from those elements alone. We do not actually need to know, or keep track of, the wavefunctions for the individual subpopulations. All we need to know are the average values of the products.

The value of matrices for these calculations should become apparent in the next step. Matrix notation is particularly useful for dealing with expressions that are linear combinations of individual terms. Matrix multiplication, in particular, is a means of expressing sums of products in a systematic way with a bare minimum of notation. The effects of a pulse on the density matrix can be expressed as two matrix multiplication steps, involving two matrices. For a pulse along the x'-axis that rotates the magnetization by an angle, a, these matrices are

$$\mathbf{R}_x(a) = \begin{bmatrix} \cos(a/2) & -i\sin(a/2) \\ -i\sin(a/2) & \cos(a/2) \end{bmatrix}$$

and

$$\mathbf{R}_x^{-1}(a) = \begin{bmatrix} \cos(a/2) & i\sin(a/2) \\ i\sin(a/2) & \cos(a/2) \end{bmatrix}$$

These two matrices are said to be the inverses of one another, as suggested by their labels. Just as two numbers are inverses of one another if their product is 1, two matrices are inverses if multiplying one by the other yields the identity matrix, a special matrix (designated $\mathbf{1}$) in which the primary diagonal elements are 1 and all other elements are 0. You should be able to show that $\mathbf{R}_x$ and $\mathbf{R}_x^{-1}$ are related to one another according to

$$\mathbf{R}_x(a)\mathbf{R}_x^{-1}(a) = \begin{bmatrix} 1 & 0 \\ 0 & 1 \end{bmatrix}$$

The density matrix following an x-pulse, which we will call $\rho(a)$, is calculated according to

$$\rho(a) = \mathbf{R}_x(a)\rho\mathbf{R}_x^{-1}(a)$$

where ρ is the starting density matrix. Although matrix multiplication is not commutative, it does satisfy the associative property, so that the product can be written as either

$$\left(\mathbf{R}_x(a)\rho\right)\mathbf{R}_x^{-1}(a)$$

implying that that first ρ is multiplied by $\mathbf{R}_x$, or as

$$\mathbf{R}_x(a)\left(\rho\mathbf{R}_x^{-1}(a)\right)$$

implying that $\mathbf{R}_x^{-1}$ is first multiplied by ρ.

To illustrate how this works, consider the specific case where $a = \pi/2$—that is, a $(\pi/2)_x$-pulse. In this case,

$$\mathbf{R}_x(\pi/2) = \begin{bmatrix} \frac{1}{\sqrt{2}} & \frac{-i}{\sqrt{2}} \\ \frac{-i}{\sqrt{2}} & \frac{1}{\sqrt{2}} \end{bmatrix} \quad \text{and} \quad \mathbf{R}_x^{-1}(\pi/2) = \begin{bmatrix} \frac{1}{\sqrt{2}} & \frac{i}{\sqrt{2}} \\ \frac{i}{\sqrt{2}} & \frac{1}{\sqrt{2}} \end{bmatrix}$$

If the initial density matrix, ρ, is defined as

$$\rho = \begin{bmatrix} \overline{c_\alpha c_\alpha^*} & \overline{c_\alpha c_\beta^*} \\ \overline{c_\beta c_\alpha^*} & \overline{c_\beta c_\beta^*} \end{bmatrix}$$

then the first multiplication is

$$\mathbf{R}_x(\pi/2)\rho = \begin{bmatrix} \frac{1}{\sqrt{2}} & \frac{-i}{\sqrt{2}} \\ \frac{-i}{\sqrt{2}} & \frac{1}{\sqrt{2}} \end{bmatrix} \begin{bmatrix} \overline{c_\alpha c_\alpha^*} & \overline{c_\alpha c_\beta^*} \\ \overline{c_\beta c_\alpha^*} & \overline{c_\beta c_\beta^*} \end{bmatrix}$$

$$= \frac{1}{\sqrt{2}} \begin{bmatrix} \overline{c_\alpha c_\alpha^*} - i\overline{c_\beta c_\alpha^*} & \overline{c_\alpha c_\beta^*} - i\overline{c_\beta c_\beta^*} \\ -i\overline{c_\alpha c_\alpha^*} + \overline{c_\beta c_\alpha^*} & -i\overline{c_\alpha c_\beta^*} + \overline{c_\beta c_\beta^*} \end{bmatrix}$$

The second multiplication is

$$\left(\mathbf{R}_x(\pi/2)\rho\right)\mathbf{R}_x^{-1}(\pi/2)$$

$$= \frac{1}{\sqrt{2}} \begin{bmatrix} \overline{c_\alpha c_\alpha^*} - i\overline{c_\beta c_\alpha^*} & \overline{c_\alpha c_\beta^*} - i\overline{c_\beta c_\beta^*} \\ -i\overline{c_\alpha c_\alpha^*} + \overline{c_\beta c_\alpha^*} & -i\overline{c_\alpha c_\beta^*} + \overline{c_\beta c_\beta^*} \end{bmatrix} \begin{bmatrix} \frac{1}{\sqrt{2}} & \frac{i}{\sqrt{2}} \\ \frac{i}{\sqrt{2}} & \frac{1}{\sqrt{2}} \end{bmatrix}$$

$$= \frac{1}{2} \begin{bmatrix} (\overline{c_\alpha c_\alpha^*} - i\overline{c_\beta c_\alpha^*}) + i(\overline{c_\alpha c_\beta^*} - i\overline{c_\beta c_\beta^*}) & i(\overline{c_\alpha c_\alpha^*} - i\overline{c_\beta c_\alpha^*}) + (\overline{c_\alpha c_\beta^*} - i\overline{c_\beta c_\beta^*}) \\ (-i\overline{c_\alpha c_\alpha^*} + \overline{c_\beta c_\alpha^*}) + i(-i\overline{c_\alpha c_\beta^*} + \overline{c_\beta c_\beta^*}) & i(-i\overline{c_\alpha c_\alpha^*} + \overline{c_\beta c_\alpha^*}) + (-i\overline{c_\alpha c_\beta^*} + \overline{c_\beta c_\beta^*}) \end{bmatrix}$$

$$= \frac{1}{2} \begin{bmatrix} \overline{c_\alpha c_\alpha^*} - i\overline{c_\beta c_\alpha^*} + i\overline{c_\alpha c_\beta^*} + \overline{c_\beta c_\beta^*} & i\overline{c_\alpha c_\alpha^*} + \overline{c_\beta c_\alpha^*} + \overline{c_\alpha c_\beta^*} - i\overline{c_\beta c_\beta^*} \\ -i\overline{c_\alpha c_\alpha^*} + \overline{c_\beta c_\alpha^*} + \overline{c_\alpha c_\beta^*} + i\overline{c_\beta c_\beta^*} & \overline{c_\alpha c_\alpha^*} + i\overline{c_\beta c_\alpha^*} + -i\overline{c_\alpha c_\beta^*} + \overline{c_\beta c_\beta^*} \end{bmatrix}$$

The important thing to note about this rather formidable-looking matrix is that each of its elements represents a distinct linear combination of each of the four elements in the starting density matrix. This reflects the fact that the $\pi/2$-pulse has the effect of mixing the relative contributions of the $|\alpha\rangle$ and $|\beta\rangle$ eigenstates in the wavefunction for each spin in the population.

For a pulse along the y'-axis, the following two matrices are used calculate the new density matrix:

$$\mathbf{R}_y(a) = \begin{bmatrix} \cos(a/2) & -\sin(a/2) \\ \sin(a/2) & \cos(a/2) \end{bmatrix}$$

$$\mathbf{R}_y^{-1}(a) = \begin{bmatrix} \cos(a/2) & \sin(a/2) \\ -\sin(a/2) & \cos(a/2) \end{bmatrix}$$

The density matrix following a pulse of angle, a, along the y'-axis is then given by

$$\rho(a) = \mathbf{R}_y(a)\rho\mathbf{R}_y^{-1}(a)$$

Thus, we can calculate the density matrix and, thereby, the x'-, y'- and z-magnetization components following any pulse applied to a system, provided that we know the density matrix before the pulse.

17.1.3   Time Evolution of the Density Matrix

In Chapter 12, we considered how the wavefunction describing a pure state changes with time in a stationary magnetic field and derived the following expression for a single spin-1/2 particle:

$$|\Psi(t)\rangle = e^{-it(E_\alpha/\hbar)}c_\alpha|\alpha\rangle + e^{-it(E_\beta/\hbar)}c_\beta|\beta\rangle$$

where c_α and c_β are the coefficients at time zero, and E_α and E_β are the energies of the two eigenstates. Thus, the coefficients at time, t, are

$$c_\alpha(t) = e^{-it(E_\alpha/\hbar)}c_\alpha \quad \text{and} \quad c_\beta(t) = e^{-it(E_\beta/\hbar)}c_\beta$$

and their complex conjugates are

$$c_\alpha^*(t) = e^{it(E_\alpha/\hbar)}c_\alpha^* \quad \text{and} \quad c_\beta^*(t) = e^{it(E_\beta/\hbar)}c_\beta^*$$

The products that are averaged to form the diagonal terms of the density matrix are

$$c_\alpha(t)c_\alpha^*(t) = e^{-it(E_\alpha/\hbar)}c_\alpha \cdot e^{it(E_\alpha/\hbar)}c_\alpha^* = c_\alpha c_\alpha^*$$

and

$$c_\beta(t)c_\beta^*(t) = e^{it(-E_\beta/\hbar)}c_\beta \cdot e^{it(E_\beta/\hbar)}c_\beta^* = c_\beta c_\beta^*$$

Thus, the diagonal terms remain constant during an evolution period in a constant magnetic field. The products that form the off-diagonal terms are given by

$$c_\alpha(t)c_\beta^*(t) = e^{-it(E_\alpha/\hbar)}c_\alpha \cdot e^{it(E_\beta/\hbar)}c_\beta^* = c_\alpha c_\beta^* e^{it(E_\beta-E_\alpha)/\hbar}$$

and

$$c_\beta(t)c_\alpha^*(t) = e^{-it(E_\beta/\hbar)}c_\beta \cdot e^{it(E_\alpha/\hbar)}c_\alpha^* = c_\beta c_\alpha^* e^{-it(E_\beta-E_\alpha)/\hbar}$$

These terms are simply the original products multiplied by a term that reflects the length of the evolution period, t, and the energy difference between the two eigenstates. These terms can also be expressed in terms of the Larmor frequency, ν:

$$c_\alpha(t)c_\beta^*(t) = c_\alpha c_\beta^* e^{-i2\pi\nu t}$$

$$c_\beta(t)c_\alpha^*(t) = c_\beta c_\alpha^* e^{i2\pi\nu t}$$

Because the exponential terms are the same for all of the molecules in the population, they can be factored out when the products are averaged to generate the terms for the density matrix:

$$\overline{c_\alpha(t)c_\beta^*(t)} = \overline{c_\alpha c_\beta^*}\, e^{-i2\pi\nu t}$$

$$\overline{c_\beta(t)c_\alpha^*(t)} = \overline{c_\beta c_\alpha^*}\, e^{i2\pi\nu t}$$

The density matrix following the evolution period, $\rho(t)$, is then given by

$$\rho(t) = \begin{bmatrix} \overline{c_\alpha c_\alpha^*} & \overline{c_\alpha c_\beta^*}\, e^{-i2\pi\nu t} \\ \overline{c_\beta c_\alpha^*}\, e^{i2\pi\nu t} & \overline{c_\beta c_\beta^*} \end{bmatrix}$$

Like the calculation for a pulse, the expression for the time evolution of the density matrix can be written as two matrix multiplications:

$$\rho(t) = \mathbf{U}_H\, \rho\, \mathbf{U}_H^{-1}$$

where $\mathbf{U}_H$ is the time-evolution matrix, and $\mathbf{U}_H^{-1}$ is its inverse. These matrices are

$$\mathbf{U}_H = \begin{bmatrix} e^{-i\pi\nu t} & 0 \\ 0 & e^{i\pi\nu t} \end{bmatrix} \quad \text{and} \quad \mathbf{U}_H^{-1} = \begin{bmatrix} e^{i\pi\nu t} & 0 \\ 0 & e^{-i\pi\nu t} \end{bmatrix}$$

The labels given here to $\mathbf{U}_H$ and $\mathbf{U}_H^{-1}$ indicate their relationship to the Hamiltonian operator and the fact that these matrices have the property of being *unitary*, a symmetry relationship that is briefly explained in Appendix D (page 649). The rotation matrices, $\mathbf{R}_x(a)$, $\mathbf{R}_y(a)$ and their inverses are also unitary, and they are sometimes represented by labels such as $\mathbf{U}_x$ and $\mathbf{U}_x^{-1}$. The unitary property of these matrices is related to the requirement that the total average magnetization described by the density matrix remains constant with rotations about the x'- and y'-axes and during precession about the z-axis (ignoring the effects of relaxation).

Provided that we know a starting density matrix, we now have the tools to do the following:

1. Calculate the effects of pulses on the density matrix.
2. Calculate the effects of a time-evolution period on the density matrix.
3. Calculate any of $\bar{I}_x$, $\bar{I}_y$, and $\bar{I}_z$, the three magnetization components, averaged over the population, from the density matrix.

In the next subsection, we consider the form of the density matrix for a population of spin-1/2 particles at equilibrium, from which we can then predict the outcome of any experiment beginning with the equilibrium state.

17.1.4   The Density Matrix at Equilibrium

The previous chapters of this book have relied on an important simplifying assumption, which is that a population of molecules at thermal equilibrium in a uniform and non-fluctuating magnetic field can be treated as if it contains only the eigenstates, with the relative numbers of molecules in the different states determined by the Boltzmann distribution. Thus, for a population of uncoupled spin-1/2 nuclei, the equilibrium population was assumed to consist of molecules in the $|\alpha\rangle$ and $|\beta\rangle$ states, with the relative number of molecules in the two states determined according to

$$\frac{N_\alpha}{N_\beta} = e^{(E_\beta - E_\alpha)/kT}$$

where N_α and N_β are the number of molecules in the $|\alpha\rangle$ and $|\beta\rangle$ states, respectively, and E_α and E_β are the corresponding energies of the states.

It is now time to reexamine the assumption that the equilibrium population can be treated as if all of the spins were in either the $|\alpha\rangle$ or $|\beta\rangle$ state, before proceeding with the density matrix treatment. There is good reason, in fact, to think that the equilibrium population will be much more complicated than suggested above. Even if each of the nuclei were to start out in either the $|\alpha\rangle$ or $|\beta\rangle$ state, the molecules are constantly rotating with respect to the static magnetic field and one another. As a consequence, each nucleus is exposed to constantly fluctuating magnetic fields, and these fluctuations will lead to changes in the wavefunction. These are the processes that lead to longitudinal relaxation. Thermal equilibrium requires that the population of eigenstates satisfy the Boltzmann distribution, but what about all of the nuclei that are likely to be in superposition states at any instant? For any macroscopic sample, there is simply no way to know what the wavefunctions of the individual nuclei will be at a given time. Fortunately, though, such complete knowledge is not necessary to predict the outcome of a macroscopic experiment. All that we really need to know are the average values of the products that make up the density matrix, as detailed in the previous sections.

The discipline that deals with the population distributions of quantum mechanical systems is known as *quantum statistical mechanics*, and one of its central results is that the density matrix for a system at equilibrium has the following properties:

1. The elements that lie on the principal diagonal have values that reflect the Boltzmann distribution of the eigenstates.
2. All other elements are zero.

As discussed further in Section 17.1.6, the off-diagonal terms reflect the average transverse magnetization components for the population, and the requirement that these terms are zero at equilibrium reflects the observation that no transverse magnetization is detected at equilibrium.

To satisfy the first condition, we calculate the average values of $c_\alpha c_\alpha^*$ and $c_\beta c_\beta^*$ for those molecules that are in either the $|\alpha\rangle$ or $|\beta\rangle$ state. For the eigenstate $|\alpha\rangle$, $c_\alpha = 1$, and for $|\beta\rangle$, $c_\alpha = 0$. Thus, for the population of eigenstates, the average product $\overline{c_\alpha c_\alpha^*}$ is given by

$$\overline{c_\alpha c_\alpha^*} = f_\alpha \cdot 1 + f_\beta \cdot 0$$

where f_α and f_β are the fraction of spins in the $|\alpha\rangle$ and $|\beta\rangle$ states at equilibrium:

$$f_\alpha = \frac{N_\alpha}{N_\alpha + N_\beta}$$

$$f_\beta = \frac{N_\beta}{N_\alpha + N_\beta}$$

For convenience of notation, it is useful to express the populations in terms of the fractional excess of spins in the $|\alpha\rangle$ state, ΔP, as introduced on page 417 in Chapter 14:

$$\Delta P = \frac{N_\alpha - N_\beta}{N_\alpha + N_\beta}$$

The fraction of spins in the $|\alpha\rangle$ state can then be written as

$$\frac{N_\alpha}{N_\alpha + N_\beta} = \frac{1}{2}(1 + \Delta P)$$

and the average of $c_\alpha c_\alpha^*$ is

$$\overline{c_\alpha c_\alpha^*} = \frac{1}{2}(1 + \Delta P)$$

Similarly, the average of $c_\beta c_\beta^*$ is calculated as

$$\overline{c_\beta c_\beta^*} = f_\alpha \cdot 0 + f_\beta \cdot 1 = \frac{1}{2}(1 - \Delta P)$$

Thus, the density matrix for the system at equilibrium is given by

$$\rho_{eq} = \begin{bmatrix} \frac{1}{2}(1 + \Delta P) & 0 \\ 0 & \frac{1}{2}(1 - \Delta P) \end{bmatrix}$$

In the section below, we will see that it this density matrix for the equilibrium state accurately predicts the observed results of a simple NMR experiment.

The equilibrium density matrix can be written as a sum of two matrices:

$$\rho_{eq} = \frac{1}{2}\begin{bmatrix} 1 & 0 \\ 0 & 1 \end{bmatrix} + \frac{\Delta P}{2}\begin{bmatrix} 1 & 0 \\ 0 & -1 \end{bmatrix} \tag{17.7}$$

(Unlike matrix multiplication, matrix addition is defined by the simple addition of the corresponding elements.) Note that the matrix in the first term is the identity matrix, **1**. One of the important properties of matrix multiplication is that it is distributive with respect to matrix addition. That is, if **A**, **B** and **C** are all matrices, then

$$\mathbf{A}(\mathbf{B}+\mathbf{C}) = \mathbf{AB}+\mathbf{AC}$$

Since all of the manipulations of the density matrix discussed previously involve multiplication steps, they can in principle be carried out separately with the two parts of the sum in Eq. 17.7, and then the final results added together to calculate the observed magnetization components. This is a very important principle that we will use again in Chapter 18.

Furthermore, the product of any matrix, **M**, and the identity matrix is simply **M**. As a consequence, the effect, for instance, of a pulse along the y'-axis on the identity matrix is simply

$$\rho(a) = \mathbf{R}_y(a)\big(\mathbf{1R}_y^{-1}(a)\big) = \mathbf{R}_y(a)\mathbf{R}_y^{-1}(a) = \mathbf{1}$$

Similarly, an evolution period in a constant magnetic field leaves the identity matrix unchanged. Finally, if any of the three magnetization operators are applied to the identity matrix, the trace of the resulting matrix is 0. The consequence of these mathematical properties is that the portion of the equilibrium density matrix that corresponds to the identity matrix makes no net contribution to any experimental observation, and we can use the following simplified form for our calculations:

$$\rho_{eq} = \frac{\Delta P}{2}\begin{bmatrix} 1 & 0 \\ 0 & -1 \end{bmatrix} \tag{17.8}$$

Earlier, on page 504, we noted that the trace of the density matrix is always equal to 1. When using this partial form of the equilibrium density matrix, or any density matrix derived from it, we must modify this statement to say that the trace is always equal to 0.

We can use this representation of the equilibrium density matrix to calculate the average magnetization components generated by the equilibrium state. For the average I_x-magnetization,

$$\bar{I}_x = \mathrm{Tr}(\mathbf{I}_x \rho_{eq})$$

$$= \mathrm{Tr}\left(\frac{1}{2}\begin{bmatrix} 0 & 1 \\ 1 & 0 \end{bmatrix}\frac{\Delta P}{2}\begin{bmatrix} 1 & 0 \\ 0 & -1 \end{bmatrix}\right)$$

$$= \mathrm{Tr}\left(\frac{\Delta P}{4}\begin{bmatrix} 0 & 0 \\ 0 & 0 \end{bmatrix}\right)$$

$$= 0$$

The same manipulations can be used to calculate the average I_y-magnetization, with the same result. These results are consistent with the experimental observation that no net transverse magnetization can be detected from a sample at thermal equilibrium. The average I_z-magnetization is calculated as

$$\bar{I}_z = \text{Tr}(\mathbf{I}_z \boldsymbol{\rho}_{\text{eq}})$$

$$= \text{Tr}\left(\frac{1}{2}\begin{bmatrix} 1 & 0 \\ 0 & -1 \end{bmatrix} \frac{\Delta P}{2} \begin{bmatrix} 1 & 0 \\ 0 & -1 \end{bmatrix}\right)$$

$$= \text{Tr}\left(\frac{\Delta P}{4}\begin{bmatrix} 1 & 0 \\ 0 & 1 \end{bmatrix}\right)$$

$$= \frac{\Delta P}{4}(1+1)$$

$$= \frac{\Delta P}{2}$$

Thus, there is a net z-magnetization component that is proportional to the population difference between the eigenstates.

Even though an NMR sample contains a very large number of individual spins (perhaps 10^{18}), each of which can exist in a limitless number of superposition states, the equilibrium density matrix (Eq. 17.8) provides enough information about the starting state to predict the outcome of any NMR experiment. It is important to note, though, that the density matrix does not uniquely define the state; there are a vast number of possible states, made up of different combinations of molecules in different spin states, that can all give rise to the same density matrix and, therefore, the same experimental observations. What the results of quantum statistical mechanics specify is that, at equilibrium, the *average* values of $c_\alpha c_\beta^*$ and $c_\beta c_\alpha^*$ are zero and that the values $c_\alpha c_\alpha^*$ and $c_\beta c_\beta^*$ reflect the Boltzmann distribution of the eigenstates. One way in which the average values of $c_\alpha c_\beta^*$ and $c_\beta c_\alpha^*$ can both be zero is if all of the molecules are in either the $|\alpha\rangle$ or $|\beta\rangle$ state. Although this state is very unlikely, it does give rise to the appropriate density matrix, and this provides the justification for the simplification that has been used throughout the earlier chapters of this book. We will consider the individual elements of the density matrix further after using it to describe a simple NMR experiment.

17.1.5   Density Matrix Description of a Single-Pulse NMR Experiment

We now have all of the mathematical tools to predict the outcome of any experiment starting with the density matrix for the equilibrium state of a population of spin-1/2 nuclei. To see how all of this works, we describe a simple one-pulse experiment. We begin with the equilibrium state, and carry out the multiplications to generate the density matrix for the state following a pulse along the y'-axis. We will call this density matrix $\boldsymbol{\rho}_{\pi/2, y}$ and calculate it as

$$\boldsymbol{\rho}_{\pi/2,y} = \mathbf{R}_y(\pi/2)\boldsymbol{\rho}_{eq}\mathbf{R}_y^{-1}(\pi/2)$$

$$= \left(\frac{1}{\sqrt{2}}\begin{bmatrix} 1 & -1 \\ 1 & 1 \end{bmatrix}\right)\left(\frac{\Delta P}{2}\begin{bmatrix} 1 & 0 \\ 0 & -1 \end{bmatrix}\right)\left(\frac{1}{\sqrt{2}}\begin{bmatrix} 1 & 1 \\ -1 & 1 \end{bmatrix}\right)$$

$$= \frac{\Delta P}{4}\begin{bmatrix} 1 & -1 \\ 1 & 1 \end{bmatrix}\begin{bmatrix} 1 & 0 \\ 0 & -1 \end{bmatrix}\begin{bmatrix} 1 & 1 \\ -1 & 1 \end{bmatrix}$$

$$= \frac{\Delta P}{4}\begin{bmatrix} 1 & -1 \\ 1 & 1 \end{bmatrix}\begin{bmatrix} 1 & 1 \\ 1 & -1 \end{bmatrix}$$

$$= \frac{\Delta P}{4}\begin{bmatrix} 0 & 2 \\ 2 & 0 \end{bmatrix}$$

$$= \frac{\Delta P}{2}\begin{bmatrix} 0 & 1 \\ 1 & 0 \end{bmatrix}$$

Notice that the pulse changes the density matrix so that the non-zero elements are moved from the diagonal to the off-diagonal positions. We can then calculate the magnetization components arising from a state described by this density matrix, starting with $\langle I_x \rangle$:

$$\bar{I}_x = \text{Tr}(\mathbf{I}_x\boldsymbol{\rho}_{\pi/2,y})$$

$$= \text{Tr}\left(\frac{1}{2}\begin{bmatrix} 0 & 1 \\ 1 & 0 \end{bmatrix}\frac{\Delta P}{2}\begin{bmatrix} 0 & 1 \\ 1 & 0 \end{bmatrix}\right)$$

$$= \text{Tr}\left(\frac{\Delta P}{4}\begin{bmatrix} 1 & 0 \\ 0 & 1 \end{bmatrix}\right)$$

$$= \frac{\Delta P}{4}(1+1)$$

$$= \frac{\Delta P}{2}$$

The average magnetization along the y'-axis is calculated as follows:

$$\bar{I}_y = \text{Tr}(\mathbf{I}_y\boldsymbol{\rho}_{\pi/2,y})$$

$$= \text{Tr}\left(\frac{1}{2}\begin{bmatrix} 0 & -i \\ i & 0 \end{bmatrix}\frac{\Delta P}{2}\begin{bmatrix} 0 & 1 \\ 1 & 0 \end{bmatrix}\right)$$

$$= \text{Tr}\left(\frac{\Delta P}{4}\begin{bmatrix} -i & 0 \\ 0 & i \end{bmatrix}\right)$$

$$= \frac{\Delta P}{4}(-i+i)$$

$$= 0$$

The analogous calculation of the z-magnetization following the pulse is left for the reader, and should show that there is no magnetization remaining along this axis. Thus, the initial z-magnetization is rotated to the x'-axis, consistent with our earlier treatments.

Next, we calculate how the density matrix changes during a period of free precession. The resulting density matrix will be called $\rho(t)$ and is calculated using the unitary time-evolution matrix and its inverse, according to

$$\rho(t) = \mathbf{U}_{\mathrm{H}}\boldsymbol{\rho}_{\pi/2,y}\mathbf{U}_{\mathrm{H}}^{-1}$$

$$= \begin{bmatrix} e^{-i\pi vt} & 0 \\ 0 & e^{i\pi vt} \end{bmatrix} \frac{\Delta P}{2} \begin{bmatrix} 0 & 1 \\ 1 & 0 \end{bmatrix} \begin{bmatrix} e^{i\pi vt} & 0 \\ 0 & e^{-i\pi vt} \end{bmatrix}$$

$$= \frac{\Delta P}{2} \begin{bmatrix} e^{-i\pi vt} & 0 \\ 0 & e^{i\pi vt} \end{bmatrix} \begin{bmatrix} 0 & e^{-i\pi vt} \\ e^{i\pi vt} & 0 \end{bmatrix}$$

$$= \frac{\Delta P}{2} \begin{bmatrix} 0 & e^{-i2\pi vt} \\ e^{i2\pi vt} & 0 \end{bmatrix}$$

The average x-magnetization is calculated as before:

$$\bar{I}_x = \mathrm{Tr}(\mathbf{I}_x\rho(t))$$

$$= \mathrm{Tr}\left(\frac{1}{2}\begin{bmatrix} 0 & 1 \\ 1 & 0 \end{bmatrix} \frac{\Delta P}{2} \begin{bmatrix} 0 & e^{-i2\pi vt} \\ e^{i2\pi vt} & 0 \end{bmatrix}\right)$$

$$= \mathrm{Tr}\left(\frac{\Delta P}{4} \begin{bmatrix} e^{i2\pi vt} & 0 \\ 0 & e^{-i2\pi vt} \end{bmatrix}\right)$$

$$= \frac{\Delta P}{4}(e^{i2\pi vt} + e^{-i2\pi vt})$$

$$= \frac{\Delta P}{2}\cos(2\pi vt)$$

Thus, the matrix treatment correctly predicts that the x'-magnetization will vary sinusoidally and that it's maximum value will be the same as the initial equilibrium z-magnetization. The analogous calculation for the y'-magnetization shows that this component fluctuates with the same frequency, but with a phase difference of $\pi/2$ rad:

$$\bar{I}_y = \mathrm{Tr}\left(\mathbf{I}_y\rho(t)\right) = \frac{\Delta P}{4}i(-e^{i2\pi vt} + e^{-i2\pi vt}) = \frac{\Delta P}{2}\sin(2\pi vt)$$

All of this is completely consistent with the simple vector picture and with the quantum mechanical calculation based on the assumption that the signal arises solely from the excess $|\alpha\rangle$ spins in an equilibrium population assumed to contain only the $|\alpha\rangle$ and $|\beta\rangle$ states.

17.1.6 Interpretation of the Density Matrix Elements: "Populations" and "Coherences"

Before proceeding to the density matrix treatment for a more complex system, with two scalar-coupled spins, it is worth pausing to consider the significance of the individual elements in the matrix. Remember, first, that the product of a complex number and its conjugate is a real number and that the square root of this product, called the modulus, reflects the magnitude of the number, irrespective of the relative contributions of its real and imaginary parts. Thus, the diagonal terms of the density matrix, $\overline{c_\alpha c_\alpha^*}$ and $\overline{c_\beta c_\beta^*}$, are real numbers and represent the relative contributions of the $|\alpha\rangle$ and $|\beta\rangle$ eigenfunctions to the wavefunctions, averaged over all of the molecules in the sample. If the sample were composed only of molecules in the two eigenstates, these two terms would simply reflect the number of molecules in each state. For this reason, the diagonal elements are sometimes referred to as "populations," even though their meaning is a a bit more subtle when applied to a more realistic distribution of states.

As shown earlier, the mean z-magnetization is calculated from the difference between the two diagonal terms, and it is relatively easy to assess the z-magnetization simply by examining the density matrix. For instance, at equilibrium, the density matrix is

$$\rho_{eq} = \frac{\Delta P}{2} \begin{bmatrix} 1 & 0 \\ 0 & -1 \end{bmatrix}$$

The difference between $\overline{c_\alpha c_\alpha^*}$ and $\overline{c_\beta c_\beta^*}$ corresponds to the net z-magnetization at equilibrium. After a $(\pi/2)_y$-pulse, the density matrix is

$$\frac{\Delta P}{2} \begin{bmatrix} 0 & 1 \\ 1 & 0 \end{bmatrix}$$

Now, the diagonal elements are equal, corresponding to no net z-magnetization. Stated another way, the contributions of the $|\alpha\rangle$ and $|\beta\rangle$ eigenfunctions are equal, when averaged over the population. If the population is subjected to a second $(\pi/2)_y$-pulse, the density matrix becomes

$$\frac{\Delta P}{2} \begin{bmatrix} -1 & 0 \\ 0 & 1 \end{bmatrix}$$

The diagonal terms have returned with the second pulse, but their signs are reversed from what they were at thermal equilibrium, so that the average z-magnetization is now negative. This is consistent with the earlier picture that a net π-pulse has the effect of interchanging the $|\alpha\rangle$ and $|\beta\rangle$ populations. If we now recognize that the equilibrium population likely includes many molecules in superposition states, we interpret the change in the density matrix as reflecting a reversal of the relative contributions of the eigenstates in each of the wavefunctions. Instead of a net excess of $|\alpha\rangle$ character, there is an excess of $|\beta\rangle$ character.

The interpretation of the off-diagonal elements, $\overline{c_\alpha c_\beta^*}$ and $\overline{c_\beta c_\alpha^*}$, is a little less obvious. At equilibrium, these terms are zero, but they take on non-zero values following a $\pi/2$-pulse and are associated with the transverse magnetization that is detected following the pulse, as indicated in our earlier expressions for calculating $\bar{I}_x$ and $\bar{I}_y$ from the average products of the complex coefficients. Note, too, that the two off-diagonal elements are not independent of one another. For any pair of complex numbers, c_1 and c_2, it can be shown that the products, $c_1 c_2^*$ and $c_2 c_1^*$, are complex conjugates of one another. That is,

$$c_1 c_2^* = (c_2 c_1^*)^*$$

This relationship can be extended to the sum or average of a series of such products, so that we may write

$$\overline{c_\alpha c_\beta^*} = (\overline{c_\beta c_\alpha^*})^*$$

Thus, the density matrix actually contains one more element than is absolutely necessary to describe the state of the system. The expressions for calculating the transverse magnetization components (Eq. 17.2 and 17.3) can be written so as to include only one or the other of these terms:

$$\bar{I}_x = \mathcal{R}e(\overline{c_\alpha c_\beta^*}) = \mathcal{R}e(\overline{c_\beta c_\alpha^*})$$

$$\bar{I}_y = \mathcal{I}m(\overline{c_\beta c_\alpha^*}) = -\mathcal{I}m(\overline{c_\alpha c_\beta^*})$$

where the symbols $\mathcal{R}e$ and $\mathcal{I}m$ indicate the real and imaginary parts of the complex numbers, respectively. Thus, the two off-diagonal terms convey the same information, and either can be readily interpreted in terms of the x'- and y'-magnetization components. For instance, earlier we found that the density matrix following a $(\pi/2)_y$-pulse to the equilibrium population is

$$\frac{\Delta P}{2} \begin{bmatrix} 0 & 1 \\ 1 & 0 \end{bmatrix}$$

Since the two off-diagonal elements are real numbers, the only transverse magnetization is along the x'-axis. A $(\pi/2)_x$-pulse to the equilibrium state gives rise to the following density matrix:

$$\frac{\Delta P}{2} \begin{bmatrix} 0 & i \\ -i & 0 \end{bmatrix}$$

In this case, the off-diagonal elements are pure imaginary numbers, and the magnetization lies along the negative y'-axis. If the off-diagonal elements contain both real and imaginary parts, then the average magnetization will include both x'- and y'-components.

Once again, it is important to emphasize that the elements of the density matrix represent the average of the products for all of the molecules in the sample, rather than an idealized pure state. To see how various subpopulations contribute to the off-diagonal elements, it is useful to express these complex numbers in the exponential form. For convenience, we will focus on the lower left term of the density matrix, $\overline{c_\beta c_\alpha^*}$, but the same information is contained in the complex conjugate of this term, $\overline{c_\alpha c_\beta^*}$. For the complex coefficients of a single wave wavefunction, we can write the product $c_\beta c_\alpha^*$ as

$$c_\beta c_\alpha^* = m_\beta e^{i\theta_\beta} m_\alpha e^{-i\theta_\alpha}$$

$$= m_\alpha m_\beta e^{i(\theta_\beta - \theta_\alpha)}$$

$$= m_\alpha m_\beta \big(\cos(\theta_\beta - \theta_\alpha) + i \, \sin(\theta_\beta - \theta_\alpha) \big)$$

where m_α and m_β are the moduli of c_α and c_β, respectively, and θ_α and θ_β are the arguments, or phases of the coefficients, which reflect the relative contributions of the real and imaginary parts to each complex number. The expression above can be simplified a bit by introducing $\Delta\theta = \theta_\beta - \theta_\alpha$ to represent the phase difference between the coefficients:

$$c_\beta c_\alpha^* = m_\alpha m_\beta e^{i\Delta\theta}$$

$$= m_\alpha m_\beta \big(\cos(\Delta\theta) + i \, \sin(\Delta\theta) \big)$$

In this representation, the product $m_\alpha m_\beta$ represents the total transverse magnetization. The coefficients in a properly normalized wavefunction must satisfy the condition that $c_\alpha c_\alpha^* + c_\beta c_\beta^* = 1$, which places constraints on the possible values of $m_\alpha m_\beta$. If the modulus of either c_α or c_β is 1, the other coefficient must be 0, and the product $c_\beta c_\alpha^*$ must be 0. Physically, this corresponds to the situation in which the magnetization is fully aligned with either the positive or negative z-axis and there is no transverse component. The maximum value of the modulus of $c_\beta c_\alpha^*$ is observed when the moduli of both coefficients are $1/\sqrt{2}$, which corresponds to the situation in which the magnetization lies entirely within the transverse plane.

The direction of any transverse magnetization component is determined by the value of the phase difference, $\Delta\theta$. If $\Delta\theta = 0$, then $c_\beta c_\alpha^*$ is a pure real number, and any transverse magnetization component is aligned with the x'-axis; if $\Delta\theta = \pi/2$, $c_\beta c_\alpha^*$ is a pure imaginary number, and the magnetization lies along the y'-axis. In general, the value of $\Delta\theta$ reflects the angle formed between the transverse projection of the magnetization and the x'-axis, with positive values representing counterclockwise rotation as viewed from the positive z-axis. Thus, for a single population, the product $m_\alpha m_\beta$ reflects the length of the projection of the magnetization onto the transverse plane, and $\Delta\theta$ represents the angle of this projection with respect to the x'-axis, as illustrated in Fig. 17.1.

If a population of molecules is made up of subpopulations, each characterized by a wavefunction, $|\Psi_j\rangle$, then the average value of $c_\beta c_\alpha^*$ can be written as

$$\overline{c_\beta c_\alpha^*} = \sum_j \frac{n_j}{N_T} c_{\beta,j} c_{\alpha,j}^* = \sum_j \frac{n_j}{N_T} m_{\alpha,j} m_{\beta,j} e^{i\Delta\theta_j}$$

Figure 17.1   A vector representation of the magnetization components for a pure spin-1/2 population with a wavefunction described by the complex coefficients c_α and c_β. The black arrow represents the average magnetization, while the gray arrow represents the projection of the magnetization vector onto the transverse plane. The length of the transverse projection is $m_\alpha m_\beta$, the product of the moduli of the complex coefficients, and the angle of this projection with the x'-axis is $\Delta\theta = \theta_\beta - \theta_\alpha$.

where n_j is again the number of molecules in subpopulation j, and N_T is the total number of molecules in the sample.

To simplify the following discussion somewhat, let's restrict ourselves to cases in which all of the magnetization lies in the transverse plane and, in addition, each of the subpopulations contains the same number of molecules, n. With these constraints, $m_{\alpha,j} m_{\beta,j} = 1/2$ and $n_j = n$ have the same values for all of the subpopulations. These terms can then be factored out of the sum to give

$$\overline{c_\beta c_\alpha^*} = \frac{n}{2N_T} \sum_j e^{i\Delta\theta_j} \tag{17.9}$$

Now, we can ask how the values of $\Delta\theta_j$ for the various subpopulations influence the average value of $c_\beta c_\alpha^*$. Because the modulus of a complex number of the form $e^{i\Delta\theta}$ is always 1, it might seem that the sum in Eq. 17.9 would have a modulus equal to the number of subpopulations (N_T/n). This would give rise a value of $\overline{c_\alpha c_\beta^*}$ such that the modulus is 1/2. This, however, will only be true if all of the subpopulations have the same value of $\Delta\theta$—that is, if the population is homogeneous. More generally, the various subpopulations will have different wavefunctions with different values of $\Delta\theta_j$, and the individual values of $e^{i\Delta\theta_j}$ will add together less constructively, leading to a smaller value for the modulus of $\overline{c_\alpha c_\beta^*}$.

For instance, suppose that there are only two subpopulations, of equal size and with wavefunctions such that the values of $\Delta\theta$ are 0 and $\pi/2$. In this case, $n/N_T = 1/2$, and the average of $c_\beta c_\alpha^*$ is given by

$$\overline{c_\beta c_\alpha^*} = \frac{1}{4}(e^{i0} + e^{i\pi/2}) = \frac{1}{4}(1 + i)$$

Since $\overline{c_\beta c_\alpha^*}$ has real and imaginary parts of equal magnitude, we can conclude that the average magnetization will have equal projections onto the x'- and y'-axes. The signs of the real and imaginary parts of $\overline{c_\beta c_\alpha^*}$ indicate that both projections are positive. The phase of $\overline{c_\beta c_\alpha^*}$, $\Delta\theta$, is $\pi/4$. The modulus of $c_\beta c_\alpha$ can be calculated as

$$\sqrt{\overline{c_\beta c_\alpha^*}(\overline{c_\beta c_\alpha^*})^*} = \sqrt{\frac{1}{4}(1+i)\cdot\frac{1}{4}(1-i)}$$

$$= \frac{1}{4}\sqrt{(1+i)(1-i)}$$

$$= \frac{\sqrt{2}}{4} = \frac{1}{2\sqrt{2}}$$

Now compare this result with the case of a homogeneous population where $\Delta\theta = \pi/4$. For this population, $\overline{c_\beta c_\alpha^*}$ can be calculated as

$$\overline{c_\beta c_\alpha^*} = \frac{1}{2}e^{i\pi/4} = \frac{1}{2\sqrt{2}}(1+i)$$

As before, the real and imaginary parts of $\overline{c_\alpha c_\beta^*}$ have equal magnitudes, indicating that the average magnetization has equal x'- and y'-components. But, the modulus of $\overline{c_\beta c_\alpha^*}$ is now

$$\sqrt{\overline{c_\beta c_\alpha^*}(\overline{c_\beta c_\alpha^*})^*} = \sqrt{\frac{1}{2\sqrt{2}}(1+i)\cdot\frac{1}{2\sqrt{2}}(1-i)} = \frac{1}{2\sqrt{2}}\sqrt{(1+i)(1-i)} = \frac{1}{2}$$

Thus, the total magnitude of the magnetization, and the strength of the signal that is generated in the spectrometer as this magnetization precesses, is greater than for the case of two populations by a factor of $1/\sqrt{2}$.

This important distinction can also be illustrated graphically. The left- and right-hand drawings in Fig. 17.2 represent, respectively, the case of a single population and the case of two populations with different values of $\Delta\theta$. The lengths of the arrows are drawn to represent the numbers of molecules in the different populations. Although the net magnetization has the same direction in both cases (as indicated by the projections drawn on the transverse plane), the length of the resultant vector for the mixed population is shorter than the vector for the pure population.

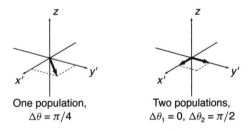

One population,
$\Delta\theta = \pi/4$

Two populations,
$\Delta\theta_1 = 0$, $\Delta\theta_2 = \pi/2$

Figure 17.2   Vector diagrams illustrating the difference between a pure spin population, on the left, and a mixture of two populations, on the right. The dashed lines indicate the total projection of the magnetization. In both cases the net magnetization lies on the transverse plane, with a direction midway between the x'- and y'-axes, but the magnitude of the magnetization is greater for the pure population.

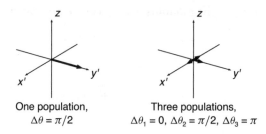

One population,
$\Delta\theta = \pi/2$

Three populations,
$\Delta\theta_1 = 0, \Delta\theta_2 = \pi/2, \Delta\theta_3 = \pi$

Figure 17.3   Vector diagrams illustrating the difference between a pure spin population, on the left, and a mixture of three populations, on the right. In both cases the net magnetization lies along the y'-axis, but the magnitude of the total magnetization for the mixed population is one-third that of the pure population.

The effect becomes more pronounced as the population is divided into more subpopulations representing a wider range of angles, $\Delta\theta$. For instance, suppose that there are three subpopulations of equal size, with $\Delta\theta = 0$, $\pi/2$ and π. Then, the average value of $c_\beta c_\alpha^*$ is given by

$$\overline{c_\beta c_\alpha^*} = \frac{1}{6}(e^{i0} + e^{i\pi/2} + e^{i\pi}) = \frac{1}{6}(1 + i - 1) = \frac{i}{6}$$

Since $\overline{c_\beta c_\alpha^*}$ is a positive pure imaginary number, the average magnetization must lie along the positive y'-axis and $\Delta\theta = \pi/2$. The modulus of $\overline{c_\beta c_\alpha^*}$ is

$$\sqrt{\overline{c_\beta c_\alpha^*}(\overline{c_\beta c_\alpha^*})^*} = \sqrt{(i/6)(-i/6)} = 1/6$$

The magnitude of the net magnetization is now only one-third of what it would be for a homogeneous population. This situation is represented in Fig. 17.3. Note that the subpopulations with magnetization along the positive and negative x'-axes cancel one another, leaving only the net contribution of the subpopulation with magnetization along the y'-axis.

From these examples, we can see that the off-diagonal elements of the density matrix are rich in information about the population of molecules and their wavefunctions. In particular these elements are influenced by the following:

- The relative contributions of the $|\alpha\rangle$ and $|\beta\rangle$ eigenfunctions. The off-diagonal elements (and the transverse magnetization) are maximized when $|\alpha\rangle$ and $|\beta\rangle$ make equal contributions.
- The difference between the phases of the complex coefficients of the $|\alpha\rangle$ and $|\beta\rangle$ eigenfunctions, expressed here as $\Delta\theta$. This difference determines the orientation of the transverse component of the magnetization.
- The extent to which the values of $\Delta\theta$ for the wavefunctions of different molecules in a population are similar to one another, corresponding to the extent to which the transverse components are aligned with one another.

This final point is particularly important and emerges only when we consider a population made up of molecules with different wavefunctions. The largest observable transverse magnetization will be created when all of the wavefunctions have the same value of $\Delta\theta$, so that the average magnetization components add together constructively. This property is called the *coherence* of a population, and this term is also sometimes applied to the off-diagonal elements of the density matrix, to distinguish them from the diagonal, or "population" elements.

As noted earlier, the equilibrium population cannot be defined uniquely and is likely to be composed of molecules in a variety of superposition states. By assuming that the off-diagonal elements of the equilibrium density matrix are zero, we are, in effect, assuming that there is no net coherence among the wavefunctions in this population. This assumption is supported by the very important observation that an NMR signal can only be detected when the sample is irradiated, thereby converting the equilibrium z-magnetization into coherent transverse magnetization. With time, this coherence is lost, contributing to transverse relaxation.

In summary, then, the diagonal elements reflect the relative contributions of the $|\alpha\rangle$ and $|\beta\rangle$ eigenfunctions to the wavefunctions represented in the population, while the off-diagonal elements are influenced by both the relative balance between the eigenstates in individual wavefunctions *and* the extent of phase coherence among these wavefunctions. As illustrated by the examples above, the diagonal, or population, terms reflect the net z'-magnetization and the off-diagonal, coherence, terms determine the magnitude and direction of the transverse magnetization.

For most practical purposes, it must now be admitted, introducing the density matrix adds relatively little to our ability to analyze the simple case of uncoupled spin-1/2 nuclei. All of the results discussed above can be obtained simply by considering the small excess population of $|\alpha\rangle$ spins in the equilibrium population. The major reason for the detailed consideration of the density matrix for this simple case is to illustrate the mechanics of the approach before introducing the density matrix for a pair of scalar-coupled spins, where the matrix manipulations really do help, by eliminating the need to keep track of four starting populations individually.

17.2   The Density Matrix for Two Scalar-Coupled Spins

As discussed in Chapter 13, the wavefunction for two weakly coupled spins can be written as a superposition of four eigenstates:

$$|\psi\rangle = c_{\alpha\alpha}|\alpha\alpha\rangle + c_{\alpha\beta}|\alpha\beta\rangle + c_{\beta\alpha}|\beta\alpha\rangle + c_{\beta\beta}|\beta\beta\rangle$$

From the four complex coefficients, there are 16 pairwise products that can be formed between these coefficients and their complex conjugates, and the density matrix for a population of molecules is composed of the average values of these products:

$$\rho = \begin{bmatrix} \overline{c_{\alpha\alpha}c_{\alpha\alpha}^*} & \overline{c_{\alpha\alpha}c_{\alpha\beta}^*} & \overline{c_{\alpha\alpha}c_{\beta\alpha}^*} & \overline{c_{\alpha\alpha}c_{\beta\beta}^*} \\ \overline{c_{\alpha\beta}c_{\alpha\alpha}^*} & \overline{c_{\alpha\beta}c_{\alpha\beta}^*} & \overline{c_{\alpha\beta}c_{\beta\alpha}^*} & \overline{c_{\alpha\beta}c_{\beta\beta}^*} \\ \overline{c_{\beta\alpha}c_{\alpha\alpha}^*} & \overline{c_{\beta\alpha}c_{\alpha\beta}^*} & \overline{c_{\beta\alpha}c_{\beta\alpha}^*} & \overline{c_{\beta\alpha}c_{\beta\beta}^*} \\ \overline{c_{\beta\beta}c_{\alpha\alpha}^*} & \overline{c_{\beta\beta}c_{\alpha\beta}^*} & \overline{c_{\beta\beta}c_{\beta\alpha}^*} & \overline{c_{\beta\beta}c_{\beta\beta}^*} \end{bmatrix}$$

Note that the diagonal elements represent the means of the moduli (squared) of the coefficients of the wavefunctions and, therefore, reflect the relative contributions of the eigenfunctions to the population as a whole. These terms determine the average magnetization along the z-axis, just as the diagonal terms for the single-spin system did. The off-diagonal terms represent the relationships between pairs of coefficients, and the significance of these terms will become apparent as we examine the density matrix representing different physical situations. As before, to utilize the density matrix we need additional matrices with which to calculate the average magnetization components and to calculate the change in the density matrix due to pulses and time evolution. We also need to know the density matrix at thermal equilibrium. Because there are now two spins involved, the number of matrices, as well as their size, increases significantly compared to the simpler case of a single spin. A full list of the matrices is provided in Appendix F, but here we will just consider a few examples to illustrate the approach and some general results.

17.2.1   The Magnetization Operator Matrices

In order to deduce the meaning of the various terms in the density matrix for a two-spin system, it is helpful to begin by considering the matrices for calculating the observable magnetization components. As with the single-spin system, the procedure is to multiply the density matrix by the appropriate operator matrix and then calculate the trace of the resulting matrix. Now, though, we need six matrices corresponding to I_x, I_y, I_z, S_x, S_y and S_z, the six magnetization components we introduced previously. Let's begin with the z-components. For the I_z-magnetization, the matrix is:

$$\mathbf{I}_z = \frac{1}{2} \begin{bmatrix} 1 & 0 & 0 & 0 \\ 0 & 1 & 0 & 0 \\ 0 & 0 & -1 & 0 \\ 0 & 0 & 0 & -1 \end{bmatrix}$$

Multiplying the general form of the density matrix by $\mathbf{I}_z$ gives

$$\frac{1}{2} \begin{bmatrix} \overline{c_{\alpha\alpha}c_{\alpha\alpha}^*} & \overline{c_{\alpha\alpha}c_{\alpha\beta}^*} & \overline{c_{\alpha\alpha}c_{\beta\alpha}^*} & \overline{c_{\alpha\alpha}c_{\beta\beta}^*} \\ \overline{c_{\alpha\beta}c_{\alpha\alpha}^*} & \overline{c_{\alpha\beta}c_{\alpha\beta}^*} & \overline{c_{\alpha\beta}c_{\beta\alpha}^*} & \overline{c_{\alpha\beta}c_{\beta\beta}^*} \\ -\overline{c_{\beta\alpha}c_{\alpha\alpha}^*} & -\overline{c_{\beta\alpha}c_{\alpha\beta}^*} & -\overline{c_{\beta\alpha}c_{\beta\alpha}^*} & -\overline{c_{\beta\alpha}c_{\beta\beta}^*} \\ -\overline{c_{\beta\beta}c_{\alpha\alpha}^*} & -\overline{c_{\beta\beta}c_{\alpha\beta}^*} & -\overline{c_{\beta\beta}c_{\beta\alpha}^*} & -\overline{c_{\beta\beta}c_{\beta\beta}^*} \end{bmatrix}$$

The effect of this operation is to divide all of the elements by 2 and, more importantly, reverse the signs of the elements in the bottom two rows. The average magnetization is then calculated as the trace of this matrix:

$$\bar{I}_z = \frac{1}{2}\left(\overline{c_{\alpha\alpha}c_{\alpha\alpha}^*} + \overline{c_{\alpha\beta}c_{\alpha\beta}^*} - \overline{c_{\beta\alpha}c_{\beta\alpha}^*} - \overline{c_{\beta\beta}c_{\beta\beta}^*} \right)$$

Thus, information about the I_z-magnetization is contained within all four of the diagonal terms in the density matrix. The expression can be rearranged slightly as the sum of two differences:

$$\bar{I}_z = \frac{1}{2}\left(\overline{c_{\alpha\alpha}c_{\alpha\alpha}^*} - \overline{c_{\beta\alpha}c_{\beta\alpha}^*}\right) + \frac{1}{2}\left(\overline{c_{\alpha\beta}c_{\alpha\beta}^*} - \overline{c_{\beta\beta}c_{\beta\beta}^*}\right)$$

Written this way, it is apparent that the total I_z-magnetization reflects the net excess of "α-character" in the population of I-spins, irrespective of the S-spins. The average S_z-magnetization is similarly calculated as follows:

$$\bar{S}_z = \mathrm{Tr}(\mathbf{S}_z\boldsymbol{\rho}) = \frac{1}{2}\left(\overline{c_{\alpha\alpha}c_{\alpha\alpha}^*} - \overline{c_{\alpha\beta}c_{\alpha\beta}^*}\right) + \frac{1}{2}\left(\overline{c_{\beta\alpha}c_{\beta\alpha}^*} - \overline{c_{\beta\beta}c_{\beta\beta}^*}\right)$$

Thus, the S_z-magnetization is determined by the same four density matrix elements as the I_z-magnetization, but now the differences reflect the net excess of α-character in the S-spins.

As you might expect, the transverse magnetization components are determined by off-diagonal elements of the density matrix. The matrix for calculating the average I_x-magnetization is

$$\mathbf{I}_x = \frac{1}{2}\begin{bmatrix} 0 & 0 & 1 & 0 \\ 0 & 0 & 0 & 1 \\ 1 & 0 & 0 & 0 \\ 0 & 1 & 0 & 0 \end{bmatrix}$$

Multiplying the density matrix by this operator matrix gives

$$\frac{1}{2}\begin{bmatrix} \overline{c_{\beta\alpha}c_{\alpha\alpha}^*} & \overline{c_{\beta\alpha}c_{\alpha\beta}^*} & \overline{c_{\beta\alpha}c_{\beta\alpha}^*} & \overline{c_{\beta\alpha}c_{\beta\beta}^*} \\ \overline{c_{\beta\beta}c_{\alpha\alpha}^*} & \overline{c_{\beta\beta}c_{\alpha\beta}^*} & \overline{c_{\beta\beta}c_{\beta\alpha}^*} & \overline{c_{\beta\beta}c_{\beta\beta}^*} \\ \overline{c_{\alpha\alpha}c_{\alpha\alpha}^*} & \overline{c_{\alpha\alpha}c_{\alpha\beta}^*} & \overline{c_{\alpha\alpha}c_{\beta\alpha}^*} & \overline{c_{\alpha\alpha}c_{\beta\beta}^*} \\ \overline{c_{\alpha\beta}c_{\alpha\alpha}^*} & \overline{c_{\alpha\beta}c_{\alpha\beta}^*} & \overline{c_{\alpha\beta}c_{\beta\alpha}^*} & \overline{c_{\alpha\beta}c_{\beta\beta}^*} \end{bmatrix}$$

The effect of this multiplication is to interchange the first and third rows of the original matrix and the second and fourth rows. One consequence of this is that terms that were previously in off-diagonal positions are now on the diagonal, where they contribute to the trace. The average I_x-magnetization is calculated as follows:

$$\bar{I}_x = \mathrm{Tr}(\mathbf{I}_x\boldsymbol{\rho}) = \frac{1}{2}\left(\overline{c_{\alpha\alpha}c_{\beta\alpha}^*} + \overline{c_{\alpha\beta}c_{\beta\beta}^*} + \overline{c_{\beta\alpha}c_{\alpha\alpha}^*} + \overline{c_{\beta\beta}c_{\alpha\beta}^*}\right)$$

From the properties of a complex number and its conjugate, the off-diagonal density matrix elements in this expression can be shown to be related to one another as follows:

$$\overline{c_{\beta\alpha}c_{\alpha\alpha}^*} = \left(\overline{c_{\alpha\alpha}c_{\beta\alpha}^*}\right)^*$$

$$\overline{c_{\beta\beta}c_{\alpha\beta}^*} = \left(\overline{c_{\alpha\beta}c_{\beta\beta}^*}\right)^*$$

We can then rewrite the expression for $\bar{I}_x$ as

$$\bar{I}_x = \frac{1}{2}\left(\overline{c_{\alpha\alpha}c_{\beta\alpha}^*} + \left(\overline{c_{\alpha\alpha}c_{\beta\alpha}^*}\right)^* + \overline{c_{\alpha\beta}c_{\beta\beta}^*} + \left(\overline{c_{\alpha\beta}c_{\beta\beta}^*}\right)^*\right)$$

$$= \frac{1}{2}\left(\left(\mathcal{R}e\left(\overline{c_{\alpha\alpha}c_{\beta\alpha}^*}\right) + i\mathcal{I}m\left(\overline{c_{\alpha\alpha}c_{\beta\alpha}^*}\right)\right) + \left(\mathcal{R}e\left(\overline{c_{\alpha\alpha}c_{\beta\alpha}^*}\right) - i\mathcal{I}m\left(\overline{c_{\alpha\alpha}c_{\beta\alpha}^*}\right)\right)\right.$$

$$\left. + \left(\mathcal{R}e\left(\overline{c_{\alpha\beta}c_{\beta\beta}^*}\right) + i\mathcal{I}m\left(\overline{c_{\alpha\beta}c_{\beta\beta}^*}\right)\right) + \left(\mathcal{R}e\left(\overline{c_{\alpha\beta}c_{\beta\beta}^*}\right) - i\mathcal{I}m\left(\overline{c_{\alpha\beta}c_{\beta\beta}^*}\right)\right)\right)$$

$$= \mathcal{R}e\left(\overline{c_{\alpha\alpha}c_{\beta\alpha}^*}\right) + \mathcal{R}e\left(\overline{c_{\alpha\beta}c_{\beta\beta}^*}\right)$$

or, equivalently, as:

$$\bar{I}_x = \mathcal{R}e\left(\overline{c_{\beta\alpha}c_{\alpha\alpha}^*}\right) + \mathcal{R}e\left(\overline{c_{\beta\beta}c_{\alpha\beta}^*}\right)$$

The elements in the density matrix that contribute to $\bar{I}_x$ are highlighted in the following representation of the matrix, in which the other elements are indicated with dashes:

$$\begin{bmatrix} - & - & \overline{c_{\alpha\alpha}c_{\beta\alpha}^*} & - \\ - & - & - & \overline{c_{\alpha\beta}c_{\beta\beta}^*} \\ \overline{c_{\beta\alpha}c_{\alpha\alpha}^*} & - & - & - \\ - & \overline{c_{\beta\beta}c_{\alpha\beta}^*} & - & - \end{bmatrix}$$

Thus, we can quickly assess the I_x-magnetization from the real parts of these positions in the density matrix.

For calculating the I_y-magnetization, the matrix is

$$\mathbf{I}_y = \frac{1}{2}\begin{bmatrix} 0 & 0 & -i & 0 \\ 0 & 0 & 0 & -i \\ i & 0 & 0 & 0 \\ 0 & i & 0 & 0 \end{bmatrix}$$

Notice that the non-zero elements of this matrix are in the same positions as the non-zero elements of $\mathbf{I}_x$, but here the values are imaginary. As a consequence, multiplication by this matrix moves the same elements of the density matrix to the diagonal, but the value of the trace is different than for calculating $\bar{I}_x$. Without going through all of the steps, the results are

$$\bar{I}_y = \mathrm{Tr}(\mathbf{I}_y \boldsymbol{\rho})$$

$$= \frac{i}{2}\left(\overline{c_{\alpha\alpha}c_{\beta\alpha}^*} + \overline{c_{\alpha\beta}c_{\beta\beta}^*} - \overline{c_{\beta\alpha}c_{\alpha\alpha}^*} - \overline{c_{\beta\beta}c_{\alpha\beta}^*}\right)$$

$$= -\mathcal{I}m\left(\overline{c_{\alpha\alpha}c_{\beta\alpha}^*}\right) - \mathcal{I}m\left(\overline{c_{\alpha\beta}c_{\beta\beta}^*}\right)$$

$$= \mathcal{I}m\left(\overline{c_{\beta\alpha}c_{\alpha\alpha}^*}\right) + \mathcal{I}m\left(\overline{c_{\beta\beta}c_{\alpha\beta}^*}\right)$$

Therefore, the average I_y-magnetization can be calculated from the imaginary parts of $\overline{c_{\alpha\alpha}c_{\beta\alpha}^*}$ and $\overline{c_{\alpha\beta}c_{\beta\beta}^*}$ or, equivalently, the imaginary parts of $\overline{c_{\beta\alpha}c_{\alpha\alpha}^*}$ and $\overline{c_{\beta\beta}c_{\alpha\beta}^*}$, while the real parts of these elements determine the average I_x-magnetization.

For the S_x-magnetization, the operator matrix is

$$S_x = \frac{1}{2} \begin{bmatrix} 0 & 1 & 0 & 0 \\ 1 & 0 & 0 & 0 \\ 0 & 0 & 0 & 1 \\ 0 & 0 & 1 & 0 \end{bmatrix}$$

Following the same set of rules to calculate the average S_x-magnetization gives

$$\bar{S}_x = \text{Tr}(S_x \boldsymbol{\rho})$$

$$= \frac{1}{2}\left(\overline{c_{\alpha\alpha}c_{\alpha\beta}^*} + \overline{c_{\alpha\beta}c_{\alpha\alpha}^*} + \overline{c_{\beta\alpha}c_{\beta\beta}^*} + \overline{c_{\beta\beta}c_{\beta\alpha}^*} \right)$$

$$= \mathcal{R}e\left(\overline{c_{\alpha\beta}c_{\alpha\alpha}^*} \right) + \mathcal{R}e\left(\overline{c_{\beta\beta}c_{\beta\alpha}^*} \right)$$

$$= \mathcal{R}e\left(\overline{c_{\alpha\alpha}c_{\alpha\beta}^*} \right) + \mathcal{R}e\left(\overline{c_{\beta\alpha}c_{\beta\beta}^*} \right)$$

The positions of the density matrix elements contributing to $\bar{S}_x$ are highlighted in the following representation:

$$\begin{bmatrix} - & \overline{c_{\alpha\alpha}c_{\alpha\beta}^*} & - & - \\ \overline{c_{\alpha\beta}c_{\alpha\alpha}^*} & - & - & - \\ - & - & - & \overline{c_{\beta\alpha}c_{\beta\beta}^*} \\ - & - & \overline{c_{\beta\beta}c_{\beta\alpha}^*} & - \end{bmatrix}$$

Note that the equation for calculating $\bar{S}_x$ has the same general form as the equation for $\bar{I}_x$, but the terms are taken from different positions of the density matrix.

By now, you should be able to guess that the average S_y-magnetization is calculated from the same terms, but using the imaginary parts. This is indeed the case, and the equations are

$$\bar{S}_y = \text{Tr}(S_y \boldsymbol{\rho})$$

$$= \frac{i}{2}\left(\overline{c_{\alpha\alpha}c_{\alpha\beta}^*} - \overline{c_{\alpha\beta}c_{\alpha\alpha}^*} + \overline{c_{\beta\alpha}c_{\beta\beta}^*} - \overline{c_{\beta\beta}c_{\beta\alpha}^*} \right)$$

$$= -\mathcal{I}m\left(\overline{c_{\alpha\alpha}c_{\alpha\beta}^*} \right) - \mathcal{I}m\left(\overline{c_{\beta\alpha}c_{\beta\beta}^*} \right)$$

$$= \mathcal{I}m\left(\overline{c_{\alpha\beta}c_{\alpha\alpha}^*} \right) + \mathcal{I}m\left(\overline{c_{\beta\beta}c_{\beta\alpha}^*} \right)$$

The eight off-diagonal elements that contribute to the expressions for the transverse magnetization components for the two-spin system are analogous to the two off-diagonal elements in the simpler density matrix for the one-spin system. In both cases, these matrix elements reflect the average relative contributions of two eigenstates, the average phase difference between the coefficients for the two eigenstates and, importantly, the coherence among the phase differences for the molecules in the population.

In the case of the one-spin system, there are only two off-diagonal elements. For the two-spin system, there are a total of 12 off-diagonal elements, but only eight of these contribute to the observable magnetization components. Four elements contribute to the I-spin transverse magnetization:

$$\overline{c_{\alpha\alpha}c_{\beta\alpha}^*}, \qquad \overline{c_{\alpha\beta}c_{\beta\beta}^*}, \qquad \overline{c_{\beta\alpha}c_{\alpha\alpha}^*} \quad \text{and} \quad \overline{c_{\beta\beta}c_{\alpha\beta}^*}$$

Each of these elements reflects the coefficients of two eigenfunctions that differ only in the state of the I-spin. Similarly, the four elements that contribute to the S-spin transverse magnetization are

$$\overline{c_{\alpha\alpha}c_{\alpha\beta}^*}, \qquad \overline{c_{\alpha\beta}c_{\alpha\alpha}^*}, \qquad \overline{c_{\beta\alpha}c_{\beta\beta}^*} \quad \text{and} \quad \overline{c_{\beta\beta}c_{\beta\alpha}^*}$$

These pairs of coefficients are associated with eigenfunctions that differ only in the state of the S-spin. The eight density matrix elements listed above are referred to as single-quantum coherence terms. Each is associated with the absorption of radiation by one of the spins, and each contributes to the transverse magnetization of that spin.

In addition to the off-diagonal density matrix elements associated with differences in *either* the I- or S-spin, there are four elements that reflect relationships between eigenstates that differ from one another with respect to both spins. These elements form the reverse, or skew, diagonal of the matrix:

$$\begin{bmatrix} - & - & - & \overline{c_{\alpha\alpha}c_{\beta\beta}^*} \\ - & - & \overline{c_{\alpha\beta}c_{\beta\alpha}^*} & - \\ - & \overline{c_{\beta\alpha}c_{\alpha\beta}^*} & - & - \\ \overline{c_{\beta\beta}c_{\alpha\alpha}^*} & - & - & - \end{bmatrix}$$

The $\overline{c_{\alpha\alpha}c_{\beta\beta}^*}$ and $\overline{c_{\beta\beta}c_{\alpha\alpha}^*}$ terms are associated with eigenstates that differ by 2 in quantum number and are, therefore, called double-quantum coherences. The other two terms are associated with eigenstates with the same net quantum number, 0, and are called zero-quantum coherences. Collectively, the zero- and double-quantum coherences are referred to as multiple-quantum coherences. These matrix elements are not associated with any *directly* observable property of the system, but they are not mere place holders either. They represent the evolution of correlations between the transverse magnetization components of the coupled nuclei—that is, the transverse correlations discussed on pages 412–414.

We have now been able to associate most, but not all, of the density matrix elements with the various magnetization components that can, in principle, be measured for the coupled two-spin system. These relationships are shown schematically below, where the elements of the density matrix are replaced with symbols indicating the magnetization components to which these elements contribute:

$$\begin{bmatrix} \bar{I}_z \,\&\, \bar{S}_z & \bar{S}_x \,\&\, \bar{S}_y & \bar{I}_x \,\&\, \bar{I}_y & - \\ \bar{S}_x \,\&\, \bar{S}_y & \bar{I}_z \,\&\, \bar{S}_z & - & \bar{I}_x \,\&\, \bar{I}_y \\ \bar{I}_x \,\&\, \bar{I}_y & - & \bar{I}_z \,\&\, \bar{S}_z & \bar{S}_x \,\&\, \bar{S}_y \\ - & \bar{I}_x \,\&\, \bar{I}_y & \bar{S}_x \,\&\, \bar{S}_y & \bar{I}_z \,\&\, \bar{S}_z \end{bmatrix}$$

Note that each of the diagonal elements contributes to both $\bar{I}_z$ and $\bar{S}_z$ and that these elements are always real numbers. The off-diagonal elements contribute to either $\bar{I}_x$

and $\bar{I}_y$ or $\bar{S}_x$ and $\bar{S}_y$, with the real parts determining the x'-components and the imaginary parts determining the y'-components. These patterns make it relatively easy to associate changes in the density matrix with changes in the observable properties of the system.

17.2.2   The Equilibrium Density Matrix

Next, we consider the density matrix that represents the state of the population at thermal equilibrium. Again, results from quantum statistical mechanics stipulate that the off-diagonal terms are zero and that the diagonal terms reflect the Boltzmann distribution of the eigenstates. Following the same approach we used for the single-spin system, the diagonal terms are given by

$$\overline{c_\alpha c_\alpha^*} = f_{\alpha\alpha} \cdot 1 + f_{\alpha\beta} \cdot 0 + f_{\beta\alpha} \cdot 0 + f_{\beta\beta} \cdot 0$$

$$\overline{c_\alpha c_\beta^*} = f_{\alpha\alpha} \cdot 0 + f_{\alpha\beta} \cdot 1 + f_{\beta\alpha} \cdot 0 + f_{\beta\beta} \cdot 0$$

$$\overline{c_\beta c_\alpha^*} = f_{\alpha\alpha} \cdot 0 + f_{\alpha\beta} \cdot 0 + f_{\beta\alpha} \cdot 1 + f_{\beta\beta} \cdot 0$$

$$\overline{c_\beta c_\beta^*} = f_{\alpha\alpha} \cdot 0 + f_{\alpha\beta} \cdot 0 + f_{\beta\alpha} \cdot 0 + f_{\beta\beta} \cdot 1$$

The fractional populations can be expressed in terms of the spin population differences, ΔP_I and ΔP_S, as defined on page 417, and the resulting equilibrium density matrix is

$$\rho_{eq} = \frac{1}{4} \begin{bmatrix} (1 + \Delta P_I + \Delta P_S) & 0 & 0 & 0 \\ 0 & (1 + \Delta P_I - \Delta P_S) & 0 & 0 \\ 0 & 0 & (1 - \Delta P_I + \Delta P_S) & 0 \\ 0 & 0 & 0 & (1 - \Delta P_I - \Delta P_S) \end{bmatrix}$$

Writing this as a sum of two matrices gives

$$\rho_{eq} = \frac{1}{4} \begin{bmatrix} 1 & 0 & 0 & 0 \\ 0 & 1 & 0 & 0 \\ 0 & 0 & 1 & 0 \\ 0 & 0 & 0 & 1 \end{bmatrix}$$

$$+ \frac{1}{4} \begin{bmatrix} (\Delta P_I + \Delta P_S) & 0 & 0 & 0 \\ 0 & (\Delta P_I - \Delta P_S) & 0 & 0 \\ 0 & 0 & (-\Delta P_I + \Delta P_S) & 0 \\ 0 & 0 & 0 & (-\Delta P_I - \Delta P_S) \end{bmatrix}$$

Applying the same arguments as were used before for the case of the population of uncoupled spins, we discard the portion of the expression corresponding to the identity matrix, to give

$$\rho_{eq} = \frac{1}{4} \begin{bmatrix} (\Delta P_I + \Delta P_S) & 0 & 0 & 0 \\ 0 & (\Delta P_I - \Delta P_S) & 0 & 0 \\ 0 & 0 & (-\Delta P_I + \Delta P_S) & 0 \\ 0 & 0 & 0 & (-\Delta P_I - \Delta P_S) \end{bmatrix}$$

Since the equilibrium density matrix contains only diagonal elements, there is no net transverse magnetization observable for either spin, consistent with all of our experience. The net z-magnetization components are calculated as

$$\bar{I}_z = \text{Tr}(\mathbf{I}_z \boldsymbol{\rho}_{\text{eq}}) = \frac{\Delta P_I}{2}$$

and

$$\bar{S}_z = \text{Tr}(\mathbf{S}_z \boldsymbol{\rho}_{\text{eq}}) = \frac{\Delta P_S}{2}$$

As expected, each of the z-magnetization components is proportional to the population difference associated with the corresponding spin.

17.2.3 Effects of Pulses on the Density Matrix

The change in the density matrix caused by a pulse along the x'- or y'-axis is calculated in the same way as described earlier for the single-spin system, using rotation matrices and their inverses. For a pulse along the y'-axis that rotates the I-magnetization by an angle, a, the matrices are

$$\mathbf{R}_{y,I}(a) = \begin{bmatrix} \cos(a/2) & 0 & -\sin(a/2) & 0 \\ 0 & \cos(a/2) & 0 & -\sin(a/2) \\ \sin(a/2) & 0 & \cos(a/2) & 0 \\ 0 & \sin(a/2) & 0 & \cos(a/2) \end{bmatrix}$$

and

$$\mathbf{R}_{y,I}^{-1}(a) = \begin{bmatrix} \cos(a/2) & 0 & \sin(a/2) & 0 \\ 0 & \cos(a/2) & 0 & \sin(a/2) \\ -\sin(a/2) & 0 & \cos(a/2) & 0 \\ 0 & -\sin(a/2) & 0 & \cos(a/2) \end{bmatrix}$$

For the case of a $(\pi/2)_{y,I}$-pulse, the matrices are

$$\mathbf{R}_{y,I}(\pi/2) = \frac{1}{\sqrt{2}} \begin{bmatrix} 1 & 0 & -1 & 0 \\ 0 & 1 & 0 & -1 \\ 1 & 0 & 1 & 0 \\ 0 & 1 & 0 & 1 \end{bmatrix}$$

and

$$\mathbf{R}_{y,I}^{-1}(\pi/2) = \frac{1}{\sqrt{2}} \begin{bmatrix} 1 & 0 & 1 & 0 \\ 0 & 1 & 0 & 1 \\ -1 & 0 & 1 & 0 \\ 0 & -1 & 0 & 1 \end{bmatrix}$$

If we begin with the equilibrium density matrix, and apply a $(\pi/2)_{y,I}$-pulse, the result is

$$\mathbf{R}_{y,I}(\pi/2)\boldsymbol{\rho}_{eq}\mathbf{R}_{y,I}^{-1}(\pi/2) = \frac{1}{4}\begin{bmatrix} \Delta P_S & 0 & \Delta P_I & 0 \\ 0 & -\Delta P_S & 0 & \Delta P_I \\ \Delta P_I & 0 & \Delta P_S & 0 \\ 0 & \Delta P_I & 0 & -\Delta P_S \end{bmatrix}$$

Notice that the terms reflecting the populations of the two spins, ΔP_I and ΔP_S, have been separated from one another. In particular, the diagonal elements include only ΔP_S, while the off-diagonal terms include only ΔP_I. This is consistent with our physical picture, in which the selective pulse rotates the I-magnetization to the transverse plane, reflected in the off-diagonal density matrix elements, while leaving the net S-magnetization aligned with the z-axis, reflected in diagonal elements. Note also that the off-diagonal elements that now have non-zero values are those corresponding to the mean products, $\overline{c_{\alpha\alpha}c_{\beta\alpha}^*}$, $\overline{c_{\alpha\beta}c_{\beta\beta}^*}$, $\overline{c_{\beta\alpha}c_{\alpha\alpha}^*}$ and $\overline{c_{\beta\beta}c_{\alpha\beta}^*}$, which we showed earlier are associated with the transverse I-magnetization. Because these terms are pure real numbers, we know that the magnetization lies along the x'-axis, just as expected. The average S_z-magnetization remains the same as before the pulse.

Next, we look at the effects of a selective pulse to the S-spin along the x'-axis, again starting with the equilibrium state. The matrices for a $(\pi/2)_{x,S}$-pulse are

$$\mathbf{R}_{x,S}(\pi/2) = \frac{1}{\sqrt{2}}\begin{bmatrix} 1 & -i & 0 & 0 \\ -i & 1 & 0 & 0 \\ 0 & 0 & 1 & -i \\ 0 & 0 & -i & 1 \end{bmatrix}$$

and

$$\mathbf{R}_{x,S}^{-1}(\pi/2) = \frac{1}{\sqrt{2}}\begin{bmatrix} 1 & i & 0 & 0 \\ i & 1 & 0 & 0 \\ 0 & 0 & 1 & i \\ 0 & 0 & i & 1 \end{bmatrix}$$

If a $(\pi/2)_{x,S}$-pulse is applied to the equilibrium state, the new density matrix is calculated as follows:

$$\mathbf{R}_{x,S}(\pi/2)\boldsymbol{\rho}_{eq}\mathbf{R}_{x,S}^{-1}(\pi/2) = \frac{1}{4}\begin{bmatrix} \Delta P_I & i\Delta P_S & 0 & 0 \\ -i\Delta P_S & \Delta P_I & 0 & 0 \\ 0 & 0 & -\Delta P_I & i\Delta P_S \\ 0 & 0 & -i\Delta P_S & -\Delta P_I \end{bmatrix}$$

This density matrix shares some similarities with the one generated by a pulse to the I-spin. As before, the terms reflecting the I- and S-spins have been separated, but now the I-spin terms are left on the diagonal, while the S-spin terms are moved to off-diagonal positions. Now, the non-zero off-diagonal elements represent the terms that contribute to the transverse magnetization of the S-spin. Since these terms are imaginary, we know that the S-magnetization is aligned with the y'-axis (in the negative direction).

To calculate the density matrix following a non-selective pulse to both spins, we simply carry out the matrix multiplications for one pulse, and then for the other. If this manipulation is carried out for a $(\pi/2)_y$-pulse to both spins, applied to the equilibrium state, the following matrix is obtained:

$$\frac{1}{4}\begin{bmatrix} 0 & \Delta P_S & \Delta P_I & 0 \\ \Delta P_S & 0 & 0 & \Delta P_I \\ \Delta P_I & 0 & 0 & \Delta P_S \\ 0 & \Delta P_I & \Delta P_S & 0 \end{bmatrix}$$

Now, the diagonal elements have been eliminated entirely, and there are four non-zero elements both above and below the diagonal. These are the same elements that were made non-zero by the individual pulses to the I- and S-spins, and they will give rise to the expected magnetization components, I_x and S_x.

17.2.4 Time Evolution of the Two-Spin Density Matrix

For two weakly coupled spins, the time-evolution matrix is given by

$$\mathbf{U_H} = \begin{bmatrix} e^{-i\pi(v_I+v_S+J/2)t} & 0 & 0 & 0 \\ 0 & e^{-i\pi(v_I-v_S-J/2)t} & 0 & 0 \\ 0 & 0 & e^{i\pi(v_I-v_S+J/2)t} & 0 \\ 0 & 0 & 0 & e^{i\pi(v_I+v_S-J/2)t} \end{bmatrix}$$

$$(17.10)$$

and its inverse is

$$\mathbf{U_H^{-1}} = \begin{bmatrix} e^{i\pi(v_I+v_S+J/2)t} & 0 & 0 & 0 \\ 0 & e^{i\pi(v_I-v_S-J/2)t} & 0 & 0 \\ 0 & 0 & e^{-i\pi(v_I-v_S+J/2)t} & 0 \\ 0 & 0 & 0 & e^{-i\pi(v_I+v_S-J/2)t} \end{bmatrix}$$

We can use these matrices to see what happens during a time-evolution period following a pulse. For instance, suppose that we start with the equilibrium state and apply a selective $\pi/2$-pulse to the I-spin along the y'-axis. From the previous section, the density matrix following the pulse is

$$\rho = \frac{1}{4}\begin{bmatrix} \Delta P_S & 0 & \Delta P_I & 0 \\ 0 & -\Delta P_S & 0 & \Delta P_I \\ \Delta P_I & 0 & \Delta P_S & 0 \\ 0 & \Delta P_I & 0 & -\Delta P_S \end{bmatrix}$$

The matrix, after time, t, is calculated as follows:

$$\rho(t) = \mathbf{U_H}\rho\mathbf{U_H^{-1}}$$

$$= \frac{1}{4}\begin{bmatrix} \Delta P_S & 0 & \Delta P_I e^{-i2\pi(v_I+J/2)t} & 0 \\ 0 & -\Delta P_S & 0 & \Delta P_I e^{-i2\pi(v_I-J/2)t} \\ \Delta P_I e^{i2\pi(v_I+J/2)t} & 0 & \Delta P_S & 0 \\ 0 & \Delta P_I e^{i2\pi(v_I-J/2)t} & 0 & -\Delta P_S \end{bmatrix}$$

From inspection of the density matrix, it should be apparent that the average S-magnetization, which lies entirely along the z-axis, is unchanged with time, and that the phases of the off-diagonal positions change. This corresponds to a change in the relative x'- and y'-magnetization components for the I-spin, which can be calculated more explicitly using the $\mathbf{I}_x$ and $\mathbf{I}_y$ operator matrices.

The x'-magnetization of the I-spin is calculated as follows:

$$\bar{I}_x(t) = \mathrm{Tr}\left(\mathbf{I}_x \rho(t)\right)$$

$$= \frac{1}{8}\,\mathrm{Tr}\left(\begin{bmatrix} \Delta P_I e^{i2\pi(\nu_I + J/2)t} & 0 & \Delta P_S & 0 \\ 0 & \Delta P_I e^{i2\pi(\nu_I - J/2)t} & 0 & -\Delta P_S \\ \Delta P_S & 0 & \Delta P_I e^{-i2\pi(\nu_I + J/2)t} & 0 \\ 0 & -\Delta P_S & 0 & \Delta P_I e^{-i2\pi(\nu_I - J/2)t} \end{bmatrix}\right)$$

$$= \frac{\Delta P_I}{8}\left(e^{i2\pi(\nu_I + J/2)t} + e^{i2\pi(\nu_I - J/2)t} + e^{-i2\pi(\nu_I - J/2)t} + e^{-i2\pi(\nu_I + J/2)t}\right)$$

With a bit of algebra and the appropriate trigonometric identities, this can be rearranged to give

$$\bar{I}_x(t) = \frac{\Delta P_I}{4}\left(\cos\left(2\pi(\nu_I + J/2)t\right) + \cos\left(2\pi(\nu_I - J/2)t\right)\right)$$

A similar calculation for $\bar{I}_y$ yields an equivalent expression, with the cosine terms replaced with sines. These are just the results we would expect from our previous analyses, with the I-magnetization precessing in the transverse plane with two frequencies separated by the coupling constant. It is worth remembering, however, that this is different than the result expected if a $(\pi/2)_{y,I}$-pulse is applied to one of the pure eigenstates, such as $|\alpha\alpha\rangle$. In that case, only one of the two possible frequencies is predicted. Only by treating the population as a mixture of the four eigenstates do we predict both frequencies. This highlights the principle of the density matrix approach, and its usefulness: All of the summation and weighting of multiple states is contained within the equilibrium density matrix. All of the subsequent manipulations yield results for the entire population, without having to deal explicitly with subpopulations again.

As a final exercise with the density matrix, we will examine the effects of a non-selective $\pi/2$-pulse, followed by an evolution period and then a second pulse, as in the homonuclear COSY experiment. As shown earlier, the density matrix following a non-selective $(\pi/2)_y$-pulse is

$$\frac{1}{4}\begin{bmatrix} 0 & \Delta P_S & \Delta P_I & 0 \\ \Delta P_S & 0 & 0 & \Delta P_I \\ \Delta P_I & 0 & 0 & \Delta P_S \\ 0 & \Delta P_I & \Delta P_S & 0 \end{bmatrix}$$

A period of time evolution, t_1, changes the phases of all of the off-diagonal elements to give the following matrix:

$$\rho(t_1) = \frac{1}{4} \begin{bmatrix} 0 & \Delta P_S e^{-i2\pi(\nu_S+J/2)t_1} & \Delta P_I e^{-i2\pi(\nu_I+J/2)t_1} & 0 \\ \Delta P_S e^{i2\pi(\nu_S+J/2)t_1} & 0 & 0 & \Delta P_I e^{-i2\pi(\nu_I-J/2)t_1} \\ \Delta P_I e^{i2\pi(\nu_I+J/2)t_1} & 0 & 0 & \Delta P_S e^{-i2\pi(\nu_S-J/2)t_1} \\ 0 & \Delta P_I e^{i2\pi(\nu_I-J/2)t_1} & \Delta P_S e^{i2\pi(\nu_S-J/2)t_1} & 0 \end{bmatrix}$$

If we were to record an FID during this time period, we would detect signals from both the I- and S-spins, with each giving rise to two frequency components.

If, after period, t_1, we apply a second non-selective $(\pi/2)_y$-pulse, the elements of the density matrix are mixed together some more. At this point, all of the elements take on non-zero values that reflect both frequencies of both spins. Applying the usual rules for calculating the three magnetization components for both spins gives

$$\bar{I}_y = \frac{\Delta P_I}{4} \left(\sin(2\pi(\nu_I + J/2)t_1) + \sin(2\pi(\nu_I - J/2)t_1) \right)$$

$$\bar{I}_z = \frac{-\Delta P_I}{4} \left(\cos(2\pi(\nu_I + J/2)t_1) + \cos(2\pi(\nu_I - J/2)t_1) \right)$$

$$\bar{S}_y = \frac{\Delta P_S}{4} \left(\sin(2\pi(\nu_S + J/2)t_1) + \sin(2\pi(\nu_S - J/2)t_1) \right)$$

$$\bar{S}_z = \frac{-\Delta P_S}{4} \left(\cos(2\pi(\nu_S + J/2)t_1) + \cos(2\pi(\nu_S - J/2)t_1) \right)$$

These results show the expected pattern; the $(\pi/2)_y$-pulse following the t_1 evolution period leaves magnetization components along the y'-axis and converts x'-magnetization to z-magnetization. The magnitudes of these components reflect the two-frequencies associated with each spin and the length of the evolution period.

At this point, the density matrix elements have become quite complicated, and it is not so easy even to write the matrix on a single piece of paper. In Chapter 18, some additional mathematics and notation will be introduced to help manage situations like this, but for now, we can proceed a little bit further by using some shorthand symbols to represent the individual elements of the matrix. We will call this matrix $\mathbf{D}(t_1)$ to indicate that its elements depend on the length of the t_1 delay, and the elements of the matrix are identified by subscripts that simply reflect their positions:

$$\mathbf{D}(t_1) = \begin{bmatrix} d_{1,1}(t_1) & d_{1,2}(t_1) & d_{1,3}(t_1) & d_{1,4}(t_1) \\ d_{2,1}(t_1) & d_{2,2}(t_1) & d_{2,3}(t_1) & d_{2,4}(t_1) \\ d_{3,1}(t_1) & d_{3,2}(t_1) & d_{3,3}(t_1) & d_{3,4}(t_1) \\ d_{4,1}(t_1) & d_{4,2}(t_1) & d_{4,3}(t_1) & d_{4,4}(t_1) \end{bmatrix}$$

For instance, the element in the upper left corner of the matrix is

$$d_{1,1}(t_1) = -\frac{1}{8}\left(\Delta P_I \cos(2\pi(\nu_I + J/2)t_1) + \Delta P_I \cos(2\pi(\nu_I - J/2)t_1)\right.$$

$$\left. + \Delta P_S \cos(2\pi(\nu_S + J/2)t_1) + \Delta P_S \cos(2\pi(\nu_S - J/2)t_1)\right)$$

One important feature of this density matrix, which we have not seen before, is that the skew-diagonal elements (i.e., the elements that lie on the diagonal from the lower left to the upper right corners of the matrix: $d_{1,4}, d_{2,3}, d_{3,2}$ and $d_{4,1}$) now have non-zero values. For instance:

$$d_{1,4}(t_1) = -\frac{i}{8}\big(\Delta P_I \sin(2\pi(v_I + J/2)t_1) - \Delta P_I \sin(2\pi(v_I - J/2)t_1)$$

$$+ \Delta P_S \sin(2\pi(v_S + J/2)t_1) - \Delta P_S \sin(2\pi(v_S - J/2)t_1)\big)$$

These elements do not contribute to any of the observable magnetization components, but they may if further manipulations are carried out. So, we should examine how these elements and the rest of the matrix change during a second evolution period, t_2.

To do this, we apply the time-evolution matrix and its inverse to generate a new density matrix, which depends on the two evolution periods:

$$\begin{bmatrix} d_{1,1}(t_1) & d_{1,2}(t_1)e^{-i2\pi(v_S+J/2)t_2} & d_{1,3}(t_1)e^{-i2\pi(v_I+J/2)t_2} & d_{1,4}(t_1)e^{-i2\pi(v_I+v_S)t_2} \\ d_{2,1}(t_1)e^{i2\pi(v_S+J/2)t_2} & d_{2,2}(t_1) & d_{2,3}(t_1)e^{-i2\pi(v_I-v_S)t_2} & d_{2,4}(t_1)e^{-i2\pi(v_I-J/2)t_2} \\ d_{3,1}(t_1)e^{i2\pi(v_I+J/2)t_2} & d_{3,2}(t_1)e^{i2\pi(v_I-v_S)t_2} & d_{3,3}(t_1) & d_{3,4}(t_1)e^{-i2\pi(v_S-J/2)t_2} \\ d_{4,1}(t_1)e^{i2\pi(v_I+v_S)t_2} & d_{4,2}(t_1)e^{i2\pi(v_I-J/2)t_2} & d_{4,3}(t_1)e^{i2\pi(v_S-J/2)t_2} & d_{4,4}(t_1) \end{bmatrix}$$

Notice that the diagonal elements, which reflect the z-magnetization components, depend only on t_1, while all of the other elements depend on both time periods. As we have come to expect, only the phases of the off-diagonal elements change during t_2, while their moduli are determined by the length of the t_1 interval. If we were to apply the appropriate operator matrices, we could calculate the fluctuations of the transverse magnetization components as a function of t_2 and, thereby, calculate the expected FIDs generated as a function of the t_1 delay time. Once again, however, simply writing out the mathematics becomes very awkward, and we will save this analysis until we have introduced a more compact notation in Chapter 18.

For now, we will just take note of the skew-diagonal terms and the frequencies associated with their phase changes during t_2. These frequencies are $(v_I + v_S)$ and $(v_I - v_S)$, corresponding to the double- and zero-quantum transitions, respectively. As noted previously, these transitions cannot be stimulated by the absorption of single photons of the corresponding energies, but these frequencies emerge here from the mixing of the wavefunction coefficients during a particular set of manipulations. Because these frequencies only appear on the skew diagonal, they do not yet influence any of the observable magnetization components. If, however, yet another pulse were applied, the additional mixing of the coefficients would move the double- and zero-quantum frequencies to the matrix elements that determine the observable magnetization components. In this way, it is possible to generate spectra with peaks corresponding to these frequencies, even though there is no direct absorption of radiation.

Summary Points for Chapter 17

- The density matrix and its associated manipulations provide a means of analyzing the behaviors of mixed spin populations, effectively combining the statistical and quantum mechanical aspects of the system.
- A given density matrix does not uniquely define a specific spin population, but rather defines the average properties of an immense number of different possible populations that all behave equivalently.
- For a population of uncoupled spins, the density matrix contains four elements, each of which is the population average of the product of one complex wavefunction coefficient and the complex conjugate of the same coefficient (for diagonal elements) or a different coefficient (for off-diagonal elements).
- The diagonal components of the density matrix represent the population average contributions of the eigenstates to the wavefunctions for the individual molecules. These terms are often referred to as "populations."
- The off-diagonal components of the density matrix represent the average phases and the degree of phase coherence among the wavefunctions for the individual molecules. These terms are referred to as "coherences."
- In the density matrix for a spin population at equilibrium, all of the off-diagonal elements are zero and the diagonal elements represent the equilibrium populations of the eigenstates.
- Each of the average magnetization components for the total population is calculated by multiplying the density matrix by an operator matrix and calculating the trace (the sum of the diagonal elements) of the resulting matrix.
- The effects of a pulse on the density matrix, ρ, are calculated by forming the product, $\mathbf{R}\rho\mathbf{R}^{-1}$, where $\mathbf{R}$ is a matrix associated with the pulse, and $\mathbf{R}^{-1}$ is the inverse of that matrix.
- The effects of an evolution period on the density matrix, ρ, are calculated by forming the product, $\mathbf{U}_H\rho\mathbf{U}_H^{-1}$, where $\mathbf{U}_H$ is a time-evolution matrix and $\mathbf{U}_H^{-1}$ is the inverse of that matrix.
- The density matrix for a population of scalar-coupled spin pairs contains 16 elements.
- The matrix manipulations used for a system of coupled spin pairs are analogous to those for the four-element matrices for uncoupled spins, but quickly become more involved.

Exercises for Chapter 17

Maxima macros for manipulating density matrices for one- and two-spin systems, along with examples, are available for download through links at http://uscibooks.com/goldenberg.htm

17.1. Confirm, for the case of a single spin population, that multiplying the density matrix by the $\mathbf{I}_y$ or $\mathbf{I}_z$ operator matrices and then taking the trace of the resulting matrix yields the expressions given in Eqs. 17.3 and 17.4 for calculating the average values of I_y and I_z.

17.2. For the case of a single spin population with a density matrix equal to the identity matrix, **1**, show that the net magnetization in any direction is zero.

17.3. Imagine a *non-equilibrium* population of spin-1/2 nuclei made up of 50% of the spins in the $|\alpha\rangle$ state and 50% in the $|\beta\rangle$ state.

   (a) Calculate the density matrix for this population.
   (b) Calculate the expected average I_x-, I_y-, and I_z-magnetization for this population.
   (c) Specify another population with the same density matrix. Your answer should include the specific wavefunctions describing the nuclei in this population and the relative number of nuclei with each wavefunction. Show that this population does, in fact, have the same density matrix.
   (d) Describe a simple set of NMR manipulations, starting with an equilibrium distribution, that would generate a population with a density matrix that is *equivalent to* the one described above. For this purpose, two density matrices are considered equivalent if they predict, within a constant of proportionality, the same observable magnetization components. How would you confirm experimentally that the population has the desired density matrix?

17.4. Beginning with the density matrix for a population of spin-1/2 particles at equilibrium,

   (a) Calculate the density matrix representing the population after applying a $\pi/4$-pulse along the y'-axis.
   (b) From the density matrix, calculate the average I_x-, I_y-, and I_z-magnetization components after the pulse.
   (c) Briefly explain the significance of the four elements of the density matrix, after the pulse, and how they are related to the average magnetization components.

17.5. For a system of two coupled spins, consider a population containing only the eigenstates, in the proportions 30% $|\alpha\alpha\rangle$, 10% $|\alpha\beta\rangle$, 25% $|\beta\alpha\rangle$ and 35% $|\beta\beta\rangle$.

   (a) Write the density matrix for this population.
   (b) From inspection of the density matrix, predict which of the possible average Cartesian magnetization components will have non-zero values. Explain your reasoning.
   (c) For the non-zero magnetization components, calculate the average values.
   (d) Calculate the resulting density matrix after applying a selective $(\pi/2)_{x,I}$-pulse to this population.

17.6. Again for a system of two coupled spins, consider a population containing equal subpopulations of molecules with the following two wavefunctions:

$$|\Psi_1\rangle = \frac{1}{\sqrt{2}}|\beta\alpha\rangle + \frac{1}{\sqrt{2}}|\beta\beta\rangle$$

and

$$|\Psi_2\rangle = \frac{1}{\sqrt{2}}|\alpha\alpha\rangle + \frac{i}{\sqrt{2}}|\alpha\beta\rangle$$

(a) Write the density matrix for this population.

(b) From inspection of the density matrix, predict which of the possible average Cartesian magnetization components will have non-zero values. Explain how the observable magnetization components arise from the two subpopulations.

(c) Describe a simple set of NMR manipulations that would generate a population with a density matrix that is equivalent (as defined in Exercise 17.3d) to the one above, starting with an equilibrium distribution of molecules containing a homonuclear spin pair.

(d) Describe how this density matrix and the observable magnetization components would evolve with time in a stationary and uniform external magnetic field aligned with the z-axis.

Further Reading

For a very thorough treatment of the density matrix and a variety of applications, see the following book. The text begins with populations of spin-1/2 particles and then develops the theory in a more general and formal way. The third edition includes new material on coupled spin pairs and the special properties of entanglement.

◇ Blum, K. (2012). *Density Matrix Theory and Applications*, 3rd ed. Springer Series on Atomic, Optical, and Plasma Physics, Vol. 64. Springer-Verlag, Berlin. http://dx.doi.org/10.1007/978-3-642-20561-3

The density matrix treatments for populations of spin-1/2 nuclei and spin pairs are presented in a more formal and concise form in the following two articles:

◇ Farrar, T. C. (1990). Density matrices in NMR spectroscopy: Part I. *Concepts Magn. Reson.*, 2, 1–12. http://dx.doi.org/10.1002/cmr.1820020102

◇ Farrar, T. C. (1990). Density matrices in NMR spectroscopy: Part II. *Concepts Magn. Reson.*, 2, 55–61. http://dx.doi.org/10.1002/cmr.1820020202

A highly influential review on the density matrix as a tool for describing mixed ensembles of quantum systems, with a range of applications:

◇ Fano, U. (1957). Description of states in quantum mechanics by density matrix and operator techniques. *Rev. Mod. Phys.*, 29, 74–93. http://dx.doi.org/10.1103/RevModPhys.29.74

For a concise derivation of the equilibrium density matrix:

◇ Sakurai, J. J. (1985). *Modern Quantum Mechanics*, pp. 174–187. Addison-Wesley, Redwood City, CA.

CHAPTER 18

Basis Sets for the Density Matrix and the Product-Operator Formalism

The density matrix and its associated manipulations provide a rigorous mechanism for predicting the outcome of any series of pulses and delay periods. Carrying out the calculations can rapidly become quite awkward, however, and it is not always easy to see the physical significance of the matrix elements once they involve several terms. In this chapter, we will introduce a mathematical method and formalism that greatly simplifies the calculations and can help restore some physical intuition to the treatment. This method is often referred to as the "product-operator formalism," and we will see that the product operators introduced in Chapter 13 play an important role in the treatment of scalar-coupled spins. The name is a bit misleading, however, because the central idea is the use of a *basis set* to represent the density matrix, as opposed to the use of the product operators per se. We will begin with describing and using a basis set for the case of a population of isolated spins, where product operators are not used at all, and then we will extend the treatment to coupled spin pairs.

18.1 A Basis Set for a Single-Spin Population

Any $n \times n$ matrix can, if we wish, be written as a linear combination of a set of n^2 basis matrices. As a rather trivial illustration of this idea, we can write the density matrix for a population of single spins as the following linear combination:

$$\rho = \begin{bmatrix} \overline{c_\alpha c_\alpha^*} & \overline{c_\alpha c_\beta^*} \\ \overline{c_\beta c_\alpha^*} & \overline{c_\beta c_\beta^*} \end{bmatrix}$$

$$= \overline{c_\alpha c_\alpha^*} \begin{bmatrix} 1 & 0 \\ 0 & 0 \end{bmatrix} + \overline{c_\alpha c_\beta^*} \begin{bmatrix} 0 & 1 \\ 0 & 0 \end{bmatrix} + \overline{c_\beta c_\alpha^*} \begin{bmatrix} 0 & 0 \\ 1 & 0 \end{bmatrix} + \overline{c_\beta c_\beta^*} \begin{bmatrix} 0 & 0 \\ 0 & 1 \end{bmatrix} \tag{18.1}$$

In this case, the basis set is composed of the four matrices in which one element is 1 and the others are 0. The linear combination is defined by the coefficients by which the matrices are multiplied. Formally, this is very similar to the way in which we can

539

write a wavefunction as a linear combination, or superposition, of basis wavefunctions. Not just any set of four matrices can be used for the basis set, however; each member of the set must be independent of the other three. That is, none of the four can be written as a linear combination of any of the other three.

The 2×2 basis set shown in Eq. 18.1 is not very useful, because the coefficients are simply the elements of the original matrix, and we don't gain any new information from them. Nor would the use of this set simplify any of the matrix multiplication required to use the basis set. There are, however, other more interesting basis sets. In particular, we can use the matrices associated with the $\hat{I}_x$-, $\hat{I}_y$- and $\hat{I}_z$-magnetization operators, along with the identity matrix as the elements of a basis set. Recall that these matrices are defined as follows:

$$I_x = \frac{1}{2} \begin{bmatrix} 0 & 1 \\ 1 & 0 \end{bmatrix}$$

$$I_y = \frac{i}{2} \begin{bmatrix} 0 & -1 \\ 1 & 0 \end{bmatrix}$$

$$I_z = \frac{1}{2} \begin{bmatrix} 1 & 0 \\ 0 & -1 \end{bmatrix}$$

$$1 = \begin{bmatrix} 1 & 0 \\ 0 & 1 \end{bmatrix}$$

Though it may not be obvious upon first inspection, any 2×2 matrix can be written as a linear combination of these four. We will refer to this set as the *operator basis set* for the single-spin population. Each pair of matrices drawn from the set is orthogonal, a condition that is satisfied when the trace of the product of the two is zero. For instance,

$$\text{Tr}(I_x I_y) = \text{Tr}\left(\frac{1}{2}\begin{bmatrix} 0 & 1 \\ 1 & 0 \end{bmatrix} \frac{i}{2}\begin{bmatrix} 0 & -1 \\ 1 & 0 \end{bmatrix}\right) = \frac{i}{4}\text{Tr}\left(\begin{bmatrix} 1 & 0 \\ 0 & -1 \end{bmatrix}\right) = 0$$

Similarly, for I_z and the identity matrix, 1,

$$\text{Tr}(I_z 1) = \text{Tr}\left(\frac{1}{2}\begin{bmatrix} 1 & 0 \\ 0 & -1 \end{bmatrix}\begin{bmatrix} 1 & 0 \\ 0 & 1 \end{bmatrix}\right) = \frac{1}{2}\text{Tr}\left(\begin{bmatrix} 1 & 0 \\ 0 & -1 \end{bmatrix}\right) = 0$$

You should be able to show that this condition is satisfied for each pair of matrices from the set. Satisfying these conditions guarantees that the basis set can be used to write any 2×2 matrix. Thus, we can write the density matrix as

$$\rho = c_x I_x + c_y I_y + c_z I_z + c_1 1 \tag{18.2}$$

where c_x, c_y, c_z and c_1 are coefficients that represent the relative contributions of the basis matrices to the sum. Since, in general, the elements of a matrix can be complex numbers, the coefficients in Eq. 18.2 can also be complex. We will see, however, that the coefficients are always real when the operator basis set is used to represent a density matrix. It is also important to keep in mind that these coefficients are *not* simply the four elements of the original 2×2 matrix, as they were in the trivial example of Eq 18.1.

18.1.1   Magnetization Components

So, what has been gained by rewriting the density matrix in this form? At first glance, it may seem that we have only made things more complicated. The utility of this representation becomes apparent when we calculate the average magnetization components for a system described by the density matrix. Recall from Chapter 17 that we calculate these quantities by first multiplying the density matrix by the appropriate operator matrix, such as I_x, and then taking the trace of the product. For instance,

$$\bar{I}_x = \text{Tr}(I_x \rho)$$

Because matrix multiplication follows the distributive property, we can now write this as

$$\bar{I}_x = \text{Tr}(c_x I_x I_x + c_y I_x I_y + c_z I_x I_z + c_1 I_x 1)$$

Since the trace operation simply involves taking a sum of the diagonal elements, we can show that the trace of a sum of matrices is equal to the sum of the traces. Thus,

$$\bar{I}_x = c_x \text{Tr}(I_x I_x) + c_y \text{Tr}(I_x I_y) + c_z \text{Tr}(I_x I_z) + c_1 \text{Tr}(I_x 1)$$

From the orthogonality condition, we know that the trace of the product of any pair of different members of the basis set is zero. Therefore, the only term that contributes to $\bar{I}_x$ is the one containing the matrix, I_x:

$$\bar{I}_x = c_x \text{Tr}(I_x I_x)$$

Calculating the average x-magnetization is then reduced to calculating the product of I_x and itself:

$$\bar{I}_x = c_x \text{Tr}\left(\frac{1}{2}\begin{bmatrix} 0 & 1 \\ 1 & 0 \end{bmatrix} \frac{1}{2}\begin{bmatrix} 0 & 1 \\ 1 & 0 \end{bmatrix}\right) = c_x \text{Tr}\left(\frac{1}{4}\begin{bmatrix} 1 & 0 \\ 0 & 1 \end{bmatrix}\right) = \frac{1}{2}c_x$$

Thus, the value of $\bar{I}_x$ is simply $1/2$ times the coefficient c_x. Notice also that the product of I_x and itself is a constant $(1/4)$ times the identity matrix, 1. This important property also holds for the I_y and I_z operator matrices. As a consequence, each of the average magnetization components can be determined by simple examination of the density matrix, provided that the matrix is written as a linear combination of the operator basis set:

$$\bar{I}_x = \frac{1}{2}c_x$$

$$\bar{I}_y = \frac{1}{2}c_y$$

$$\bar{I}_z = \frac{1}{2}c_z$$

Because the average values of the magnetization components must be real numbers, the coefficients must also be real. One cost of this method of representing the density matrix in this form is that we have introduced a potential source for confusion:

The operator matrices are now playing two roles in the formalism, first as the matrices used to calculate the magnetization components from the density matrix and second as the basis set for writing the density matrix itself. As a consequence, some care is required to distinguish the density matrix from the operators. The other catch is that the density matrix must be written in the appropriate form. If the density matrix is, instead, written as a single matrix, generating the coefficients of the linear combination reverts back to the problem of calculating the magnetization components, so nothing is really gained. In order to make calculating the magnetization components as simple as possible, we must begin with the operator basis-set representation of the density matrix and then retain this form through the various manipulations.

18.1.2   Pulses

To make practical use of the operator basis set, we need a set of rules to describe how various elements in the linear combination change with pulses and evolution periods. We begin by considering a pulse along the x'-axis. From Chapter 17, the pulse changes the density matrix according to the following matrix multiplications:

$$\rho(a) = \mathbf{R}_x(a)\rho\mathbf{R}_x^{-1}(a)$$

where ρ is the starting density matrix, $\rho(a)$ is the density matrix following a pulse with angle, a, and the rotation matrix and its inverse are given, respectively, by

$$\mathbf{R}_x(a) = \begin{bmatrix} \cos(a/2) & -i\,\sin(a/2) \\ -i\,\sin(a/2) & \cos(a/2) \end{bmatrix}$$

$$\mathbf{R}_x^{-1}(a) = \begin{bmatrix} \cos(a/2) & i\,\sin(a/2) \\ i\,\sin(a/2) & \cos(a/2) \end{bmatrix}$$

With the density matrix written as a linear combination of the operator basis set, we can write

$$\rho(a) = \mathbf{R}_x(a)\Big(c_x\mathbf{I}_x + c_y\mathbf{I}_y + c_z\mathbf{I}_z + c_1\mathbf{1}\Big)\mathbf{R}_x^{-1}(a)$$

Again taking advantage of the fact that matrix multiplication is distributive, this can be rewritten as

$$\rho(a) = c_x\mathbf{R}_x(a)\mathbf{I}_x\mathbf{R}_x^{-1}(a) + c_y\mathbf{R}_x(a)\mathbf{I}_y\mathbf{R}_x^{-1}(a) + c_z\mathbf{R}_x(a)\mathbf{I}_z\mathbf{R}_x^{-1}(a) + c_1\mathbf{R}_x(a)\mathbf{1}\mathbf{R}_x^{-1}(a)$$

With the expression written in this way, it becomes apparent that if the matrix products are calculated just once, we can use the results to calculate the result for any density matrix. For the term containing $\mathbf{I}_x$,

$$\mathbf{R}_x(a)\mathbf{I}_x\mathbf{R}_x^{-1}(a) = \begin{bmatrix} \cos(a/2) & -i\sin(a/2) \\ -i\sin(a/2) & \cos(a/2) \end{bmatrix} \frac{1}{2} \begin{bmatrix} 0 & 1 \\ 1 & 0 \end{bmatrix} \begin{bmatrix} \cos(a/2) & i\sin(a/2) \\ i\sin(a/2) & \cos(a/2) \end{bmatrix}$$

$$= \frac{1}{2} \begin{bmatrix} -i\sin(a/2) & \cos(a/2) \\ \cos(a/2) & -i\sin(a/2) \end{bmatrix} \begin{bmatrix} \cos(a/2) & i\sin(a/2) \\ i\sin(a/2) & \cos(a/2) \end{bmatrix}$$

$$= \frac{1}{2} \begin{bmatrix} 0 & \sin^2(a/2) + \cos^2(a/2) \\ \cos^2(a/2) + \sin^2(a/2) & 0 \end{bmatrix}$$

From the identity, $\cos^2\theta + \sin^2\theta = 1$, this reduces to

$$\mathbf{R}_x(a)\mathbf{I}_x\mathbf{R}_x^{-1}(a) = \frac{1}{2}\begin{bmatrix} 0 & 1 \\ 1 & 0 \end{bmatrix} = \mathbf{I}_x$$

Therefore, a rotation about the x'-axis, of any angle, leaves the $\mathbf{I}_x$-component unchanged.

What about the $\mathbf{I}_y$-component? Using the same approach,

$$\mathbf{R}_x(a)\mathbf{I}_y\mathbf{R}_x^{-1}(a) = \begin{bmatrix} \cos(a/2) & -i\sin(a/2) \\ -i\sin(a/2) & \cos(a/2) \end{bmatrix} \frac{i}{2} \begin{bmatrix} 0 & -1 \\ 1 & 0 \end{bmatrix} \begin{bmatrix} \cos(a/2) & i\sin(a/2) \\ i\sin(a/2) & \cos(a/2) \end{bmatrix}$$

$$= \frac{1}{2} \begin{bmatrix} \sin(a/2) & -i\cos(a/2) \\ i\cos(a/2) & -\sin(a/2) \end{bmatrix} \begin{bmatrix} \cos(a/2) & i\sin(a/2) \\ i\sin(a/2) & \cos(a/2) \end{bmatrix}$$

$$= \frac{1}{2} \begin{bmatrix} 2\sin(a/2)\cos(a/2) & i\sin^2(a/2) - i\cos^2(a/2) \\ i\cos^2(a/2) - i\sin^2(a/2) & -2\sin(a/2)\cos(a/2) \end{bmatrix}$$

Using a few trigonometric identities, this can be rewritten as

$$\mathbf{R}_x(a)\mathbf{I}_y\mathbf{R}_x^{-1}(a) = \frac{1}{2} \begin{bmatrix} \sin(a) & -i\cos(a) \\ i\cos(a) & -\sin(a) \end{bmatrix}$$

which, in turn, can be written as a linear combination of $\mathbf{I}_y$ and $\mathbf{I}_z$:

$$\mathbf{R}_x(a)\mathbf{I}_y\mathbf{R}_x^{-1}(a) = \cos(a)\frac{i}{2} \begin{bmatrix} 0 & -1 \\ 1 & 0 \end{bmatrix} + \sin(a)\frac{1}{2} \begin{bmatrix} 1 & 0 \\ 0 & -1 \end{bmatrix}$$

$$= \cos(a)\mathbf{I}_y + \sin(a)\mathbf{I}_z$$

Thus, a pulse along the x'-axis converts the $\mathbf{I}_y$-component of the density matrix into a mixture of $\mathbf{I}_y$ and $\mathbf{I}_z$, depending on the pulse angle, a.

Using the same manipulations, it can be shown that an x-pulse causes the following effect on the $\mathbf{I}_z$-component:

$$\mathbf{R}_x(a)\mathbf{I}_z\mathbf{R}_x^{-1}(a) = -\sin(a)\mathbf{I}_y + \cos(a)\mathbf{I}_z$$

All of this should seem rather familiar, since it corresponds exactly to the vector picture that we developed much earlier. For a pulse along the y'-axis, the corresponding relationships are

$$\mathbf{R}_y(a)\mathbf{I}_x\mathbf{R}_y^{-1}(a) = \cos(a)\mathbf{I}_x - \sin(a)\mathbf{I}_z$$

$$\mathbf{R}_y(a)\mathbf{I}_y\mathbf{R}_y^{-1}(a) = \mathbf{I}_y$$

$$\mathbf{R}_y(a)\mathbf{I}_z\mathbf{R}_y^{-1}(a) = \sin(a)\mathbf{I}_x + \cos(a)\mathbf{I}_z$$

18.1.3   Time Evolution

Next, consider how each of the components is affected by an evolution period in a constant magnetic field. From Chapter 17, this change is given by the following relationship:

$$\rho(t) = \mathbf{U}_H\rho\mathbf{U}_H^{-1}$$

where ρ is the starting matrix and $\mathbf{U}_H$ and $\mathbf{U}_H^{-1}$ are the time-evolution matrix and its inverse, respectively. These matrices are given by

$$\mathbf{U}_H = \begin{bmatrix} e^{-i\pi vt} & 0 \\ 0 & e^{i\pi vt} \end{bmatrix} \quad \text{and} \quad \mathbf{U}_H^{-1} = \begin{bmatrix} e^{i\pi vt} & 0 \\ 0 & e^{-i\pi vt} \end{bmatrix}$$

Applying these matrices to $\mathbf{I}_x$ gives

$$\mathbf{U}_H\mathbf{I}_x\mathbf{U}_H^{-1} = \begin{bmatrix} e^{-i\pi vt} & 0 \\ 0 & e^{i\pi vt} \end{bmatrix} \frac{1}{2}\begin{bmatrix} 0 & 1 \\ 1 & 0 \end{bmatrix}\begin{bmatrix} e^{i\pi vt} & 0 \\ 0 & e^{-i\pi vt} \end{bmatrix}$$

$$= \frac{1}{2}\begin{bmatrix} 0 & e^{-i\pi vt} \\ e^{i\pi vt} & 0 \end{bmatrix}\begin{bmatrix} e^{i\pi vt} & 0 \\ 0 & e^{-i\pi vt} \end{bmatrix}$$

$$= \frac{1}{2}\begin{bmatrix} 0 & e^{-i2\pi vt} \\ e^{i2\pi vt} & 0 \end{bmatrix}$$

where v is the Larmor frequency. The exponentials can be written in the following trigonometric form,

$$\mathbf{U}_H\mathbf{I}_x\mathbf{U}_H^{-1} = \frac{1}{2}\begin{bmatrix} 0 & \cos(2\pi vt) - i\sin(2\pi vt) \\ \cos(2\pi vt) + i\sin(2\pi vt) & 0 \end{bmatrix}$$

which can be written as a linear combination of $\mathbf{I}_x$ and $\mathbf{I}_y$:

$$\mathbf{U}_H\mathbf{I}_x\mathbf{U}_H^{-1} = \cos(2\pi vt)\frac{1}{2}\begin{bmatrix} 0 & 1 \\ 1 & 0 \end{bmatrix} + \sin(2\pi vt)\frac{i}{2}\begin{bmatrix} 0 & -1 \\ 1 & 0 \end{bmatrix}$$

$$= \cos(2\pi vt)\mathbf{I}_x + \sin(2\pi vt)\mathbf{I}_y$$

Thus, the $\mathbf{I}_x$-component evolves with time into a mixture of $\mathbf{I}_x$ and $\mathbf{I}_y$, with the relative contributions changing cyclically. The analogous relationship for the evolution of the $\mathbf{I}_y$-component of the density matrix is

$$\mathbf{U}_H\mathbf{I}_y\mathbf{U}_H^{-1} = -\sin(2\pi vt)\mathbf{I}_x + \cos(2\pi vt)\mathbf{I}_y$$

Applying the same procedure to the I_z-component reveals, as you might expect, that it remains constant with time:

$$U_H I_z U_H^{-1} = I_z$$

We should also consider what happens to the identity matrix, **1**, under the influence of the pulses and evolution periods, since this matrix is required to form a complete basis set for the family of 2×2 matrices and could, in principle, contribute to the density matrix. As detailed above, each of the changes in the density matrix is calculated by carrying out a two-step matrix multiplication:

$$\rho' = A\rho A^{-1}$$

where ρ and ρ' are the matrices before and after, respectively, the process defined by the matrix A and its inverse, A^{-1}. If ρ is the identity matrix, then

$$\rho' = A1A^{-1} = AA^{-1} = 1$$

Therefore, none of the operations has any effect on the component of the density matrix corresponding to **1**. Furthermore, multiplying the identity matrix by any of the magnetization operator matrices yields just the operator matrix. Since the traces of these matrices are all zero, the identity matrix makes no contribution to any of the magnetization components. As a consequence, we can discard this component of the density matrix, safe in the knowledge that it will remain unchanged and will not contribute to any observable parameter.

We now have what amounts to a set of rules for calculating the density matrix following any combination of pulses and evolution periods. If we begin with an arbitrary density matrix written as a linear combination of the magnetization operator matrices, then the effects of a pulse of angle, a, along the x'-axis on the individual terms in the linear combination will be

$$I_x \rightarrow I_x$$
$$I_y \rightarrow \cos(a)I_y + \sin(a)I_z$$
$$I_z \rightarrow -\sin(a)I_y + \cos(a)I_z$$

For a pulse along the y'-axis, the rules are as follows:

$$I_x \rightarrow \cos(a)I_x - \sin(a)I_z$$
$$I_y \rightarrow I_y$$
$$I_z \rightarrow \sin(a)I_x + \cos(a)I_z$$

For an evolution period of time t and Larmor frequency of v:

$$I_x \rightarrow \cos(2\pi vt)I_x + \sin(2\pi vt)I_y$$
$$I_y \rightarrow -\sin(2\pi vt)I_x + \cos(2\pi vt)I_y$$
$$I_z \rightarrow I_z$$

Suppose, then, that we consider a $(\pi/2)_x$-pulse applied to a system described by an arbitrary density matrix written in the following general form:

$$\rho = c_x\mathbf{I}_x + c_y\mathbf{I}_y + c_z\mathbf{I}_z$$

Following the pulse, the density matrix will be

$$\rho = c_x\mathbf{I}_x + c_y\big(\cos(\pi/2)\mathbf{I}_y + \sin(\pi/2)\mathbf{I}_z\big) + c_z\big(-\sin(\pi/2)\mathbf{I}_y + \cos(\pi/2)\mathbf{I}_z\big)$$

$$= c_x\mathbf{I}_x + c_y(0 \cdot \mathbf{I}_y + 1 \cdot \mathbf{I}_z) + c_z(-1 \cdot \mathbf{I}_y + 0 \cdot \mathbf{I}_z)$$

$$= c_x\mathbf{I}_x - c_z\mathbf{I}_y + c_y\mathbf{I}_z$$

From the new coefficients of the matrix components, we instantly know the average magnetization components in all three directions.

18.1.4   Equilibrium and a Single-Pulse Experiment

The final ingredient needed to predict the outcome of an NMR experiment is a representation of the density matrix at thermal equilibrium. In Chapter 17, the equilibrium density matrix was given as

$$\rho_{eq} = \begin{bmatrix} \frac{1}{2}(1+\Delta P) & 0 \\ 0 & \frac{1}{2}(1-\Delta P) \end{bmatrix}$$

where $\Delta P = h\nu/(2kT)$ is the fractional population difference. The density matrix can be rewritten as

$$\rho_{eq} = \frac{1}{2}\begin{bmatrix} 1 & 0 \\ 0 & 1 \end{bmatrix} + \frac{\Delta P}{2}\begin{bmatrix} 1 & 0 \\ 0 & -1 \end{bmatrix}$$

$$= \frac{1}{2}\mathbf{1} + \Delta P\mathbf{I}_z$$

From the arguments made on page 545, we know that the identity matrix, $\mathbf{1}$, may be dropped from further consideration, leaving us with a term proportional to $\mathbf{I}_z$ as the representation of the density matrix. This makes sense, because only z-magnetization is present at equilibrium. For most purposes, we can also drop the initial coefficient for the matrix, keeping in mind that all of the subsequent forms of the density matrix are implicitly multiplied by this term.

Suppose, then, that we begin with the equilibrium state, apply a $(\pi/2)_y$-pulse and then record an FID. The rule governing the pulse is

$$\mathbf{I}_z \rightarrow \sin(a)\mathbf{I}_x + \cos(a)\mathbf{I}_z \tag{18.3}$$

with $a = \pi/2$. Thus, the new density matrix can be represented as $\mathbf{I}_x$. Using the rule for a time-evolution period of length t,

$$\mathbf{I}_x \rightarrow \cos(2\pi\nu t)\mathbf{I}_x + \sin(2\pi\nu t)\mathbf{I}_y$$

Figure 18.1
Vector diagram representing a spin population with the density matrix, $\rho = c_x\mathbf{I}_x + c_y\mathbf{I}_y + c_z\mathbf{I}_z$.

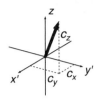

Thus, the initial pulse converts z-magnetization to positive x-magnetization, which is then cyclically converted to y-magnetization and back to x-magnetization. This, of course, is exactly what we expect from simple vector diagrams, and one could argue convincingly that it is easier to remember the rules for graphically manipulating the vectors than it is to write the operator rules. We will see, though, that the benefits are considerably greater when this approach is applied to a system of two coupled spins.

We can also think of this rather involved exercise as providing a more formal interpretation, based on quantum and statistical mechanics, of the graphical vector diagrams. In this view, the projections of a vector onto the x'-, y'- and z-axes represent the contributions of the operator basis-set matrices to the density matrix. Thus, for the general density matrix given in the form,

$$\rho = c_x\mathbf{I}_x + c_y\mathbf{I}_y + c_z\mathbf{I}_z$$

the vector diagram is as shown in Fig. 18.1.

18.2   A Basis Set for a Population of Scalar-Coupled Spin Pairs

From Chapter 17, the density matrix for a population of coupled spin pairs has 16 elements and is written as

$$\rho = \begin{bmatrix} \overline{c_{\alpha\alpha}c^*_{\alpha\alpha}} & \overline{c_{\alpha\alpha}c^*_{\alpha\beta}} & \overline{c_{\alpha\alpha}c^*_{\beta\alpha}} & \overline{c_{\alpha\alpha}c^*_{\beta\beta}} \\ \overline{c_{\alpha\beta}c^*_{\alpha\alpha}} & \overline{c_{\alpha\beta}c^*_{\alpha\beta}} & \overline{c_{\alpha\beta}c^*_{\beta\alpha}} & \overline{c_{\alpha\beta}c^*_{\beta\beta}} \\ \overline{c_{\beta\alpha}c^*_{\alpha\alpha}} & \overline{c_{\beta\alpha}c^*_{\alpha\beta}} & \overline{c_{\beta\alpha}c^*_{\beta\alpha}} & \overline{c_{\beta\alpha}c^*_{\beta\beta}} \\ \overline{c_{\beta\beta}c^*_{\alpha\alpha}} & \overline{c_{\beta\beta}c^*_{\alpha\beta}} & \overline{c_{\beta\beta}c^*_{\beta\alpha}} & \overline{c_{\beta\beta}c^*_{\beta\beta}} \end{bmatrix}$$

For this system, there are six magnetization operators, three each for the I- and S-spins. The matrices corresponding to these operators are

$$\mathbf{I}_x = \frac{1}{2}\begin{bmatrix} 0 & 0 & 1 & 0 \\ 0 & 0 & 0 & 1 \\ 1 & 0 & 0 & 0 \\ 0 & 1 & 0 & 0 \end{bmatrix} \qquad \mathbf{S}_x = \frac{1}{2}\begin{bmatrix} 0 & 1 & 0 & 0 \\ 1 & 0 & 0 & 0 \\ 0 & 0 & 0 & 1 \\ 0 & 0 & 1 & 0 \end{bmatrix}$$

$$\mathbf{I}_y = \frac{i}{2}\begin{bmatrix} 0 & 0 & -1 & 0 \\ 0 & 0 & 0 & -1 \\ 1 & 0 & 0 & 0 \\ 0 & 1 & 0 & 0 \end{bmatrix} \qquad \mathbf{S}_y = \frac{i}{2}\begin{bmatrix} 0 & -1 & 0 & 0 \\ 1 & 0 & 0 & 0 \\ 0 & 0 & 0 & -1 \\ 0 & 0 & 1 & 0 \end{bmatrix}$$

$$\mathbf{I}_z = \frac{1}{2}\begin{bmatrix} 1 & 0 & 0 & 0 \\ 0 & 1 & 0 & 0 \\ 0 & 0 & -1 & 0 \\ 0 & 0 & 0 & -1 \end{bmatrix} \qquad \mathbf{S}_z = \frac{1}{2}\begin{bmatrix} 1 & 0 & 0 & 0 \\ 0 & -1 & 0 & 0 \\ 0 & 0 & 1 & 0 \\ 0 & 0 & 0 & -1 \end{bmatrix}$$

Now, we are faced with a problem that did not arise in the case of the single spin: For a 4×4 matrix, we need a total of 16 matrices to form a basis set, but we only have six magnetization operators so far, plus the identity matrix. We need another nine matrices, and they can't be just any nine 4×4 matrices; each must be orthogonal to all of the operator matrices shown above, and to one another. We would also like these matrices to carry some physical meaning.

The other nine matrices for the basis set can be generated by forming products between the matrices for the I and S operators:

$$\mathbf{I}_x\mathbf{S}_x = \frac{1}{4}\begin{bmatrix} 0 & 0 & 0 & 1 \\ 0 & 0 & 1 & 0 \\ 0 & 1 & 0 & 0 \\ 1 & 0 & 0 & 0 \end{bmatrix} \quad \mathbf{I}_x\mathbf{S}_y = \frac{i}{4}\begin{bmatrix} 0 & 0 & 0 & -1 \\ 0 & 0 & 1 & 0 \\ 0 & -1 & 0 & 0 \\ 1 & 0 & 0 & 0 \end{bmatrix} \quad \mathbf{I}_x\mathbf{S}_z = \frac{1}{4}\begin{bmatrix} 0 & 0 & 1 & 0 \\ 0 & 0 & 0 & -1 \\ 1 & 0 & 0 & 0 \\ 0 & -1 & 0 & 0 \end{bmatrix}$$

$$\mathbf{I}_y\mathbf{S}_x = \frac{i}{4}\begin{bmatrix} 0 & 0 & 0 & -1 \\ 0 & 0 & -1 & 0 \\ 0 & 1 & 0 & 0 \\ 1 & 0 & 0 & 0 \end{bmatrix} \quad \mathbf{I}_y\mathbf{S}_y = \frac{1}{4}\begin{bmatrix} 0 & 0 & 0 & -1 \\ 0 & 0 & 1 & 0 \\ 0 & 1 & 0 & 0 \\ -1 & 0 & 0 & 0 \end{bmatrix} \quad \mathbf{I}_y\mathbf{S}_z = \frac{i}{4}\begin{bmatrix} 0 & 0 & -1 & 0 \\ 0 & 0 & 0 & 1 \\ 1 & 0 & 0 & 0 \\ 0 & -1 & 0 & 0 \end{bmatrix}$$

$$\mathbf{I}_z\mathbf{S}_x = \frac{1}{4}\begin{bmatrix} 0 & 1 & 0 & 0 \\ 1 & 0 & 0 & 0 \\ 0 & 0 & 0 & -1 \\ 0 & 0 & -1 & 0 \end{bmatrix} \quad \mathbf{I}_z\mathbf{S}_y = \frac{i}{4}\begin{bmatrix} 0 & -1 & 0 & 0 \\ 1 & 0 & 0 & 0 \\ 0 & 0 & 0 & 1 \\ 0 & 0 & -1 & 0 \end{bmatrix} \quad \mathbf{I}_z\mathbf{S}_z = \frac{1}{4}\begin{bmatrix} 1 & 0 & 0 & 0 \\ 0 & -1 & 0 & 0 \\ 0 & 0 & -1 & 0 \\ 0 & 0 & 0 & 1 \end{bmatrix}$$

Although matrix multiplication is in general not commutative, the magnetization operator matrices have properties such that these particular multiplications can be carried out in either order.

The use of these matrix products as part of the basis set for the two-spin system is the reason that this approach to the density matrix is often referred to as the product-operator formalism. Recall from Section 13.1.4 that the products of the magnetization operators have an important physical significance, providing information about the correlation between the magnetization components of the two coupled spins. We will return to the physical interpretation of the product operators shortly. For now, the important point is that the six magnetization operator matrices, the nine product-operator matrices and the identity matrix form a complete basis set that can be used to represent the density matrix for the two-spin system. As before, the identity matrix does not make any contribution to any observable properties and can be ignored from here on. Just as with the simpler one-spin case, representing the density matrix in this way greatly simplifies all of the manipulations.

The reader might be wondering about other products that could be formed, such as I_xI_y or I_yI_z. Could these be used to form a basis set? The answer is no, at least not if the individual I-magnetization operators are also to be used. Consider the product, I_xI_y:

$$I_xI_y = \frac{1}{2}\begin{bmatrix} 0 & 0 & 1 & 0 \\ 0 & 0 & 0 & 1 \\ 1 & 0 & 0 & 0 \\ 0 & 1 & 0 & 0 \end{bmatrix} \frac{i}{2}\begin{bmatrix} 0 & 0 & -1 & 0 \\ 0 & 0 & 0 & -1 \\ 1 & 0 & 0 & 0 \\ 0 & 1 & 0 & 0 \end{bmatrix} = \frac{i}{4}\begin{bmatrix} 1 & 0 & 0 & 0 \\ 0 & 1 & 0 & 0 \\ 0 & 0 & -1 & 0 \\ 0 & 0 & 0 & -1 \end{bmatrix}$$

Though this is a perfectly good matrix, notice that it is exactly the same as I_z, multiplied by a constant, $i/2$. Thus, the product, I_xI_y, doesn't bring anything new to the table. This reflects the physical principle that the I_x- and I_y-magnetization components cannot be measured independently, as demonstrated by the Stern–Gerlach experiments described in Chapter 11. The following relationships hold for the products that can be formed among the I-magnetization operator matrices:

$$I_xI_y = \frac{i}{2}I_z$$

$$I_yI_z = \frac{i}{2}I_x \tag{18.4}$$

$$I_zI_x = \frac{i}{2}I_y$$

These relationships can be visualized as forming a circle, illustrated in Fig. 18.2. Multiplying any pair of operator matrices in the order indicated by the arrows results in the third (multiplied by $i/2$). These multiplications are *not* commutative, but the products taken in the opposite order (counterclockwise in Fig. 18.2) are the negative of those given in Eq. 18.4. For S-magnetization operators, the analogous

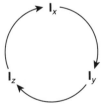

Figure 18.2   Cyclic relationship among the pairwise products of I_x, I_y and I_z. The product of each pair, taken in the indicated clockwise direction, is $i/2$ times the third operator matrix. If the product is taken in the counterclockwise direction, the result is $= -i/2$ times the third operator matrix.

relationships are

$$S_x S_y = \frac{i}{2} S_z$$

$$S_y S_z = \frac{i}{2} S_x$$

$$S_z S_x = \frac{i}{2} S_y$$

Again, taking the products in the opposite order gives the same results, but multiplied by −1. Although these products are not used to form a basis set for the density matrix, they do arise in the manipulations described later, and knowing these relationships will be quite useful.

Just as with the single-spin system, the virtues of using a basis-set representation of the density matrix for the two-spin system are that (1) the individual components of the density matrix can be analyzed individually, using relatively simple rules, and (2) the physically observable magnetization components are derived from simple inspection of the coefficients. However, the relationships between the coefficients and average observable properties are slightly different in this case.

If we write the density matrix as a linear combination of the basis-set matrices,

$$\rho = c_{I_x} I_x + c_{I_y} I_y + c_{I_z} I_z$$
$$+ c_{S_x} S_x + c_{S_y} S_y + c_{S_z} S_z$$
$$+ c_{I_x S_x} I_x S_x + c_{I_x S_y} I_x S_y + c_{I_x S_z} I_x S_z$$
$$+ c_{I_y S_x} I_y S_x + c_{I_y S_y} I_y S_y + c_{I_y S_z} I_y S_z$$
$$+ c_{I_z S_x} I_z S_x + c_{I_z S_y} I_z S_y + c_{I_z S_z} I_z S_z \qquad (18.5)$$

then we can, for instance, calculate the average I_z-magnetization as

$$\bar{I}_z = \text{Tr}(I_z \rho)$$

Using the same arguments as before, we can discard all of the terms in the linear combination except for the one corresponding to the operator of interest, which is I_z in this case:

$$\bar{I}_z = \text{Tr}(I_z c_{I_z} I_z)$$

For the two-spin system, the product, $I_z I_z$, is equal to $(1/4)\mathbf{1}$, where $\mathbf{1}$ is the 4×4 identity matrix. The trace of the identity matrix is 4, so we have the following result:

$$\bar{I}_z = c_{I_z} \frac{1}{4} \text{Tr}(\mathbf{1}) = c_{I_z}$$

Thus, the average magnetization is equal to the coefficient in the expansion (rather than 1/2 times the coefficient, as seen for the single-spin case).

The result is slightly different for the case of average values calculated using the product operators, reflecting a factor of 1/4 that is introduced when the product matrices are generated by multiplication of the magnetization operator matrices. For instance, the average value of the product, $I_x S_y$, is calculated as follows:

$$\overline{I_x S_y} = \mathrm{Tr}(\mathbf{I}_x \mathbf{S}_y c_{I_x S_y} \mathbf{I}_x \mathbf{S}_y)$$

Here, the product, $\mathbf{I}_x \mathbf{S}_y \mathbf{I}_x \mathbf{S}_y$, equals $(1/16)$ times the identity matrix, so

$$\overline{I_x S_y} = c_{I_x S_y} \frac{1}{16} \mathrm{Tr}(\mathbf{1}) = \frac{1}{4} c_{I_x S_y}$$

Thus, the average value of the product is obtained as 1/4 times the coefficient.[1] The same result is obtained for each of the other eight product operators. In practice, it is not possible to actually measure these averages experimentally. However, these terms can be converted into observable magnetization components, as discussed in Chapter 14, so that it is important to keep track of their relative magnitudes.

18.2.1   Vector Representations and Their Interpretation

For the case of a single spin, we saw that the effects of pulses and time evolution on the operator basis-set representation were entirely consistent with our earlier description using magnetization vectors. We can also use a vector representation for the two-spin system, though some care is required, especially in the interpretation of the vectors used to represent the product-operator components.

For the components of the density matrix corresponding to the simple magnetization operators, we use the rather obvious representations shown in Fig. 18.3. In previous chapters, the magnetization vectors for the two spins were often placed on the same diagram. Now, however, the vectors representing the operator components of the density matrix components will be kept separate, to avoid confusion with the vectors used to represent the product-operator components, described below.

Recall from Chapters 13 and 14 that the product operators are used with the wavefunction, Ψ, to calculate the correlations between the magnetization components of two spins that arise when the spins are coupled. For instance, the operator $\hat{I}_z \hat{S}_z$ is used to calculate the average value of the product of the z-magnetization components of the two spins:

$$\langle I_z S_z \rangle = \langle \Psi | \hat{I}_z \hat{S}_z | \Psi \rangle$$

For an individual spin pair, I_z and S_z can, in principle, be measured simultaneously with a Stern–Gerlach type apparatus, and each measurement can yield the two possible values, 1/2 and −1/2. The possible products of these values are 1/4 and −1/4, and

---

[1] In many texts and papers, the product-operator matrices are multiplied by 2 when used in the basis set (e.g., as $2\mathbf{I}_x \mathbf{S}_y$). When written in this form, the average products are one-half of the coefficients.

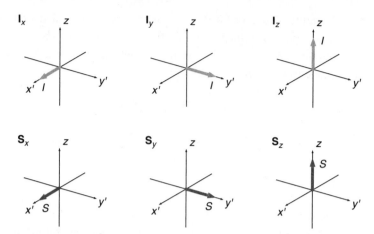

Figure 18.3   Vector representations of the operator components of the density matrix corresponding to the observable magnetization components.

the average value of the product, when the measurements are applied to a population of spin pairs, all with the same wavefunction, depends on the particular wavefunction. For the wavefunctions $|\alpha\alpha\rangle$ and $|\alpha\beta\rangle$, the average products are 1/4 and −1/4, respectively. For other wavefunctions, however, the z-magnetization components may be uncorrelated, leading to an average product of 0.

In Chapter 14 (pages 407–409), an extension of the standard vector diagrams was introduced to represent the correlations calculated by the product operators. For instance, Fig. 18.4 shows the diagram representing a correlation between I_z and S_z. In this drawing, the vectors representing positive I_z- and S_z-magnetizations are linked together to indicate that, for an individual spin pair, a positive value of I_z is always associated with a positive value of S_z. Conversely, a negative value of I_z is correlated with a negative value of S_z. The correlation between two magnetization components can also have a negative sign, implying that a positive value of one is associated with a negative value of the other. This is represented in Fig. 18.5 for $-I_zS_z$. What, then, is the significance of the I_zS_z product-operator matrix, for instance, when it appears as a component of the density matrix? As we just saw, the coefficient for a product-operator matrix in the expansion shown in Equation 18.5 corresponds to four times the value of the corresponding product, in this case the product I_zS_z, averaged over the population represented by the density matrix. It is important to

Figure 18.4

Vector representation of the I_zS_z correlation.

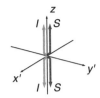

Figure 18.5

Vector representation of a negative $I_z S_z$ correlation.

stress that this calculation involves, conceptually, measuring the product for each of the individual spin pairs and then calculating the average, rather than the product of the average magnetization components for the population. Thus, the value of $\overline{I_z S_z}$ contains information about the correlations between spins in individual molecules, averaged over the population, rather than the bulk magnetization. For instance, $\bar{I}_z$ and $\bar{S}_z$ might both be zero, but $\overline{I_z S_z}$ could still have a positive or negative value.

Drawings to represent each of the product-operator components of the density matrix are shown in Fig. 18.6. In each of these diagrams, a positive I-magnetization is correlated with a positive S-magnetization, and the negative magnetization components are similarly correlated. Another set of diagrams can be drawn to represent the negative correlations. The information conveyed in these diagrams could be represented more simply using just a single pair of connected I- and S-vectors, such as

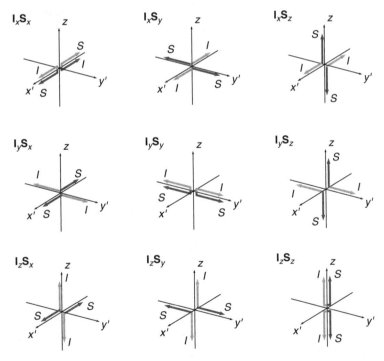

Figure 18.6   Vector representations of the operator components of the density matrix corresponding to spin correlations.

the positive I_x- and S_x-vectors for I_xS_x. However, the use of both pairs serves as a reminder of the underlying physics, such that measurements for the spins in individual coupled pairs may yield either of the quantized results, 1/2 or –1/2, but are correlated as indicated in the diagrams. Another important feature of the diagrams in Fig. 18.6 is that, in each case, the two vectors associated with each spin point in opposite directions, reflecting the fact that the operators they represent do not, in themselves, imply any observable magnetization.

18.2.2   Representation of the Equilibrium Density Matrix

Before considering the effects of pulses and precession periods, it is useful to examine the density matrix representing the two-spin system at thermal equilibrium, which provides some further insights into the significance of the operator matrices. From Chapter 17, the equilibrium density matrix can be represented as

$$\rho_{eq} = \frac{1}{4} \begin{bmatrix} (\Delta P_I + \Delta P_S) & 0 & 0 & 0 \\ 0 & (\Delta P_I - \Delta P_S) & 0 & 0 \\ 0 & 0 & (-\Delta P_I + \Delta P_S) & 0 \\ 0 & 0 & 0 & (-\Delta P_I - \Delta P_S) \end{bmatrix}$$

where ΔP_I and ΔP_S are the fractional excess populations of the I- and S-spins at equilibrium. Comparison with the operator matrices shows that this can be written as the following sum:

$$\rho_{eq} = \Delta P_I \frac{1}{4} \begin{bmatrix} 1 & 0 & 0 & 0 \\ 0 & 1 & 0 & 0 \\ 0 & 0 & -1 & 0 \\ 0 & 0 & 0 & -1 \end{bmatrix} + \Delta P_S \frac{1}{4} \begin{bmatrix} 1 & 0 & 0 & 0 \\ 0 & -1 & 0 & 0 \\ 0 & 0 & 1 & 0 \\ 0 & 0 & 0 & -1 \end{bmatrix}$$

$$= \frac{\Delta P_I}{2} I_z + \frac{\Delta P_S}{2} S_z$$

This corresponds with our expectation that, at equilibrium, the I- and S-spin populations will each give rise to a net positive z-magnetization and that there will be no detectable transverse magnetization components. The magnitudes of the I_z- and S_z-components depend on the equilibrium distribution of spin states, which depends, in turn, on the gyromagnetic ratios of the nuclei, the magnetic field strength and the temperature.

Note that the equilibrium density matrix does *not* include a contribution from the product-operator matrix, I_zS_z. Even though the average z-magnetization components of both spins are positive, any correlations between the spins of individual pairs average out to zero. It is worthwhile to examine this important result in a bit more detail.

As noted previously, we can treat the equilibrium distribution *as if* it were composed only of spin pairs in the states represented by the $|\alpha\alpha\rangle$, $|\alpha\beta\rangle$, $|\beta\alpha\rangle$ and $|\beta\beta\rangle$

wavefunctions, with the fractional populations of the four eigenstates written as

$$f_{\alpha\alpha} = \frac{N_{\alpha\alpha}}{N_T} = \frac{1}{4}\left(1 + \Delta P_I + \Delta P_S\right)$$

$$f_{\alpha\beta} = \frac{N_{\alpha\beta}}{N_T} = \frac{1}{4}\left(1 + \Delta P_I - \Delta P_S\right)$$

$$f_{\beta\alpha} = \frac{N_{\beta\alpha}}{N_T} = \frac{1}{4}\left(1 - \Delta P_I + \Delta P_S\right)$$

$$f_{\beta\beta} = \frac{N_{\beta\beta}}{N_T} = \frac{1}{4}\left(1 - \Delta P_I - \Delta P_S\right) \tag{18.6}$$

where N_T is the total number of spin pairs in the total population. For the individual spin states making up this population, the values determined by the product operators are *not* all zero. Applying the operators to the four wavefunctions gives the following results:

$$\text{for } |\alpha\alpha\rangle, \ \langle I_z S_z\rangle = 1/4$$

$$\text{for } |\alpha\beta\rangle, \ \langle I_z S_z\rangle = -1/4$$

$$\text{for } |\beta\alpha\rangle, \ \langle I_z S_z\rangle = -1/4$$

$$\text{for } |\beta\beta\rangle, \ \langle I_z S_z\rangle = 1/4$$

All of the other product averages are zero. To calculate the total average value of $I_z S_z$, we weight each of these values by the corresponding fraction of the total population and sum the results:

$$\langle I_z S_z\rangle = f_{\alpha\alpha}\frac{1}{4} + f_{\beta\alpha}\frac{-1}{4} + f_{\beta\alpha}\frac{-1}{4} + f_{\beta\beta}\frac{1}{4}$$

When this sum is calculated, the result is zero, just as expected from the analysis of the density matrix. Thus, the individual spin pairs do, in fact, show correlations, but these correlations average to zero when the entire population is considered. As discussed in Chapter 17, this particular distribution is consistent with the equilibrium density matrix, but is only one of a vast number of such distributions. Importantly, however, any distribution that satisfies the requirements for an equilibrium distribution will have the same properties, including the cancellation of any net correlation between the I- and S-magnetization components of individual spin pairs.

18.2.3 Pulses

Now we need rules to describe the behavior of the individual components during pulses and evolution periods. As you might expect, pulses applied selectively to the I- or S-spin affect the matrix components representing observable magnetization components in a relatively simple way, as illustrated in Fig. 18.7 for the $\mathbf{I}_x$-, $\mathbf{I}_y$- and $\mathbf{I}_z$-components. The corresponding rules for pulses to the S-spin are exactly the same, with the symbol $\mathbf{S}$ replacing $\mathbf{I}$ in the expressions.

Pulse of angle *a* along the *x'*-axis

$I_x \rightarrow I_x$

$I_y \rightarrow \cos(a)I_y + \sin(a)I_z$

$I_z \rightarrow \cos(a)I_z - \sin(a)I_y$

Pulse of angle *a* along the *y'*-axis

$I_x \rightarrow \cos(a)I_x - \sin(a)I_z$

$I_y \rightarrow I_y$

$I_z \rightarrow \cos(a)I_z + \sin(a)I_x$

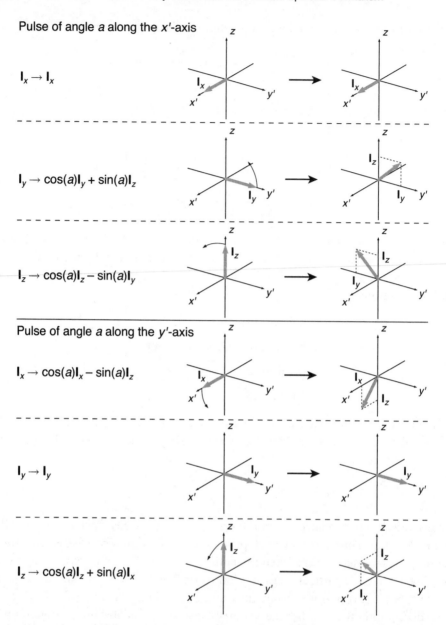

Figure 18.7   Rules for calculating the effects of pulses on the operators representing the *I*-magnetization components. Adapted with permission from Goldenberg, D.P. (2010) *Concepts. Magn. Reson.* **36A**, 49-83. Copyright © 2010 by John Wiley and Sons.

The effects of a non-selective pulse are calculated by first applying the rules to one set of matrix components and then to the other. For instance, if we start with the equilibrium state for a homonuclear pair, the effects of a $(\pi/2)_y$-pulse to both spins is

$$I_z + S_z \rightarrow I_x + S_x$$

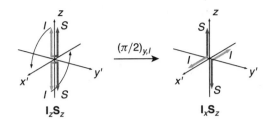

Figure 18.8
Vector representation of the conversion of an I_zS_z correlation into an I_xS_z correlation by a selective $(\pi/2)_{y,I}$-pulse.

The effects of pulses on matrix components representing correlations are calculated using the same rules, applied to each of the terms. For instance, the effect of a $(\pi/2)_{y,I}$-pulse on the matrix, I_zS_z, is calculated as

$$I_zS_z \rightarrow I_xS_z$$

which is visualized using the correlation diagrams shown in Fig. 18.8. The pulse changes the net correlation between I_z and S_z into a correlation between I_x and S_z. It is important to emphasize that these results reflect only the changes in the correlations, rather than the net magnetization for the population. The pulses convert the contribution of one product-operator matrix into a contribution of a different product operator. The effects on the net magnetization components arise from the effects of the pulses on the magnetization operator components, I_z, S_z, etc., as calculated earlier.

18.2.4   Time Evolution in the Absence of Scalar Coupling

As we have seen previously, starting in Chapter 14, the interesting phenomena involving scalar coupling emerge during periods in which the coupled spins undergo precession. For the purposes of calculations, it is convenient to separate the time-dependent effects of chemical-shift differences from those due to scalar coupling. This results in two sets of time-evolution matrices, which can be represented as two sets of rules for the operator basis representation and are applied sequentially, in either order. The rules for evolution due to chemical-shift differences are derived from the time-evolution matrix, $\mathbf{U}_H$ (page 532), evaluated with $J = 0$. The results are just as expected for the magnetization components, as shown in Fig. 18.9 for the I-components. Analogous rules are applied to the S-magnetization components.

As seen before with pulses, the chemical-shift evolution rules are also applied to the individual components of the product-operator terms. For instance, I_xS_z evolves according to

$$I_xS_z \rightarrow \left(\cos(2\pi \nu_I t)I_x + \sin(2\pi \nu_I t)I_y \right)S_z = \cos(2\pi \nu_I t)I_xS_z + \sin(2\pi \nu_I t)I_yS_z$$

The initial I_xS_z-component is converted into an I_yS_z-component and back again, with the frequency, ν_I. This result is also consistent with a vector representation, as shown in Fig. 18.10.

Evolution due to chemical shift (CS)

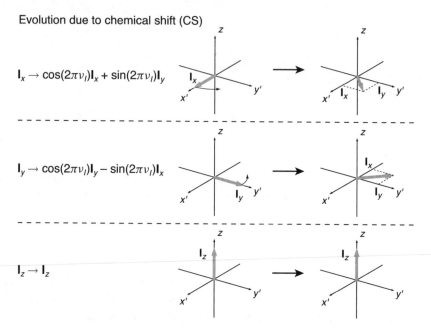

$$\mathbf{I}_x \rightarrow \cos(2\pi\nu_I)\mathbf{I}_x + \sin(2\pi\nu_I)\mathbf{I}_y$$

$$\mathbf{I}_y \rightarrow \cos(2\pi\nu_I)\mathbf{I}_y - \sin(2\pi\nu_I)\mathbf{I}_x$$

$$\mathbf{I}_z \rightarrow \mathbf{I}_z$$

Figure 18.9 Rules for calculating the effects of chemical-shift evolution on the operators representing the I-magnetization components. Adapted with permission from Goldenberg, D.P. (2010) *Concepts. Magn. Reson.* **36A**, 49-83. Copyright © 2010 by John Wiley and Sons.

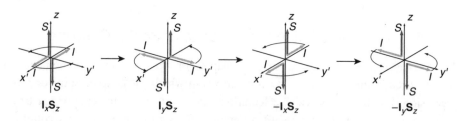

Figure 18.10 Vector representation of the interconversion of $\mathbf{I}_x\mathbf{S}_z$ and $\mathbf{I}_y\mathbf{S}_z$ correlations due to chemical-shift evolution, in the absence of scalar coupling.

The situation is somewhat more complex when there is a correlation between transverse magnetization components for the two spins, such as when starting from $\mathbf{I}_x\mathbf{S}_x$. When both the $\mathbf{I}$- and $\mathbf{S}$-components evolve with time, the rules for each spin can be applied sequentially. First, we apply the rule for the $\mathbf{I}_x$-component,

$$\mathbf{I}_x\mathbf{S}_x \rightarrow \cos(2\pi\,\nu_I t)\mathbf{I}_x\mathbf{S}_x + \sin(2\pi\,\nu_I t)\mathbf{I}_y\mathbf{S}_x$$

followed by the same rule applied to $\mathbf{S}_x$:

$$\cos(2\pi\,\nu_I t)\mathbf{I}_x\mathbf{S}_x + \sin(2\pi\,\nu_I t)\mathbf{I}_y\mathbf{S}_x \rightarrow \cos(2\pi\,\nu_I t)\mathbf{I}_x\big(\cos(2\pi\,\nu_S t)\mathbf{S}_x + \sin(2\pi\,\nu_S t)\mathbf{S}_y\big)$$

$$+ \sin(2\pi\,\nu_I t)\mathbf{I}_y\big(\cos(2\pi\,\nu_S t)\mathbf{S}_x + \sin(2\pi\,\nu_S t)\mathbf{S}_y\big)$$

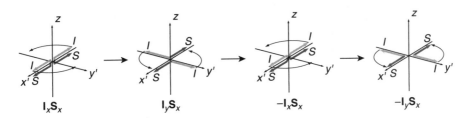

Figure 18.11   Vector representation of the interconversion of the I_xS_x and I_yS_x correlations due to chemical-shift evolution, in the absence of scalar coupling and with v_S set to zero.

Collecting the operator terms gives

$$I_xS_x \rightarrow \cos(2\pi v_I t)\cos(2\pi v_S t)I_xS_x + \cos(2\pi v_I t)\sin(2\pi v_S t)I_xS_y$$

$$+ \sin(2\pi v_I t)\cos(2\pi v_S t)I_yS_x + \sin(2\pi v_I t)\sin(2\pi v_S t)I_yS_y \quad (18.7)$$

Though this looks rather formidable, there is an underlying simplicity. To see the physical significance of this result, it is helpful to set one of the frequencies, say v_S, to zero. This is equivalent to setting the rotating-frame frequency to v_S, and there is no loss in generality from this manipulation. The expression in Eq. 18.7 then simplifies to

$$I_xS_x \rightarrow \cos(2\pi v_I t)I_xS_x + \sin(2\pi v_I t)I_yS_x$$

Now, the S-component of the correlation remains aligned with the x'-axis, and the I-component precesses as expected. The correlation between I_x and S_x is converted cyclically into a correlation between I_y and S_x, and back again, as shown in Fig. 18.11. If the S-spin Larmor frequency does not match the rotating-frame frequency, then both the S- and I-vectors in the diagram precess, representing the evolution of all four of the transverse correlation components in Eq. 18.7.

18.2.5   Evolution Under the Influence of Scalar Coupling

Finally, we consider the effect of scalar coupling on the various operator matrices during an evolution period. The matrix for calculating these effects is obtained from the overall time-evolution matrix, U_H (page 532), by setting both v_I and v_S to zero. When applied to the individual **I**-magnetization components, the rules illustrated in Fig. 18.12 are obtained. In these diagrams, net transverse magnetization components (I_x and I_y) are represented by two vector pairs showing a coupling between the I- and S-spins, but with the two vectors representing the I-spin both pointing in the same direction, along the x'- or y'-axis. These diagrams, which were introduced earlier in Chapter 14 (page 411), indicate a net I-magnetization and the *absence* of a correlation between the two spins. For instance, in the representation of I_x, positive I_x-magnetization is associated equally with positive and negative S_z-magnetization. As the system evolves, two frequency components are generated, separated by the

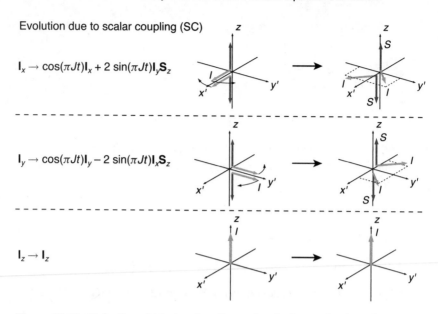

Evolution due to scalar coupling (SC)

$$I_x \rightarrow \cos(\pi Jt)I_x + 2\sin(\pi Jt)I_yS_z$$

$$I_y \rightarrow \cos(\pi Jt)I_y - 2\sin(\pi Jt)I_xS_z$$

$$I_z \rightarrow I_z$$

Figure 18.12 Rules for calculating the effects of evolution under the influence of scalar coupling on the operators representing the I-magnetization components. Adapted with permission from Goldenberg, D.P. (2010) *Concepts. Magn. Reson.* **36A**, 49-83. Copyright © 2010 by John Wiley and Sons.

scalar coupling constant, J. Since we are considering only the effect of scalar coupling and assuming, in essence, that v_I and v_S are zero, the direction of the *average* magnetization remains stationary in the rotating frame. In the case of an initial I_x-component, as the two I-vectors precess in opposite directions away from the x'-axis, the y'-component of each is correlated with an S_z-component of the same sign. When $t = 1/(2J)$, the I_x-component of the density matrix disappears, and the correlation term, I_yS_z, reaches a maximum.

The physical interpretation of the S_z-vectors in these diagrams can be rather subtle. In Chapter 14, we saw that the appearance of split magnetization vectors can reflect two rather different physical situations. The easiest situation to visualize is a population composed of spin pairs with different wavefunctions, each with a distinct precession frequency, with the S-spin in either the α or β state. In this case, the vector pairs in Fig. 18.12 represent spin pairs in which the I-magnetization lies in the transverse plane and the S-magnetization is aligned with either the positive or negative z-axis. However, the two I-frequency components can also arise from a pure population with a single wavefunction with a transverse S-magnetization component. In this latter case, the S-spin is in a state that represents a superposition of α and β states, and the positive and negative S_z-vectors must be interpreted as a sort of virtual magnetization that is only expressed when the I-spin precesses under the influence of scalar coupling. A virtue of the density-matrix treatment, and its representation using the operator basis set, is that it allows us to ignore the details of the population and wavefunctions and simply carry out the appropriate manipulations to predict the outcome of an experiment. At the same time, however, the physical origins of the phenomena are sometimes obscured by these more abstract treatments.

Evolution due to scalar coupling (SC)

$$\mathbf{I}_x\mathbf{S}_z \rightarrow \cos(\pi Jt)\mathbf{I}_x\mathbf{S}_z + \frac{1}{2}\sin(\pi Jt)\mathbf{I}_y$$

$$\mathbf{I}_y\mathbf{S}_z \rightarrow \cos(\pi Jt)\mathbf{I}_y\mathbf{S}_z - \frac{1}{2}\sin(\pi Jt)\mathbf{I}_x$$

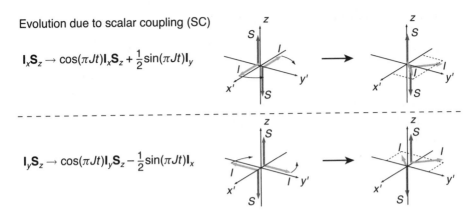

Figure 18.13   Rules for calculating the effects of evolution under the influence of scalar coupling on the operators representing the $\mathbf{I}_x\mathbf{S}_z$ and $\mathbf{I}_y\mathbf{S}_z$ correlations. Adapted with permission from Goldenberg, D.P. (2010) *Concepts. Magn. Reson.* **36A**, 49-83. Copyright © 2010 by John Wiley and Sons.

The diagrams in Fig. 18.12 also suggest how the $\mathbf{I}_x\mathbf{S}_z$ and $\mathbf{I}_y\mathbf{S}_z$ product-operator components will change with time, since these processes simply represent the continued precession of the transverse $\mathbf{I}$-components, as shown in Fig. 18.13. Notice that the vector representation of the behavior of $\mathbf{I}_x\mathbf{S}_z$ and $\mathbf{I}_y\mathbf{S}_z$ under the influence of scalar coupling is very similar to that shown in Fig. 18.10 for the evolution due to chemical shift, *except* that now the vectors representing the I-magnetization are precessing in opposite directions. During chemical-shift evolution, the two vectors precess in the same direction and at the same frequency, so that a product operator never evolves into a magnetization operator, or vice versa. The interconversion of the two types of density-matrix components only occurs when scalar coupling is involved.

Just as the z-components of the $\mathbf{I}_x\mathbf{S}_z$, $\mathbf{I}_y\mathbf{S}_z$, $\mathbf{I}_z\mathbf{S}_x$, and $\mathbf{I}_z\mathbf{S}_y$ density-matrix components are unaffected during evolution under the influence of scalar coupling, the matrix representing the correlation between z-components, $\mathbf{I}_z\mathbf{S}_z$, is unaffected by scalar-coupling evolution.

There are four other product-operator matrices to be considered, those representing correlations between the x- and y-magnetization components for the two spins: $\mathbf{I}_x\mathbf{S}_x$, $\mathbf{I}_x\mathbf{S}_y$, $\mathbf{I}_y\mathbf{S}_x$ and $\mathbf{I}_y\mathbf{S}_y$. The effects of scalar-coupling evolution can be calculated either by carrying out the matrix multiplication or by using the results for the individual components and substitution. Since this latter procedure is easier and generally used in the application of the product-operator formalism, we will employ it here, taking $\mathbf{I}_x\mathbf{S}_x$ as an example. First, the rule for the evolution of $\mathbf{I}_x$ is applied

$$\mathbf{I}_x\mathbf{S}_x \rightarrow \big(\cos(\pi Jt)\mathbf{I}_x + 2\sin(\pi Jt)\mathbf{I}_y\mathbf{S}_z\big)\mathbf{S}_x$$

followed by the rule for $\mathbf{S}_x$:

$$\big(\cos(\pi Jt)\mathbf{I}_x + 2\sin(\pi Jt)\mathbf{I}_y\mathbf{S}_z\big)\mathbf{S}_x$$

$$\rightarrow \big(\cos(\pi Jt)\mathbf{I}_x + 2\sin(\pi Jt)\mathbf{I}_y\mathbf{S}_z\big)\big(\cos(\pi Jt)\mathbf{S}_x + 2\sin(\pi Jt)\mathbf{I}_z\mathbf{S}_y\big)$$

Expanding this result and collecting terms gives

$$\mathbf{I}_x\mathbf{S}_x \rightarrow \cos^2(\pi J t)\mathbf{I}_x\mathbf{S}_x + 2\cos(\pi J t)\sin(\pi J t)\mathbf{I}_x\mathbf{I}_z\mathbf{S}_y$$

$$+ 2\cos(\pi J t)\sin(\pi J t)\mathbf{I}_y\mathbf{S}_z\mathbf{S}_x + 4\sin^2(\pi J t)\mathbf{I}_y\mathbf{S}_z\mathbf{I}_z\mathbf{S}_y$$

We now have some matrix products with three or even four terms. The products can be simplified using some of the special multiplication rules that have been introduced earlier on page 549. For instance,

$$\mathbf{I}_x\mathbf{I}_z = -\mathbf{I}_z\mathbf{I}_x = \frac{-i}{2}\mathbf{I}_y,$$

so that we can write

$$\mathbf{I}_x\mathbf{I}_z\mathbf{S}_y = -\frac{i}{2}\mathbf{I}_y\mathbf{S}_y$$

Similarly, $\mathbf{S}_z\mathbf{S}_x = \frac{i}{2}\mathbf{S}_y$, so that we can simplify $\mathbf{I}_y\mathbf{S}_z\mathbf{S}_x$ as

$$\mathbf{I}_y\mathbf{S}_z\mathbf{S}_x = \frac{i}{2}\mathbf{I}_y\mathbf{S}_y$$

For the term containing four matrices, we use the fact that matrices corresponding to different spins *do* commute with multiplication:

$$\mathbf{I}_y\mathbf{S}_z\mathbf{I}_z\mathbf{S}_y = \mathbf{I}_y\mathbf{I}_z\mathbf{S}_z\mathbf{S}_y$$

Because $\mathbf{I}_y\mathbf{I}_z = \frac{i}{2}\mathbf{I}_x$ and $\mathbf{S}_z\mathbf{S}_y = -\mathbf{S}_y\mathbf{S}_z = -\frac{i}{2}\mathbf{S}_x$,

$$\mathbf{I}_y\mathbf{I}_z\mathbf{S}_z\mathbf{S}_y = \frac{i}{2}\mathbf{I}_x\frac{-i}{2}\mathbf{S}_x = \frac{1}{4}\mathbf{I}_x\mathbf{S}_x$$

Substituting these simplified matrix products back into the expression for the change of $\mathbf{I}_x\mathbf{S}_x$ gives

$$\mathbf{I}_x\mathbf{S}_x \rightarrow \cos^2(\pi J t)\mathbf{I}_x\mathbf{S}_x - i\cos(\pi J t)\sin(\pi J t)\mathbf{I}_y\mathbf{S}_y$$

$$+ i\cos(\pi J t)\sin(\pi J t)\mathbf{I}_y\mathbf{S}_y + \sin^2(\pi J t)\mathbf{I}_x\mathbf{S}_x$$

$$= \left(\cos^2(\pi J t) + \sin^2(\pi J t)\right)\mathbf{I}_x\mathbf{S}_x$$

$$= \mathbf{I}_x\mathbf{S}_x$$

Therefore, $\mathbf{I}_x\mathbf{S}_x$ is unaffected by scalar coupling. This result may be surprising at first glance, since it seems not to follow the pattern we have seen with other matrices representing transverse components. It should be noted, however, that in all of the earlier cases in which the transverse components precessed due to scalar coupling, the transverse components were associated with z-components of the other spin. It is only magnetization along the z-axis that influences the precession frequency in the transverse plane, and so only z-correlations are detectable during precession. Remember also that we are only considering the contribution of scalar coupling here. As shown earlier, the transverse components of all of the product operators do undergo precession under the influence of chemical-shift evolution, which reflects the influence of the external field. The same result is obtained for the other three product operators that involve only transverse components: $\mathbf{I}_x\mathbf{S}_y$, $\mathbf{I}_y\mathbf{S}_x$ and $\mathbf{I}_y\mathbf{S}_y$ are all unaffected by the scalar coupling interaction.

18.3   Some Examples

The rules presented in the previous sections provide the basis for analyzing any experiment involving a scalar-coupled spin pair, provided that the weak-coupling condition is satisfied. As we have seen in earlier chapters, most multi-pulse NMR experiments can be broken down into a set of basic blocks that serve common roles in a variety of experiments. The product-operator analysis of complex experiments can often be simplified by knowing how the various operator components are manipulated by the common pulse-sequence elements. In these final pages, the general approach to applying the product-operator formalism will be illustrated with a few examples, including refocusing pulses and the important INEPT sequence introduced in Chapter 16, followed by the HSQC and COSY two-dimensional experiments.

18.3.1   Refocusing Pulses

Refocusing pulses, in which a π-pulse separates two evolution periods of equal length, are one of the most common elements in modern NMR experiments and serve many functions. In heteronuclear experiments, the pulse may be applied to either or both of the scalar-coupled spins, with quite different results depending on the situation.

A Decoupling Pulse Sequence

One common application of refocusing pulses is the "decoupling" of the influence of one spin when the other precesses, such as during the data acquisition period. As discussed in Chapter 16, this procedure does not really eliminate the scalar coupling between spins, but rather hides the effect on the precession of the spins. In the treatment below, we will use both vector diagrams and the algebraic rules to follow the operator components for the case of a $\pi_{x,I}$-refocusing pulse applied to an evolving transverse S-component.

The periods before and after the pulse are of length $\tau/2$, and the precession during these periods reflects both chemical-shift evolution and scalar coupling. In calculating the net effect of each evolution period, the product-operator rules for chemical-shift and scalar-coupling evolution can be calculated in either order. The diagram in Fig. 18.14 starts with an $\mathbf{S}_x$-component and shows the chemical-shift evolution followed by scalar-coupling evolution for the first period. Chemical-shift evolution during the first period converts $\mathbf{S}_x$ into a mixture of $\mathbf{S}_x$ and $\mathbf{S}_y$:

$$\mathbf{S}_x \rightarrow \cos(\pi\nu_S\tau)\mathbf{S}_x + \sin(\pi\nu_S\tau)\mathbf{S}_y$$

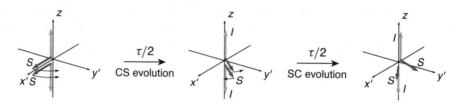

Figure 18.14   Vector diagrams representing the evolution of an $\mathbf{S}_x$-component under the influence of chemical shift (CS) and scalar coupling (SC).

The two resulting components are then treated independently to describe the scalar-coupling evolution during the same period:

$$\cos(\pi \nu_S \tau)\mathbf{S}_x \rightarrow \cos(\pi \nu_S \tau)\left(\cos(\pi J\tau/2)\mathbf{S}_x + 2\sin(\pi J\tau/2)\mathbf{I}_z\mathbf{S}_y \right)$$

$$= \cos(\pi \nu_S \tau)\cos(\pi J\tau/2)\mathbf{S}_x + 2\cos(\pi \nu_S \tau)\sin(\pi J\tau/2)\mathbf{I}_z\mathbf{S}_y$$

$$\sin(\pi \nu_S \tau)\mathbf{S}_y \rightarrow \sin(\pi \nu_S \tau)\left(\cos(\pi J\tau/2)\mathbf{S}_y - 2\sin(\pi J\tau/2)\mathbf{I}_z\mathbf{S}_x \right)$$

$$= \sin(\pi \nu_S \tau)\cos(\pi J\tau/2)\mathbf{S}_y - 2\sin(\pi \nu_S \tau)\sin(\pi J\tau/2)\mathbf{I}_z\mathbf{S}_x$$

Combining these gives the total effect of the first evolution period, which is to convert $\mathbf{S}_y$ into a mixture of two magnetization components and two correlations:

$$\mathbf{S}_x \xrightarrow{t/2} \cos(\pi \nu_S \tau)\cos(\pi J\tau/2)\mathbf{S}_x$$

$$+ \sin(\pi \nu_S \tau)\cos(\pi J\tau/2)\mathbf{S}_y$$

$$- 2\sin(\pi \nu_S \tau)\sin(\pi J\tau/2)\mathbf{I}_z\mathbf{S}_x$$

$$+ 2\cos(\pi \nu_S \tau)\sin(\pi J\tau/2)\mathbf{I}_z\mathbf{S}_y$$

The selective $\pi_{x,I}$-pulse leaves the S-components unchanged, but reverses the signs of the $\mathbf{I}_z$-components of the two correlations, as shown in Fig. 18.15. The density matrix is now represented as

$$\cos(\pi \nu_S \tau)\cos(\pi J\tau/2)\mathbf{S}_x + \sin(\pi \nu_S \tau)\cos(\pi J\tau/2)\mathbf{S}_y$$

$$+ 2\sin(\pi \nu_S \tau)\sin(\pi J\tau/2)\mathbf{I}_z\mathbf{S}_x - 2\cos(\pi \nu_S \tau)\sin(\pi J\tau/2)\mathbf{I}_z\mathbf{S}_y$$

The chemical-shift and scalar coupling processes during the second evolution period can also be treated in either order, but it is more convenient to treat scalar coupling first, as shown in Fig. 18.16.

Because the $\mathbf{I}_z$-components of the two correlations have changed signs, the relative directions of the scalar-coupling evolution are reversed after the pulse. This is

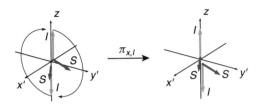

Figure 18.15 Vector diagrams representing the effects of a selective $\pi_{x,I}$-pulse on the magnetization and correlation components generated by the evolution period, $\tau/2$, of Fig. 18.14.

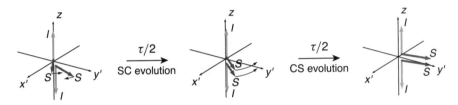

Figure 18.16 Vector diagrams representing evolution during the second interval of the I-decoupling pulse sequence.

expressed in the operators as

$$\cos(\pi \nu_S \tau) \cos(\pi J \tau/2)\mathbf{S}_x$$
$$\rightarrow \cos(\pi \nu_S \tau) \cos(\pi J \tau/2)\big(\cos(\pi J \tau/2)\mathbf{S}_x + 2 \sin(\pi J \tau/2)\mathbf{I}_z\mathbf{S}_y\big)$$
$$= \cos(\pi \nu_S \tau) \cos^2(\pi J \tau/2)\mathbf{S}_x + 2 \cos(\pi \nu_S \tau) \cos(\pi J \tau/2) \sin(\pi J \tau/2)\mathbf{I}_z\mathbf{S}_y$$

$$\sin(\pi \nu_S \tau) \cos(\pi J \tau/2)\mathbf{S}_y$$
$$\rightarrow \sin(\pi \nu_S \tau) \cos(\pi J \tau/2)\big(\cos(\pi J \tau/2)\mathbf{S}_y - 2 \sin(\pi J \tau/2)\mathbf{I}_z\mathbf{S}_x\big)$$
$$= \sin(\pi \nu_S \tau) \cos^2(\pi J \tau/2)\mathbf{S}_y - 2 \sin(\pi \nu_S \tau) \cos(\pi J \tau/2) \sin(\pi J \tau/2)\mathbf{I}_z\mathbf{S}_x$$

$$2 \sin(\pi \nu_S \tau) \sin(\pi J \tau/2)\mathbf{I}_z\mathbf{S}_x$$
$$\rightarrow 2 \sin(\pi \nu_S \tau) \sin(\pi J \tau/2)\big(\cos(\pi J \tau/2)\mathbf{I}_z\mathbf{S}_x + (1/2) \sin(\pi J \tau/2)\mathbf{S}_y\big)$$
$$= 2 \sin(\pi \nu_S \tau) \sin(\pi J \tau/2) \cos(\pi J \tau/2)\mathbf{I}_z\mathbf{S}_x + \sin(\pi \nu_S \tau) \sin^2(\pi J \tau/2)\mathbf{S}_y$$

$$- 2 \cos(\pi \nu_S \tau) \sin(\pi J \tau/2)\mathbf{I}_z\mathbf{S}_y$$
$$\rightarrow -2 \cos(\pi \nu_S \tau) \sin(\pi J \tau/2)\big(\cos(\pi J \tau/2)\mathbf{I}_z\mathbf{S}_y - (1/2) \sin(\pi J \tau/2)\mathbf{S}_x\big)$$
$$= -2 \cos(\pi \nu_S \tau) \sin(\pi J \tau/2) \cos(\pi J \tau/2)\mathbf{I}_z\mathbf{S}_y + \cos(\pi \nu_S \tau) \sin^2(\pi J \tau/2)\mathbf{S}_x$$

When all of these components are added together, the two terms for $\mathbf{I}_z\mathbf{S}_x$ cancel one another, as do the terms for $\mathbf{I}_z\mathbf{S}_y$, leaving only the $\mathbf{S}_x$ and $\mathbf{S}_y$ terms. Combining the terms for $\mathbf{S}_x$ and applying the identity, $\cos^2(x) + \sin^2(x) = 1$, gives:

$$\cos(\pi \nu_S \tau) \cos^2(\pi J \tau/2)\mathbf{S}_x + \cos(\pi \nu_S \tau) \sin^2(\pi J \tau/2)\mathbf{S}_x = \cos(\pi \nu_S \tau)\mathbf{S}_x$$

Similarly,

$$\sin(\pi \nu_S \tau) \cos^2(\pi J \tau/2)\mathbf{S}_y + \sin(\pi \nu_S \tau) \sin^2(\pi J \tau/2)\mathbf{S}_y = \sin(\pi \nu_S \tau)\mathbf{S}_y$$

At this point, we have

$$\mathbf{S}_x \xrightarrow[\text{CS}]{\tau/2} \xrightarrow[\text{SC}]{\tau/2} \xrightarrow{\pi_{x,I}} \xrightarrow[\text{SC}]{\tau/2} \cos(\pi \nu_S \tau)\mathbf{S}_x + \sin(\pi \nu_S \tau)\mathbf{S}_y$$

The final step is the chemical-shift evolution during the second delay period, which affects both of the remaining terms:

$$\cos(\pi \nu_S \tau)\mathbf{S}_x \rightarrow \cos(\pi \nu_S \tau)\big(\cos(\pi \nu_S \tau)\mathbf{S}_x + \sin(\pi \nu_S \tau)\mathbf{S}_y \big)$$

$$= \cos^2(\pi \nu_S \tau)\mathbf{S}_x + \cos(\pi \nu_S \tau) \sin(\pi \nu_S \tau)\mathbf{S}_y$$

$$\sin(\pi \nu_S \tau)\mathbf{S}_y \rightarrow \sin(\pi \nu_S \tau)\big(\cos(\pi \nu_S \tau)\mathbf{S}_y - \sin(\pi \nu_S \tau)\mathbf{S}_x \big)$$

$$= \sin(\pi \nu_S \tau) \cos(\pi \nu_S \tau)\mathbf{S}_y - \sin^2(\pi \nu_S \tau)\mathbf{S}_x$$

Combining the terms for $\mathbf{S}_x$ and applying the identity, $\cos^2(x) - \sin^2(x) = \cos(2x)$, gives

$$(\cos^2(\pi \nu_S \tau) - \sin^2(\pi \nu_S \tau))\mathbf{S}_x = \cos(2\pi \nu_S \tau)\mathbf{S}_x$$

Similarly, combining the terms for $\mathbf{S}_y$ and applying $\cos(x) \sin(x) = \frac{1}{2} \sin(2x)$ gives:

$$2 \sin(\pi \nu_S \tau) \cos(\pi \nu_S \tau)\mathbf{S}_y = \sin(2\pi \nu_S \tau)\mathbf{S}_y$$

The total effect of the refocusing sequence is

$$\mathbf{S}_x \xrightarrow{\tau/2} \xrightarrow{\pi_{x,I}} \xrightarrow{\tau/2} \cos(2\pi \nu_S \tau)\mathbf{S}_x + \sin(2\pi \nu_S \tau)\mathbf{S}_y$$

Thus, we obtain the result for chemical-shift evolution alone, showing that the effect of scalar coupling is suppressed. If a rapid series of pulses are applied as the signal from the S-spin is recorded, only the single frequency will be observed. A decoupling pulse can also be applied during an evolution period in a multidimensional experiment, thereby eliminating splitting in an indirectly observed dimension of the final spectrum. In addition to suppressing the splitting of the precession frequency, the decoupling pulse also prevents the conversion of an observable magnetization component into a correlation.

This example also shows that the algebraic manipulations of the operators can become quite involved, even for a relatively simple pulse sequence. The vector diagrams can help visualize the various transformations, however. With practice and care, the diagrams can be used without carrying out all of the algebraic manipulations, but it is critical that all of the components be carried through the calculations, whatever method is used. In the following examples, the diagrams will be used without the algebra.

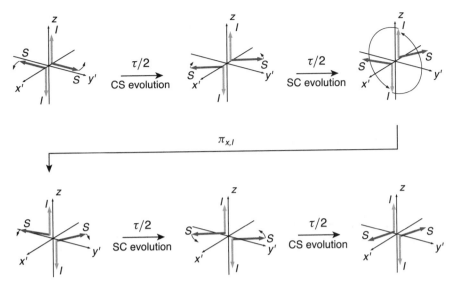

Figure 18.17   Vector diagrams representing the evolution of an I_zS_y correlation component during a I-decoupling pulse sequence.

A Decoupling Pulse Applied to I_zS_y

The result illustrated above for the S_x-magnetization component also applies to a transverse component when it is part of a product, such as S_y as a component of the I_zS_y correlation, as illustrated in Fig. 18.17. As before, the key feature is that the $\pi_{x,I}$-pulse reverses the direction of precession of the transverse S-components due to scalar coupling. In this case, the net effect is to change the I_zS_y correlation into a mixture of I_zS_y and I_zS_x, according to

$$I_zS_y \rightarrow -\cos(2\pi\nu_S\tau)I_zS_y + \sin(2\pi\nu_S\tau)I_zS_x$$

Again, the effect is to hide the influence of scalar coupling during the overall evolution period, so there is no conversion of the correlation into an observable magnetization component. Notice, however, that the π-pulse to the I-spin also has the effect of reversing the sign of the I_zS_x- and I_zS_y-components from what they would otherwise be.

In summary, anytime that we see a transverse component (I_x, I_y, S_x or S_y) in a product-operator calculation, we know that the effect of an evolution period split by a π-pulse to the *other* spin can be calculated by considering only the chemical-shift evolution, along with the change in sign of the terms representing the irradiated spin.

A Refocusing Pulse Applied to Transverse Components

Consider now the effect of a refocusing pulse applied selectively to an evolving transverse component, as opposed to the associated z-component of the coupled spin discussed above. The effects on an evolving S_x-component are illustrated in Fig. 18.18. The coupled I_z-components are unaffected by the pulse, so the direction of the J-evolution for each S-component is unchanged. However, the $\pi_{x,S}$-pulse reverses the

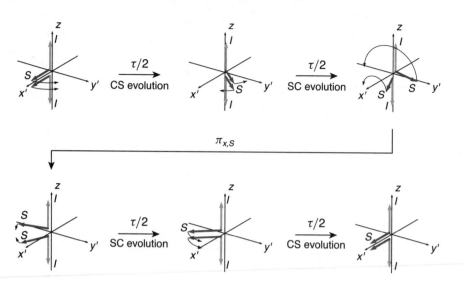

Figure 18.18 Vector diagrams representing the evolution of an S_x-component with refocusing.

relative positions of the two S-components with respect to the x-axis, so the two components have both returned to the y'-axis at the end of the second evolution period. Thus, the effects of both chemical-shift and scalar-coupling evolution are suppressed.

The same pulse sequence has a similar effect on an initial S_y-component, but with a change in sign:

$$S_y \rightarrow -S_y$$

A second iteration will return the $-S_y$-component back to S_y. Conversely, if the π-pulse is applied along the y'-axis, an S_y-component will be refocused to the y'-axis, while an S_x-component will be converted to $-S_x$.

A Non-Selective Refocusing Pulse

Finally, consider the case when a refocusing pulse is applied to both spins, again beginning with an S_x-component, illustrated in Fig. 18.19. The $\pi_{x,I}$-pulse has the effect of reversing the correlation between the I_z-components and the transverse S-components and, therefore, the directions of the scalar-coupling evolution before and after the pulse. However, the positions of the transverse components are also reversed by the $\pi_{x,S}$-pulse. As a consequence, the relative displacements due to scalar coupling during the two periods are additive, while the net chemical-shift evolution is zero. Thus, this pulse sequence allows the effects of scalar coupling to be manifest, without any precession (relative to the rotating frame) due to chemical shift. This leads to an interconversion between S_x and I_zS_y:

$$S_x \rightarrow \cos(\pi J \tau)S_x + 2\sin(\pi J \tau)I_zS_y$$

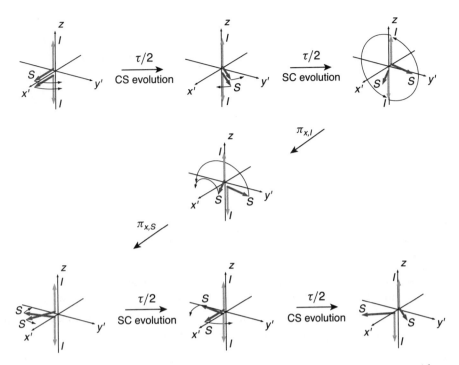

Figure 18.19 Vector diagrams representing the evolution of an S_x component with refocusing of both spins.

If τ is set to $1/(2J)$, S_x is completely converted to $I_z S_y$. This forms the basis for magnetization transfer in many heteronuclear experiments, as illustrated below for the INEPT pulse sequence.

18.3.2 INEPT

In Chapter 16, we discussed the INEPT pulse sequence, which is illustrated again in Fig. 18.20. In the earlier treatment, the experiment was analyzed by following each of the starting eigenstates present at equilibrium, $|\alpha\alpha\rangle$, $|\alpha\beta\rangle$, $|\beta\alpha\rangle$ and $|\beta\beta\rangle$. Now, we will reexamine it using the product operators and vector diagrams to follow the full population at once.

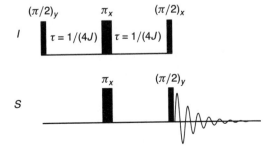

Figure 18.20
Pulse sequence for the INEPT experiment.

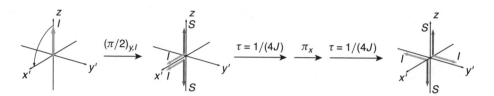

Figure 18.21   Vector diagrams illustrating the effects of the first part of the INEPT pulse sequence on the initial I_z-magnetization.

The starting equilibrium state is given by

$$\rho_{eq} = \frac{\Delta P_I}{2}\mathbf{I}_z + \frac{\Delta P_S}{2}\mathbf{S}_z$$

The initial $(\pi/2)_{y,I}$-pulse converts the starting $\mathbf{I}_z$-component into an $\mathbf{I}_x$-component. For this component, the effects of the subsequent delays, with refocusing pulses applied to both spins, are exactly analogous to the situation described in the previous section for a starting $\mathbf{S}_x$-component, with the overall result,

$$\frac{\Delta P_I}{2}\mathbf{I}_z \xrightarrow{(\pi/2)_{y,I}} \xrightarrow{1/(4J)} \xrightarrow{\pi_x} \xrightarrow{1/(4J)} \Delta P_I \mathbf{I}_y \mathbf{S}_z$$

which is drawn as shown in Fig. 18.21. At the end of the second delay period, a $\pi/2$-pulse along the x'-axis is applied to the I-spin, and a $\pi/2$-pulse is applied to the S-spin along the y'-axis. Together, these two pulses convert the $\mathbf{I}_y\mathbf{S}_z$ correlation into an $\mathbf{I}_z\mathbf{S}_x$ correlation, as diagrammed in Fig. 18.22. Thus, the full INEPT sequence has the following effect on the initial equilibrium I-magnetization:

$$\frac{\Delta P_I}{2}\mathbf{I}_z \xrightarrow{\text{INEPT}} \Delta P_I \mathbf{I}_z \mathbf{S}_x$$

By comparison, the effects of the INEPT sequence on the S-spin are quite simple, because the magnetization remains aligned with the positive or negative z-axis until the final pulse. The π_x-pulse inverts the initial z-magnetization, and the final $(\pi/2)_y$-pulse rotates the magnetization to the $-x'$-axis, as shown in Fig. 18.23. Expressed

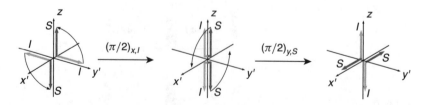

Figure 18.22   Vector diagrams illustrating the effects of the final two pulses of the INEPT pulse sequence on the components arising from the initial I_z-magnetization.

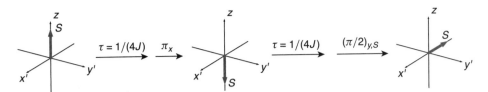

Figure 18.23 Vector diagrams illustrating the effects of the INEPT pulse sequence on the initial S_z-magnetization.

mathematically, the effect on the initial S-magnetization is

$$\frac{\Delta P_S}{2} S_z \xrightarrow{\text{INEPT}} -\frac{\Delta P_S}{2} S_x$$

Next, we must consider the evolution of the $I_z S_x$- and S_x-components during the data acquisition period, which is influenced by chemical shift and scalar coupling. Although both terms interconvert with magnetization components and correlations, it is only the S-magnetization components that contribute to the signal. Note that the magnitude of the $I_z S_x$-component is determined by the initial equilibrium population difference for the I-spin, ΔP_I, and it is this feature that leads to enhancement of the signal from the S-spin. To fully explain the signal, it is necessary to examine the evolution of both $I_z S_x$ and S_x.

The diagrams in Fig. 18.24 show in parallel the evolution from the two components for an arbitrary time interval, t, with the scalar-coupling evolution preceding the chemical-shift process. Immediately after the final pulse, there is no net magnetization from the $I_z S_x$-component. However, scalar-coupling evolution interconverts this correlation with an observable S_y-magnetization, while chemical-shift evolution simultaneously leads to precession of the net magnetization component—that is, the interconversion of S_y and S_x. At any instant, the net S-magnetization represents the sum of the two transverse S-components.

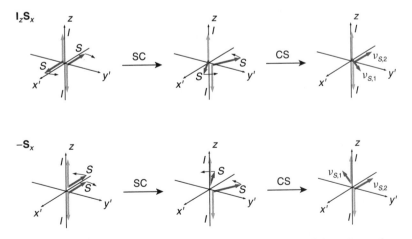

Figure 18.24 Vector diagrams illustrating the evolution of the $I_z S_x$- and S_x-components during the data acquisition period of the INEPT experiment.

The $-\mathbf{S}_x$-component present at the beginning of the acquisition period also evolves, as shown in the lower part of Fig. 18.24. Scalar coupling causes the net magnetization to decrease as it is converted into an $\mathbf{I}_z\mathbf{S}_y$ correlation, while chemical-shift evolution interconverts $\mathbf{S}_x$ with $\mathbf{S}_y$ and $\mathbf{I}_z\mathbf{S}_y$ with $\mathbf{I}_z\mathbf{S}_x$.

Examination of the diagrams in Fig. 18.24 shows that the two starting components, $\mathbf{I}_z\mathbf{S}_x$ and $-\mathbf{S}_x$, generate two frequencies each. For one of the frequencies ($\nu_{S,2} = \nu_S - J/2$), represented by the S-vectors associated with the negative I_z-vector, the corresponding vectors in the two rows always point in the same direction, so they combine to create a signal in the spectrometer coils. For the other frequency ($\nu_{S,1} = \nu_S + J/2$), associated with the positive I_z-vector, the two vectors point in opposite directions, so they partially cancel out.

The final signal intensities for the two frequencies depend on the population differences associated with $\mathbf{I}_z\mathbf{S}_x$ and $-\mathbf{S}_x$ at the beginning of the acquisition period. Recall that the $\mathbf{I}_z\mathbf{S}_x$-component is derived from the equilibrium population difference of the I-spin, while the $-\mathbf{S}_x$-component comes from the population difference for the S-spin. For the frequency, $\nu_{S,2}$, the signal intensity is proportional to the sum of starting populations differences:

$$\Delta P_I + \Delta P_S$$

The intensity of the other frequency component is proportional to the difference:

$$\Delta P_I - \Delta P_S$$

The two frequency components initially differ in phase by π rad, so the peaks in the final spectrum have opposite signs. If the I-spin is ^{1}H, and the S-spin is ^{15}N, then ΔP_I will be nearly 10-fold greater than ΔP_S, and both of the recorded frequency components will be much greater than if the spectrum were generated by a simple $\pi/2$-pulse to the S-spin. If ^{13}C is the S-spin, the increase in sensitivity is about four-fold.[2]

18.3.3   HSQC

In addition to being useful for one-dimensional spectra of insensitive nuclei, the INEPT pulse sequence is widely used as the first stage of many multidimensional heteronuclear experiments, including the HSQC (Heteronuclear Single-Quantum Coherence) experiment discussed in Section 16.5 and diagrammed in Fig. 18.25. As shown above, the operator basis-set representation of the density matrix following the INEPT sequence is

$$-\frac{\Delta P_S}{2}\mathbf{S}_x + \Delta P_I\mathbf{I}_z\mathbf{S}_x$$

---

[2] The analysis presented here assumes that the gyromagnetic ratios of both spins and the coupling constant, J, are all positive, as is the case for a directly bonded ^{1}H–^{13}C pair. For a directly bonded ^{1}H–^{15}N pair, however, the gyromagnetic ratio of one of the spins (^{15}N) and the coupling constant are both negative. This will alter some of the sign relationships in the analysis, but not the overall effect on the signal intensity.

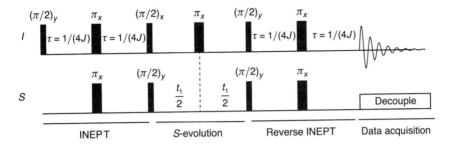

Figure 18.25   Pulse sequence for the HSQC experiment.

Figure 18.26
Vector diagrams representing the evolution of the $-\mathbf{S}_x$-magnetization component during the t_1 evolution period of the HSQC experiment.

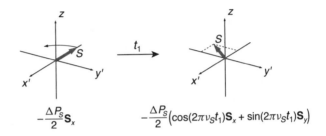

Consider, first, the fate of the $-\mathbf{S}_x$-component during the S-evolution and reverse INEPT stages. During the evolution period, with incremented total duration of t_1, the I-spin is refocused, so the effects of scalar coupling are hidden, leaving just the chemical-shift evolution of the S-magnetization. The result is a mixture of $\mathbf{S}_x$- and $\mathbf{S}_y$-components, determined by the evolution period and precession frequency, as shown in the vector diagrams of Fig. 18.26.

The first step of the reverse INEPT portion of the experiment is a non-selective $(\pi/2)_y$-pulse, which converts the $\mathbf{S}_x$-component to $\mathbf{S}_z$, as shown in Fig. 18.27. The rest of the reverse INEPT portion is a delay period of $1/(2J)$ with a π_x-refocusing pulse to both spins, which eliminates the effect of chemical-shift evolution. Scalar coupling

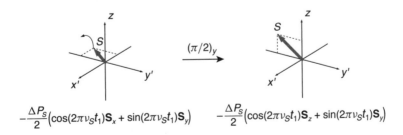

Figure 18.27   Vector diagrams representing the effect of the $(\pi/2)_y$-pulse of the reverse INEPT segment in the HSQC experiment on the magnetization component that begins as $-\mathbf{S}_x$ at the start of the evolution period.

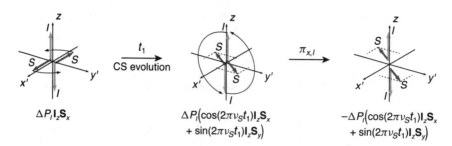

Figure 18.28   Vector diagrams representing the evolution of the I_zS_x correlation during the t_1 period of the HSQC experiment.

converts the S_y-component into an I_zS_x correlation component. During the FID, only I-magnetization is recorded, with decoupling applied to the S-spin. Thus, the I_zS_x-component does not make any detectable contribution. The S_z-component present after the $(\pi/2)_y$-pulse (at the beginning of the reverse INEPT segment) is inverted by the refocusing pulse, but also does not contribute to the recorded FID. Therefore, the $-S_x$-component that is generated by the INEPT portion of the experiment is entirely hidden in the final FID.

At the end of the INEPT sequence, there is also an I_zS_x-component. During the S-evolution period, which includes a $\pi_{x,I}$-refocusing pulse, the I_zS_x-component is influenced only by chemical-shift evolution and the effect of the pulse, generating a mixture of I_zS_x and I_zS_y correlation components, as shown in Fig. 18.28. At this point in the analysis, it is convenient to treat the I_zS_x- and I_zS_y-components separately, resolving the vector diagram as shown in Fig. 18.29.

First, we consider the fate of the $-I_zS_y$-component. The initial $(\pi/2)_y$-pulse of the reverse INEPT sequence converts this correlation into $-I_xS_y$, as shown in Fig. 18.30. Recall that transverse correlations such as I_xS_y are unaffected by scalar-coupling evolution. Because the effects of chemical-shift are also suppressed by the refocusing pulse of the reverse INEPT portion, the only change in the $-I_xS_y$ correlation is a change in sign due to the π_x-pulse, as diagrammed in Fig. 18.30. During the data acquisition phase, the I_xS_y and I_xS_x correlations evolve, but never interconvert with observable magnetization components, so this component is also invisible in the FID.

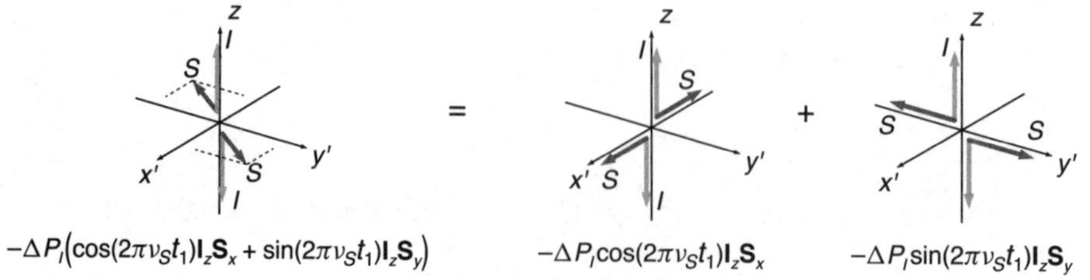

Figure 18.29   Resolution of the I_zS_x- and I_zS_y-components generated during the t_1 evolution period of the HSQC experiment.

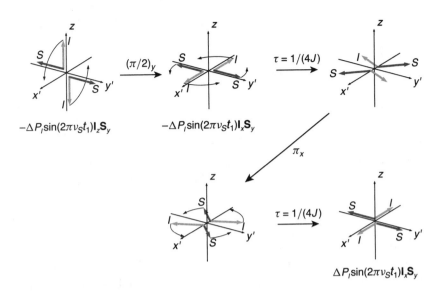

Figure 18.30 Vector diagrams representing the effect of the reverse INEPT segment of the HSQC experiment on the $-\mathbf{I}_z\mathbf{S}_y$ correlation present at the end of the t_1 evolution period. The lower-right diagram represents the final state of this component before the data acquisition period.

Finally, we consider the $-\mathbf{I}_z\mathbf{S}_x$-component that is present at the end of the t_1 evolution period. As shown in Fig. 18.31, the $(\pi/2)_{y,I}$- and $(\pi/2)_{y,S}$-pulses of the reverse INEPT sequence convert this component into $\mathbf{I}_x\mathbf{S}_z$. During the refocused evolution period of the reverse INEPT segment, scalar coupling converts the $\mathbf{I}_x\mathbf{S}_z$ correlation into an $\mathbf{I}_y$-magnetization component, the sign of which is changed by the π_x-pulse, as shown in Fig. 18.32. This $-\mathbf{I}_y$-component is what finally gives rise to the FID during the data acquisition period. For each increment in the two-dimensional experiment, the intensity of the FID is modulated by the chemical-shift evolution of the S-spin during the t_1 period. Decoupling is usually applied to the S-spin during data acquisition, resulting in a single frequency for the I-spin. When the two-dimensional data

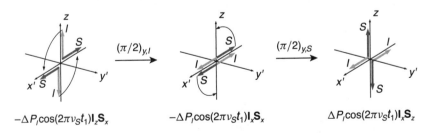

Figure 18.31 Vector diagrams representing the effects of the $(\pi/2)_y$-pulses of the reverse INEPT segment in the HSQC experiment on the $-\mathbf{I}_z\mathbf{S}_x$ correlation present at the end of the t_1 evolution period.

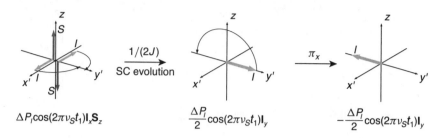

$$\Delta P_I \cos(2\pi \nu_S t_1) \mathbf{I}_x \mathbf{S}_z \qquad \frac{\Delta P_I}{2} \cos(2\pi \nu_S t_1) \mathbf{I}_y \qquad -\frac{\Delta P_I}{2} \cos(2\pi \nu_S t_1) \mathbf{I}_y$$

Figure 18.32   Vector diagrams representing the conversion of the $\mathbf{I}_x \mathbf{S}_z$ correlation into an $\mathbf{I}_y$-magnetization component during the refocused delay period of the reverse INEPT segment of the HSQC experiment.

set is Fourier transformed in both dimensions, a spectrum with I- and S-frequency axes is generated, with a cross peak for each coupled I–S spin pair.

Because the signal intensity is determined by both the initial I-spin population difference and the precession of the I-spin during the data acquisition period, the HSQC experiment makes maximum use of this spin's larger gyromagnetic ratio. However, the conventional HSQC experiment described above suffers from an inherent inefficiency, because the $\mathbf{I}_z \mathbf{S}_y$-component that is present at the end of the S-evolution period is converted to an $\mathbf{I}_x \mathbf{S}_y$ correlation that does not contribute to the final signal (Fig. 18.30). A slightly more elaborate experiment, usually described as a *sensitivity-enhanced* HSQC, has been developed by Mark Rance, Lewis L. Kay and their co-workers to capture both the $\mathbf{I}_z \mathbf{S}_x$- and $\mathbf{I}_z \mathbf{S}_y$-components in a way that contributes to the signal, and this is now the more commonly used version.

18.3.4   Homonuclear COSY

As a final example, we return to the homonuclear COSY experiment described in Chapter 15, with the two-pulse sequence shown again in Fig. 18.33. Although the pulse sequence is the simplest possible for a two-dimensional experiment, the resulting density matrices are remarkably complex, as we began to see in Chapter 17. Using the product-operator formalism greatly simplifies the treatment and can help provide a better physical understanding of the experiment.

As in the previous examples, the starting equilibrium density matrix is given by

$$\rho_{eq} = \frac{\Delta P_I}{2} \mathbf{I}_z + \frac{\Delta P_S}{2} \mathbf{S}_z$$

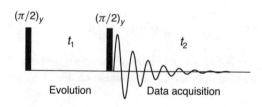

Figure 18.33
Pulse diagram for the homonuclear COSY experiment.

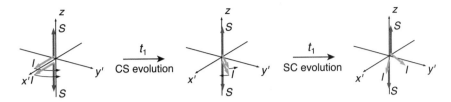

Figure 18.34 Vector diagrams representing the conversion of the I_x-component into I_y-magnetization and the I_xS_z and I_yS_z correlations, during the t_1 evolution period of the homonuclear COSY experiment.

In this case, however, the two spins are of the same type (usually ^{1}H), and the equilibrium population differences are essentially identical. As a consequence, it is not necessary to keep track of the population differences, and these terms are dropped in the following treatment. Following the initial $(\pi/2)_y$-pulse, I_z is converted to I_x and S_z is converted to S_x. Except for the chemical shifts, the subsequent evolution of the I_x and S_x terms are identical through the experiment, and it is sufficient to follow just one of them.

Starting with the I_x term, the evolution during the t_1 period is influenced by both the chemical shift and scalar coupling, as illustrated in the vector diagrams of Fig. 18.34. Following the evolution period, the density matrix contains terms representing both observable magnetization (I_x and I_y) and the correlations, I_xS_z and I_yS_z. In the example illustrated in Fig. 18.34, the I-vector correlated with the positive S_z-vector has a more positive y'-component than does the I-vector associated with the negative S_z-vector, corresponding to a positive I_yS_z correlation. At the same time, there is a negative correlation between I_x and S_z.

Before considering the effects of the second pulse, it is convenient to break down the vector representation into the four components shown in Fig. 18.35. As indicated in the figure, the relative contributions of the four components are determined by the duration of the t_1 evolution period, the chemical shifts, expressed as v_I and v_S, and the scalar coupling constant, J. The two frequencies associated with the I-spin are shown as $v_{I,1} = v_I + J/2$ and $v_{I,2} = v_I - J/2$. We can then consider the effects of the second pulse and the evolution during the data acquisition period separately for each of these components.

For the two I-magnetization components, the subsequent steps are easy to visualize. The second $(\pi/2)_y$-pulse converts the I_x-component to $-I_z$. During the data acquisition period, the $-I_z$ component will relax to positive I_z, but will not contribute to any observable magnetization. On the other hand, the I_y-component is unaffected by the pulse, and will evolve during the t_2 period, as shown in Fig. 18.36. As well as the interconversion between I_x- and I_y-magnetization components, this evolution will generate I_xS_z and I_yS_z correlations, but these will not contribute to observable magnetization. The observable components will be

$$-\frac{1}{4}\Big(\sin(2\pi v_{I,1}t_1) + \sin(2\pi v_{I,2}t_1)\Big)\Big(\sin(2\pi v_{I,1}t_2) + \sin(2\pi v_{I,2}t_2)\Big)I_x$$

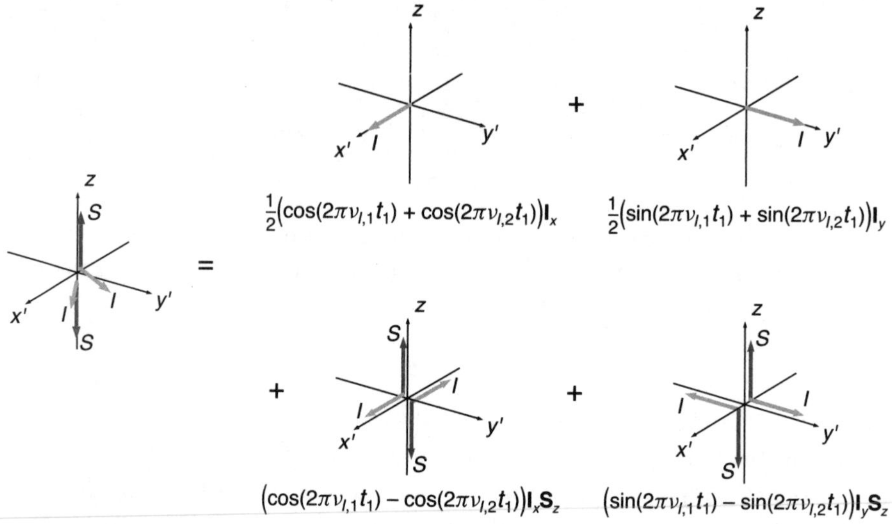

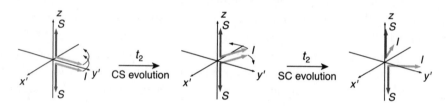

Figure 18.35 Vector diagrams representing the decomposition of the magnetization components and correlations present at the end of the t_1 evolution period of the homonuclear COSY experiment.

Figure 18.36 Vector diagrams representing evolution of the I_y-magnetization component during the data acquisition period of the homonuclear COSY experiment.

and

$$\frac{1}{4}\left(\sin(2\pi\nu_{I,1}t_1) + \sin(2\pi\nu_{I,2}t_1) \right)\left(\cos(2\pi\nu_{I,1}t_2) + \cos(2\pi\nu_{I,2}t_2) \right)I_y \quad (18.8)$$

The terms on the left in Eq. 18.8 represent the evolution of the I-spin during the t_1 period, with two frequencies, which determine the amplitudes of the magnetization components detected during the t_2 period. The terms on the right represent the evolution during t_2, which has the same two frequency components. After the two-dimensional Fourier transform, these components give rise to a multiplet of four peaks on the diagonal, as shown in Figure 15.2.

For the I_yS_z-component present after the t_1 evolution period, the effect of the second pulse is to convert this into an I_yS_x correlation (Fig. 18.37A). During the t_2 period, I_yS_x interconverts with the other three transverse correlations (I_xS_x, I_xS_y and I_yS_y), but none of these components contributes to the observable signal.

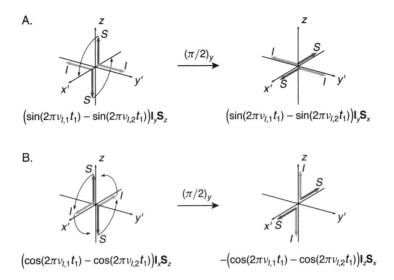

Figure 18.37 Vector diagrams representing the conversion of the $I_y S_z$ (A) and $I_x S_z$ (B) correlations present at the end of the evolution period into $I_y S_x$ and $-I_z S_x$, respectively, by the second pulse of the COSY experiment.

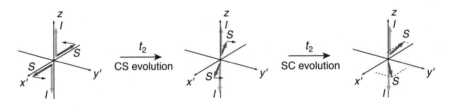

Figure 18.38 Vector diagrams representing evolution of the $-I_z S_x$ correlation to generate observable magnetization during the data acquisition period of the homonuclear COSY experiment.

The most important components of the COSY spectrum arise from the $I_x S_z$ correlation that is present at the end of t_1 (and the corresponding $I_z S_x$ correlation that is derived from the initial S_z-magnetization). The second $(\pi/2)_y$-pulse rotates both the I- and S-components of this correlation, to generate a $-I_z S_x$-component, as shown in Fig. 18.37B. During the t_2 period, this correlation evolves to generate S_x- and S_y-magnetization components, as well as the $I_z S_x$ correlation, as shown in Fig. 18.38. The important feature here is that the S-magnetization component is precessing and generating a signal in the FID, but the amplitude of this signal was determined by the precession of the I-spin during the t_1 evolution period.

The observable magnetization components derived from $I_z S_x$ are given by

$$-\frac{1}{4}\left(\cos(2\pi \nu_{I,1} t_1) - \cos(2\pi \nu_{I,2} t_1) \right)\left(\cos(2\pi \nu_{S,1} t_2) - \cos(2\pi \nu_{S,2} t_2) \right)S_x$$

and

$$-\frac{1}{4}\big(\cos(2\pi \nu_{I,1}t_1) - \cos(2\pi \nu_{I,2}t_1)\big)\big(\sin(2\pi \nu_{S,1}t_2) - \sin(2\pi \nu_{S,2}t_2)\big)\mathbf{S}_y \quad (18.9)$$

As in Eq. 18.8, the left-hand terms represent the evolution during t_1, and the terms on the right represent evolution during the data acquisition period. Notice, however, that the sine terms in Eq. 18.8 are replaced with cosine terms, and vice versa. The consequence of this is that the cross peaks in the final two-dimensional spectrum have dispersive, rather than absorptive, shapes, as shown in Fig. 15.2. The phase of the spectrum can be changed to generate absorptive cross peaks, but this will also change the diagonal multiplets to a dispersion shape, as in Fig. 15.3. In addition, the cross-peak multiplets in the re-phased spectrum have the anti-phase pattern of signs that is characteristic of COSY spectra.

18.4   Final Comments

We thus end this chapter, and the book, with a final description of the most basic and prototypical two-dimensional NMR experiment, using the power of the density matrix and the convenience of the product-operator representation. This treatment allows us to derive a description of the experiment and the resulting spectrum with a relatively limited number of mathematical manipulations, and the vector diagrams used here provide a visual guide to the evolution of the various magnetization components and correlations. It is important to remember, however, that the operator terms represent population averages over states that, themselves, correspond to quantum mechanical superpositions of eigenstates. This allows for efficiency in calculation, but the operator components and the vector diagrams demand careful interpretation.

It is also important to note what is missing from this treatment. Although the density matrix can be used to describe systems of strongly coupled spins, the time-evolution matrix we have used is based on an assumption of weak coupling (page 532). As a consequence, the expressions used to describe the evolution of the operator terms cannot be extended to strongly coupled spins. Although this is not a problem for describing heteronuclear experiments or most homonuclear experiments performed at high field strengths, experiments such as TOCSY, where scalar-coupling evolution takes place in the rotating frame, cannot be treated this way. Furthermore, we have limited ourselves to systems composed of only two coupled spins. Systems of three or more spins are not at all uncommon, and the density matrix treatment can be extended to those as well. The size of the matrices grow quite rapidly, however, and the manipulations become increasingly cumbersome. Fortunately, consideration of the two-spin system is usually adequate to understand how most experiments work and to design new ones.

Perhaps the most important omission from this presentation of the product-operator formalism is the effect of relaxation, which occurs during all of the delay periods in multi-pulse experiments, as well as during the data acquisition period.

Not only do the observable magnetization components undergo relaxation, but so do the operator terms representing correlations, and these relaxation processes can have important consequences for practical experiments (and can be measured to help characterize molecular motions). The theories introduced in Chapter 8 have been extended to the product-operator components, allowing for a more complete description.

In spite of these limitations, the product-operator formalism has proven to be a remarkably powerful tool, and proficiency in its use will greatly aid your future study and practice of NMR spectroscopy.

Summary Points for Chapter 18

- Any $n \times n$ matrix can be represented as a linear combination of the members of a basis set composed of n^2 mutually orthogonal matrices. Two matrices are orthogonal if the trace of their product is zero.
- A convenient basis set for the 2×2 density matrix representing a population of uncoupled spin-1/2 nuclei is formed by the matrices corresponding to the three magnetization components (I_x, I_y and I_z), along with the 2×2 identity matrix, 1.
- When the density matrix for a population of uncoupled spins is represented as described above, the coefficients of the operator matrices (c_x, c_y and c_z) are simply the average values of the magnetization components, multiplied by the constant, 2.
- The changes in the density matrix for a population of uncoupled spins, due to pulses or evolution periods, can be calculated using simple rules that define the changes in the coefficients in the linear combination. These rules are easily visualized with vector diagrams.
- To represent the density matrix for a population of scalar-coupled spin pairs, a basis set of 16 4×4 matrices is required:
  - The operator matrices for the individual magnetization components (I_x, I_y, I_z, S_x, S_y and S_z) provide six of the matrices required for the basis set.
  - Another nine matrices can be formed from the products of an I-spin operator and an S-spin operator—for instance, $I_x S_x$ or $I_y S_z$.
  - The 4×4 identity matrix serves as the final member of the basis set but can be ignored in calculations.
- When the density matrix for a population of coupled spin pairs is represented as a linear combination of the basis set described above, the coefficients of the single-operator matrices are equal to the corresponding observable magnetization components.
- The coefficients of the product-operator matrices in the linear combination are equal to four times the population averages of the corresponding magnetization

products. These coefficients reflect the correlations between the magnetization components of individual spin pairs, averaged over the population.

- The correlation components of the density matrix cannot be directly observed, but they can be converted into observable magnetization components by appropriate pulses and evolution periods.

- The correlation components, and their interconversions with other correlations or with observable magnetization, can be represented using special vector diagrams.

- Complex NMR experiments can be analyzed by following the magnetization and correlation components of the density matrix through the pulses and evolution periods of the experiment.

- Only the observable magnetization components will contribute to the final signal from an experiment, but all of the components must be followed through the analysis up to the data acquisition period.

Exercises for Chapter 18

The Maxima macros and examples available through links at http://uscibooks.com/goldenberg.htm include functions that are useful for some of these exercises.

18.1. Show that the I_x, I_y and I_z operator matrices for a single spin-1/2, along with the identity matrix, form an orthogonal set.

18.2. In Exercise 17.5 (page 537), a density matrix was constructed for a hypothetical population of coupled spins composed of 30% $|\alpha\alpha\rangle$, 10% $|\alpha\beta\rangle$, 25% $|\beta\alpha\rangle$ and 35% $|\beta\beta\rangle$. Write this density matrix as a linear combination of magnetization and product operator matrices.

18.3. As discussed in Section 18.2.2, the thermal equilibrium state of a population of coupled spin pairs displays no net correlation between the magnetization components of the two spins, even though there is a net z-magnetization component from each spin. This represents an important distinction between the equilibrium state and a pure population of spins in the $|\alpha\alpha\rangle$ state. Devise a simple pulse sequence that would convert the equilibrium state into one that represents only a net $I_z S_z$ correlation.

  Hint: Recall from Chapter 10 that pulsed field gradients can be used to eliminate transverse magnetization components, if necessary.

18.4. Like the thermal equilibrium state, the $I_z S_z$ state prepared in Exercise 18.2 would not contain any transverse magnetization components and would not generate an FID signal without further manipulation. Suppose that you were presented with a sample and told that it represented either the equilibrium state or the $I_z S_z$ state. You then apply either a non-selective $(\pi/2)_y$-pulse or a selective $(\pi/2)_{y,I}$-pulse. Describe the different results you would expect from the two types of sample, for each pulse type.

18.5. In Chapters 14, 15 and 16, we analyzed a variety of different experiments by treating the four eigenstates for a coupled spin pair individually and then combining the observable magnetization components according to the relative populations of the eigenstates in the initial population. With the density matrix and the product-operator formalism, we can now treat the equilibrium population in one set of calculations. Using the product-operator formalism and correlation diagrams, reexamine the following experiments:

(a) The three demonstration experiments discussed in Section 14.7 (page 414).

(b) The heteronuclear COSY experiment diagrammed in Fig. 15.2 (page 426).

(c) The HMQC experiment diagrammed in Fig. 16.38 (page 483).

Further Reading

The idea of representing the density matrix for coupled spins as a linear combination of magnetization operator matrices and their products was developed independently by three groups in the early 1980s. The first of the three articles listed below is generally considered the definitive reference, but the other two also offer useful perspectives. Note, however, that the paper by Sørensen et al. follows the convention of right-hand rotations, as used in this book, while the other two papers employ a left-hand convention.

◇ Sørensen, O. W., Eich, G. W., Levitt, M. H., Bodenhausen, G. & Ernst, R. R. (1983). Product-operator formalism for the description of NMR pulse experiments. *Prog. Nuci. Mag. Res. Sp.*, 16, 163–192. http://dx.doi.org/10.1016/0079-6565%2884%2980005-9

◇ van de Ven, F. J. M. & Hilbers, C. W. (1983). A simple formalism for the description of multiple-pulse experiments: Application to a weakly-coupled two-spin ($I = 1/2$) system. *J. Magn. Reson.*, 54, 512–520. http://dx.doi.org/10.1016/0022-2364(83)90331-1

◇ Wang, P.-K. & Slichter, C. P. (1986). A pictorial operator formalism for NMR coherence phenomenon. *Bull. Magn. Reson.*, 8, 3–16.

A particularly clear review and tutorial on the product-operator formalism:

◇ Shriver, J. (1992). Product operators and coherence transfer in multiple-pulse NMR experiments. *Concepts Magn. Reson.*, 4, 1–33. http://dx.doi.org/10.1002/cmr.1820040102

A Mathematica notebook for carrying out product operator calculations, described in this paper, is available from http://daffy.uah.edu/nmr/software/DownloadSoftware.html

A description of another Mathematica notebook for product operator calculations:

◇ Güntert, P., Schaefer, N., Otting, G. & Wüthrich, K. (1993). POMA: A complete *Mathematica* implementation of the NMR product-operator formalism. *J. Magn. Reson. A.*, 101, 103–105. http://dx.doi.org/10.1006/jmra.1993.1016

The notebook is available for download at http://www.bpc.uni-frankfurt.de/guentert/wiki/index.php/POMA

List of Symbols, Numerical Constants, and Abbreviations

A.1 Roman Letter Symbols

$\hat{A}$ Quantum mechanical operator associated with the measurement of quantity, A.

B Magnetic field vector.

B Magnetic flux density, the magnitude of the magnetic field vector, with units of tesla (SI) or gauss (cgs), commonly referred to as the magnetic field strength.

$\mathbf{B}_0$ Stationary magnetic field vector.

B_0 Magnitude of the stationary magnetic field vector.

$\mathbf{B}_1$ Oscillating transverse magnetic field vector

B_1 Magnitude of the oscillating magnetic field vector.

B_g Magnitude of field vector at a specific position in a spatial field gradient.

B_{rot} Magnitude of the stationary magnetic field vector in the rotating frame of reference.

c Chemical shift anisotropy coefficient, with dimension of inverse time.

c Also used to represent complex coefficients in representations of quantum mechanical wavefunctions as linear superpositions, with subscripts indicating the basis wavefunctions (e.g., c_α and c_β).

c^* Conjugate of the complex number or function c.

$|c|$ Modulus of the complex number or function c; $|c| = \sqrt{cc^*}$.

d Dipolar coupling constant, with dimension of inverse time.

D Diffusion coefficient, with units of m^2/s.

E Energy, with units of joule.

f_1, f_2 and f_3 Frequency axes in two- or three-dimensional NMR spectra. The subsripts indicate the corresponding time periods in the pulse sequence, with the largest subscript (f_2 or f_3) identifying the directly detected dimension, which is derived from Fourier transformation of the free induction decays.

f_c Critical, or Nyquist, frequency.

$\hat{F}^+$ and $\hat{F}^-$ Combined raising and lowering operators for a coupled spin pair. $\hat{F}^+ = \hat{I}^+ + \hat{S}^+$ and $\hat{F}^- = \hat{I}^- + \hat{S}^-$.

$\mathcal{F}(g(x))$ Fourier transform of the function $g(x)$.

$g(\Delta t)$ Autocorrelation function evaluated for the time interval Δt.

G Magnetic field gradient strength, with units of T/m or g/cm, often written with a subscript (x, y, or z) indicating the axis.

h The Planck constant, $\approx 6.62606957 \times 10^{-34}$ J· s. ($1\,\text{J·s} = 1\,\text{m}^2$ kg/s.)

$\hbar$ The Planck constant divided by 2π, $\approx 1.05457172 \times 10^{-34}$ J· s

$\mathcal{H}$ Quantum mechanical Hamiltonian operator.

$\mathcal{H}_{sc}$ Hamiltonian operator for the scalar coupling interaction.

$\mathcal{H}_T$ Total Hamiltonian operator for the scalar coupling interaction in the presence of a stationary external magnetic field.

$\mathbf{I}$ Angular momentum vector.

I Identifier of a specific spin.

I_x, I_y, and I_z Cartesian angular momentum components of spin I, also commonly identified with magnetization components.

$\hat{I}_x$, $\hat{I}_y$, and $\hat{I}_z$ Quantum mechanical operators for the corresponding magnetization components of spin I.

$\mathbf{I}_x$, $\mathbf{I}_y$, and $\mathbf{I}_z$ Matrix representations of operators for the corresponding magnetization components of spin I.

$\hat{I}^+$ and $\hat{I}^-$ Raising and lowering operators, respectively, for the spin quantum number of spin I.

$\langle I_x \rangle$, $\langle I_y \rangle$, and $\langle I_z \rangle$ Quantum mechanical averages, or expectations, of the corresponding magnetization components of spin I.

$\bar{I}_x$, $\bar{I}_y$, and $\bar{I}_z$ Population averages of the corresponding magnetization components of spin I.

$\bar{I}_z^0$ Equilibrium population average of I_z-magnetization.

$\mathcal{Im}(c)$ Imaginary part of the complex number c.

J Scalar coupling constant, with units of Hz.

^nJ Scalar coupling constant, with the superscript $n = 1$, 2 or 3 indicating the number of covalent bonds separating the coupled nuclei.

$J(\omega)$ Spectral density function evaluated at angular frequency, ω, with dimensions of time.

k The Boltzmann constant, $\approx 1.380658 \times 10^{-23}$ J/K.

k_{ab} and k_{ba} First-order rate constants for the interconversion of alternative states in the treatment of chemical or conformational exchange (i.e., changes giving rises to fluctuations in chemical shift), with dimensions of inverse time.

k_{ex} Total rate constant for chemical or conformational exchange, $k_{ex} = k_{ab} + k_{ba}$, with dimensions of inverse time.

M   Matrix.

N The number of particles or molecules in a particular state, usually identified with a subscript.

p_a and p_b Fractional populations of alternative states in the treatment of chemical or conformational exchange; dimensionless.

ΔP Population difference, usually expressed as a fractional value.

$\mathbf{R}_x(a)$, $\mathbf{R}_x^{-1}(a)$ Rotation matrix and its inverse, respectively, for calculating the change in the density matrix due to a pulse of angle, a, about the x'-axis. The subscript may include a designation of a specific spin.

$\mathbf{R}_y(a)$, $\mathbf{R}_y^{-1}(a)$ Rotation matrix and its inverse, respectively, for calculating the change in the density matrix due to a pulse of angle, a, about the y'-axis.

R_1 Longitudinal relaxation rate constant, with dimensions of inverse time.

R_1^{csa} Portion of R_1 due to chemical shift anisotropy.

R_1^{dd} Portion of R_1 due to dipolar coupling interactions.

$R_{1\rho}$ Rate constant for longitudinal relaxation in the rotating frame, with dimensions of inverse time.

R_2 Transverse relaxation rate constant, with dimensions of inverse time.

R_2^{csa} Portion of R_2 due to chemical shift anisotropy.

R_2^{dd} Portion of R_2 due to dipolar coupling interactions.

R_{ex} Portion of R_2 due to chemical or conformational exchange.

$\mathcal{R}e(c)$ Real part of the complex number c.

RMS(x)   Root-mean-square average of quantity x.

S Identifier of a specific spin, usually paired with spin I. Symbols representing operators and average magnetization components are analogous to those used for spin I.

S^2 Order parameter in the Lipari–Szabo treatment of heteronuclear relaxation; dimesionless.

$s(t)$ Time-domain signal, inverse Fourier transform of $S(f)$.

$S(f)$ Frequency-domain signal, Fourier transform of $s(t)$.

t_1, t_2 and t_3 Variable delay or data acquisition times in two- or three-dimensional NMR experiments. The subsripts indicate the order of the time periods in the pulse sequence, with the largest subscript (t_2 or t_3) corresponding to the data acquisition time.

T Temperature.

T_1 Longitudinal relaxation time.

T_2 Transverse relaxation time.

$\mathbf{U}_H(t)$ and $\mathbf{U}_H^{-1}(t)$ Unitary matrix and its inverse, respectively, for calculating the time evolution of the density matrix in the presence of a stationary magnetic field.

$\mathbf{v}$ Vector.

W Quantum transition rate constant, with dimensions of inverse time. Subscripts indicate a change in quantum number (usually 0, 1 or 2), and superscripts are used to identify spins (e.g., I or S).

x' and y' Cartesian coordinates in a frame of reference rotating about the z-axis.

$\langle x \rangle$ Average value of quantity x.

A.2   Greek Letter Symbols

$|\alpha\rangle$ Dirac ket representation of the eigenfunction for the z-magnetization operator for a single spin-1/2, with the magnetic dipole aligned parallel with the external field.

$|\beta\rangle$ Dirac ket representation of the eigenfunction for the z-magnetization operator for a single spin-1/2, with the magnetic dipole aligned anti-parallel to the external field.

γ Gyromagnetic ratio, or magnetogyric ratio, with units of $T^{-1}s^{-1}$.

δ Chemical shift; dimensionless but usually expressed in parts per million (ppm).

δ_{11}, δ_{22} and δ_{33} Orientation-specific chemical shifts in the directions of principal axes 1, 2 and 3.

δ_{iso} Isotropic chemical shift.

$\Delta\delta$ Chemical shift anisotropy.

$\Delta\omega$ Absolute value of the difference in angular frequencies of alternate states in the treatment of chemical or conformational exchange.

η Viscosity, with units of Pa· s, equivalent to N· m^{-2} s.

η Relaxation interference rate constant, with units of inverse time.

$\eta_I(S)$ Steady-state nuclear Overhauser effect expressed as the relative change in signal itensity from spin I due to saturation of spin S.

λ Wavelength.

λ_i The ith eigenvalue of an eigenfunction.

$\boldsymbol{\mu}$ Magnetic moment vector.

μ_0 The magnetic permeability of free space; $4\pi \times 10^{-7}$ T^{2}m^3/J $\approx 1.25664 \times 10^{-6}$ T^{2}m^3/J.

ν Frequency, with units of Hz or s^{-1}.

ν_0 Reference frequency.

ν_{rot} Precession frequency in the rotating frame of reference.

ν' Frequency of the component of a heterodyned signal with the lower absolute frequency, generated by mixing with a reference signal and removing the high-frequency component.

π_x and π_y Pulses to rotate magnetization by 1/2 rotation about the x'- and y'-axes, respectively, in the rotating frame.

$\pi_{x,I}$ and $\pi_{x,S}$ Selective pulses to rotate the magnetization of I- or S-spins, respectively, by 1/2 rotation about the x'-axis, respectively, in the rotating frame.

$(\pi/2)_x$ and $(\pi/2)_y$ Pulses to rotate magnetization by 1/4 rotation about the x'- and y'-axes, respectively, in the rotating frame.

$(\pi/2)_{x,I}$ and $(\pi/2)_{x,S}$ Selective pulses to rotate the magnetization of I- or S-spins, respectively, by 1/4 rotation about the x'-axis, in the rotating frame.

Ψ General representation of a quantum wavefunction.

$|\Psi\rangle$ Dirac ket notation for the wavefunction, Ψ.

$\langle\Psi|$ Dirac bra notation for the complex conjugate of the wavefunction, Ψ.

$|\psi_j\rangle$ Dirac ket notation for the jth eigenvalue of the Hamiltonian operator, $\mathcal{H}$.

ρ Density matrix.

σ Shielding coefficient; dimensionless.

σ_{11}, σ_{22} and σ_{33} Orientation-specific shielding coefficients in the directions of principal axes 1, 2 and 3.

σ_{iso} Isotropic shielding coefficient.

$\Delta\sigma$ Shielding anisotropy.

$\sigma_{\parallel}$ Parallel shielding coefficient component, equal to σ_{11}.

$\sigma_{\perp}$ Perpendicular shielding coefficient component, equal to $(\sigma_{22} + \sigma_{33})/2$.

τ Correlation time or duration of a fixed delay time in an NMR pulse sequence.

τ_c Correlation time for overall molecular tumbling.

τ_e Correlation time for internal motions faster than molecular tumbling.

τ_{cp} Delay time in the Carr–Purcell–Meiboom–Gill pulse sequence.

τ_m Duration of mixing time in a pulse sequence. (Also sometimes used as the correlation time for overall molecular tumbling.)

ω Angular frequency, with units rad/s or s^{-1}.

A.3   Abbreviations

CSA   Chemical shift anisotropy.

COSY   COrrelation SpectroscopY.

CPMG   Carr–Purcell–Meiboom–Gill pulse sequence for measuring transverse relaxation.

CS   Chemical shift.

CW   Continuous wave.

Da   Dalton unit of molecular mass, equivalent to the atomic mass unit.

DFT   Discrete Fourier transform.

FID   Free induction decay.

FT   Fourier transform.

HMQC   Heteronuclear multiple-quantum correlation.

HNCA   A three-dimensional NMR experiment used to establish correlations between amide-^{1}H, amide-^{15}N and α-^{13}C nuclei in proteins.

HN(CO)CA   A three-dimensional NMR experiment used to establish correlations between amide-^{1}H, amide-^{15}N and α-^{13}C nuclei in proteins, with the magnetization transfer mediated by the carbonyl-^{13}C nucleus separating an amide group and the α-^{13}C of the preceding residue.

HSQC   Heteronuclear single-quantum correlation.

kDa   Unit of molecular mass equal to 10^3 Da.

NOE   Nuclear Overhauser effect (or enhancement).

NOESY   Nuclear Overhauser Effect SpectroscopY.

PFG   Pulsed field gradient.

RDC   Residual dipolar coupling.

SC   Scalar coupling.

rf   Radio frequency.

TOCSY   Total COrrelation SpectroscopY.

TROSY   Transverse Relaxation Optimized SpectroscopY.

Trigonometric Functions and Complex Numbers

Appendices B, C, and D provide brief introductions to some of the mathematical tools that are frequently used in the theory and practice of NMR spectroscopy. Much of this material is also included in the main text, where it is introduced as needed for various applications. It is presented again here, sometimes with additional explanation or extensions, so as to provide a concise reference. This material is not meant to be either comprehensive or particularly rigorous. For more information on these topics, the reader is referred to standard textbooks and references, as well as the online resources Mathworld (http://mathworld.wolfram.com) and Wikipedia (http://www.wikipedia .org). Another excellent resource on basic mathematical relationships is Chapter 22 of the Feynman Lectures, http://www.feynmanlectures.caltech.edu/I_22.html.

B.1   The Basic Trigonometric Functions

Using the conventional $x-y$ coordinate system, as shown in Fig. B.1, the basic trigonometric functions are defined as

$$\sin \theta = \frac{y}{r}$$

$$\cos \theta = \frac{x}{r}$$

$$\tan \theta = \frac{y}{x}$$

where $r = \sqrt{x^2 + y^2}$.

In using the trigonometric functions in physics and mathematics, it is often convenient to express angles in the units of radians (rad), rather than the more commonly used degrees. The radian is defined, as illustrated in Fig. B.1, as the contour length, l, of the arc subsumed by the angle divided by the radius, r. Since l is 2π when the angle

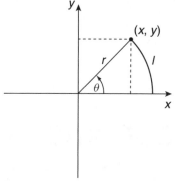

Figure B.1

Cartesian coordinate system used to define the trigonometric functions and the radian unit of angular measurement, $\theta = l/r$ rad, where l is the contour length of the arc subsumed by the angle θ.

is 360°, radians can be converted to degrees according to the following:

$$radians = degrees\frac{2\pi}{360}$$

When the angle, θ, is expressed in radians, the following important relationships hold:

$$\frac{d \sin \theta}{d\theta} = \cos \theta$$

$$\frac{d \cos \theta}{d\theta} = -\sin \theta$$

$$\lim_{\theta \to 0} \frac{\sin \theta}{\theta} = 1$$

Using radians is also particularly convenient for calculations involving the exponentials of imaginary numbers, as discussed in the next section. A function closely related to the sine is the sine cardinal, often called the sinc function, which is discussed on page 607.

The unit of the radian (along with its analog for the measurement of solid angles, the steradian) are rather special among the standard units of measurement. If the radian is expressed in terms of the base SI (Système International) units, it is simply a ratio of lengths, meter/meter. Thus, it is a unit with no units! For a period, from 1960 to 1995, the SI placed the radian and steradian in a special class of their own, called supplementary units. In 1995, however, it was decided that this special classification didn't really clarify anything, and the radian and steradian were moved to the class of derived units, which are those (such as the newton) made up of the seven base units (the meter, kilogram, second, ampere, kelvin, candela and mole). In its recommendations for the use of the radian and steradian, the SI states that their names and symbols "may, but need not, be used in expressions for other SI derived units, as is convenient." (See http://www.bipm.org/en/CIPM/db/1980/1/ and http://www.bipm.org/en/CGPM/db/20/8/.) In practice, it is common to include the radian when specifying angles or angular frequencies, but it is often dropped, for instance, in correlation times and the spectral density function. Though the latter practice may seem sloppy, it is, in fact, sanctioned by the highest authorities on such matters!

B.2   Trigonometric Identities

There are a number of relationships among the trigonometric functions that are referred to as identities. The following are some of the identities that frequently come in handy:

$$\cos^2\theta + \sin^2\theta = 1$$
$$\cos(-\theta) = \cos\theta$$
$$\sin(-\theta) = -\sin\theta$$
$$\sin^2\theta = \frac{1}{2}\left(1 - \cos(2\theta)\right)$$
$$\cos^2\theta = \frac{1}{2}\left(1 + \cos(2\theta)\right)$$
$$\cos\theta\cos\phi = \frac{1}{2}\left(\cos(\theta - \phi) + \cos(\theta + \phi)\right)$$
$$\sin\theta\sin\phi = \frac{1}{2}\left(\cos(\theta - \phi) - \cos(\theta + \phi)\right)$$
$$\sin\theta\cos\phi = \frac{1}{2}\left(\sin(\theta + \phi) + \sin(\theta - \phi)\right) \tag{B.1}$$

B.3   Imaginary and Complex Numbers

An imaginary number is composed of a real number multiplied by the unit imaginary number, i, which is defined as the square root of -1. Thus,

$$i \cdot i = -1$$

The square of any imaginary number will be a negative number, in contrast to real numbers, for which the square is always positive.

A complex number, c, is the sum of a real number and an imaginary number and can be written in the form,

$$c = a + ib \tag{B.2}$$

where a and b are both real numbers. For every complex number, c, there is a complex conjugate, c^*:

$$c^* = a - ib$$

The product of c and $c*$ will always be a positive real number:

$$cc^* = (a + ib)(a - ib) = a^2 - iab + iab - i^2b^2 = a^2 + b^2$$

The square root of this product is called the *modulus* and is represented as

$$|c| = \sqrt{cc^*} = \sqrt{a^2 + b^2}$$

The modulus represents the "magnitude" of a complex number, much as the absolute value of a real number, which is equal to the square root of the square ($|a| = \sqrt{a^2}$), is a positive number that represents the magnitude of the real number.

Exponential functions of imaginary numbers, of the form e^{ib}, are used widely in physics. The complex conjugate of e^{ib} is e^{-ib}. There is a very special relationship between the exponential of an imaginary number and the trigonometric functions:

$$e^{ib} = \cos b + i \sin b$$

This is known as *Euler's formula* (actually just one of many Euler's formulae) and turns up in lots of interesting places, especially whenever cyclic or periodic processes are described. (Rather ironically, imaginary numbers are used extensively in some very real applications, such as the mathematics of alternating current, upon which electrical power systems are based.) A special case of this relationship is

$$e^{i\pi} + 1 = 0$$

This equation is remarkable because it brings together five of the most fundamental numbers in mathematics. It has even been used as an argument for the existence of a deity! For an entertaining and illuminating derivation of Euler's formula, see Chapter 22 of the Feynman Lectures (Vol I, http://www.feynmanlectures.caltech.edu/I_22.html). Feynman describes this equation as "one of the most remarkable, almost outstanding, formulas in all of mathematics" and "our jewel"!

From Euler's formula, the following useful relationships can be derived:

$$e^{i\theta} + e^{-i\theta} = 2 \cos \theta$$
$$e^{i\theta} - e^{-i\theta} = i \cdot 2 \sin \theta$$

Also, Euler's formula is often the easiest way to derive relationships among the trigonometric functions, such as those listed in the previous section.

The product of any complex number of the form e^{ib} and its complex conjugate, e^{-ib}, is unity. Thus, the modulus of any such number is 1. More generally, any complex number can be written as

$$c = me^{i\theta} = m(\cos \theta + i \sin \theta)$$

where m is the modulus. Comparison with the other representation of a complex number (Eq. B.2) shows that

$$a = m \cos \theta$$
$$b = m \sin \theta$$

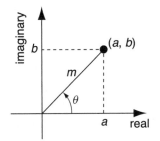

Figure B.2
Representation of a complex
number as a point on a plane
defined by perpendicular real
and imaginary axes.

This suggests that complex numbers can be thought of as points on a "complex plane" defined by real and imaginary axes, as illustrated in Fig. B.2. The modulus can then be thought of as the length of a line segment, and the angle, θ, represents the relative contributions of the real and imaginary parts. This angle is sometimes called the *argument* or *phase* of the complex number.

Fourier Series and Transforms

In the early 1800s, Jean Baptiste Joseph Fourier, a French mathematician, physicist, archeologist and government official (just some of his vocations), addressed the problem of heat flow in solids of various shapes and developed an approach to this problem that was based on infinite sums of sine and cosine functions of increasing frequencies. Although the idea of such series was not entirely new, Fourier's generalization of it was a major advance and met some initial resistance, largely because of a perceived lack of rigor in his treatment. In the centuries since, the mathematics of Fourier series and their related operations were extensively developed and have become central to a wide range of subjects, including modern NMR spectroscopy. The language (and notation) of this subject can be somewhat confusing, and it is useful to distinguish among the following:

- Fourier series
- Continuous Fourier transforms
- Discrete Fourier transforms

These three topics are closely related and can be developed from one another in a variety of ways. In Chapter 6, Fourier series and transforms were introduced to deal specifically with the conversion of the signal from a pulse NMR experiment (the free induction decay) to a frequency spectrum. Fourier transforms were also used in the same chapter in the context of shaped pulses, as well as in Chapter 8 when discussing the autocorrelation and spectral density functions. In this appendix, Fourier series are presented in a more general way, followed by sections covering continuous and discrete transforms. A closely related topic, convolution, is also discussed here.

C.1 Fourier Series

First we consider the use of Fourier series to represent a general periodic function, $g(t)$. Because so many applications of Fourier series and transforms involve time-dependent signals characterized by frequency components, and are relatively easy to visualize in these terms, it is convenient to express the mathematics in terms of a frequency variable, f, and a time variable, t, even though the applications are much wider.

A periodic function can be described as having a repeat period, T, and a frequency, $f = 1/T$. Many, but not all, such functions can be expressed in terms of an infinite series of cosine and sine terms of the form

$$g(t) = a_0 + \sum_{n=1}^{\infty} \left(a_n \cos(2\pi nft) + b_n \sin(2\pi nft) \right) \tag{C.1}$$

where a_n and b_n are constants that represent the contributions of the cosine and sine terms, respectively, to $g(t)$. Functions that are not periodic can also be represented this way, but only over a defined interval, usually specified as $-T/2$ to $T/2$. Both real- and complex-valued functions can be represented as Fourier series. For complex functions, the coefficients, a_0, a_n and b_n, will, in general, be complex numbers. If the following integrals exist, then the coefficients can be evaluated as

$$a_0 = \frac{1}{T} \int_{-T/2}^{T/2} g(t)dt \tag{C.2a}$$

$$a_n = \frac{2}{T} \int_{-T/2}^{T/2} g(t) \cos(2\pi nft)dt \tag{C.2b}$$

$$b_n = \frac{2}{T} \int_{-T/2}^{T/2} g(t) \sin(2\pi nft)dt \tag{C.2c}$$

A Fourier series, representing either a real- or complex-valued function, can also be expressed in an exponential form:

$$g(t) = \sum_{n=-\infty}^{\infty} c_n e^{i2\pi nft} \tag{C.3}$$

where c_n are, in general, complex coefficients, which are evaluated as

$$c_n = \frac{1}{T} \int_{-T/2}^{T/2} g(t)e^{-i2\pi nft}dt \tag{C.4}$$

The integrals in Eq. C.2 (or equivalently C.4) represent the degree to which the various terms in the Fourier series contribute to the function, $g(t)$, a relationship that is best illustrated with an example, as discussed below.

The left panel of Fig. C.1 shows the graph of a function, $g(t)$, over the interval $t = -0.5$ to 0.5. For this example, $T = 1$ and $f = 1$. The first coefficient of the Fourier series is calculated as the integral of $g(t)$ divided by T (Eq. C.2a). This term represents the average value of the function and the portion of $g(t)$ that does not vary with t.

The coefficients for the cosine terms of the Fourier series are calculated by multiplying $g(t)$ by the corresponding cosine functions, $\cos(2\pi nft)$, and integrating the resulting product function, as illustrated in Fig. C.2 for $n = 1$. Graphs of $g(t)$ and the $n = 1$ cosine function are shown in the left-hand panel, and the product of the

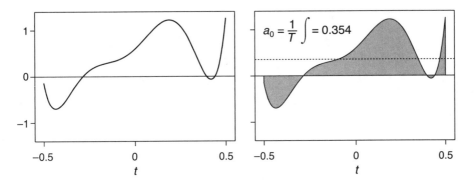

Figure C.1   A function and the integral for the constant term, a_0, in the Fourier series. The left-hand panel shows the graph of an arbitrary function for which a Fourier series is to be calculated. The same function is shown in the right-hand panel, with the area representing the integral of the function shaded. The dashed line represents the average value of the function and, from Eq. C.2a, a_0.

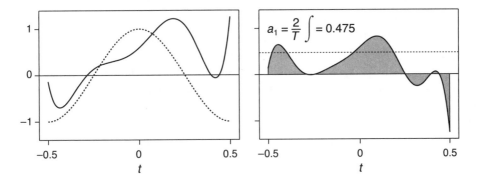

Figure C.2   A function and the integral for the first cosine coefficient, a_1, in the Fourier series. The left-hand panel shows the graph of the function, $g(t)$ (solid curve), and the cosine term $\cos(2\pi n f t)$, with $n = 1$. The product of $g(t)$ and the cosine term are graphed in the right-hand panel, with the area representing the integral of the product shaded. The dashed line represents twice the average value of the product and, from Eq. C.2b, a_1.

two functions is shown on the right. The product of the two functions represents the degree to which they match one another: For values of t where both functions are positive or both are negative, the product is positive. Where one function is positive and one is negative, the product is negative. In this case, there is a substantial positive correlation between the two functions, and the integral is positive. Thus, the term $\cos(2\pi f t)$ has a positive coefficient in the Fourier series.

Fig. C.3 shows the product of $g(t)$ and the second cosine term, $\cos(4\pi f t)$, for the same example. For this term, the integral is negative, largely due to the region between $t \approx 0.15$ and $t \approx 0.4$, where $g(t)$ is positive and the cosine is negative. This term thus makes a net negative contribution to the Fourier series.

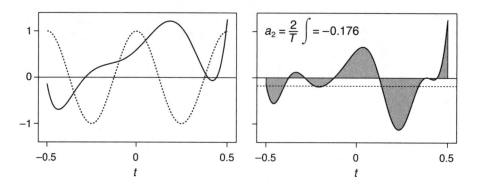

Figure C.3   The integral for the second cosine coefficient, a_2, in the Fourier series. Details as in Fig. C.2.

The coefficients for the sine terms are calculated in the same way, according to Eq. C.2c and illustrated in Fig. C.4 for b_1. Although this term has the same frequency as the one illustrated in Fig. C.2, its phase is shifted by $\pi/2$ rad and represents a different component of $g(t)$. In this case, the lowest-frequency cosine and sine terms make roughly equal contributions to the Fourier series.

The first five terms of the Fourier series for the function $g(t)$ illustrated in Figs. C.1–C.4 are graphed individually in the left-hand panel of Fig. C.5. The sum of the five terms is represented by the dashed curve in the right-hand panel, along with the graph of the original function. In this case, the first five Fourier terms do moderately well at approximating the function, $g(t)$, except for the sharp upturns near the extreme values, $t = -0.5$ and 0.5.

Successively better approximations can be made by adding cosine and sine terms with higher frequencies, as shown in Fig. C.6. The dashed curves represent the sums of the constant term, a_0, and the first $n = 1, 3$ and 10 cosine and sine terms. With a total of 21 terms, the sum very closely approximates $g(t)$ for $-0.4 \lesssim t \lesssim 0.4$. Even with the highest-frequency terms, however, there are significant deviations at the edges of the defined interval. These deviations reflect the discontinuity between the values of

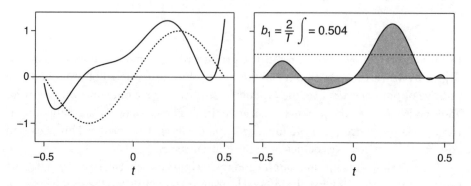

Figure C.4   The integral for the first sine coefficient, b_1, in the Fourier series. Details as in Fig. C.2.

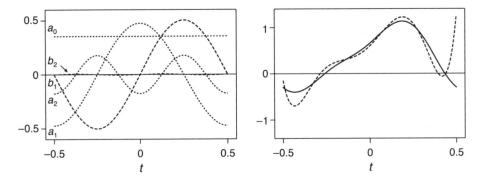

Figure C.5 The first five terms of the Fourier series representing the function, $g(t)$, for the example illustrated in Figs. C.1–C.4. In the left-hand panel, the constant value, a_0, and the first two cosine terms, labeled with the coefficients a_1 and a_2, are drawn with short dashes, and longer dashes are used to represent the sine terms, with the coefficients b_1 and b_2. The second sine term has a very small amplitude and is barely distinguishable from the $y = 0$ axis. The sum of the five terms is represented by the dashed curve in the right-hand panel, along with the original function, $g(t)$ (solid curve).

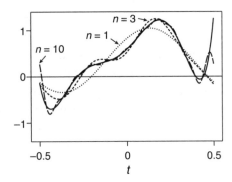

Figure C.6 Successive approximations of a function by increasing the number of terms from the Fourier series. The dashed curves represent sums of the Fourier terms, with the labels indicating the index of the last cosine and sine terms in the sum. The original function, $g(t)$, is represented by the solid curve.

the function at the two ends of the defined interval and will not approach zero with any finite number of terms in the series, though the deviations will move closer and closer to the point of the discontinuity as the number of terms increases. The same effect will be seen when discontinuities appear at interior points of the function. This behavior is known as the Gibbs phenomenon, named for the same J. Willard Gibbs renowned for his contributions to thermodynamics.

An important feature of the terms making up the Fourier series is that they are orthogonal, implying that none of them can be expressed as linear combinations of any of the others. In this context, orthogonality means that the integrals, over the interval $-T/2$ to $T/2$, of the pairwise products formed between distinct terms are all zero. For the cosine terms, for instance, the following is true for any n and m that are not equal to one another:

$$\int_{-T/2}^{T/2} \cos(2\pi n f t) \cos(2\pi m f t) dt = 0$$

Similarly, for $n \neq m$,

$$\int_{-T/2}^{T/2} \sin(2\pi n f t) \sin(2\pi m f t) dt = 0$$

and, for any n and m,

$$\int_{-T/2}^{T/2} \sin(2\pi n f t) \cos(2\pi m f t) dt = 0$$

These relationships can be confirmed using some of the trigonometric identities from page 593 and ensure that there is a unique Fourier series for a given function that can be represented in this way.

Since evaluating the coefficients requires knowledge of the function, $g(t)$, over the interval of integration, it might not appear that much is gained by representing the function as a Fourier series. But, in many situations, it proves easier to carry out other mathematical operations on the function when expressed as a Fourier series. In Fourier's original analysis of heat flow, solutions of the relevant differential equation were already known for the cosine and sine functions, and Fourier recognized that these solutions could be applied to many other cases, provided that the problem of interest could be represented as a series of the trigonometric functions. Although an exact solution might require an infinite number of cosine and sine terms, any specified degree of precision can be obtained with a finite number of terms (provided that the function does not contain any discontinuities, such as seen at the extremes of our example).

Beyond the practical advantages that a series representation may have, the coefficients of the Fourier series convey information about the frequency components of a function, particularly when a function represents a signal made up of a fundamental frequency and integer multiples, or harmonics, of that frequency. Physically, such functions arise in describing vibrational modes, such as the strings or air column of a musical instrument. Often, however, the cosine and sine terms of a Fourier series will not have any particular physical significance. For instance, we can represent the FID from a pulse NMR experiment (over a defined time period) as a Fourier series, but it is very unlikely that the Larmor frequencies of individual nuclei in the molecule of interest will be related as simple multiples. There will, however, be a frequency spectrum of the Larmor frequencies, and the function describing the spectrum is related to the FID by a Fourier transform, which brings us to the next subject.

C.2   Continuous Fourier Transforms

Very generally, the Fourier transform of a function, $g(x)$, is another function, which we will express in terms of another variable, s, and will write as $G(s)$. To express the relationship between the two functions, we write

$$G(s) = \mathcal{F}\big(g(x)\big)$$

The function, $G(s)$, is defined in terms of $g(x)$ as

$$G(s) = \int_{-\infty}^{\infty} g(x)e^{-i2\pi xs}dx$$

$$= \int_{-\infty}^{\infty} g(x)\cos(2\pi xs)dx - i \int_{-\infty}^{\infty} g(x)\sin(2\pi xs)dx \qquad (C.5)$$

Both x and s are assumed to be real, but the functions, $g(x)$ and $G(s)$, can be real- or complex-valued. The integration from $-\infty$ to ∞ is with respect to x, so this variable disappears with the integration, leaving only the new variable, s. Put another way, evaluation of the function, $G(s)$, for a given value of s requires integration over all values of x. The Fourier transform of a function only exists if this integral converges, but it fortunately does for many functions describing physical phenomena.

Notice, also, that the variables, x and s, appear as a product in the exponent of Eq. C.5, and the other terms in the exponent are pure, dimensionless numbers. Because the exponent must be dimensionless, this implies that if x and s are quantities with dimensions, their units must be reciprocals of one another.[1] Thus, if g is a function of time, G must be a function of frequency, and vice versa. If $g(x)$ is a function of length, then s must have units of inverse length in $G(s)$. Though inverse length might appear to be a strange concept, it does appear in just this form in the use of Fourier transforms to describe diffraction in x-ray crystallography and related techniques, where the spacing of atoms is described in "real space," typically with units of Å , and the spacing of diffraction peaks is described in "reciprocal space," with units of Å^{-1}. In NMR, we are generally interested in functions of time and frequency, however, and the relationship between the units of the arguments of $g(x)$ and $G(s)$ provides a clue to understanding the physical meaning of a Fourier transform of, for instance, an FID.

An additional clue to understanding the meaning of the Fourier transform of $g(x)$ comes from comparing the definition of the transform (Eq. C.5) with the very similar equations for evaluating the coefficients of a Fourier series, in either trigonometric or exponential form (Eqs. C.2 and C.4, p. 598). As illustrated by the examples in the previous section, the integral of a product of two functions represents the degree to which one matches, or follows the trends of, the other. For a given value of s, the

---

[1] In the third edition of his treatise on heat flow (*Théorei Analytique de la Chaleur*, 1822), Joseph Fourier formulated some of the basic concepts of what is now referred to as dimensional analysis— that is, the treatment of units as algebraic entities that should balance or cancel out in a reasonable mathematical description of a physical phenomenon.

integral in Eq. C.5 represents the extent to which the function,

$$e^{-i2\pi xs} = \cos(2\pi xs) - i\,\sin(2\pi xs)$$

contributes to $g(x)$. Depending on the value of $G(s)$, this contribution can be real, imaginary or complex and can represent a cosine or sine function, or both, with period $1/s$. As discussed earlier, when used to describe a time-domain signal, the coefficients in a Fourier series represent the relative contributions of sine and co-sine functions with frequencies that are integer multiples of a fundamental frequency, $f = 1/T$. In the same context, the continuous Fourier transform represents an ex-tension of this idea, with the function, $G(s)$, representing the contributions to the time-domain signal, $g(t)$, of sine and cosine functions with frequencies that are not restricted to integer multiples. For the specific case of $s = 0$, the value of the trans-form, $G(0)$, is the integral of $g(x)$ over all values of x, in approximate analogy to the a_0 term of the Fourier series, which represents the average value of the function, $g(t)$, over the range $-T/2 \le t \le T/2$.

The relationship between a function and its Fourier transform is *almost* recipro-cal, so if $G(s)$ is the transform of $g(x)$, then the transform of $G(s)$ is given by

$$\mathcal{F}\big(G(s)\big) = \int_{-\infty}^{\infty} G(s)e^{-i2\pi xs}ds = g(-x) \tag{C.6}$$

The change in sign of x is a consequence of the term, $-i$, in the exponent in the integral defining the transform. Because of this relationship, functions and their transforms generally come in reciprocal pairs, a few of which are discussed below. We can also define an inverse Fourier transform, so that if $G(s)$ is the transform of $g(x)$, then the inverse transform of $G(s)$ is $g(x)$. The inverse transform is given by the following integral:

$$\mathcal{F}^{-1}\big(G(s)\big) = g(x) = \int_{-\infty}^{\infty} G(s)e^{i2\pi xs}ds \tag{C.7}$$

This integral differs from the one for the forward transform (Eq. C.5) only by the sign of the exponential term. The choice of the negative sign in the forward transform and positive in the inverse, rather than the opposite, is arbitrary, but reflects current convention.

The reader should also be aware of differences in convention regarding the term, 2π, in the exponential (or trigonometric) terms. In some disciplines it is common to remove this factor and write the forward transform as

$$G(s) = \int_{-\infty}^{\infty} g(x)e^{-ixs}dx$$

$$= \int_{-\infty}^{\infty} g(x)\cos(xs)dx - i \int_{-\infty}^{\infty} g(x)\sin(xs)dx$$

If x represents a time domain, then s is an angular frequency with units of rad/s, versus s^{-1} or Hz when the definition of Eq. C.5 is used. Also, with the factor 2π

removed from the exponential, the inverse transform of $G(s)$ becomes

$$\mathcal{F}^{-1}\big(G(s)\big) = g(x) = \frac{1}{2\pi} \int_{-\infty}^{\infty} G(s)e^{ixs}ds$$

Yet another convention removes the factor 2π from the exponential, but splits it between the forward and inverse transforms. The resulting forward transform is

$$G(s) = \frac{1}{\sqrt{2\pi}} \int_{-\infty}^{\infty} g(x)e^{-ixs}dx$$

and the corresponding inverse transform is

$$\mathcal{F}^{-1}\big(G(s)\big) = g(x) = \frac{1}{\sqrt{2\pi}} \int_{-\infty}^{\infty} G(s)e^{ixs}ds$$

While all of these conventions are valid if used consistently, the one defined by Eqs. C.5 and C.7 often leads to the simplest mathematical expressions. Be aware, however, that other texts and computer programs may use other conventions. Some computer programs (including Mathematica, from Wolfram Research) offer options to specify the constants in the different parts of the equations.

Another important and general property of Fourier transforms is that they are linear operations. If $g(x)$ and $h(x)$ are both functions of x with transforms $G(s)$ and $H(s)$, respectively, then

$$\mathcal{F}\big(g(x) + h(x)\big) = \mathcal{F}\big(g(x)\big) + \mathcal{F}\big(h(x)\big) = G(s) + H(s) \qquad \text{(C.8)}$$

Linearity also implies that if c is a constant, then

$$\mathcal{F}\big(c \cdot g(x)\big) = c\mathcal{F}\big(g(x)\big) = c \cdot G(s) \qquad \text{(C.9)}$$

There are a number of other important and useful properties of Fourier transforms, some of which will be discussed after examining a few specific examples.

C.2.1   Some Examples of Fourier Transform Pairs

Some of the important properties of Fourier transforms are best illustrated by considering specific examples, a few of which are shown in graphical form in Fig. C.7, without proof. Each row of the figure shows the graph of a function, $g(x)$, in the left-hand panel, and that of its transform, $G(s)$, in the right-hand panel.

The continuous Fourier transform can only be calculated for a relatively limited set of functions, since the integral of Eq. C.5 only converges if the function, $g(x)$, equals or approaches zero as x approaches $-\infty$ from the right and ∞ from the left. An example of a function that is non-zero for all values but approaches zero when $|x| \to \infty$ is the Gaussian function, one form of which is written as

$$g(x) = e^{-\pi x^2} \qquad \text{(C.10)}$$

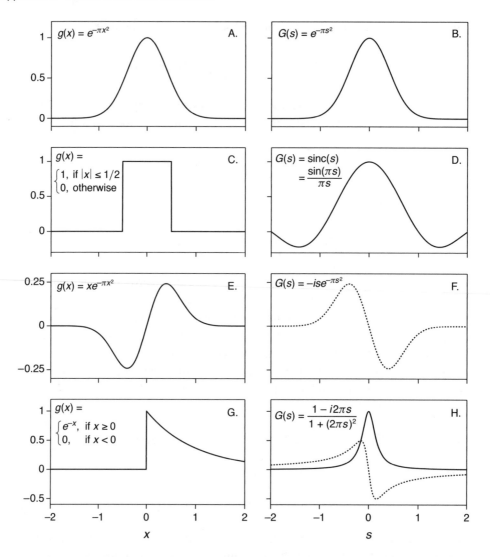

Figure C.7 Graphical representations of some Fourier transform pairs. The left-hand panels contain graphs of four functions, $g(x)$, and the right-hand panels show graphs of the corresponding Fourier transforms, $G(s)$. Solid and dashed curves represent the real and imaginary components of the transforms, respectively.

This function describes the classic "bell curve" that arises frequently in probability and statistics and is graphed in panel A of Fig. C.7. The Gaussian function in the form given in Eq. C.10 has the special property of being its own Fourier transform, as graphed in panel B of Fig. C.7.

A simple example of a function that does not have a Fourier transform is $g(x) = a$, where a is a constant other than zero. If, however, the function is redefined so that it is equal to a constant value over a finite range of x-values and zero elsewhere, then the integral of Eq. C.5 takes on a finite value and the transform is defined. Such functions are commonly described as rectangle or top-hat functions, and a special case is the

Heaviside Π function, which is defined so that

$$\Pi(x) = \begin{cases} 1 & \text{if } |x| \leq 1/2 \\ 0 & \text{otherwise} \end{cases}$$

A graph of $\Pi(x)$ is shown in Fig. C.7C. The Fourier transform of $\Pi(x)$ is the sine cardinal, or sinc, function, defined as[2]

$$\text{sinc}(s) = \begin{cases} \dfrac{\sin(\pi s)}{\pi s} & \text{if } s \neq 0 \\ 1 & \text{if } s = 0 \end{cases} \tag{C.11}$$

The sinc function, graphed in panel D of Fig. C.7, is characterized by a maximum at $s = 0$ and oscillations between positive and negative values that extend indefinitely in both positive and negative directions, but with ever-decreasing amplitude. These oscillations, often referred to as "sinc wiggles," are associated with the Fourier transforms of functions with sharp boundaries and reflect the requirement of terms with indefinitely high frequency to account for these boundaries. In NMR, sinc wiggles are frequently seen when an FID is truncated because of time or other constraints and can be minimized by smoothing, or apodizing, the signal before transformation (Section 6.8.2, p. 146).

Both the Gaussian function (panel A) and the Heaviside Π function (panel C) are examples of "even" functions, those for which $g(-x)$ is equal to $g(x)$. The Fourier transforms of even functions are also always even functions. Another property of even functions is that their transforms are fully reciprocal, so that if $G(s)$ is the transform of an even function, $g(x)$, then the transform of $G(s)$ is $g(x)$. Furthermore, if $g(x)$ is even and has only real values, then the transform, $G(s)$, is also real-valued, as in the examples shown in panels B and D of Fig. C.7.

Functions for which $g(-x) = -g(x)$ are described as "odd." An example of an odd function, for which the Fourier transform exists, is the product of a Gaussian function and its argument:

$$g(x) = xe^{-\pi x^2}$$

A graph of this function is shown in panel C of Fig. C.7, with its transform in panel E. For a real-valued odd function, the Fourier transform has only an imaginary component, which is also an odd function, so that $G(-s) = -G(s)$. (In Fig. C.7, the imaginary components, when present, are shown as dashed curves.) Note also that for this and other odd functions, both $g(0)$ and $G(0)$ are zero, reflecting the fact that the integrals of both are zero. For an odd function, $g(x)$, with transform, $G(s)$, the transform of $G(s)$ is $g(-x) = -g(x)$.

Most functions are neither even nor odd, so $g(-x)$ is, in general, not equal to either $g(x)$ or $-g(x)$. An example of this class of functions, graphed in Fig. C.7G,

---

[2] In some contexts, the sinc function is defined as $\sin(x)/x$ for $x \neq 0$ and 1 for $x = 0$. Especially when using computer software in which the sinc function is predefined, it is important to check just which definition is being used.

describes an exponential decay for positive values of x:

$$g(x) = \begin{cases} e^{-x} & \text{if } x \geq 0 \\ 0 & \text{if } x < 0 \end{cases}$$

The Fourier transform of e^{-x} for all values of x does not exist, because that function increases without bound as x becomes more negative, and the transformation integral does not converge. But, by specifying that $g(x) = 0$ for negative x, the transform integral is made to converge. The transforms of real-valued functions that are neither even nor odd are composed, in general, of both real and imaginary components, as shown for the example in panels G and H of Fig. C.7. For this case, the Fourier transform is

$$G(s) = \frac{1 - i2\pi s}{1 + (2\pi s)^2} \tag{C.12}$$

which is a special case of the Lorentzian function introduced in Eq. 6.4 (p. 129), with the frequency offset, ν', equal to zero and r, the decay rate, equal to 1. To keep things compact, we will refer to this case of the Lorentzian as $L(s)$. The real and imaginary components represent the absorptive and dispersive line shapes, respectively, that are characteristic of NMR spectra.

Although the sine, cosine and complex exponential (e^{ix}) functions play central roles in Fourier series and transforms, their own transforms do not, strictly speaking, exist. Transforms for these functions can, however, be described as limiting cases, by introducing another function that, again strictly speaking, is not a true function. This not-quite function is variously called the "δ function," the "Dirac δ" or the "impulse symbol," and is usually written $\delta(x)$. One way of visualizing $\delta(x)$ is to consider a variation on the Heaviside Π function, which we will call Π_τ and is defined in terms of the constant $\tau > 0$ so that

$$\Pi_\tau(x) = \begin{cases} 1/\tau & \text{if } |x| \leq \tau/2 \\ 0 & \text{otherwise.} \end{cases}$$

This defines a "top-hat" function with width τ and height $1/\tau$. The integral of this function over all values of x, for any value of τ, is 1. The δ function, $\delta(x)$, can then be defined as the limit of $\Pi_\tau(x)$ as $\tau \to 0$. In this limit, Π_τ is zero for all values of x except zero, and $\Pi_\tau(0) \to \infty$. A not-very-proper definition of $\delta(x)$ can be written as

$$\delta(x) = \begin{cases} \infty & \text{if } x = 0 \\ 0 & \text{otherwise} \end{cases}$$

The impropriety here comes from the fact that ∞ is not really a number and the function is not defined for the one value of x where it is non-zero. Nonetheless, a common graphical representation of $\delta(x)$ is an upward-pointing arrow, or impulse, located at $x = 0$, as shown in Fig. C.8. The δ function, $\delta(x)$, has the property that, for any function, $g(x)$, that is continuous at $x = 0$,

$$\int_{-\infty}^{\infty} \delta(x)g(x)dx = g(0)$$

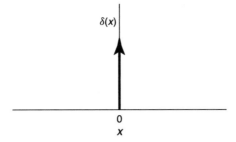

Figure C.8
Graph of the Dirac δ function,
or impulse symbol, $\delta(x)$.

Intuitively, this means that when the function, $g(x)$, is multiplied by $\delta(x)$ over the infinitesimal range of values for which $\delta(x)$ is non-zero, (and the integral of $\delta(x)$ is 1), the integral of the product is just $1 \times g(0) = g(0)$. More generally, if we consider a shifted δ function, $\delta(x - a)$, where a is a real constant, then:

$$\int_{-\infty}^{\infty} \delta(x - a)g(x)dx = g(a)$$

if $g(x)$ is continuous at $x = a$. This, then, allows us to define the Fourier transform of $\delta(x - a)$:

$$\mathcal{F}\big(\delta(x - a)\big) = \int_{-\infty}^{\infty} \delta(x - a)e^{-i2\pi xs}dx = e^{-i2\pi as}$$

For the special case of $a = 0$, this means that the Fourier transform of $\delta(x)$ is just equal to 1, and, by the reciprocal relationship (Eq. C.6), that the transform of the constant 1 is $\delta(x)$, as illustrated in Figs. C.9A and B.

Earlier, however, we argued that a function with a constant value does not have a transform, since the integral of Eq. C.5 does not converge. The two conclusions can be reconciled by noting, again, that $\delta(x)$ represents a limiting case of a rectangle function (Fig. C.7C) that becomes very narrow and very tall. Its transform is the limiting case of a very broad sinc function (Fig. C.7D), which approaches a horizontal line. Conversely, the transform of a very broad sinc function is a very narrow rectangle function, which, in the limiting case, is $\delta(x)$.

Two other cases allow us to derive the limiting Fourier transforms for the cosine and sine functions, as well as for $e^{i2\pi x}$. For the case where $a = 1$,

$$\mathcal{F}\big(\delta(x - 1)\big) = \int_{-\infty}^{\infty} \delta(x - 1)e^{-i2\pi xs}dx = e^{-i2\pi s} = \cos(2\pi s) - i\,\sin(2\pi s)$$

From the reciprocal relationship (Eq. C.6, p. 604), we can write

$$\mathcal{F}(e^{-i2\pi x}) = \delta(-s - 1) = \delta(s + 1) \tag{C.13}$$

For $a = -1$, we have

$$\mathcal{F}\big(\delta(x + 1)\big) = e^{i2\pi s} = \cos(2\pi s) + i\,\sin(2\pi s)$$

and

$$\mathcal{F}(e^{i2\pi x}) = \delta(-s + 1) = \delta(1 - s) \qquad\qquad (C.14)$$

By applying the linearity property of Fourier transforms (Eqs. C.8 and C.9, p. 605) to the results of Eqs. C.13 and C.14, the following Fourier transforms can be derived:

$$\mathcal{F}\big(\cos(2\pi x)\big) = \frac{1}{2}\big(\delta(s + 1) + \delta(s - 1)\big)$$

$$\mathcal{F}\big(\sin(2\pi x)\big) = \frac{i}{2}\big(\delta(s + 1) - \delta(s - 1)\big)$$

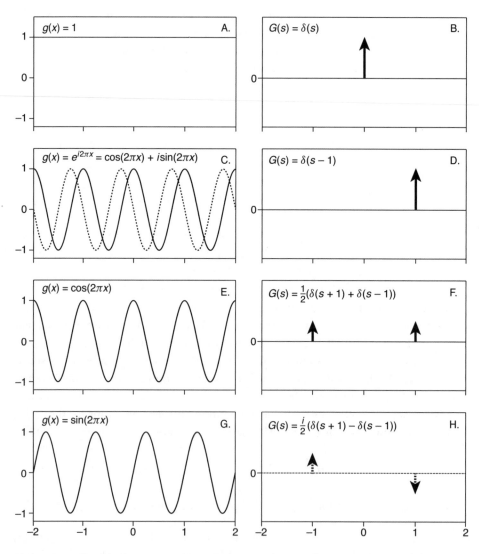

Figure C.9 Graphical representations of some Fourier transform pairs that are defined only as limiting cases and are expressed in terms of the Dirac δ function, or impulse symbol. The left-hand panels contain graphs of four functions, $g(x)$, and the right-hand panels show graphs of the corresponding Fourier transforms, $G(s)$. Solid and dashed curves represent the real and imaginary components of the functions, respectively.

These and other Fourier transform pairs based on the Dirac δ function are represented graphically in Fig. C.9. The result for the Fourier transform of the cosine function makes intuitive sense, since the function is composed of a single frequency component, but is the same for either a positive or negative frequency. Similarly, the function, $\sin(2\pi x)$ represents a single frequency, but is identical to $-\sin(-2\pi x)$. That the transform of the sine function is imaginary reflects the fact that $\sin(x)$ is an odd function.

C.2.2 Some Additional Fourier Transform Relationships

In the previous subsection, we made use of two important properties of Fourier transforms, the reciprocal relationship (Eq. C.6) and the linearity of the transform operation (Eqs. C.8 and C.9). There are several additional theorems that make it relatively easy to extend the result from one function to related ones. In addition to providing a means of deriving Fourier transforms for new functions, the relationships defined by these theorems arise frequently in many applications and can provide important insights into physical phenomena. A few of these theorems are stated here and illustrated with examples. Although proofs are not provided, these relationships can be derived from the definition of the transform (Eq. C.5).

The Similarity Theorem
If $G(s)$ is the Fourier transform of $g(x)$, then the transform of $g(ax)$, where a is a constant, is

$$\mathcal{F}\big(g(ax)\big) = \frac{1}{|a|} G(s/a) \tag{C.15}$$

This theorem represents the basic idea that increased spacings in one domain (time, for instance) are associated with decreased spacings in the transform domain (such as frequency). An example of this relationship is shown in Fig. C.10, for the case of a rectangular function and its transform, the sinc function.

In the case illustrated, x is replaced with $2x$ in $g(x)$, which reduces the width of the rectangle by $1/2$ and increases the width of the sinc function. In the context of a pulse in an NMR experiment, this represents a doubling in the spectrum of frequencies generated. In addition, the height of the transform is reduced by a factor of two, reflecting the fact that energy is now spread over a wider frequency range. Note that the narrowing of the rectangle, without a change in its height, reduces the integral of $g(x)$, so the value of $G(0)$ is reduced proportionally.

A special case of the similarity theorem describes the effect of a change in sign of x:

$$\mathcal{F}\big(g(-x)\big) = G(-s)$$

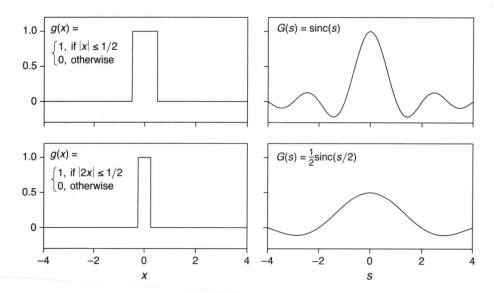

Figure C.10   An example of the application of the similarity theorem, Eq. C.15. Replacing x with $2x$ in $g(x)$ reduces the width of the rectangle functions and increases the width of the transform, as well as reducing the scale of $G(s)$.

For an even function, the transform, which is also even, is unaffected. For other functions, reversing the direction of g along the x-axis reverses the direction of the transform, G, along the s-axis.

The Shift Theorem

If $G(s)$ is the Fourier transform of $g(x)$, then the transform of $g(x - a)$, where a is a constant, is

$$\mathcal{F}\big(g(x - a)\big) = e^{-i2\pi as} G(s) \tag{C.16}$$

$$= \big(\cos(2\pi as) - i\,\sin(2\pi as)\big)G(s) \tag{C.17}$$

The effect of subtracting a positive constant from x in $g(x)$ is to shift the function to the right along the x-axis, whereas adding a positive constant shifts $g(x)$ to the left. A shift in either direction has what may seem a surprisingly dramatic effect on both the real and imaginary components of the transform, as illustrated in Fig. C.11, again for the case of the rectangle and sinc functions.

In this case, x is replaced by $x - 0.5$, shifting the rectangle to the right. As noted earlier, the transform of a real-valued even function is always another even real-valued function. Shifting an even function eliminates its symmetry about $x = 0$ and leads to an imaginary component in the transform.

Recall that any complex number written in the form $e^{i\theta}$ has a modulus of one and that the modulus represents the total magnitude of the real and imaginary parts. Multiplying the original transform, $G(s)$, by the factor $e^{-i2\pi as}$ does not change the modulus of $G(s)$ for any value of s, but does change the relative contributions of the real and imaginary parts, and importantly, this change depends on both a and s.

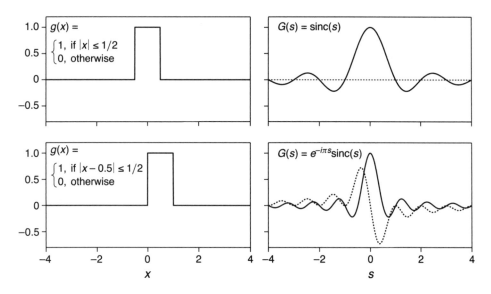

Figure C.11   An example of the application of the shift theorem, Eq. C.17. Replacing x with $x - 0.5$ in $g(x)$ shifts the rectangle function to the right by half its width and modulates the real component of the transform by $\cos(2\pi as)$. The shift also introduces an imaginary component modulated by $\sin(2\pi as)$.

Although it might seem that a constant shift along the x-axis would lead to a simple parallel shift along s, the effect is clearly more complicated, and it may be more easily visualized by considering the specific case of a time-domain function and its frequency-domain transform. A fixed shift in the timing of a series of repeating events will have a larger proportional effect on a high-frequency process than on one with a lower frequency. As a consequence, the phase shift in the transform changes as the frequency moves from zero in either direction.

Amidst all of the changes, however, notice that, in the case illustrated in Fig. C.11, the value of $G(0)$ is unchanged, since the integral of the rectangle function is the same. Conversely, the total integral of the transform is the same as before, since the value of $g(0)$ is still 1. If, however, $g(x)$ is shifted further to the right, so that $g(0) = 0$, then the integrals of both the real and imaginary parts of the transform will become 0.

The Cosine Modulation Theorem

If $G(s)$ is the Fourier transform of $g(x)$, then the transform of $\cos(2\pi vx)g(x)$, where v is a constant, is

$$\mathcal{F}\big(\cos(2\pi vx)g(x)\big) = \frac{1}{2}G(s - v) + \frac{1}{2}G(s + v) \qquad \text{(C.18)}$$

An example of this theorem is illustrated in Fig. C.12, with the exponential decay function, e^{-x}, for positive x. As discussed earlier (page 608), the transform of this function is a Lorentzian function, written here as $L(s)$, with both real and imaginary components.

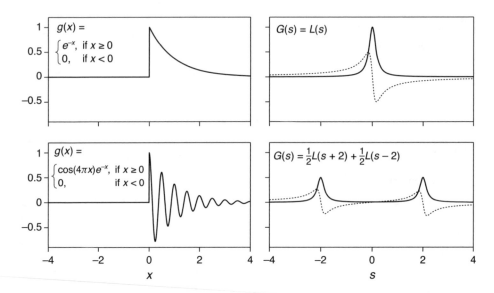

Figure C.12 An example of the application of the cosine modulation theorem, Eq. C.18. The transform of an exponential decay, for $x > 0$, is the Lorentzian function centered at $s = 0$. Multiplying the exponential function by a cosine function in the x-domain splits the Lorentzian peak, $L(s)$, into two peaks in the s-domain, each shifted from $s = 0$ by the frequency of the cosine function.

Multiplying the decay function by $\cos(2\pi \nu x)$ results in a damped oscillation, as seen in the free induction decay signal from a single spin population. The transform of the damped cosine is the sum of two Lorentzian functions, each with real and imaginary parts, with midpoints at $\pm\nu$. The two peaks in the frequency domain reflect the fact that the same function in the time domain can arise from either the positive or negative frequency, since $\cos(2\pi \nu x) = \cos(-2\pi \nu x)$.

The modulation theorem, which treats the product of a general function, $g(x)$, and the cosine function, $\cos(2\pi \nu x)$, is one example of a much more broadly applicable theorem regarding the transforms of products, referred to as the convolution theorem. Convolutions and their relationship to Fourier transforms are discussed next.

C.3   Convolution

Convolution is an operation that takes two mathematical functions, designated here as $f(x)$ and $g(x)$, and generates a third, $h(x)$. As the name suggests, the two starting functions are merged in a way that can be a bit difficult to describe in words, but is often more easily visualized graphically. The symbol used to indicate the operation of convolution is the asterisk, so the relationship between $f(x)$, $g(x)$ and $h(x)$ is written as

$$h(x) = f(x) * g(x)$$

Figure C.13
Graphs of the Lorentzian
(Eq. C.20, solid curve) and
rectangular (Eq. C.21, dashed
curve) functions, both centered
at zero. The values of the
functions have been scaled
arbitrarily for convenience of
illustration.

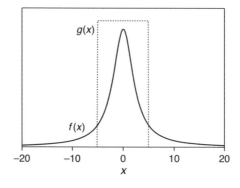

Importantly, convolution is *not* equivalent to simple multiplication. Rather, the operation is defined by an integral:

$$h(x) = \int_{-\infty}^{\infty} f(u)g(x-u)du \qquad (C.19)$$

The expression in Eq. C.41 introduces another variable, u, but this variable is used only for the integration and disappears when the convolution is evaluated for a specific value of x.

Convolution is most readily explained using some simple examples. We start with $f(x)$ defined as the real component of a Lorentzian function centered at zero:

$$f(x) = \frac{r}{r^2 + 4\pi^2 x^2} \qquad (C.20)$$

where r determines the width of the Lorentzian peak. And we choose for $g(x)$ a rectangular function of width w, also centered at zero. The rectangle function can be written as $\Pi(x/w)$, where $\Pi(x)$ is the Heaviside function introduced earlier, so that

$$g(x) = \Pi(x/w) = \begin{cases} 1 & \text{if } |x| \leq w/2 \\ 0 & \text{otherwise.} \end{cases} \qquad (C.21)$$

The functions $f(x)$ and $g(x)$, are plotted in Fig. C.13 for $r = 15$ and $w = 10$.

To evaluate $h(x)$ at a specific value of x, following Eq. C.19, we plot the two functions, $f(u)$ and $g(x-u)$, as shown in Fig. C.14 for the case of $x = 10$. Note that $f(u)$ plotted as a function of u is just like $f(x)$ plotted as a function of x, but the function, $g(x-u)$, for a given value of x is shifted to the right by the amount x, relative to the plot of $f(x)$ versus x in Fig. C.13. We then form the product, $f(u)g(x-u)$, and integrate this function from $-\infty$ to ∞. In this case, the integral has a simple form, because $g(x-u)$ is unity between $u = x - w/2$ and $u = x + w/2$ and zero outside of this range. The integrated region is shaded in Fig. C.14, and the convolution is given as

$$h(x) = \int_{x-w/2}^{x+w/2} f(u)du$$

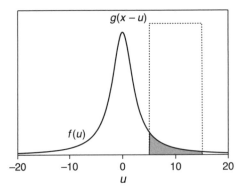

Figure C.14

Graphs of the functions, $f(u)$ (solid curve) and $g(x - u)$, evaluated for $x = 10$ (dashed curve). The shaded region represents the integral of $f(u)g(x - u)$ with respect to u. This integral is the value of $h(x)$ for $x = 10$.

The process of evaluating the convolution for all values of x can be envisioned as sliding the function $g(u)$ along the u-axis and calculating the integral of $f(u)g(x - u)$ for each point, as illustrated for a few values of x in Fig. C.15. The right-hand panels in Fig. C.15 show plots of $h(x)$ as a function of x, with the specific values of x corresponding to the adjacent left-hand panels indicated by a filled circle. From these graphs, it should be apparent that the value of $h(x)$ for each x is proportional to the average of $f(x)$ over the range from $x - w/2$ to $x + w/2$. Thus, the convolution represents a kind of moving, or rolling average. For this particular convolution (using the rectangle function), the values of $f(x)$ are weighted equally in the average, but convolution with other functions will lead to other weightings, as illustrated in the next example.

Another feature of the convolution can be appreciated by comparing the shapes of $g(x)$, $f(x)$ and $h(x)$: The convolution represents a blending of the shapes of the input functions. The function, $h(x)$, is both more squat, or boxlike, than the Lorentzian function and more curvaceous and center-weighted than the rectangle function.

Other features of convolution become apparent when we look at functions that are not so symmetrical as the Lorentzian and rectangle functions. As an example, consider a function with the shape of a sawtooth, such as the one plotted in Fig. C.16, along with the Lorentzian function. The sawtooth function is defined as:

$$g(x) = \begin{cases} x/w + 1/2 & \text{if } |x| \leq w/2 \\ 0 & \text{otherwise} \end{cases} \tag{C.22}$$

where w again represents the width of the function.

As before, the convolution is calculated by considering the functions, $f(u)$ and $g(x - u)$, as plotted in Fig. C.17. The first thing to notice in Fig. C.17 is that the sawtooth function is not only displaced, but its shape is reversed from the way it was when $g(x)$ was plotted in Fig. C.16. This is because of the negative sign introduced in $g(x - u)$. The sign change is somewhat arbitrary, since the convolution could be defined in terms of $g(u - x)$, preserving the orientation of asymmetric functions, but the standard definition leads to some convenient features.

Also note that the shaded region shown in Fig. C.17, which represents the integral of $f(u)g(x - u)$ with respect to u, no longer matches the shape of $f(u)$ where the two

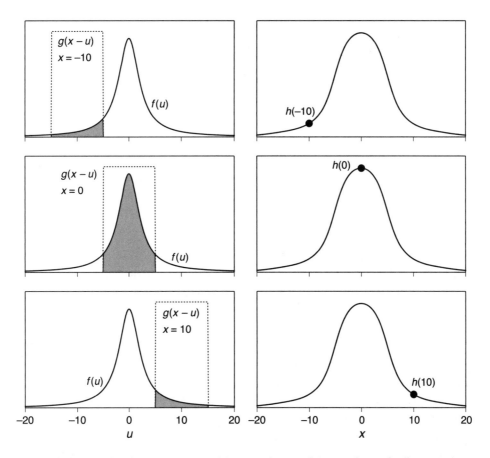

Figure C.15   Graphical representation of the convolution of $f(x)$ and $g(x)$ for the case where $f(x)$ and $g(x)$ are Lorentzian and rectangle functions, respectively. The left-hand panels show plots of $f(u)$ (solid curves) and $g(x - u)$ (dashed curve) for x equal to -10, 0 and 10, as indicated. The shaded regions represent the integrals of $f(u)g(x - u)$ with respect to u. The right-hand panels show plots of $h(x)$ (solid curves). In each right-hand panel, a filled circle indicates the value of $h(x)$ for the value of x in the corresponding left-hand panel.

Figure C.16

Graphs of the Lorentzian (solid curve) and sawtooth functions (Eq. C.22 with $w = 10$, dashed curve) functions, both centered at zero. The values of the functions have been scaled arbitrarily for convenience of illustration.

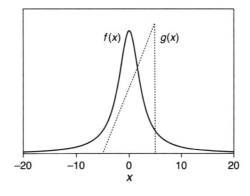

Figure C.17
Graphs of the functions,
$f(u)$ (solid curve) and
$g(x - u)$, evaluated for $x = 7.5$
(dashed curve), with $g(x)$
representing the sawtooth
function (Eq. C.22). The
shaded region represents the
integral of $f(u)g(x - u)$ with
respect to u.

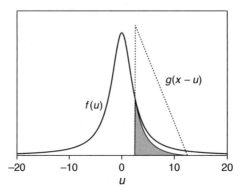

functions overlap. Instead, $f(u)g(x - u)$ in this region decreases more rapidly than does $f(u)$, following the leading edge of the reversed sawtooth. The integral again is proportional to an average of the values of $f(u)$ within the overlap region, but now it is a weighted average, in which the values of $f(u)$ towards the left are weighted more heavily.

In Fig. C.18, convolution integrals and the resulting convolution function, $h(x)$, are shown for three values of x, for the case of the sawtooth function. As in the first case, the convolution represents a merging of the shapes represented by the two functions. The convolution, $h(x)$, is again broader than the original Lorentzian function and has also has acquired some of the asymmetry of the sawtooth function. The stronger weighting of $f(u)$ values from the left-hand side of the overlap region of each integral tends to "push" the mass of the curve to the right, causing the sharp fall off of the original sawtooth function to reappear, in a smoother form, in the convolution.

Note that the smoothed sawtooth appears in $h(x)$ in the same orientation as the original function, with the steeper fall off to the right. This is due to the reversal of $g(x)$ in the convolution integral and is one reason for defining the convolution in this way. A more important reason is that the reversal of $g(x)$ ensures that the convolution operator is commutative—that is,

$$f(x) * g(x) = g(x) * f(x)$$

Convolution also satisfies the distributive property with respect to addition, so that

$$f(x) * \big(g(x) + h(x)\big) = f(x) * g(x) + f(x) * h(x)$$

where $h(x)$, in this case, is another independent function. It should be noted at this point, however, that convolution is not defined for all function pairs. The convolution can only be calculated if the integral of Eq. C.19 converges to a finite value.

Convolutions of functions based on the Dirac δ function, such as those illustrated in Fig. C.9, arise frequently and have particularly simple forms. Consider, for instance, the function $\delta(x - a)$, which represents an impulse at $x = a$. The convolution of this

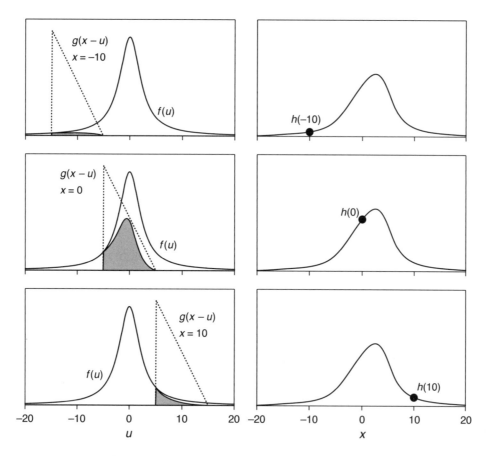

Figure C.18 Graphical representation of the convolution of the Lorentzian and sawtooth functions ($f(x)$ and $g(x)$, respectively). The left-hand panels show plots of $f(u)$ (solid curves) and $g(x-u)$ (dashed curves) for x equal to -10, 0 and 10, as indicated. The shaded regions represent the integrals of $f(u)g(x-u)$ with respect to u. The right-hand panels show plots of $h(x)$ (solid curves). In each right-hand panel, a filled circle indicates the value of $h(x)$ for the value of x in the corresponding left-hand panel.

function with another function, $f(x)$, can be written as

$$f(x) * \delta(x-a) = \int_{-\infty}^{\infty} f(u)\delta(x-a-u)du \qquad (C.23)$$

Recall from the earlier discussion of the δ function that

$$\int_{-\infty}^{\infty} g(x)\delta(x-a)dx = g(a) \qquad (C.24)$$

Replacing $g(x)\delta(x-a)dx$ in Eq. C.24 with $f(u)\delta(x-a-u)du$, and recognizing that $\delta(x-a-u) = \delta(u-(x-a))$, yields the following result:

$$f(x) * \delta(x-a) = \int_{-\infty}^{\infty} f(u)\delta(u-(x-a))du = f(x-a) \qquad (C.25)$$

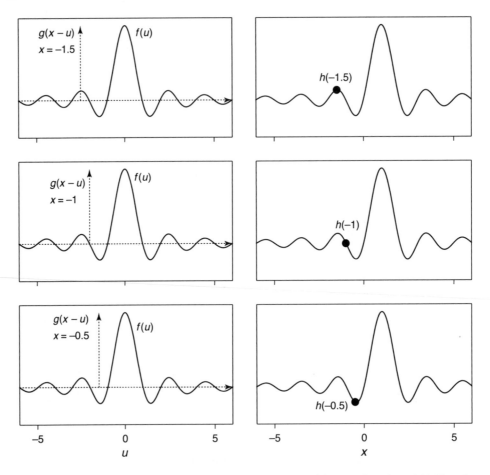

Figure C.19 Graphical representation of the convolution of the sinc functions ($f(x)$) and a δ function, $g(x) = \delta(x - 1)$. The left-hand panels show plots of $f(u)$ (solid curves) and $g(x - u)$ (dashed curves) for x equal to -1.5, -1 and -0.5, as indicated. In this case, the integral of $f(u)g(x - u)$ with respect to u is $g(x - 1)$. The right-hand panels show plots of $h(x) = f(x) * g(x)$ (solid curves). In each right-hand panel, a filled circle indicates the value of $h(x)$ for the value of x in the corresponding left-hand panel.

Thus, the convolution of $g(x)$ by $\delta(x - a)$ has the simple effect of shifting $f(x)$ to the right by the distance, a.

An example of a convolution with a δ function is shown graphically in Fig. C.19. In this example, $f(x)$ is the sinc function and $g(x) = \delta(x - 1)$. As the δ function is moved along the u-axis, its only non-zero value is where $u = x - 1$, and the integral of $f(u)g(x - u)$ with respect to u is $f(x - 1)$. The convolution, $h(x)$, is therefore $f(x - 1) = \text{sinc}(x - 1)$.

The Convolution Theorem

One reason that convolutions arise frequently, particularly in signal-processing applications, is an intimate relationship with Fourier transformations. If, as before, we use the symbol $\mathcal{F}$ to indicate the Fourier transform of a function, and both $f(x)$ and

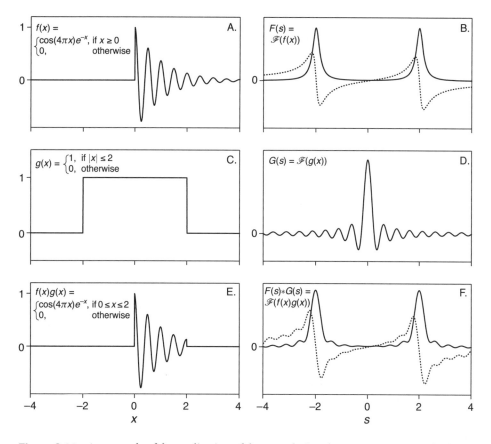

Figure C.20 An example of the application of the convolution theorem, Eq. C.26. Multiplying a damped cosine function ($f(x)$, panel A) by a rectangle function ($g(x)$, panel C) results in truncation of the damped cosine function (panel E). The transform of the truncated function (panel F) can be calculated as the convolution of the transforms of the damped cosine and rectangle functions ($F(s)$ and $G(s)$, plotted in panels B and D, respectively).

$g(x)$ have Fourier transforms, $F(s)$ and $G(s)$, the convolution theorem states that

$$\mathcal{F}\big(f(x)g(x)\big) = F(s) * G(s) \qquad (C.26)$$

That is, the Fourier transform of the product, $f(x)g(x)$, is the convolution of the transforms of the individual functions. Conversely, the transform of a convolution of two functions can be calculated as the product of the transforms:

$$\mathcal{F}\big(f(x) * g(x)\big) = F(s)G(s)$$

In a given case, it may be easier to calculate the Fourier transform of a product of functions as a convolution of the individual transforms, or vice versa. Furthermore, knowledge of a few transforms can be readily extended, using the convolution theorem, to gain insight into the transforms of more complex functions.

An application of the convolution theorem is illustrated in Fig. C.20, which shows the effect of multiplying a damped cosine function, with frequency $v = 2$, by a rectangle function. As shown earlier, the transform of the damped cosine function is the sum of two Lorentzian functions, centered at $\pm \nu$ (panel B), and the transform of the rectangle function is a sinc function (panel D). The transform of the truncated damped cosine function is the convolution of the transforms plotted in panels B and D, as shown in panel F of the figure.

As in the earlier examples, convolution has the effect of merging the characteristics of the two functions. The convolved transform, $F(s) * G(s)$, resembles $F(s)$ with respect to the positions of the two sets of peaks, as well as the approximate shapes of the absorptive and dispersive components, and also has the "wiggles" characteristic of the sinc function. Note also that truncation of the damped cosine function increases the width of the peaks. If $f(x)$ represents a time-domain signal, such as an idealized FID from a single spin population in an NMR experiment, the broadening of the peaks can be understood as the consequence of recording the signal for a shorter time, which reduces the precision with which the frequency is defined. The sinc wiggles, on the other hand, are due to the introduction of a sharp cutoff in the signal, which leads to additional frequency components that slowly diminish as the absolute values of $\nu - s$ and $\nu + s$ increase.

Another important application of the convolution theorem in NMR is its use in calculating the frequency spectra generated by rf pulses with different shapes, as described in Section 6.9. In this application, the pulse is described as the product of the rf signal (a cosine or sine function) and a function that defines the overall shape of the pulse, such as a rectangle or a smoother shape designed to minimize the presence of high-frequency components.

C.4   Discrete Fourier Transforms

The continuous Fourier transform discussed in the previous sections is useful for describing general properties of mathematical functions important in NMR and other fields, but most day-to-day work is done with experimental data measured at discrete intervals. Even when considering continuous functions, evaluating the Fourier integrals analytically may be difficult or impossible, so numerical methods must be used. For these purposes, the Fourier transform must be reformulated somewhat to deal with discrete values. Although most of the concepts introduced above for the continuous Fourier transform carry over to the discrete approximation, additional considerations arise when a finite number of values separated by discrete intervals are introduced.

C.4.1   Definition of the Discrete Fourier Transform

Suppose that instead of the continuous function, $g(x)$, we have a series of N evenly spaced values of x, which we will call x_j, and a corresponding set of values called g_j, with $j = 0$ to $N - 1$. The values of g_j might be measurements from an instrument, such as an NMR spectrometer, or they might be calculated from a mathematical function. If the first value of x is x_0 and the interval between x-values is Δx, then $x_j = x_0 + j \Delta x$. Without imposing other restrictions for the moment, we can approximate the integral of Eq. C.5 as follows:[3]

$$G(s) = \int_{-\infty}^{\infty} g(x)e^{-i2\pi xs}dx$$

$$\approx \sum_{j=0}^{N-1} g_j e^{-i2\pi x_j s}\Delta x$$

$$= \sum_{j=0}^{N-1} g_j e^{-i2\pi(x_0 + j\Delta x)s}\Delta x$$

$$= e^{-i2\pi x_0 s}\Delta x \sum_{j=0}^{N-1} g_j e^{-i2\pi j\Delta xs} \tag{C.27}$$

Note that we have gone from a integral over an infinite range to a finite sum. Not only are the values of x_j finite, but there are a finite number of them.

The sum shown in Eq. C.53 is relatively straightforward to evaluate (at least in principle, for a modest value of N), and it might seem that we could use it for any value of s, just as there are no explicit boundaries on s for the continuous transform. In fact, however, by limiting the number of values of x, we have placed a limit on the number of *meaningful* values of s for which the sum can be evaluated. One constraint on the acceptable values for s comes from the sampling theorem, introduced in Chapter 6 (page 133). The theorem is usually presented in terms of time-domain signals and frequencies, but it can be applied to other areas, such as digital imaging, as well. For a specified sampling interval in the x-domain, Δx, the theorem defines a critical frequency, sometimes called the Nyquist frequency, f_c:

$$f_c = \frac{1}{2\Delta x} \tag{C.28}$$

The sampling theorem states that signals that contain only frequencies that lie within the range $-f_c$ to f_c can be fully described by samples separated by Δx. If, however, there are components with frequencies outside of this range, they will not be properly reconstructed by a Fourier transform (or any other method). So, it makes little sense to calculate the transform for frequencies outside of this range. Within the constraints of the sampling theorem, it might still seem that we could evaluate Eq. C.27 for values of s spaced as closely as we like, but if N values of g are used as inputs, there should only be N parameters derived from them, so the spacing is also limited.

Rather than dealing explicitly with values of, for instance, x and s, the standard definition of the discrete Fourier transform (DFT) considers only the discrete input values, g_j, and an equal number of output values of the transform, G_k. For a series of

---

[3] This is not generally the best way to numerically evaluate an integral, but it is the simplest and leads us towards the discrete Fourier transform.

N values of g_j, the discrete values of G_k are calculated as

$$G_k = \sum_{j=0}^{N-1} g_j e^{-i2\pi jk/N} \tag{C.29}$$

Although not absolutely necessary, it is convenient for purposes of notation to make N an even number. A further restriction arises from the algorithm most commonly used to calculate DFTs, the fast Fourier transform (FFT), which is most efficient when N is a power of two.

The inverse discrete Fourier transform will regenerate the sequence g_j from its transform, G_k, and is defined as

$$g_j = \frac{1}{N} \sum_{k=0}^{N-1} G_k e^{i2\pi jk/N} \tag{C.30}$$

As discussed further below, the values of the index, k, can be set either to 0 to $N-1$, as indicated in the equation above, or, equivalently, to $-N/2$ to $N/2 - 1$, as is more common. The factor $1/N$ in the inverse transform is necessary to keep g_j and G_k from growing larger with each cycle of transformation, but some definitions include this term in the forward transform and some in the inverse.

The definition of the DFT given by Eq. C.29 bears some obvious similarity to the approximation to the continuous Fourier transform given in Eq. C.27, but it's not identical. First, the factor, $e^{-i2\pi x_0 s} \Delta x$, preceding the sum has been dropped. The first part of this factor, $e^{-i2\pi x_0 s}$, is a phase term, with modulus always equal to one. Thus, it reflects the relative contributions of real and imaginary components in the transform. Dropping this term is equivalent to assuming that the first value of x, x_0, is zero. The factor Δx is a normalization factor that makes Eq. C.27 approximately equal to the continuous transform, irrespective of the sampling interval in the x-domain. Because of its absence in Eq. C.29, G_k is not normalized and its absolute value generally increases when the sampling interval is decreased. The other important difference between Eqs. C.27 and C.29 is that the exponents in the formal definition of the discrete transform do not contain explicit values of x and s, but are represented by the indices j and k. Thus, the DFT can be calculated without reference to variables defining the domains of g and G, such as x and s, but these variables can be reintroduced as appropriate for specific applications, as discussed below.

The discrete Fourier transform is also very closely related to the Fourier series representation of a continuous function, as defined in Eq. C.3 (p. 598), except that it is composed of a finite number of terms. Like the Fourier series, the DFT is a sum of cosine and sine functions with discrete frequencies and can be viewed as an approximation of a continuous function, but it need not be directly related to one.

An important feature of Eq. C.29 is that it defines a periodic sequence of values for G_k, if k is not constrained to N sequential values. Consider what happens if we replace k with $k + mN$, where m is a positive or negative integer:

$$G_{k+mN} = \sum_{j=0}^{N-1} g_j e^{-i2\pi j(k+mN)/N} = \sum_{j=0}^{N-1} g_j e^{-i2\pi jk/N} e^{-i2\pi jm}$$

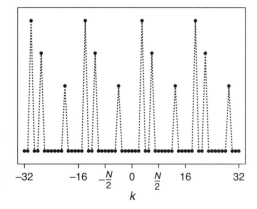

Figure C.21

Periodicity of a hypothetical discrete Fourier transform with $N = 16$. For any value of k, G_k, calculated from Eq. C.29, is equal to G_{k+mN}, for any integer value of m.

Because j and m are both integers, their product is an integer, and $e^{-i2\pi jm}$ is always equal to one. Thus, G_{k+mN} is always equal to G_k. The periodicity of G_k is illustrated for a hypothetical example, with $N = 16$, in Fig. C.21.

As a consequence of this periodicity, there are only N unique values of G_k, and the choice of which values of k to use is somewhat arbitrary. G_0, however, always has special significance. Examination of Eq. C.29 shows that G_0 is the simple sum of g_0 through g_{N-1}, much as $G(0)$ in the continuous transform is the integral of $g(x)$ over all values of x. In the case of a time-domain signal, G_0 represents the zero-frequency, or time-independent, component. Although 0 to $N - 1$ might seem like the obvious choice for the range of k, a consideration of physical applications in which both positive and negative frequencies are relevant leads to the more common practice of placing G_0 near the center of the sequence with the range of k usually defined as

$$-\frac{N}{2}, -\frac{N}{2} + 1, -\frac{N}{2} + 2, \ldots, -1, 0, 1, \ldots, \frac{N}{2} - 2, \frac{N}{2} - 1$$

This sequence is illustrated in Fig. C.22, for the same hypothetical example shown in Fig. C.21. Note that $k = -N/2$ (–8 in the example) is included, but $k = N/2$ (8) is *not*, since $G_{N/2}$ would be term $N + 1$ in the sequence, and $G_{N/2} = G_{-N/2}$.

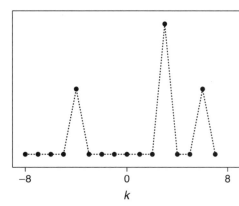

Figure C.22

A hypothetical discrete Fourier transform with $N = 16$ and k restricted to the values between $-N/2$ and $(N/2 - 1)$, for the same values as plotted in Fig. C.21.

The discrete Fourier transform also implies a periodic series of values g_j for j less than zero or greater than $N-1$, even though it is based on the finite range from 0 to $N-1$. Using the same logic as above, if we replace j with $j+mN$, where m is an integer, Eq. C.29 becomes

$$G_k = \sum_{j=0}^{N-1} g_{(j+mN)} e^{-i2\pi(j+mN)k/N}$$

$$= \sum_{j=0}^{N-1} g_{(j+mN)} e^{-i2\pi jk/N} e^{-i2\pi km}$$

$$= \sum_{j=0}^{N-1} g_{(j+mN)} e^{-i2\pi jk/N}$$

Only if g_j is periodic, so that $g_{(j+mN)} = g_j$, will the output of the transform be the same for any integer, m. However, all that the transform algorithm knows about are the actual input values, so that we can think of the output as the transform of either a unique sequence or a periodic function based on this sequence.

When specific x-values are to be associated with the input values to the transform, g_j, they are, by convention, defined so that

$$x_j = j\Delta x$$

where Δx is the sampling interval. By this definition, the values of x range from 0 to $(N-1)\Delta x$. If the x-values in the original data do not begin with zero, they may be shifted appropriately, but this will introduce a change in the relative magnitudes of the real and imaginary components, as per the shift theorem, a point that we will return to later. (The phase shift corresponds to the term, $e^{-i2\pi x_0 s}$, which was factored out of Eq. C.27 and is not included in Eq. C.29.)

Values of s associated with G_k for a specified value of Δx are given by

$$s_k = \frac{k}{N\Delta x}$$

The values of s_k then range from $-1/(2\Delta x)$ to $(N/2-1)/(N\Delta x)$, and the interval between sequential values of s_k is $\Delta s = 1/(N\Delta x)$.

C.4.2   Some DFT Examples

As an example of a discrete Fourier transform of a periodic function, Fig. C.23 shows the result of applying the DFT to a series of 32 values sampled from the cosine function, $\cos(2\pi x)$, for $x = 0$ to (almost) 4. For this sequence, Δx was $4/32 = 0.125$, so the function was sampled for $x = 0$ to $(N-1)\Delta x = 3.875$. The transform also contains 32 values, with $k = -16$ to 15, corresponding to the range in the s-domain from $-1/(2\Delta x) = -4$ to $(N/2-1)/(N\Delta x) = 3.75$, and the interval between points in the transform is $\Delta s = 0.25$.

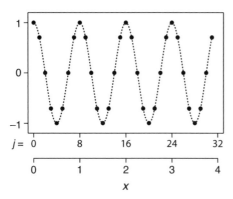 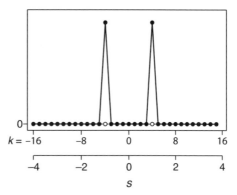

Figure C.23   Discrete Fourier transform of values sampled from the function, $\cos(2\pi x)$. The cosine function is sampled at $N = 32$ discrete values, leading to 32 values for the transform, calculated according to Eq. C.29. The left-hand panel shows the cosine function as a dashed curve, with the sampled values indicated by the filled circles. The right-hand panel shows the DFT of the sampled values, with the real components shown as filled circles and connected with solid line segments, and the imaginary components shown as open circles connected with dashed lines. The horizontal axes are labeled with both the indices of the discrete values, j and k, and with the corresponding values of the variables, x and s.

Recall that the continuous Fourier transform of the function, $\cos(2\pi x)$, is represented by Dirac δ functions at $s = -1$ and 1 (Fig. C.9F), and that the transform has only a real component because the cosine function is even, in the sense that $g(-x) = g(x)$. These features are recapitulated in the DFT, but the values of the DFT at $s = \pm 1$ are finite, and the resolution in the s-domain is limited by the discrete values at which the transform is evaluated.

Fig. C.24 shows the DFT of the same function, but sampled at smaller intervals. In this case, there are 64 samples covering the same range of x, with $\Delta x = 4/64$. The smaller interval between sampled x-values leads to a wider range of values in the s-domain, from -8 to 7.75, even though the total range of x-values has remained the same. In this case, there are no higher-frequency components in the original function, and the transform is equal to zero in regions covering the extra frequency range. Also, the spacing between values in the s-domain, $\Delta s = 1/(N\Delta x)$, is unchanged by the smaller sampling interval in the x-domain, so that if the s-scale of Fig. C.24 were to be expanded to cover only the range from -4 to 3.75, it would be indistinguishable from the one shown in Fig. C.23.

To increase the resolution in the s-domain, the total *range* of x-values must be increased, as illustrated in Fig. C.25, where the same function is sampled over the range of $x = 0$ to 7.75. The number of samples is $N = 32$, as in Fig. C.23, so that Δx is now 0.25. Increasing the range of x-values sampled, while keeping N the same, reduces both the range of s-values covered in the transform and the spacing between these values. In terms of a frequency spectrum, this represents higher resolution, at the cost of a narrower frequency range.

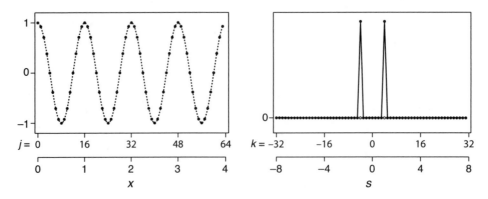

Figure C.24 Discrete Fourier transform of values sampled from the function, $\cos(2\pi x)$, as in Fig. C.23, but with the cosine function sampled at $N = 64$ discrete values.

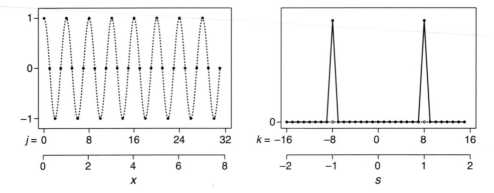

Figure C.25 Discrete Fourier transform of values sampled from the function, $\cos(2\pi x)$, for $x = 0$ to 7.75. As in Fig. C.23, $N = 32$, but Δx is increased from 4/32 to 8/32 in order to cover twice the range of x-values. The spacing of values in the s-domain is decreased from $1/(N\Delta x) = 1/4$ to 1/8, representing an increase in resolution, but the range of s values in the transform is reduced.

As another, somewhat more complicated, example of a discrete Fourier transform, consider the rectangle function, $\Pi(x)$, introduced earlier. Recall that the continuous Fourier transform of this function is a real function, $\text{sinc}(s)$ (Fig. C.7C and D). The left-hand panel of Fig. C.26 shows a graph of $\Pi(x)$ for $x = 0$ to 4, sampled at 32 points, and the right-hand panel shows the DFT. In this case, the DFT is not quite what we might have expected by analogy with the continuous Fourier transform. Although the real component of the transform approximates the sinc function, there is an imaginary component that is not found in the continuous transform.

Recall that the transform of an even real-valued function, such as either the cosine function or the rectangle function, is always an even real-valued function. The imaginary component of the transform shown in Fig. C.26 arises because sampling the rectangle function at only positive values of x eliminates its symmetry about $x = 0$. This problem does not arise with the cosine function discussed above, and the

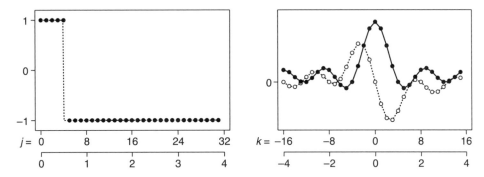

Figure C.26   Discrete Fourier transform of values sampled from the rectangle function, $\Pi(x)$, for $x = 0$ to 2. The function is sampled at $N = 32$ discrete values. The left-hand panel shows the rectangle function as a dashed curve, with the sampled values indicated by the filled circles. The right-hand panel shows the DFT of the sampled values, with the real components shown as filled circles linked with solid lines, and the imaginary components shown as open circles connected with dashed lines.

difference between the two examples is a consequence of the implicit periodicity of g_j and how this is expressed in the two cases. This difference is illustrated in Fig. C.27. In both panels of the figure, the shaded rectangle indicates the values for x (and j) for which the functions are sampled (as indicated by the filled circles), and the implied periodic functions extending beyond the sampled regions are indicated by dashed curves. In each case, the periodicity is described by the relationship between points that lie outside of the sampled region ($j < 0$ or $j \geq N$) and those within the sampled region:

$$g_{(j+mN)} = g_j$$

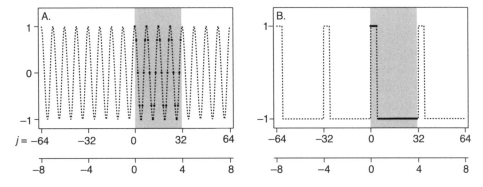

Figure C.27   Implicit periodicity of discrete data sampled from the cosine function (panel A) and the rectangle function (panel B), as in Figs. C.23 and C.26, with the range of j extended from $j = -2N$ to $2N - 1$. In each case, $N = 32$ and $\Delta x = 1/8$. The sampled values are indicated by filled circles and the implicit periodic function are represented by dashed curves. The sampled regions of the x-domain are highlighted by the shaded boxes.

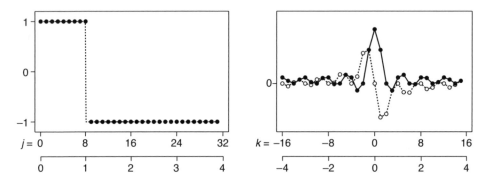

Figure C.28   Discrete Fourier transform of values sampled from a rectangle function for which the function is 1 for $x = 0$ to 1 and 0 elsewhere. As in Fig. C.26, the function is sampled at $N = 32$ discrete values.

where m is an integer. The value g_{-1}, for instance, is equal to g_{N-1}. For the cosine function, this relationship leads to the result for an even function, $g_{-1} = g_1$, because $g_{N-1} = g_1$. For the sampled rectangle function, however, $g_1 = 1$, whereas $g_{-1} = g_{N-1} = 0$.

There are two ways (which are formally equivalent) in which the asymmetry of the sampled rectangle function can be addressed, so as to generate a result consistent with the continuous Fourier transform of the Heaviside Π function. One approach involves manipulating the transform, while the other changes the input data. In either approach, we also need to address the fact that the width of the rectangle sampled for the DFT is only 1/2, representing $0 \le x \le 1/2$, whereas the width of the function used for the continuous transform is 1, for $-1/2 \le x \le 1/2$. In the first method, we will begin by making the function 1 for x between 0 and 1, as shown in Fig. C.28. As expected from the similarity theorem for the continuous Fourier transform, doubling the width of the non-zero portion of g_j has the effect of narrowing the features in the transform. However, the imaginary component of the transform still distinguishes the DFT from the continuous form. In fact, this DFT represents an approximation of the continuous transform shown in Fig. C.11 for the rectangle function generated by shifting $\Pi(x)$ to the right by 1/2 unit.

What we now need to do is create the transform expected if $g(x)$ were shifted to the left along the x-axis by 1/2 unit. For this we can utilize the shift theorem discussed on page 612, but rewritten for the discrete transform. If the sequence G_k is the transform of the sequence g_j, and the first x-value is to be shifted from 0 to x_0, then the DFT of the shifted function is given by

$$G_k e^{-i2\pi x_0 k/(N\Delta x)} = e^{-i2\pi x_0 k/(N\Delta x)} \sum_{j=0}^{N-1} g_j e^{-i2\pi jk/N} \tag{C.31}$$

Note that if $x_0 = 0$ or an integer multiple of $N\Delta x$, then the exponential equals 1, and the shift has no effect. In this instance, we want to make $x_0 = -1/2$, and $N\Delta x = 32/8 = 4$. Multiplying the transform illustrated in the right-hand panel of Fig. C.28

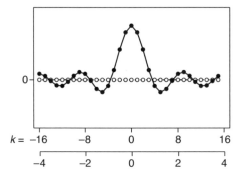

Figure C.29   Discrete Fourier transform of the rectangle function, from Fig. C.28, after shifting x_0 from 0 to $-1/2$ to center the rectangle at $x = 0$. The original DFT was multiplied by the phase factor indicated in Eq. C.31. The real components of the transform are indicated by filled circles and the imaginary components by open circles.

by the phase term, $e^{-i2\pi(-1/2)k/4}$, yields the result shown in Fig. C.29. After the phase correction corresponding to the shift in the x-domain, the DFT has only a real component and approximates the sinc function, as expected from the continuous Fourier transform.

The second approach to "fixing" the DFT of the rectangle function, so that it resembles the Fourier transform of a rectangle centered at $x = 0$, is to modify the input sequence, g_j, so that it correctly represents a periodic rectangle function centered at $j = mN$, where m is an integer. This modification is illustrated in Fig. C.30. The dashed curves in both panels of Fig. C.30 represent implicit periodic functions

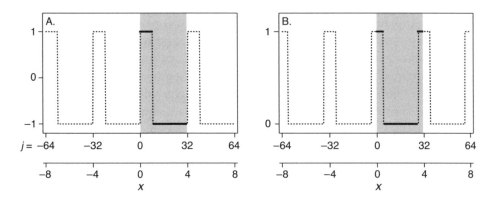

Figure C.30   Implicit periodicity of discrete data sampled from rectangle functions of width 1. In panel A, the data are sampled from a rectangle centered at $x = 0.5$, and the implied periodic function is composed of rectangles centered at points where $j = 4 + mN$, where m is an integer. In panel B, the periodic function is shifted along the x-axis by 0.5 units to the left, so that the rectangles are centered a points where $j = mN$. Data sampled from the periodic function in panel A corresponds to the input for the DFT illustrated in Fig. C.28. The sampled regions of the x-domain are highlighted by the shaded boxes.

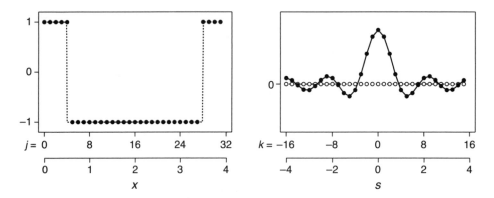

Figure C.31   Discrete Fourier transform of values sampled from a shifted periodic rectangle function, as shown in panel B of Fig. C.30. The rectangle of width 1 is split between the left- and right-hand sides of the sampled region. The right-hand panel shows the DFT of the sampled values, with the real components shown as filled circles linked with solid lines, and the imaginary components shown as open circles connected with dashed lines.

corresponding to discrete data sampled for $j = 0$ to $N - 1$. Panel A of the figure corresponds to the rectangle function of width 1 sampled to generate the DFT shown in Fig. C.28. In panel B, the implicit periodic function has been shifted to the left by 1/2 unit, so that periodic rectangles are centered at $x = 0$ and $x = mN\Delta x$, where m is an integer. If we now examine the data to be sampled for the discrete transform (shaded in the figure), the left-hand part of this region is the same as in Fig. C.26, where $g_j = 1$ for the region corresponding to $x \leq 1/2$, but the right-hand region now corresponds to the left-hand side of the rectangle centered at $x = 0$.

Fig. C.31 shows the DFT of the data sampled from the shifted rectangle function. As shown in the figure, the DFT of the shifted function contains only the real component, an approximation to the sinc function, and is exactly the same as the result of applying a phase correction to the DFT of the unshifted function (Fig. C.29).

All of the examples shown so far involve real-valued input sequences, and the real components of the transforms are symmetrical about $s = 0$, so that $\mathcal{R}e(G_{-k}) = \mathcal{R}e(G_k)$. Where the transform contains an imaginary component, $\mathcal{I}m(G_{-k}) = -\mathcal{I}m(G_k)$. This pattern is general: For any real-valued sequence, g_k, the elements of the DFT (G_k) for positive and negative k are complex conjugates of one another, so that $G_{-k} = G_k^*$. A practical consequence of this relationship is that the computational cost of calculating the DFT of a real-valued sequence can be cut nearly in half by calculating G_k for only $k = -N/2$ to 0, or $= 0$ to $N/2 - 1$.

The influence of an imaginary component in the input sequence for a discrete Fourier transform is illustrated by examples shown in Fig. C.32. The data plotted in panel A were sampled from an exponentially damped cosine function of the form,

$$g(x) = e^{-rx} \cos(2\pi \nu x) \tag{C.32}$$

with $r = 0.5$ and $\nu = 1$. The DFT of these data, plotted in panel B, approximates the sum of two Lorentzian functions with peaks at $s = \pm 1$. The presence of the two peaks

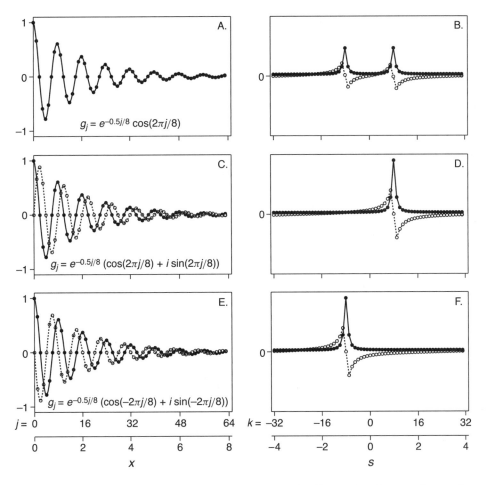

Figure C.32 Influence of an imaginary component in the discrete Fourier transforms of sequences representing a damped oscillation. The left-hand panels show values sampled from a damped cosine function with only a real component (A) and from the same function with imaginary components introduced (C and E), as described in the text. The right-hand panels shows the DFT of the sampled values in the left-hand panels. Real components are indicated by filled circles linked with solid curves, and the imaginary components are shown as open circles connected with dashed curves.

reflects the fact that the same values of g_j, and therefore the same transform, can be generated if $\nu = -1$. Panels C and E show data sampled from a complex function with an imaginary component that is phase shifted from the real component by $\pi/2$ rad:

$$g(x) = e^{-rx}\big(\cos(2\pi \nu x) + i \, \sin(2\pi \nu x)\big) \qquad (C.33)$$

The data plotted in panel C were sampled from the function above with $\nu = 1$, whereas the data in panel E correspond to the same function with $\nu = -1$. As shown in panels D and F, introduction of the imaginary component eliminates the symmetry of the transform shown in panel B, and leads to a single peak in each transform, with positions correctly indicating the sign of ν in the original data. This example simulates the use of quadrature detection in pulse NMR experiments, as discussed in detail in Sections 6.1 and 6.7.

C.4.3   Alias Peaks in Discrete Fourier Transforms

The sampling, or Nyquist theorem, discussed on page 623 (and in Section 6.5), defines a critical frequency in terms of the sampling interval, $f_c = 1/(2\Delta x)$, and stipulates that any signal composed only of frequency components that lie within the range $-f_c$ to f_c can be accurately reconstructed from data sampled at intervals of Δx. The signal can also be described in the s-domain by the DFT calculated from the sampled data, with $-f_c < s < f_c$. If, however, the signal contains components with frequencies that lie outside of this range, these components will still contribute to the DFT and will generate features in the transform that appear to correspond to frequencies within the allowed range. These features are usually referred to as aliases, because they appear as something other than what they really are.

The behavior of alias peaks depends in part on whether the data input to the DFT are real-valued or complex. In Fig. C.33, the effect of aliasing is shown for a set of real-valued data sets, representing a damped cosine function of the form described by Eq. C.32 with increasing frequencies, v. For clarity, only the real components of the transforms are shown in the panels. For this example, the number of sampled values in the x-domain is 128 and $\Delta x = 0.1$. The critical frequency, therefore, is $f_c = 1/0.2 = 5$. For the plots shown in the top row, the condition of the sampling theorem is satisfied, with $-f_c \leq v \leq f_c$, and two peaks appear in each of the transforms, with $s = \pm v$. For the examples in the bottom row, however, v is greater than f_c, leading

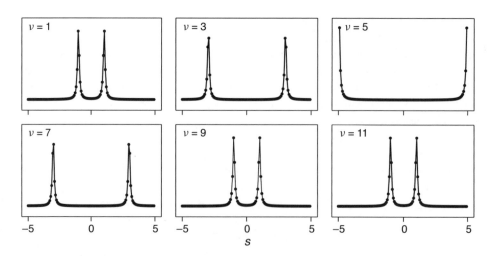

Figure C.33   True and alias peaks in the discrete Fourier transform of a real-valued function. Each panel represents the real component of a discrete Fourier transform of data sampled from a damped cosine function, as defined by Eq. C.32, with the frequency, v, as indicated. The sampling interval, Δx, is 0.1 and the total number of data points is 128. The critical frequency is $f_c = 1/(2\Delta x) = 5$.

Figure C.34

Positions, s_{app}, of true and alias peaks in the DFT of a real-valued signal as a function of the signal frequency, v, for $s \geq 0$. The dashed vertical lines and curved arrows indicate how the graph segments can be folded onto one another and superimposed.

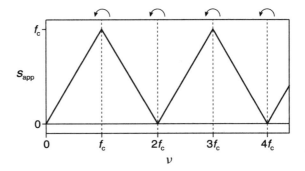

to alias peaks. For $v = 7, 9$ and 11, the alias peaks appear at $s = \pm 3, \pm 1$ and ± 1, respectively.

The pattern describing the positions of the alias peaks may not be obvious from the examples illustrated in Fig. C.33, but it can be shown that the positions of the apparent peaks, s_{app}, are related to the true frequency, v, according to

$$s_{app} = |(v + f_c) \text{ modulo } (2f_c) - f_c| \qquad \text{(C.34)}$$

where (a modulo b) is the remainder of the division of a by b, defined so that the remainder is non-negative.[4] A graph of this relationship is shown in Fig. C.34.

One way to visualize the pattern described by Eq. C.34 and graphed in Fig. C.34, is to imagine folding the graph along the dotted lines accordion-style, so that each segment corresponding to $v = nf_c$ to $(n + 1)f_c$ is layered on top of the region corresponding to $v = 0$ to f_c. All of the segments in the graph of s_{app} versus v will then be superimposed. The same folding pattern can be used to describe the relationship between the positions of alias peaks and their "true" positions—that is, the positions they would occupy if the transform were not limited to the range $s = 0$ to f_c. This pattern is illustrated in Fig. C.35, where each panel represents a peak corresponding to a different true frequency component, all of which create a peak at the same position in a DFT with a given critical frequency. Because their locations can be described by this pattern, alias peaks in discrete transforms are often described as being "folded."

For the DFT of a complex function, the situation is somewhat different, because positive and negative frequencies are distinguishable, as illustrated by the examples shown in Fig. C.36. For this set of examples, the input data are sampled from a damped complex oscillation function, such as illustrated earlier in Fig. C.32B, with different frequencies. As in Fig. C.35, only the real components of the transforms are shown.

For the examples shown in the top row of Fig. C.36, the absolute value of the signal frequency, v, is less than $f_c = 5$, and the peaks appear at their expected positions.

---

[4] The modulo, or remainder, operator is often defined differently in different contexts, and in different computer languages. The operator is defined here so that if a and b are real numbers, (a modulo b) $= a - n \cdot b$, where n is the positive or negative integer such that $0 \leq (a \text{ modulo } b) < |b|$.

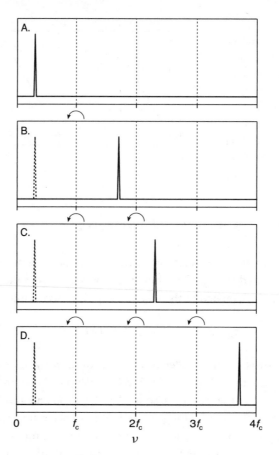

Figure C.35
Folding of alias peak positions in the DFT of a real-valued signal. In panel A, the signal frequency, ν, lies within the range 0 to f_c and the peak appears at its expected position. In each of the other panels, ν has a value that leads to an alias peak at the same position in the DFT. The relationship between the true frequencies, ν, and the peak position in the DFT can be described by folding the graphs along the vertical dotted lines, as indicated by the curved arrows.

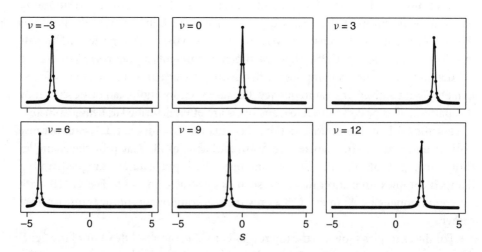

Figure C.36 True and alias peaks in the discrete Fourier transform of a complex-valued function. Each panel represents the real component of a discrete Fourier transform of data sampled from a function describing a damped complex oscillation, as defined by Eq. C.33 and the frequency, ν is as indicated. The sampling interval, Δx, is 0.1 and the total number of data points is 128. The critical frequency is $f_c = 1/(2\Delta x) = 5$.

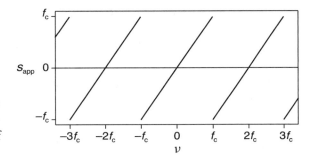

Figure C.37

Positions, s_{app}, of true and alias peaks in the DFT of a complex-valued function as a function of the signal frequency, ν.

In the lower panels, however, signals with frequencies of $\nu = 6$, 9 and 12 are found to generate peaks at positions $s = -4$, -1, and 2 in the transform. This pattern is described by the expression,

$$s_{app} = (\nu + f_c) \text{ modulo } (2f_c) - f_c$$

which is graphed in Fig. C.37.

Unlike the relationship for the positions of alias peaks in discrete transforms of real-valued functions, the pattern shown in Fig. C.37 does not correspond to a folding of the graph. Rather, the aliases follow a cyclic pattern as ν increases—disappearing when s exceeds f_c and reappearing at $-f_c$. Nonetheless, alias peaks in the transforms of complex data, such as a quadrature-detected NMR signal, are commonly referred to as folded peaks.

Although it is straightforward to predict the location of an alias peak from the signal frequency and critical frequency, identifying an alias peak and deducing the corresponding true frequency from the position of the alias is much less so. The simplest approaches to dealing with the potential problem of aliases are to either filter out unwanted signals with frequencies that are greater than f_c or to use sufficiently high sampling frequencies to prevent any of the signals from creating aliased peaks. With recent advances in electronics, computational power and data storage, increased sampling frequencies are now much more practical than they were a decade or two ago. Nonetheless, in some applications, such as multidimensional NMR spectroscopy where the total data acquisition time may be a limitation, alias peaks may not be easily avoided. Such peaks can usually be identified by recording and transforming the same signal using different sampling frequencies, which will usually cause the positions of the alias peaks to shift (unless the two sampling frequencies happen to be unfortunately chosen).

C.4.4   Summary of a Few Key Relationships for Discrete Fourier Transforms

1. For an input sequence, g_j, with $j = 0$ to $N - 1$, the DFT generates an output sequence, G_k, calculated according to Eq. C.29. The indices of the output

sequence are most commonly chosen as

$$k = -\frac{N}{2}, -\frac{N}{2} + 1, -\frac{N}{2} + 2, \ldots, -1, 0, 1, \cdots, \frac{N}{2} - 2, \frac{N}{2} - 1$$

Outside this range, the values of G_k will follow a periodic pattern, such that

$$G_{k+mN} = G_k$$

where m is an integer.

2. The value of G_0 is the sum of g_0 through g_{N-1}.

3. If the input sequence, g_j, represents values of a function or signal, $g(x)$, evaluated at discrete values of x, G_k can be interpreted as a function, $G(s)$, evaluated at discrete values of s. The values of the variables, x and s, are related to the indices, j and k, according to

$$x_j = j\Delta x$$

$$s_k = \frac{k}{N\Delta x}$$

4. A shift in the x-domain so that $x_0 = a$ can be represented in the DFT by multiplying the values of G_k by the phase factor,

$$e^{-i2\pi x_0 k/(N\Delta x)}$$

5. The resolution in the s-domain of the transform is given by

$$\Delta s = \frac{1}{N\Delta x}$$

6. The critical, or Nyquist, frequency is defined by the sampling interval,

$$f_c = \frac{1}{2\Delta x}$$

A signal that contains only components with frequencies limited to the range from $-f_c/2$ to $f_c/2$ can be fully described by samples recorded at intervals of Δx. The values of s represented in the DFT cover the range from $-f_c/2$ to $f_c/2 - \Delta s$. Signal components with frequencies outside of the range from $-f_c/2$ to $f_c/2$ will give rise to alias peaks in the DFT.

7. The DFT of a real-valued function is symmetrical about G_0 such that

$$G_k = G^*_{-k}$$

Further Reading

Suggested references on Fourier transforms and their applications are included at the end of Chapter 6 (page 158).

D

Vectors and Matrices

D.1 Vectors

Vectors are mathematical objects that convey information about both magnitude and direction in two or more dimensions. (Strictly speaking, there are also one-dimensional vectors, but these amount to being simply real or complex numbers.) In two or three, dimensions, it is common to represent a vector containing real numbers as an arrow, usually with its "tail" positioned at the origin of the coordinate system, as shown in Fig. D.1 for a two-dimensional vector. The vectors used to represent bulk magnetization components in three dimensions are examples of this type of vector. More generally, and more formally, a vector in n dimensions is defined as a list of n real or complex numbers. For real-valued vectors, these numbers represent the coordinates of the head of the vector when the tail is placed at the origin, as for the case of the two-dimensional vector, $[x, y]$, shown in Fig. D.1.

In this book, vectors are identified with bold symbols, such as $\mathbf{A}$ or $\boldsymbol{\mu}$. Vectors are also often labeled with an arrow drawn over the symbol, as $\vec{A}$ or $\vec{\mu}$.

Vector addition is defined so that the sum of two n-dimensional vectors is a vector of the same dimension composed of the sums of the corresponding elements of the two vectors. Thus, if $\mathbf{A} = [A_1, A_2]$ and $\mathbf{B} = [B_1, B_2]$, then the sum, $\mathbf{A} + \mathbf{B}$, is given by

$$\mathbf{A} + \mathbf{B} = [A_1 + B_1, A_2 + B_2]$$

For two- and three-dimensional real vectors, this can be visualized as lining up the vectors head-to-tail, as shown in Fig. D.2. The sum is then the vector from the tail of the first vector to the head of the second. The sum of two or more vectors is often called the *resultant*. Similarly, vector subtraction is defined so that

$$\mathbf{A} - \mathbf{B} = [A_1 - B_1, A_2 - B_2]$$

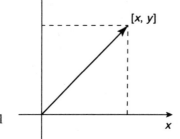

Figure D.1
Representation of a vector as
an arrow on a two-dimensional
plane.

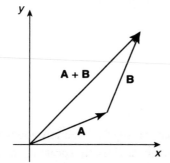

Figure D.2
Graphical representation
of vector addition in two
dimensions.

Vector multiplication is a somewhat more complicated, with several different kinds of multiplication involving vectors. Vectors can be multiplied by real or complex numbers, which are often referred to as *scalars* when vectors are also involved. When a vector is multiplied by a scalar, each element of the vector is multiplied by the scalar, to generate a new vector. Thus, if $\mathbf{A} = [A_1, A_2]$ and c is a scalar, then the product is given by

$$c\mathbf{A} = [cA_1, cA_2]$$

Multiplication by a scalar is commutative, so $c\mathbf{A} = \mathbf{A}c$.

Two n-dimensional vectors can also be multiplied, but there is more than one kind of product defined. The most commonly used product is variously called the *dot product*, *scalar product*, or *inner product*. The dot product is a special case of the *inner product*, which can be defined for a more general class of mathematical objects, including wavefunctions. (Note that the scalar product is *not* the result of multiplication of a vector by a scalar, a potential source of confusion.) The dot product is a scalar and is calculated by taking the sum of the pairwise products of the corresponding vector elements. For the two-dimensional vectors, $\mathbf{A}$ and $\mathbf{B}$, the dot product is given by

$$\mathbf{A} \cdot \mathbf{B} = A_1 B_1 + A_2 B_2$$

Figure D.3

Graphical representation of the dot product of two vectors, in two dimensions. A_B is the length of the projection of vector **A** onto the direction of vector **B**. The dot product, $\mathbf{A} \cdot \mathbf{B}$, is equal to $A_B |\mathbf{B}| = \cos \theta |\mathbf{A}| |\mathbf{B}|$, where $|\mathbf{A}|$ and $|\mathbf{B}|$ are the lengths of vectors **A** and **B**, respectively.

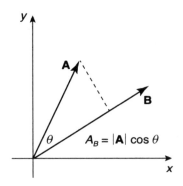

The dot product can also be defined geometrically, as illustrated in Fig. D.3. With the tail of the two vectors placed at the same point, the length of the projection of **A** onto the direction of **B** is given by

$$A_B = \cos \theta \, |\mathbf{A}|$$

where θ is the angle between the two vectors, and $|\mathbf{A}|$ is the length of vector **A**. The dot product is equal to the product of A_B and $|\mathbf{B}|$, the length of vector **B**:

$$\mathbf{A} \cdot \mathbf{B} = A_B \, |\mathbf{B}| = \cos \theta \, |\mathbf{A}| \, |\mathbf{B}|$$

The length of a vector is given by the square root of the dot product of the vector with itself:

$$|\mathbf{A}| = \sqrt{\mathbf{A} \cdot \mathbf{A}}$$

When two vectors are perpendicular, $\cos \theta = 0$, and the dot product is zero. When the two vectors are parallel, the dot product is simply the product of their lengths. If two vectors point in opposite directions, then $\cos \theta = -1$, and $\mathbf{A} \cdot \mathbf{B} = -|\mathbf{A}||\mathbf{B}|$. Thus, the dot product is a measure of the extent to which two vectors are aligned with one another, as well as the product of their lengths.

Another commonly used vector product is called the *cross product* and is only defined for three-dimensional vectors. This product is another three-dimensional vector that is perpendicular to the other two, with their tails placed at a common point. The length of the cross product is equal to the area of the parallelogram defined by the two vectors. This vector product is not used in this text, and interested readers are referred to http://en.wikipedia.org/wiki/Cross_product and http://mathworld.wolfram.com/CrossProduct.html.

Yet another kind of vector multiplication is defined by the *outer product*, which is, in a sense, the opposite of the dot (or inner) product. Whereas the dot product is a scalar created from two vectors, multiplication to form the outer product creates a matrix from two vectors. The explanation of this product is left until page 644, after matrix multiplication has been introduced.

D.2   Matrices

Matrices can be thought of as an extension of vectors, which are one-dimensional lists of numbers, into two dimensions. For instance, an $n \times m$ matrix is composed of n rows and m columns of real or complex numbers:

$$\mathbf{A} = \begin{bmatrix} A_{1,1} & A_{1,2} & A_{1,3} & \cdots & A_{1,m} \\ A_{2,1} & A_{2,2} & A_{2,3} & \cdots & A_{2,m} \\ \vdots & \vdots & \vdots & \ddots & \vdots \\ A_{n,1} & A_{n,2} & A_{n,3} & \cdots & A_{n,m} \end{bmatrix}$$

By convention, the indices of a two-dimensional matrix element define the row and column numbers of the element, in that order.

For square matrices—that is, those with the same number, n, of rows and columns—it is sometimes useful to distinguish the diagonal elements. The elements lying on the diagonal from the upper-left to lower-right corners are referred to as making up the main diagonal, or alternatively, the major, principal or primary diagonal, as highlighted below:

$$\begin{bmatrix} \boxed{A_{1,1}} & A_{1,2} & \cdots & A_{1,n} \\ A_{2,1} & \boxed{A_{2,2}} & \cdots & A_{2,n} \\ \vdots & \vdots & \ddots & \vdots \\ A_{n,1} & A_{n,2} & \cdots & \boxed{A_{n,n}} \end{bmatrix} \tag{D.1}$$

The elements along the diagonal from lower left to upper right, highlighted below, are referred to as the skew diagonal, the minor diagonal, the secondary diagonal or the antidiagonal:

$$\begin{bmatrix} A_{1,1} & \cdots & A_{1,n-1} & \boxed{A_{1,n}} \\ A_{2,1} & \cdots & \boxed{A_{2,n-1}} & A_{2,n} \\ \vdots & \ddots & \vdots & \vdots \\ \boxed{A_{n,1}} & \cdots & A_{n,n-1} & A_{n,n} \end{bmatrix} \tag{D.2}$$

D.2.1   Matrix addition and multiplication

Matrices of the same size and shape can be added or subtracted in the same way as vectors. For instance, the sum of two 2×2 matrices is given by

$$\begin{bmatrix} A_{1,1} & A_{1,2} \\ A_{2,1} & A_{2,2} \end{bmatrix} + \begin{bmatrix} B_{1,1} & B_{1,2} \\ B_{2,1} & B_{2,2} \end{bmatrix} = \begin{bmatrix} A_{1,1} + B_{1,1} & A_{1,2} + B_{1,2} \\ A_{2,1} + B_{2,1} & A_{2,2} + B_{2,2} \end{bmatrix}$$

Also like vectors, matrices can be multiplied by scalars:

$$c \begin{bmatrix} A_{1,1} & A_{1,2} \\ A_{2,1} & A_{2,2} \end{bmatrix} = \begin{bmatrix} cA_{1,1} & cA_{1,2} \\ cA_{2,1} & cA_{2,2} \end{bmatrix}$$

Matrix addition and scalar multiplication satisfy the commutative, associative and distributive properties, so the following relationships hold:

$$\mathbf{A} + \mathbf{B} = \mathbf{B} + \mathbf{A}$$

$$(\mathbf{A} + \mathbf{B}) + \mathbf{C} = \mathbf{A} + (\mathbf{B} + \mathbf{C})$$

$$c(\mathbf{A} + \mathbf{B}) = c\mathbf{A} + c\mathbf{B}$$

Matrix multiplication is more complicated and is restricted to cases where the number of columns in the first matrix is equal to the number of rows in the second. If the first matrix has n rows and m columns, and the second matrix has m rows and p columns, the result of the multiplication is a matrix with n rows and p columns. For instance, if $\mathbf{A}$ is a 2×3 matrix, and $\mathbf{B}$ is a 3×3 matrix, the product,

$$\mathbf{AB} = \mathbf{C}$$

will be a 2×3 matrix. The element $C_{i,j}$ in the product is calculated from row i of $\mathbf{A}$ and column j of $\mathbf{B}$, as indicated below:

$$\mathbf{AB} = \begin{bmatrix} A_{1,1} & A_{1,2} & \cdots & A_{1,m} \\ \vdots & \vdots & \ddots & \vdots \\ \boxed{A_{i,1} \quad A_{i,2} \quad \cdots \quad A_{i,m}} \\ \vdots & \vdots & \ddots & \vdots \\ A_{n,1} & A_{n,2} & \cdots & A_{n,m} \end{bmatrix} \begin{bmatrix} B_{1,1} & \cdots & B_{1,j} & \cdots & B_{1,p} \\ B_{2,1} & \cdots & B_{2,j} & \cdots & B_{2,p} \\ \vdots & \ddots & \vdots & \ddots & \vdots \\ B_{m,1} & \cdots & B_{m,j} & \cdots & B_{m,p} \end{bmatrix}$$

$$= \begin{bmatrix} C_{1,1} & \cdots & C_{1,j} & \cdots & C_{1,p} \\ \vdots & \ddots & \vdots & \ddots & \vdots \\ C_{i,1} & \cdots & \boxed{C_{i,j}} & \cdots & C_{i,p} \\ \vdots & \ddots & \vdots & \ddots & \vdots \\ C_{n,1} & \cdots & C_{n,j} & \cdots & C_{n,p} \end{bmatrix} \tag{D.3}$$

Element $C_{i,j}$ is calculated as the dot product of row i from the first matrix and column j from the second:

$$C_{i,j} = A_{i,1}B_{1,j} + A_{i,2}B_{2,j} + \cdots + A_{i,m}B_{m,j}$$

Square matrices of the same dimension (i.e., $n \times n$ matrices) can be multiplied in both orders, and the result of either multiplication is another $n \times n$ matrix. An $n \times m$ matrix (with $n \neq m$) and an $m \times n$ matrix can also be multiplied in either order, resulting in an $m \times m$ or $n \times n$ matrix, depending on the order. Even for square matrices, however, matrix multiplication is not, in general, commutative, so that

$$\mathbf{AB} \neq \mathbf{BA}$$

For dimension n, there is an $n \times n$ identity matrix, **1**, which contains 1 for each diagonal element, and 0 for all other elements. For instance, the three-dimensional identity matrix is

$$\mathbf{1} = \begin{bmatrix} 1 & 0 & 0 \\ 0 & 1 & 0 \\ 0 & 0 & 1 \end{bmatrix}$$

Multiplication of any square matrix by the identity matrix of the same dimension yields the same matrix:

$$\mathbf{1A} = \mathbf{A}$$

This is one of the special cases where matrix multiplication is commutative, so that

$$\mathbf{1A} = \mathbf{A1} = \mathbf{A}$$

Matrices composed of a single row or a single column can be thought of as vectors, and are sometimes referred to as row vectors and column vectors, respectively, which returns us to the subject of vector multiplication. Consider the 1×3 matrix, **A**, and the 3×1 matrix, **B**:

$$\mathbf{A} = [\, A_{1,1} \quad A_{1,2} \quad A_{1,3} \,] \qquad \mathbf{B} = \begin{bmatrix} B_{1,1} \\ B_{2,1} \\ B_{3,1} \end{bmatrix}$$

Since the matrix product is defined as long as the number of columns in the first matrix is equal to the number of rows in the second, both **AB** (three columns and three rows) and **BA** (one column and one row) are defined. The product, **AB**, is a one-element matrix:

$$\mathbf{AB} = [\, A_{1,1} \quad A_{1,2} \quad A_{1,3} \,] \begin{bmatrix} B_{1,1} \\ B_{2,1} \\ B_{3,1} \end{bmatrix} = [\, A_{1,1}B_{1,1} + A_{1,2}B_{2,1} + A_{1,3}B_{3,1} \,]$$

The single element of this matrix is simply the dot product that would be formed by treating **A** and **B** as vectors. Although a bit sloppy, it is common to ignore the distinction between row vectors and column vectors, especially when only vectors (and scalars) are involved. Thus, if **A** and **B** are both vectors with n elements, the product written as **AB** is taken to be the dot (or scalar, or inner) product, and the order of multiplication does not matter. To minimize confusion, though, it is probably best to write the product as $\mathbf{A} \cdot \mathbf{B}$

If we take **A** and **B** to be matrices, and we multiply them in the opposite order, we get a 3×3 matrix:

$$\mathbf{BA} = \begin{bmatrix} B_{1,1} \\ B_{2,1} \\ B_{3,1} \end{bmatrix} [\, A_{1,1} \quad A_{1,2} \quad A_{1,3} \,] = \begin{bmatrix} B_{1,1}A_{1,1} & B_{1,1}A_{1,2} & B_{1,1}A_{1,3} \\ B_{2,1}A_{1,1} & B_{2,1}A_{1,2} & B_{2,1}A_{1,3} \\ B_{3,1}A_{1,1} & B_{3,1}A_{1,2} & B_{3,1}A_{1,3} \end{bmatrix}$$

This matrix contains all nine of the possible pairwise products of the elements of **A** and **B** and corresponds to another kind of vector product, mentioned earlier, called the outer product.

The outer product of vectors **A** and **B** is written as $\mathbf{A} \otimes \mathbf{B}$. In this operation, **A** is taken to be a column vector and **B** is a row vector, even if they are both written as rows or columns. The outer product is then the matrix multiplication:

$$\mathbf{A} \otimes \mathbf{B} = \mathbf{AB}$$

For the vectors, $\mathbf{A} = [A_1,\ A_2,\ A_3]$ and $\mathbf{B} = [B_1,\ B_2,\ B_3]$,

$$\mathbf{A} \otimes \mathbf{B} = \begin{bmatrix} A_1 \\ A_2 \\ A_3 \end{bmatrix} [\,B_1 \quad B_2 \quad B_3\,] = \begin{bmatrix} A_1B_1 & A_1B_2 & A_1B_3 \\ A_2B_1 & A_2B_2 & A_2B_3 \\ A_3B_1 & A_3B_2 & A_3B_3 \end{bmatrix}$$

Although the inner product is commutative, the outer product is not:

$$\mathbf{B} \otimes \mathbf{A} = \begin{bmatrix} B_1 \\ B_2 \\ B_3 \end{bmatrix} [\,A_1 \quad A_2 \quad A_3\,] = \begin{bmatrix} B_1A_1 & B_1A_2 & B_1A_3 \\ B_2A_1 & B_2A_2 & B_2A_3 \\ B_3A_1 & B_3A_2 & B_3A_3 \end{bmatrix}$$

Notice that the rows of $\mathbf{B} \otimes \mathbf{A}$ are equivalent to the columns of $\mathbf{A} \otimes \mathbf{B}$ and vice versa. The two matrices are said to be transposes of one another, a relationship discussed further on page 648.

In the context of quantum mechanics, the outer product of two vectors representing wavefunctions defines an operator.

D.2.2   Matrix inversion

For *some* square matrices, **A**, there is a matrix inverse, $\mathbf{A}^{-1}$, such that the product of **A** and $\mathbf{A}^{-1}$ is the identity matrix:

$$\mathbf{AA}^{-1} = \mathbf{1}$$

This multiplication is also commutative, so that

$$\mathbf{A}^{-1}\mathbf{A} = \mathbf{1}$$

The relationship between the two matrices, **A** and $\mathbf{A}^{-1}$, is reciprocal, so that

$$(\mathbf{A}^{-1})^{-1} = \mathbf{A}$$

A matrix that does have an inverse is said to be *invertible*, while one that does not is called *noninvertible*, *singular* or *degenerate*. The general requirement for a matrix to be invertible is that the columns (or, equivalently, the rows) are linearly independent of one another—that is, none of the columns can be written as linear combinations of the others.

Finding the inverse of a matrix is closely related to finding the solution to n linear equations in n unknowns. For instance, if we have three equations in which x, y and

z are unknown variables, we could write the equations as

$$a_1 x + a_2 y + a_3 z = a_4$$
$$b_1 x + b_2 y + b_3 z = b_4$$
$$c_1 x + c_2 y + c_3 z = c_4$$

where the coefficients, a_i, b_i and c_i, are specified real numbers. This can also be written as a matrix multiplication, with the following definitions:

$$\mathbf{A} = \begin{bmatrix} a_1 & a_2 & a_3 \\ b_1 & b_2 & b_3 \\ c_1 & c_2 & c_3 \end{bmatrix}$$

$$\mathbf{X} = \begin{bmatrix} x \\ y \\ z \end{bmatrix}$$

and

$$\mathbf{B} = \begin{bmatrix} a_4 \\ b_4 \\ c_4 \end{bmatrix}$$

The system of equations is then written as

$$\mathbf{AX} = \mathbf{B}$$

Multiplying both sides of this equation by $\mathbf{A}^{-1}$ gives

$$\mathbf{A}^{-1}\mathbf{AX} = \mathbf{A}^{-1}\mathbf{B}$$
$$\mathbf{1X} = \mathbf{A}^{-1}\mathbf{B}$$
$$\mathbf{X} = \mathbf{A}^{-1}\mathbf{B}$$

In general, finding the inverse of a matrix requires a good deal of numerical computation (which goes up dramatically with the number of dimensions), and this is usually not the most efficient way to solve a system of linear equations.

D.2.3 Matrix exponentials

Like matrix multiplication, taking the exponential of a matrix is not quite as simple as one might guess; the exponential of matrix, $\mathbf{A}$, is not simply the one composed of elements in which the corresponding element of $\mathbf{A}$ is made the exponent of e. Instead, the matrix exponential function is defined by analogy to the Taylor's series for the exponential function of a number, which is

$$\exp(x) = e^x = 1 + x + \frac{x^2}{2!} + \frac{x^3}{3!} + \cdots = \sum_{k=0}^{\infty} \frac{x^k}{k!}$$

where $k! = k(k-1)(k-2)\cdots 1$. For a matrix, $\mathbf{A}$, the analogous expression is

$$\exp(\mathbf{A}) = e^{\mathbf{A}} = \sum_{k=0}^{\infty} \frac{1}{k!} \mathbf{A}^k$$

Here, the term $\mathbf{A}^k$ represents matrix multiplication of $\mathbf{A}$ by itself $k - 1$ times. This sum will always converge, for both real and complex matrices, but carrying out the calculation can involve a lot of matrix multiplication! For the special case of a matrix in which all of the non-diagonal elements are zero (called a diagonal matrix), however, the result has a simple form. If $\mathbf{A}$ is diagonal, then the exponential of $\mathbf{A}$ is

$$\exp(\mathbf{A}) = \exp \left(\begin{bmatrix} A_{1,1} & 0 & 0 & \cdots & 0 \\ 0 & A_{2,2} & 0 & \cdots & 0 \\ 0 & 0 & A_{3,3} & \cdots & 0 \\ \vdots & \vdots & \vdots & \ddots & 0 \\ 0 & 0 & 0 & \cdots & A_{n,n} \end{bmatrix} \right)$$

$$= \begin{bmatrix} e^{A_{1,1}} & 0 & 0 & \cdots & 0 \\ 0 & e^{A_{2,2}} & 0 & \cdots & 0 \\ 0 & 0 & e^{A_{3,3}} & \cdots & 0 \\ \vdots & \vdots & \vdots & \ddots & 0 \\ 0 & 0 & 0 & \cdots & e^{A_{n,n}} \end{bmatrix}$$

In some cases, an efficient way of calculating the exponential of a non-diagonal matrix is to transform it into one that is diagonal, calculate the exponential, and then reverse the transformation. In quantum mechanics, the matrix used to calculate the time evolution of a wavefunction is derived by exponentiation of the matrix for the Hamiltonian operator.

D.2.4   Eigenvectors and eigenvalues

For a square $n \times n$ matrix, $\mathbf{A}$, there may be one or more associated $n \times 1$ matrices, $\mathbf{X_i}$, such that the product $\mathbf{AX_i}$ is equal to $\mathbf{X_i}$ times a constant, λ_i:

$$\mathbf{AX_i} = \lambda_i \mathbf{X_i}$$

As noted above, an $n \times 1$ matrix can be thought of as a column vector, and $\mathbf{X_i}$ is commonly referred to as an eigenvector of the matrix, $\mathbf{A}$, and λ_i is called an eigenvalue. The eigenvectors of a matrix are not usually considered to include the trivial example of a vector containing only zeros. For a given $n \times n$ matrix, there can be from 0 to n eigenvalues and the same number of linearly independent eigenvectors. If there are n linearly independent eigenvectors, then any other n-dimensional vector can be written as a linear combination of the eigenvectors.

Geometrically, the multiplication of a real n-dimensional vector by a real $n \times n$ matrix can be thought of as a rotation and scaling of the vector into another vector in the n-dimensional space. For an eigenvector, the transformed vector is parallel to the original vector, and the eigenvalue represents the length of the new vector relative to the old one.

Eigenvectors and eigenvalues arise in a number of areas in mathematics, including finding the solutions to polynomials. In quantum mechanics, operators are frequently represented as square matrices, with associated eigenvectors that represent the eigenfunctions of the operators.

D.2.5 Matrix transposition

Any $n \times m$ matrix has associated with it a transpose $m \times n$ matrix, in which the rows and columns are transposed. For instance, if $\mathbf{A}$ is the 3×2 matrix given by

$$\mathbf{A} = \begin{bmatrix} A_{1,1} & A_{1,2} \\ A_{2,1} & A_{2,2} \\ A_{3,1} & A_{3,2} \end{bmatrix}$$

then the transpose of $\mathbf{A}$ is (using the labels from the original matrix)

$$\mathbf{A}^{\mathrm{T}} = \begin{bmatrix} A_{1,1} & A_{2,1} & A_{3,1} \\ A_{1,2} & A_{2,2} & A_{3,2} \end{bmatrix}$$

In general, the elements of the transpose are

$$A^{\mathrm{T}}_{i,j} = A_{j,i}$$

If matrix, $\mathbf{A}$, is square, then so is its transpose, and the diagonal elements in the two matrices are the same. A square matrix is said to be symmetric if it is equal to its own transpose. For instance, a symmetric 3×3 matrix contains at most six unique elements, and both it and its transpose can be written as

$$\mathbf{A} = \mathbf{A}^{\mathrm{T}} = \begin{bmatrix} A_{1,1} & A_{1,2} & A_{1,3} \\ A_{1,2} & A_{2,2} & A_{2,3} \\ A_{1,3} & A_{2,3} & A_{3,3} \end{bmatrix}$$

A special case of a symmetric matrix is one in which only the diagonal elements are non-zero, which is referred to as a diagonal matrix. A 3×3 diagonal matrix and its transpose have the following form:

$$\mathbf{A} = \mathbf{A}^{\mathrm{T}} = \begin{bmatrix} A_{1,1} & 0 & 0 \\ 0 & A_{2,2} & 0 \\ 0 & 0 & A_{3,3} \end{bmatrix}$$

If an $n \times m$ matrix contains complex numbers as elements, then its *conjugate transpose* is an $m \times n$ matrix in which the rows and elements are transposed, as described above, and then the individual elements are replaced with their complex conjugates. If the matrix, $\mathbf{A}$, contains elements, $A_{i,j}$, then the elements of the conjugate transpose matrix, $\mathbf{A}^{*}$, are given by

$$A^{*}_{i,j} = (A_{j,i})^{*}$$

where $(A_{j,i})^*$ is the complex conjugate of $A_{j,i}$. For instance, if **A** is a 3×3 matrix composed of elements of the form $a_{i,j} + ib_{i,j}$, where $a_{i,j}$ and $b_{i,j}$ are real

$$\mathbf{A} = \begin{bmatrix} a_{1,1} + ib_{1,1} & a_{1,2} + ib_{1,2} & a_{1,3} + ib_{1,3} \\ a_{2,1} + ib_{2,1} & a_{2,2} + ib_{2,2} & a_{2,3} + ib_{2,3} \\ a_{3,1} + ib_{3,1} & a_{3,2} + ib_{3,2} & a_{3,3} + ib_{3,3} \end{bmatrix}$$

then the conjugate transpose is given by

$$\mathbf{A}^* = \begin{bmatrix} a_{1,1} - ib_{1,1} & a_{2,1} - ib_{2,1} & a_{3,1} - ib_{3,1} \\ a_{1,2} - ib_{1,2} & a_{2,2} - ib_{2,2} & a_{3,2} - ib_{3,2} \\ a_{1,3} - ib_{1,3} & a_{2,3} - ib_{2,3} & a_{3,3} - ib_{3,3} \end{bmatrix}$$

The conjugate transpose matrix is also commonly called the *adjoint* matrix, though that term is also sometimes used with a different meaning, leading to possible confusion.

There is a rather special class of square matrices, which have the property that they are their own conjugate transposes—that is,

$$\mathbf{A}^* = \mathbf{A}$$

Matrices of this class are said to be *self-adjoint* or *Hermitian*. A 3×3 Hermitian matrix can be written as

$$\mathbf{A} = \mathbf{A}^* = \begin{bmatrix} a_{1,1} & a_{1,2} + ib_{1,2} & a_{1,3} + ib_{1,3} \\ a_{1,2} - ib_{1,2} & a_{2,2} & a_{2,3} + ib_{2,3} \\ a_{1,3} - ib_{1,3} & a_{2,3} - ib_{2,3} & a_{3,3} \end{bmatrix}$$

Notice that each of the diagonal elements of a Hermitian matrix must be a real number, for which the complex conjugate is itself. A matrix containing only real-valued elements can be Hermitian only if it is symmetric. One of the special properties of an $n \times n$ Hermitian matrix is that it possesses n eigenvalues and n linearly independent eigenvectors. These eigenvectors can then be linearly combined to represent any other n-dimensional vector. These properties make Hermitian matrices particularly important in quantum mechanics, where they are used to represent measurement operators, as discussed in Appendices E and F.

Another special class of square matrices is described as being *unitary*. Matrices of this class have the property that their conjugate transpose matrices are also the inverse of the unitary matrix. Thus, if **U** is a unitary matrix, then

$$\mathbf{U}^{-1} = \mathbf{U}^*$$

and

$$\mathbf{U}\mathbf{U}^* = 1$$

The matrices used to calculate the effects on the density matrix of rotations about the x'- and y'-axes (e.g., $\mathbf{R}_x$ and $\mathbf{R}_x^{-1}$) are examples of unitary matrices, as are the matrices that determine the time evolution of the density matrix in a stationary external field, $\mathbf{U}_\mathrm{H}$ and $\mathbf{U}_\mathrm{H}^{-1}$.

D.2.6   Matrix calculations

A wide variety of scientific calculations can be expressed in matrix form, which offers both efficiency in notation and a common language for computation. Matrix calculations can be carried out either numerically, for specific matrices, or algebraically to yield general results. Both types of calculations can become quite challenging as the sizes of the matrices increase. Inverting matrices, exponentiation and finding eigenvectors can be particularly difficult for larger systems. However, very efficient numerical methods for solving these problems have been developed and have been implemented as libraries in a variety of computer languages and programs. The commercial program Matlab is particularly well suited for numerical matrix calculations, as are the open-source programs Octave and Scilab, which are designed to be compatible with Matlab, and the NumPy package for the Python language. There are also programs referred to as computer algebra systems (CAS), which include capabilities for symbolic manipulation of matrices, as well as numerical calculations. These include the commercial programs Mathematica and Maple, as well as open-source alternatives, including Maxima and Xcas.

Mathematica and Maxima libraries of macros for manipulating the vectors and matrices associated with spin-1/2 systems are available through the web site http://uscibooks.com/goldenberg.htm.

E

Mathematics for Uncoupled Spin-1/2 Particles

E.1 Wavefunctions and Eigenstates

For an uncoupled spin-1/2 particle, the general form of the wavefunction can be written as a linear combination of the eigenfunctions for the z-magnetization operator ($\hat{I}_z$, defined below), $|\alpha\rangle$ and $|\beta\rangle$:

$$|\Psi\rangle = c_\alpha|\alpha\rangle + c_\beta|\beta\rangle$$

where c_α and c_β are complex coefficients. The wavefunctions are written as kets in Dirac notation. The complex conjugate of the wavefunction, or bra, is given by

$$\langle\Psi| = c_\alpha^*\langle\alpha| + c_\beta^*\langle\beta|$$

where c_α^* and c_β^* are the complex conjugates of the coefficients in the wavefunction, and $\langle\alpha|$ and $\langle\beta|$ are the complex conjugates of $|\alpha\rangle$ and $|\beta\rangle$, respectively. In Dirac notation, the product of a wavefunction and the complex conjugate of the same or a different wavefunction are written as brackets, such as

$$\langle\Psi|\Psi\rangle$$

The eigenfunctions of the $\hat{I}_z$ operator are orthonormal, so that the following relationships hold:

$$\langle\alpha|\alpha\rangle = 1$$
$$\langle\beta|\beta\rangle = 1$$
$$\langle\alpha|\beta\rangle = 0$$
$$\langle\beta|\alpha\rangle = 0$$

The z-magnetization eigenfunctions, $|\alpha\rangle$ and $|\beta\rangle$, are also the stationary states, or eigenstates, of the spin in a stationary magnetic field aligned with the z-axis, so that the magnetization components of these states remain constant in the stationary field. The eigenstates are also the eigenfunctions of the Hamiltonian operator, with eigenvalues corresponding to their energies. The energies are given by

$$E_\alpha = -\frac{h\gamma B}{4\pi}$$

$$E_\beta = \frac{h\gamma B}{4\pi}$$

where h is Planck's constant, γ is the gyromagnetic ratio of the particle and B is the strength of the external field.

E.2   Magnetization Operators

The operator associated with magnetization along the z-axis for the wavefunction, $|\Psi\rangle = c_\alpha|\alpha\rangle + c_\beta|\beta\rangle$, is given by

$$\hat{I}_z|\Psi\rangle = \hat{I}_z\left(c_\alpha|\alpha\rangle + c_\beta|\beta\rangle\right) = \frac{1}{2}(c_\alpha|\alpha\rangle - c_\beta|\beta\rangle)$$

The two eigenfunctions for this operator are $|\alpha\rangle$ and $|\beta\rangle$, with eigenvalues $-1/2$ and $1/2$, respectively. Thus,

$$\hat{I}_z|\alpha\rangle = \frac{1}{2}|\alpha\rangle$$

$$\hat{I}_z|\beta\rangle = -\frac{1}{2}|\alpha\rangle$$

The operator associated with magnetization along the x-axis for the wavefunction, $|\Psi\rangle$, is given by

$$\hat{I}_x|\Psi\rangle = \hat{I}_x\left(c_\alpha|\alpha\rangle + c_\beta|\beta\rangle\right) = \frac{1}{2}(c_\beta|\alpha\rangle + c_\alpha|\beta\rangle)$$

The eigenfunctions for this operator are

$$\frac{1}{\sqrt{2}}|\alpha\rangle + \frac{1}{\sqrt{2}}|\beta\rangle$$

and

$$-\frac{1}{\sqrt{2}}|\alpha\rangle + \frac{1}{\sqrt{2}}|\beta\rangle$$

with eigenvalues of $1/2$ and $-1/2$, respectively. Like the eigenvectors of $\hat{I}_z$, these are also orthonormal with one another.

The operator associated with magnetization along the y-axis for the wavefunction, $|\Psi\rangle$, is

$$\hat{I}_y|\Psi\rangle = \hat{I}_y\left(c_\alpha|\alpha\rangle + c_\beta|\beta\rangle\right) = \frac{i}{2}(-c_\beta|\alpha\rangle + c_\alpha|\beta\rangle)$$

Like the other operators, $\hat{I}_y$ has two eigenfunctions:

$$\frac{1}{\sqrt{2}}|\alpha\rangle + \frac{i}{\sqrt{2}}|\beta\rangle$$

and

$$\frac{1}{\sqrt{2}}|\alpha\rangle - \frac{i}{\sqrt{2}}|\beta\rangle$$

and the corresponding eigenvalues are $1/2$ and $-1/2$, respectively. These eigenfunctions are also orthonormal to one another.

E.3 Equations to Calculate Average Magnetization Components

Following the standard procedure of quantum mechanics to calculate the average value of an observation, the average I_z-magnetization for a population of spin-1/2 particles, all with the wavefunction, $|\Psi\rangle$, is calculated as

$$\langle I_z \rangle = \langle \Psi | \hat{I}_z | \Psi \rangle = \left(c_\alpha^* \langle \alpha | + c_\beta^* \langle \beta | \right) \hat{I}_z \left(c_\alpha |\alpha\rangle + c_\beta |\beta\rangle \right)$$

$$= \left(c_\alpha^* \langle \alpha | + c_\beta^* \langle \beta | \right) \left(\frac{c_\alpha}{2} |\alpha\rangle - \frac{c_\beta}{2} |\beta\rangle \right)$$

$$= \frac{1}{2} \left(c_\alpha^* c_\alpha \langle \alpha | \alpha \rangle - c_\alpha^* c_\beta \langle \alpha | \beta \rangle + c_\beta^* c_\alpha \langle \beta | \alpha \rangle - c_\beta^* c_\beta \langle \beta | \beta \rangle \right)$$

$$= \frac{1}{2} \left(c_\alpha^* c_\alpha - c_\beta^* c_\beta \right)$$

Analogous expressions for calculating the average x- and y-magnetization components are

$$\langle I_x \rangle = \langle \Psi | \hat{I}_x | \Psi \rangle = \frac{1}{2} \left(c_\alpha c_\beta^* + c_\beta c_\alpha^* \right)$$

$$\langle I_y \rangle = \langle \Psi | \hat{I}_y | \Psi \rangle = \frac{i}{2} \left(c_\alpha c_\beta^* - c_\beta c_\alpha^* \right)$$

E.4 Shift Operators

The raising and lowering operators for a single spin-1/2 particle are defined as

$$\hat{I}^+ = \hat{I}_x + i\hat{I}_y$$

$$\hat{I}^- = \hat{I}_x - i\hat{I}_y$$

The shift operators can also be expressed in terms of their action on an arbitrary wavefunction, $|\Psi\rangle = c_\alpha |\alpha\rangle + c_\beta |\beta\rangle$:

$$\hat{I}^+ |\Psi\rangle = c_\beta |\alpha\rangle$$

$$\hat{I}^- |\Psi\rangle = c_\alpha |\beta\rangle$$

Unlike the magnetization operators, the shift operators are not associated with any experimental observable, and they do not have any eigenvectors or eigenvalues. Instead, these operators are used to calculate the probabilities of a transition between two states with defined wavefunctions. The raising operator is used to calculate the probability of a transition in which the quantum number increases (e.g., the transition from $|\beta\rangle$ to $|\alpha\rangle$), and the lowering operator is used for transitions in which the quantum number decreases. For instance, the probability of an upward transition from $|\Psi_a\rangle$ to $|\Psi_b\rangle$ is calculated as

$$P_{a,b}^+ = |\langle \Psi_b | \hat{I}^+ | \Psi_a \rangle|^2$$

The calculation is performed by first applying the operator to the initial wavefunction and then multiplying the result by the complex conjugate of the final wavefunction.

E.5   Evolution of the Wavefunction in a Stationary Magnetic Field

The evolution of a wavefunction over time is calculated from the time-dependent Schrödinger equation:

$$\frac{\partial}{\partial t} |\Psi(t)\rangle = \frac{-i}{\hbar} \mathcal{H} |\Psi(t)\rangle$$

where $|\Psi(t)\rangle$ is the wavefunction at time, t, $\hbar$ is Planck's constant divided by 2π and $\mathcal{H}$ is the Hamiltonian operator. For a spin-1/2 particle in a stationary magnetic field, the Hamiltonian is given by

$$\mathcal{H}|\Psi\rangle = -\hbar \gamma B \hat{I}_z |\Psi\rangle$$

The eigenfunctions of $\mathcal{H}$ are also the eigenfunctions of $\hat{I}_z$, with eigenvalues equal to the energies of the $|\alpha\rangle$ and $|\beta\rangle$ states, as given above. For the case of a spin-1/2 particle in a stationary field, if the initial wavefunction is given by $\Psi(0) = c_\alpha |\alpha\rangle + c_\beta |\beta\rangle$, then integration of the Schrödinger equation gives

$$|\Psi(t)\rangle = e^{-it E_\alpha / \hbar} c_\alpha |\alpha\rangle + e^{-it E_\beta / \hbar} c_\beta |\beta\rangle$$

The quantities $E_\alpha / \hbar$ and $E_\beta / \hbar$ can also be written as $\nu\pi$ and $-\nu\pi$, respectively, where ν is the Larmor frequency (and is negative for a positive gyromagnetic ratio). The evolution of the wavefunction is then given by

$$|\Psi(t)\rangle = e^{-i\pi \nu t} c_\alpha |\alpha\rangle + e^{i\pi \nu t} c_\beta |\beta\rangle$$

The complex conjugate of the wavefunction evolves as

$$\langle \Psi(t)| = e^{i\pi vt} c_\alpha^* \langle\alpha| + e^{-i\pi vt} c_\beta^* \langle\beta|$$

E.6   Pulse Operators to Rotate Magnetization

For an initial wavefunction given by $\Psi(0) = c_\alpha|\alpha\rangle + c_\beta|\beta\rangle$, the change in the wavefunction due to an x'-pulse in the rotating frame is given by

$$|\Psi(a)\rangle = \big(c_\alpha \cos(a/2) - i c_\beta \sin(a/2)\big)|\alpha\rangle$$
$$+ \big(-i c_\alpha \sin(a/2) + c_\beta \cos(a/2)\big)|\beta\rangle \tag{E.1}$$

where a is the angle of rotation in the rotating frame. In units of radians, a is given by

$$a = t|\gamma|B_1$$

where t is the duration of the pulse and B_1 is the strength of the oscillating magnetic field. The absolute value of the gyromagnetic ratio is used so that the directions of pulses are defined by the right-handed coordinate system and are the same for nuclei with positive or negative gyromagnetic ratios.

For a pulse along the y'-axis, the change in wavefunction is

$$|\Psi(a)\rangle = \big(c_\alpha \cos(a/2) - c_\beta \sin(a/2)\big)|\alpha\rangle$$
$$+ \big(c_\beta \cos(a/2) + c_\alpha \sin(a/2)\big)|\beta\rangle \tag{E.2}$$

The wavefunctions calculated from Eqs. E.1 and E.2 assume a rotating frame. Thus, the magnetization components calculated from the wavefunctions and the $\hat{I}_x$, $\hat{I}_y$, and $\hat{I}_z$ operators refer to the rotating-frame coordinates.

E.7   Vector and Matrix Representations of Wavefunctions and Operators

The mathematical relationships expressed above using Dirac notation can also be written in matrix form, providing a more compact representation and simplifying calculations. The general form of the wavefunction, $\Psi = c_\alpha|\alpha\rangle + c_\beta|\beta\rangle$, is written as a 2×1 matrix, or column vector, containing the coefficients c_α and c_β:

$$\Psi = \begin{bmatrix} c_\alpha \\ c_\beta \end{bmatrix} \tag{E.3}$$

The matrix serves as a sort of shorthand, in which the eigenstates, $|\alpha\rangle$ and $|\beta\rangle$, are implicit. The matrix form can be expanded into the Dirac notation by forming the dot (or inner) product of the column vector in Eq. E.3 and a row vector containing the kets for the eigenstates:

$$|\Psi\rangle = [\ |\alpha\rangle \quad |\beta\rangle\ ] \cdot \begin{bmatrix} c_\alpha \\ c_\beta \end{bmatrix} = c_\alpha|\alpha\rangle + c_\beta|\beta\rangle$$

The complex conjugate of the wavefunction is written as a 1×2 matrix, or row vector:

$$\Psi^* = [\ c_\alpha^* \quad c_\beta^*\ ]$$

Magnetization Operators

The magnetization operators are represented as 2×2 matrices, referred to as the Pauli spin matrices:

$$\mathbf{I}_x = \frac{1}{2}\begin{bmatrix} 0 & 1 \\ 1 & 0 \end{bmatrix}$$

$$\mathbf{I}_y = \frac{i}{2}\begin{bmatrix} 0 & -1 \\ 1 & 0 \end{bmatrix}$$

$$\mathbf{I}_z = \frac{1}{2}\begin{bmatrix} 1 & 0 \\ 0 & -1 \end{bmatrix}$$

The operations are then expressed as matrix multiplications. For instance, applying the $\hat{I}_z$ operator is equivalent to multiplying the wavefunction vector by $\mathbf{I}_z$:

$$\mathbf{I}_z\Psi = \frac{1}{2}\begin{bmatrix} 1 & 0 \\ 0 & -1 \end{bmatrix}\begin{bmatrix} c_\alpha \\ c_\beta \end{bmatrix} = \frac{1}{2}\begin{bmatrix} c_\alpha \\ -c_\beta \end{bmatrix}$$

The average z-magnetization is then calculated as the dot product of this product and the complex conjugate of Ψ:

$$\langle I_x\rangle = \Psi^*\mathbf{I}_z\Psi = [\ c_\alpha^* \quad c_\beta^*\ ]\frac{1}{2}\begin{bmatrix} c_\alpha \\ -c_\beta \end{bmatrix} = \frac{1}{2}(c_\alpha c_\alpha^* - c_\beta c_\beta^*)$$

Notice that the matrices representing the magnetization operators are all Hermitian—that is they are equal to their own transpose conjugates (or adjoints). This ensures that they have eigenvectors, corresponding to the eigenfunctions of the operators, and that the eigenvalues are real numbers, as they should be for a measurement.

Shift Operators

The shift operators can also be written as matrices:

$$\mathbf{I}^+ = \mathbf{I}_x + i\mathbf{I}_y = \frac{1}{2}\begin{bmatrix} 0 & 1 \\ 1 & 0 \end{bmatrix} + i\frac{i}{2}\begin{bmatrix} 0 & -1 \\ 1 & 0 \end{bmatrix} = \begin{bmatrix} 0 & 1 \\ 0 & 0 \end{bmatrix}$$

and

$$\mathbf{I}^- = \mathbf{I}_x - i\mathbf{I}_y = \frac{1}{2}\begin{bmatrix} 0 & 1 \\ 1 & 0 \end{bmatrix} - i\frac{i}{2}\begin{bmatrix} 0 & -1 \\ 1 & 0 \end{bmatrix} = \begin{bmatrix} 0 & 0 \\ 1 & 0 \end{bmatrix}$$

These matrices are *not* Hermitian, reflecting the fact that they do not represent operators for observables.

The Hamiltonian Operator and Time Evolution

The Hamiltonian operator for a single spin is written in matrix form as

$$\mathcal{H} = -\hbar\gamma\, B\mathbf{I}_z = h\nu\mathbf{I}_z = \frac{h\nu}{2}\begin{bmatrix} 1 & 0 \\ 0 & -1 \end{bmatrix}$$

The unitary time-evolution matrix, $\mathbf{U}_H$, is related to the Hamiltonian matrix as follows:

$$\mathbf{U}_H = \exp\left(\frac{-it}{\hbar}\mathcal{H}\right) = \begin{bmatrix} e^{-i\pi\nu t} & 0 \\ 0 & e^{i\pi\nu t} \end{bmatrix}$$

The time evolution of the wavefunction is calculated by multiplying the starting wavefunction vector by the time-evolution matrix:

$$\boldsymbol{\Psi}(t) = \mathbf{U}_H\boldsymbol{\Psi} = \begin{bmatrix} e^{-i\pi\nu t} & 0 \\ 0 & e^{i\pi\nu t} \end{bmatrix}\begin{bmatrix} c_\alpha \\ c_\beta \end{bmatrix} = \begin{bmatrix} e^{-i\pi\nu t}c_\alpha \\ e^{i\pi\nu t}c_\beta \end{bmatrix}$$

The inverse of the time-evolution matrix, which is used in density-matrix calculations (Chapter 17), is

$$\mathbf{U}_H^{-1} = \begin{bmatrix} e^{i\pi\nu t} & 0 \\ 0 & e^{-i\pi\nu t} \end{bmatrix}$$

Rotation by Pulses

The matrix for rotation about the x'-axis by the angle, a, is

$$\mathbf{R}_x(a) = \begin{bmatrix} \cos(a/2) & -i\,\sin(a/2) \\ -i\,\sin(a/2) & \cos(a/2) \end{bmatrix}$$

For rotation about the y'-axis, the matrix is

$$\mathbf{R}_y(a) = \begin{bmatrix} \cos(a/2) & -\sin(a/2) \\ \sin(a/2) & \cos(a/2) \end{bmatrix}$$

The inverses of these matrices are

$$\mathbf{R}_x^{-1}(a) = \begin{bmatrix} \cos(a/2) & i\,\sin(a/2) \\ i\,\sin(a/2) & \cos(a/2) \end{bmatrix}$$

and

$$\mathbf{R}_y^{-1}(a) = \begin{bmatrix} \cos(a/2) & \sin(a/2) \\ -\sin(a/2) & \cos(a/2) \end{bmatrix}$$

As with the time-evolution operator, the effects of the rotation operators are calculated by multiplying the vector for the wavefunction by the rotation matrices.

For instance, the effect of a $(\pi/2)_x$-pulse is calculated as

$$
\mathbf{R}_x(\pi/2)\mathbf{\Psi} = \begin{bmatrix} \cos(\pi/4) & -i\,\sin(\pi/4) \\ -i\,\sin(\pi/4) & \cos(\pi/4) \end{bmatrix} \begin{bmatrix} c_\alpha \\ c_\beta \end{bmatrix}
$$

$$
= \frac{1}{\sqrt{2}} \begin{bmatrix} 1 & -i \\ -i & 1 \end{bmatrix} \begin{bmatrix} c_\alpha \\ c_\beta \end{bmatrix}
$$

$$
= \frac{1}{\sqrt{2}} \begin{bmatrix} c_\alpha - ic_\beta \\ -ic_\alpha + c_\beta \end{bmatrix}
$$

With the set of matrices given above, any calculations for a pure population of spin-1/2 particles can be carried out as a set of matrix multiplications. The operator and rotation matrices are also used in the manipulations of density matrices representing mixed populations of spins, as described in Chapter 17.

Mathematics of the Two-Spin System

F.1   Wavefunctions and Eigenstates

The wavefunction for an arbitrary state composed of two scalar-coupled spins can be expressed as a linear combination of the eigenfunctions for the $\hat{I}_z$ and $\hat{S}_z$ operators (defined below):

$$|\alpha\alpha\rangle, \ |\alpha\beta\rangle, \ |\beta\alpha\rangle, \ |\beta\beta\rangle$$

The general expression for a linear combination of these is

$$|\Psi\rangle = c_{\alpha\alpha}|\alpha\alpha\rangle + c_{\alpha\beta}|\alpha\beta\rangle + c_{\beta\alpha}|\beta\alpha\rangle + c_{\beta\beta}|\beta\beta\rangle$$

where $c_{\alpha\alpha}$, $c_{\alpha\beta}$, $c_{\beta\alpha}$ and c_β are complex coefficients. The complex conjugate of the wavefunction, or bra, is given by

$$\langle\Psi| = c_{\alpha\alpha}^*\langle\alpha\alpha| + c_{\alpha\beta}^*\langle\alpha\beta| + c_{\beta\alpha}^*\langle\beta\alpha| + c_{\beta\beta}^*\langle\beta\beta|$$

where $c_{\alpha\alpha}^*$, $c_{\alpha\beta}^*$, $c_{\beta\alpha}^*$ and $c_{\beta\beta}^*$ are the complex conjugates of the coefficients in the expression for the wavefunction, and $\langle\alpha\alpha|$, $\langle\alpha\beta|$, $\langle\beta\alpha|$ and $\langle\beta\beta|$ are the complex conjugates of the eigenfunctions.

The functions $|\alpha\alpha\rangle$, $|\alpha\beta\rangle$, $|\beta\alpha\rangle$, and $|\beta\beta\rangle$ are also the stationary states (eigenstates) in the limit of weak coupling—that is, when $J^2 \ll (v_I - v_S)^2$, where v_I and v_S are the Larmor frequencies of the two spins in the absence of coupling and J is the scalar coupling constant. The energies of the eigenstates in the weak-coupling limit are

$$E_{\alpha\alpha} = h(v_I + v_S)/2 + hJ/4$$

$$E_{\alpha\beta} = h(v_I - v_S)/2 - hJ/4$$

$$E_{\beta\alpha} = -h(v_I - v_S)/2 - hJ/4$$

$$E_{\beta\beta} = -h(v_I + v_S)/2 + hJ/4$$

More generally, without imposing the weak-coupling limit, the eigenstates are given by

$$|\alpha\alpha\rangle$$
$$|\Psi_A\rangle = -\sin\theta|\alpha\beta\rangle + \cos\theta|\beta\alpha\rangle$$
$$|\Psi_B\rangle = \cos\theta|\alpha\beta\rangle + \sin\theta|\beta\alpha\rangle$$
$$|\beta\beta\rangle$$

where θ is defined so that

$$\cos(2\theta) = (\nu_I - \nu_S)/D$$

and

$$D = \sqrt{J^2 + (\nu_I - \nu_S)^2}$$

The energies of the four eigenstates are

$$E_{\alpha\alpha} = h(\nu_I + \nu_S)/2 + hJ/4$$

$$E_A = -h\sqrt{J^2 + (\nu_I - \nu_S)^2}/2 - hJ/4$$

$$E_B = h\sqrt{J^2 + (\nu_I - \nu_S)^2}/2 - hJ/4$$

$$E_{\beta\beta} = -h(\nu_I + \nu_S)/2 + hJ/4$$

F.2   Magnetization Operators

For the I-spin,

$$\hat{I}_x|\Psi\rangle = \frac{1}{2}\left(c_{\beta\alpha}|\alpha\alpha\rangle + c_{\beta\beta}|\alpha\beta\rangle + c_{\alpha\alpha}|\beta\alpha\rangle + c_{\alpha\beta}|\beta\beta\rangle\right)$$

$$\hat{I}_y|\Psi\rangle = \frac{i}{2}\left(-c_{\beta\alpha}|\alpha\alpha\rangle - c_{\beta\beta}|\alpha\beta\rangle + c_{\alpha\alpha}|\beta\alpha\rangle + c_{\alpha\beta}|\beta\beta\rangle\right)$$

$$\hat{I}_z|\Psi\rangle = \frac{1}{2}\left(c_{\alpha\alpha}|\alpha\alpha\rangle + c_{\alpha\beta}|\alpha\beta\rangle - c_{\beta\alpha}|\beta\alpha\rangle - c_{\beta\beta}|\beta\beta\rangle\right)$$

For the S-spin,

$$\hat{S}_x|\Psi\rangle = \frac{1}{2}\left(c_{\alpha\beta}|\alpha\alpha\rangle + c_{\alpha\alpha}|\alpha\beta\rangle + c_{\beta\beta}|\beta\alpha\rangle + c_{\beta\alpha}|\beta\beta\rangle\right)$$

$$\hat{S}_y|\Psi\rangle = \frac{i}{2}\left(-c_{\alpha\beta}|\alpha\alpha\rangle + c_{\alpha\alpha}|\alpha\beta\rangle - c_{\beta\beta}|\beta\alpha\rangle + c_{\beta\alpha}|\beta\beta\rangle\right)$$

$$\hat{S}_z|\Psi\rangle = \frac{1}{2}\left(c_{\alpha\alpha}|\alpha\alpha\rangle - c_{\alpha\beta}|\alpha\beta\rangle + c_{\beta\alpha}|\beta\alpha\rangle - c_{\beta\beta}|\beta\beta\rangle\right)$$

F.3   Equations to Calculate Average Magnetization Components

For the I-spin,

$$\langle I_x \rangle = \langle \Psi | \hat{I}_x | \Psi \rangle = \frac{1}{2} \left(c_{\beta\alpha} c_{\alpha\alpha}^* + c_{\beta\beta} c_{\alpha\beta}^* + c_{\alpha\alpha} c_{\beta\alpha}^* + c_{\alpha\beta} c_{\beta\beta}^* \right)$$

$$\langle I_y \rangle = \langle \Psi | \hat{I}_y | \Psi \rangle = \frac{i}{2} \left(-c_{\beta\alpha} c_{\alpha\alpha}^* - c_{\beta\beta} c_{\alpha\beta}^* + c_{\alpha\alpha} c_{\beta\alpha}^* + c_{\alpha\beta} c_{\beta\beta}^* \right)$$

$$\langle I_z \rangle = \langle \Psi | \hat{I}_z | \Psi \rangle = \frac{1}{2} \left(c_{\alpha\alpha} c_{\alpha\alpha}^* + c_{\alpha\beta} c_{\alpha\beta}^* - c_{\beta\alpha} c_{\beta\alpha}^* - c_{\beta\beta} c_{\beta\beta}^* \right)$$

For the S-spin,

$$\langle S_x \rangle = \langle \Psi | \hat{S}_x | \Psi \rangle = \frac{1}{2} \left(c_{\alpha\beta} c_{\alpha\alpha}^* + c_{\alpha\alpha} c_{\alpha\beta}^* + c_{\beta\beta} c_{\beta\alpha}^* + c_{\beta\alpha} c_{\beta\beta}^* \right)$$

$$\langle S_y \rangle = \langle \Psi | \hat{S}_y | \Psi \rangle = \frac{i}{2} \left(-c_{\alpha\beta} c_{\alpha\alpha}^* + c_{\alpha\alpha} c_{\alpha\beta}^* - c_{\beta\beta} c_{\beta\alpha}^* + c_{\beta\alpha} c_{\beta\beta}^* \right)$$

$$\langle S_z \rangle = \langle \Psi | \hat{S}_z | \Psi \rangle = \frac{1}{2} \left(c_{\alpha\alpha} c_{\alpha\alpha}^* - c_{\alpha\beta} c_{\alpha\beta}^* + c_{\beta\alpha} c_{\beta\alpha}^* - c_{\beta\beta} c_{\beta\beta}^* \right)$$

F.4   Shift Operators

Raising and lowering operators for the individual spins:

$$\hat{I}^+ | \Psi \rangle = c_{\beta\alpha} | \alpha\alpha \rangle + c_{\beta\beta} | \alpha\beta \rangle$$

$$\hat{I}^- | \Psi \rangle = c_{\alpha\alpha} | \beta\alpha \rangle + c_{\alpha\beta} | \beta\beta \rangle$$

$$\hat{S}^+ | \Psi \rangle = c_{\alpha\beta} | \alpha\alpha \rangle + c_{\beta\beta} | \beta\alpha \rangle$$

$$\hat{S}^- | \Psi \rangle = c_{\alpha\alpha} | \alpha\beta \rangle + c_{\beta\alpha} | \beta\beta \rangle$$

Combined raising and lowering operators for both spins:

$$\hat{F}^+ | \Psi \rangle = (c_{\beta\alpha} + c_{\alpha\beta}) | \alpha\alpha \rangle + c_{\beta\beta} | \alpha\beta \rangle + c_{\beta\beta} | \beta\alpha \rangle$$

$$\hat{F}^- | \Psi \rangle = c_{\alpha\alpha} | \alpha\beta \rangle + c_{\alpha\alpha} | \beta\alpha \rangle + (c_{\beta\alpha} + c_{\alpha\beta}) | \beta\beta \rangle$$

F.5   Product Operators

$$\hat{I}_x\hat{S}_x|\Psi\rangle = \frac{1}{4}(c_{\beta\beta}|\alpha\alpha\rangle + c_{\beta\alpha}|\alpha\beta\rangle + c_{\alpha\beta}|\beta\alpha\rangle + c_{\alpha\alpha}|\beta\beta\rangle)$$

$$\hat{I}_x\hat{S}_y|\Psi\rangle = \frac{i}{4}(-c_{\beta\beta}|\alpha\alpha\rangle + c_{\beta\alpha}|\alpha\beta\rangle - c_{\alpha\beta}|\beta\alpha\rangle + c_{\alpha\alpha}|\beta\beta\rangle)$$

$$\hat{I}_x\hat{S}_z|\Psi\rangle = \frac{1}{4}(c_{\beta\alpha}|\alpha\alpha\rangle - c_{\beta\beta}|\alpha\beta\rangle + c_{\alpha\alpha}|\beta\alpha\rangle - c_{\alpha\beta}|\beta\beta\rangle)$$

$$\hat{I}_y\hat{S}_x|\Psi\rangle = \frac{i}{4}(-c_{\beta\beta}|\alpha\alpha\rangle - c_{\beta\alpha}|\alpha\beta\rangle + c_{\alpha\beta}|\beta\alpha\rangle + c_{\alpha\alpha}|\beta\beta\rangle)$$

$$\hat{I}_y\hat{S}_y|\Psi\rangle = \frac{1}{4}(-c_{\beta\beta}|\alpha\alpha\rangle + c_{\beta\alpha}|\alpha\beta\rangle + c_{\alpha\beta}|\beta\alpha\rangle - c_{\alpha\alpha}|\beta\beta\rangle)$$

$$\hat{I}_y\hat{S}_z|\Psi\rangle = \frac{i}{4}(-c_{\beta\alpha}|\alpha\alpha\rangle + c_{\beta\beta}|\alpha\beta\rangle + c_{\alpha\alpha}|\beta\alpha\rangle - c_{\alpha\beta}|\beta\beta\rangle)$$

$$\hat{I}_z\hat{S}_x|\Psi\rangle = \frac{1}{4}(c_{\alpha\beta}|\alpha\alpha\rangle + c_{\alpha\alpha}|\alpha\beta\rangle - c_{\beta\beta}|\beta\alpha\rangle - c_{\beta\alpha}|\beta\beta\rangle)$$

$$\hat{I}_z\hat{S}_y|\Psi\rangle = \frac{i}{4}(-c_{\alpha\beta}|\alpha\alpha\rangle + c_{\alpha\alpha}|\alpha\beta\rangle + c_{\beta\beta}|\beta\alpha\rangle - c_{\beta\alpha}|\beta\beta\rangle)$$

$$\hat{I}_z\hat{S}_z|\Psi\rangle = \frac{1}{4}(c_{\alpha\alpha}|\alpha\alpha\rangle - c_{\alpha\beta}|\alpha\beta\rangle - c_{\beta\alpha}|\beta\alpha\rangle + c_{\beta\beta}|\beta\beta\rangle)$$

F.6   Evolution of the Wavefunction Over Time

Under the conditions of weak coupling, where the eigenfunction of the Hamiltonian are approximated by the set, $|\alpha\alpha\rangle$, $|\alpha\beta\rangle$, $|\beta\alpha\rangle$ and $|\beta\beta\rangle$, the time dependence of the wavefunction in a stationary magnetic field is given by

$$|\Psi(t)\rangle = e^{-itE_{\alpha\alpha}/\hbar}c_{\alpha\alpha}|\alpha\alpha\rangle + e^{-itE_{\alpha\beta}/\hbar}c_{\alpha\beta}|\alpha\beta\rangle$$
$$+ e^{-itE_{\beta\alpha}/\hbar}c_{\beta\alpha}|\beta\alpha\rangle + e^{-itE_{\beta\beta}/\hbar}c_{\beta\alpha}|\beta\beta\rangle$$

where $c_{\alpha\alpha}$, $c_{\alpha\beta}$, $c_{\beta\alpha}$ and $c_{\beta\beta}$ are the initial coefficients of the eigenstates at time $t = 0$.

With the energies of the eigenstates expressed in terms of the Larmor frequencies, ν_I and ν_S, and the coupling constant, J, the time evolution of the wavefunction is

$$|\Psi(t)\rangle = e^{-i\pi t(\nu_I+\nu_S+J/2)}c_{\alpha\alpha}|\alpha\alpha\rangle + e^{-i\pi t(\nu_I-\nu_S-J/2)}c_{\alpha\beta}|\alpha\beta\rangle$$
$$+ e^{i\pi t(\nu_I-\nu_S+J/2)}c_{\beta\alpha}|\beta\alpha\rangle + e^{i\pi t(\nu_I+\nu_S-J/2)}c_{\beta\alpha}|\beta\beta\rangle$$

F.7   Pulse Operators to Rotate Magnetization

The initial wavefunction is given by $|\Psi\rangle = c_{\alpha\alpha}|\alpha\alpha\rangle + c_{\alpha\beta}|\alpha\beta\rangle + c_{\beta\alpha}|\beta\alpha\rangle + c_{\beta\beta}|\beta\beta\rangle$
Rotation of I-magnetization about x'-axis:

$$|\Psi(a)\rangle = \left(c_{\alpha\alpha}\cos(a/2) - ic_{\beta\alpha}\sin(a/2)\right)|\alpha\alpha\rangle$$
$$+ \left(c_{\alpha\beta}\cos(a/2) - ic_{\beta\beta}\sin(a/2)\right)|\alpha\beta\rangle$$
$$+ \left(c_{\beta\alpha}\cos(a/2) - ic_{\alpha\alpha}\sin(a/2)\right)|\beta\alpha\rangle$$
$$+ \left(c_{\beta\beta}\cos(a/2) - ic_{\alpha\beta}\sin(a/2)\right)|\beta\beta\rangle$$

Rotation of I-magnetization about y'-axis:

$$|\Psi(a)\rangle = \left(c_{\alpha\alpha}\cos(a/2) - c_{\beta\alpha}\sin(a/2)\right)|\alpha\alpha\rangle$$
$$+ \left(c_{\alpha\beta}\cos(a/2) - c_{\beta\beta}\sin(a/2)\right)|\alpha\beta\rangle$$
$$+ \left(c_{\beta\alpha}\cos(a/2) + c_{\alpha\alpha}\sin(a/2)\right)|\beta\alpha\rangle$$
$$+ \left(c_{\beta\beta}\cos(a/2) + c_{\alpha\beta}\sin(a/2)\right)|\beta\beta\rangle$$

Rotation of S-magnetization about x'-axis:

$$|\Psi(a)\rangle = \left(c_{\alpha\alpha}\cos(a/2) - ic_{\alpha\beta}\sin(a/2)\right)|\alpha\alpha\rangle$$
$$+ \left(c_{\alpha\beta}\cos(a/2) - ic_{\alpha\alpha}\sin(a/2)\right)|\alpha\beta\rangle$$
$$+ \left(c_{\beta\alpha}\cos(a/2) - ic_{\beta\beta}\sin(a/2)\right)|\beta\alpha\rangle$$
$$+ \left(c_{\beta\beta}\cos(a/2) - ic_{\beta\alpha}\sin(a/2)\right)|\beta\beta\rangle$$

Rotation of S-magnetization about y'-axis:

$$|\Psi(a)\rangle = (c_{\alpha\alpha}\cos(a/2) - c_{\alpha\beta}\sin(a/2))|\alpha\alpha\rangle$$
$$+ (c_{\alpha\beta}\cos(a/2) + c_{\alpha\alpha}\sin(a/2))|\alpha\beta\rangle$$
$$+ (c_{\beta\alpha}\cos(a/2) - c_{\beta\beta}\sin(a/2))|\beta\alpha\rangle$$
$$+ (c_{\beta\beta}\cos(a/2) + c_{\beta\alpha}\sin(a/2))|\beta\beta\rangle$$

F.8   Vector and Matrix Representations of Wavefunctions and Operators

Wavefunctions

The wavefunction for a scalar-coupled spin pair can be represented as a column vector,

$$\Psi = \begin{bmatrix} c_{\alpha\alpha} \\ c_{\alpha\beta} \\ c_{\beta\alpha} \\ c_{\beta\beta} \end{bmatrix}$$

where the elements of the vector are the complex coefficients of the z-magnetization eigenfunctions, $|\alpha\alpha\rangle$, $|\alpha\beta\rangle$, $|\beta\alpha\rangle$ and $|\beta\beta\rangle$. The complex conjugate of the wavefunction is written as a row vector:

$$\mathbf{\Psi}^* = [\, c^*_{\alpha\alpha} \quad c^*_{\alpha\beta} \quad c^*_{\beta\alpha} \quad c^*_{\beta\beta} \,]$$

Magnetization Operators

The magnetization operators for the I-spin:

$$\mathbf{I}_x = \frac{1}{2}\begin{bmatrix} 0 & 0 & 1 & 0 \\ 0 & 0 & 0 & 1 \\ 1 & 0 & 0 & 0 \\ 0 & 1 & 0 & 0 \end{bmatrix} \quad \mathbf{I}_y = \frac{i}{2}\begin{bmatrix} 0 & 0 & -1 & 0 \\ 0 & 0 & 0 & -1 \\ 1 & 0 & 0 & 0 \\ 0 & 1 & 0 & 0 \end{bmatrix} \quad \mathbf{I}_z = \frac{1}{2}\begin{bmatrix} 1 & 0 & 0 & 0 \\ 0 & 1 & 0 & 0 \\ 0 & 0 & -1 & 0 \\ 0 & 0 & 0 & -1 \end{bmatrix}$$

The magnetization operators for the S-spin:

$$\mathbf{S}_x = \frac{1}{2}\begin{bmatrix} 0 & 1 & 0 & 0 \\ 1 & 0 & 0 & 0 \\ 0 & 0 & 0 & 1 \\ 0 & 0 & 1 & 0 \end{bmatrix} \quad \mathbf{S}_y = \frac{i}{2}\begin{bmatrix} 0 & -1 & 0 & 0 \\ 1 & 0 & 0 & 0 \\ 0 & 0 & 0 & -1 \\ 0 & 0 & 1 & 0 \end{bmatrix} \quad \mathbf{S}_z = \frac{1}{2}\begin{bmatrix} 1 & 0 & 0 & 0 \\ 0 & -1 & 0 & 0 \\ 0 & 0 & 1 & 0 \\ 0 & 0 & 0 & -1 \end{bmatrix}$$

The operators are applied to the wavefunction by multiplying the wavefunction vector by the operator matrix. For instance,

$$\mathbf{I}_x\mathbf{\Psi} = \frac{1}{2}\begin{bmatrix} 0 & 0 & 1 & 0 \\ 0 & 0 & 0 & 1 \\ 1 & 0 & 0 & 0 \\ 0 & 1 & 0 & 0 \end{bmatrix}\begin{bmatrix} c_{\alpha\alpha} \\ c_{\alpha\beta} \\ c_{\beta\alpha} \\ c_{\beta\beta} \end{bmatrix} = \frac{1}{2}\begin{bmatrix} c_{\beta\alpha} \\ c_{\beta\beta} \\ c_{\alpha\alpha} \\ c_{\alpha\beta} \end{bmatrix}$$

The average I_x-magnetization is then calculated as the dot product of this product and the complex conjugate of $\mathbf{\Psi}$:

$$\langle I_x\rangle = \mathbf{\Psi}^*\mathbf{I}_x\mathbf{\Psi}$$

$$= [\, c^*_{\alpha\alpha} \quad c^*_{\alpha\beta} \quad c^*_{\beta\alpha} \quad c^*_{\beta\beta} \,]\frac{1}{2}\begin{bmatrix} c_{\beta\alpha} \\ c_{\beta\beta} \\ c_{\alpha\alpha} \\ c_{\alpha\beta} \end{bmatrix}$$

$$= \frac{1}{2}(c_{\beta\alpha}c^*_{\alpha\alpha} + c_{\beta\beta}c^*_{\alpha\beta} + c_{\alpha\alpha}c^*_{\beta\alpha} + c_{\alpha\beta}c^*_{\beta\beta})$$

Shift Operators

The shift operators are written in matrix form as

$$\mathbf{I}^+ = \mathbf{I}_x + i\mathbf{I}_y = \begin{bmatrix} 0 & 0 & 1 & 0 \\ 0 & 0 & 0 & 1 \\ 0 & 0 & 0 & 0 \\ 0 & 0 & 0 & 0 \end{bmatrix} \quad \mathbf{I}^- = \mathbf{I}_x - i\mathbf{I}_y = \begin{bmatrix} 0 & 0 & 0 & 0 \\ 0 & 0 & 0 & 0 \\ 1 & 0 & 0 & 0 \\ 0 & 1 & 0 & 0 \end{bmatrix}$$

$$\mathbf{S}^+ = \mathbf{S}_x + i\mathbf{S}_y = \begin{bmatrix} 0 & 1 & 0 & 0 \\ 0 & 0 & 0 & 0 \\ 0 & 0 & 0 & 1 \\ 0 & 0 & 0 & 0 \end{bmatrix} \qquad \mathbf{S}^- = \mathbf{S}_x - i\mathbf{S}_y = \begin{bmatrix} 0 & 0 & 0 & 0 \\ 1 & 0 & 0 & 0 \\ 0 & 0 & 0 & 0 \\ 0 & 0 & 1 & 0 \end{bmatrix}$$

$$\mathbf{F}^+ = \mathbf{I}^+ + \mathbf{S}^+ = \begin{bmatrix} 0 & 1 & 1 & 0 \\ 0 & 0 & 0 & 1 \\ 0 & 0 & 0 & 1 \\ 0 & 0 & 0 & 0 \end{bmatrix} \qquad \mathbf{F}^- = \mathbf{I}^- + \mathbf{S}^- = \begin{bmatrix} 0 & 0 & 0 & 0 \\ 1 & 0 & 0 & 0 \\ 1 & 0 & 0 & 0 \\ 0 & 1 & 1 & 0 \end{bmatrix}$$

Product Operators

The product operators are constructed by the appropriate multiplication of the magnetization operator matrices. For instance,

$$\mathbf{I}_x\mathbf{S}_x = \frac{1}{2}\begin{bmatrix} 0 & 0 & 1 & 0 \\ 0 & 0 & 0 & 1 \\ 1 & 0 & 0 & 0 \\ 0 & 1 & 0 & 0 \end{bmatrix} \frac{1}{2}\begin{bmatrix} 0 & 1 & 0 & 0 \\ 1 & 0 & 0 & 0 \\ 0 & 0 & 0 & 1 \\ 0 & 0 & 1 & 0 \end{bmatrix} = \frac{1}{4}\begin{bmatrix} 0 & 0 & 0 & 1 \\ 0 & 0 & 1 & 0 \\ 0 & 1 & 0 & 0 \\ 1 & 0 & 0 & 0 \end{bmatrix}$$

The full set of nine product operator matrices are as follows:

$$\mathbf{I}_x\mathbf{S}_x = \frac{1}{4}\begin{bmatrix} 0 & 0 & 0 & 1 \\ 0 & 0 & 1 & 0 \\ 0 & 1 & 0 & 0 \\ 1 & 0 & 0 & 0 \end{bmatrix} \quad \mathbf{I}_x\mathbf{S}_y = \frac{i}{4}\begin{bmatrix} 0 & 0 & 0 & -1 \\ 0 & 0 & 1 & 0 \\ 0 & -1 & 0 & 0 \\ 1 & 0 & 0 & 0 \end{bmatrix} \quad \mathbf{I}_x\mathbf{S}_z = \frac{1}{4}\begin{bmatrix} 0 & 0 & 1 & 0 \\ 0 & 0 & 0 & -1 \\ 1 & 0 & 0 & 0 \\ 0 & -1 & 0 & 0 \end{bmatrix}$$

$$\mathbf{I}_y\mathbf{S}_x = \frac{i}{4}\begin{bmatrix} 0 & 0 & 0 & -1 \\ 0 & 0 & -1 & 0 \\ 0 & 1 & 0 & 0 \\ 1 & 0 & 0 & 0 \end{bmatrix} \quad \mathbf{I}_y\mathbf{S}_y = \frac{1}{4}\begin{bmatrix} 0 & 0 & 0 & -1 \\ 0 & 0 & 1 & 0 \\ 0 & 1 & 0 & 0 \\ -1 & 0 & 0 & 0 \end{bmatrix} \quad \mathbf{I}_y\mathbf{S}_z = \frac{i}{4}\begin{bmatrix} 0 & 0 & -1 & 0 \\ 0 & 0 & 0 & 1 \\ 1 & 0 & 0 & 0 \\ 0 & -1 & 0 & 0 \end{bmatrix}$$

$$\mathbf{I}_z\mathbf{S}_x = \frac{1}{4}\begin{bmatrix} 0 & 1 & 0 & 0 \\ 1 & 0 & 0 & 0 \\ 0 & 0 & 0 & -1 \\ 0 & 0 & -1 & 0 \end{bmatrix} \quad \mathbf{I}_z\mathbf{S}_y = \frac{i}{4}\begin{bmatrix} 0 & -1 & 0 & 0 \\ 1 & 0 & 0 & 0 \\ 0 & 0 & 0 & 1 \\ 0 & 0 & -1 & 0 \end{bmatrix} \quad \mathbf{I}_z\mathbf{S}_z = \frac{1}{4}\begin{bmatrix} 1 & 0 & 0 & 0 \\ 0 & -1 & 0 & 0 \\ 0 & 0 & -1 & 0 \\ 0 & 0 & 0 & 1 \end{bmatrix}$$

The Hamiltonian Operator and Time Evolution

The Hamiltonian operator for a scalar-coupled spin pair in a stationary external magnetic field along the z-axis is written in matrix form as

$$\mathcal{H} = h\nu_I\mathbf{I}_z + h\nu_S\mathbf{S}_z + hJ(\mathbf{I}_x\mathbf{S}_x + \mathbf{I}_y\mathbf{S}_y + \mathbf{I}_z\mathbf{S}_z)$$

$$= \frac{h}{2}\begin{bmatrix} (\nu_I + \nu_S + J/2) & 0 & 0 & 0 \\ 0 & (\nu_I - \nu_S - J/2) & J & 0 \\ 0 & J & (-\nu_I + \nu_S - J/2) & 0 \\ 0 & 0 & 0 & (-\nu_I - \nu_S + J/2) \end{bmatrix}$$

In the limit of weak coupling, the Hamiltonian simplifies to

$$\mathcal{H} = h\nu_I I_z + h\nu_S S_z + h J I_z S_z$$

$$= \frac{h}{2} \begin{bmatrix} (\nu_I + \nu_S + J/2) & 0 & 0 & 0 \\ 0 & (\nu_I - \nu_S - J/2) & 0 & 0 \\ 0 & 0 & (-\nu_I + \nu_S - J/2) & 0 \\ 0 & 0 & 0 & (-\nu_I - \nu_S + J/2) \end{bmatrix}$$

The unitary time-evolution matrix is then derived by integrating the Schrödinger equation, which gives:

$$\mathbf{U}_H = \exp\left(\frac{-it}{\hbar}\mathcal{H}\right)$$

In the weak-coupling limit (where the Hamiltonian is diagonal and matrix exponentiation is easy)

$$\mathbf{U}_H = \begin{bmatrix} e^{-i\pi(\nu_I+\nu_S+J/2)t} & 0 & 0 & 0 \\ 0 & e^{-i\pi(\nu_I-\nu_S-J/2)t} & 0 & 0 \\ 0 & 0 & e^{i\pi(\nu_I-\nu_S+J/2)t} & 0 \\ 0 & 0 & 0 & e^{i\pi(\nu_I+\nu_S-J/2)t} \end{bmatrix}$$

The inverse of $\mathbf{U}_H$ is

$$\mathbf{U}_H^{-1} = \begin{bmatrix} e^{i\pi(\nu_I+\nu_S+J/2)t} & 0 & 0 & 0 \\ 0 & e^{i\pi(\nu_I-\nu_S-J/2)t} & 0 & 0 \\ 0 & 0 & e^{-i\pi(\nu_I-\nu_S+J/2)t} & 0 \\ 0 & 0 & 0 & e^{-i\pi(\nu_I+\nu_S-J/2)t} \end{bmatrix}$$

The time dependence of the wavefunction is calculated as

$$\mathbf{\Psi}(t) = \mathbf{U}_H \mathbf{\Psi}(0)$$

where $\mathbf{\Psi}(0)$ is the vector representation of the wavefunction at time zero. If $\mathbf{\Psi}(0)$ is written as a column vector,

$$\mathbf{\Psi}(0) = \begin{bmatrix} c_{\alpha\alpha}(0) \\ c_{\alpha\beta}(0) \\ c_{\beta\alpha}(0) \\ c_{\beta\beta}(0) \end{bmatrix}$$

then the wavefunction after time, t, is

$$\mathbf{\Psi}(t) = \begin{bmatrix} e^{-i\pi(\nu_I+\nu_S+J/2)t} & 0 & 0 & 0 \\ 0 & e^{-i\pi(\nu_I-\nu_S-J/2)t} & 0 & 0 \\ 0 & 0 & e^{i\pi(\nu_I-\nu_S+J/2)t} & 0 \\ 0 & 0 & 0 & e^{i\pi(\nu_I+\nu_S-J/2)t} \end{bmatrix} \begin{bmatrix} c_{\alpha\alpha}(0) \\ c_{\alpha\beta}(0) \\ c_{\beta\alpha}(0) \\ c_{\beta\beta}(0) \end{bmatrix}$$

$$= \begin{bmatrix} c_{\alpha\alpha}(0)e^{-i\pi(\nu_I+\nu_S+J/2)t} \\ c_{\alpha\beta}(0)e^{-i\pi(\nu_I-\nu_S-J/2)t} \\ c_{\beta\alpha}(0)e^{i\pi(\nu_I-\nu_S+J/2)t} \\ c_{\beta\beta}(0)e^{i\pi(\nu_I+\nu_S-J/2)t} \end{bmatrix}$$

The time-evolution matrix can be divided into two matrices to represent separately the evolution due to chemical-shift differences (CS) and scalar coupling (SC):

$$\mathbf{U}_H = \mathbf{U}_{H,cs}\mathbf{U}_{H,sc}$$

$$\mathbf{U}_{H,cs} = \begin{bmatrix} e^{-i\pi(\nu_I+\nu_S)t} & 0 & 0 & 0 \\ 0 & e^{-i\pi(\nu_I-\nu_S)t} & 0 & 0 \\ 0 & 0 & e^{i\pi(\nu_I-\nu_S)t} & 0 \\ 0 & 0 & 0 & e^{i\pi(\nu_I+\nu_S)t} \end{bmatrix}$$

$$\mathbf{U}_{H,sc} = \begin{bmatrix} e^{-i\pi(J/2)t} & 0 & 0 & 0 \\ 0 & e^{i\pi(J/2)t} & 0 & 0 \\ 0 & 0 & e^{i\pi(J/2)t} & 0 \\ 0 & 0 & 0 & e^{-i\pi(J/2)t} \end{bmatrix}$$

The wavefunction vector can be multiplied by $\mathbf{U}_{H,cs}$ and $\mathbf{U}_{H,sc}$ in either order:

$$\mathbf{\Psi}(t) = \mathbf{U}_{H,cs}\mathbf{U}_{H,sc}\mathbf{\Psi}(0) = \mathbf{U}_{H,sc}\mathbf{U}_{H,cs}\mathbf{\Psi}(0)$$

Rotation by Pulses

The rotation matrices for the I-spin and their inverses:

$$\mathbf{R}_{x,I}(a) = \begin{bmatrix} \cos(a/2) & 0 & -i\sin(a/2) & 0 \\ 0 & \cos(a/2) & 0 & -i\sin(a/2) \\ -i\sin(a/2) & 0 & \cos(a/2) & 0 \\ 0 & -i\sin(a/2) & 0 & \cos(a/2) \end{bmatrix}$$

$$\mathbf{R}_{y,I}(a) = \begin{bmatrix} \cos(a/2) & 0 & -\sin(a/2) & 0 \\ 0 & \cos(a/2) & 0 & -\sin(a/2) \\ \sin(a/2) & 0 & \cos(a/2) & 0 \\ 0 & \sin(a/2) & 0 & \cos(a/2) \end{bmatrix}$$

$$\mathbf{R}_{x,I}^{-1}(a) = \begin{bmatrix} \cos(a/2) & 0 & i\sin(a/2) & 0 \\ 0 & \cos(a/2) & 0 & i\sin(a/2) \\ i\sin(a/2) & 0 & \cos(a/2) & 0 \\ 0 & i\sin(a/2) & 0 & \cos(a/2) \end{bmatrix}$$

$$\mathbf{R}_{y,I}^{-1}(a) = \begin{bmatrix} \cos(a/2) & 0 & \sin(a/2) & 0 \\ 0 & \cos(a/2) & 0 & \sin(a/2) \\ -\sin(a/2) & 0 & \cos(a/2) & 0 \\ 0 & -\sin(a/2) & 0 & \cos(a/2) \end{bmatrix}$$

The rotation matrices for the S-spin and their inverses:

$$R_{x,S}(a) = \begin{bmatrix} \cos(a/2) & -i\sin(a/2) & 0 & 0 \\ -i\sin(a/2) & \cos(a/2) & 0 & 0 \\ 0 & 0 & \cos(a/2) & -i\sin(a/2) \\ 0 & 0 & -i\sin(a/2) & \cos(a/2) \end{bmatrix}$$

$$R_{y,S}(a) = \begin{bmatrix} \cos(a/2) & -\sin(a/2) & 0 & 0 \\ \sin(a/2) & \cos(a/2) & 0 & 0 \\ 0 & 0 & \cos(a/2) & -\sin(a/2) \\ 0 & 0 & \sin(a/2) & \cos(a/2) \end{bmatrix}$$

$$R_{x,S}^{-1}(a) = \begin{bmatrix} \cos(a/2) & i\sin(a/2) & 0 & 0 \\ i\sin(a/2) & \cos(a/2) & 0 & 0 \\ 0 & 0 & \cos(a/2) & i\sin(a/2) \\ 0 & 0 & i\sin(a/2) & \cos(a/2) \end{bmatrix}$$

$$R_{y,S}^{-1}(a) = \begin{bmatrix} \cos(a/2) & \sin(a/2) & 0 & 0 \\ -\sin(a/2) & \cos(a/2) & 0 & 0 \\ 0 & 0 & \cos(a/2) & \sin(a/2) \\ 0 & 0 & -\sin(a/2) & \cos(a/2) \end{bmatrix}$$

The effects of the rotation operators are calculated by multiplying the vector for the wavefunction by the appropriate rotation matrix. For instance, the wavefunction generated when a selective $(\pi/2)_x$-pulse is applied to the I-spin of a spin pair with the initial wavefunction, Ψ, is calculated as follows:

$$\Psi\big((\pi/2)_x\big) = R_{x,I}(\pi/2)\Psi$$

$$= \begin{bmatrix} \cos(\pi/4) & 0 & -i\sin(\pi/4) & 0 \\ 0 & \cos(\pi/4) & 0 & -i\sin(\pi/4) \\ -i\sin(\pi/4) & 0 & \cos(\pi/4) & 0 \\ 0 & -i\sin(\pi/4) & 0 & \cos(\pi/4) \end{bmatrix} \begin{bmatrix} c_{\alpha\alpha} \\ c_{\alpha\beta} \\ c_{\beta\alpha} \\ c_{\beta\beta} \end{bmatrix}$$

$$= \frac{1}{\sqrt{2}} \begin{bmatrix} 1 & 0 & -i & 0 \\ 0 & 1 & 0 & -i \\ -i & 0 & 1 & 0 \\ 0 & -i & 0 & 1 \end{bmatrix} \begin{bmatrix} c_{\alpha\alpha} \\ c_{\alpha\beta} \\ c_{\beta\alpha} \\ c_{\beta\beta} \end{bmatrix}$$

$$= \frac{1}{\sqrt{2}} \begin{bmatrix} c_{\alpha\alpha} - ic_{\beta\alpha} \\ c_{\alpha\beta} - ic_{\beta\beta} \\ -ic_{\alpha\alpha} + c_{\beta\alpha} \\ -ic_{\alpha\beta} + c_{\beta\beta} \end{bmatrix}$$

The time evolution and rotation matrices, along with their inverses, are also used to calculate the effects of pulses on the density matrix, as described in Chapter 17.

Index

Page numbers ending with f indicate a figure; ending with n indicate a footnote; ending with r indicate a bibliographic reference; ending with t indicate a table.